Lecture Notes in Computer Science 15960

The series Lecture Notes in Computer Science (LNCS), including its subseries Lecture Notes in Artificial Intelligence (LNAI) and Lecture Notes in Bioinformatics (LNBI), has established itself as a medium for the publication of new developments in computer science and information technology research, teaching, and education.

LNCS enjoys close cooperation with the computer science R & D community, the series counts many renowned academics among its volume editors and paper authors, and collaborates with prestigious societies. Its mission is to serve this international community by providing an invaluable service, mainly focused on the publication of conference and workshop proceedings and postproceedings. LNCS commenced publication in 1973.

James C. Gee · Daniel C. Alexander ·
Jaesung Hong · Juan Eugenio Iglesias ·
Carole H. Sudre · Archana Venkataraman ·
Polina Golland · Jong Hyo Kim · Jinah Park
Editors

Medical Image Computing and Computer Assisted Intervention – MICCAI 2025

28th International Conference
Daejeon, South Korea, September 23–27, 2025
Proceedings, Part I

Springer

Editors
James C. Gee
University of Pennsylvania
Philadelphia, PA, USA

Daniel C. Alexander
University College London
London, UK

Jaesung Hong
DGIST
Daegu, Korea (Republic of)

Juan Eugenio Iglesias
Massachusetts General Hospital and Harvard Medical School
Charlestown, MA, USA

Carole H. Sudre
University College London
London, UK

Archana Venkataraman
Boston University
Boston, MA, USA

Polina Golland
MIT
Cambridge, MA, USA

Jong Hyo Kim
Seoul National University
Seoul, Korea (Republic of)

Jinah Park
KAIST
Daejeon, Korea (Republic of)

ISSN 0302-9743 ISSN 1611-3349 (electronic)
Lecture Notes in Computer Science
ISBN 978-3-032-04926-1 ISBN 978-3-032-04927-8 (eBook)
https://doi.org/10.1007/978-3-032-04927-8

This Springer imprint is published by the registered company Springer Nature Switzerland AG
The registered company address is: Gewerbestrasse 11, 6330 Cham, Switzerland

Preface

We are pleased to present the proceedings of the 28th International Conference on Medical Image Computing and Computer-Assisted Intervention (MICCAI 2025). For the first time in its history, the MICCAI conference was held in South Korea, taking place at the Daejeon Convention Center in Daejeon from September 23 to 27, 2025. The conference included a variety of Satellite Events held on September 23 and 27, featuring 99 workshops, tutorials, and challenges. This year marked a significant milestone for the Clinical Translation in Medical Image Computing and Computer-Assisted Intervention (CLINICCAI) conference, which held its 5th edition on September 25. Unlike in the past when it was organized independently, this year's CLINICCAI was prepared in close collaboration with the MICCAI Organizing Committee for the first time, further strengthening the conference's emphasis on clinical translation and engagement. Open science was another continuing focus for advancement at the meeting as the conference hosted the 2nd Open Data Session on September 26.

MICCAI 2025 set a new record for submissions, with a 28% increase compared to 2024. We received 3677 full submissions, of which 3447 advanced to full review. One hundred fifty-nine papers were desk-rejected as they did not comply with submission guidelines, and 71 papers were withdrawn, including 60 that violated the double submission rules. The number of papers accepted for presentation after review was 1027, which represents an acceptance rate of about 30%. These papers comprise the 16 volumes of these Lecture Notes in Computer Science (LNCS) proceedings as follows:

- Part I, LNCS Volume 15960: Multimodal Fusion and Contextual Reasoning in Medical Imaging
- Part II, LNCS Volume 15961: Surgical Navigation, Scene Understanding, and Video Modeling
- Part III, LNCS Volume 15962: Learning and Augmented Reality for Surgical and Endoscopic Applications (I)
- Part IV, LNCS Volume 15963: Learning and Augmented Reality for Surgical and Endoscopic Applications (II)
- Part V, LNCS Volume 15964: Graph-Based Methods in Medical Imaging
- Part VI, LNCS Volume 15965: Datasets and Methods for Image Quality Enhancement
- Part VII, LNCS Volume 15966: Trustworthy and Responsible AI for Medical Imaging
- Part VIII, LNCS Volume 15967: Multimodal Learning for Diagnosis, Risk Prediction, and Survival Analysis
- Part IX, LNCS Volume 15968: Core Techniques in Medical Imaging: Segmentation, Registration, Synthesis, Reconstruction, and Other Emerging Methods (I)
- Part X, LNCS Volume 15969: Core Techniques in Medical Imaging: Segmentation, Registration, Synthesis, Reconstruction, and Other Emerging Methods (II)
- Part XI, LNCS Volume 15970: Core Techniques in Medical Imaging: Segmentation, Registration, Synthesis, Reconstruction, and Other Emerging Methods (III)
- Part XII, LNCS Volume 15971: Core Techniques in Medical Imaging: Segmentation, Registration, Synthesis, Reconstruction, and Other Emerging Methods (IV)

- Part XIII, LNCS Volume 15972: Adapting Foundation Models for Medical Imaging: LLMs, VLMs, and Cross-Domain Generalization (I)
- Part XIV, LNCS Volume 15973: Adapting Foundation Models for Medical Imaging: LLMs, VLMs, and Cross-Domain Generalization (II)
- Part XV, LNCS Volume 15974: Adapting Foundation Models for Medical Imaging: LLMs, VLMs, and Cross-Domain Generalization (III)
- Part XVI, LNCS Volume 15975: Statistical Techniques in Medical Imaging: Causality, Imputation, Weak Supervision, and Other Methods

The papers for the proceedings were selected after a rigorous double-blind peer-review process with three reviews per paper on average. The MICCAI 2025 Program Committee (PC) consisted of 187 area chairs (ACs), over 2400 scientific reviewers, and 12 dual-submissions reviewers, with representation from 61 countries across all major continents. About a quarter (22%) of area chairs self-identified as women. Each area chair managed the review of roughly 34 papers, and the average workload was four papers per reviewer. Area chairs and reviewers were selected following public calls to the community, and were vetted by the program chairs. As in the past, the Microsoft Conference Management Toolkit (CMT) was used for paper management and organization of the overall review process. The Toronto Paper Matching System (TPMS) was, again, employed to help ensure papers were evaluated by area chairs and reviewers with the appropriate expertise. While special effort was made this year to improve the quality of paper matches by reworking both TPMS formulae and CMT categorization of papers and PC member profiles, we recognize the ongoing challenge of securing high-quality reviews in the face of ever-increasing numbers of submissions.

Based on the double-blinded reviews, area chairs' recommendations, and program chairs' decisions, 9% (310) of papers were provisionally accepted without rebuttal, 48% (1663) were rejected, and 43% (1474) proceeded to the rebuttal stage. Final decisions were made based on all available information, including paper screening, adherence to the author submission guidelines, reviews, rebuttals, 2–3 independent meta-reviews, communications during the review process, and, in some cases, additional assessment by the program chairs. A dedicated team was formed to enforce dual submissions policy. The program committee scrutinized papers flagged during reviews, manuscripts of special interest, potential plagiarism, and appeals from authors. The committee was particularly attentive to suggestions of misuse of LLMs both in submissions and reviews - an increasingly vexing issue not only at MICCAI but in science in general.

The 1027 accepted papers that constitute this year's scientific program originated from 43 countries. Seventy-nine percent of these papers are from the following five countries, as identified by a paper's first author: China, including Hong Kong SAR (501), USA (118), Germany (66), South Korea (64), and UK (62). The next five countries in descending order of the number of accepted papers are France (26), Australia (22), Canada (22), Switzerland (17), and Singapore (15). Nineteen papers were from lower and middle-income countries in Asia, a region highlighted by this MICCAI edition toward the conference's continuing efforts to broaden its global impact.

To build the oral program of 76 presentations, accepted papers were first clustered by technical theme. For each theme, oral presentation candidates were chosen from the papers that received overwhelming enthusiasm among the area chairs. These candidates were then narrowed down to the final list by ensuring the program was representative of the range of clinical applications covered by the papers and reflected both geographic and demographic inclusivity among its presenters. Additionally, 36 posters were highlighted through rapid talks ahead of their presentation at subsequent poster sessions. These pieces of work were recommended by ACs for their likelihood to provoke discussion and for specific contributions to the community in the form of new open-source datasets.

MICCAI 2025 centered around two overarching themes. The first, "From Concept to Clinical," emphasized the translational journey of MICCAI methodologies: from foundational research to real-world clinical application. To support this theme, we actively encouraged industry engagement through dedicated panels and presentations, and featured mentoring programs with industry leaders in collaboration with the MICCAI Student Board.

The second theme, "Pan-Asian Connections," highlighted the growing collaboration within Asian research communities and reinforced MICCAI's ongoing commitment to inclusion and outreach. Initiatives under this theme included targeted mentorship programs for young researchers, from writing technical papers to navigating the submission and rebuttal process. Another key initiative involved real-time broadcasts of the conference by selected young researchers across Asia who facilitated local community discussions, particularly in countries historically underrepresented at MICCAI.

Furthermore, MICCAI 2025 spotlighted focused discussions on healthcare challenges across Asia, with special attention to underrepresented countries in Area 3. These efforts culminated in a special plenary session titled "Challenges and Opportunities of Medical AI in Asia." With over half of the world's population residing in Asia, the region represents immense potential and need for developing scalable and generalizable healthcare AI solutions. The recent increase in cross-border research collaboration further positions Asia as a fertile ground for piloting innovative medical technologies. This highlight session brought together leading experts from South Korea, Japan, China, Singapore, Mongolia, and Nepal to share insights on the current landscape, region-specific challenges, and the future of AI integration in healthcare. These collective efforts were made possible through the unique platform provided by the MICCAI meeting.

To further the mission of enhancing inclusion at MICCAI, we offered travel grants to candidates from lower-to-middle-income countries as well as from countries historically underrepresented in the MICCAI community, including participants from Africa, Asia, and Latin America. We also provided partial support to additional participants through the MICCAI Society Registration Grants. Funding was made possible by generous support from the MICCAI Society, donations from individual groups within the MICCAI community, and financial support from non-commercial sponsors. Support for US-based students and early-career investigators was also provided in part by the National Institutes of Health (NIH) through conference grant R13EB035922 from the National Institute of Biomedical Imaging and Bioengineering. The views expressed in written conference materials or publications and by speakers and moderators do not necessarily reflect the official policies of the NIH; nor does mention by trade names, commercial practices, or organizations imply endorsement by the U.S. Government. We additionally supported young Asian researchers through a conference outreach program that provided registration and accommodation support, made possible by the generous support of the City of Daejeon.

A meeting of MICCAI's scale and impact is only possible through the collective dedication and tireless efforts of many individuals, most notably, the members of this year's organizing committees. While every member played a vital role, we would like to especially recognize several key contributors. Luping Zhou, Fernando Bello, and Nicholas Heller led the Workshops, Tutorials, and Challenges teams, respectively, in reviewing a total of 191 Satellite Event proposals and assembling a diverse, high-quality program. We are particularly grateful to our Satellite Program Chair, Michal Kozubek, for skillfully coordinating the 99 accepted events into a cohesive and well-structured program. We also recognize Lena Maier-Hein for her primary role in launching the Lighthouse Challenges this year. The Industry Engagement Committee, headed by Stephen Aylward, established a strong industry presence that brought our conference themes to life. Tanveer Syeda-Mahmood coordinated both the industry and clinical panel sessions. The Young Researchers Initiative Committee, under the leadership of Minjeong Kim, introduced a new and impactful program that further enriched the participation and engagement of early-career researchers.

We also appreciate the exemplary work of our Sponsorship team, Kwang Gi Kim, Mohammad Yaqub, Jihoon Cho, and Maria Valdés Hernández, for their efforts in securing conference financial support despite the challenging economic climate. Joonbeom Seo and Jonghye Woo, together with the CLINICCAI Program Committee, ensured the continued success of CLINICCAI as a distinctive flagship of MICCAI. Nassir Navab and Seong Tae Kim provided consummate oversight of the Outreach Program, helping to further expand MICCAI's reach into underrepresented regions. Martijn Starmans, Apostolia Tsirikoglou, and Namkug Kim led with distinction in advancing the Open Data Initiative.

A tremendous amount of work was carried out by the Local Organizing Committee, chaired by Sanghyun Park, whose commitment and meticulous attention to every detail were critical to the smooth execution of all parts of the meeting. We are also grateful to Jaeil Kim, Finance Chair, as well as our Communications and Signage Chairs, Keri Kim, Won Hwa Kim, and Sangrok Jin, whose expert support in managing various elements of the local venue helped ensure a pleasant experience for all participants. Seungryong Cho directed the Student Volunteers with enthusiasm, while Hyunjoo Jenny Lee successfully organized KAIST Day, creating a memorable and festive pre-conference event. Jae Sung Lee, Ilwoo Lyu, and Joonho Seo provided impeccable support to facilitate networking and foster meaningful connections and engagement among the community throughout the event.

We are deeply grateful to our local supporting institutions, KAIST (Korea Advanced Institute of Science and Technology) and KSIIM (Korea Society of Imaging Informatics in Medicine), for their invaluable organizational and logistical support. We also wish to extend our heartfelt thanks to the City of Daejeon and the Daejeon Tourism Organization for their generous contributions, which made it possible to bring MICCAI 2025 to South Korea.

The MICCAI Student Board, chaired by Naren Akash, contributed significantly by organizing highly popular networking and social events, including the Ph.D. Thesis

Madness as well as the Doctoral Consortium, which was held for the first time and offered a valuable forum for recent graduates to present their work and engage with the broader community. Likewise, the Women in MICCAI and RISE communities, under the guidance of their respective presidents, Ruogu Fang and Esther Puyol Anton, presented topical events that further strengthened the inclusivity and depth of our technical program.

We would also like to express our sincere gratitude to the MICCAI Society Board of Directors and in particular to its Chair, Marius George Linguraru, and former Chair, Caroline Essert, for their support and constructive feedback. We gratefully acknowledge, too, the contributions of the MICCAI Society secretariat personnel, Janette Wallace and Johanne Langford, who oversaw logistics and budgets, and Diana Cunningham, Anna Van Vliet, and John Baxter for the timely conference announcements in the MICCAI Society newsletters and MICCAI Morning issues. The proficient on-site organization of the conference, including budget management and event operations, was the commendable work of our Professional Conference Organizer, Dekon Congress & Tourism, led by Mehmet Eldegez. Dekon's essential efforts were carried out in partnership with Lorraine Han and her team at the MICE Bureau of the Daejeon Tourism Organization, to whom we offer our wholehearted thanks.

The proceedings of MICCAI 2025 are the fruit of the extraordinary work of our entire community. Kitty Wong, the MICCAI Submission System Manager, deserves special mention and our profound thanks for her crucial support throughout the paper submission, review, program planning, and proceedings preparation process. We are particularly grateful, too, to the members of the Program Committee for their dedicated and diligent work in the reviewer assignments and final paper selection, and to all the reviewers for their commitment and support during the paper evaluation process. Congratulations to the outstanding area chairs and reviewers! Finally, and most importantly, we reserve our deepest appreciation and gratitude for all the authors for their effort in preparing their manuscripts and supporting the conference through their submissions. The success of MICCAI 2025 is a direct reflection of the high-quality work presented at the meeting. Our warmest congratulations to all!

In closing, we would like to thank once more every individual and team involved in the organization of MICCAI 2025. Their inspiring dedication, professionalism, and collaborative spirit made this year's conference a truly outstanding and memorable event. We hope you enjoyed MICCAI 2025 in Daejeon, the conference's first meeting in South Korea. We now look forward to seeing you next year in Abu Dhabi at MICCAI 2026.

September 2025

James C. Gee
Daniel Alexander
Jaesung Hong
Juan Eugenio Iglesias
Carole Sudre
Archana Venkataraman
MICCAI 2025 Program Chairs

Jinah Park
Polina Golland
Jonghyo Kim
MICCAI 2025 General Chairs

Organization

General Chairs

Jinah Park	Korea Advanced Institute of Science and Technology, South Korea
Polina Golland	Massachusetts Institute of Technology, USA
Jonghyo Kim	Seoul National University, South Korea

Program Committee Chairs

James C. Gee	University of Pennsylvania, USA
Daniel C. Alexander	University College London, UK
Juan Eugenio Iglesias	Massachusetts General Hospital and Harvard Medical School, USA
Jaesung Hong	Daegu Gyeongbuk Institute of Science and Technology, South Korea
Carole H. Sudre	University College London, UK
Archana Venkataraman	Boston University, USA

Satellite Events Chairs

Michal Kozubek	Masaryk University, Czech Republic
Masahiro Oda	Nagoya University, Japan

Workshops

Luping Zhou (Chair)	University of Sydney, Australia
Masahiro Oda	Nagoya University, Japan
Heung-Il Suk	Korea University, South Korea
Guorong Wu	University of North Carolina at Chapel Hill, USA

Challenges

Lena Maier-Hein (Chair)	German Cancer Research Center (DKFZ), Germany
Nicholas Heller (Co-chair)	Cleveland Clinic, USA
Suyash Awate	Indian Institute of Technology Bombay, India
Sophia Bano	University College London, UK
Jun Ma	University Health Network, Canada
Annika Reinke	German Cancer Research Center (DKFZ), Germany

Tutorials

Fernando Bello (Chair)	Imperial College London, UK and Duke-NUS Medical School, Singapore
Diana Mateus	Centrale Nantes LS2N, France
Daniel H. Pak	Yale University, USA
Chang Min Park	Seoul National University College of Medicine, South Korea
Hoo-Chang Shin	NVIDIA, USA

MICCAI-Clinical (CLINICCAI)

Joon Beom Seo (Chair)	University of Ulsan College of Medicine, South Korea
Steven Wong	Sengkang General Hospital, Singapore
Jonghye Woo	Massachusetts General Hospital and Harvard Medical School, USA

Industry Engagement

Stephen Aylward (Chair)	NVIDIA, USA
Kyu-Hwan Jung (Co-chair)	Sungkyunkwan University, South Korea
Alex Lemm	Amazon Web Services, Germany
Ma Luo	Medtronic, USA
Brendon Lutnick	Johnson & Johnson Innovative Medicine, USA

Sponsorship

Kwang Gi Kim (Co-chair)	Gachon University, South Korea
Mohammad Yaqub (Co-chair)	Mohamed bin Zayed University of Artificial Intelligence, UAE
Jihoon Cho	Samsung Research, South Korea
Maria Valdés Hernández	University of Edinburgh, UK

Young Scholars Initiative

Minjeong Kim (Chair)	University of North Carolina at Greensboro, USA
Ehsan Adeli	Stanford University, USA
Seong Jae Hwang	Yonsei University, South Korea
Anees Kazi	Massachusetts General Hospital and Harvard Medical School, USA
Islem Rekik	Imperial College London, UK

Outreach

Nassir Navab (Co-chair)	Technical University of Munich, Germany
Seong Tae Kim (Co-chair)	Kyung Hee University, South Korea
Shadi Albarqouni	Universität Bonn and Helmholtz Munich, Germany
Shoji Kido	Osaka University, Japan
Namkug Kim	University of Ulsan College of Medicine, South Korea
Anna Seo	SEEANN Solution Co., South Korea

Open Data Chairs

Martijn Starmans	Erasmus University Medical Center, Netherlands
Apostolia Tsirikoglou	Karolinska Institutet, Sweden
Namkug Kim	University of Ulsan College of Medicine, South Korea
Lidia Garrucho Moras	Universitat de Barcelona, Spain
Kaouther Mouheb	Erasmus University Medical Center, Netherlands

Panels Chair

Tanveer Syeda-Mahmood	IBM Research, USA

Local Organization Committee

Sang Hyun Park (Chair)	Daegu Gyeongbuk Institute of Science and Technology, South Korea

Finance Chair

Jaeil Kim	Kyungpook National University, South Korea

Communications and Signage Chairs

Keri Kim	Korea Institute of Science and Technology, South Korea
Won Hwa Kim	Pohang University of Science and Technology, South Korea
Sangrok Jin	Pusan National University, South Korea

Student Volunteers Chair

Seungryong Cho	Korea Advanced Institute of Science and Technology, South Korea

KAIST Day Chair

Hyunjoo Jenny Lee	Korea Advanced Institute of Science and Technology, South Korea

Network and Arrangements

Jae Sung Lee	Seoul National University College of Medicine, South Korea
Ilwoo Lyu	Pohang University of Science and Technology, South Korea
Joonho Seo	Korea Institute of Machinery & Materials, South Korea

Convention Liaison

Lorrain Han	Daejeon Tourism Organization, South Korea
Jinny Seo	Daejeon Tourism Organization, South Korea
Hazel Seok	Daejeon Tourism Organization, South Korea

Women in MICCAI President

Ruogu Fang	University of Florida, USA

RISE Committee President

Esther Puyol-Antón	Heartflow and King's College London, UK

Submission Platform Manager

Kitty Wong	MICCAI Society, Canada

Program Committee

Pablo Arbelaez	Universidad de los Andes, Colombia
Angelica I. Aviles-Rivero	Tsinghua University, China
Wenjia Bai	Imperial College London, UK
Yaël Balbastre	University College London, UK
Abhirup Banerjee	University of Oxford, UK
Omri Bar	Theator, Israel
Neslihan Bayramoglu	University of Oulu, Finland

Cosmin I. Bercea	Technical University Munich and Helmholtz AI and Helmholtz Center Munich, Germany
Ryoma Bise	Kyushu University, Japan
Esther E. Bron	Erasmus University Medical Center, Netherlands
Mariano Cabezas	University of Sydney, Australia
Weidong Cai	University of Sydney, Australia
Chen Chen	University of Sheffield, UK
Cheng Chen	Chinese University of Hong Kong, China
Elvis C. S. Chen	Robarts Research Institute, Canada
Hao Chen	Hong Kong University of Science and Technology, China
Jianan Chen	University College London, UK
Zhen Chen	Yale University, USA
Jang-Hwan Choi	Ewha Womans University, South Korea
Zhiming Cui	ShanghaiTech University, China
Kang Dang	Xi'an Jiaotong-Liverpool University, China
Eleonora D'Arnese	University of Edinburgh, UK
Charles Delahunt	Global Health Labs, USA
Ruining Deng	Weill Cornell Medicine, USA
Christian Desrosiers	ÉTS Montréal, Canada
Blake E. Dewey	Johns Hopkins University, USA
Jose Dolz	ÉTS Montréal, Canada
Niharika S. D'Souza	IBM Research, USA
Nicha Dvornek	Yale University, USA
Dmitry V. Dylov	Skoltech, Russia
Mostafa El Habib Daho	University of Western Brittany, France
Deng-Ping Fan	Nankai University, China
Ruogu Fang	University of Florida, USA
Azade Farshad	Technical University of Munich, Germany
Aasa Feragen	Technical University of Denmark, Denmark
Huazhu Fu	A*STAR Institute of High Performance Computing, Singapore
Jin Kyu Gahm	Pusan National University, South Korea
Adrian Galdran	Universitat Pompeu Fabra, Spain
Mingchen Gao	University at Buffalo, SUNY, USA
Elodie Germani	Universitätsklinikum Bonn, Germany
Syed Zulqarnain Gilani	Edith Cowan University, Australia
Karthik Gopinath	Massachusetts General Hospital and Harvard Medical School, USA
Pietro Gori	Télécom Paris, France
Yun Gu	Imperial College London, UK
Xiaoqing Guo	University of Oxford, UK

Yao Sui	Peking University, China
Kyung Hyun Sung	University of California, Los Angeles, USA
Tanveer Syeda-Mahmood	IBM Research, USA
Lin Tian	Massachusetts General Hospital and Harvard Medical School, USA
Aleksei Tiulpin	University of Oulu, Finland
Martin Urschler	Medical University Graz, Austria
Marta Varela	Imperial College London, UK
Harini Veeraraghavan	Memorial Sloan Kettering Cancer Center, USA
Chengjia Wang	Heriot-Watt University, UK
Guotai Wang	University of Electronic Science and Technology of China, China
Hu Wang	Mohamed bin Zayed University of Artificial Intelligence, United Arab Emirates
Li Wang	University of North Carolina at Chapel Hill, USA
Liansheng Wang	Xiamen University, China
Qian Wang	ShanghaiTech University, China
Renzhen Wang	Xi'an Jiaotong University, China
Shujun Emma Wang	Hong Kong Polytechnic University, China
Shuo Wang	Fudan University, China
Xiaosong Wang	Shanghai AI Laboratory, China
Yan Wang	East China Normal University, China
Donglai Wei	Boston College, USA
Matthias Wilms	University of Calgary, Canada
Jelmer Wolterink	University of Twente, Netherlands
Ye Wu	Nanjing University of Science and Technology, China
Yicheng Wu	Monash University, Australia
Yong Xia	Northwestern Polytechnical University, China
Yiming Xiao	Concordia University, Canada
Yutong Xie	Mohamed bin Zayed University of Artificial Intelligence, United Arab Emirates
Fuyong Xing	University of Colorado Denver, USA
Xiaohan Xing	Stanford University, USA
Chenchu Xu	Anhui University, China
Xuanang Xu	Rensselaer Polytechnic Institute, USA
Yanwu Xu	South China University of Technology, China
Yanyu Xu	Shandong University, China
Zhe Xu	Chinese University of Hong Kong, China
Wufeng Xue	Shenzhen University, China
Zengqiang Yan	Huazhong University of Science and Technology, China

Jiancheng Yang	Swiss Federal Technology Institute of Lausanne, Switzerland
Qiushi Yang	City University of Hong Kong, China
Yanwu Yang	University of Tübingen, Germany
Jiawen Yao	DAMO Academy, Alibaba Group, China
Qingsong Yao	Stanford University, USA
Chuyang Ye	Beijing Institute of Technology, China
Dong Hye Ye	Georgia State University, USA
Zhaozheng Yin	Stony Brook University, USA
Chenyu You	Stony Brook University, USA
Fatemeh Zabihollahy	University of Toronto, Canada
Jianpeng Zhang	Alibaba DAMO Academy, China
Jiong Zhang	Cixi Institute of Biomedical Engineering, Ningbo Institute of Materials Technology and Engineering, Chinese Academy of Sciences, China
Lichi Zhang	Shanghai Jiao Tong University, China
Yanfu Zhang	William and Mary, USA
Yi Zhang	Sichuan University, China
Can Zhao	Nvidia, USA
Fenqiang Zhao	DAMO Academy, Alibaba Group, China
Yitian Zhao	Cixi Institute of Biomedical Engineering, Ningbo Institute of Industrial Technology, Chinese Academy of Sciences, China
Yue Zhao	Chongqing University of Posts and Telecommunications, China
Yushan Zheng	Beihang University, China
Zichun Zhong	Wayne State University, USA
Qin Zhou	East China University of Science and Technology, China
S. Kevin Zhou	University of Science and Technology of China, China
Tao Zhou	Nanjing University of Science and Technology, China
Tianfei Zhou	Beijing Institute of Technology, China
Dajiang Zhu	University of Texas at Arlington, USA
Lei Zhu	Hong Kong University of Science and Technology (Guangzhou), China
Xiaofeng Zhu	University of Electronic Science and Technology of China, China
Ke Zou	Sichuan University, China

Outstanding Area Chairs

Angelica I. Aviles-Rivero	Tsinghua University, China
Mariano Cabezas	University of Sydney, Australia
Martin Urschler	Medical University Graz, Austria
Fuyong Xing	University of Colorado Denver, USA

Outstanding Reviewers

Syed Farhan Abbas	Norwegian University of Science and Technology, Norway
Tajamul Ashraf	Indian Institute of Technology Delhi, India
Matthew Baugh	Imperial College London, UK
Benjamin Billot	Inria, France
Katharina Breininger	University of Würzburg, Germany
Sebastian Doerrich	University of Bamberg, Germany
Catalin Fetita	Télécom SudParis, France
Isabel Funke	National Center for Tumor Diseases (NCT) Dresden, Germany
Zdravko Marinov	Karlsruhe Institute of Technology, Germany
Chinedu Nwoye	University of Strasbourg, France
Gasper Podobnik	University of Ljubljana, Slovenia
Isabella Poles	Politecnico di Milano, Italy
Hatem Rashwan	Rovira i Virgili University, Spain
Amirhossein Sabour	McMaster University, Canada
Weiyi Zhang	Hong Kong Polytechnic University, China
Lianrui Zuo	Vanderbilt University, USA

Honorable Mentions (Reviewers)

Seyed-Ahamad Ahmadi	Technical University of Munich, Germany
Yasmina Al Khalil	Eindhoven University of Technology, Netherlands
Mehdi Astaraki	Karolinska Institutet, Sweden
Mohammad Saber Azimi	Shahid Beheshti University, Iran
Joao Barreto	University of Coimbra, Portugal
Víctor M. Batlle	I3A, Universidad de Zaragoza, Spain
Federico Bolelli	Università degli Studi di Modena e Reggio Emilia, Italy
Ester Bonmati Coll	University of Westminster, UK
Yi Hao Chan	Nanyang Technological University, Singapore

Shengjia Chen	Icahn School of Medicine at Mount Sinai, USA
Ziyang Chen	Northwestern Polytechnical University, China
Philip Chikontwe	Harvard University, USA
Allison Clement	University of Oxford, UK
Francesco Cremonesi	Inria, France
Meghane Decroocq	RIKEN Center for Brain Science, Japan
Yu Deng	King's College London, UK
Kerol Djoumessi	University of Tübingen, Germany
Reuben Dorent	Inria, France
Peiyu Duan	Yale University, USA
Benoit Dufumier	CEA, France
Felix Dülmer	Technical University of Munich, Germany
Jim Duncan	Yale University, USA
Marawan Elbatel	Hong Kong University of Science and Technology, China
Huihui Fang	Nanyang Technological University, Singapore
Greg Fleishman	Howard Hughes Medical Institute, USA
Shangqi Gao	University of Cambridge, UK
Yifan Gao	University of Science and Technology of China, China
Negin Ghamsarian	University of Bern, Switzerland
Catalina Gomez	Johns Hopkins University, USA
Alvaro Gonzalez-Jimenez	University Hospital of Basel, Switzerland
Pengfei Gu	University of Notre Dame, USA
Ida Häggström	Chalmers Tekniska Högskola, Sweden
Shahad Hardan	Mohamed bin Zayed University of Artificial Intelligence, United Arab Emirates
Hideaki Hayashi	Osaka University, Japan
Ruobing Huang	Shenzhen University, China
Juntao Jiang	Zhejiang University, China
Amelia Jiménez-Sánchez	IT University of Copenhagen, Denmark
Shantanu Joshi	University of California, Los Angeles, USA
Bernhard Kainz	Imperial College London, UK & FAU Erlangen-Nürnberg, Germany
Yunsoo Kim	University College London, UK
Tobias Klinder	Philips, Germany
Nicholas Konz	Duke University, USA
Daniel Lang	Helmholtz Center Munich, Germany
Moon Hwan Lee	Daegu Gyeongbuk Institute of Science and Technology, South Korea
Won Hee Lee	Kyung Hee University, South Korea
Fan Li	Shanghai Jiao Tong University, China

Francisco Vasconcelos	University College London, UK
Hooman Vaseli	University of British Columbia, Canada
Trinh Thi Le Vuong	Korea University, South Korea
Thomas Wendler	Technical University of Munich, Germany
Markus Wenzel	Fraunhofer Institute for Digital Medicine MEVIS, Germany
Julia Wolleb	Yale University, USA
McKell Woodland	University of Texas MD Anderson Cancer Center, USA
Fuping Wu	University of Oxford, UK
Jianghao Wu	University of Electronic Science and Technology of China, China
Yunheng Wu	Nagoya University, Japan
Zijian Wu	University of British Columbia, Canada
Geng-Xin Xu	Shenzhen Institute of Advanced Technology, Chinese Academy of Sciences, China
Zikang Xu	University of Science and Technology of China, China
Yingyu Yang	University of Oxford, UK
Chenfei Ye	Harbin Institute of Technology at Shenzhen, China
Siyeop Yoon	Massachusetts General Hospital and Harvard Medical School, USA
Zhen Yu	Monash University, Australia
Ziqi Yu	Fudan University, China
Paul Yushkevich	University of Pennsylvania, USA
Bodong Zhang	University of Utah, USA
Dong Zhang	University of British Columbia, Canada
Tianyi Zhang	Agency for Science, Technology and Research, Singapore
Weihang Zhang	Beijing Institute of Technology, China
Yushan Zheng	Beihang University, China

Mentorship Program (Mentors)

Sèmèvo A. R. M. Ahouandjinou	University of Abomey-Calavi, Benin
Joseph Akinyemi	University of York, UK
Shadi Albarqouni	University of Bonn & Helmholtz AI, Germany
Sharib Ali	University of Leeds, UK
Adel Alii	University of Sfax, Tunisia
Amine Amyar	Harvard Medical School, USA

Syed Anwar	Children's National Hospital, USA
Zhen Chen	Yale University, USA
Kathleen Curran	University College Dublin, Ireland
Idy Diop	Cheikh Anta Diop University, Senegal
Mostafa El Habib Daho	University of Western Brittany, France
Mohammed El Hassouni	Mohammed V University in Rabat, Morocco
Lamiaa Elfadaly	October University for Modern Science and Arts, Egypt
Parastoo Farnia	Tehran University of Medical Sciences, Iran
Huazhu Fu	A*STAR Institute of High Performance Computing, Singapore
Florin-Cristian Ghesu	Siemens Healthineers, Germany
Sanaa Ghouzali	International University of Rabat, Morocco
Syed Zulqarnain Gilani	Edith Cowan University, Australia
Prashnna Gyawali	West Virginia University, USA
Kaiyuan Hu	McGill University, Canada
Uzair Iqbal	University of Malaya, Malaysia
Mobarakol Islam	University College London, UK
Eric Jagwara	Tufuna Technologies, Uganda
Amal Jlassi	University of Tunis El Manar, Tunisia
Wonsik Jung	Korea University, South Korea
Manjunath K. N.	Manipal Institute of Technology, India
Dmitrii Kaplun	China University of Mining and Technology, China
Baiying Lei	Shenzhen University, China
Jianming Liang	Arizona State University, USA
Meharban M. S.	Sheridian, Canada
Michail Mamalakis	University of Cambridge, UK
Subhamoy Mandal	Indian Institute of Technology Kharagpur, India
Mounia Mikram	École des Sciences de l'Information, Morocco
Ahmed Nebli	Heirich Heine University, Düsseldorf, Germany
Akash Parvatikar	HistoWiz Inc., USA
Mengyun Qiao	Imperial College London, UK
Rizwan Qureshi	University of Central Florida, USA
Tariq Rahim	Kingston University London, UK
Maryem Rhanoui	Université Claude Bernard Lyon 1, France
Juan A. Sánchez-Margallo	Jesús Usón Minimally Invasive Surgery Centre, Spain
Sahar Selim Soussa	Nile University, Egypt
Keewon Shin	Korea University College of Medicine, South Korea
Kyung Sung	University of California, Los Angeles, USA

Jiacheng Wang	Manteia Technology Co. Ltd, China
Ying Weng	University of Nottingham Ningbo China, China
Ye Wu	Nanjing University of Science and Technology, China
Bôa Djavph Yoganguina	École Supérieure Polytechnique de Dakar, Senegal
S. Kevin Zhou	University of Science and Technology of China, China
Zongwei Zhou	Johns Hopkins University, USA
Lei Zhu	HKUST (GZ) and HKUST, China

Reviewers

Syed Farhan Abbas
Hamid Abbasi
Zeinab Abboud
Lamiaa Abdel-Hamid
Sanne Abeln
Kumar Abhishek
Purang Abolmaesumi
Burak Acar
Hamidreza Aftabi
Vinti Agarwal
Harshit Agrawal
Antonio Agudo
Seyed-Ahamad Ahmadi
Ninad Aithal
Naren Akash
Faranak Akbarifar
Emre Akbas
Zeynettin Akkus
Fatih Aksu
Dawood Al Chanti
Wafa Al Ghallabi
Yasmina Al Khalil
Khaoula Alaoui Belghiti
Julia Alekseenko
Hassan Alhajj
Shahzad Ali
Sharib Ali
Sikandar Ali
Omar Al-Kadi
Simeon Allmendinger
Mohammed Al-masni
Pablo Alvarez
Charlems Alvarez-Jimenez
Jennifer Alvén
Victor Alves
Lidia Al-Zogbi
Dayeong An
Haill An
Elsa Angelini
Anna Anikina
André Anjos
Sameer Antani
Michela Antonelli
Michel Antunes
Syed Anwar
Yoshimitsu Aoki
Dooman Arefan
Ignacio Arganda-Carreras
Mohammad Ali Armin
Abu Bakor Hayat Arnob
Chetan Arora
Dewa Made Sri Arsa
Muhammad Arsalan
Janan Arslan
Marius Arvinte
Md Ashikuzzaman
Tajamul Ashraf
Mehdi Astaraki
Arda Atalık
Chinmayee Athalye
Thomas Athey
Benjamin Aubert
Marc Aubreville
Chloé Audigier
Ignacio Avellino
Akash Awasthi
Suyash Awate
Ishak Ayad
Nicolás Ayobi
Sana Ayromlou
Md Abulkalam Azad
Mohammad Farid Azampour
Mehran Azimbagirad
Mohammad Saber Azimi
Abu Zahid Bin Aziz
Britty Baby
Hyeon-Min Bae
Seung Baek
Ulas Bagci
Harrison Bai
Jieyun Bai
Long Bai
Muhammad Naseer Bajwa
Eduard Bakstein
Guha Balakrishnan
Abdourahmane Balde
Abhirup Banerjee
Imon Banerjee
Soumyanil Banerjee

Chandan Ganesh Bangalore Yogananda
Jordan Bannister
Jaume Banus
Jamal Banzi
Shunxing Bao
D. M. Bappy
Joao Barreto
Hritam Basak
Tamer Basha
Deepti R. Bathula
Víctor M. Batlle
Matthew Baugh
Christian Baumgartner
John Baxter
Nourhan Bayasi
Siming Bayer
Roza Bayrak
Roozbeh Bazargani
Clément Beitone
Soufiane Belharbi
Simone Bendazzoli
Sean Benson
Mariana Bento
Guy Ben-Yosef
Christos Bergeles
Gabriel Bernardino
Riddhish Bhalodia
Nithya Bhasker
Moinak Bhattacharya
Binod Bhattarai
Lei Bi
Ning Bi
Qi Bi
Yuan Bi
Zhangxing Bian
Benjamin Billot
Manuel Birlo
George Biros
Pinar Bisgin
Koushik Biswas
Sebastian Bodenstedt
Lea Bogensperger
Patrick Bolan
Federico Bolelli
Mahyar Bolhassani
Fabian Bongratz
Ester Bonmati Coll
Sébastien Bougleux
Sylvain Bouix
Emmanuelle Bourigault
Alexandre Bousse
Arnaud Boutillon
Maria Boyko
Afshin Bozorgpour
Nadia Brancati
Kit Bransby
Biagio Brattoli
Diego Bravo
Katharina Breininger
Clara Brémond Martin
Michael Briden
Christopher Bridge
Tom Brosch
Michael Brudno
Michal Brzus
Nhat-Tan Bui
Phuoc-Nguyen Bui
Junghyun Bum
Arianna Bunnell
Stefano Buoso
Ninon Burgos
Nikolay Burlutskiy
Joshua Butke
Yunsu Byeon
Guido Caccianiga
Noa Cahan
Jiangdong Cai
Linghan Cai
Lishan Cai
Ming Cai
Shangyan Cai
Tongan Cai
Zhiyuan Cai
Michael Camilleri
Gabriele Campanella
Victor Campello
Liane Canas
Dongliang Cao
Guogang Cao
Jianguo Cao
Weiguo Cao
Xu Cao
Yankun Cao
Yuzhu Cao
Daniel Capellán-Martín
Gianluca Carloni
Owen Carmichael
Patrick Carnahan
Jacob Carse
Alessandro Casella
Eduardo Castañeda
Philippe C. Cattin
Stefano Cerri
Kate Cevora
Richard Cha
Matthieu Chabanas
Dong-Kyu Chae
Krishna Chaitanya
Souradeep Chakraborty
Tapabrata Chakraborty
Arunava Chakravarty
Kwan Ho Ryan Chan
S.-H. Gary Chan
Yi Hao Chan
Shekhar Chandra
Yao-Jen Chang
Young Hwan Chang
Yuchou Chang
Simon Chatelin
Soumitri Chattopadhyay
Muhammad Faizyab Ali Chaudhary
Shivesh Chaudhary
Joël Chavas
Yiming Che
Antong Chen
Boqi Chen
Chuangquan Chen
Dongdong Chen
Geng Chen
Guannan Chen
Han Chen
Haoyuan Chen
Hongbo Chen

Jiahe Chen
Jianxu Chen
Jiayi Chen
Jiazhou Chen
Jie Chen
Jieneng Chen
Jun Chen
Junxiang Chen
Junzhou Chen
Kai Chen
Kanghao Chen
Kevin Chen
Lei Chen
Li Chen
Lixuan Chen
Pengcheng Chen
Pingjun Chen
Pingyi Chen
Qi Chen
Qiang Chen
Qishen Chen
Shanxiong Chen
Shaobin Chen
Shengchao Chen
Shengjia Chen
Tao Chen
Tingting Chen
Tong Chen
Wanwen Chen
Wenting Chen
Wenxuan Chen
Xiang Chen
Xin Chen
Xinrong Chen
Xiumei Chen
Xueli Chen
Yaxi Chen
Yifei Chen
Yiliang Chen
Yinda Chen
Yinran Chen
Yixiong Chen
Yufei Chen
Yu-Jen Chen
Yunjie Chen
Zhihao Chen
Zhu Chen
Zhuangzhuang Chen
Zifan Chen
Zihao Chen
Zijian Chen
Zijiao Chen
Ziyang Chen
Chang-Chieh Cheng
Chi-Tung Cheng
Jiale Cheng
Jing Cheng
Jun Cheng
Kun Cheng
Pujin Cheng
Teja Krishna Cherukuri
Aichi Chien
Mark Chiew
Philip Chikontwe
Debarghya China
Ching-Hao Chiu
Seungryong Cho
Sue Min Cho
Woo-Jin Cho Kim
Tomáš Chobola
Changyong Choi
Hyoungshin Choi
Kyu Sung Choi
Min-Kook Choi
Sang-Il Choi
Yoonseok Choi
Young Sang Choi
Jakub Chojnacki
Kyobin Choo
Yu-Cheng Chou
Argyrios Christodoulidis
Hung Chu
Lingyang Chu
Miao Chu
Yuetan Chu
Yusung Chu
Kai-Cheng Chuang
Hyungjin Chung
Minyoung Chung
Yunsung Chung
Ozkan Cigdem
Davide Ciraolo
Allison Clement
Bryan Clifford
Robert Cobb
Dana Cobzas
Jaume Coll-Font
Olivier Colliot
Marco Colussi
Cong Cong
Runmin Cong
Laura Connolly
William Consagra
Pierre-Henri Conze
Tim Cootes
Valentina Corbetta
Teresa Correia
Aurélien Corroyer-Dulmont
Baris Coskunuzer
Stephane Cotin
Francesco Cremonesi
Alex Crimi
Ines Alejandro Cruz Guerrero
Hengfei Cui
Jianan Cui
Jiaqi Cui
Xiaoxiao Cui
Xiaoyu Cui
Yue Cui
Zhuo-Xu Cui
Sasa Cukovic
Tolga Cukur
Ariel Curiale
Tobias Czempiel
Duwei Dai
Pengyu Dai
Wei Dai
Sixtus Dakurah
Tingting Dan
Maik Dannecker
Eleonora D'Arnese
Dhritiman Das
Debashis Das Chakladar

Ananyananda Dasari
Aigerim Dautkulova
Palak Dave
Tareen Dawood
Riccardo De Feo
Jonas De Vylder
Bram de Wilde
Meghane Decroocq
Andreas Deistung
Herve Delingette
Liang-Jian Deng
Tianpeng Deng
Yu Deng
Zengtian Deng
Zhipeng Deng
Zhongying Deng
Felix Denzinger
Adrien Depeursinge
Hrishikesh Deshpande
Thomas DeSilvio
Neel Dey
Rohan Raju Dhanakshirur
Maxime Di Folco
Daniel Di Giovanni
Francesco Di Salvo
Songhui Diao
Alina Dima
Hao Ding
Hongxu Ding
Jiaqi Ding
Jun-En Ding
Kexin Ding
Xinpeng Ding
Ying Ding
Zhengyao Ding
Nicola Dinsdale
Nico Disch
Kerol Djoumessi
Konstantin Dmitriev
Daniel Dobkin
Sebastian Doerrich
Melanie Dohmen
Mischa Dombrowski
Ines Domingues
Caixia Dong
Di Dong
Haoyu Dong
Hexin Dong
Mengjin Dong
Min Dong
Shoubin Dong
Siyuan Dong
Xingbo Dong
Zijian Dong
Reuben Dorent
Gianfranco Doretto
Sven Dorkenwald
Haoran Dou
Yimeng Dou
Jason Dowling
Simon Drouin
Karen Drukker
Niharika D'Souza
Chenhe Du
Guodong Du
Honghui Du
Jie Du
Jing Du
Shaoyi Du
Siyi Du
Xia Du
Yuexi Du
Bin Duan
Peiyu Duan
Wenting Duan
Shikha Dubey
Nicolas Duchateau
David Dueñas Gaviria
Benoit Dufumier
Felix Dülmer
Jim Duncan
Joshua Durso-Finley
Pallabi Dutta
Oleh Dzyubachyk
Jan Egger
Gudmundur Einarsson
Favour Ekong
Nafie El Amrani
Mohammed El Hassouni
Ahmed Elazab
Marawan Elbatel
Mohamed El Amine Elforaici
Shireen Elhabian
David Ellis
Mohamed Elsharkawy
Daniel Elson
Solayman Hossain Emon
Sandy Engelhardt
Justin Engelmann
Andinet Enquobahrie
Ertunc Erdil
Marius Erdt
Ayhan Can Erdur
Selman Erel
Meltem Esengönül
Hooman Esfandiari
Nazila Esmaeili
Marco Esposito
Oscar Esteban
Alfonso Estudillo-Romero
Alessio Fagioli
Di Fan
Jianan Fan
Jiansong Fan
Jingfan Fan
Lei Fan
Liang Fan
Xiaoyao Fan
Xin Fan
Huihui Fang
Xi Fang
Yingying Fang
Yuqi Fang
Zhenghan Fang
Zhengqing Fang
Benito Farina
Aiman Farooq
Syed Hasib Akhter Faruqui
Jamil Fayyad
Botond Fazekas
Manman Fei
Paula Feldman
Lina Felsner
Anqi Feng

Mengling Feng
Wei Feng
Weiwei Feng
Zishun Feng
Zunlei Feng
Philipp Feodorovici
Abdesselam Ferdi
André Ferreira
Catalin Fetita
Jean Feydy
Leo Fillioux
Martina Finocchiaro
Alessio Fiorin
Ahmad Firjani
Lukas Fischer
Paul Fischer
Stefan Fischer
Greg Fleishman
Kevin Flores
Kyriakos Flouris
Stefano Fogarollo
Bernhard Föllmer
Nils Daniel Forkert
Jean-Rassaire Fouefack
Kevin François-Bouaou
Jannik Franzen
Moti Freiman
Friedhelm Schwenker
Magda Friedjungová
Paul Friedrich
Yannik Frisch
Jia Fu
Jingru Fu
Qiang Fu
Suzhong Fu
Wei Fu
Xianghua Fu
Xueming Fu
Xueyang Fu
Yu Fu
Yunguan Fu
Moritz Fuchs
Gareth Funka-Lea
Isabel Funke
Pedro Furtado
Ryo Furukawa
Lucas Gago
Mélanie Gaillochet
Gurur Gamgam
Fan Gan
Yu Gan
Melanie Ganz
Bingchen Gao
Chang Gao
Cong Gao
Fei Gao
Linlin Gao
Ruiliang Gao
Shangqi Gao
Shuo Gao
Xiaoru Gao
Xiaoyi Gao
Yibo Gao
Yifan Gao
Yuan Gao
Yue Gao
Zeyu Gao
Zheyao Gao
Sumita Garai
Utpal Garain
Piotr Garbat
Carles Garcia Cabrera
Lin Ge
Yuyan Ge
Zongyuan Ge
Eloy Geenjaar
Manuela Geiss
Elodie Germani
Tijana Geroski
Salah Ghamizi
Negin Ghamsarian
Yazan Gharaibeh
Ghazal Ghazaei
Behnaz Gheflati
Florin Ghesu
Shantanu Ghosh
Yosr Ghozzi
Yongha Gi
Diana Giraldo
Yannik Glaser
Ben Glocker
Bharti Goel
Wesley Gohn
Negar Golestani
Mayank Golhar
Alberto Gomez
Catalina Gomez
Santiago Gómez Hernández
Estibaliz Gómez-de-Mariscal
Mikhail Goncharov
Andrea Gondova
Haifan Gong
Xun Gong
Yongyi Gong
Ricardo Gonzales
Camila Gonzalez
German Gonzalez
Miguel Angel González Ballester
Alvaro Gonzalez-Jimenez
Anisleidy Gonzalez-Mitjans
Sharath Gopal
Vivek Gopalakrishnan
Philippe Gottfrois
Michael Götz
Maged Goubran
Robert Graf
Alejandro Granados
Eric Granger
Vicente Grau
Michael Green
Thomas Greer
Thomas Grenier
Fabian Gröger
Michal Grzeszczyk
Hanxue Gu
Leslie Gu
Pengfei Gu
Ran Gu
Yi Gu
Zaiwang Gu
Francesco Guarnera

Malitha Gunawardhana
Changlu Guo
Fumin Guo
Jintao Guo
Jixiang Guo
Jun Guo
Song Guo
Ting Guo
Xiaojuan Guo
Xueqi Guo
Yi Guo
Yuanbo Guo
Zhanqiang Guo
Gagan Gupta
Danna Gurari
Simon Gutwein
Lisa Guzzi
Sung Min Ha
Benjamin Haeffele
Zahra Hafezi Kafshgari
Paul Hager
Ida Häggström
Justin Haldar
Ghassan Hamarneh
Hua Han
Jinyoung Han
Jiyong Han
John Han
Kangfu Han
Luyi Han
Wenchao Han
Xiangmin Han
Yoseob Han
Lasse Hansen
Degan Hao
Jing Hao
Luoying Hao
Qin Hao
You Hao
Nazim Haouchine
Shota Harada
Shahad Hardan
Michael Hardisty
Mohamed Harmanani
Jeffry Hartanto
Septian Hartono
Md Kamrul Hasan
Ravi Hassanaly
Hideaki Hayashi
Dongdong He
Hongliang He
Huiguang He
Jianzhong He
Mengshen He
Tiantian He
Xiaoxiao He
Xinwei He
Xinzi He
Yifei He
Zhenqi He
Zhibin He
Moein Heidari
Mattias Heinrich
Nicholas Heller
Anja Hennemuth
Pit Henrich
Julius Hense
Hwan Heo
Keun-Soo Heo
Alessa Hering
Monica Hernandez
Daniel Hieber
John Hipwell
David Ho
Edmond S. L. Ho
Matthew Holden
Felix Holm
David Holmes III
Gregory Holste
Barmak Honarvar Shakibaei Asli
Jaeseong Hong
Jaesung Hong
Qingqi Hong
Nicolas Honnorat
Fabian Hörst
Tonmoy Hossain
Golriz Hosseinimanesh
Peyman Hosseinzadeh Kassani
Benjamin Hou
Junlin Hou
Qingshan Hou
Xuege Hou
Franko Hrzic
Dan Hu
Dianlin Hu
Jihong Hu
Kyle Hu
Minhao Hu
Qiang Hu
Qingqiao Hu
Renjiu Hu
Shishuai Hu
Xinrong Hu
Yan Hu
Yang Hu
Zhehao Hu
Shengyi Hua
Mengdi Huai
Baoru Huang
Bin Huang
Cheng Huang
Huifang Huang
Jiahao Huang
Jiashuang Huang
Jiaxing Huang
Kai Huang
Kuan Huang
Liqin Huang
Qinwen Huang
Ruobing Huang
Shuo Huang
Weitian Huang
Wenhui Huang
Wentao Huang
Xinrui Huang
Yanyan Huang
Yi-Jie Huang
Yijin Huang
Yiming Huang
Ying Huang
Yongsong Huang
Yuhao Huang
Yuliang Huang

Zengan Huang
Zhe Huang
Zhen Huang
Zheng Huang
Zhenxing Huang
Arnaud Huaulmé
Alex Hung
Jiayu Huo
Mohammad Arafat Hussain
Alexander Hustinx
Yasin Ibrahim
Ilknur Icke
Tali Ilovitsh
Abdullah-Al-Zubaer Imran
Veera Venkata Raghunath Indugu
Samra Irshad
Aisha Isa
Kh Tohidul Islam
Md Shazid Islam
Mohammad Tariqul Islam
Lukman Ismaila
Koichi Ito
Hayato Itoh
Krithika Iyer
Nabil Jabareen
Mohammad Jafari
Bahram Jafrasteh
Clément Jailin
Amoon Jamzad
Ananya Jana
Ikbeom Jang
Joseph Janizek
Pierre Jannin
Uditha Jarayathne
Ronnachai Jaroensri
Chathurangi Shyalika Jayakody Kankanamalage
Nivetha Jayakumar
Alissa Jell
Alexander Jenke
Mark Jenkinson
Rachid Jennane
Duhee Jeon
Wangsu Jeon
Hwihun Jeong
Ji-Hoon Jeong
Jiwoong Jeong
Won-Ki Jeong
Ranjeet Ranjan Jha
Zexin Ji
Zexuan Ji
Dengqiang Jia
Haozhe Jia
Bailiang Jian
Baichuan Jiang
Hao Jiang
Haotian Jiang
Hongxu Jiang
Juntao Jiang
Ning Jiang
Shancheng Jiang
Songhan Jiang
Xiaoyi Jiang
Xixi Jiang
Yufeng Jiang
Yuncheng Jiang
Zhencun Jiang
Zhongliang Jiang
Yining Jiao
Amelia Jiménez-Sánchez
Qiangguo Jin
Qiuye Jin
Qunchao Jin
Ruinan Jin
Yinzhu Jin
Jie Jing
Yongcheng Jing
Pierre-Marc Jodoin
Craig Jones
Anand Joshi
Shantanu Joshi
Smriti Joshi
Samuel Joutard
Ce Ju
Lie Ju
Vincent Jugnon
Gao Jun
Yohan Jun
Jaehee Jung
Wonsik Jung
Manjunath K. N.
Ali Kafaei Zad Tehrani
Joshua Kaggie
Dagmar Kainmueller
Bernhard Kainz
Mohammad Mahdi Kalantari
Niveditha Kalavakonda
John Kalkhof
Gabriel Kalweit
Maria Kalweit
Tae-Eui Kam
Sreeram Kamabattula
Naoki Kamiya
Michael C. Kampffmeyer
Renuga Kanagavelu
Dongwoo Kang
Junghwa Kang
Ming Kang
Mingon Kang
Qingbo Kang
Sohyun Kang
Wenjie Kang
Arunkumar Kannan
Dmitrii Kaplun
Saarthak Kapse
Siddhartha Kapuria
Tamás Karácsony
Mokshagna Sai Teja Karanam
Neerav Karani
Mert Asim Karaoglu
Turkay Kart
Tushar Kataria
Eytan Kats
Alexander Katzmann
Gurwinder Kaur
Matthias Keicher
Elif Keles
Erwan Kerrien
Ali Keshavarzi
Sara Ketabi

Mohsen Khadem
Firas Khader
Abbas Khan
Hassan Khan
Izhar Ahmed Khan
Ufaq Khan
Bidur Khanal
Bishesh Khanal
Benjamin Killeen
Daeseung Kim
Daewoon Kim
Dohoon Kim
Dohyun Kim
Hyeongyu Kim
Hyun Gi Kim
Hyun-Woo Kim
Jeongchul Kim
Jin Kim
Jinhee Kim
Jonghun Kim
JongWoo Kim
Kiduk Kim
Kyungsu Kim
Mansu Kim
MinWoo Kim
Sangwook Kim
Seong Tae Kim
Soopil Kim
Suhyeon Kim
Tae Soo Kim
Young-Min Kim
Yunsoo Kim
Alvin Kimbowa
Benjamin Kimia
Andrew King
Atilla Kiraly
Ayesha Kiran
Yannick Kirchhoff
Gabriel Kiss
Andreas Kist
Itaru Kitahara
Yoshiro Kitamura
Jens Kleesiek
Stefan Klein
Tobias Klinder
Kazuma Kobayashi
Lisa Koch
Yeong Jun Koh
Taha Koleilat
Anudeep Konda
Satoshi Kondo
Ming Kong
Ender Konukoglu
Aishik Konwer
Nicholas Konz
Mohamad Koohi-Moghadam
Steven Korevaar
Kivanc Kose
Sebastian Kozerke
Anna Kreshuk
Axel Krieger
Murali Krishnamoorthy
AnithaPriya Krishnan
Florian Kromp
Frithjof Kruggel
Henry Krumb
Hulin Kuang
Nanyu Kuang
David Kügler
Aaron Kujawa
Pranav Kulkarni
Amar Kumar
Ashwin Kumar
Nishant Kumar
Logiraj Kumaralingam
Kassymzhomart Kunanbayev
Soumya Kundu
Manuela Kunz
Po-Chih Kuo
Jan Kybic
Charlotte Kyeremah
Aymen Laadhari
Dmitrii Lachinov
Chaymaa Lahmar
Bolin Lai
Shang-Hong Lai
Songning Lai
Yingxin Lai
Brejesh Lall
Bennett Landman
Daniel Lang
Frank Langbein
Georg Langs
Qicheng Lao
Othmane Laousy
Catherine Laporte
Carole Lartizien
Shlomi Laufer
Max-Heinrich Laves
Aonghus Lawlor
Ana Lawry Aguila
An Le
Lawrence Le
Loic Le Folgoc
Bumshik Lee
Dong-Gyu Lee
Eungjoo Lee
Hong Joo Lee
Hyunyeol Lee
Jangho Lee
Jemyoung Lee
Jong-Hwan Lee
Jong-Min Lee
Jungwook Lee
Kyungsu Lee
Moon Hwan Lee
Sangmin Lee
Sangseon Lee
Seoyoung Lee
Seungeun Lee
Seungjun Lee
Sing Chun Lee
Soochahn Lee
Tiarna Lee
Won Hee Lee
Woonghee Lee
Youngjoon Lee
Antoine Legouhy
Long Lei
Minglong Lei
Wen Hui Lei
Yiming Lei
Rogers Jeffrey Leo John

Riccardo Levi
Michael Levine
Caizi Li
Chao Li
Chen Li
Cheng Li
Chengyin Li
Cynthia Xinran Li
Dongyang Li
Fan Li
Fangda Li
Ganping Li
Guangyao Li
Guoshi Li
Haiqing Li
Haitao Li
Han Li
Hanyu Li
Hao Li
Haofeng Li
Haoshen Li
Haoyang Li
Heng Li
Hongming Li
Hongyu Li
Huiqi Li
Jiawen Li
Jingming Li
Jingxiong Li
Jinpeng Li
Jun Li
Juncheng Li
Junjian Li
Jupeng Li
Kang Li
Lei Li
Lin Li
Miao Li
Qi Li
Qian Li
Qiyue Li
Rui Li
Ruizhe Li
Shiwei Li
Shuaibo Li
Shulong Li
Shumeng Li
Shuo Li
Siqi Li
Tiancheng Li
Wen Li
Wenxue Li
Xiangyu Li
Xiao-Xin Li
Xiaoxu Li
Xingjian Li
Xingyu Li
Xinyue Li
Xuesong Li
Yamin Li
Yang Li
Ye Li
Yi Li
Yicong Li
Yingshu Li
Yingtai Li
Yitong Li
Yuexiang Li
Yuheng Li
Zeju Li
Zhe Li
Zhiyuan Li
Zhjin Li
Zhongyu Li
Zhuoan Li
Zi Li
Zihan Li
Chunfeng Lian
Sheng Lian
Dong Liang
Libin Liang
Ruixing Liang
Yaqian Liang
Ziyun Liang
Guiqiu Liao
Hongen Liao
Junda Liao
Wei Liao
Zehui Liao
Marta Ligero
Chern Hong Lim
Gilbert Lim
Ailiang Lin
Chieh-Te Lin
Fangzhou Lin
Huei-Yung Lin
Jianxin Lin
Li Lin
Manxi Lin
Mingquan Lin
Runjia Lin
Xian Lin
Yi Lin
Yiqun Lin
Yiyang Lin
Zehui Lin
Claudia Lindner
Haibin Ling
Xitong Ling
Anran Liu
Baodi Liu
Bo Liu
Chang Liu
Che Liu
Chen Liu
Chenyu Liu
Chi Liu
Chuanbin Liu
Dichao Liu
Fei Liu
Feihong Liu
Guiran Liu
Guoyang Liu
Han Liu
Hao Liu
Hengjie Liu
Huafeng Liu
Huaxiang Liu
Jianan Liu
Jianming Liu
Jiawei Liu
Jia-Wei Liu
Jiaxiang Liu
Jie Liu
Jinduo Liu

Jingsong Liu
Jingxin Liu
Jinhua Liu
Jinyang Liu
Juan Liu
Jun Liu
Junyi Liu
Kang Liu
Meng Liu
Mengting Liu
Mianxin Liu
Mingxin Liu
Mingxuan Liu
Mingyuan Liu
Peng Liu
Shengyuan Liu
Shuting Liu
Sidong Liu
Tiange Liu
Ting Liu
Tingwei Liu
Wei Liu
Wenjie Liu
Xianyuan Liu
Xinyu Liu
Yan Liu
Yang Liu
Yanhong Liu
Yanzhen Liu
Yi Liu
Yifan Liu
Yihao Liu
Yikang Liu
Yue Liu
Yunyi Liu
Yusheng Liu
Yuxiao Liu
Zelong Liu
Zeyu Liu
Zhentao Liu
Zhi Liu
Ziteng Liu
Zixuan Liu
Lisette Lockhart
Andrea Loddo
Herve Lombaert
Nicolas Loménie
Yonghao Long
Giovanni Lonia
Miguel López-Pérez
Marco Lorenzi
Ange Lou
Meng Lou
Wei Lou
Chenggang Lu
Chun-Shien Lu
Guangming Lu
Jiaxuan Lu
Jingfeng Lu
Mengkang Lu
Shengnan Lu
Shuai Lu
Xing Lu
Yao Lu
Yu Lu
Yuhuan Lu
Ziru Lu
Xiao Luan
Ruben Lucassen
Miguel Luna
Guibo Luo
Haozhe Luo
Huoling Luo
Jiaqi Luo
Ma Luo
Mingyuan Luo
Xiangde Luo
Xiaoling Luo
Xinzhe Luo
Yong Luo
Zhiming Luo
Ngoc Ly
Ilwoo Lyu
Mengye Lyu
Qing Lyu
Yiheng Lyu
Deeksha M. Shama
Benteng Ma
Chenglong Ma
Da Ma
DongAo Ma
Hehuan Ma
Jiquan Ma
Jun Ma
Lei Ma
Ming Ma
Qiang Ma
Xinke Ma
Yingliang Ma
Yuanzheng Ma
Yuhui Ma
Zhuoqi Ma
Diego Machado Reyes
Pedro Macias Gordaliza
Anderson Maciel
Derek Magee
Dwarikanath Mahapatra
S. Sara Mahdavi
Ahsan Mahmood
Mohammed Mahmoud
Huayu Mai
Jennifer Maier
Klaus H. Maier-Hein
Lena Maier-Hein
Abrar Majeedi
Roman Malashin
Hashmat Shadab Malik
Mayur Mallya
Norberto Malpica
Constantina Maltezou-Papastylianou
Jan Emily Mangulabnan
Arooba Maqsood
Radu Marculescu
Jan Margeta
Umar Marikkar
Zdravko Marinov
Iman Marivani
Pablo Márquez Neila
Carsten Marr
Robert Marti
Nicolas Martin
Josué Martínez-Martínez
Queralt Martín-Saladich
Najmeh Mashhadi

Martin Maška
Tejas Sudharshan Mathai
Sasan Matinfar
Shinnosuke Matsuo
Hiroki Matsuzaki
Thomas Paul Matthews
Petr Matula
Satyajit Maurya
Dimitrios Mavroeidis
Voichita Maxim
Arnaldo Mayer
Perla Mayo
Evangelos Mazomenos
Badhan Mazumder
Ashery Mbilinyi
Jamie McClelland
Julian McGinnis
Pau Medrano-Gracia
Mostafa Mehdipour Ghazi
Deval Mehta
Raghav Mehta
Liye Mei
Felix Meister
Alaa Melek
Steve Mendoza
Mingyuan Meng
Runqi Meng
Yan Meng
Vivek Menon
Martin Menten
Talha Meraj
Dorit Merhof
Odyssée Merveille
Adrien Meyer
Leo Milecki
Wenwen Min
Zhe Min
Duy Minh Ho Nguyen
Golrokh Mirzaei
Nahal Mirzaie
Virendra Mishra
Ashish Mittal
Tadashi Miyamoto
Pooya Mobadersany
Sovesh Mohapatra
Ramtin Mojtahedi
Tony C. W. Mok
Javier A. Montoya-Zegarra
Mehdi Moradi
José Morano
Kensaku Mori
Shohei Mori
Elena Morotti
Lia Morra
Luca Mossina
Ana M. Mota
Lei Mou
Harold Mouchère
Kaouther Mouheb
Seyed Amir Mousavi
Daniel Moyer
Pritam Mukherjee
Anirban Mukhopadhyay
Johanna Müller
Philip Müller
Sarah Müller
Asbjørn Munk
Aditya Murali
Vittorio Murino
Saif Musluh
Yasmine Mustafa
Inye Na
Shahabedin Nabavi
Sheida Nabavi
Roman Naeem
Manas Nag
Enamundram Naga Karthik
Noor Nakhaei
Sahil Nalawade
Jakub Nalepa
Khashayar Namdar
Yang Nan
Nandakishor Nandakishor
Ahmed Nebli
Fnu Neha
Debanga Raj Neog
Favour Nerrise
Gia Ngo
Binh Nguyen
Hang Nguyen
Huy Nguyen
Khoa Nguyen
Tan Nguyen
Phi Nguyễn Văn
Duong Nhu
Dong Ni
Haomiao Ni
Brennan Nichyporuk
Hannes Nickisch
Xinyu Nie
Joshua Niemeijer
Aditya Nigam
Lipeng Ning
Xiaolin Ning
Hareem Nisar
Kazuya Nishimura
Sijie Niu
Usma Niyaz
Fuad Noman
Mehrdad Noori
Chinedu Nwoye
Emeka Obi
Gilberto Ochoa-Ruiz
Marzieh Oghbaie
Baris Oguz
Hyun-Jic Oh
Esther Okafor
Ajibola Oladokun
Laura Oliva Maza
Arlindo Oliveira
Bruno Miguel Gomes Oliveira
Antoine Olivier
Juan Olmos
Jimena Olveres
Yu Hon On
Doruk Oner
John Onofrey
Hiroko Oshibe
Andreas Østvik
Chubin Ou
Yafei Ou
Muhammad Owais
Utku Ozbulak

Caner Ozer
Ege Özsoy
Kutsev Ozyoruk
Martina Paccini
Nicolas Padoy
José Blas Pagador Carrasco
Suraj Pai
Daniel Pak
Ashkan Pakzad
Basudha Pal
Christoph Palm
Delin Pan
Hongyi Pan
Jiazhen Pan
Liangrui Pan
Wentao Pan
Xiang Pan
Yimu Pan
Zhifang Pan
Fengqian Pang
Haowen Pang
Jiaxuan Pang
Yan Pang
Hsing-Kuo Pao
Tasos Papastylianou
Bartlomiej Papiez
Jay Paranjape
Abhijeet Parida
Hyunjin Park
Joon Sik Park
Robin Park
Sang Hyun Park
Sung-Hong Park
Yongkeun Park
Mario Parreno-Centeno
Srivathsa Pasumarthi
Ashay Patel
Sushant Patkar
Mayank Patwari
Angshuman Paul
Riti Paul
Rasmus Paulsen
Thierry Pécot
João Pedrosa
Paraskevas Pegios
Jialun Pei
Yuchen Pei
Yuru Pei
Himashi Peiris
Chantal Pellegrini
Danni Peng
Linkai Peng
Yige Peng
Yunsong Peng
Zelin Peng
Matteo Pennisi
Tobias Penzkofer
Jacob Peoples
Sergio Pereira
Alejandra Perez
Fernando Pérez-García
Matthias Perkonigg
Denis Peruzzo
Eike Petersen
Jens Petersen
Simon Pezold
Elisabeth Pfaehler
Ashish Phophalia
Thomas Pinetz
Sergey Plis
Sergei Pnev
Gasper Podobnik
Kilian Pohl
Isabella Poles
Dirk H. J. Poot
Mirela Popa
Peter Pott
Alison Pouch
Pranav Poudel
Chinmay Prabhakar
Ferran Prados
Manish Prajapati
Rishav Pramanik
Raphael Prevost
Federica Proietto Salanitri
Sergey Protserov
Lemuel Puglisi
Sergi Pujades
Behrus Puladi
Lorenzo Putzu
Elodie Puybareau
Kehan Qi
Liangce Qi
Lin Qi
Yaolei Qi
Buyue Qian
Yan Qiang
Mengyun Qiao
Yuchuan Qiao
Fangbo Qin
Jiahao Qin
Yiming Qin
Zengyi Qin
Huaqi Qiu
Jie Qiu
Jingna Qiu
Peijie Qiu
Shihan Qiu
Linhao Qu
Tingwei Quan
Sandro Queiros
Sandro Queirós
Rizwan Qureshi
Prashanth R.
Daniel Racoceanu
Mehdi Rahim
Md Masudur Rahman
Md Mostafijur Rahman
Ankita Raj
Ajitha Rajan
Jagath Rajapakse
João Ramalhinho
Sirisha Rambhatla
Jayroop Ramesh
Xuming Ran
Amin Ranem
Bryan Ranger
Aneesh Rangnekar
Carolina Raposo
Hatem Rashwan
Sebastian Rassmann
Aditya Rastogi
Daniele Ravì
Sadhana Ravikumar

Vikram Ravindra
Sudarshan Regmi
Simon Reiß
Samuel Remedios
Qin Ren
Zhihang Ren
Mauricio Reyes
Abel Reyes-Angulo
Hadrien Reynaud
Helena Ribeiro Torres
Daniel Riccio
Anne-Marie Rickmann
Bastian Rieck
Brandalyn Riedel
Tammy Riklin Raviv
Morgan Ringel
Harry Robertshaw
Lucas Robinet
Alfie Roddan
Jessica Rodgers
Ranga Rodrigo
Javier Rodriguez-Puigvert
Robert Rohling
Melanie Roschewitz
Pooneh Roshanitabrizi
Florence Rossant
Alberto Rota
Ayush Roy
Kaushiki Roy
Saikat Roy
Dan Ruan
Jiacheng Ruan
Daniel Rueckert
Daniela Ruiz
Dewinda Rumala
Josef Lorenz Rumberger
Mirabela Rusu
Patryk Rygiel
Kanghyun Ryu
Muralikrishna S. N.
Amirhossein Sabour
Maziar Sabouri
Ario Sadafi
Numan Saeed
Farhang Sahba
Heba Sailem
Hideo Saito
Simone Saitta
Gulfam Saju
Md Sirajus Salekin
Mostafa Salem
Benjamin Salmon
Thomas Sanchez
Yudi Sang
Sara Sangalli
Sivaramakrishnan Sankarapandian
Alice Santilli
Bikash Santra
Nishchal Sapkota
Duygu Sarikaya
Gary Sarwin
Olivier Saut
Ricky Savjani
Cian Scannell
Yoni Schirris
Alexander Schlaefer
Maja Schlereth
Adam Schmidt
Linda-Sophie Schneider
Steffen Schneider
Daniel Scholz
Thomas Schultz
Michael Schwier
Michaël Sdika
Eric Seibel
Matthias Seibold
Jothiraj Selvaraj
Sourya Sengupta
Thilo Sentker
Hyewon Seo
Sumin Seo
Nazmus Shakib Shadin
Ahmed Shaffie
Hina Shah
Keyur Devendra Shah
Nisarg Shah
Nagur Shareef Shaik
S. Shailja
Fereshteh Shakeri
Kun Shang
Wei Shao
Maxim Sharaev
Mostafa Sharifzadeh
Saurav Sharma
Vanshali Sharma
Vatsala Sharma
Bingzhi Shen
Hualei Shen
Linlin Shen
Xiongri Shen
Yiqing Shen
Zhenrong Shen
Zhiqiang Shen
Victor S. Sheng
Enze Shi
Fashun Shi
Ji Shi
Jian Shi
Jun Shi
Junjie Shi
Kangdi Shi
Lulin Shi
Pengcheng Shi
Ruohua Shi
Xiaoshuang Shi
Yaying Shi
Yi Shi
Yinghuan Shi
Yonggang Shi
Zhan Shi
Shu-Fu Shih
Kaito Shiku
Jitae Shin
Yejee Shin
Isaac Shiri
Boris Shirokikh
Hongchao Shu
Yucheng Shu
Zhongyi Shui
Amita Shukla-Dave
Weixin Si
Vasiliki Sideri-Lampretsa
Alberto Signoroni
Monica Sigovan

Saurabh Sihag
Wilson Silva
Julio Silva-Rodríguez
Margarida Silveira
Marco Simoes
Ivor Simpson
Malay Singh
Vivek Singh
Abhishek Singh Sambyal
Rohit Singla
Ashish Sinha
Ayushi Sinha
Vasilis Siomos
Raviteja Sista
Elena Sizikova
Luka Škrlj
Gregory Slabaugh
Paddy Slator
Dane Smith
Kevin Smith
Tiffany So
Roger Soberanis-Mukul
Elanchezhian Somasundaram
Young-Han Son
Sifan Song
Tao Song
Xinrui Song
Tananun Songdechakraiwut
Kristine Sørensen
Moein Sorkhei
Aristeidis Sotiras
Mazen Soufi
Mayssa Soussia
Raissa Souza
Roberto Souza
Georgia Sovatzidi
Michal Španěl
Rachel Sparks
Tyler Spears
William Speier
Ziga Spiclin
Jon Sporring
Arvind Krishna Sridhar
Lawrence Staib
Marc Stamminger
Emma Stanley
Johannes Stegmaier
Jan Steinbrener
Nil Stolt-Ansó
Skylar Stolte
Robin Strand
Moritz Sturm
Martin Styner
Weifeng Su
Yao Su
Zhe Su
Julian Suk
Emese Sükei
Chao Sun
Hongfu Sun
Jianqi Sun
Kaicong Sun
Liyan Sun
Poly Z. H. Sun
Susu Sun
Xiaowu Sun
Xiaoxiao Sun
Yan Sun
Yihua Sun
Yiran Sun
Zhaobin Sun
Zhe Sun
Zhichao Sun
Vaanathi Sundaresan
Wonmo Sung
Mohammod Naimul Islam Suvon
Michele Svanera
Numi Sveinsson Cepero
Tomasz Szczepański
Behrad Taghibeyglou
Pablo Tahoces
Divyanshu Tak
Shumpei Takezaki
Estefania Talavera
Jeremy Tan
Jing Wei Tan
Shan Tan
Tao Tan
Wenjun Tan
Zhiwei Tan
Feilong Tang
Fenghe Tang
Haoteng Tang
He Tang
Jiankai Tang
Xiaoqin Tang
YuCheng Tang
Zhenyu Tang
Zhi-Ri Tang
Zihao Tang
Qian Tao
Tianli Tao
Austin Tapp
Elias Tappeiner
Mickael Tardy
Giacomo Tarroni
Wallapak Tavanapong
Thomas Tavolara
Jacopo Teneggi
Lin Teng
Paul Terrassin
Kaveri Thakoor
Franz Thaler
Antoine Theberge
Belvin Thomas
Sarina Thomas
Elinor Thompson
Stephen Thompson
Felix Thomsen
Alexander Thorley
Jacob Thrasher
Xin Tian
Xuanyu Tian
Yapeng Tian
Yun Tian
Matthew Tivnan
Minh Nguyen Nhat To
Santiago Toledo-Cortés
Lyuyang Tong
Marwan Torki
Nicolas Toussaint
Henry Tregidgo

Jakob Troidl
Yen Nhi Truong Vu
Apostolia Tsirikoglou
Mingxiao Tu
Oezguen Turgut
Aayush Tyagi
Georgios Tziritas
Kwang-Hyun Uhm
Mathias Unberath
Balagopal Unnikrishnan
Hristina Uzunova
Sami Väänänen
Valentina Vadori
Theodoros Vagenas
Anurag Vaidya
Maria Vakalopoulou
Vanya Valindria
Juan Miguel Valverde
Fons van der Sommen
Louis van Harten
Tom van Sonsbeek
Gijs van Tulder
Bino Varghese
Francisco Vasconcelos
Hooman Vaseli
Rajat Vashistha
S. Swaroop Vedula
Matthew Velazquez
Vikram Venkatraghavan
Mitko Veta
Vibujithan Vigneshwaran
Veronica Vilaplana
Pierre-Frederic Villard
Thanaporn Viriyasaranon
Christiaan Viviers
Hung Vo
Wolf-Dieter Vogl
Emiliano Votta
Hung Vu
Milos Vukadinovic
Trinh Thi Le Vuong
Christian Wachinger
Martin Wagner
Jia Wan
Liang Wan
Shouhong Wan
Xiaohua Wan
An Wang
Bin Wang
Changmiao Wang
Chen Wang
Cheng Wang
Chenhui Wang
Chong Wang
Congcong Wang
Diwei Wang
Dongang Wang
Fan Wang
Fu Wang
Fuying Wang
Haifeng Wang
Haolin Wang
Haoran Wang
Hongxiao Wang
Hongyi Wang
Hongyu Wang
Hongzhi Wang
Hualiang Wang
Huayu Wang
Janet Wang
Jiacheng Wang
Jiamu Wang
Jian Wang
Jiaqi Wang
Jiazhen Wang
Jinfeng Wang
Jixin Wang
Jiyao Wang
Jueqi Wang
Jui-Kai Wang
Jun Wang
Junchen Wang
Lehan Wang
Lei Wang
Lin Wang
Linwei Wang
Lyuyang Wang
Manning Wang
Meng Wang
Nan Wang
Peiqi Wang
Qi Wang
Qiuli Wang
Rongfang Wang
Rui Wang
Ruixuan Wang
Runmin Wang
Sen Wang
Shiyi Wang
Shuxian Wang
Tao Wang
Tianyu Wang
Weiming Wang
Xi Wang
Xiaofei Wang
Xin Wang
Xingce Wang
Xinliang Wang
Xu Wang
Yan Wang
Yi Wang
Yifan Wang
Yifeng Wang
Yinsong Wang
Yipei Wang
Yiqing Wang
Yixin Wang
Yiyang Wang
Yiying Wang
Yonghao Wang
Yuheng Wang
Yulin Wang
Yun Wang
Yuxi Wang
Zeyi Wang
Zhaoyang Wang
Zhewei Wang
Zhong Wang
Zhu Wang
Zichen Wang
Ziyi Wang
Ziyue Wang
Zuhui Wang
Valentine Wargnier Dauchelle

Mohammad Wasih
Dong Wei
Guangshun Wei
Jia Wei
Jie Wei
Meng Wei
Ruofeng Wei
Shuwen Wei
Wen Wei
Jonas Weidner
Wolfgang Wein
Oliver Weinheimer
Xuyun Wen
Zheyu Wen
Thomas Wendler
Nina Weng
Wenhai Weng
Markus Wenzel
Julia Werner
Bryan Williams
Brady Williamson
Paul Wilson
Alexander Winkler
Eric Wisotzky
Adam Wittek
Marek Wodzinski
Ivo Wolf
Georg Wölflein
Julia Wolleb
Dong-Ok Won
Ka-Chun Wong
Ken C. L. Wong
McKell Woodland
Stefan Wörz
Binghong Wu
Chengchuan Wu
Chenwei Wu
Fuping Wu
Fuzhi Wu
Ji Wu
Jianghao Wu
Jie Ying Wu
Junde Wu
Junyang Wu
Linshan Wu
Mengqi Wu
Qian Wu
Qianyu Wu
Qing Wu
Ruoyou Wu
Weiyi Wu
Wenbin Wu
Wenxuan Wu
Xian Wu
Xiao Wu
Xinyao Wu
Yihang Wu
Yinhao Wu
Yinzhe Wu
Yunheng Wu
Zhan Wu
Zhengwang Wu
Zhiying Wu
Zihao Wu
Zijian Wu
Julian Wyatt
Magdalena Wysocki
Pengcheng Xi
Chao Xia
Jing Xia
Jun Xia
Peng Xia
Tian Xia
Weihao Xia
Yihao Xia
Jinhai Xiang
Lei Xiang
Nan Xiang
Bin Xiao
Deqiang Xiao
Li Xiao
Qing Xiao
Sa Xiao
Xiaojiao Xiao
Bin Xie
Chi Xie
Hao Xie
Jianyang Xie
Kangxian Xie
Kunzi Xie
Long Xie
Qihang Xie
Xingyu Xie
Yiting Xie
Bowen Xin
Fangxu Xing
Jiarui Xing
Shuwei Xing
Zhaohu Xing
Conghao Xiong
Honglin Xiong
Juming Xiong
Yiheng Xiong
Yujian Xiong
Chi Xu
Di Xu
Dongrong Xu
Fangqiang Xu
Feng Xu
Gelei Xu
Geng-Xin Xu
Hongming Xu
Hongyan Xu
Huihui Xu
Jialang Xu
Jiangchang Xu
Jiaqi Xu
Jiaxing Xu
Jilan Xu
Meilong Xu
Min Xu
Moucheng Xu
Rui Xu
Senyan Xu
Weixin Xu
Wentian Xu
Xiao Xu
Xiaowei Xu
Xiuyuan Xu
Yongchao Xu
Yunqiu Xu
Zhoubing Xu
Ziang Xu
Zikang Xu
Ziyue Xu

Zunnan Xu
Cheng Xue
Jie Xue
Yuyang Xue
Zhong Xue
Fang Yan
Ke Yan
Kun Yan
Rui Yan
Siyuan Yan
Wen Yan
Xiangyi Yan
Yige Yan
Yuguang Yan
Yunlu Yan
Zipei Yan
Baoyao Yang
Carl Yang
Changchun Yang
Chen Yang
Gang Yang
Guanyu Yang
Heran Yang
Huijuan Yang
Jianhua Yang
Jiaqi Yang
Jie Yang
Junlin Yang
Lin Yang
Peng Yang
Qianye Yang
Sejung Yang
Shu Yang
Shuojue Yang
Su Yang
Xiaochen Yang
Xiaofei Yang
Xikai Yang
Xinyi Yang
Xuan Yang
Yan Yang
Yang Yang
Yijun Yang
Yingyu Yang
Zefan Yang
Zhicheng Yang
Zhikai Yang
Zhiwei Yang
Zhiwen Yang
Zixin Yang
Ziyuan Yang
Jiachen Yao
Linlin Yao
Tianyuan Yao
Xiaohui Yao
Xing Yao
Yongcheng Yao
Zhao Yao
Choon Hwai Yap
Ziya Ata Yazıcı
Chenfei Ye
Jian Ye
Jiayu Ye
Johan Ye
Kai Ye
Shuchang Ye
Xujiong Ye
Yiwen Ye
Chun-Hsiao Yeh
Srikar Yellapragada
Si Yong Yeo
Joe Yeong
Pak Hei Yeung
Murong Yi
Weixi Yi
Yugen Yi
Hanlong Yin
Yunting Yin
Zijin Yin
Haochao Ying
Wenhao Ying
Tatsuya Yokota
Jaejun Yoo
Seok Bong Yoo
Youngjin Yoo
Jee Seok Yoon
Jihun Yoon
Kyungho Yoon
Siyeop Yoon
Xin You
Alexandra Young
Boxiao Yu
Fanyang Yu
Hanyi Yu
Jianxun Yu
Jinhua Yu
Minhui Yu
Qinji Yu
Renping Yu
Sigang Yu
Weihao Yu
Weimin Yu
Yang Yu
Zhen Yu
Ziqi Yu
Cheng Yuan
Kun Yuan
Mingze Yuan
Xiaohan Yuan
Hailin Yue
Jiaxin Yue
Xiaodong Yue
Yutao Yue
Paul Yushkevich
Anna Zapaishchykova
Masoud Zarepisheh
Kamilia Zaripova
Tal Zeevi
Rachid Zeghlache
Ramy Zeineldin
Bolun Zeng
Dewen Zeng
Dong Zeng
Hongye Zeng
Qingjie Zeng
Tianyi Zeng
Yuwen Zeng
Hasnae Zerouaoui
Oliver Zettinig
Ruyi Zha
Baochang Zhang
Bodong Zhang
Chenxi Zhang
Chuyan Zhang
Daoqiang Zhang

Dong Zhang
Fan Zhang
Guoning Zhang
Hanxiao Zhang
Hanyun Zhang
Haochen Zhang
Haoyue Zhang
Hongliang Zhang
Jianhai Zhang
Jianjia Zhang
Jianping Zhang
Jiansong Zhang
Jianwei Zhang
Jiaqi Zhang
Jiaqing Zhang
Jiazhen Zhang
Jingyang Zhang
Jinwei Zhang
Kaiyu Zhang
Ke Zhang
Le Zhang
Lei Zhang
Li Zhang
Lu Zhang
Meng ze Zhang
Miaomiao Zhang
Minghui Zhang
Minqing Zhang
Molin Zhang
Pingping Zhang
Rongzhao Zhang
Ruisi Zhang
Shu Zhang
Shuai Zhang
Shuaitong Zhang
Song Zhang
Tian Tian Zhang
Tianjing Zhang
Tianyi Zhang
Tuo Zhang
Wei Zhang
Weihang Zhang
Weiyi Zhang
Wenhua Zhang
Xiaoming Zhang
Xiaoran Zhang
Xiaoshan Zhang
Xiaotong Zhang
Xin Zhang
Xuhong Zhang
Xukun Zhang
Xuzhe Zhang
Yanfu Zhang
Yanteng Zhang
Yi Zhang
Yichi Zhang
Yike Zhang
Yilan Zhang
Yizhe Zhang
You Zhang
Yuan Zhang
Yuanwang Zhang
Yue Zhang
Yuhan Zhang
Yundi Zhang
Yundong Zhang
Yuqi Zhang
Yuxin Zhang
Zeyu Zhang
Zheng Zhang
Zhenxuan Zhang
Zheyu Zhang
Zhicheng Zhang
An Zhao
Baoliang Zhao
Beidi Zhao
Chen Zhao
Cheng Zhao
Chongyue Zhao
Haoyu Zhao
He Zhao
Jichao Zhao
Jun Zhao
Kun Zhao
Mingyang Zhao
Puyang Zhao
Qingyu Zhao
Shen Zhao
Shensheng Zhao
Weibing Zhao
Weiqin Zhao
Weiren Zhao
Wenzhao Zhao
Xiangyu Zhao
Xingyue Zhao
Xinkai Zhao
Yidong Zhao
Yizhou Zhao
Yue Zhao
Zhihao Zhao
Zihao Zhao
Ziheng Zhao
Ziwei Zhao
Zixu Zhao
Xiantong Zhen
Boyun Zheng
Fudan Zheng
Hao Zheng
Haoxin Zheng
Qiaoyu Zheng
Qingqing Zheng
Shaoming Zheng
Shuai Zheng
Tianshu Zheng
Weihao Zheng
Yalin Zheng
Yefeng Zheng
Yijiang Zheng
Yushan Zheng
Zhou Zheng
Jiayang Zhong
Lanfeng Zhong
Yi Zhong
Yuan Zhong
Binggui Zhou
Congcong Zhou
Guangquan Zhou
Haoyin Zhou
Jun Zhou
Li Zhou
Meng Zhou
Qian Zhou
Rong Zhou
Shuo Zhou
Sijin Zhou

Wei Zhou
Wu Zhou
Xiaogen Zhou
Xinrui Zhou
Yaxuan Zhou
Yi Zhou
Yuan Zhou
Yuanpin Zhou
Yukun Zhou
Yuyue Zhou
Zhen Zhou
Zhengbo Zhou
Zongwei Zhou
Hancan Zhu
Jiayuan Zhu
Junchao Zhu
Lei Zhu
Lingting Zhu
Meilu Zhu
Qikui Zhu
Yanjie Zhu
Yazhou Zhu
Yitao Zhu
Yuemin Zhu
Zhanshi Zhu
Zhenyang Zhu
Jia-Xin Zhuang
Jun Zhuang
Yupeng Zhuo
Arif Hassan Zidan
Alexander Ziller
Veronika Zimmer
David Zimmerer
Fangrong Zong
Jing Zou
Xinrui Zou
Jian Zu
Wenqiang Zu
Maria A. Zuluaga
Lianrui Zuo
Gerald Zwettler

Contents – Part I

Multimodal Fusion and Contextual Reasoning in Medical Imaging

A Composite Alignment-Aware Framework for Myocardial Lesion Segmentation in Multi-sequence CMR Images

Yifan Gao[1,2], Shaohao Rui[2,3,5], Haoyang Su[2,4,5], Jinyi Xiang[6], Lianming Wu[6], and Xiaosong Wang[2,5](✉)

[1] School of Biomedical Engineering (Suzhou), Division of Life Science and Medicine, University of Science and Technology of China, Hefei, China
240107100014@sii.edu.cn

[2] Shanghai Innovation Institute, Shanghai, China
wangxiaosong@pjlab.org.cn

[3] Shanghai Jiao Tong University, Shanghai, China

[4] Fudan University, Shanghai, China

[5] Shanghai Artificial Intelligence Laboratory, Shanghai, China

[6] Department of Radiology, Renji Hospital, Shanghai Jiao Tong University, Shanghai, China

Abstract. Accurate segmentation of myocardial lesions from multi-sequence cardiac magnetic resonance imaging is essential for cardiac disease diagnosis and treatment planning. However, achieving optimal feature correspondence is challenging due to intensity variations across modalities and spatial misalignment caused by inconsistent slice acquisition protocols. We propose CAA-Seg, a composite alignment-aware framework that addresses these challenges through a two-stage approach. First, we introduce a selective slice alignment method that dynamically identifies and aligns anatomically corresponding slice pairs while excluding mismatched sections, ensuring reliable spatial correspondence between sequences. Second, we develop a hierarchical alignment network that processes multi-sequence features at different semantic levels, i.e., local deformation correction modules address geometric variations in low-level features, while global semantic fusion blocks enable semantic fusion at high levels where intensity discrepancies diminish. We validate our method on a large-scale dataset comprising 397 patients. Experimental results show that our proposed CAA-Seg achieves superior performance on most evaluation metrics, with particularly strong results in myocardial infarction segmentation, representing a substantial 5.54% improvement over state-of-the-art approaches. The code is available at https://github.com/yifangao112/CAA-Seg.

Keywords: Multi-sequence CMR · Myocardial lesion segmentation · Alignment

J. C. Gee et al. (Eds.): MICCAI 2025, LNCS 15960, pp. 3–13, 2026.
https://doi.org/10.1007/978-3-032-04927-8_1

1 Introduction

Cardiovascular diseases remain a leading global health burden, with myocardial pathologies such as infarction and edema requiring precise quantification for effective clinical management [1,2]. While manual segmentation of these pathological regions is time-consuming and subject to significant inter-observer variability [3,4], automatic segmentation presents substantial challenges due to the small size and subtle appearance of infarction regions. Multi-sequence cardiac magnetic resonance (CMR) imaging has emerged as a cornerstone for comprehensive myocardial assessment, where late gadolinium enhancement (LGE) delineates lesion with high contrast [5], while complementary T1 and T2 mapping sequences provide additional pathological indicators for both infarction and edema [6,7].

Leveraging these imaging sequences offers promising opportunities to enhance segmentation performance by combining distinct pathological markers from each modality [8,9]. However, the integration of these complementary sequences faces two significant technical barriers, as shown in Fig. 1. First, the substantial difference in slice sampling between sequences creates severe anatomical mismatches. LGE typically employs 6–8 slices to capture localized pathologies, whereas T1/T2 mapping sequences use only 2–3 slices for rapid quantification. When these slices are acquired at different anatomical positions, direct registration becomes anatomically inconsistent and clinically unreliable (Fig. 1b.2). Second, attempting to resolve this mismatch through aggressive resampling introduces significant artifacts. Resampling T1/T2 mapping from 3 to 8 slices forces interpolation across large gaps, creating artificial structural continuity and distorting the very pathological features we aim to segment (Fig. 1b.1). Thus, the challenge lies in establishing both inter-sequence semantic correspondence and intra-sequence spatial consistency across these heterogeneously sampled data streams.

Existing methods struggle to address these challenges effectively. Conventional approaches predominantly focus on single-modality analysis [10–12], particularly LGE, sacrificing complementary pathological information from other sequences. While recent multi-modal fusion attempts [13–16] demonstrate promise on small datasets (25 patients) [17] with pre-aligned inputs, such idealized conditions rarely reflect the misalignment prevalent in routine clinical acquisitions. This discrepancy forces networks to process anatomically inconsistent features, particularly in mid-ventricular regions where interpolation artifacts dominate.

To address these challenges, we propose CAA-Seg, a composite alignment-aware framework for myocardial lesion segmentation in multi-sequence CMR images. The framework consists of two sequential stages: selective slice alignment (SSA) and hierarchical feature fusion. First, we propose a selective slice alignment method that handles *resampling distortion* and *anatomical mismatch* through dynamic slice matching. Instead of forcing alignment between all slices, SSA employs sliding-window optimization to identify anatomically corresponding slice pairs while excluding mismatched regions. This generates aligned multi-sequence regions for lesion fusion and single-sequence regions for stan-

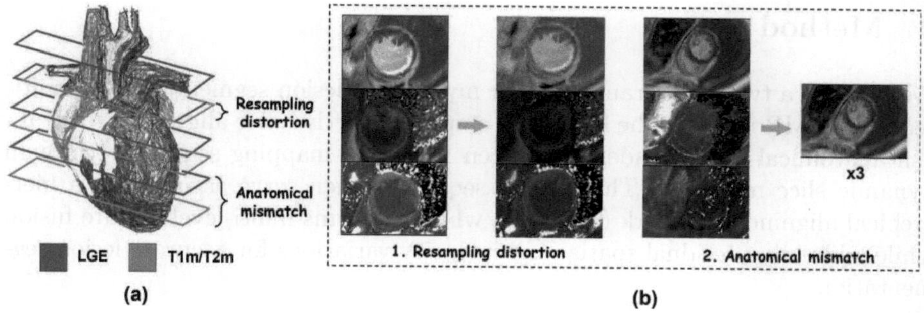

Fig. 1. (a) Visualization of misalignment challenges in multi-sequence CMR. (b.1) **Resampling distortion:** Aggressive resampling of T1/T2 mapping (3 to 8 slices) introduces interpolation artifacts and structural distortion. (b.2) **Anatomical mismatch:** LGE and T1/T2 mapping slices were acquired at different anatomical positions, causing misalignment. These challenges motivate our selective slice alignment strategy that preserves only reliable correspondences while avoiding forced matching of incompatible slices. Our slice-level alignment significantly reduces resampling-induced misregistration. Additionally, we use only duplicated LGE slices to avoid anatomical mismatch.

dard segmentation, which are processed through task-specific channels. Second, we develop a hierarchical alignment network (HA-Net) that addresses multi-sequence feature fusion at different semantic levels. At low-level features where local geometric variations dominate, we employ deformable convolutions and pixel-wise modulation to correct spatial misalignments. High-level features, which encode pathological patterns with reduced intensity variations, are aligned through cascaded cross-attention modules to capture semantic relationships between LGE and mapping sequences. A task-aware controller at the bottleneck level adaptively modulates features based on input characteristics, ensuring effective fusion of multi-sequence information.

Our comprehensive validation on a large-scale CMR dataset demonstrates significant improvements over state-of-the-art methods. CAA-Seg achieves 78.09% Dice score for myocardium segmentation and 65.49% for myocardial edema delineation. Notably, for myocardial infarction segmentation, our method achieves 51.11% Dice score, outperforming the second-best approach by a substantial margin of 5.54%. The main technical contributions are:

1. We propose a novel framework for multi-sequence CMR segmentation that effectively integrates heterogeneous cardiac imaging sequences through composite alignment.
2. We propose a selective slice alignment approach that establishes reliable inter-sequence correspondence through dynamic slice matching and mutual information optimization.
3. We develop a hierarchical alignment network that combines local deformation correction and global semantic fusion for effective multi-sequence feature fusion.

2 Method

We propose a two-stage framework for myocardial lesion segmentation in multi-sequence CMR images. The first stage employs selective slice alignment to establish anatomical correspondence between LGE and mapping sequences through dynamic slice matching. The aligned sequences then serve as input to a hierarchical alignment network (HA-Net), which performs multi-level feature fusion while addressing residual spatial and intensity variations for accurate lesion segmentation.

2.1 Selective Slice Alignment

Multi-sequence CMR acquisition results in varying slice counts and positions across sequences, making direct registration between LGE and mapping sequences challenging. Rather than forcing alignment between all slices, which can introduce artifacts and misleading information, we propose the SSA approach that dynamically identifies and aligns only the most reliable slice pairs.

Given a series of 2D slices, let $I_f^k \in \mathbb{R}^{H\times W}$ denote the k-th fixed LGE slice and $I_m^j \in \mathbb{R}^{H\times W}$ represent the j-th moving slice from either T1m or T2m sequence. We formulate the slice-wise registration as an optimization problem:

$$\phi_k^* = \arg\min_{\phi_k} \underbrace{\mathcal{L}_{MMI}(I_f^k, I_m^j \circ \phi_k)}_{\text{Multi-sequence similarity}} + \lambda \underbrace{\mathcal{R}(\phi_k)}_{\text{Regularization}}, \quad k \in [1, K] \tag{1}$$

where K is the total number of slices. To ensure robust multi-modal alignment, we adopt Mattes mutual information (MMI) as the similarity metric:

$$\mathcal{L}_{MMI}(I_f^k, I_m^j \circ \phi_k) = -\sum_{x,y} p_k(x,y) \log\left(\frac{p_k(x,y)}{p_f^k(x)p_m^j(y)}\right) \tag{2}$$

where $p_k(x,y)$ represents the joint intensity distribution. Our transformation model accommodates both global and local deformations through a hierarchical composition:

$$\phi_k = \phi_{k,rigid} \circ \phi_{k,affine} \circ \phi_{k,diff} \tag{3}$$

The key innovation of our approach lies in its selective slice matching strategy. Instead of a fixed window, we employ an adaptive search to identify the most plausible anatomical correspondence for each slice. First, we compute a similarity score for each potential slice pair:

$$S(k,j) = MMI(I_{LGE}^k, \phi_k(I_{T1m}^j)) + MMI(I_{LGE}^k, \phi_k(I_{T2m}^j)) \tag{4}$$

For the k-th LGE slice, we then identify the best-matching moving slice, j_k^*, by searching within a dynamically defined range. This range is constrained by

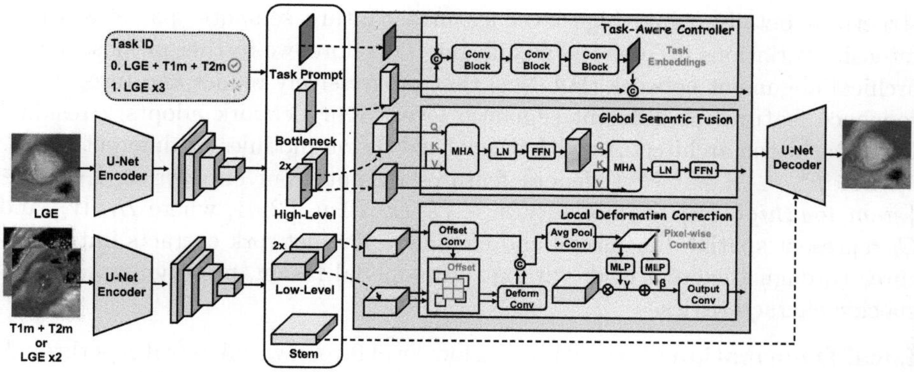

Fig. 2. Overview of the proposed HA-Net. The framework consists of dual encoders for multi-sequence CMR processing, with specialized alignment modules at different feature levels. Local deformation correction modules address spatial misalignment in low-level features, while cascaded cross-attention blocks enable semantic fusion at high-level features. A task-aware controller at the bottleneck adaptively modulates features based on different input images.

the position of the previously matched slice (j_{k-1}) to maintain sequential order and ensure enough subsequent slices are available for future matches:

$$j_k^* = \underset{j \in [j_{k-1}, N-M+k]}{\arg\min} \quad S(k, j) \tag{5}$$

where j_{k-1} is the matching position of the previous slice, N is the total number of moving image slices, and M is the remaining number of slices from the current position k to the end of the LGE sequence. This ensures that:

$$\begin{cases} j_k^* > j_{k-1} & \text{maintain sequence order} \\ j_k^* \leq N - M + k & \text{reserve sufficient slices} \end{cases} \tag{6}$$

This selective mechanism naturally handles through-plane motion while maintaining anatomical consistency within each 2D plane. The final registered volumes are reconstructed only from the reliably matched and transformed slice pairs:

$$V_{reg} = \{I_m^{j_k^*} \circ \phi_k^*\}_{k=1}^{K} \tag{7}$$

This method enables our framework to focus on high-quality alignments while avoiding error propagation from forced matching of incompatible slices (Fig. 2).

2.2 Hierarchical Alignment Network

While SSA provides reliable inter-sequence alignment, minor local deformations and intensity variations may persist at the feature level. The low-level features primarily encode local tissue textures where geometric misalignments affect structural details, while high-level features capture semantic patterns where intensity variations become less significant. Therefore, we further propose a hierarchical alignment network (HA-Net) that progressively refines and fuses multi-sequence features at different semantic levels. Our network adopts a residual UNet backbone architecture with specialized fusion modules at different scales.

Let $F_m^l \in \mathbb{R}^{H_l \times W_l \times C_l}$ denote features output from encoder at level $l \in \{stem, low, high, bn\}$ for modality $m \in \{LGE, T1m, T2m\}$, where H_l, W_l, and C_l represent spatial dimensions and channels. The network extracts initial features through the stem layer without cross-modal fusion, preserving modality-specific characteristics.

Local Deformation Correction: At low-level layers, we align features through deformable convolution and pixel-wise modulation. For each modality pair, the offset field Δp and aligned features $F_{aligned}$ are computed as:

$$\begin{aligned} \Delta p &= f_{offset}(F_{LGE}^{low}) \\ F_{aligned} &= \text{DeformConv}(F_{T1m/T2m}^{low}, \Delta p) \end{aligned} \tag{8}$$

The features are then modulated through learned parameters:

$$\begin{aligned} &= f_{mod}(\text{GlobalPool}([F_{LGE}^{low}, F_{aligned}])) \\ F_{out}^{low} &= \gamma \cdot F_{LGE}^{low} + \beta \end{aligned} \tag{9}$$

where f_{mod} maps the global context to modulation parameters γ and β for pixel-wise feature adjustment.

Global Semantic Fusion: High-level features undergo cascaded cross-attention for semantic alignment. The process consists of two sequential attention blocks:

$$\begin{aligned} F_1^{high} &= \text{Attn}(Q = F_{LGE}^{high}, K, V = F_{T1m}^{high}) \\ F_{out}^{high} &= \text{Attn}(Q = F_1^{high}, K, V = F_{T2m}^{high}) \end{aligned} \tag{10}$$

where the attention operation follows the standard scaled dot-product mechanism with multi-head projection.

Task-Aware Controller: Inspired by [18], we incorporate a task-aware controller design that adaptively processes features based on different inputs. At the bottleneck, we integrate task information through a learned prompt P_{task}:

$$F_{out}^{bn} = \text{MLP}(\text{LayerNorm}([F_{LGE}^{bn}; P_{task}])) \tag{11}$$

where $[\cdot;\cdot]$ represents channel-wise concatenation. This controller adapts the feature processing based on the input characteristics from different modalities.

3 Experiments and Results

Dataset and Experimental Setup: We evaluate our method on an IRB-approved, large-scale in-house dataset comprising 397 patients with confirmed

Table 1. Quantitative evaluation results of different methods in myocardial lesion segmentation dataset. The best and second best results are highlighted in bold and underlined respectively. Myo: myocardium, ME: myocardial edema, MI: myocardial infarction.

Methods	Myo		ME		MI		Overall	
	Dice (%)	HD95 (pixel)	Dice (%)	HD95 (pixel)	Dice (%)	HD95 (pixel)	Dice (%)	HD95 (pixel)
nnU-Net [19]	76.46	10.45	62.68	9.52	45.30	12.78	61.48	10.92
UNet++ [20]	77.51	9.84	65.03	9.77	44.26	14.03	62.27	11.21
TransUNet [21]	77.05	9.87	63.00	11.08	43.25	13.21	61.10	11.39
Swin UNETR [22]	75.23	11.31	61.39	11.21	35.31	14.94	57.30	12.49
UTNet [23]	76.99	10.25	62.16	10.15	42.06	14.23	60.40	11.54
UMamba [24]	77.58	10.56	64.93	**9.23**$^{\sim}$	45.57	13.11	62.69	10.97
MFU-Net [13]	74.84	10.89	59.22	12.53	36.68	14.34	56.91	12.59
AWSNet [14]	76.50	10.71	63.06	9.72	40.00	14.71	59.85	11.71
MyoPS-Net [15]	77.20	9.88	63.90	9.55	44.15	14.56	61.75	11.33
CAA-Seg (ours)	**78.09**$^{\sim}$	**9.00**‡	**65.49**‡	9.25	**51.11**‡	**11.43**‡	**64.89**‡	**9.89**‡

$\sim$: p-value > 0.05, $\ddagger$: p-value < 0.05, in comparison between the best and second-best results.

myocardial pathologies, randomly split into training (278 cases), validation (39 cases), and testing (80 cases) sets. Among these, all cases included annotations for myocardium (Myo) and myocardial edema (ME), while 208 cases presented with myocardial infarction (MI).

We use seven encoder stages with feature dimensions of {32, 64, 128, 256, 512, 512, 512} channels. The deformable convolution modules use 3×3 kernels with 18 sampling points. Input images are resampled and standardized to 384×384 pixels via crop or pad. The features naturally separate into low-level (64–256 channels) and high-level (512 channels) representations. The cross-attention blocks employ 8 attention heads with a hidden dimension of 256. The task controller generates a 128-dimensional task embedding that is projected to match feature dimensions.

Our comparisons include general medical segmentation frameworks (nnU-Net [19], UNet++ [20], TransUNet [21], Swin UNETR [22], UTNet [23], UMamba [24]) and cardiac-specific approaches (MFU-Net [13], AWSNet [14], MyoPS-Net [15]). General frameworks were evaluated using only LGE images due to performance degradation with multi-sequence input, while cardiac-specific approaches utilized multi-sequence data with their dedicated fusion mechanisms. For fair comparison, all methods follow identical preprocessing and postprocessing protocols as nnU-Net, and are trained for 200 epochs with an initial learning rate of 0.01 using cosine decay scheduler and batch size of 16 on an NVIDIA RTX 4090 GPU. The network was trained using a combination of Dice loss and Cross-Entropy loss, consistent with the standard nnU-Net framework. The results are assessed using Dice coefficient (%) and 95% Hausdorff Distance (HD95, in pixels) across three targets: Myo, ME, and MI. Statistical significance is assessed using the Wilcoxon signed rank test between our method and other approaches.

Results: The experimental results comparing different methods for myocardial pathology segmentation are presented in Table 1. Our CAA-Seg achieves consis-

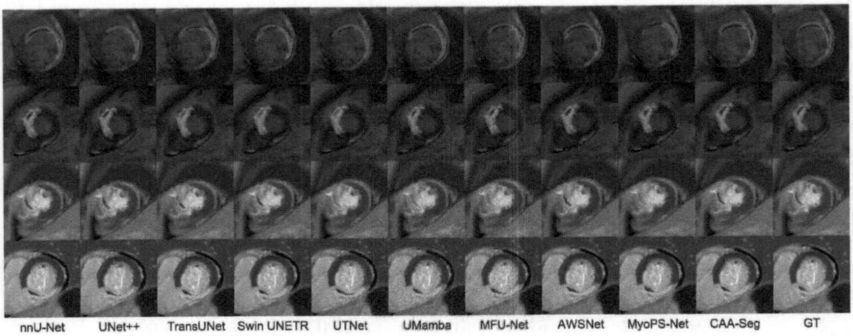

Fig. 3. Visual comparison of myocardial lesion segmentation results across different methods with myocardium shown in red, myocardial edema in green, and myocardial infarction in blue. (Color figure online)

Table 2. Ablation study of CAA-Seg, analyzing the impact of different registration methods, network architectures, and input sequences on segmentation performance.

Settings	Registration			Network		Input		Overall	
	Baseline	MvMM [25]	SSA	nnU-Net	HA-Net	LGE	Multi	Dice (%)	HD95 (pixel)
1	✓			✓			✓	59.63	12.17
2	✓				✓		✓	60.12	11.92
3		✓			✓		✓	<u>62.98</u>	<u>10.83</u>
4			✓		✓	✓		61.51	11.85
5 (proposed)			✓		✓		✓	**64.89**	**9.89**

tent improvements across all metrics, with notable gains in the challenging MI segmentation task where it reaches a Dice score of 51.11% ($p<0.05$ improvement over the second-best result). The performance advantage is particularly evident in overall metrics, where CAA-Seg obtains a mean Dice of 64.89% and HD of 9.89 pixels, representing statistically significant improvements over previous methods. Interestingly, while UMamba shows competitive performance in Myo and ME segmentation, its effectiveness diminishes in MI detection, highlighting the advantages of our alignment-aware design. In terms of computational efficiency, CAA-Seg remains competitive. The average inference time per case was 0.71 s with 1.8 GB of GPU memory, compared to nnU-Net (0.52 s, 1.1 GB) and UMamba (0.62 s, 1.3 GB). While slightly more intensive due to the explicit alignment, our framework offers substantial performance gains for a modest increase in computational cost.

The qualitative results visualized in Fig. 3 further demonstrate how our method effectively handles complex cases where traditional approaches struggle with multi-sequence alignment and feature fusion. CAA-Seg achieves more accurate target localization, especially in regions with minor pathological changes that are typically challenging to identify.

Ablation Study: We further conduct ablation experiments to validate the effectiveness of each component in our framework. The results in Table 2 demonstrate that our SSA strategy combined with the HA-Net achieves the best performance, significantly outperforming both traditional registration methods and single-modality approaches. Notably, while the multivariate mixture model (MvMM) [25] registration shows competitive results, the integration of SSA with multi-sequence input further improves performance by handling spatial misalignment more effectively.

4 Discussion and Conclusion

In this work, we presented CAA-Seg, a composite alignment-aware framework for myocardial lesion segmentation in multi-sequence CMR images. Our method effectively handles the heterogeneous nature of multi-sequence cardiac imaging data, where misalignments stem from both acquisition protocols and tissue characteristics. The selective slice alignment strategy successfully establishes reliable correspondence between LGE and mapping sequences, while the hierarchical alignment network enables effective feature fusion across modalities with varying intensity distributions and spatial resolutions. Experimental validation on a large-scale dataset of 397 cases demonstrates significant performance improvements across all lesion regions.

We acknowledge that the performance of CAA-Seg may degrade in cases with extreme inter-sequence misalignment beyond the capture range of our alignment module, or for very small and indistinct lesions that are challenging even for expert analysis. Future work could focus on extending the framework to handle more complex clinical scenarios and incorporating temporal information from cine sequences. The composite alignment strategy demonstrated in this work shows promise for application in other multi-sequence medical imaging tasks that require cross-modality feature fusion.

Disclosure of Interests. The authors have no competing interests.

References

1. Vaduganathan, M., Mensah, G.A., Turco, J.V., Fuster, V., Roth, G.A.: The global burden of cardiovascular diseases and risk: a compass for future health (2022)
2. Rui, S., Su, H., Xiang, J., Wu, L.-M., Wang, X.: CardioCoT: hierarchical reasoning for multimodal survival analysis. arXiv preprint arXiv:2505.19195 (2025)
3. Gao, Y., Xia, W., Dingdu, H., Wang, W., Gao, X.: DeSAM: decoupled segment anything model for generalizable medical image segmentation. In: Linguraru, M.G., et al. (eds.) MICCAI 2024. LNCS, vol. 15012, pp. 509–519. Springer, Cham (2024). https://doi.org/10.1007/978-3-031-72390-2_48
4. Gao, Y., Xia, W., Wang, W., Gao, X.: MBA-net: sam-driven bidirectional aggregation network for ovarian tumor segmentation. In: Linguraru, M.G., et al. (eds.) MICCAI 2024. LNCS, vol. 15011, pp. 437–447. Springer, Cham (2024)
5. Unger, A., et al.: Prognostic value of cardiac MRI late gadolinium enhancement granularity in participants with ischemic cardiomyopathy. Radiology **314**(1), e240806 (2025)

6. Liu, X., et al.: Cardiac magnetic resonance t1 mapping for evaluating myocardial fibrosis in patients with type 2 diabetes mellitus: correlation with left ventricular longitudinal diastolic dysfunction. Eur. Radiol. **32**(11), 7647–7656 (2022)
7. Warnica, W.: Clinical impact of cardiac MRI t1 and t2 parametric mapping in patients with suspected cardiomyopathy. Radiology **305**(2), 319–326 (2022)
8. Gao, Y., Dai, Y., Liu, F., Chen, W., Shi, L.: An anatomy-aware framework for automatic segmentation of parotid tumor from multimodal MRI. Comput. Biol. Med. **161**, 107000 (2023)
9. Dai, Y., Gao, Y., Liu, F.: Transmed: transformers advance multi-modal medical image classification. Diagnostics **11**(8), 1384 (2021)
10. Yue, Q., Luo, X., Ye, Q., Xu, L., Zhuang, X.: Cardiac segmentation from LGE MRI using deep neural network incorporating shape and spatial priors. In: Shen, D., et al. (eds.) MICCAI 2019. LNCS, vol. 11765, pp. 559–567. Springer, Cham (2019). https://doi.org/10.1007/978-3-030-32245-8_62
11. Xinhua, Yu., Chen, J., Fang, B., Wang, W., Zhang, L., Lv, Z.: Cardiac LGE MRI segmentation with cross-modality image augmentation and improved u-net. IEEE J. Biomed. Health Inform. **27**(2), 588–597 (2021)
12. Li, L., Zimmer, V.A., Schnabel, J.A., Zhuang, X.: Medical image analysis on left atrial LGE MRI for atrial fibrillation studies: a review. Med. Image Anal. **77**, 102360 (2022)
13. Jiang, H., Wang, C., Chartsias, A., Tsaftaris, S.A.: Max-fusion U-net for multi-modal pathology segmentation with attention and dynamic resampling. In: Zhuang, X., Li, L. (eds.) MyoPS 2020. LNCS, vol. 12554, pp. 68–81. Springer, Cham (2020). https://doi.org/10.1007/978-3-030-65651-5_7
14. Wang, K.-N., et al.: AWSNet: an auto-weighted supervision attention network for myocardial scar and edema segmentation in multi-sequence cardiac magnetic resonance images. Med. Image Anal. **77**, 102362 (2022)
15. Qiu, J., et al.: MYOPS-net: myocardial pathology segmentation with flexible combination of multi-sequence CMR images. Med. Image Anal. **84**, 102694 (2023)
16. Xu, X., Chen, J., Thakur, D., Hong, D.: Multi-modal disease segmentation with continual learning and adaptive decision fusion. Inf. Fus. 102962 (2025)
17. Li, L., et al.: MYOPS: a benchmark of myocardial pathology segmentation combining three-sequence cardiac magnetic resonance images. Med. Image Anal. **87**, 102808 (2023)
18. Ye, Y., Xie, Y., Zhang, J., Chen, Z., Xia, Y.: UniSeg: a prompt-driven universal segmentation model as well as a strong representation learner. In: Greenspan, H., et al. (eds.) MICCAI 2023. LNCS, vol. 14222, pp. 508–518. Springer, Cham (2023). https://doi.org/10.1007/978-3-031-43898-1_49
19. Isensee, F., Jaeger, P.F., Kohl, S.A.A., Petersen, J., Maier-Hein, K.H.: NNU-Net: a self-configuring method for deep learning-based biomedical image segmentation. Nat. Methods **18**(2), 203–211 (2021)
20. Zhou, Z., Rahman Siddiquee, M.M., Tajbakhsh, N., Liang, J.: UNet++: a nested u-net architecture for medical image segmentation. In: Stoyanov, D., et al. (eds.) DLMIA/ML-CDS -2018. LNCS, vol. 11045, pp. 3–11. Springer, Cham (2018). https://doi.org/10.1007/978-3-030-00889-5_1
21. Chen, J., et al.: TransuNet: rethinking the U-Net architecture design for medical image segmentation through the lens of transformers. Med. Image Anal. **97**, 103280 (2024)
22. Hatamizadeh, A., Nath, V., Tang, Y., Yang, D., Roth, H.R., Xu, D.: Swin unetr: Swin transformers for semantic segmentation of brain tumors in MRI images. In:

Crimi, A., Bakas, S. (eds.) BrainLes 2021. LNCS, vol. 12962, pp. 272–284. Springer, Cham (2021). https://doi.org/10.1007/978-3-031-08999-2_22

23. Gao, Y., Zhou, M., Metaxas, D.N.: UTNet: a hybrid transformer architecture for medical image segmentation. In: de Bruijne, M., et al. (eds.) MICCAI 2021. LNCS, vol. 12903, pp. 61–71. Springer, Cham (2021). https://doi.org/10.1007/978-3-030-87199-4_6
24. Ma, J., Li, F., Wang, B.: U-mamba: enhancing long-range dependency for biomedical image segmentation. arXiv preprint arXiv:2401.04722 (2024)
25. Zhuang, X.: Multivariate mixture model for myocardial segmentation combining multi-source images. IEEE Trans. Pattern Anal. Mach. Intell. **41**(12), 2933–2946 (2018)

A Curvature-Guided Diffeomorphic Mesh Deformation Framework for Lifespan Brain Cortical Surface Reconstruction

Lin Teng[1], Shen Zhao[2], Feng Shi[2(✉)], and Dinggang Shen[1,2,3(✉)]

[1] School of Biomedical Engineering and State Key Laboratory of Advanced Medical Materials and Devices, ShanghaiTech University, Shanghai 201210, China

[2] Department of Research and Development, United Imaging Intelligence, Shanghai 200230, China

feng.shi@uii-ai.com

[3] Shanghai Clinical Research and Trial Center, Shanghai 201210, China

dgshen@shanghaitech.edu.cn

Abstract. Accurate and automatic lifespan brain cortical surface reconstruction (CSR) is crucial for analyzing brain development and aging. Traditional pipelines involve multiple processing steps, which are time-intensive and inefficient for handling larger datasets. While deep learning-based methods can accelerate reconstruction speed and produce high-quality meshes compared to traditional approaches, they are often constrained to a single time point. The limitation arises from the significant variations in cortical surfaces across age groups, particularly in folding patterns. In this paper, we propose a novel curvature-guided diffeomorphic mesh deformation framework for lifespan brain CSR. Specifically, to preserve correct topology structure and uniformity, the framework employs multiple deformation blocks to gradually warp a simple smooth template mesh to a complex target surface with high folding. Considering that curvature is closely associated with folding patterns, we introduce curvature map prediction as an auxiliary task to guide the deformation process, enhancing the anatomical accuracy to facilitate subsequent cortical morphometry. Notably, incorporating curvature can also expedite model convergence. Our method is evaluated on a large-scale brain dataset with **2,132** subjects spanning ages 0 to 100 years. Experimental results show that our reconstructed surfaces have fewer geometric errors and optimal mesh regularity while being several orders of magnitude faster than traditional pipelines. Our code is available at https://github.com/TL9792/CCF.

Keywords: Lifespan · 3D cortical surface reconstruction · Curvature-guided diffeomorphic mesh deformation · Deep learning

J. C. Gee et al. (Eds.): MICCAI 2025, LNCS 15960, pp. 14–23, 2026.
https://doi.org/10.1007/978-3-032-04927-8_2

1 Introduction

The human brain undergoes dramatic changes in size, shape, and tissue intensity distribution throughout the lifespan[1], particularly during the first postnatal year [1]. These changes are especially prominent in cortical surfaces[2] where cortical shape and thickness vary with age. Compared to volumetric data, the cortical surface provides more detailed geometric representations of the brain, enabling the visualization of cortical structures and supporting various downstream analyses by calculating cortical properties (*e.g.*, cortical curvature and thickness) [23]. Consequently, accurate brain cortical surface reconstruction (CSR) from volumetric data is crucial for understanding brain development, aging, and associated neurodegenerative disease diagnosis.

Traditional surface analysis packages designed for a single time point, such as Infant FreeSurfer [24], FreeSurfer [8], and FastSurfer [10], typically involve multiple time-consuming steps. This makes it challenging to efficiently analyze large datasets. To address this, deep learning-based CSR approaches have emerged, and these can be broadly categorized into implicit and explicit methods depending on the 3D shape representation they operate on. Implicit methods leverage deep neural network (DNN) to learn an implicit surface representation, such as an occupancy field and a signed distance function (SDF) [4,7,8,10,16]. For example, Cruz *et al.* [4] proposed DeepCSR, a model that predicts implicit surface representations for points and subsequently applies an isosurface extraction method to reconstruct adult brain cortical surfaces. However, these implicit methods often fail to ensure the quality of reconstructed meshes with topology correctness. For explicit methods [13,14,18,22], a DNN is trained in an end-to-end manner to directly generate an explicit mesh by deforming an initial mesh into a target mesh. For example, Ma *et al.* [14] proposed a PialNN to generate outer surfaces from inner surfaces as input through multiple deformation blocks. Similarly, Lebrat *et al.* [18] proposed CorticalFlow^{++}, which employs diffeomorphic mesh deformation (DMD) modules to learn a series of stationary velocity fields (SVF) defining diffeomorphic deformations from an initial template to inner and outer surfaces. These explicit methods have gained attention for their capability to preserve topological accuracy while achieving exceptional computational efficiency. However, they often ignore the deformation of key regions with high folding, making it challenging to reconstruct brain cortical surfaces with significant variations across large-scale age groups.

Here, we propose a curvature-guided diffeomorphic mesh deformation framework for lifespan brain CSR. Specifically, we build upon CorticalFlow^{++} as our baseline due to its efficiency in preserving topology details. Beyond CorticalFlow^{++}, we use brain tissue segmentation maps predicted from our

[1] **Lifespan** refers to the duration of an organism's life, from birth to old age.
[2] **Brain cortical surface**, consists of inner/white and outer/pial cerebral cortex in each hemisphere. The inner surface refers to the boundary between white matter (WM) and gray matter (GM), and the outer surface is the interface between GM and cerebrospinal fluid (CSF) [5].

proposed segmentation model [19,20] rather than brain MR images as input. This allows the model to focus directly on interfaces between different tissues, thereby alleviating the partial volume effect (PVE) problem commonly encountered in brain MR images. Moreover, we introduce curvature as prior knowledge to guide the diffeomorphic deformation of high-folding regions through a multi-task learning strategy. This enhances reconstruction accuracy, particularly in regions with rich anatomical information, such as sulci and gyri. Notably, introducing curvature not only improve accuracy but also accelerate convergence without increasing GPU consumption. The proposed method is evaluated on a large-scale lifespan brain dataset, and experimental results demonstrate that our mesh outperforms other state-of-the-art (SOTA) methods in terms of geometry and regularity.

2 Methodology

Figure 1 illustrates our proposed framework for lifespan brain CSR, which consists of a series of curvature-guided deformation blocks that are trained sequentially (detailed in Sect. 2.1). Each block comprises one UNet-like architecture [3] (detailed in Sect. 2.3) and a DMD module [13]. The UNet-like model predicts a curvature-aware flow field that estimates per-vertex displacement vectors on the current state of deformation and a curvature map. The DMD module computes a diffeomorphic mapping to transform an initial template mesh into the target mesh, such as inner or outer surfaces, shown in Sect. 2.2.

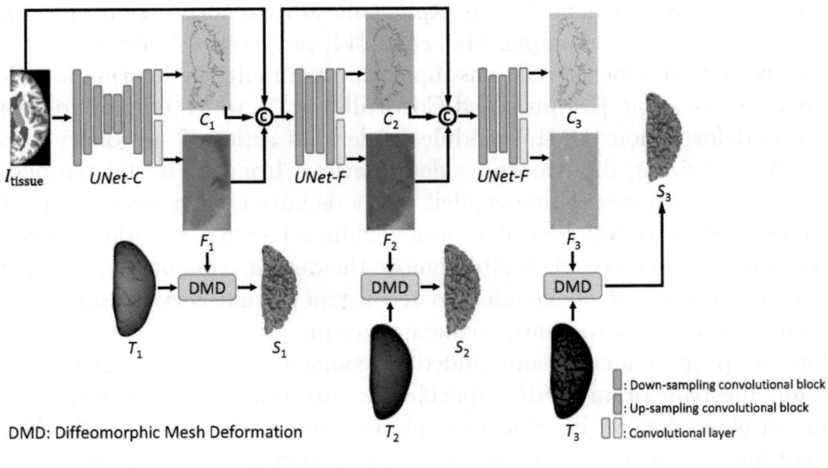

Fig. 1. An overview of our proposed curvature-guided diffeomorphic mesh deformation framework for lifespan brain CSR. The framework consists of multiple curvature-guided deformation blocks to sequentially deform an initial template mesh to the target mesh through a coarse-to-fine learning strategy (*i.e.*, inner or outer surfaces).

2.1 Overview

To mitigate the challenges associated with large deformations from a smooth template mesh to a complex brain cortical surface, we adopt a coarse-to-fine learning strategy. This approach decomposes a substantial deformation into multiple simpler deformations, enhancing both accuracy and efficiency. The framework includes three sequential deformation stages for reconstructing inner and outer surfaces of each brain hemisphere. Specifically, in the initial stage, a 3D brain tissue segmentation map (I_{tissue}) with dimension $\mathrm{H} \times \mathrm{W} \times \mathrm{D} \times 1$ is fed into the first deformation model (UNet-C), and the output includes a curvature-aware flow field (F_1) of dimension $\mathrm{H} \times \mathrm{W} \times \mathrm{D} \times 3$ and a curvature map (C_1) of dimension $\mathrm{H} \times \mathrm{W} \times \mathrm{D} \times 1$ (detailed in Sect. 2.2). Subsequently, F_1 and a smooth template mesh comprising 40k vertices (T_1) are processed through a DMD module to produce a coarse cortical surface (S_1). In the second stage, the concatenation of I_{tissue}, F_1, and C_1 serve as input to the second deformation model UNet-F, and the output includes a curvature-aware flow field (F_2) and curvature map (C_2). Both F_2 and T_2 with 140k vertices are fed into DMD module to yield a refined cortical surface (S_2). The third deformation stage mirrors the second, differing only in the use of a template mesh with 380k vertices (T_3). This hierarchical approach ensures a progressive refinement of the cortical surface, effectively capturing intricate anatomical details. The total deformation equation is as follows:

$$\begin{aligned} F_1, C_1 &= \text{UNet-C}(I_{tissue}), \\ S_1 &= \text{DMD}\left(F_1, T_1\right), \\ F_{i+1}, C_{i+1} &= \text{UNet-F}_{i+1}\left(F_1^\frown C_1 \cdots F_i C_i I_{tissue}\right), \quad \text{for } i \geq 1. \\ S_{i+1} &= \text{DMD}\left(F_{i+1}, T_{i+1}\right), \end{aligned} \tag{1}$$

where F_i, C_i, and S_i are i-th predicted curvature-aware flow field, i-th predicted curvature map, and i-th reconstructed surface, respectively. i is the number of deformation iteration, *i.e.*, $i = 3$. $\frown$ represents the concatenation operator.

2.2 Curvature-Guided Diffeomorphic Mesh Deformation

Cortical surfaces exhibit significant variations across different age groups, particularly in detailed sulcal and gyral patterns. These variations pose challenges in accurately aligning a template with each individual's unique cortical structure while preserving the fine details of the surface topology. To solve this issue, we introduce curvature as prior knowledge to guide the deformation process. Curvature effectively quantifies the folding patterns of cortical surfaces, capturing both local variations in surface shape and the overall structure of the brain's gyral and sulcal regions.

Curvature Map Generation. Due to the lack of a one-to-one correspondence between the vertices of the template mesh and the pseudo-ground truth, we generate the curvature map within volumetric space rather than directly on the surface. Initially, we compute the mean curvature value at each vertex of the pseudo-ground-truth cortical surface using the PyVista library. To align them with brain MR image, these vertex-specific curvature values are then mapped into 3D volumetric space by correlating vertex coordinates to voxel positions. The process yields a curvature map as C with dimension of $\mathrm{H} \times \mathrm{W} \times \mathrm{D}$, where H, W, and D are the length, width, and depth of brain MR image, respectively. Finally, the curvature values are normalized to a range between -1 and 1, where negative values represent sulcal regions while positive values are gyral regions.

Multi-Task Learning. To effectively utilize curvature to guide the deformation of critical regions, we adopt a multi-task learning strategy. Each deformation model is designed with two branches: one for predicting a curvature-aware flow field and the other for predicting a curvature map. By using curvature map prediction as an auxiliary task, the model becomes more anatomy-aware, enabling it to better capture features in high-folded regions. This approach enhances the model's ability to reconstruct cortical surfaces with high accuracy, particularly in regions with complex anatomical structures.

Curvature-Aware Diffeomorphic Deformation. To accurately deform a template mesh to the target mesh, the predicted curvature-aware flow field (F) and a template mesh (T) are fed into a DMD module, computing a diffeomorphic mapping ϕ as an ordinary differential equation (ODE) for each vertex position $x \in \mathbb{R}^3$, to obtain a new mesh that is closer to the target mesh. Formally, the procedure of solving the flow ODE

$$\frac{\partial \phi_i\left(s; x_{i,k}\right)}{\partial s} = U_i\left(\phi_i\left(s; x_{i,k}\right)\right), \phi_i\left(0; x_{i,k}\right) = x_{i-1,k}, \tag{2}$$

using the fourth-order Runge-Kutta (RK4) with 30 time-steps [17]. ϕ_i, U_i, and $x_{i,k}$ are the i-th diffeomorphic mapping, the i-th curvature-aware flow field, and the k-th vertex of i-th surface, respectively. The condition is that $x_{i,k}$ is the same as $x_{i-1,k}$ when s=0. Formulating a deformation as an ODE significantly reduces the occurrence of self-intersecting faces, thereby more accurately approximating the target surface.

2.3 Network Architecture

The framework comprises three sub-networks with a 3D UNet-shape architecture: UNet-C and two UNet-F. UNet-C consists of four down-/up-sampling convolutional blocks and two $3 \times 3 \times 3$ convolutional layers, and UNet-F has two down/up-sampling convolutional blocks and two convolutional layers. Down-sampling convolutional block includes a $3 \times 3 \times 3$ convolutional layer and a LeakyReLU activation layer. Up-sampling convolutional block includes a $3 \times 3 \times 3$ convolutional layer, a LeakyReLU activation layer, and a transposed convolutional layer. The rationale behind this architectural change is to facilitate the learning of coarse-to-fine deformations [13].

2.4 Loss Function

The training process is supervised by a composite loss function comprising Chamfer distance loss L_{cd} [22], edge length loss L_{edge} [22], and curvature loss $L_{curvature}$, each weighted by hyperparameters λ_1, λ_2, and λ_3, respectively.

$$L(S_p, S_g, C_p, C_g) = \lambda_1 L_{cd}(S_p, S_g) + \lambda_2 L_{edge}(S_p) + \lambda_3 L_{curvature}(C_p, C_g), \quad (3)$$

where L_{cd} and L_{edge} are computed on 150k points from uniformly sampling at random on the predicted (S_p) and pseudo-ground-truth surfaces (S_g). $L_{curvature}$ calculates the mean absolute error between the predicted (C_p) and ground truth curvature map (C_g). λ_1, λ_2, and λ_3 are set to 1.0, 1.0, and 0.001 based on the magnitude of loss values, respectively.

3 Experiments and Results

3.1 Datasets and Metrics

We have compiled a comprehensive lifespan brain dataset encompassing individuals aged 0 to 100 years, comprising **2,132** T1-weighted (T1w) MR images sourced from 18 repositories. These include 341 subjects (0–6 years) from Baby Connectome Project (BCP) [11], 30 subjects (9–11 years) from Adolescent Brain Cognitive Development (ABCD) [2], 295 subjects (18–81 years) from CBMFM [9], 22 subjects (18–94 years) from Open Access Series of Imaging Studies (OASIS) [15], 129 subjects (56–96 years) from Alzheimer's Disease Neuroimaging Initiative (ADNI) [12], 69 subjects (6–56 years) from Autism Brain Imaging Data Exchange (ABIDE) [6], and 875 subjects (2–95 years) from private hospitals. Each subject has a T1w MR image, a tissue segmentation map generated by our previously developed segmentation model mentioned above, and pseudo-ground-truth surfaces derived using Infant FreeSurfer [24] and FreeSurfer pipelines with different smoothing parameters set, based on our tissue segmentation maps. The T1w MR image is preprocessed with skull stripping, intensity normalization, and bias field correction [21]. All volume data are standardized to a resolution of $193 \times 229 \times 193$ with voxel spacing of $1.0 \times 1.0 \times 1.0$ mm^3. Each dataset is randomly partitioned into training, validation, and testing in a ratio of 8:1:1 for each age group [1]. Performance is evaluated using Chamfer distance (CD), 95th-percentile Hausdorff distance (HD), and normal consistency (NC) calculated for point clouds of 200k points with uniform sampling.

3.2 Implementation Details

Experiments are implemented on one NVIDIA A100 GPU with 40GB memory. Each deformation model is trained with 70K iterations for inner surface reconstruction. After that, the estimated inner surfaces serve as the initial template mesh to reconstruct outer surfaces, in which each deformation model is trained with 70K iterations while freezing the parameters of the white model. This ensures a one-to-one mapping between the vertices of the inner and outer surfaces, facilitating brain analysis. An Adam optimizer with an initial learning rate of e^{-4} is used. Batch size is 3.

Table 1. Quantitative comparison of brain cortical surfaces from different SOTA methods in terms of CD, HD, and NC (↑ and ↓ denote larger and smaller values, respectively. \ is no result. The best results are in **bold**).

Method	Left Inner Surface			Right Inner Surface		
	CD (mm) ↓	HD (mm) ↓	NC ↑	CD (mm) ↓	HD (mm) ↓	NC ↑
PialNN	\	\	\	\	\	\
DeepCSR	4.98±21.49	1.71±2.80	0.88±0.06	3.00±10.77	1.45±1.98	0.89±0.06
CorticalFlow^{++}	1.31±1.22	1.15±0.47	0.88±0.06	1.02±0.93	1.01±0.39	0.89±0.05
Ours	**0.46±0.22**	**0.69±0.10**	**0.95±0.01**	**0.53±0.18**	**0.76±0.10**	**0.94±0.01**
	Left Outer Surface			Right Outer Surface		
	CD (mm) ↓	HD (mm) ↓	NC ↑	CD (mm) ↓	HD (mm) ↓	NC ↑
PialNN	1.36±0.54	1.27±0.29	0.87±0.02	1.26±0.44	1.21±0.24	0.87±0.02
DeepCSR	4.91±4.19	3.00±1.44	0.84±0.04	5.37±4.56	3.18±1.41	0.84±0.04
CorticalFlow^{++}	1.25±1.44	1.06±0.56	0.87±0.05	1.10±1.16	1.04±0.52	0.91±0.04
Ours	**0.47±0.18**	**0.68±0.09**	**0.94±0.01**	**0.51±0.09**	**0.77±0.06**	**0.95±0.01**

3.3 Results

Comparison with SOTA Methods. We evaluated our method against several representative SOTA methods in brain CSR, including PialNN, DeepCSR, and CorticalFlow^{++}. To ensure a fair comparison, we reproduced all methods using their official code and applied the same dataset and partition strategy.

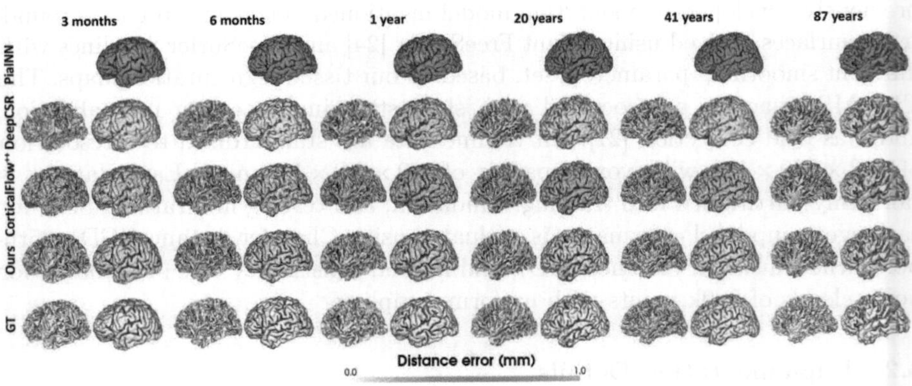

Fig. 2. Reconstructed left inner and outer surfaces across different age groups from different SOTA methods, which are color-coded with distance errors relative to pseudo-ground-truth surfaces.

Table 1 shows the quantitative comparison of brain CSR from different SOTA methods in terms of CD, HD, and NC. As we can see, our method outperforms other SOTA methods on all metrics. For example, PialNN shows relatively strong performance in outer surface reconstruction, although it is not suitable for reconstructing inner surfaces. DeepCSR exhibits the poorest performance in all metrics, likely due to its implicit representation learning approach that does not ensure topological correctness. In contrast, our method achieves significant improvements over the baseline (*i.e.*, CorticalFlow^{++}), with reductions of 0.78 mm in CD, 0.38 mm in HD, and 0.11 in NC for left outer surface reconstruction, respectively. Figure 2 visualizes predicted left cortical surfaces across multiple age groups, color-coded to indicate the distance to pseudo-ground-truth surfaces from different SOTA methods. Our surfaces have smaller geometric errors than other SOTA methods, as evidenced by the distance errors overlaid on the surfaces. For example, reconstructed surfaces from CorticalFlow^{++} exhibit significant errors in highly folded regions (*e.g.*, sulci and gyri). Besides, our method is more robust than other methods. Compared to DeepCSR and PialNN, our reconstructed surfaces across multiple age groups are more consistent.

Ablation Study. We conduct an ablation study to assess the impact of using the tissue segmentation map as input and adding the curvature prediction branch on the performance of lifespan brain CSR.

Tissue Segmentation Map as Input. We conduct two experiments based on CorticalFlow^{++} using either brain MR image or tissue segmentation map as input to verify the benefit of incorporating tissue-specific information, respectively. Experimental results show that CorticalFlow^{++}+I_{tissue} achieves CD, HD, and NC of 0.63/0.67 mm, 0.83/0.85 mm, and 0.90/0.90 for left inner/outer surface reconstruction, and 0.72/0.73 mm, 0.86/0.86 mm, and 0.91/0.93 for right inner/outer surface reconstruction, respectively. Compared to CorticalFlow^{++}, CorticalFlow^{++}+I_{tissue} brings improvements of 68% in CD and 32% in HD for left inner surface reconstruction. This demonstrates that tissue maps provide clearer boundary information, effectively mitigating the PVE effects present in brain MR images.

Curvature Prediction Branch. Building upon CorticalFlow^{++}+I_{tissue}, we introduce a curvature prediction branch using a multi-task learning strategy. Compared to CorticalFlow^{++}+I_{tissue}, curvature prior brings improvements across all metrics, including 18%/21% (CD) and 12%/13% (HD) for inner/outer surface reconstruction. This highlights that curvature prediction provides additional anatomical features for deformation models, leading to enhanced reconstruction accuracy and structural fidelity. Furthermore, incorporating curvature reduces the convergence time from 3.5 days to 3 days.

4 Conclusion

In this paper, we have presented a novel curvature-guided flow-based approach for efficient lifespan brain CSR. To preserve accurate topology and anatomical properties for cortical surfaces with large variations across multiple age groups, the framework consists of several curvature-guided deformation blocks that gradually transform a smooth template mesh to the inner surface and subsequently into the outer surface using a coarse-to-fine strategy. Experiments conducted on extensive lifespan brain datasets demonstrate that our method achieves more precise mesh regularity and better preservation of cortical folding patterns compared to other SOTA methods. Therefore, this method will be integrated into our developed lifespan brain analysis pipeline to support clinical studies of brain development and aging.

Acknowledgments. This work was supported in part by National Natural Science Foundation of China (82441023, U23A20295, 62131015, 82394432), National Key Technologies R&D Program of China (82027808), the China Ministry of Science and Technology (S20240085, STI2030-Major Projects-2022ZD0209000, STI2030-Major Projects-2022ZD0213100), Shanghai Municipal Central Guided Local Science and Technology Development Fund (No. YDZX20233100001001), The Key R&D Program of Guangdong Province, China (2023B0303040001), and HPC Platform of ShanghaiTech University. This research was supported in part by the National Key Research and Development Program of China (2022YFE0205700).

Disclosure of Interests. S.Z., F.S., and D.S. are employees of United Imaging Intelligence. The company has no role in designing and performing the surveillances and analyzing and interpreting the data. All other authors report no conflicts of interest relevant to this article.

References

1. Bethlehem, R.A., et al.: Brain charts for the human lifespan. Nature **604**(7905), 525–533 (2022)
2. Casey, B.J., et al.: The adolescent brain cognitive development (ABCD) study: imaging acquisition across 21 sites. Dev. Cogn. Neurosci. **32**, 43–54 (2018)
3. Çiçek, Ö., Abdulkadir, A., Lienkamp, S.S., Brox, T., Ronneberger, O.: 3D U-Net: learning dense volumetric segmentation from sparse annotation. In: Ourselin, S., Joskowicz, L., Sabuncu, M.R., Unal, G., Wells, W. (eds.) MICCAI 2016. LNCS, vol. 9901, pp. 424–432. Springer, Cham (2016). https://doi.org/10.1007/978-3-319-46723-8_49
4. Cruz, R.S., Lebrat, L., Bourgeat, P., Fookes, C., Fripp, J., Salvado, O.: Deepcsr: a 3D deep learning approach for cortical surface reconstruction. In: Proceedings of the IEEE/CVF Winter Conference on Applications of Computer Vision, pp. 806–815 (2021)
5. Dale, A.M., Fischl, B., Sereno, M.I.: Cortical surface-based analysis: I. segmentation and surface reconstruction. Neuroimage **9**(2), 179–194 (1999)
6. Di Martino, A., et al.: The autism brain imaging data exchange: towards a large-scale evaluation of the intrinsic brain architecture in autism. Mol. Psychiatry **19**(6), 659–667 (2014)
7. Fan, J., Cao, X., Wang, Q., Yap, P.T., Shen, D.: Adversarial learning for mono-or multi-modal registration. Med. Image Anal. **58**, 101545 (2019)
8. Fischl, B.: Freesurfer. Neuroimage **62**(2), 774–781 (2012)

9. Gu, D., et al.: An artificial-intelligence-based age-specific template construction framework for brain structural analysis using magnetic resonance images. Hum. Brain Mapp. **44**(3), 861–875 (2023)
10. Henschel, L., Conjeti, S., Estrada, S., Diers, K., Reuter, M.: Fastsurfer - a fast and accurate deep learning based neuroimaging pipeline. Neuroimage **219**, 117012 (2020)
11. Howell, B.R., et al.: The UNC/UMN baby connectome project (BCP): an overview of the study design and protocol development. Neuroimage **185**, 891–905 (2019)
12. Jack Jr, C.R., et al.: The Alzheimer's disease neuroimaging initiative (ADNI): MRI methods. J. Mag. Resonan. Imaging Off. J. Int. Soc. Magnet. Resonan. Med. **27**(4), 685–691 (2008)
13. Lebrat, L., et al.: Corticalflow: a diffeomorphic mesh transformer network for cortical surface reconstruction. Adv. Neural. Inf. Process. Syst. **34**, 29491–29505 (2021)
14. Ma, Q., Robinson, E.C., Kainz, B., Rueckert, D., Alansary, A.: PialNN: a fast deep learning framework for cortical Pial surface reconstruction. In: Abdulkadir, A., et al. (eds.) MLCN 2021. LNCS, vol. 13001, pp. 73–81. Springer, Cham (2021). https://doi.org/10.1007/978-3-030-87586-2_8
15. Marcus, D.S., Wang, T.H., Parker, J., Csernansky, J.G., Morris, J.C., Buckner, R.L.: Open access series of imaging studies (oasis): cross-sectional MRI data in young, middle aged, nondemented, and demented older adults. J. Cogn. Neurosci. **19**(9), 1498–1507 (2007)
16. Park, J.J., Florence, P., Straub, J., Newcombe, R., Lovegrove, S.: DeepSDF: learning continuous signed distance functions for shape representation. In: Proceedings of the IEEE/CVF Conference on Computer Vision and Pattern Recognition, pp. 165–174 (2019)
17. Press, W., Flannery, B., Teukolsky, S., Vetterling, W.: Runge-Kutta method in numerical recipes in Fortran: the art of scientific computing (1992)
18. Santa Cruz, R., et al.: Corticalflow++: boosting cortical surface reconstruction accuracy, regularity, and interoperability. In: Wang, L., Dou, Q., Fletcher, P.T., Speidel, S., Li, S. (eds.) MICCAI 2022. LNCS, vol. 13435, pp. 496–505. Springer, Cham (2022)
19. Teng, L., He, Y., Cao, Z., Hua, R., Han, Y., Feng, Q., Shi, F., Shen, D.: Unified model for children's brain image segmentation with co-registration framework guided by longitudinal MRI. IEEE J. Biomed. Health Inform. (2024)
20. Teng, L., et al.: Knowledge-guided prompt learning for lifespan brain MR image segmentation. In: Linguraru, M.G., et al. (eds.) MICCAI 2024. LNCS, vol. 15002, pp. 238–248. Springer, Cham (2024). https://doi.org/10.1007/978-3-031-72069-7_23
21. Tustison, N.J., Avants, B.B., Cook, P.A., Zheng, Y., Egan, A., Yushkevich, P.A., Gee, J.C.: N4itk: improved n3 bias correction. IEEE Trans. Med. Imaging **29**(6), 1310–1320 (2010)
22. Wickramasinghe, U., Remelli, E., Knott, G., Fua, P.: Voxel2Mesh: 3D mesh model generation from volumetric data. In: Martel, A.L., Abolmaesumi, P., Stoyanov, D., Mateus, D., Zuluaga, M.A., Zhou, S.K., Racoceanu, D., Joskowicz, L. (eds.) MICCAI 2020. LNCS, vol. 12264, pp. 299–308. Springer, Cham (2020). https://doi.org/10.1007/978-3-030-59719-1_30
23. Zhao, F., Wu, Z., Li, G.: Deep learning in cortical surface-based neuroimage analysis: a systematic review. Intell. Med. **3**(1), 46–58 (2023)
24. Zöllei, L., Iglesias, J.E., Ou, Y., Grant, P.E., Fischl, B.: Infant freesurfer: an automated segmentation and surface extraction pipeline for t1-weighted neuroimaging data of infants 0–2 years. Neuroimage **218**, 116946 (2020)

A Holistic Time-Aware Classification Model for Multimodal Longitudinal Patient Data

Tobias Susetzky[1(✉)], Huaqi Qiu[1], Rickmer Braren[2], and Daniel Rueckert[1,3,4]

[1] Chair for AI in Healthcare and Medicine, Technical University of Munich (TUM) and TUM University Hospital, Munich, Germany
{tobias.susetzky,huaqi.qiu,daniel.rueckert}@tum.de

[2] TUM University Hospital, Munich, Germany
rbraren@tum.de

[3] Department of Computing, Imperial College London, London, UK

[4] Munich Center for Machine Learning (MCML), Munich, Germany

Abstract. Current prognostic and diagnostic AI models for healthcare often limit informational input capacity by being time-agnostic and focusing on single modalities, therefore lacking the holistic perspective clinicians rely on. To address this, we introduce a **T**ime-**A**ware **M**ulti**M**odal Transformer **E**ncoder (TAMME) for longitudinal medical data. Unlike most state-of-the-art models, TAMME integrates longitudinal imaging, textual, numerical, and categorical data together with temporal information. Each element is represented as the sum of embeddings for high-level categorical type, further specification of this type, time-related data, and value. This composition overcomes limitations of a closed input vocabulary, enabling generalization to novel data. Additionally, with temporal context including the delta to the preceding element, we eliminate the requirement for evenly sampled input sequences. For long-term EHRs, the model employs a novel summarization mechanism that processes sequences piecewise and prepends recent data with history representations in end-to-end training. This enables balancing recent information with historical signals via self-attention. We demonstrate TAMME's capabilities using data from 431k+ hospital stays, 73k ICU stays, and 425k Emergency Department (ED) visits from the MIMIC dataset for clinical classification tasks: prediction of triage acuity, length of stay, and readmission. We show superior performance over state-of-the-art approaches especially gained from long-term data. Overall, our approach provides versatile processing of entire patient trajectories as a whole to enhance predictive performance on clinical tasks. Code is available at github.com/go31glX57/tamme.

Keywords: Multimodal Learning · Temporal Modeling · Longitudinal Health Records · Clinical AI

J. C. Gee et al. (Eds.): MICCAI 2025, LNCS 15960, pp. 24–33, 2026.
https://doi.org/10.1007/978-3-032-04927-8_3

1 Introduction

In clinical decision-making, practitioners exploit the full range of information available for an individual patient, comprising medical imaging, textual reports, prescriptions, vitals, previously applied procedures, and more. This multimodal data is typically used to put the current health status of a patient into the context

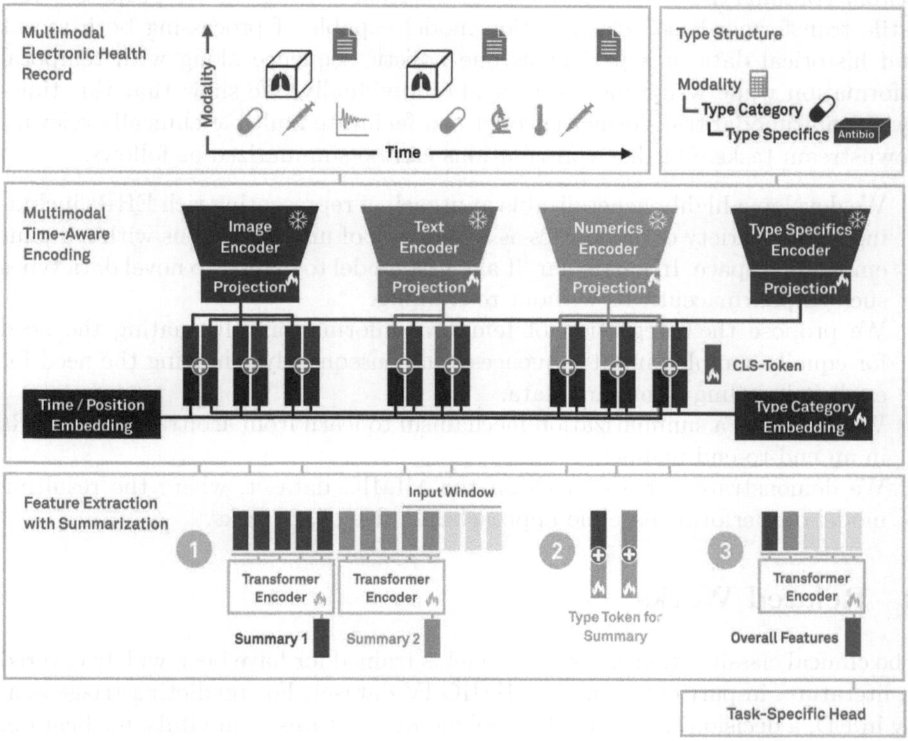

Fig. 1. TAMME is a transformer encoder that takes in a sequence of EHR elements of different modalities comprising numerics, categoricals, imaging and text. Each element is represented as a token resulting from the sum of embeddings for value, position/time, type specifics, and type category. The latter are learned during training; numerical values are represented by Fourier Features; for other values, external pre-trained models are used as encoders. All embeddings are projected into a shared space by trainable projection modules. By this, a uniform sequence of tokens is obtained that can be processed by a regular transformer with a task-specific head. In order to cope with long sequences beyond regular capacity in an end-to-end fashion, we add a summarization mechanism. By this, features for overflowing subsequences are extracted piece-wise. Subsequently, a dedicated trainable type token is added to those summaries before pre-pendeding them to the most recent items. The transformer encoder then extracts the final features from this resulting sequence. Throughout the whole diagram, the flame icon indicates trainable components while a snowflake symbol is attached to modules that are frozen during the whole training process.

of their individual records and disease progression. In particular, incorporated temporal information for each entry allows doctors to compare items longitudinally when assessing individual relevancy. Recent AI models in healthcare also incorporate multiple modalities but are mostly limited to a few modalities [5,20] or focus on vision-language tasks [1]. Moreover, they often omit longitudinal dynamics or temporal information and isolate recent single-time events. However, in extensive real-world datasets such as MIMIC [7–9,11], the patient's history contains the vast majority of information (cf. Fig. 3). We propose a versatile transformer-based classification model capable of processing both recent and historical data of a patient as one holistic sequence along with temporal information while weighting its elements individually. We show that this time-aware multimodal transformer encoder can facilitate multiple clinically relevant downstream tasks. Our key contributions can be summarized as follows:

1. We develop a highly-generalizable approach of representing rich EHRs including a wide variety of modalities as a sequence of uniform tokens within a joint embedding space. In particular, it allows a model to extend to novel data types such as pharmaceuticals without re-training.
2. We propose the integration of temporal information, eliminating the need for equally sampled input sequences and consequently removing the need for explicit handling of missing data.
3. We introduce a summarization mechanism to learn from arbitrary long EHRs in an end-to-end manner.
4. We demonstrate our approach on the MIMIC dataset, where the resulting model outperforms baseline approaches in five of six tasks.

2 Related Works

The clinical classification tasks our model is trained for have been widely covered in literature, in particular for the MIMIC-IV dataset. For predicting triage acuity in ED, a decision tree trained on polynomial features from vitals, medication, complaints, and metadata achieves SOTA results [23]. MeTra [12], a modified vision transformer, is the first use of a transformer for ICU survival prediction. It prepends patch projections of one image per patient with tokens for selected numerical data from a limited timespan. On MIMIC-IV, XGBoost succeeded in ICU survival prediction [16] from crafted numerical features, while MedFuse [5] combines numerical EHR elements with images to predict in-hospital mortality. ICU length of stay classification has been done using BERT-based models [19], Random Forest Classifier [6], or an LSTM model [3]. The duration of hospital stay and readmission have been successfully predicted, for example, by LSTM models [3] or a so-called Hierarchical Attention Network [4], again using numerical clinical data. There is also a considerable amount of work leveraging the longitudinal data structure of EHRs, some also incorporating temporal information. MOTOR [21] and BEHRT [15] both use a sequence of categorical EHR events timestamped with the current patient age to train a transformer for prognosis, albeit with highly different architecture and training strategies. Another proposed option for time integration is explicitly scaling the transformer's self-attention to prioritize recent over historic data [13]. There have also been experiments with both, time-aware attention scaling and time embeddings that are added to the input tokens for longitudinal imaging data [14]. Early work on multimodal sequence modeling including temporal aspects uses a transformer to process numeric and categorical ICU data together with patient metadata and history statistics [20]. We significantly go beyond this with a highly versatile input representation that can include all kinds of numeric, categorical, and meta data as well as texts and images, while still being extendable to novel data types. Further, our model incorporates all relevant relative temporal information and is capable of processing long-term history restricted in length only by the availability of computational power.

3 Methods

Modular Token Representation

We propose a novel approach for holistic token representations of multi-modal EHRs. In our setting, a temporally ordered sequence of numerical, categorical, textual, and imaging EHR elements from a single patient constitutes a sample. Each element consists of a high-level type category (e.g., drug, procedure), free text type specifics (e.g., drug name, description), temporal information, and a value (image, text, number, or none for categoricals). TAMME obtains a uniform sequence where each token representation is the sum of embeddings for type category, type specifics, time, and value (cf. Figure 1). To obtain the value embeddings, numerics are encoded as Fourier Features [22] while for images and texts, external domain-specific pre-trained models are used, namely RAD-DINO [17] and BioLORD 2023 [18]. All value embeddings remain frozen throughout the whole training process. For temporal information, we use a sinusoidal representation of patient age, time delta to the preceding EHR element, element position within the sequence, and an optional time delta to an anchor event such as the latest hospital admission. For type information, a large-scale closed vocabulary would limit extendability of a trained model to novel datasets and EHR elements such as unseen drugs. We overcome this by splitting type information into a high-level category and specifics. In close consultation with experienced clinicians, we define a minimal yet comprehensive set of 32 categories which are represented by randomly initialized and fully trainable embeddings. These categories are further specified by free-text details which are statically encoded by the same domain-specific pre-trained model as used for text values. Figure 1 includes an illustration. By this, we obtain semantically valid embeddings which are robust to most data inconsistencies in spelling, terminology, etc. and technically enable the zero-shot application of a trained model to novel data. Overall, we propose a versatile approach for joint learning from most kinds of patient data with longitudinal structure and temporal information.

Model Architecture

TAMME employs a well-known transformer architecture with minor modifications: input sequences are being encoded as described above; static embeddings for type specifics as well as for numeric, textual, and image values are projected to a shared space by linear layers that are trained end-to-end along with the other building blocks of the model. The projections are combined with the trainable type category embedding via addition, resulting in a sequence of uniform tokens. Instead of using standard positional embeddings, we leverage the temporal information of each token represented as sinusoidal frequencies and add them to the corresponding tokens. A transformer encoder is used as a feature extractor on these token sequences together with a CLS token. We attach a specific head and train the whole model separately for each one of a selection of classification tasks relevant to clinical decision making. To enable the full processing of individual EHRs far beyond the regular input length n of the model, TAMME features piecewise summarization: the remainder of an input sequence exceeding specified capacity by m tokens is processed piecewise. Each of the $s := \lceil \frac{m}{n} \rceil$ overflowing subsequences of maximal length n is being replaced by the feature vector produced by the model for this isolate subsequence. The most recent $n-s$ tokens of the input sequence are kept unchanged and prepended with the s summary tokens. An illustration is included in Fig. 1. To mark a summary token as such, a learnable type embedding is applied via addition. The entire setup remains trainable in an end-to-end manner and enables the complete processing of individual patient journeys in the form of long-term EHRs.

Self-Supervised Pre-training

We also explore a variant where we first pre-train TAMME using Masked Sequence Modelling in a self-supervised fashion. Respecting imbalance among high-level type categories, we apply biased-random sampling when selecting the tokens to mask. Only the unmasked tokens are processed by the transformer encoder and a transformer decoder model is added to reconstruct the masked tokens from context. We task the model to recover the type category as well the embeddings for type specifics and value for each of the masked tokens. In a second stage, we omit the decoder and fine-tune the transformer encoder module for each of our tasks individually.

4 Experiments and Results

To investigate our approach we merge MIMIC-IV [8], MIMIC-ED [7], MIMIC-Note [9], as well as MIMIC-CXR [10,11], all at version 2.2 from PhysioNet [2]. From MIMIC-Note, we use the findings section of radiology reports and all contents of discharge summaries. We add information such as patient age, precompute static embeddings, and create subsets centered around hospital, ED, and ICU stays. All data is split randomly into training, validation, and test set using a 70-15-15 split. We ensure all data from one patient is only included in one split. The dataset contains 425M+ single events from 303k+ patients, 431k+ hospital stays, 73k+ ICU stays, and 425k+ ED visits. We train and evaluate our model on ED visits, stationary hospital stays, and ICU stays for the following classification tasks, where applicable: length of stay, survival, readmission, and triage acuity (i.e. assigning a criticality level to patients at ED, see [7]). Table 1 provides an overview. For each stay, we label a patient as survived, if there is no death charted within the respective stay nor within 24h after discharge, i.e. the end of the stay. This temporal tolerance accounts for precision shortcomings in the original charting and data collection process of MIMIC. To prevent information leakage, we exclude events from the model's input that might reveal the prediction target, such as the current discharge summary. We constrain the model's input data based on the covered time span. By default, it ranges from the onset of the current stay to the earliest of the following: prediction target event, end of stay, or a fixed maximum of 12 h from the beginning of the stay. For readmission prediction, the 12-hour cap is omitted, allowing the model to leverage all data of the most recent stay. For all tasks, we experiment with extending this time window to historic records dating back up to one week, six months, or one year prior to the beginning of the current stay. The results detailed in Table 2 generally indicate performance improvements by historic data, albeit subject to the specific task. In comparison to training task-specific models from scratch, we also evaluate fine-tuning a task-agnostic model that has been pre-trained in a self-supervised manner (see Sect. 3). We assume that uniform pre-training affects performance equally for all context windows used in fine-tuning. Thus, we only use to best configuration from Table 2 to reduce computational cost. Pre-training is done on 400k+ multimodal EHR subsequences obtained by a sliding window of size 3000. At a second stage, we then fine-tune the resulting model on the exact same data and parameter set as we did when training from scratch. Further, we benchmark against the state of the art baseline. Overall, our trainings are done on 2–8 Nvidia A100 GPUs for up to 103 h, highly depending on the length of input sequences. As summarized in Table 3, we overall observe slight gains from pre-training, while TAMME exceeds baseline performance in five of six tasks. For the triage acuity task, we follow [23] using a subsampled dataset to tackle heavy class imbalance. We still observe inferior performance of our general approach when compared to the specialized baseline. We hypothesize that short-term signals relevant for triage might be easily cluttered by long-term data. We suspect a similar signal-vs-noise tradeoff explains the minor performance drops when using an extended EHR window in some tasks. However, this negative impact from long-term information is marginal, whereas its positive impact is significant (Tasks 2, 3, 5). Overall, we discover that the ideal EHR input window varies among tasks (Figs. 2 and 4).

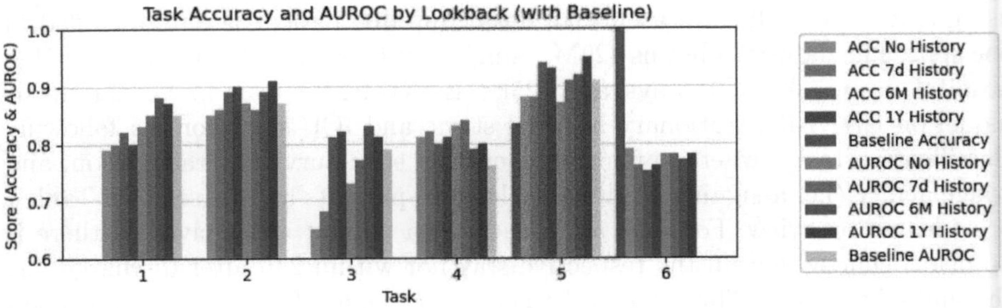

Fig. 2. Comparison of classification performance of TAMME using different extents of included EHR history versus the performance of the baseline model.

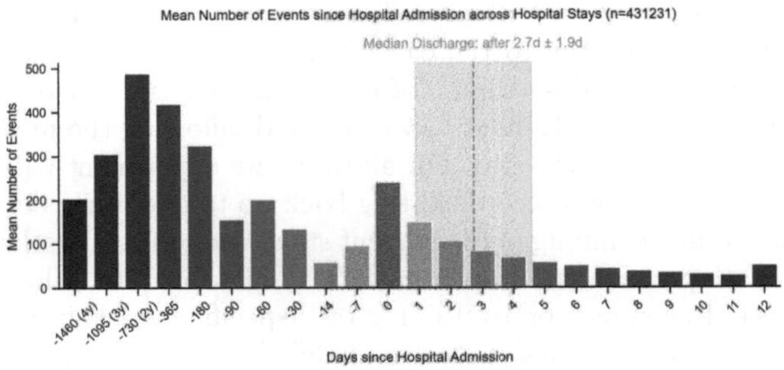

Fig. 3. The mean number of EHR entries (events) per day before and after admission, across all patients in the MIMIC dataset. For readability, historic data points are aggregated into bars covering time spans larger than one day. For instance, on average, patients have around 300 EHR entries charted 90 to 180 days before admission. Overall, this illustrates that patient history is much richer on information compared to the current stay.

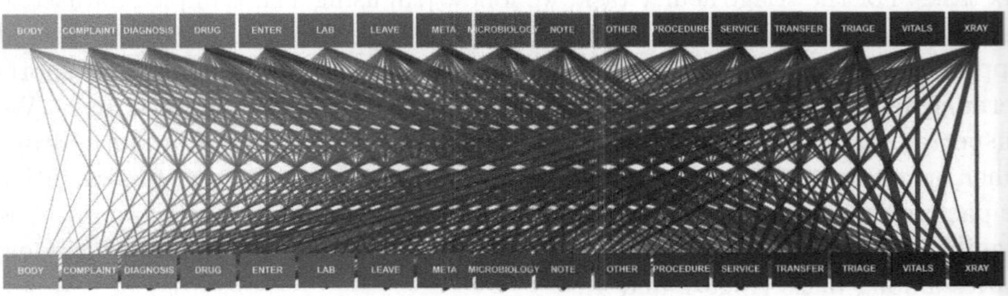

Fig. 4. Average attention among high-level type categories of a model trained for prediction of hospital length of stay. Higher line thickness indicates increased attention. We observe that, for instance, vitals appear to be crucial for interpreting most other data, especially for triage scores.

Table 1. Task targets and classes.

Task ID	Target	Classes
1	Hospital Length of Stay	short (≤3d), long (>3d)
2	In-Hospital Survival	True, False
3	Hospital Readmission	within 30 days, after 30 days/never
4	ICU Length of Stay	short (≤3d), long (>3d)
5	In-ICU Survival	True, False
6	ED Triage Acuity	1, 2, 3, 4, 5

Table 2. Classification performance for different extents of included historic EHRs.

Task ID	Target	No History		7 Days		6 Months		1 year	
		ACC	AUC	ACC	AUC	ACC	AUC	ACC	AUC
1	Hospital Length of Stay	0,79	0,83	0,80	0,85	**0,82**	**0,88**	0,80	0,87
2	In-Hospital Survival	0,85	0,87	0,86	0,86	0,89	0,89	**0,90**	**0,91**
3	Hospital Readmission	0,65	0,73	0,68	0,77	0,81	**0,83**	**0,82**	0,81
4	ICU Length of Stay	0,81	0,83	**0,82**	**0,84**	0,80	0,79	0,81	0,80
5	In-ICU Survival	0,88	0,87	0,88	0,91	**0,94**	0,92	0,93	**0,95**
6	ED Triage Acuity	**0,79**	**0,78**	0,76	0,78	0,75	0,77	0,76	0,78

Table 3. Classification performance of our model trained from scratch vs. a fine-tuned variant based on self-supervised pre-training, all compared against the baseline. For each task, the best configuration according to Table 2 has been used.

Task ID	Target	Ours				Baseline		
		pre-trained		from scratch		result		ref
		ACC	AUC	ACC	AUC	ACC	AUC	
1	Hospital Length of Stay	0.83	**0.89**	0.82	0.88	–	0.85	[3]
2	In-Hospital Survival	0.96	**0.95**	0.90	0.91	–	0.87	[5]
3	Hospital Readmission	0.81	**0.84**	0.81	0.83	–	0.79	[3]
4	ICU Length of Stay	0.82	0.83	0.82	**0.84**	–	0.76	[3]
5	In-ICU Survival	0.94	**0.95**	0.93	**0.95**	0.83	0.91	[16]
6	ED Triage Acuity	0.83	0.81	0.79	0.78	**0.99**	–	[23]

5 Discussion and Conclusion

We introduced TAMME, a time-aware transformer encoder for multimodal longitudinal EHRs. Unlike existing models, TAMME is extendable, versatile, and integrates most clinically relevant data types while preserving longitudinal structure and temporal information. Using piecewise summarization, it can process long patient histories effectively. While still being limited by the availability of computational power, this enhances control on resource usage and contributes to the model's ability to distinguish between recent and historical signals. Our experiments demonstrate that leveraging extended patient histories can indeed improve predictive performance on clinical tasks, surpassing state-of-the-art baselines in five of six tasks. This highlights the importance of capturing temporal dependencies and multimodal information for targeted decision-making. Further evaluation on other datasets and additional clinical tasks, such as personalized treatment recommendations, might help to assess the model's broader applicability as well as a clinical trial, exploring the model's reasoning, and an extensive ablation study beyond the length of historic context. In conclusion, TAMME offers a scalable approach for processing multimodal longitudinal EHRs, advancing clinical predictive modeling and decision support.

Acknowledgments. This work benefited from resources provided by the joint project "Open Medical Inference", a Module 3 project of the Medical Informatics Initiative of the Federal Government, funded by the German Federal Ministry for Research, Technology, and Space (Grant Number 01ZZ2315B).

Disclosure of Interests. The authors have no competing interests to declare that are relevant to the content of this article.

References

1. Bannur, S., et al.: Learning to exploit temporal structure for biomedical vision-language processing. In: Proceedings of the IEEE/CVF Conference on Computer Vision and Pattern Recognition, pp. 15016–15027 (2023)
2. Goldberger, A.L., et al.: Physiobank, physiotoolkit, and physionet: components of a new research resource for complex physiologic signals. Circulation **101**(23), e215–e220 (2000). https://doi.org/10.1161/01.CIR.101.23.e215, https://www.ahajournals.org/doi/10.1161/01.CIR.101. 23.e215, [Online]
3. Gupta, M., Gallamoza, B., Cutrona, N., Dhakal, P., Poulain, R., Beheshti, R.: An extensive data processing pipeline for mimic-iv. In: Machine Learning for Health, pp. 311–325. PMLR (2022)
4. Harerimana, G., Kim, J.W., Jang, B.: A deep attention model to forecast the length of stay and the in-hospital mortality right on admission from ICD codes and demographic data. J. Biomed. Inform. **118**, 103778 (2021)
5. Hayat, N., Geras, K.J., Shamout, F.E.: Medfuse: multi-modal fusion with clinical time-series data and chest x-ray images. In: Machine Learning for Healthcare Conference, pp. 479–503. PMLR (2022)

6. Hempel, L., Sadeghi, S., Kirsten, T.: Prediction of intensive care unit length of stay in the mimic-iv dataset. Appl. Sci. **13**(12), 6930 (2023)
7. Johnson, A., Bulgarelli, L., Pollard, T., Celi, L.A., Mark, R., Horng, S.: Mimic-iv-ed (version 2.2) (2023). https://doi.org/10.13026/5ntk-km72
8. Johnson, A., Bulgarelli, L., Pollard, T., Horng, S., Celi, L.A., Mark, R.: Mimic-iv (version 2.2) (2023). https://doi.org/10.13026/6mm1-ek67
9. Johnson, A., Pollard, T., Horng, S., Celi, L.A., Mark, R.: Mimic-iv-note: deidentified free-text clinical notes (version 2.2) (2023). https://doi.org/10.13026/1n74-ne17
10. Johnson, A.E.W., et al.: Mimic-cxr, a de-identified publicly available database of chest radiographs with free-text reports. Sci. Data **6**, 317 (2019). https://doi.org/10.1038/s41597-019-0322-0
11. Johnson, A.E.W., Pollard, T.J., Mark, R., Berkowitz, S.J., Horng, S.: Mimic-cxr database (version 2.0.0) (2019). https://doi.org/10.13026/C2JT1Q, https://doi.org/10.13026/C2JT1Q
12. Khader, F., et al.: Medical transformer for multimodal survival prediction in intensive care: integration of imaging and non-imaging data. Sci. Rep. **13**(1), 10666 (2023)
13. Li, T.Z., et al.: Longitudinal multimodal transformer integrating imaging and latent clinical signatures from routine EHRS for pulmonary nodule classification. In: Greenspan, H., et al. (eds.) MICCAI 2023. LNCS, vol. 14221, pp. 649–659. Springer, Cham (2023). https://doi.org/10.1007/978-3-031-43895-0_61
14. Li, T.Z., et al.: Time-distance vision transformers in lung cancer diagnosis from longitudinal computed tomography. In: Medical Imaging 2023: Image Processing, vol. 12464, pp. 229–238. SPIE (2023)
15. Li, Y., et al.: BEHRT: transformer for electronic health records. Sci. Rep. **10**(1), 7155 (2020)
16. Pang, K., Li, L., Ouyang, W., Liu, X., Tang, Y.: Establishment of ICU mortality risk prediction models with machine learning algorithm using mimic-IV database. Diagnostics **12**(5), 1068 (2022)
17. Pérez-García, F., et al.: Exploring scalable medical image encoders beyond text supervision. Nat. Mach. Intell. (2025). https://doi.org/10.1038/s42256-024-00965-w
18. Remy, F., Demuynck, K., Demeester, T.: BioLORD-2023: semantic textual representations fusing large language models and clinical knowledge graph insights. J. Am. Med. Inf. Assoc. ocae029 (024). https://doi.org/10.1093/jamia/ocae029
19. Röhr, T., et al.: Revisiting clinical outcome prediction for mimic-iv. In: Proceedings of the 6th Clinical Natural Language Processing Workshop, pp. 208–217 (2024)
20. Shickel, B., et al.: Multi-dimensional patient acuity estimation with longitudinal EHR tokenization and flexible transformer networks. Front. Dig. Health **4**, 1029191 (2022)
21. Steinberg, E., Fries, J., Xu, Y., Shah, N.: Motor: a time-to-event foundation model for structured medical records. arXiv preprint arXiv:2301.03150 (2023)
22. Tancik, M., et al.: Fourier features let networks learn high frequency functions in low dimensional domains. Adv. Neural. Inf. Process. Syst. **33**, 7537–7547 (2020)
23. Yaddaden, Y., Benahmed, Y., Rioux, M.D., Kallel, M.: Machine learning-based pre-diagnosis tools in emergency departments: predicting hospitalization, mortality and triage acuity. In: 2023 IEEE Third International Conference on Signal, Control and Communication (SCC), pp. 1–6. IEEE (2023)

Accurate and Efficient Fetal Birth Weight Estimation from 3D Ultrasound

Jian Wang[1,2,3], Qiongying Ni[1,3], Hongkui Yu[5], Ruixuan Yao[6], Jinqiao Ying[7], Bin Zhang[5], Xingyi Yang[5], Jin Peng[1,3], Jiongquan Chen[1,3], Junxuan Yu[1,3], Wenlong Shi[8], Chaoyu Chen[1,3], Zhongnuo Yan[1,3], Mingyuan Luo[1,3], Gaocheng Cai[8], Dong Ni[1,2,3], Jing Lu[4(✉)], and Xin Yang[1,3(✉)]

[1] National-Regional Key Technology Engineering Laboratory for Medical Ultrasound, School of Biomedical Engineering, Shenzhen University, Shenzhen, China
xinyang@szu.edu.cn
[2] College of Computer Science and Software Engineering, Shenzhen University, Shenzhen, China
[3] Medical Ultrasound Image Computing (MUSIC) Lab, School of Biomedical Engineering, Medical School, Shenzhen University, Shenzhen, China
[4] The First Affiliated Hospital of Xiamen University, School of Medicine, Xiamen University, Xiamen, China
lujing810@foxmail.com
[5] Shenzhen Baoan Women's and Children's Hospital, Shenzhen, China
[6] University of California, Berkeley, Berkeley, CA, USA
[7] Yongkang Maternal and Child Health Hospital, Yongkang, China
[8] Shenzhen RayShape Medical Technology Co., Ltd., Shenzhen, China

Abstract. Accurate fetal birth weight (FBW) estimation is essential for optimizing delivery decisions and reducing perinatal mortality. However, clinical methods for FBW estimation are inefficient, operator-dependent, and challenging to apply in cases of complex fetal anatomy. Existing deep learning methods are based on 2D standard ultrasound (US) images or videos that lack spatial information, limiting their prediction accuracy. In this study, we propose the first method for directly estimating FBW from 3D fetal US volumes. Our approach integrates a multi-scale feature fusion network (MFFN) and a synthetic sample-based learning framework (SSLF). The MFFN effectively extracts and fuses multi-scale features under sparse supervision by incorporating channel attention, spatial attention, and a ranking-based loss function. SSLF generates synthetic samples by simply combining fetal head and abdomen data from different fetuses, utilizing semi-supervised learning to improve prediction performance. Experimental results demonstrate that our method achieves superior performance, with a mean absolute error of 166.4 ± 155.9 g and a mean absolute percentage error of $5.1 \pm 4.6\%$, outperforming existing methods and approaching the accuracy of a senior doctor. Code is available at: https://github.com/Qioy-i/EFW.

Keywords: Fetal birth weight · 3D ultrasound volumes · Semi-supervised learning

J. Wang, Q. Ni and H. Yu—Contribute equally to this work.

J. C. Gee et al. (Eds.): MICCAI 2025, LNCS 15960, pp. 34–44, 2026.
https://doi.org/10.1007/978-3-032-04927-8_4

1 Introduction

Accurate estimation of fetal birth weight (FBW) prior to delivery is crucial for making optimal delivery decisions (i.e., vaginal birth or cesarean section) and reducing perinatal mortality [14]. As shown in Fig. 1, the standard clinical method for FBW assessment involves a cumbersome procedure: obtaining 2D standard ultrasound (US) planes, measuring biometric parameters, and applying an empirical formula to estimate it [6]. However, each steps poses significant challenges. First, acquiring standard US planes requires radiologists to deeply understand fetal anatomy and accurately locate the 2D plane in 3D space, which is technically demanding and time-consuming [4]. Second, both standard plane acquisition and biometric measurements depend heavily on the radiologist's experience, leading to intra- and inter-observer variability, which affects the accuracy and reproducibility [10]. Finally, the empirical formulas currently in use are based on regression analysis of a limited set of biometric parameters, failing to account for diverse and complex anatomical variations among different fetuses [12].

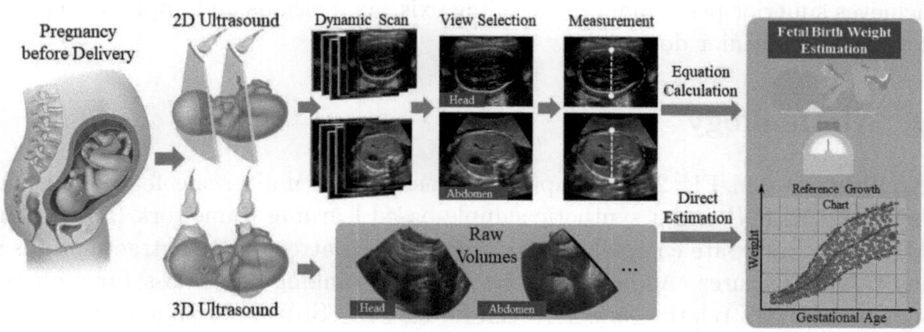

Fig. 1. Workflow for conventional 2D US-based and our 3D US-based FBW estimation.

To address these clinical challenges, several artificial intelligence-based approaches have been proposed, broadly categorized into three types: (1) Some studies use manually measured parameters and maternal information, applying machine learning methods to predict FBW [3,11,21]. (2) Other studies leverage deep learning to predict parameters from standard US planes or videos, followed by applying empirical formulas to estimate FBW [1,17,18]. (3) A few studies employ deep learning models to directly predict FBW from standard US videos combined with clinical information [15,16,19]. However, to date, no approaches have been developed to comprehensively address all three key challenges. Furthermore, due to the lack of spatial information in 2D US imaging, methods based on it struggle to accurately capture fetal anatomical structures, limiting prediction accuracy. In contrast, as shown in Fig. 1, 3D fetal US imaging provides more comprehensive anatomical information and reduces reliance on the operator's technique, making it a promising alternative for estimating FBW [7,9,13].

However, its relatively lower resolution and higher noise levels pose challenges for manual biometric measurements and empirical formula-based estimation. Moreover, while 3D US has not yet achieved widespread clinical adoption due to factors like cost and accessibility, it is increasingly available in advanced obstetric settings. This trend highlights the growing need for automated methods that can fully leverage 3D US data and bypass the limitations of handcrafted features and empirical modeling.

In this study, we propose the first method for directly estimating FBW from 3D fetal US volumes. The key challenges lie in extracting informative features from large, anatomically complex 3D data under sparse supervision, and addressing data scarcity given the wide prediction space. To tackle these, we introduce a multi-scale feature fusion network (MFFN) that integrates head and abdomen volumes using channel attention and Mamba-based spatial attention. We further design a ranking-based loss based on inter-fetal FBW relationships to improve generalization. To alleviate limited sample size, we propose a synthetic sample-based learning framework (SSLF), which generates new fetal samples by mixing head and abdomen volumes from different fetuses and learns from them via semi-supervised training. The experimental results demonstrate that our method achieves superior performance, exceeding existing methods and approaching the accuracy of a senior doctor.

2 Methodlogy

As illustrated in Fig. 2, our approach consists of a multi-scale feature fusion network (MFFN) and a synthetic sample-based learning framework (SSLF). In MFFN, we integrate channel attention and spatial attention to extract and fuse multi-scale features. Additionally, we design a ranking-based loss function to model the FBW relationships between fetuses. In SSLF, we design a simple yet effective sample synthesis method to expand the dataset and then apply semi-supervised learning to enhance FBW estimation.

2.1 Multi-scale Feature Fusion Network

Due to the significant variations in the spatial scales of fetal anatomical structures, features extracted from a single scale are inadequate for accurate FBW prediction. Thus, we design a multi-scale feature fusion network (Fig. 2(a)).

Problem Definition: Let $\{(x_i^h, x_i^a, s_i^h, s_i^a, g_i, y_i) \,|i \in 1, 2, ...N\}$ denote a mini-batch randomly sampled from the 3D fetal US dataset. Here, $x_i^h, x_i^a \in \mathbb{R}^{1\times D\times H\times W}$ represent the 3D US volumes of the fetal head and abdomen, where D, H, and W are the depth, height, and width of the volumes, respectively. $s_i^h, s_i^a \in \mathbb{R}^3$ are the voxel spacing in the depth, height, and width directions for the head and abdomen. $g_i \in \{0, 1, 2, 3\}$ is the time interval (in days) from the 3D US scan to delivery. $y_i \in [0, 1]$ denotes the normalized true FBW, and N is the batch size. During clinical inference, g_i is provided by the doctor as the time interval from the 3D US scan to the estimated delivery. To incorporate spacing

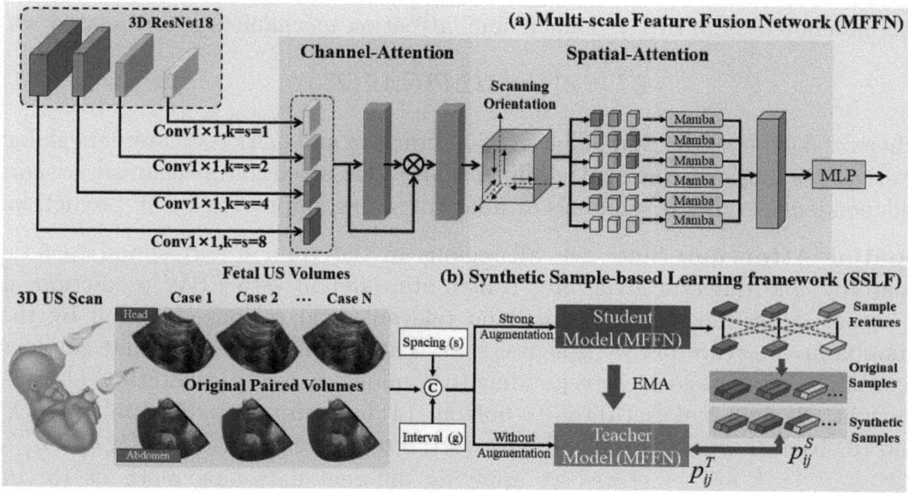

Fig. 2. Overview of the proposed FBW estimation method.

and time interval information, we expand the dimensions of s_i^h and g_i and concatenate them along the channel dimension with x_i^h, resulting in a five-channel head volume $X_i^h \in \mathbb{R}^{5\times D\times H\times W}$. Similarly, we obtain abdominal volume data X_i^a. The objective of this study is to predict FBW y_i given the input volumes (X_i^h, X_i^a).

Feature Fusion: We designed a shared network for both head and abdomen data, so their forward computation processes are identical. For simplicity, we describe the computation process using head data as an example. 3DResNet18 [23] is adopted as the backbone for feature extraction. The features from four intermediate layers, denoted as $Z_i^{h^1}, Z_i^{h^2}, Z_i^{h^3}, Z_i^{h^4}$, are extracted. These features are spatially downsampled by factors of 4, 8, 16, and 32, with corresponding channel dimensions of 64, 128, 256, and 512. To enable multi-scale feature fusion at a uniform spatial resolution, we first apply $1\times 1\times 1$ convolutions to project all feature maps to a consistent channel dimension of 128. Then, four sets of convolutions are applied to downsample the features to the same spatial resolution, with kernel sizes and strides set to 8, 4, 2, and 1, respectively. This process aligns features across all four scales both spatially and channel-wise. Finally, we concatenate the features along the channel dimension, resulting in the fused representation $Z_i^h \in \mathbb{R}^{512\times \frac{D}{32}\times \frac{H}{32}\times \frac{W}{32}}$.

Channel Attention: High-level features typically capture global semantics but with lower resolution, while low-level features focus on fine-grained details at higher resolution. Given their different roles, it is necessary to weigh them appropriately. To achieve this, we apply channel attention to the fused features Z_i^h. We use a bottleneck MLP with the same input and output dimensions, where the activation function is ReLU and the dimensionality reduction factor is set to 16, denoted as $MLP(\cdot)$. The channel attention mechanism is formulated as:

$$Z_i^h = Z_i^h \cdot \sigma(MLP(GAP(Z_i^h))), \tag{1}$$

where $\sigma(\cdot)$ denotes the sigmoid activation function and $GAP(\cdot)$ represents global average pooling over the spatial dimensions. This attention mechanism assigns independent weights to channels at different scales, improving FBW prediction.

Spatial Attention: Since not all regions in the volume correspond to fetal anatomy, and different structures contribute differently to FBW prediction, it is essential for the model to focus on relevant fetal regions. Inspired by the Mamba architecture [5], we design a spatial attention mechanism that employs six Mamba models, each corresponding to a unique scanning orientation: (1) left-to-right, (2) right-to-left, (3) top-to-bottom, (4) bottom-to-top, (5) front-to-back, and (6) back-to-front. Specifically, six spatial sequences $Z_i^{h_1}, Z_i^{h_2}, Z_i^{h_3}, Z_i^{h_4}, Z_i^{h_5}, Z_i^{h_6} \in \mathbb{R}^{512 \times L}$ are generated by applying different flattening methods to the input features, where $L = \frac{D}{32} \times \frac{H}{32} \times \frac{W}{32}$. The sequences are processed by their corresponding Mamba models, and their outputs are fused using an averaging operation, yielding the final feature map $Z_i^h \in R^{512 \times L}$. This fusion captures spatial information from multiple perspectives, improving FBW prediction.

Loss Function: After the above three steps, we obtain the head and abdomen features Z_i^h and Z_i^a. These features are concatenated along the channel dimension and subjected to global average pooling over the spatial dimensions, resulting in the fused head-abdomen representation $Z_{i,i}^{ha} \in \mathbb{R}^{1024}$. A single fully connected layer followed by a sigmoid activation function is applied to predict the FBW p_i. The regression loss is defined as:

$$Loss_{reg} = \frac{1}{N} \sum_{i=1}^{N} MSE(p_i, y_i), \tag{2}$$

where $MSE(\cdot)$ denotes the mean squared error between the predicted FBW p_i and the ground truth y_i. To further enhance the model's ability to capture relative weight differences among fetuses, we introduce a ranking-based loss function. This loss encourages the predicted order of FBWs to align with the ground truth order. The ranking loss is formulated as:

$$Loss_{rank} = \frac{1}{N^2} \sum_{i=1}^{N} \sum_{j=1}^{N} max[0, -(p_i - p_j)] \cdot \tau(y_i - y_j), \tag{3}$$

where $\tau(\cdot)$ equal to 1 if the argument is greater than 0 and 0 otherwise.

2.2 Synthetic Sample-Based Learning Framework

In this study, a key challenge we face is the conflict between limited sample size and nearly infinite possible predictions, making conventional supervised learning models prone to severe overfitting. To address this issue, we propose an synthetic sample-based learning framework, as shown in Fig. 2(b).

Sample Synthesis: Here, the sample refers to the fetus rather than the head or abdominal volume. We hypothesize that the model predicts FBW by estimating fetal size from both head and abdominal volumes. Due to the anatomical independence between head and abdominal regions, there is little or no direct correlation between their volumes within the same fetus. Based on this observation, we propose a simple method to synthesize new samples from different fetuses. Specifically, we create synthetic samples by combining the head volume of one fetus with the abdominal volume of another. Let $\{(X_i^h, X_j^a)\,|i, j \in 1, 2, ...N, i \neq j\}$ denote the set of synthetic samples within a mini-batch, where N is the batch size. For each mini-batch, this method can generate $N(N-1)$ synthetic samples, greatly expanding the dataset without any complicated operations.

Semi-supervised Learning: Although the previous steps synthesize a large number of new samples to expand the dataset, they lack real FBW labels. To address this, we leverage semi-supervised learning by treating synthetic samples as unlabeled data. Inspired by the MeanTeacher framework [22], we design a semi-supervised learning setup consisting of two models: the teacher model and the student model, both of which share the same architecture (i.e., MFFN). During training, the student model's parameters are updated via backpropagation, while the teacher model's parameters are updated using an exponential moving average (EMA) of the student model's parameters, as follows:

$$\theta_T^t = m\theta_T^{t-1} + (1-m)\theta_S^t, \tag{4}$$

where θ_T and θ_S denote the parameters of the teacher and student models, respectively. t denotes the iteration step, and m is the momentum coefficient that controls the update rate of the teacher model.

In each training batch, there are N head and N abdomen data samples, resulting in N^2 possible combinations. Of these, N combinations correspond to real fetal samples (with labels), while the remaining $N(N-1)$ combinations correspond to synthetic fetal samples (without labels). For these combinations, let $\{p_{ij}^T|i, j \in 1, 2, ..., N\}$ and $\{p_{ij}^S|i, j \in 1, 2, ..., N\}$ represent the predictions from the teacher and student models, respectively. We treat the teacher Model's predictions as pseudo-labels to supervise the student Model. The loss function for the semi-supervised learning phase is defined as:

$$Loss_{semi} = \frac{1}{N(N-1)} \sum_{i=1}^{N} \sum_{j=1}^{N} MSE(p_{ij}^S, p_{ij}^T), i \neq j. \tag{5}$$

Finally, the total loss function is the weighted sum of three components:

$$Loss_{total} = Loss_{reg} + \alpha Loss_{rank} + \beta Loss_{semi}, \tag{6}$$

where α and β are the loss weighting coefficients. With this approach, our model can effectively learn from synthetic samples generated by combining fetal head data with abdomen data from different fetuses.

3 Experiments and Results

3.1 Datasets and Implementations

With ethics committee approval, we collected a dataset of 491 pregnancies by two senior doctors, each with 17 years of experience. Each case includes one head, abdomen, and femur ultrasound scan acquired within 72 h before delivery, using a GE HealthCare E8 system with an RAB6-D probe. Prior to delivery, a senior doctor estimated fetal birth weight (FBW) using the Hadlock formula based on manual biometric measurements. Post-delivery, the true FBW was recorded as ground truth. The mean true FBW was 3229.3 ± 467.8 g, ranging from 1000 g to 4610 g. Voxel spacing was extracted from metadata. The dataset was randomly split into training, validation, and test sets at a 7:1:2 ratio, and all volumes were resized to $160 \times 128 \times 96$ while preserving aspect ratio.

We trained the model for 200 epochs with a batch size of 16 using the Adam optimizer. The learning rate started at $1e^{-4}$, warmed up for 5 epochs, and then decayed via cosine annealing. Data augmentation included rotation, flipping, scaling, and contrast and brightness adjustments to improve generalizability. Hyperparameters were set as follows: $\alpha = 0.001$, β increased linearly from 0 to 0.2, and momentum m from 0.99 to 0.9999. These values were selected via grid search on the validation set. All experiments ran on an RTX 4090 GPU server.

3.2 Results

To evaluate our method, we compared it with several existing approaches, including Hadlock [6], INTERGROWTH-21st [20], BabyNet [16], BabyNet++ [15], 3DCNN-1 [24], 3DCNN-2 [2], and DSCNN [8]. Hadlock and INTERGROWTH-21st are widely used clinical methods based on manual biometric measurements and empirical formulas; we used the pre-delivery estimate by a senior doctor as the Hadlock reference. BabyNet and BabyNet++ predict FBW from 2D ultrasound videos, with the latter also incorporating biometric and maternal data; we adopted the reported results from their original studies. 3DCNN-1, 3DCNN-2, and DSCNN are recent 3D regression models for other tasks, which we re-implemented following their published protocols. All methods were evaluated using mean absolute error (MAE), root mean square error (RMSE), and mean absolute percentage error (MAPE).

Table 1 summarizes the comparison results. Our method achieves strong performance, with a MAE of 166.4 ± 154.9 g, RMSE of 227.3 ± 227.3 g, and MAPE of $5.1 \pm 4.6\%$. Notably, it is highly competitive with Hadlock—widely used in clinical practice—and INTERGROWTH-21st, a global fetal growth standard. The closest method, BabyNet++, leverages standard US videos, manual biometric measurements, and maternal information. In contrast, our approach relies solely on 3D US volumes, requiring minimal scanning expertise and no manual input. We also evaluated the two key components of our method. MFFN alone achieved a MAE of 172.0 ± 164.6 g, RMSE of 238.1 ± 237.1 g, and MAPE of $5.3 \pm 4.9\%$, demonstrating its ability to extract features effectively from 3D volumes under

Table 1. Comparison results of different methods.

Method	MAE (g)↓	RMSE (g)↓	MAPE (%)↓
Hadlock [6]	160.2 ± 118.5	199.3 ± 197.9	5.0 ± 3.7
INTERGROWTH-21st [20]	182.9 ± 137.0	228.5 ± 220.9	5.7 ± 4.2
BabyNet [16]*	254.0 ± 230.0	341.0 ± 215.0	7.5 ± 6.6
BabyNet++ [15]*	179.0 ± 19.0	203.0	5.1 ± 0.6
3DCNN-1 [24]	268.7 ± 245.2	363.8 ± 358.7	8.2 ± 7.1
3DCNN-2 [2]	253.1 ± 222.1	336.8 ± 336.2	7.9 ± 6.9
DSCNN [8]	264.7 ± 239.1	356.7 ± 355.0	8.3 ± 7.8
$MFFN_{(Ours)}$	172.0 ± 164.6	238.1 ± 237.1	5.3 ± 4.9
3DCNN-1 + SSLF	264.7 ± 227.8	349.3 ± 348.9	8.2 ± 7.1
3DCNN-2 + SSLF	248.3 ± 212.2	326.6 ± 325.1	8.1 ± 6.7
DSCNN + SSLF	220.0 ± 218.3	310.0 ± 309.1	6.8 ± 6.5
$MFFN + SSLF_{(Ours)}$	166.4 ± 154.9	227.3 ± 227.3	5.1 ± 4.6

*Results reported from original publication; model not retrained on our dataset due to input modality mismatch.

Table 2. Experimental results of ablation study with different configurations of key components of our model. 3DR18: 3DResNet18, WS: Weight-sharing, FF: Feature-fusion, CA: Channel attention, SA: Spatial attention, RL: Ranking-based loss

Model Architecture							Input		Results	
3DR18	WS	FF	CA	SA	RL	SSLF	Head	Abdomen	MAE (g)	MAPE (%)
✓							✓		283.1 ± 229.0	9.0 ± 7.5
✓								✓	252.8 ± 191.8	8.0 ± 6.4
✓							✓	✓	222.4 ± 193.2	6.7 ± 5.1
✓	✓						✓	✓	200.6 ± 178.1	6.1 ± 5.1
✓	✓	✓					✓	✓	189.7 ± 173.9	5.9 ± 5.4
✓	✓	✓	✓				✓	✓	191.0 ± 166.1	5.9 ± 5.0
✓	✓	✓	✓	✓			✓	✓	177.2 ± 175.2	5.9 ± 5.0
✓	✓	✓	✓	✓	✓		✓	✓	172.0 ± 164.6	5.3 ± 4.9
✓	✓	✓	✓	✓	✓	✓	✓	✓	166.4 ± 154.9	5.1 ± 4.6

sparse supervision. When combined with SSLF, multiple 3D regression models showed improved performance, indicating its adaptability and potential for broader application.

To evaluate the effectiveness of each component in our method, we conducted a series of ablation experiments. Specifically, we assessed the impact of single-site and multi-site volumes, weight-sharing in multi-site volume fusion, feature fusion, channel attention, spatial attention, ranking-based loss function, and SSLF. The results, presented in Table 2, demonstrate that weight-sharing in multi-site volume fusion is crucial for accurate FBW estimation. Additionally, simple multi-scale feature fusion enhances accuracy, confirming its necessity. While channel attention increases the average MAE by 1.3 g, it reduces the standard deviation by 7.8 g, making it beneficial overall. Spatial attention further improves performance, decreasing the average MAE by 13.8 g. The ranking-based loss function aligns the FBW relationships between fetuses with the groundtruth, effectively boosting prediction accuracy. Finally, SSLF significantly expands the dataset and improves performance through a semi-supervised learning framework.

4 Conclusion

In this study, we present a novel method for predicting FBW directly from 3D fetal US volumes. To address challenges in feature extraction and sample size limitations, our method designs a multi-scale feature fusion network and a synthetic sample-based learning framework. Our results outperform existing methods and close to clinical accuracy. Rather than replacing 2D workflows, our method offers a complementary tool for enhanced decision-making. Future work will extend this framework to jointly estimate biometric parameters and FBW, improving automation and interpretability.

Acknowledgments. This work was supported by the grant from National Natural Science Foundation of China (82201851, 12326619, 62171290); Guangxi Province Science Program (2024AB17023); Yunnan Major Science and Technology Special Project Program (No. 202402AA310052); Yunnan Key Research and Development Program (202503AP140014); Science and Technology Planning Project of Guangdong Province (2023A0505020002); Frontier Technology Development Program of Jiangsu Province (No. BF2024078).

Disclosure of Interests. The authors have no competing interests to declare that are relevant to the content of this article.

References

1. Bano, S., et al.: AutoFB: automating fetal biometry estimation from standard ultrasound planes. In: de Bruijne, M., et al. (eds.) MICCAI 2021, Part VII. LNCS, vol. 12907, pp. 228–238. Springer, Cham (2021). https://doi.org/10.1007/978-3-030-87234-2_22
2. Cali, R.J., et al.: The influence of brain MRI defacing algorithms on brain-age predictions via 3d convolutional neural networks. In: 2023 45th Annual International Conference of the IEEE Engineering in Medicine & Biology Society (EMBC), pp. 1–6 (2023). https://doi.org/10.1109/EMBC40787.2023.10340740
3. Camargo-Marín, L., Guzmán-Huerta, M., Piña-Ramirez, O., Perez-Gonzalez, J.: Multimodal early birth weight prediction using multiple kernel learning. Sensors **24**(1), 2 (2023)
4. Chen, H., et al.: Automatic fetal ultrasound standard plane detection using knowledge transferred recurrent neural networks. In: Navab, N., Hornegger, J., Wells, W.M., Frangi, A.F. (eds.) MICCAI 2015, Part I. LNCS, vol. 9349, pp. 507–514. Springer, Cham (2015). https://doi.org/10.1007/978-3-319-24553-9_62

5. Gu, A., Dao, T.: Mamba: linear-time sequence modeling with selective state spaces. arXiv preprint arXiv:2312.00752 (2023)
6. Hadlock, F.P., Harrist, R., Sharman, R.S., Deter, R.L., Park, S.K.: Estimation of fetal weight with the use of head, body, and femur measurements–a prospective study. Am. J. Obstet. Gynecol. **151**(3), 333–337 (1985)
7. Hamza, H.A., Shabana, A.A.E., Abu Shar, A.M., Egiz, M.N.: Accuracy of fetal thigh circumference and fractional thigh volume in prediction of fetal birth weight. Egypt. J. Fertil. Steril. **29**(1), 3–13 (2025)
8. Jafrasteh, B., Lubián-López, S.P., Benavente-Fernández, I.: A deep sift convolutional neural networks for total brain volume estimation from 3d ultrasound images. Comput. Methods Programs Biomed. **242**, 107805 (2023)
9. Kang, L., Wu, Q.Q., Sun, L.J., Gao, F.Y., Wang, J.J.: Predicting fetal weight by three-dimensional limb volume ultrasound (AVol/TVol) and abdominal circumference. Chin. Med. J. **134**(9), 1070–1078 (2021)
10. Li, J., Gao, Z., Wang, C., Pu, B., Li, K.: A rule-guided interpretable lightweight framework for fetal standard ultrasound plane capture and biometric measurement. Neurocomputing **621**, 129290 (2025)
11. Lu, Y., Fu, X., Chen, F., Wong, K.K.: Prediction of fetal weight at varying gestational age in the absence of ultrasound examination using ensemble learning. Artif. Intell. Med. **102**, 101748 (2020)
12. Ma, J., Cheng, D., Zhang, Z., Cai, B., Xu, X.: Evaluating the accuracy of sonographic fetal weight estimations using the Hadlock IV formula in a Chinese population. Quant. Imaging Med. Surg. **13**(6), 3726 (2023)
13. Metwally, S., Alsherbiny, M., Elsayed, M., Abuelezz, S.: Three-versus two-dimensional sonographic biometry for predicting birth weight at term. Benha J. Appl. Sci. **7**(10), 79–83 (2022)
14. Milner, J., Arezina, J.: The accuracy of ultrasound estimation of fetal weight in comparison to birth weight: a systematic review. Ultrasound **26**(1), 32–41 (2018)
15. Płotka, S., et al.: BabyNet++: fetal birth weight prediction using biometry multimodal data acquired less than 24 hours before delivery. Comput. Biol. Med. **167**, 107602 (2023)
16. Płotka, S., et al.: BabyNet: residual transformer module for birth weight prediction on fetal ultrasound video. In: Wang, L., Dou, Q., Fletcher, P.T., Speidel, S., Li, S. (eds.) MICCAI 2022. LNCS, vol. 13434, pp. 350–359. Springer, Cham (2022). https://doi.org/10.1007/978-3-031-16440-8_34
17. Płotka, S., et al.: Deep learning fetal ultrasound video model match human observers in biometric measurements. Phys. Med. Biol. **67**(4), 045013 (2022)
18. Płotka, S., Włodarczyk, T., Klasa, A., Lipa, M., Sitek, A., Trzciński, T.: FetalNet: multi-task deep learning framework for fetal ultrasound biometric measurements. In: Mantoro, T., Lee, M., Ayu, M.A., Wong, K.W., Hidayanto, A.N. (eds.) ICONIP 2021. CCIS, vol. 1517, pp. 257–265. Springer, Cham (2021). https://doi.org/10.1007/978-3-030-92310-5_30
19. Płotka, S.S., et al.: Deep learning for estimation of fetal weight throughout the pregnancy from fetal abdominal ultrasound. Am. J. Obstetr. Gynecol. MFM **5**(12), 101182 (2023)
20. Stirnemann, J., Cheikh Ismail, L., et al.: International estimated fetal weight standards of the intergrowth-21st project. Ultrasound Obstetr. Gynecol. **49**(4), 478–486 (2017)
21. Tao, J., Yuan, Z., Sun, L., Yu, K., Zhang, Z.: Fetal birthweight prediction with measured data by a temporal machine learning method. BMC Med. Inform. Decis. Mak. **21**, 1–10 (2021)

22. Tarvainen, A., Valpola, H.: Mean teachers are better role models: weight-averaged consistency targets improve semi-supervised deep learning results. Adv. Neural Inf. Process. Syst. **30** (2017)
23. Tran, D., Wang, H., Torresani, L., Ray, J., LeCun, Y., Paluri, M.: A closer look at spatiotemporal convolutions for action recognition. In: Proceedings of the IEEE Conference on Computer Vision and Pattern Recognition, pp. 6450–6459 (2018)
24. Wang, Y., Wen, J., Xin, J., Zhang, Y., Xie, H., Tang, Y.: 3DCNN predicting brain age using diffusion tensor imaging. Med. Biol. Eng. Comput. **61**(12), 3335–3344 (2023)

Adaptive Spatial Transcriptomics Interpolation via Cross-modal Cross-slice Modeling

Ningfeng Que[1,2], Xiaofei Wang[3], Jingjing Chen[1,2], Yixuan Jiang[1,2], and Chao Li[1,3,4,5](✉)

[1] School of Science and Engineering, University of Dundee, Dundee, UK
[2] College of Medicine and Biological Information Engineering, Northeastern University, Boston, China
[3] Department of Clinical Neurosciences, University of Cambridge, Cambridge, UK
cl647@cam.ac.uk
[4] School of Medicine, University of Dundee, Dundee, UK
[5] Department of Applied Mathematics and Theoretical Physics, University of Cambridge, Cambridge, UK

Abstract. Spatial transcriptomics (ST) is a promising technique that characterizes the spatial gene profiling patterns within the tissue context. Comprehensive ST analysis depends on consecutive slices for 3D spatial insights, whereas the missing intermediate tissue sections and high costs limit the practical feasibility of generating multi-slice ST. In this paper, we propose C2-STi, the first attempt for interpolating missing ST slices at arbitrary intermediate positions between adjacent ST slices. Despite intuitive, effective ST interpolation presents significant challenges, including 1) limited continuity across heterogeneous tissue sections, 2) complex intrinsic correlation across genes, and 3) intricate cellular structures and biological semantics within each tissue section. To mitigate these challenges, in C2-STi, we design 1) a distance-aware local structural modulation module to adaptively capture cross-slice deformations and enhance positional correlations between ST slices, 2) a pyramid gene co-expression correlation module to capture multi-scale biological associations among genes, and 3) a cross-modal alignment module that integrates the ST-paired hematoxylin and eosin (H&E)-stained images to filter and align the essential cellular features across ST and H&E images. Extensive experiments on the public dataset demonstrate our superiority over state-of-the-art approaches on both single-slice and multi-slice ST interpolation. Codes are available at https://github.com/XiaofeiWang2018/C2-STi.

Keywords: Spatial transcriptomics · Interpolation · Deep learning · Multimodal modeling

N. Que and X. Wang—Equal contribution.

J. C. Gee et al. (Eds.): MICCAI 2025, LNCS 15960, pp. 45–54, 2026.
https://doi.org/10.1007/978-3-032-04927-8_5

1 Introduction

Spatial transcriptomics (ST) has transformed the understanding of tissue function by linking gene expression with spatial context [12,18,20]. However, in real-world scenarios, tissue sections are often lost or discarded during pre-testing, staining, or quality control, making continuous tissue profiling challenging for current ST platforms [15]. Additionally, the high cost of repeated experiments to generate consecutive ST slices limits continuous ST analysis, especially for methods relying on tissue continuity, causing distortions in 3D reconstruction [24]. This highlights the need for computational methods to overcome the challenges.

Existing approaches focus on super-resolving or imputing regional gene expression within a single slice [19], while few address interpolating the missing ST slices. A recent study [11] proposed a semi-supervised method that first interpolates a missing H&E slice and then used it to predict the ST. Despite success, the two-step approach may cause bias in leveraging the in-depth spatial information of tissue. In particular, it can only interpolate one single slice at the middle position, assuming the missing tissue is spatially smooth, thus impractical for real-world scenarios that require multiple slices interpolations and reflect the tissue heterogeneity. Therefore, it is necessary to develop methods for effectively interpolating ST slices at arbitrary locations between available slices.

In other medical domains, approaches have been proposed to capture the dynamic changes between adjacent slices for estimating the deformation of under-interpolated slices, thereby generating high-quality interpolation. For instance, Frakes *et al.* [2] employed a multi-resolution optical flow estimation approach to interpolate 3D anisotropic MRI. However, this method lacks a coarse-to-fine process, making it susceptible to errors for low-quality input. Similarly, Li *et al.* [7] developed an interpolation network leveraging joint space-frequency feature extraction and cross-view fusion to achieve CT reconstruction. However, the alignment errors between different views can introduce artefacts during fusion.

Challenges remain in interpolating the missing ST slices. **Firstly,** ST slices are often affected by factors such as tissue heterogeneity and experimental procedures, resulting in limited continuity and high deformation between slices. However, traditional interpolation algorithms such as bilinear methods assume strong continuity and fail to capture the nonlinear deformations in tissue structure [6]. To generate arbitrary ST slices, tailored methods are demanded for learning deformation features at given positions. **Secondly**, ST maps contain pixel-wise expression of thousands of genes, with sparse expression in most genes [8]. However, existing methods are not tailored to model the spatial gene co-expression in ST maps. **Lastly**, biological systems are complex with not only rich gene expression information but also fine cellular structures and biological semantics. Existing methods fail to model gene expression within the context of tissue structures, often resulting in blurred cell boundaries and distorted interpolation.

To address these challenges, we propose a bespoke framework for ST interpolation at arbitrary positions via cross-modal cross-slice modeling (C2-STi).

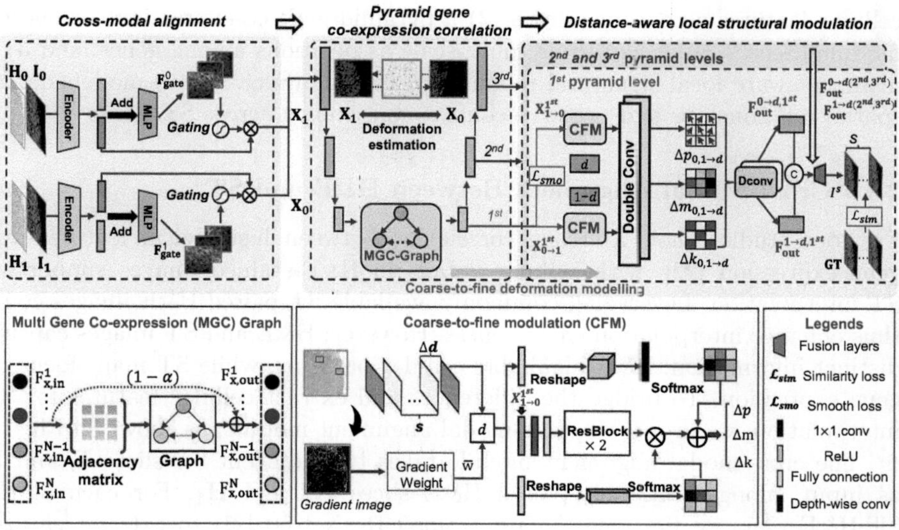

Fig. 1. Framework of the C2-STi, including three modules of cross-modal alignment, pyramid gene co-expression correlation and distance-aware local structural modulation.

Firstly, we propose a distance-aware local structural modulation (DLSM) module that leverages distance perception and local tissue structure to adaptively capture cross-slice deformations between ST slices, promising to achieve high-quality interpolation at arbitrary positions. **Secondly**, we emphasize capturing multi-scale deformation information across slices by modeling gene co-expression. Specifically, We design a pyramid gene co-expression correlation module that integrates a pyramid encoder, a multi-gene co-expression graph (MGC-Graph), and a multi-scale decoder, capturing deformation information between slices at various scales while reflecting gene co-expression relationships. **Lastly**, we propose a novel cross-modal alignment module that leverages gated attention weight to integrate H&E and ST features for preserving key cellular information [23]. To our best knowledge, this is the first algorithm to interpolate arbitrary number of missing sections between adjacent ST slices. Experiments demonstrate that our method outperforms other state-of-the-art (SOTA) methods.

2 Methodology

Figure 1 illustrates the proposed C2-STi. Our model achieves St interpolation by adaptively interpolating the missing slices through cross-modal cross-section modeling, leveraging H&E images and ST maps from adjacent slices. C2-STi takes as input two adjacent ST slices of N genes, $\mathbf{I}_0$ and $\mathbf{I}_1$, along with their corresponding H&E images $\mathbf{H}_0$ and $\mathbf{H}_1$. It generates s imputed intermediate ST slices, $\{\mathbf{I}^i_{\text{out}}\}^s_{i=1}$. In general, three key modules are proposed in C2-STi, including **1)** a cross-modal alignment module (Sect. 2.1) to filter and match the essential cellular features across modalities, **2)** a pyramid gene co-expression correlation module (Sect. 2.2) for capturing multi-scale associations among genes, and **3)** a distance-aware local structural modulation module (Sect. 2.3) to model the the spatial relationships and preserve context consistency across ST slices.

2.1 Cross-Modal Alignment Between H&E and ST

Previous studies show a strong correlation between histological features and gene expression [22], with cellular details in H&E-stained images supporting ST interpolation. Leveraging commonly available ST-paired H&E images could thus enhance interpolation effectiveness. However, H&E and ST images capture distinct information: H&E highlights cellular patterns, while ST maps focus on gene expression. To bridge the differences and extract cellular features for ST interpolation, we propose a cross-modal alignment module, as shown in Fig. 1.

The cross-modal alignment module takes two adjacent ST slices $\mathbf{I}_0$ and I_1 as input, along with their paired H&E slices $\mathbf{H}_0$ and $\mathbf{H}_1$. For each of the ST-H&E pair, we first use the pre-trained ResNet-50 [21] to extract features $\mathbf{M_h} \in \mathbb{R}^{C_h \times H \times W}$ (with C_h, H and W denoting the channel number, height and width of the features) and $\mathbf{M_s} \in \mathbb{R}^{C_s \times H \times W}$ for the H&E and ST image, respectively. Then, to mitigate the impact of inherent noise from each modality on the overall fusion, we further conduct gated-attention-based cross-modal alignment on the concatenated multimodal features $\mathbf{F}_{\text{cat}} = [\mathbf{M_h}; \mathbf{M_s}] \in \mathbb{R}^{(C_s+C_h) \times H \times W}$. Specifically, we used two 1×1 convolutional layers coupled with activations to generate the attention weights $\mathbf{G}$ for retaining useful ST features with rich cellular patterns. Formally, $\mathbf{G}$ is defined as:

$$G = \text{Sigmoid}\left(\text{Conv}_{1\times1}(\mathbf{F}_{\text{gate}})\right), \text{where } \mathbf{F}_{\text{gate}} = \text{ReLU}\left(\text{Conv}_{1\times1}(\mathbf{F}_{\text{cat}})\right) \quad (1)$$

Finally, we obtained the cross-modal aligned features $\mathbf{X} = \mathbf{M_s} \odot \mathbf{G}$, where $\odot$ represents element-wise multiplication, for each of the two ST maps.

2.2 Pyramid Gene Co-expression Correlation

As shown above, effective ST slice interpolation requires both tissue-level and cellular-level deformations while considering gene-specific variations . To address this, we propose the pyramid gene co-expression correlation module, shown in Fig. 1, which integrates a pyramid encoder-decoder structure with a multi-gene co-expression graph (MGC-Graph). Details follow the processing flow below.

Pyramid Feature Encoder. For each cross-modal fused feature $\mathbf{X}_0$ and $\mathbf{X}_1$ from adjacent ST slices, we design a weight-shared extractor to capture contextual information using pyramid features. At each pyramid level $L \in \{1, 2, 3\}$, four convolutional layers perform spatial downsampling, producing the final multi-scale features $\mathbf{C}_0^L, \mathbf{C}_1^L$, each with the feature channel of C.

MGC-Graph. To capture the co-expression relationships among genes in adjacent ST slices, we propose MGC-Graph, denoted as $\mathbf{g} = (\mathcal{V}, \mathcal{E})$, where $\mathcal{V}$ represents nodes (genes) and $\mathcal{E}$ represents edges based on co-expression intensity.

Specifically, given the output features of the pyramid feature encoder $\{\mathbf{C}_0^L, \mathbf{C}_1^L\}$ as input nodes, we construct a multi-gene co-expression based correlation matrix $\mathbf{A} \in \mathbb{R}^{N \times N}$ to reflect the relationships among each node feature, with a weight matrix $\mathbf{W}_n \in \mathbb{R}^{C \times C}$ to update the value of $\mathbf{C}_0^L$ and $\mathbf{C}_1^L$. Formally, the correlation matrix $\mathbf{A}$ can be defined as:

$$\mathbf{A} = [\rho_{ik}]_{i,k=1}^N, \text{where } \rho_{ik} = \tilde{\mathbf{E}}_i \cdot \tilde{\mathbf{E}}_k, \text{and } \tilde{\mathbf{E}}_i = \frac{\mathbf{E}_i - \bar{\mathbf{E}}_i}{\|\mathbf{E}_i - \bar{\mathbf{E}}_i\|_2^2}. \tag{2}$$

where $\cdot$ denotes the calculation of Pearson correlation coefficient (PCC), and $\bar{\mathbf{E}}_i$ represents the average expression intensity of i-th gene across ST slices. Based on the correlation matrix $\mathbf{A}$, the output nodes $\mathbf{F}_0^L$ of $\mathbf{C}_0^L$ are formulated by a single graph convolutional network(GCN) layer as

$$\mathbf{F}_0^L = \lambda\, \mathbf{C}_{0,graph}^L + (1-\lambda)\, \mathbf{C}_0^L, \text{where } \mathbf{C}_{0,graph}^L = \delta(\mathbf{P}\, \mathbf{C}_0^L\, W_n). \tag{3}$$

where $\delta(\cdot)$ is an activation function, and λ is a graph balancing hyper-parameter Similarly, the output nodes $\mathbf{F}_1^L$ of $\mathbf{C}_1^L$ are obtained using the same GCN layer.

Multi-deformation Estimation Decoder. After extracting hierarchical information, we add a separate decoder $D^{(L)}$ for the feature maps $\mathbf{F}_0^L$ and $\mathbf{F}_1^L$ of multiple scales. Specifically, the coarse deformation features between $\mathbf{F}_0$ and $\mathbf{F}_1$ are processed as follows: $\{\mathbf{F}_{0\to1}^n, \mathbf{F}_{1\to0}^n\}_{n=1}^L = D^{(L)}(\mathbf{F}_0^L, \mathbf{F}_1^L)$.

2.3 Distance-Aware Local Structural Modulation Across ST Slices

To enhance structural consistency across input ST slices and enable interpolation at arbitrary positions, we propose the DLSM module, shown in Fig. 1. It refines coarse deformation features $\{\mathbf{F}_{0\to1}^n, \mathbf{F}_{1\to0}^n\}_{n=1}^L$ based on slice positions. The module comprises three key components: cross-section distance modeling, coarse-to-fine modulation, and deformable convolutional fusion layers.

Cross-Section Distance Modeling. As shown in Fig. 1, we model the distance between adjacent ST slices to to quantify biological structure similarity and enable effective interpolation. Given s under-interpolated slices and interval Δd interval between two adjacent ST slices, the distance between any two slices $\mathbf{I}_i$ and $\mathbf{I}_j$, across the entire set of s slices can then be expressed as $d_{i,j} = (s+1)\Delta d$.

Next, the cross-slice distance can be further refined via weighted on the amplitude of the gradient map[1] of the H&E images to emphasize regions with significant structural changes. Specifically, the Sobel operator [4] is used to calculate the gradient-based weight of i-th slice as $\overline{w}_i = \text{Norm}(1+\alpha|\nabla \mathbf{H}_i|)$, where α is a gradient scaling parameter. Thus, the position set of the under-interpolated slices is obtained as $\mathbf{P} = \{\text{p}_i \mid \text{p}_i = \frac{i}{s} \cdot \overline{w}_i \in (0,1),\ i \text{ values from } 1 \text{ to } s-1\}$.

Coarse-to-Fine Modulation. Based on the position set $\mathbf{P}$, we design a coarse-to-fine modulation to generate multiple deformation features at different positions, adaptively adjusting the coarse deformation features $\{\mathbf{F}_{0\to1}^n, \mathbf{F}_{1\to0}^n\}_{n=1}^L$.

[1] The gradient map can capture the structural information of an image [9].

The forward feature $\mathbf{F}_{0\to1}$ is modulated by $\{\text{p}_i\}_{i=1}^{s-1}$, while the backward feature $\mathbf{F}_{1\to0}$ is modulated by $\{1-\text{p}_i\}_{i=1}^{s-1}$ for symmetry. The process is detailed below.

First, we apply global average pooling to $\mathbf{F}_{0\to1}$ to obtain a statistical vector $\bar{\mathbf{f}}$, while the positions $\mathbf{P} = \{\text{p}_i\}_{i=1}^{s-1}$ are mapped to the embedding vector $\mathbf{e}_p = \phi(\mathbf{P}) \in \mathbb{R}^d$. Concatenating $\bar{\mathbf{f}}$ and $\mathbf{e}_p$ to obtain global context and positional information. Further, the concatenated features are fed into fully-connected layers to generate channel weight $\mathbf{w}_c$, used for obtaining the channel-adaptive feature $\mathbf{F}_{\text{ca}} = \mathbf{w}_c \odot \mathbf{F}_{0\to1}$, where $\odot$ denotes Kronecker product. Moreover, to extract local details, a 3×3 convolutional is applied to $\mathbf{F}_{0\to1}$ to obtain local structural features $\mathbf{F}_{\text{loc}}$. The positions $\mathbf{P} = \{\text{p}_i\}_{i=1}^{s-1}$ are mapped by another embedding function to $\mathbf{e}'_p = \psi(\mathbf{P}) \in \mathbb{R}^{d'}$ and expanded into a tensor $\mathbf{E}_p \in \mathbb{R}^{d' \times H \times W}$.

Next, we use 1×1 convolution and multilayer perceptron (MLP) to obtain the fused feature $\mathbf{F}_{\text{fused}}$. Then, taking $\mathbf{E}'_p$, a small MLP is utilized to generate a adjustable kernel k applied to $\mathbf{F}_{\text{fused}}$ through depth-wise convolution, further generating the spatial-adaptive feature $\mathbf{F}_{\text{sp}}$. Further, the output of the channel branch $\mathbf{F}_{\text{ca}}$ and the spatial branch $\mathbf{F}_{\text{sp}}$ are combined using preset weights β and γ to obtain the final fully modulated deformable feature $\mathbf{F}_{\text{out}} = \beta \mathbf{F}_{\text{ca}} + \gamma \mathbf{F}_{\text{sp}}$.

Finally, three estimators predict adjustments for $\mathbf{F}_{\text{out}}^{0\to1}$ with its position-related kernel $\Delta k_{0\to d}$, offsets $\Delta p_{0\to d}$, and masks $\Delta m_{0\to d}$ as follows:

$$\begin{cases} \Delta k_{0\to d} &= \text{KernelEstimator}(\mathbf{F}_{\text{out}}) \in \mathbb{R}^{C \times k_h \times k_w}, \\ \Delta p_{0\to d} &= \text{OffsetEstimator}(\mathbf{F}_{\text{out}}) \in \mathbb{R}^{\text{offset\_dim}}, \\ \Delta m_{0\to d} &= \text{MaskEstimator}(\mathbf{F}_{\text{out}}) \in \mathbb{R}^{\text{mask\_dim}}. \end{cases} \tag{4}$$

Similarly, the fully modulated deformable feature $\mathbf{F}_{\text{out}}^{1\to0}$ for $\mathbf{F}_{1\to0}$ can also be obtained via the above process.

Deformable Convolution Fusion Layers: Based on the learned offsets, masks, and kernels, we further use the deformable convolution [17], denoted as $\mathcal{F}_{\text{DConv}}$, to generate the intermediate interpolation feature maps:

$$\begin{cases} \mathbf{F}_{\text{out}}^{0\to d} = \mathcal{F}_{\text{DConv}}(\mathbf{F}_{0\to1}, \Delta p_{0\to d}, \Delta m_{0\to d}, \Delta k_{0\to d}), \\ \mathbf{F}_{\text{out}}^{1\to d} = \mathcal{F}_{\text{DConv}}(\mathbf{F}_{1\to0}, \Delta p_{1\to d}, \Delta m_{1\to d}, \Delta k_{1\to d}). \end{cases} \tag{5}$$

At each pyramid level, $\mathbf{F}_{\text{out}}^{0\to d}$ and $\mathbf{F}_{\text{out}}^{1\to d}$ are fused and upsampled with a pixel-shuffle layer to obtain the imputed s intermediate ST slices $\{\mathbf{I}_{\text{out}}^i\}_{i=1}^s$.

2.4 Loss Function

We use two loss functions to align the interpolated slice $\mathbf{I}^s$ and ground truth $\mathbf{I}^{gt}$: $\mathcal{L}_{\text{sim}} = \sum_s \|\mathbf{I}^{gt} - \mathbf{I}^s\|_1$ measuring the pixel-wise discrepancy, and $\mathcal{L}_{\text{smo}} = \|\nabla F_{0\to1}\|_1 + \|\nabla F_{1\to0}\|_1$ regularizing pixel variations in the interpolated slice. The final loss function is expressed as:

$$\mathcal{L} = \lambda_{\text{sim}} \mathcal{L}_{\text{sim}} + \lambda_{\text{smo}} \mathcal{L}_{\text{smo}}, \tag{6}$$

where $\lambda_{\text{sim}} = 1$ and $\lambda_{\text{smo}} = 1$, based on prior knowledge that similarity and smoothness are equally important. These weights control the balance between prediction accuracy and spatial consistency.

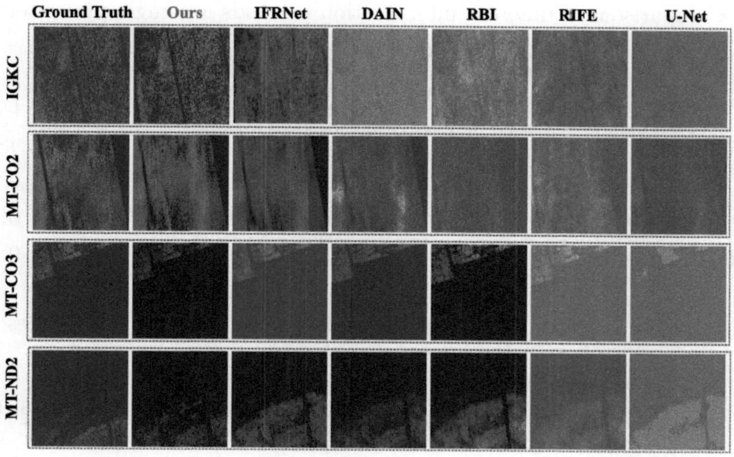

Fig. 2. Visual comparisons for single slice interpolation result. Note that IGKC, MT-CO2, MT-CO3, and MT-ND2 denote different genes.

3 Experiments and Results

Dataset. we use the largest 3D H&E-ST dataset, i.e., HNSCC [16], which includes 1,216 aligned H&E-ST image pairs from 19 consecutive slices spanning 350 mm. The patches are with the size of 512 × 512 and splitted into the training (852), validation (100), and test (264) sets. For the single-slice interpolation task, for each of the three consecutive patches, we use the $1st$ and $3rd$ slices as input, and the $2nd$ slice as the targeted slice for interpolation. Similarly, for N-slice interpolation tasks, we further construct the input-output pair using each of the $N+2$ consecutive patches of the whole dataset.

Implementation Details. Our method is optimized using AdamW [10] and implemented with PyTorch. We trained our model for 40 epochs on seven NVIDIA RTX V100 32 GB GPUs, with a weight decay of 10^{-4}. The initial learning rate is set to 10^{-4}, which is decreased to 10^{-6} using cosine annealing. The batch size for each training step is set to 6.

3.1 Performance Evaluation

Single Slice Interpolation. We compared our model with five other SOTA methods, i.e., IFRNet [5], RBI [14], DAIN [1], RIFE [3] and U-Net [13]. Figure 2 shows the visual results of the single-slice interpolation task, comparing our method with other SOTA methods. Specifically, it presents visualizations for four representative genes: IGKC, MT-CO2, MT-CO3, and MT-ND2.Of note, IFRNet, DAIN and RIFE are SOTA video frame interpolation approaches, while RBI is MR image interpolation method, and U-Net serves as the baseline. The left panel of Table 1 shows that C2-STi outperforms all competing methods in single-slice interpolation. Compared to the other five methods, our method improves the

Table 1. Comparisons of different interpolation methods for single slice interpolation.

Method	Interpolation network						Ablation (*w/o*)		
	RIFE	RBI	DAIN	IFRNet	Unet	Ours	Cross-modal	MGC-Graph	DLSM
PSNR↑	41.02	41.47	44.48	51.07	40.21	**54.11**	48.72	50.15	48.80
SSIM↑	0.76	0.71	0.76	0.73	0.61	**0.81**	0.77	0.78	0.77
PCC↑	0.17	0.23	0.33	0.38	0.15	**0.49**	0.42	0.47	0.47
RMSE↓	0.16	0.19	0.15	0.14	0.23	**0.11**	0.13	0.12	0.12

Table 2. Comparisons of different interpolation methods for multi-slices task

Method	Position-based	2 slices				3 slices				4 slices			
		PSNR	RMSE	PCC	SSIM	PSNR	RMSE	PCC	SSIM	PSNR	RMSE	PCC	SSIM
RIFE	Pre-defined	41.50	0.29	0.34	0.68	38.30	0.27	0.25	0.58	39.80	0.29	0.21	0.57
DAIN	Pre-defined	42.30	0.28	0.42	0.72	39.10	0.27	0.25	0.64	40.50	0.27	0.24	0.51
IFRNet	Pre-defined	40.75	0.31	0.45	0.74	**42.95**	0.21	0.28	0.65	39.30	0.22	0.27	0.74
RBI	Adaptive	42.80	0.25	0.17	0.61	35.50	0.30	0.15	0.59	33.30	0.33	0.17	0.47
Unet	Pre-defined	43.10	0.27	0.15	0.59	33.00	0.28	0.14	0.55	33.20	0.36	0.13	0.44
w/o Cross-modal	Adaptive	43.30	0.18	0.43	0.72	42.60	0.20	0.28	0.72	41.80	0.19	0.34	0.70
w/o MGC-Graph	Adaptive	43.70	0.16	0.40	0.78	42.40	0.22	0.26	0.72	41.90	0.19	0.35	0.69
w/o DLSM	Pre-defined	43.00	0.14	0.44	0.77	42.20	0.18	0.28	0.73	41.20	0.19	0.33	0.72
Ours	Adaptive	**43.90**	**0.11**	**0.46**	**0.78**	42.77	**0.18**	**0.34**	**0.77**	**42.00**	**0.19**	**0.35**	**0.76**

peak signal-to-noise ratio (PSNR), structural similarity index measure (SSIM), and PCC by at least 6.0%, 6.6%, and 29.0%, respectively, while also advancing the root mean square error (RMSE) by at least 21.4%. These results demonstrate that our method effectively recovers slice information, enhances image quality, and preserves structural consistency.

Multi-slice Interpolation. The right panel of Table 2 shows the performance of various interpolation methods under multi-slice settings (2, 3, and 4 slices). Our model consistently outperforms all competitors. Specifically, in the 2-slice scenario, our approach achieves a PSNR of 43.90 dB, which is at least 3.2% higher than the other methods. Similarly, for the 3-slice case, C2-STi attains the best SSIM of 0.77, representing an improvement of at least 2.5% over the alternatives. Finally, in the 4-slice setting, our method demonstrates robust performance by achieving a PSNR that is at least 3.1% greater than those of the compared approaches. These results highlight that our model not only excels in single-slice interpolation but also in handling more complex multi-slice scenarios.

3.2 Ablation Experiments

We assess the contribution of three key components in our model: 1) *w/o* Cross-modal – using only the ST modality; 2) *w/o* MGC-Graph – removing gene co-expression modeling; and 3) *w/o* DLSM – inserting slices at predefined positions without considering structural correlations. The ablation results of single-slice and multi-slice interpolation tasks are tabulated in Table 1 and Table 2, respectively. As shown in both tables, all three ablated models perform worse than the complete model, suggesting that these components substantially enhance overall performance. Notably, the *w/o* Cross-modal variant performs the worst, consistent with our hypothesis that conventional fusion strategies fail to fully leverage complementary multi-modal information.

4 Conclusion

In this study, we introduced C2-STi, the first work for interpolating missing ST slices at arbitrary intermediate positions. Addressing key challenges in ST interpolation, our approach ensures structural continuity across heterogeneous tissue sections, models gene correlations, and preserves intricate cellular structures. By integrating a distance-aware local structural modulation module, a pyramid gene co-expression correlation module, and a cross-modal alignment module, C2-STi effectively captures cross-slice deformations, enhances positional correlations, and leverages H&E images for refinement. Extensive experiments on a public dataset demonstrate its superiority over SOTA methods in both single-slice and multi-slice ST interpolation.

Disclosure of Interests. XW is grateful for the China Scholarship Council (CSC) Cambridge scholarship. CL is funded by Guarantors of Brain Fellowship.

References

1. Bao, W., Lai, W.S., Ma, C., Zhang, X., Gao, Z., Yang, M.H.: Depth-aware video frame interpolation (2019). https://arxiv.org/abs/1904.00830
2. Frakes, D.H., et al.: A new method for registration-based medical image interpolation. IEEE Trans. Med. Imaging **27**(3), 370–377 (2008). https://doi.org/10.1109/TMI.2007.907324
3. Huang, Z., Zhang, T., Heng, W., Shi, B., Zhou, S.: Real-time intermediate flow estimation for video frame interpolation (2022). https://arxiv.org/abs/2011.06294
4. Kanopoulos, N., Vasanthavada, N., Baker, R.L.: Design of an image edge detection filter using the Sobel operator. IEEE J. Solid-State Circuits **23**(2), 358–367 (1988)
5. Kong, Let al.: IFRNet: intermediate feature refine network for efficient frame interpolation (2022). https://arxiv.org/abs/2205.14620
6. Li, M., Que, N., Zhang, J., Du, P., Dai, Y.: VTCNet: a feature fusion dl model based on CNN and VIT for the classification of cervical cells. Int. J. Imaging Syst. Technol. **34**(5), e23161 (2024)
7. Li, W., et al.: SFCLI-Net: spatial-frequency collaborative learning interpolation network for computed tomography slice synthesis. Expert Syst. Appl. **272**, 126602 (2025). https://doi.org/10.1016/j.eswa.2025.126602. https://www.sciencedirect.com/science/article/pii/S0957417425002246
8. Liang, Y., et al.: Prost: quantitative identification of spatially variable genes and domain detection in spatial transcriptomics. Nat. Commun. **15**(1), 600 (2024). https://doi.org/10.1038/s41467-024-44835-w

9. Liu, A., Lin, W., Narwaria, M.: Image quality assessment based on gradient similarity. IEEE Trans. Image Process. **21**(4), 1500–1512 (2011)
10. Loshchilov, I., Hutter, F.: Decoupled weight decay regularization. arXiv preprint arXiv:1711.05101 (2017)
11. Monjo, T., Koido, M., Nagasawa, S., Suzuki, Y., Kamatani, Y.: Efficient prediction of a spatial transcriptomics profile better characterizes breast cancer tissue sections without costly experimentation. Sci. Rep. **12**(1), 4133 (2022)
12. Moses, L., Pachter, L.: Museum of spatial transcriptomics. Nat. Methods **19**(5), 534–546 (2022)
13. Park, S., Han, D., Kwak, N.: The U-Net based glow for optical-flow-free video interframe generation (2021). https://arxiv.org/abs/2103.09576
14. Penney, G., Schnabel, J., Rueckert, D., Viergever, M., Niessen, W.: Registration-based interpolation. IEEE Trans. Med. Imaging **23**(7), 922–926 (2004). https://doi.org/10.1109/TMI.2004.828352
15. Qiu, X., et al.: Spatiotemporal modeling of molecular holograms. Cell **187**(26), 7351–7373 (2024)
16. Schott, M., et al.: Open-ST: high-resolution spatial transcriptomics in 3d. Cell **187**(15), 3953–3972.e26 (2024). https://doi.org/10.1016/j.cell.2024.05.055
17. Shi, Z., Liu, X., Shi, K., Dai, L., Chen, J.: Video frame interpolation via generalized deformable convolution (2021). https://arxiv.org/abs/2008.10680
18. Tian, L., Chen, F., Macosko, E.Z.: The expanding vistas of spatial transcriptomics. Nat. Biotechnol. **41**(6), 773–782 (2023)
19. Wang, X., Huang, X., Price, S., Li, C.: Cross-modal diffusion modelling for super-resolved spatial transcriptomics. In: Linguraru, M.G., et al. (eds.) MICCAI 2024. LNCS, vol. 15003, pp. 98–108. Springer, Cham (2024). https://doi.org/10.1007/978-3-031-72384-1_10
20. Wang, X., Price, S., Li, C.: Multi-task learning of histology and molecular markers for classifying diffuse glioma. arXiv preprint arXiv:2303.14845 (2023)
21. You, Y., Zhang, Z., Hsieh, C.J., Demmel, J., Keutzer, K.: ImageNet training in minutes. In: Proceedings of the 47th International Conference on Parallel Processing, ICPP 2018. Association for Computing Machinery, New York (2018). https://doi.org/10.1145/3225058.3225069
22. Zhang, D., et al.: Inferring super-resolution tissue architecture by integrating spatial transcriptomics with histology. Nat. Biotechnol. **42**(9), 1372–1377 (2024). https://doi.org/10.1038/s41587-023-02019-9
23. Zhang, Y., Wang, X., Meng, F., Tang, J., Li, C.: Knowledge-driven subspace fusion and gradient coordination for multi-modal learning. In: Linguraru, M.G., et al. (eds.) MICCAI 2024LNCS, vol. 15004, pp. 263–273. Springer, Cham (2024). https://doi.org/10.1007/978-3-031-72083-3_25
24. Zhao, Q., Zhang, Q.: High-dimensional Bayesian model for disease-specific gene detection in spatial transcriptomics (2024). https://arxiv.org/abs/2409.02397

Anatomy-Aware Frequency-Attention Transformer Networks for Liver Couinaud CT/MR Segmentation

Wenkang Fan[1], Hao Fang[1,2], Rui Li[3], Yanduan Lin[3], Chao An[4], and Xiongbiao Luo[1,2,5](✉)

[1] Department of Computer Science and Engineering, Xiamen University, Xiamen 361102, China
[2] National Institute for Data Science in Health and Medicine, School of Medicine, Xiamen University, Xiamen 361102, China
[3] Zhongshan Hospital, Xiamen University, Xiamen 361004, China
[4] Department of Interventional Ultrasound, Chinese PLA General Hospital, Beijing 100853, China
[5] Discipline of Intelligent Instrument and Equipment, Xiamen University, Xiamen 361102, China

xiongbiao.luo@gmail.com

Abstract. Accurate Couinaud segmentation of liver CT/MR is essential in helping surgeons perceive the positional relationship between liver anatomy and intrahepatic lesions to make surgical planning. Unfortunately, current conventional and deep-learning based methods remain challenges in accurate Couinaud segmentation since the segmentation boundaries of different categories depending on hepatic vascular information are hard to predict. This work proposes a new deeply learned framework called anatomy-aware frequency-attention transformer networks (AFATN) for Couinaud segmentation of liver anatomy which contains the hybrid anatomy-aware preprocessing and frequency-attention transformer networks (FATN). Specifically, our framework first uses hybrid anatomy-aware preprocessing to integrate the hybrid cues of liver contour and hepatic venous centerline, then effectively utilizes hybrid cues for accurate Couinaud segmentation through the frequency-attention transformer networks with omission re-detected loss function. Our segmentation model FATN uses transformers to extract local structure and global semantic features and further focus on the hybrid cues with frequency-attention mechanisms. The proposed method was evaluated on clinical CT data and compared with currently available deep learning approaches, with the experimental results demonstrating that our method outperforms other approaches especially in accurately segmenting the Couinaud boundaries.

Keywords: Couinaud segmentation · Liver segmentation · Hepatic vessels · Transformer · Frequency attention

1 Introduction

The Couinaud segmentation standard is widely used to describe the functional anatomy of the liver, which divides the liver into eight distinct units based on its umbilical fis-

J. C. Gee et al. (Eds.): MICCAI 2025, LNCS 15960, pp. 55–65, 2026.
https://doi.org/10.1007/978-3-032-04927-8_6

sure and vascular supply, with each unit having its own inflow and outflow hepatic vessels [6]. Generating an accurate three-dimensional visualization of the Couinaud segmentation from Computed Tomography (CT) and Magnetic Resonance (MR) can help surgeons intuitively perceive the positional relationship between hepatic segments and intrahepatic lesions, and determine suitable treatment between radiofrequency ablation and hepatectomy procedures [19]. Since manual annotation in clinics is time-consuming and labor-intensive, automatic and accurate Couinaud segmentation for surgical planning is certainly beneficial to surgeons and patients.

Conventional Couinaud segmentation contains two categories: voxel-based and plane-based. Voxel-based methods allocate liver voxels to these segments by considering the distances between their locations and a specific branch of the hepatic vein [23], which does not follow the standard of Couinaud segmentation. Plane-based approaches first detect anatomical geometric elements like landmarks and lines from the liver surface and vessels, then reconstruct hepatic segments by the planes made up of these elements. For instance, Pamulapati et al. [1] built skeletonization veins and depend on the orientation between points and skeleton to construct the functional segment, while Wang and Arya et al. [2,20] employed key point detection for Couinaud classification.

Deep-learning based methods are more popular for automatic medical image segmentation. Convolutional neural networks (CNN) such as U-Net [5,11,12] and ResUnet [24] have achieved remarkable results for many segmentation tasks. CNN works well in extracting the local texture, while they have the limited ability to extract the global features because convolution only focuses on the surrounding regions instead of capturing long-distance relationships. To solve this problem, some works propose attention mechanisms to focus on crucial information of images [15,21,25,26]. Recently, transformer-based models or integrated CNN with transformer models have become more popular and achieved great performance on many liver-related segmentation tasks [4,9,10,16,22].

However, existing methods still remain challenging for accurate Couinaud segmentation because the boundary of the hepatic segments depends on the location of the hepatic veins and portal veins and there is no significant contrast between the boundaries of the different categories. This work considers the importance of utilizing the hepatic vessels for Couinaud segmentation and proposes a new deeply learned anatomy-aware frequency-attention transformer framework (AFATN) to accurately obtain hepatic segments. The highlights of our work are clarified as follows. First, we propose a hybrid anatomy-aware preprocessing to extract hybrid cues of liver contour and hepatic vessels and highlight these cues on CT/MR images to help networks perform feature extraction, which applies the advantages of plane-based approaches to deep learning methods. Second, we propose the frequency-attention transformer networks (FATN) for anatomy-aware Couinaud segmentation which uses a transformer-based encoder to extract sufficient local texture and global semantic features and employs global frequency-attention to enhance the structural boundary, especially the hepatic vascular information.

2 Methods

The proposed anatomy-aware frequency-attention transformer framework (Fig. 1) generally consists of two parts: (1) hybrid anatomy-aware preprocessing, and (2)

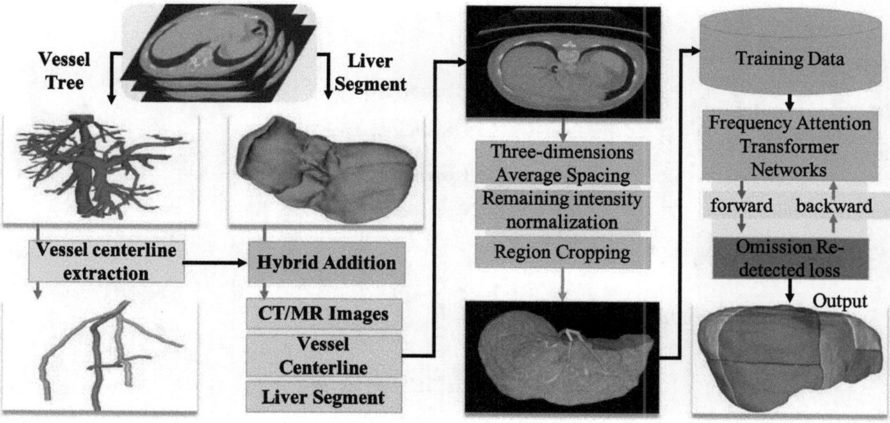

Fig. 1. Our new framework consists of hybrid anatomy-aware preprocessing and frequency-attention transformer networks (FATN) with omission re-detected loss.

frequency-attention transformer networks (Fig. 2) with omission re-detected loss function.

2.1 Hybrid Anatomy-Aware Preprocessing

The Couinaud segmentation boundary is based on the five planes consisting of the left umbilical fissure and four venous trunks (middle and right of the hepatic vein, left and right of the portal vein). Therefore, our framework aims to extract the vascular trunks and highlight this information in training CT/MR data, so that the segmentation network can better focus on this information to extract accurate hepatic segment boundaries.

For the input CT/MR images $\mathrm{I}(x, y, z)$, we first segment the liver mask $\mathrm{L}(x, y, z)$ through nnUNet [11] and segment hepatic and portal veins $\widehat{\mathrm{V}}(x, y, z)$ through [22]. Then, we employ the vessel centerline extraction approach from VTK [13] to extract hepatic venous trunks $\mathrm{V}(x, y, z)$ for segmented veins $\widehat{\mathrm{V}}(x, y, z)$. Note that all points on the extracted centerline will be selected according to their obtained vessel diameters, and points with relatively small vessel diameters are removed.

Now we have three intermediate outputs: the CT/MR input $\mathrm{I}(x, y, z)$, liver mask $\mathrm{L}(x, y, z)$ and intra-hepatic veins centerline $\mathrm{V}(x, y, z)$. The new CT/MR data $\mathrm{H}(x, y, z)$ can be obtained by a hybrid addition operation which integrates the original input with the extracted hybrid cues of liver mask and hepatic venous trunks

$$\mathrm{H}(x, y, z) = \begin{cases} 0, & \mathrm{L}(x, y, z) = False, \\ \text{maximum HU}, & \mathrm{L}(x, y, z) = True, \mathrm{V}(x, y, z) = True, \\ \mathrm{I}(x, y, z), & \mathrm{L}(x, y, z) = True, \mathrm{V}(x, y, z) = False, \end{cases} \tag{1}$$

where HU represents the intensity value range of $\mathrm{I}(x, y, z)$. After the hybrid addition operation, we averagely space the three-dimensions voxel (x, y, z) to (0.78, 0.78, 4.0) for all cases and use Z-score normalization to normalize the intensity value of $\mathrm{H}(x, y, z)$

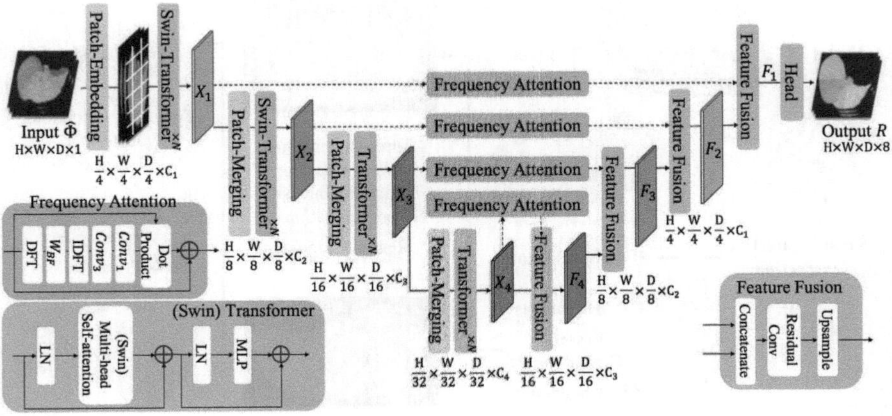

Fig. 2. Frequency-attention transformer networks (FATN) for Couinaud segmentation.

to narrow the variability from different patients according to [11]

$$\Phi(x, y, z) = \psi_{(0.78,0.78,4.0)}(\mathrm{Norm}_z^{x,y}(\mathrm{H}(x, y, z))), \tag{2}$$

where ψ is uniform spacing operation and $\mathrm{Norm}_z^{x,y}$ describes Z-score normalization.

Eventually, to reduce the size of input images and decrease unnecessary distractions of irrelevant regions, we perform region cropping operation $\varphi(\cdot)$ for all cases to exclude regions without liver to obtain the final training data $\hat{\Phi}$

$$\hat{\Phi} = \varphi(\Phi(x, y, z)) \text{ , } if \ \mathrm{L}(x, y, z) = False. \tag{3}$$

2.2 Frequency-Attention Transformer Networks

We explore and analyze the performance of CNN and the transformer for medical image segmentation. The CNN modules such as residual block [24] can extract sufficient local texture and structure features while they have the limited ability to extract the global semantic features. On the contrary, the transformer [7] using multi-head self-attention can capture long-distance relationships to extract more global semantic features, and the sliding window-based attention [14] can also capture sufficient local features.

Therefore, we design a transformer-based feature extraction model that integrates the swin-transformer [14] and standard transformer [7] to extract both local and global features. Additionally, we propose a frequency-attention module to enhance the edge structural information of feature maps, which makes networks pay more attention to vessel characteristics of anatomy-aware CT/MR data to obtain more accurate boundaries of hepatic segments. The architecture of our FATN is shown in Fig. 2.

Encoder. The transformer-based encoder contains four stages. The first stage will first divide the input training data $\hat{\Phi} \in \mathbb{R}^{H \times W \times D \times 1}$ into 3-D patches $\in \mathbb{R}^{M \times (P^3 \times 1)}$, where P is the patch size we set to 4 and $M = (H \times W \times D)/P^3$ represents the number of patches. Then, each 3-D patch will be flattened and projected into C_1 dimension token through a trainable linear projection. The first stage finally employs N swin-transformer layers [14] with shifted-window multi-head self-attention (MSA) and multi-layer perceptron (MLP) to extract local texture features $X_1 \in \mathbb{R}^{M \times C_1}$ from these tokens. After feature extraction, local features X_1 will be assembled into image-like feature maps $\in \mathbb{R}^{H/4 \times W/4 \times D/4 \times C_1}$ for the next feature-extraction stage and feature-fusion decoder. The operation of the second to the fourth feature-extraction stage is almost similar to the first stage. All the patch-merging modules set the patch size to 2 for downsampling the feature map of the previous stage. Note that the second stage also uses swin-transformer layers [14] for local feature extraction, while the third and fourth use standard transformer layers [7] to extract multi-scale global features. After that, we obtain two local feature maps and two global feature maps with different sizes $\{X_1, X_2, X_3, X_4\}$ from the transformer-based encoder.

Decoder. The feature-fusion decoder can be divided into two steps. It first uses the global frequency-attention module to further highlight the noteworthy or essential boundary structural information on each feature map, then fuses these feature maps through the feature-fusion module from coarse to fine and predicts the final segmentation result by a head module. The operation of the frequency attention module on feature maps $X_i \in \mathbb{R}^{H_i \times W_i \times D_i \times C_i}$ can be formulated as

$$\hat{X}_i = X_i + W_{FA}^i \circ X_i, i = \{1, 2, 3, 4\}, \tag{4}$$

$$W_{FA}^i = Conv_1(Conv_3(f_{IDFT}(W_{BF} \circ (f_{DFT}(X_i))))), \tag{5}$$

$$W_{BF}(x, y, z) = 1 - exp(-\alpha(\frac{\sqrt{(x - x_c)^2 + (y - y_c)^2 + (z - z_c)^2}}{\sqrt{(H_i/2)^2 + (W_i/2)^2 + (D_i/2)^2}})^{\beta}) \tag{6}$$

where f_{DFT} and f_{IDFT} are 3-D discrete Fourier transform (DFT) and 3-D inverse discrete Fourier transform (IDFT), which are applied to each channel of the feature maps, W_{BF} is our designed bandpass filter weighted map which is used to filter out low-frequency information in the spectrum and preserve high-frequency edge structural information, x_c, y_c, and z_c represents the center position of map, α and β are hyper-parameters used to control the degree of low-frequency information filtering,"$\circ$" represents dot product operation, and $Conv_3$ and $Conv_1$ are $3 \times 3 \times 3$ and $1 \times 1 \times 1$ convolution operations to obtain frequency attentive map W_{FA}.

After feature maps $\{\hat{X}_1, \hat{X}_2, \hat{X}_3, \hat{X}_4 X_4\}$ are obtained through the transformer-based encoder and the frequency attention module, four feature fusion modules corresponding to four stages of the encoder are used to fuse these features

$$F_4 = \mathcal{U}(\Theta(\otimes(X_4, \hat{X}_4))), \tag{7}$$

$$F_i = \mathcal{U}(\Theta(\otimes(F_{i+1}, \hat{X}_i))), i = 3, 2, 1, \tag{8}$$

where $\otimes$, Θ, and $\mathcal{U}$ represent channel-based concatenation, residual block [24], and upsample, respectively. Eventually, the head module with a residual block, upsample operation, and a $1 \times 1 \times 1$ convolution is to obtain the segmentation results

$$R = Conv_1(\mathcal{U}(\Theta(F_1))) \tag{9}$$

2.3 Omission Re-Detected Loss

In the multi-class segmentation task, topological interactions between different foreground classes generally include two types of Containment and Exclusion according to [8]. Because the relationship between different hepatic segments is excluded in the Couinaud segmentation task, we employ the convolution-based topological interaction module [8] to go through all pairs of adjacent voxels of the result R and identify the positions that violate the desired constraint

$$\mathrm{E}(x, y, z) = \begin{cases} 0, & \Pi_{R,G}(x, y, z) = False, \\ 1, & \Pi_{R,G}(x, y, z) = True, \end{cases} \tag{10}$$

where R and G describe the segmentation result and ground truth, respectively. $\Pi_{R,G} = True$ denotes the prediction violate the desired constraint, while $\Pi_{R,G} = False$ conversely. We re-detect the omission region $\mathrm{E}(x, y, z)$ and calculate the loss as follows.

We adopt a hybrid loss containing cross-entropy loss and dice coefficient loss

$$\mathcal{S}_L = \omega_{Dice}(R, G) + \omega_{BCE}(R, G), \tag{11}$$

$$\omega_{Dice}(R, G) = 1 - \frac{2\sum x_i \hat{x}_k}{\sum(x_k) + \sum(\hat{x}_k)}, \tag{12}$$

$$\omega_{BCE}(R, G) = -\sum(x_k \log(\hat{x}_k) + (1 - x_k)\log(1 - \hat{x}_k)), \tag{13}$$

where x_k and $\hat{x}_k$ are each pixel in G and R, respectively. Finally, the total omission re-detected Loss $\mathcal{OR}_L$ can be describe as:

$$\mathcal{OR}_L = (1 + \lambda_{\mathcal{E}}\mathrm{E}(x, y, z))\mathcal{S}_L(R, G) \tag{14}$$

where $\lambda_{\mathcal{E}}$ is the coefficient to measure the penalty level for the omission area.

3 Experimental Settings

We collected 330 CT cases from Task08_HepaticVessel [17] in the Medical segmentation Decathlon study (180 cases with hepatic vessel labels and Couinaud labels), 3D-IRCADb-01 [18] dataset (20 cases with liver and hepatic vessel labels) and Liver Tumor Segmentation Benchmark (LiTS) [3] (130 cases with liver labels). Couinaud labels of all 150 cases from 3D-IRCADb-01 and LiTS were annotated manually by three physicians.

We used Task08_HepaticVessel and 3D-IRCADb-01 to train our FATN to segment hepatic and portal veins of LiTS, and used 3D-IRCADb-01 and LiTS to train the networks [11] to obtain liver labels of Task08_HepaticVessel. The average spacing of all cases is around (0.78,0.78,4.0)mm^3 and we uniformed the size of 320 × 256 × 96 for all cases after region-cropping. For pre-processing and data augmentation of training data, we refer to nnU-Net [11]. The transformer layers N was 4 and the α, β, and $\lambda_{\mathcal{E}}$ were 5, 2, and 0.4, respectively. We set the learning rate from 10^{-4} to 10^{-3} and used a stochastic gradient descent algorithm as an optimizer with a momentum of 0.9 during

Table 1. Quantitative comparison of the hepatic vessel and liver Couinaud segmentation.

Methods	Liver Couinaud				Hepatic Vessels			
	DSC	IoU	Precision	Recall	DSC	IoU	Precision	Recall
APVN [25]	0.858	0.752	0.855	0.862	–	–	–	–
3D UNet [5]	0.887	0.799	0.895	0.882	0.814	0.687	0.818	0.811
UnetR++ [16]	0.910	0.836	0.928	0.894	0.819	0.694	0.822	0.817
TransUNnet [4]	0.935	0.879	0.942	0.930	0.837	0.720	**0.839**	0.836
FATN-(ours)	0.927	0.864	0.931	0.924	0.829	0.709	0.832	0.828
FATN(ours)	0.949	0.888	0.941	0.940	**0.840**	**0.724**	**0.839**	**0.842**
AFATN(ours)	**0.952**	**0.909**	**0.956**	**0.949**	–	–	–	–

training. We divided the collected data into training and testing data according to 7:3. The batch size, epoch, and iterations were set to 1, 500, and 1000, respectively.

We employ four popular metrics to evaluate the Couinaud segmentation of different methods: Dice similarity coefficient (DSC), intersection over union (IoU), precision, and recall. We compare our method anatomy-aware attention transformer networks (AFATN) with several 3-D segmentation methods: (1) Anatomical-aware point-voxel network (APVN) [25] which also highlights intra-hepatic veins for Couinaud segmentation, (2) 3-D UNet [5] which is a convolution-based model, (3) UnetR++ [16] which uses U-shape transformers to extract global features, (4) TransUNnet [4] which integrates convolution and transformer to extract local and global features with attention. Then, we conduct ablation studies to evaluate the effectiveness of our proposed anatomy-aware processing and frequency attention module. Therefore, we compare (5) FATN and (6) FATN-: without frequency attention. Additionally, we compare the effects of different methods on both hepatic vessels (hepatic veins and portal veins) and liver Couinaud segmentation.

4 Results and Discussion

Results. Figure 3 visually compares the 3-D Couinaud segmentation results of different methods and Fig. 4 illustrates the selected 2-D slices of results. Although UnetR++ [16] and TransUNnet [4] predict little misclassification within hepatic segments, the boundaries of different categories are not accurate, especially near the middle and right hepatic veins (according to *Row* 2 in Fig. 3 and *Columns* 1-2 Fig. 4) and right portal veins (according to *Rows* 4-5 in Fig. 3). We can see our FATN performs better than UnetR++ [16] and TransUNnet [4] in boundaries prediction and our AFATN can predict more accurate hepatic segment boundaries that better fit the location of the left umbilical fissure and the hepatic and portal veins. Table 1 summarizes the quantitative results of different models. For both hepatic vessels and liver Couinaud segmentation, transformer-based models UnetR++ [16], TransUNnet [4], and our models perform better than convolution-based or local feature extraction models APVN [25] and 3D

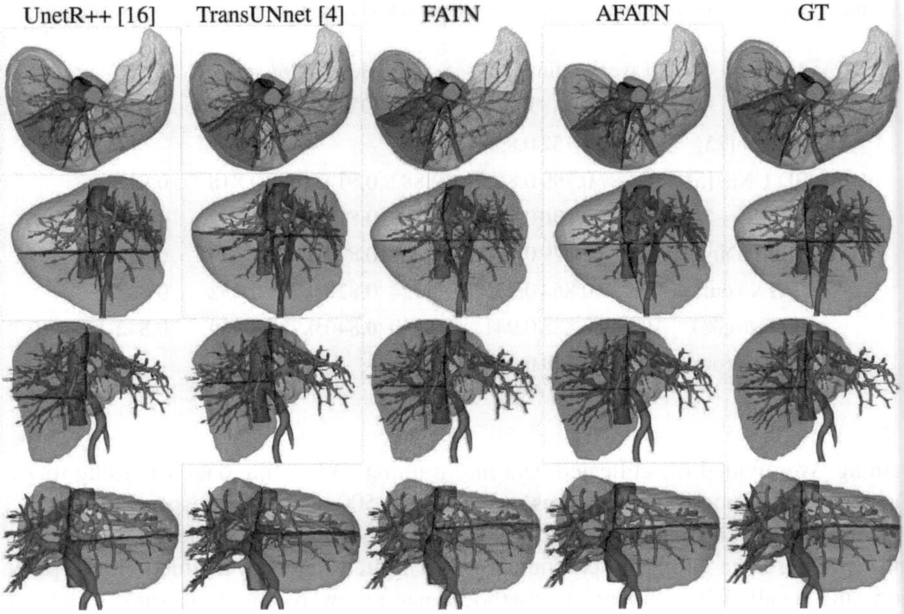

Fig. 3. Visual comparison of the Couinaud segmentation. *Rows* 2–5 shows a vertical view of the result with hepatic veins, the results near the right hepatic veins, near the middle hepatic veins and left portal veins, and near the right portal veins, respectively.

UNet [5]. Our FATN predicts higher DSC, IoU, and Recall than UnetR++ [16], TransUNnet [4], and FATN-. Meanwhile, our AFATN achieves better performance across various metrics compared to other methods, especially in terms of precision and recall.

Discussion. This work proposes a new deep-learning framework for Couinaud segmentation. The effectiveness of our method is as follows. First, our FATN extracts sufficient local texture and global structure features through transformers and employs global frequency attention to enhance the structural boundary, which is well-suited for segmenting objects with complex boundaries. Moreover, our AFATN uses hybrid anatomy-aware processing to extract and highlight the hybrid cues of liver contour and intra-hepatic veins, which can be effectively utilized for more accurate boundary segmentation. Transformer-based methods perform much better than APVN [25] and 3D UNet [5] in our experiments since APVN uses small volumes for prediction and 3-D UNet only extracts local features. FATN- performs better than UnetR++ [16] because it integrates local and global features. Since both FATN and TransUNnet [4] employ attention modules to enhance extracted features, they achieve better results than FATN-. Our FATN achieves better results than TransUNnet [4] due to the effective frequency-attention mechanism. Our approach has several limitations. First, it is difficult to accurately segment some cases with irregular lesions. Moreover, the effectiveness of our method depends on the accurate hepatic vessel segmentation. However, vessel segmentation remains challenging due to the irregular distribution. Additionally, the proposed

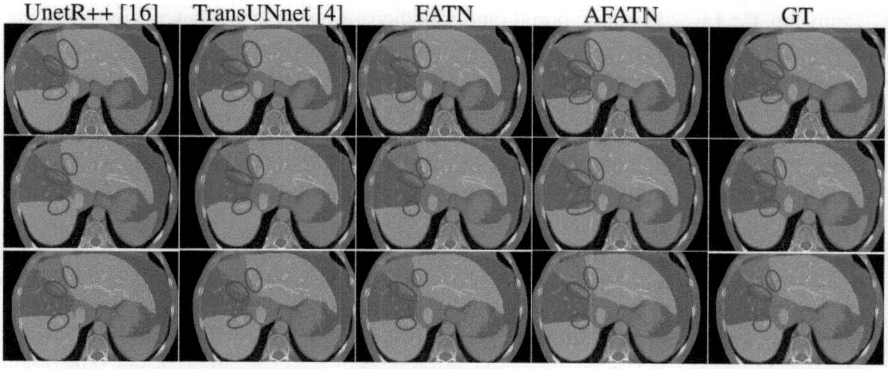

Fig. 4. Visual comparison of the Couinaud segmentation results of 2D slices using the four methods. Note that the *red circles* on the figures from top to down represent the position of the left umbilical fissure, and the middle and right hepatic veins, respectively. (Color figure online)

segmentation model has a large number of parameters and a high computational cost. We will optimize it using lightweight convolutional and self-attention modules.

Conclusion. In summary, this work proposes anatomy-aware frequency-attention transformer networks (AFATN) using the information of liver contour and hepatic vessels to achieve accurate Couinaud segmentation.

Acknowledgments. This work was supported in part by the National Natural Science Foundation of China under Grant 82272133, in part by the High-Quality Development Science and Technology Major Project of Xiamen Health Commission under Grant 2024GZL-ZD03, and in part by Ningbo 2035 Key Research and Development Program under Grant 2024Z127.

Disclosure of Interests. The authors have no competing interests to declare that are relevant to the content of this article.

References

1. Alirr, O.I., Abd. Rahni, A.A.: Automatic atlas-based liver segmental anatomy identification for hepatic surgical planning. Int. J. Comput. Assist. Radiol. Surg. **15**, 239–248 (2020)
2. Arya, Z., Ridgway, G., Jandor, A., Aljabar, P.: Deep learning-based landmark localisation in the liver for Couinaud segmentation. In: Papież, B.W., Yaqub, M., Jiao, J., Namburete, A.I.L., Noble, J.A. (eds.) MIUA 2021. LNCS, vol. 12722, pp. 227–237. Springer, Cham (2021). https://doi.org/10.1007/978-3-030-80432-9_18
3. Bilic, P., et al.: The liver tumor segmentation benchmark (LiTS), 1–43. arXiv preprint arXiv:1901.04056 (2019)
4. Chen, J., et al.: TransUNet: rethinking the u-net architecture design for medical image segmentation through the lens of transformers. Med. Image Anal. **97**, 103280 (2024)
5. Çiçek, Ö., Abdulkadir, A., Lienkamp, S.S., Brox, T., Ronneberger, O.: 3D U-Net: learning dense volumetric segmentation from sparse annotation. In: Ourselin, S., Joskowicz, L., Sabuncu, M.R., Unal, G., Wells, W. (eds.) MICCAI 2016, Part II. LNCS, vol. 9901, pp. 424–432. Springer, Cham (2016). https://doi.org/10.1007/978-3-319-46723-8_49

6. Couinaud, C.: Liver anatomy: portal (and suprahepatic) or biliary segmentation. Dig. Surg. **16**(6), 459–467 (1999)
7. Dosovitskiy, A., et al.: An image is worth 16x16 words: transformers for image recognition at scale. arXiv preprint arXiv:2010.11929 (2020)
8. Gupta, S., et al.: Learning topological interactions for multi-class medical image segmentation. In: Avidan, S., Brostow, G., Cissé, M., Farinella, G.M., Hassner, T. (eds.) ECCV 2022. LNCS, vol. 13689, pp. 701–718. Springer, Cham (2022). https://doi.org/10.1007/978-3-031-19818-2_40
9. He, Y., Nath, V., Yang, D., Tang, Y., Myronenko, A., Xu, D.: SwinUNETR-V2: stronger swin transformers with stagewise convolutions for 3d medical image segmentation. In: Greenspan, H., et al. (eds.) MICCAI 2023. LNCS, vol. 14223, pp. 416–426. Springer, Cham (2023). https://doi.org/10.1007/978-3-031-43901-8_40
10. Heidari, M., et al.: HiFormer: hierarchical multi-scale representations using transformers for medical image segmentation. In: Proceedings of the IEEE/CVF Winter Conference on Applications of Computer Vision, pp. 6202–6212 (2023)
11. Isensee, F., Jaeger, P.F., Kohl, S.A., Petersen, J., Maier-Hein, K.H.: nnU-Net: a self-configuring method for deep learning-based biomedical image segmentation. Nat. Methods **18**(2), 203–211 (2021)
12. Isensee, F., et al.: nnU-Net revisited: a call for rigorous validation in 3d medical image segmentation. In: Linguraru, M.G., et al. (eds.) MICCAI 2024. LNCS, vol. 15009, pp. 488–498. Springer, Cham (2024). https://doi.org/10.1007/978-3-031-72114-4_47
13. Izzo, R., Steinman, D., Manini, S., Antiga, L.: The vascular modeling toolkit: a python library for the analysis of tubular structures in medical images. J. Open Source Softw. **3**(25), 745 (2018)
14. Liu, Z., et al.: Swin transformer: hierarchical vision transformer using shifted windows. In: Proceedings of the IEEE/CVF International Conference on Computer Vision, pp. 10012–10022 (2021)
15. Oktay, O., et al.: Attention U-Net: learning where to look for the pancreas. arxiv 2018. arXiv preprint arXiv:1804.03999 (1804)
16. Shaker, A.M., Maaz, M., Rasheed, H., Khan, S., Yang, M.H., Khan, F.S.: UNETR++: delving into efficient and accurate 3d medical image segmentation. IEEE Trans. Med. Imag. (2024)
17. Simpson, A.L., et al.: A large annotated medical image dataset for the development and evaluation of segmentation algorithms. arXiv preprint arXiv:1902.09063 (2019)
18. Soler, L., et al.: 3D image reconstruction for comparison of algorithm database (2010). https://www.ircad.fr/research/data-sets/liver-segmentation-3d-ircadb-01
19. Wakabayashi, G., et al.: The Tokyo 2020 terminology of liver anatomy and resections: updates of the Brisbane 2000 system. J. Hepatobiliary Pancreat. Sci. **29**(1), 6–15 (2022)
20. Wang, M., Jin, R., Lu, J., Song, E., Ma, G.: Automatic CT liver Couinaud segmentation based on key bifurcation detection with attentive residual hourglass-based cascaded network. Comput. Biol. Med. **144**, 105363 (2022)
21. Woo, S., Park, J., Lee, J.-Y., Kweon, I.S.: CBAM: convolutional block attention module. In: Ferrari, V., Hebert, M., Sminchisescu, C., Weiss, Y. (eds.) ECCV 2018. LNCS, vol. 11211, pp. 3–19. Springer, Cham (2018). https://doi.org/10.1007/978-3-030-01234-2_1
22. Wu, Y., Shen, D., Jin, J., Xu, G., Chen, Y., Luo, X.: Local-global progressive u-transformers for accurate hepatic and portal veins segmentation in abdominal MR images. In: ICASSP 2023-2023 IEEE International Conference on Acoustics, Speech and Signal Processing (ICASSP), pp. 1–5. IEEE (2023)
23. Yang, X., et al.: Segmentation of liver and vessels from CT images and classification of liver segments for preoperative liver surgical planning in living donor liver transplantation. Comput. Methods Programs Biomed. **158**, 41–52 (2018)

24. Yu, W., Fang, B., Liu, Y., Gao, M., Zheng, S., Wang, Y.: Liver vessels segmentation based on 3D residual U-Net. In: 2019 IEEE International Conference on Image Processing (ICIP), pp. 250–254. IEEE (2019)
25. Zhang, X., et al.: Anatomical-aware point-voxel network for Couinaud segmentation in liver CT. In: Greenspan, H., et al. (eds.) MICCAI 2023. LNCS, vol. 14222, pp. 465–474. Springer, Cham (2023). https://doi.org/10.1007/978-3-031-43898-1_45
26. Zhou, W., et al.: HENet: hierarchical enhancement network for pulmonary vessel segmentation in non-contrast CT images. In: Greenspan, H., et al. (eds.) MICCAI 2023. LNCS, vol. 14222, pp. 551–560. Springer, Cham (2023). https://doi.org/10.1007/978-3-031-43898-1_53

Attention-Guided Vector Quantized Variational Autoencoder for Brain Tumor Segmentation

Danish Ali[1(✉)], Ajmal Mian[1], Naveed Akhtar[1,2], and Ghulam Mubashar Hassan[1]

[1] Computer Science and Software Engineering, The University of Western Australia, Perth, Australia
danish.ali@research.uwa.edu.au

[2] School of Computing and Information Systems, The University of Melbourne, Melbourne, Australia

Abstract. Precise brain tumor segmentation is critical for effective treatment planning and radiotherapy. Existing methods rely on voxel-level supervision and often struggle to accurately delineate tumor boundaries, increasing potential surgical risks. We propose an Attention-Guided Vector Quantized Variational Autoencoder (AG-VQ-VAE)—a two-stage network specifically designed for boundary-focused tumor segmentation. Stage 1 comprises a VQ-VAE which learns a compact, discrete latent representation of segmentation masks. In stage 2, a conditional network extracts contextual features from MRI scans and aligns them with discrete mask embeddings to facilitate precise structural correspondence and improved segmentation fidelity. Additionally, we propose an attention scaling module to reinforce discriminative feature learning and a soft masking module to refine attention in uncertain tumor regions. Comprehensive evaluations on BraTS 2021 demonstrate that our AG-VQ-VAE sets a new benchmark, improving the HD95 metric by 4.83 mm (Whole Tumor), 2.14 mm (Tumor Core), and 2.39 mm (Enhancing Tumor), compared to state-of-the-art methods, while achieving a 0.23% improvement in Dice score for whole tumor. Furthermore, our qualitative results and ablation study demonstrate that feature-level supervision significantly enhances boundary delineation compared to voxel-level approaches. The code is available at https://github.com/danishali6421/AG-VQVAE-MICCAI.

Keywords: Tumor boundary · Vector quantization · Self attention

1 Introduction

Brain tumor, one of the most common and life-threatening cancers, is caused by the abnormal growth of glial cells. Approximately 300,000 cases are reported annually, with 75% being malignant [24]. Magnetic resonance imaging (MRI) is

J. C. Gee et al. (Eds.): MICCAI 2025, LNCS 15960, pp. 66–76, 2026.
https://doi.org/10.1007/978-3-032-04927-8_7

frequently used to capture the detailed structure of brain tissues. Four major MRI modalities/sequences (T1, $T1_{CE}$, T2, and FLAIR) are used to capture heterogeneous and complementary information from the brain tissues. Expert radiologists segment tumor regions by carefully analyzing each MRI slice across these modalities. However, manual segmentation is a labor-intensive process due to the substantial heterogeneity in tumor morphology, including size, shape, and spatial location within the brain. Furthermore, low contrast between the healthy and tumorous regions in 3D multimodal MRI scans pose additional challenges in precise boundary delineation.

Advancement in deep learning frameworks has driven researchers to explore their potential for automatic tumor segmentation. Preliminary research [14,18] adopted a 2D U-Net architecture where each MRI volume was split into 2D slices, and each slice was segmented independently. Several CNN-based 2D U-Net models such as Znet [19], LeU-Net [21], and BU-Net [23], have demonstrated promising results. While 2D U-Net models made considerable progress in advancing brain tumor segmentation, their slice-wise processing of MRI data inherently disrupts spatial continuity across adjacent slices, leading to potential performance degradation [9]. To avoid this, various techniques such as 3DFCNN [1], ERV-Net [34], VAT [20], and MBANet [5] have been proposed that exploit full 3D MRI volumes. In addition, Zhang et al. [32] introduced a hierarchical multi-scale network (HMNET) to capture the tumor structures at various scales. Similarly, in the multi-scale residual U-Net (mResU-Net) [15], dilated convolutions of varying scales are employed to expand the receptive field, leading to improved segmentation accuracy for targets of varying sizes. Although these methods have excellent representation ability, they struggle to establish explicit long-range dependencies due to the limited local receptive fields of convolution kernels.

Recently, attention-based models, particularly Transformers [27], have gained significant recognition for their effectiveness in modeling global contextual relationships. Numerous Transformer models [4,7,26] have been proposed for medical image segmentation, achieving favorable results. The TransBTS [28] marks the pioneering attempt to integrate Transformer model into 3D CNN for brain tumor segmentation. This architecture extends the encoder-decoder framework by incorporating a Transformer in the bottleneck layer, leading to significantly improved segmentation. In contrast to TransBTS, UNETR [12] used a Transformer encoder to derive sequence representations of the input volume, effectively capturing global multiscale information. Jia et al. proposed a refined version of TransBTS called BiTr-Unet [13] which employs two sets of ViT layers to extract global information at two different scales. This solution surpasses the bottleneck approach (TransBTS) and offers computational efficiency compared to UNETR.

While transformer models have shown remarkable success in brain tumor segmentation, their attention mechanism involves computationally expensive operations with quadratic cost. To enchance computational efficiency, several window-based transformer models [3,8,11,35] have been proposed, with attention computed within localized windows. Dual-branch vision transformer (DBTrans) [31] models relationship between non-adjacent windows, significantly improving the segmentation accuracy. Despite considerable progress made by these methods, the challenge of accurately delineating tumor boundaries remains unsolved. Most existing methods emphasize maximizing the overlap between predicted and ground truth masks using voxel-level supervision, often relying on low-level spatial features passed through skip connections. However, they struggle with precise tumor boundary delineation [29], which is crucial for effective treatment planning and tumor resection while preserving surrounding healthy tissues [25].

To address the above limitations, we propose a novel two-stage brain tumor segmentation network that not only enhances tumor boundary delineation but also eliminates the reliance on low-level spatial features for segmentation tasks, thereby improving computational efficiency. This two-stage design is loosely inspired by recent advances in generative modeling, particularly DALL-E [22] and VQ-Diffusion [10], which leverage discrete latent representations for more structured generation. The first stage of our proposed model employs a Vector Quantized Variational Autoencoder (VQ-VAE) trained on segmentation masks to learn discrete latent representation, effectively capturing fine-grained boundary details. The second stage introduces a cross-latent alignment network, integrating convolutional layers with a Transformer-based bottleneck to effectively map the continuous latent representations of MRI modalities to the discrete representation of the segmentation mask. Our contributions are summarized below:

1. We propose a two-stage Attention-Guided Vector Quantized Variational Autoencoder (AG-VQ-VAE) to focus on structural coherence and accurate boundary delineation across all tumor regions using feature-level supervision, without relying on low-level spatial features from skip connections.
2. We propose a novel attention scaling mechanism that dynamically modulates the contribution of each attention head during training, prioritizing those that are most relevant for feature learning, resulting in improved generalization.
3. We propose a soft masking module to adaptively assign weights to spatial features based on their uncertainty following the attention calculation. This enforces the model to focus more on learning features associated with tumor boundaries that are inherently uncertain [17].

Extensive evaluation on the BraTS 2021 dataset demonstrates that our method significantly improves tumor boundary delineation compared to existing state-of-the-art models [16,31]. Our method improves the HD95 metric by 4.83 mm, 2.14 mm and 2.39 mm for whole, core and enchancing tumors, respectively.

2 Methodology

We develop a two stage AG-VQ-VAE model for brain tumor segmentation, see Fig. 1. The first stage employs a VQ-VAE to learn a discrete latent representation of the ground truth segmentation masks. In the second stage, MRI conditional network processes the four stacked MRI modalities, mapping their continuous high-dimensional latent features to the discrete indices of the pre-trained VQ-VAE codebook. Constituent components of our model are explained below.

2.1 Vector Quantized Variational Autoencoder

We cast brain tumor segmentation as a MRI-conditioned mask generation problem. To achieve this, three separate VQ-VAE networks are trained, each dedicated to a specific tumor sub-region: Whole Tumor (WT), Tumor Core (TC), and Enhancing Tumor (ET), using their respective 3D ground truth segmentation masks. Figure 1 shows that Stage 1 comprises a hierarchical VQ-VAE network, where the encoder (E_{vq}) processes the ground truth input masks and maps them to continuous structural features ($Z_e(x)$) that capture essential details of the tumor shape, size, and boundaries. These continuous features ($Z_e(x)$) are then passed through the vector quantization layer, which maps them to a finite set of discrete latent features. The vector quantizer consists of $K = 512$ embedding vectors, each of dimensionality $D = 32$ in its codebook E. The quantizer assigns each continuous feature to the nearest embedding vector in the codebook based on Euclidean distance, defined as:

$$Z_q(x) = \arg\min_{e_k \in E} \|Z_e(x) - e_k\|_2\,, \tag{1}$$

where $Z_e(x)$ and $Z_q(x)$ are the latent feature maps before and after quantization respectively, and e_k denotes the k-th embedding vector in the codebook E. Finally, the decoder (D_{vq}) reconstructs the mask from these quantized latent features. Dice loss $\mathcal{L}_{\text{Dice}}$ is employed as the reconstruction loss for VQ-VAE training. Straight-through estimator is applied to allow gradients to propagate

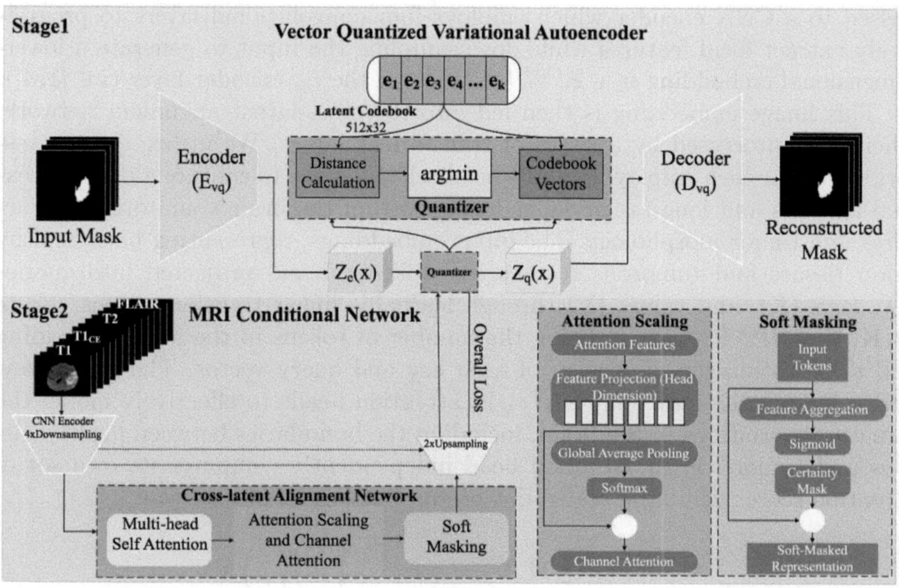

Fig. 1. The overall architecture of the proposed AG-VQ-VAE method. Stage 1 consists of VQ-VAE, while Stage 2 incorporates an attention-based MRI conditional network.

through the vector quantization step. Codebook vectors are updated using the exponential moving average of the encoder output. Additionally, a commitment loss

$$\mathcal{L}_{\text{ct}} = \|Z_e(x) - s_g[Z_q(x)]\|_2^2, \tag{2}$$

where s_g denotes the stop-gradient operator, is used to align the encoder's output with the codebook embeddings, to ensure stable training. The overall loss is:

$$\mathcal{L}_{\text{overall}} = \alpha_d \mathcal{L}_{\text{Dice}} + \alpha_c \mathcal{L}_{\text{ct}}, \tag{3}$$

where α_d and α_c are weights to balance Dice loss $\mathcal{L}_{\text{Dice}}$ and commitment loss $\mathcal{L}_{\text{ct}}$, respectively. By learning a discrete latent space, VQ-VAE ensures a structured representation of segmentation masks, where the quantized latent embeddings effectively capture essential tumor characteristics. The organization of similar tumor structural patterns into distinct codebook embeddings allows for a robust encoding of tumor morphology, which can help in precise boundary delineation during reconstruction. Moreover, the adopted discrete encoding discourages the model from memorizing specific instances, thereby enhancing its ability to handle diverse anatomical variations.

2.2 MRI Conditional Network

The MRI conditional network (Fig. 1 stage 2) comprises a CNN encoder, cross-latent alignment network and upsampling block. The input MRI of shape (4, H, W, D), where four modalities are stacked along the channel dimension, is passed to a CNN encoder which employs four convolutional layers to progressively extract local features while downsampling the input to generate a lower-dimensional embedding $\mathbf{e}_i \in \mathbb{R}^{C_i \times H_i \times W_i \times D_i}$ in the i_{th} encoder layer ($i \in [1, 4]$).

This image embedding is then fed into the cross-latent alignment network, where it is processed by a series of transformer layers. We utilize eight transformer layers, each employing multi-head self attention to capture global contextual features and long-range dependencies within the brain's anatomical structures and tumor morphology. The input embeddings, representing both healthy brain tissues and tumorous regions, are tokenized and projected into queries (Q), keys (K), and values (V) through learnable linear transformations, where $\mathbf{Q}, \mathbf{K}, \mathbf{V} \in \mathbb{R}^{N \times d_k}$, with N being the number of tokens in the MRI embedding and d_k denoting the dimension of each key and query vector. The query, key, and value matrices are split into eight attention heads to effectively model the structural variations in the brain, including the boundaries between healthy tissues and tumors. Each attention head independently computes its own set of attention scores using the scaled dot-product attention mechanism:

$$\text{Attention}(\mathbf{Q}, \mathbf{K}, \mathbf{V}) = \text{softmax}\left(\frac{\mathbf{Q}\mathbf{K}^T}{\sqrt{d_k}}\right)\mathbf{V}. \tag{4}$$

Following attention computation across all heads, the resulting features are concatenated and passed through residual connections and layer normalization, to ensure stable gradient propagation. A multi-layer perceptron (MLP) with two fully connected layers and GELU2 activation is then applied to further refine the learned representations. The refined tokenized embeddings undergo attention scaling and soft masking before being transformed into spatial embeddings. Finally, these embeddings are upsampled using two convolution layers to match their feature dimensions with those of the pre-trained VQ-VAE codebook vectors, with each feature vector assigned to the closest codebook entry. To align the MRI latent embeddings with the corresponding segmentation mask, two loss functions are employed: A cross-entropy loss that ensures each MRI embedding vector is assigned to the codebook vector index that corresponds to its ground truth (GT) mask feature vector, and a focal loss to prioritize the ambiguous features. The overall loss for training the MRI conditional network is

$$\mathcal{L}_{\text{overall}} = \left[(-\sum_i y_i \log(\hat{y}_i)) + (-\alpha(1-\hat{y}_i)^\gamma \log(\hat{y}_i)) \right], \tag{5}$$

where, y_i is the target codebook index, $\hat{y}_i$ is the predicted probability, α is a balancing factor, and γ controls the focus on hard examples.

Attention Scaling Module: The multi-head attention mechanism captures diverse features across multiple attention heads, but not all heads contribute equally to the task of discriminative feature extraction. To optimize the learning process, we introduce an attention scaling module that dynamically modulates the contribution of each head based on its relevance in extracting discriminative features. The module works by first projecting the attention features in the head dimension, followed by global average pooling (GAP) within each head to derive a compact descriptor that captures its overall significance. A softmax function is applied across the attention heads to compute a set of normalized weights, allowing the model to prioritize more relevant heads while suppressing less informative ones. The attention scaling process can be formulated as:

$$\begin{aligned} Y &= \text{Proj}(X) \quad \text{where} \quad Y \in \mathbb{R}^{B \times L \times N_{\text{head}} \times D_{\text{head}}}, \\ S &= \text{Softmax}\left(\text{GAP}_{D_{\text{head}}}(Y)\right), \\ X_{\text{scaled}} &= X_{resh} \odot S \quad \text{where} \quad X_{resh} \in R^{B \times L \times N_{\text{head}} \times D_{\text{head}}}, \end{aligned} \tag{6}$$

where "$\odot$" represents element wise rescaling and X_{resh} denotes the reshaped attention features in the head dimension. After attention scaling, the model incorporates squeeze and excitation to refine channel-wise attention, enhancing feature representation and improving model generalization by emphasizing key activations while attenuating redundant information.

Soft Masking: The Soft Masking Module, introduced after the first five transformer layers, aggregates the attention features into a single channel using a learnable transformation, followed by a sigmoid to compute a certainty map. This map generates a mask to prioritize uncertain features in the attention computation of subsequent transformer layers. This results in a more precise and reliable representation of complex tumor structures.

3 Experiments

Setup: We evaluate our method on the BraTS 2021 dataset [2], comprising MRI data of four modalities (T_1, T_2, $T1_{CE}$, FLAIR) from 1,251 patients. Following recent studies [16,31], we split the dataset into 834, 208, and 209 for training, validation and test sets respectively. Ground truth is annotated by radiologists into four regions: background, peritumoral edema (ED), necrosis (NCR), and enhancing tumor (ET). Segmentation results are evaluated across three regions: whole tumor (WT), Tumor Core (TC), and enhancing tumor (ET) using Hausdorff distance (HD95) and Dice similarity score. Our model is implemented in PyTorch and trained on each MRI conditional network for 150 epochs on NVIDIA RTX 3090 (24 GB) with a batch size of 3. We use GELU2 activation, the Adam optimizer with learning rate of $1e^{-4}$, and a weight decay of $1e^{-4}$. We apply data augmentations, including random flipping across three planes, rotation, intensity shifting, and scaling, with probabilities of 0.5, 0.5, 0.1, and 0.1, respectively.
Results and Analysis: Table 1 compares the performance of our model with state-of-the-art methods including 3D U-Net [6], TransBTS [28], UNETR [12], DBTrans [31], NestedFormer [30], and Cascaded Causal Intervention [16] on the HD95 and Dice score metrics for three tumor regions: WT, TC, and ET.

Compared to nearest competitor DBTrans [31], our boundary-focused feature-level supervision approach improves HD95 by 4.83 mm (WT), 2.14 mm (TC), and 2.39 mm (ET), and the Dice score of WT by 0.23%. Furthermore, our proposed AG-VQ-VAE architecture, comprising three dedicated networks

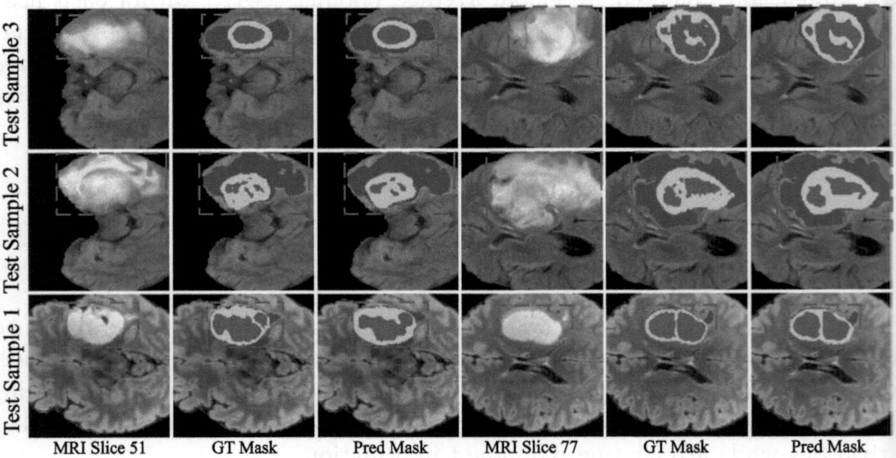

Fig. 2. Qualitative results of our method on three test samples. Red box represents the tumor region, with green, yellow, and red colors representing peritumoral edema (ED), enhancing tumor (ET), and necrotic tumor (NCR) respectively. Test Sample 1 has Avg HD95 (1.24 mm) and Avg Dice (91%), Sample 2 has Avg HD95 (1.41 mm) and Avg Dice (94%), and Sample 3 has Avg HD95 (1.27 mm) and Avg Dice (95%). (Color figure online)

Table 1. Comparison on BraTS 2021 dataset across three tumor regions: Whole Tumor (WT), Tumor Core (TC), and Enhancing Tumor (ET). Best results are in bold.

Methods	HD95 (mm) ↓				Dice Score (%) ↑			
	WT	TC	ET	AVG	WT	TC	ET	AVG
3D U-Net [6]	11.49	6.18	6.15	7.94	89.59	86.28	83.39	86.42
TransBTS [28]	15.12	8.21	7.83	10.38	89.25	85.35	80.35	84.99
UNETR [12]	15.99	10.01	9.72	11.90	90.10	83.66	79.78	84.51
NestedFormer [30]	10.23	6.43	6.08	7.58	90.12	88.18	85.62	87.97
DBTrans [31]	9.84	6.24	6.13	7.40	92.41	90.26	86.70	89.79
Causal Intervention [16]	13.92	5.85	6.43	8.73	92.32	**91.19**	**87.21**	**90.24**
Ours	**5.01**	**4.10**	**3.74**	**4.28**	**92.64**	89.05	82.25	87.98

(one per tumor region), results in a total of 441G FLOPs and 40M parameters. In contrast, DBTrans [31] require 146G FLOPs and 25M parameters. Although our model requires approximately 780 ms to process a full 3D MRI volume ($240 \times 240 \times 155$) on NVIDIA RTX 3090 (24 GB) compared to 254 ms for DBTrans [31], it still produces segmentation results in less than one second, with memory usage below 8 GB, and achieves superior HD95 scores, highlighting a favorable trade-off between accuracy and computational cost.

Note that accurate tumor boundary delineation is crucial for surgical decision-making, as minor inconsistencies in inner voxel predictions of tumor regions are more interpretable for clinicians than errors in boundary predictions [33]. Qualitative results in Fig. 2 show that, despite some inconsistencies in the prediction of necrotic regions (red segmentation mask), the boundaries of tumor sub-regions ($WT = ED + ET + NCR$, $TC = ET + NCR$, and ET) remain accurately delineated. This makes boundary-focused approaches, such as our AG-VQ-VAE, more clinically relevant than the models that prioritize inner voxel overlap but struggle with boundary precision.

Table 2. Ablation study on Attention Scaling (AS) and Soft Masking (SM) to evaluate the performance of the single-stage AG-UNet and two-stage AG-VQ-VAE.

Models	AS	SM	HD95 (mm) ↓			Dice Score (%) ↑		
			WT	TC	ET	WT	TC	ET
AG-UNet (Single-stage)	✓	✓	6.15	5.39	4.37	**93.19**	88.26	**85.09**
AG-VQ-VAE (Two-stage)	✓	✓	**5.01**	**4.10**	**3.74**	92.64	**89.05**	82.25
AG-VQ-VAE (Two-stage)	✓	×	5.87	4.64	4.20	91.35	88.51	80.27
AG-VQ-VAE (Two-stage)	×	✓	5.57	4.99	4.43	91.74	87.88	80.41

Ablation Study: We first evaluate the segmentation performance of the single-stage AG-UNet with voxel-level supervision, incorporating both Attention Scaling (AS) and Soft Masking (SM), as shown in Table 2. While this model achieves better Dice scores for WT (93.19%) and ET (85.09%), its higher HD95 values indicate less precise boundary delineation. In contrast, our two-stage AG-VQ-VAE achieves significantly lower HD95 scores, demonstrating superior tumor boundary delineation, even when using only one of the two modules. The best HD95 scores are achieved when using both AS and SM and excluding either one leads to performance degradation, highlighting the importance of both modules for accurate segmentation. These results validate that our two-stage AG-VQ-VAE offers superior tumor boundary delineation compared to single-stage AG-UNet.

4 Conclusion

We proposed a novel two-stage network for brain tumor segmentation with a strong emphasis on precise boundary delineation. By integrating attention scaling and soft masking, our approach enhances feature representation while effectively handling uncertainty in tumor regions. Evaluations on BraTS 2021 demonstrate the superiority of our method. We showed that our feature-level supervision enables superior tumor boundary delineation compared to voxel-level supervision commonly used in other methods. These advancements have significant clinical implications for surgical planning.

Acknowledgments. This research was supported by the Australian Government. Professor Ajmal Mian is the recipient of an ARC Future Fellowship Award (project #FT210100268), funded by the Australian Government. Dr. Naveed Akhtar is a recipient of the ARC Discovery Early Career Researcher Award (project #DE230101058), funded by the Australian Government.

Disclosure of Interests. The authors have no competing interests to declare that are relevant to the content of this article.

References

1. Anand, V.K., et al.: Brain tumor segmentation and survival prediction using automatic hard mining in 3D CNN architecture. In: Crimi, A., Bakas, S. (eds.) BrainLes 2020. LNCS, vol. 12659, pp. 310–319. Springer, Cham (2021). https://doi.org/10.1007/978-3-030-72087-2_27
2. Baid, U., et al.: The RSNA-ASNR-MICCAI BraTS 2021 benchmark on brain tumor segmentation and radiogenomic classification. arXiv preprint arXiv:2107.02314 (2021)
3. Cai, Y., et al.: Swin UNet3D: a 3-dimensional medical image segmentation network combining vision transformer and convolution. BMC Med. Inform. Decis. Making **23**(1) (2023)
4. Cao, H., et al.: Swin-Unet: Unet-like pure transformer for medical image segmentation. In: Karlinsky, L., Michaeli, T., Nishino, K. (eds.) ECCV 2022. LNCS, vol. 13803, pp. 205–218. Springer, Cham (2022). https://doi.org/10.1007/978-3-031-25066-8_9

5. Cao, Y., Zhou, W., Zang, M., An, D., Feng, Y., Yu, B.: MBANet: a 3D convolutional neural network with multi-branch attention for brain tumor segmentation from MRI images. Biomed. Signal Process. Control **80**, 104296 (2023)
6. Çiçek, Ö., Abdulkadir, A., Lienkamp, S.S., Brox, T., Ronneberger, O.: 3D U-Net: learning dense volumetric segmentation from sparse annotation. In: Ourselin, S., Joskowicz, L., Sabuncu, M.R., Unal, G., Wells, W. (eds.) MICCAI 2016, Part II. LNCS, vol. 9901, pp. 424–432. Springer, Cham (2016). https://doi.org/10.1007/978-3-319-46723-8_49
7. Gao, Y., Zhou, M., Metaxas, D.N.: UTNet: a hybrid transformer architecture for medical image segmentation. In: de Bruijne, M., et al. (eds.) MICCAI 2021, Part III. LNCS, vol. 12903, pp. 61–71. Springer, Cham (2021). https://doi.org/10.1007/978-3-030-87199-4_6
8. Ghazouani, F., Vera, P., Ruan, S.: Efficient brain tumor segmentation using swin transformer and enhanced local self-attention. Int. J. Comput. Assist. Radiol. Surg. **19**(2), 273–281 (2024)
9. Gholami, A.: A novel domain adaptation framework for medical image segmentation. In: Crimi, A., Bakas, S., Kuijf, H., Keyvan, F., Reyes, M., van Walsum, T. (eds.) BrainLes 2018. LNCS, vol. 11384, pp. 289–298. Springer, Cham (2019). https://doi.org/10.1007/978-3-030-11726-9_26
10. Gu, S., et al.: Vector quantized diffusion model for text-to-image synthesis. In: CVPR, pp. 10696–10706 (2022)
11. Hatamizadeh, A., Nath, V., Tang, Y., Yang, D., Roth, H.R., Xu, D.: Swin UNETR: swin transformers for semantic segmentation of brain tumors in MRI images. In: Crimi, A., Bakas, S. (eds.) BrainLes 2021. LNCS, vol. 12962, pp. 272–284. Springer, Cham (2021). https://doi.org/10.1007/978-3-031-08999-2_22
12. Hatamizadeh, A., et al.: UNETR: transformers for 3D medical image segmentation. In: Proceedings of the IEEE/CVF WACV, pp. 574–584 (2022)
13. Jia, Q., Shu, H.: BiTr-Unet: a CNN-transformer combined network for MRI brain tumor segmentation. In: Crimi, A., Bakas, S. (eds.) BrainLes 2021. LNCS, vol. 12963, pp. 3–14. Springer, Cham (2021). https://doi.org/10.1007/978-3-031-09002-8_1
14. Kermi, A., Mahmoudi, I., Khadir, M.T.: Deep convolutional neural networks using U-Net for automatic brain tumor segmentation in multimodal MRI volumes. In: Crimi, A., Bakas, S., Kuijf, H., Keyvan, F., Reyes, M., van Walsum, T. (eds.) BrainLes 2018. LNCS, vol. 11384, pp. 37–48. Springer, Cham (2019). https://doi.org/10.1007/978-3-030-11726-9_4
15. Li, P., Li, Z., Wang, Z., Li, C., Wang, M.: mResU-Net: multi-scale residual U-net-based brain tumor segmentation from multimodal MRI. Med. Biol. Eng. Comput. **62**(3), 641–651 (2024)
16. Liu, H., Li, Q., Nie, W., Xu, Z., Liu, A.: Causal intervention for brain tumor segmentation. In: Linguraru, M.G., et al. (eds.) MICCAI 2024. LNCS, vol. 15009, pp. 160–170. Springer, Cham (2024). https://doi.org/10.1007/978-3-031-72114-4_16
17. McKinley, R., Rebsamen, M., Dätwyler, K., Meier, R., Radojewski, P., Wiest, R.: Uncertainty-driven refinement of tumor-core segmentation using 3D-to-2D l. In: Crimi, A., Bakas, S. (eds.) BrainLes 2020. LNCS, vol. 12658, pp. 401–411. Springer, Cham (2021). https://doi.org/10.1007/978-3-030-72084-1_36
18. Munir, K., Frezza, F., Rizzi, A.: Brain tumor segmentation using 2D-Unet convolutional neural network. In: Deep Learning for Cancer Diagnosis, pp. 239–248 (2021)
19. Ottom, M.A., Rahman, H.A., Dinov, I.D.: Znet: deep learning approach for 2D MRI brain tumor segmentation. IEEE Transl. Eng. Health Med. (2022)

20. Peiris, H., Chen, Z., Egan, G., Harandi, M.: Reciprocal adversarial learning for brain tumor segmentation: a solution to brats challenge 2021 segmentation task. In: Crimi, A., Bakas, S. (eds.) BrainLes 2021. LNCS, vol. 12962, pp. 171–181. Springer, Cham (2021). https://doi.org/10.1007/978-3-031-08999-2_13
21. Rai, H.M., Chatterjee, K.: 2D MRI image analysis and brain tumor detection using deep learning CNN model LeU-Net. Multimedia Tools Appl. **80**(28) (2021)
22. Ramesh, A., et al.: Zero-shot text-to-image generation. In: ICML, pp. 8821–8831 (2021)
23. Rehman, M.U., Cho, S., Kim, J.H., Chong, K.T.: BU-Net: brain tumor segmentation using modified U-Net architecture. Electronics **9**(12), 2203 (2020)
24. Reynoso-Noverón, N., Mohar-Betancourt, A., Ortiz-Rafael, J.: Epidemiology of brain tumors. In: Principles of Neuro-oncology: Brain & Skull Base, pp. 15–25 (2021)
25. Scoccianti, S., et al.: Organs at risk in the brain and their dose-constraints in adults and in children: a radiation oncologist's guide for delineation in everyday practice. Radiother. Oncol. **114**(2) (2015)
26. Tragakis, A., Kaul, C., Murray-Smith, R., Husmeier, D.: The fully convolutional transformer for medical image segmentation. In: Proceedings of the IEEE/CVF Winter Conference on Applications of Computer Vision, pp. 3660–3669 (2023)
27. Vaswani, A.: Attention is all you need. Adv. Neural Inf. Process. Syst. (2017)
28. Wang, W., Chen, C., Ding, M., Yu, H., Zha, S., Li, J.: TransBTS: multimodal brain tumor segmentation using transformer. In: de Bruijne, M., et al. (eds.) MICCAI 2021. LNCS, vol. 12901, pp. 109–119. Springer, Cham (2021). https://doi.org/10.1007/978-3-030-87193-2_11
29. Xiao, Q., Nie, D.: Blurry boundary segmentation with semantic-aware feature learning. In: Yap, M.H., Kendrick, C., Behera, A., Cootes, T., Zwiggelaar, R. (eds.) MMIUA 2024. LNCS, vol. 14860, pp. 101–111. Springer, Cham (2024). https://doi.org/10.1007/978-3-031-66958-3_8
30. Xing, Z., Yu, L., Wan, L., Han, T., Zhu, L.: NestedFormer: nested modality-aware transformer for brain tumor segmentation. In: Wang, L., Dou, Q., Fletcher, P.T., Speidel, S., Li, S. (eds.) MICCAI 2022. LNCS, vol. 13435, pp. 140–150. Springer, Cham (2022). https://doi.org/10.1007/978-3-031-16443-9_14
31. Zeng, X., Zeng, P., Tang, C., Wang, P., Yan, B., Wang, Y.: DBTrans: a dual-branch vision transformer for multi-modal brain tumor segmentation. In: Greenspan, H., et al. (eds.) MICCAI 2023. LNCS, vol. 14223, pp. 502–512. Springer, Cham (2023). https://doi.org/10.1007/978-3-031-43901-8_48
32. Zhang, R., Jia, S., Adamu, M.J., Nie, W., Li, Q., Wu, T.: HMNet: hierarchical multi-scale brain tumor segmentation network. Clin. Med. **12**(2) (2023)
33. Zhang, Z.Z., Shields, L.B., Sun, D.A., Zhang, Y.P., Hunt, M.A., Shields, C.B.: The art of intraoperative glioma identification. Front. Oncol. **5**, 175 (2015)
34. Zhou, X., Li, X., Hu, K., Zhang, Y., Chen, Z., Gao, X.: ERV-Net: an efficient 3d residual neural network for brain tumor segmentation. Expert Syst. Appl. **170**, 114566 (2021)
35. Zhu, Z., He, X., Qi, G., Li, Y., Cong, B., Liu, Y.: Brain tumor segmentation based on the fusion of deep semantics and edge information in multimodal MRI. Inf. Fusion **91**, 376–387 (2023)

Blood Pressure Assisted Cerebral Microbleed Segmentation via Meta-matching

Junmo Kwon[1], Jonghun Kim[1,2], Taehyeon Kim[3], Sang Won Seo[4], Hwan-ho Cho[5(✉)], and Hyunjin Park[1,2(✉)]

[1] Department of Electrical and Computer Engineering, Sungkyunkwan University, Suwon 16419, South Korea
skenfn1231@skku.edu

[2] Center for Neuroscience Imaging Research, Institute for Basic Science, Suwon 16419, South Korea
hyunjinp@skku.edu

[3] Department of Computer and Information Technology, Purdue University, West Lafayette, IN 47907, USA

[4] Department of Neurology, Samsung Medical Center, Sungkyunkwan University School of Medicine, Seoul 06351, South Korea

[5] Department of Electronics Engineering, Incheon National University, Incheon 22012, South Korea
hwanho@inu.ac.kr

Abstract. Cerebral microbleeds (CMBs) are small hemorrhagic lesions that pose significant challenges for accurate segmentation due to the high rate of false positives and false negatives. CMBs have two subtypes: lobar and deep microbleeds (MBs). Motivated by the strong association between deep MBs and hypertension, we propose a **b**lood **p**ressure-driven **nnU-Net** (**BP-nnUNet**) that integrates blood pressure (BP) prompt into the state-of-the-art nnU-Net framework through three key strategies. First, we estimate BP using the pre-trained Meta-matching model, that requires only MRI images. This allows our method to be successfully applied to public datasets with missing clinical demographics. Second, we categorize CMBs into lobar and deep MB, enriching input text prompts with multiple classes while constraining the BP effect to deep MBs. Lastly, we introduce a novel anatomically-aware joint prompt fusion module that combines lobar and deep MB prompts. Experiments on both in-house and public datasets demonstrate that our BP-nnUNet outperforms existing CMB segmentation models and universal models incorporating medical prompts. Ablation studies validate the effectiveness of integrating subtype-level and case-level prompts, as well as our fusion module. Our method paves the way for the incorporation of clinically relevant information into a segmentation framework. Our code is available at https://github.com/junmokwon/BP-nnUNet

Keywords: Cerebral microbleeds · Blood pressure · Meta-matching · Prompt-driven medical image segmentation

J. C. Gee et al. (Eds.): MICCAI 2025, LNCS 15960, pp. 77–86, 2026.
https://doi.org/10.1007/978-3-032-04927-8_8

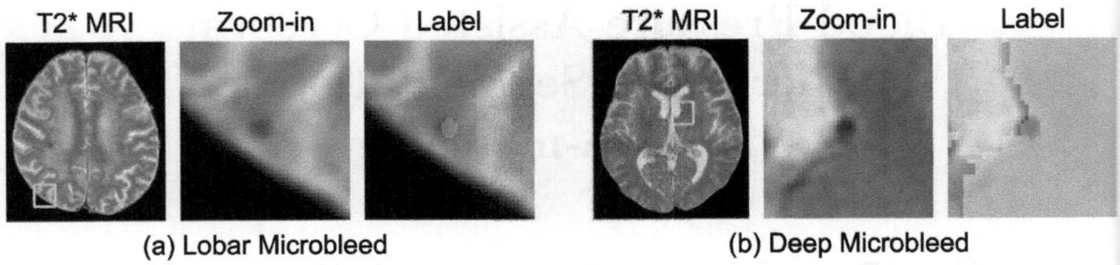

Fig. 1. Visualization of two subtypes of cerebral microbleeds: (a) Lobar microbleed and (b) deep microbleed. Red denotes microbleeds, blue denotes the lobar brain region, and green denotes the deep supratentorial region. (Color figure online)

1 Introduction

Cerebral microbleeds (CMBs) are small brain hemorrhages that appear as hypointense lesions on T2*-weighted MRI [6,21]. The term *micro* indicates their small size, typically ranging from 2 to 10 mm in diameter [21]. Hemorrhages exceeding this size threshold are no longer classified as microbleeds (MBs) but as larger intracerebral hemorrhages [4]. Therefore, CMB segmentation is a challenging task that requires accurate localization of small lesions while dealing with diverse voxel resolutions, varying from a fine-grained spacing of 0.2 mm^3 to a coarse-grained spacing of 3.0 mm^3 [23].

Segmenting small lesions requires additional modifications to tune state-of-the-art (SOTA) segmentation backbones, ranging from modifying loss functions [14,19] to applying task-specific false positive reduction modules [15]. Existing studies on CMB segmentation have proposed various false positive reduction algorithms to overcome these challenges [15,16]. AG-nnUNet [15] extends the original binary segmentation of CMBs to multi-label segmentation across lobar, supratentorial, and infratentorial regions and CMB lesions using an nnU-Net [10] backbone. This multi-label approach enhances the identification of CMB lesions surrounded by brain parenchyma, thereby reducing false positives [5]. CMB-UNETR++ [16] adapted the masked image modeling [7] technique to pre-train both the encoder and decoder components of UNETR++ [22], demonstrating superior performance over plain nnU-Net and UNETR++ for CMB segmentation.

Recently, leveraging text prompts has shown remarkable improvements in medical image segmentation [17,18,26]. The CLIP-driven Universal Model [17,18] uses both image and CLIP-driven text embeddings as convolutional weights and biases in the final convolutional layers. UniSeg [26] partitions a prompt into universal and task-specific parts, achieving superior performance in both single and multiple tasks. These studies emphasize that fusing imaging features and text embeddings is particularly beneficial for medical image segmentation.

In this study, we propose BP-nnUNet, a blood pressure-driven nnU-Net framework for CMB segmentation. CMB lesions are pathologically categorized into lobar and deep MBs (Fig. 1). High blood pressure (BP) is associated with deep MB [6]. Our framework leverages medical text prompts from the pre-trained

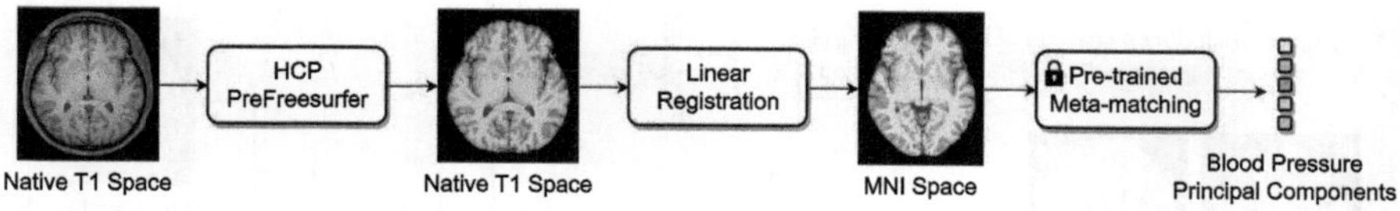

Fig. 2. Overview of the blood pressure (BP) estimation pipeline. T1-weighted MRI scans are processed using (1) HCP PreFreesurfer pipeline, (2) FMRIB's Linear Registration Tool to register T1 images to a 1 mm^3 MNI space, and (3) a pre-trained Meta-matching model, which outputs five phenotypes representing the principal components of BP measurements.

text encoder of BiomedCLIP [27]. In addition, we use the estimated BP from a pre-trained phenotype prediction model called Meta-matching [2,8,25], trained on large datasets. This BP estimation procedure only requires MRI data and thus can be applied not only to in-house datasets, but also to public datasets where clinical demographics are missing. Furthermore, we introduce a novel anatomically-aware joint prompt fusion module to effectively integrate our lobar and deep MB prompts into the segmentation network. We collected 621 cases of T1- and T2*-weighted MRI as an in-house dataset and evaluated our framework using the public MICCAI VALDO 2021 challenge dataset [23]. Our BP-nnUNet was evaluated against comparative methods, including task-specific models as well as SOTA universal models.

Our main contributions are three-fold:

1. We are the first to incorporate BP data, reflecting the pathology of CMB lesions, into the CMB segmentation framework. Unlike the task-level prompts used in universal models, BP reflects the underlying subject-level features.
2. The BP estimation procedure can be applied to anonymized public datasets without requiring access to clinical demographics.
3. We introduce a novel anatomically-aware joint prompt fusion module that achieves superior performance by leveraging a joint prompt of medical text and BP data.

2 Methods

2.1 Blood Pressure Estimation

Figure 2 summarizes the overall procedure for BP estimation using a pre-trained Meta-matching framework [2,8,25]. Meta-matching was trained using 265 phenotypes from the UK Biobank dataset [1], including diastolic BP, systolic BP, and intraocular pressure [8]. While the original Meta-matching [8] was proposed using resting-state fMRI inputs, the subsequent study [25] extended its application to T1 MRI inputs, leveraging the simple fully convolutional network as proposed in [20]. Following the pre-processing pipeline for anatomical Meta-matching [3], each T1 MRI scan yields five distinct blood pressure components, referred to as *BP eye* C2 through *BP eye* C6.

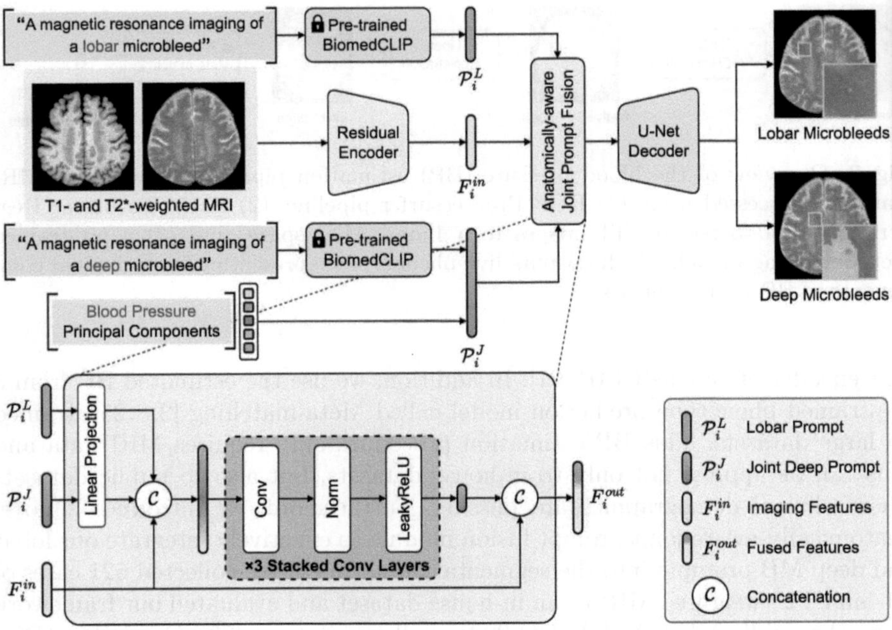

Fig. 3. Overview of our **BP-nnUNet**. Built upon nnU-Net ResEnc, we integrate lobar and deep microbleed prompts along with imaging features at the final layer of the residual encoder to predict binary segmentation masks for lobar and deep microbleeds.

2.2 Cerebral Microbleed Segmentation

We aim to develop a fully automated CMB segmentation framework by leveraging a joint prompt of medical text and BP data. Our proposed network, **BP-nnUNet**, is illustrated in Fig. 3. To seamlessly integrate imaging features and joint prompts, we propose an anatomically-aware joint prompt fusion module.

Subtyping Approach. We categorize CMB lesions based on their pathology as lobar and deep MBs, as illustrated in Fig. 1, applying different strategies for each subtype. The process is automated and followed by proxy label generation as proposed in AG-nnUNet [15]. MBs in lobar brain regions are classified as lobar MBs, while those in infratentorial and deep supratentorial regions are categorized as deep MBs. We follow the standardized medical prompt of "A magnetic resonance imaging of [CLS]" as used in the CLIP-driven Universal Model [17,18], obtaining 512-dimensional sentence embeddings using BiomedCLIP [27]. For lobar MBs, we use only a medical prompt as an input prompt. By contrast, for deep MBs, we use both a medical prompt and the estimated BP as a joint prompt, motivated by the fact that hypertension is the primary cause of deep MBs [6]. Compared with a plain task-level prompt, our approach can enrich medical prompts based on their subtypes.

Anatomically-Aware Joint Prompt Fusion. We introduce a novel fusion approach, *anatomically-aware joint prompt fusion*, for CMB segmentation. As illustrated in Fig. 3, our module integrates all input imaging features and joint prompts, and processes them through stacked convolutional layers:

$$F_i^{out} = \text{cat}(F_i^{in}, \text{StackedConv}(\text{MLP}_L(\mathcal{P}_i^L), \text{MLP}_D(\mathcal{P}_i^J), F_i^{in})) \quad (1)$$

where F_i^{in} denotes the imaging features from the two MRI modalities (i.e., T1 and T2*) of the i-th subject, obtained from the last convolutional layer of the residual encoder. cat($\cdot$) denotes concatenation along the channel dimension, assuming given inputs have the same spatial dimensions. StackedConv($\cdot$) first concatenates all inputs and then processes them through a stacked convolutional layer, followed by instance normalization and LeakyReLU layers. There are two separate linear projection layers: MLP_L and MLP_D, each projecting lobar prompt $\mathcal{P}_i^L \in \mathbb{R}^{512}$ and joint deep prompt $\mathcal{P}_i^J = \text{cat}(\mathcal{P}_i^D, \mathcal{P}_i^B)$ into spatial vectors to ensure that both prompts have the same spatial dimensions as the imaging feature F_i^{in}. $\mathcal{P}_i^D \in \mathbb{R}^{512}$ denotes the medical prompt for deep MBs and $\mathcal{P}_i^B \in \mathbb{R}^5$ represents the estimated BP principal components.

3 Experiments and Results

3.1 Datasets

We collected 621 MRI scans as an in-house dataset and internally divided them into 430 training, 87 validation, and 104 test cases. External validation was conducted using the MICCAI VALDO 2021 challenge dataset [23], utilizing 38 of the 72 cases. We excluded 34 cases obtained using 1.5T MRI because the Meta-matching framework requires 3T MRI inputs [25].

3.2 Implementation Details

All models except the CLIP-driven-UNet [17,18] were trained using the nnU-Net framework [10,11]. The input shape of 40 × 192 × 160 and the voxel resolution of 1.0 × 1.0 × 3.0 mm^3 were automatically determined during the nnU-Net pre-processing steps. For the CMB-UNETR++ [16], an input shape of 192 × 160 and a pixel resolution of 1 mm^2 were used. The CLIP-driven-UNet was trained for 2000 epochs, employing cosine annealing with a warm restart for the initial 100 epochs, and an initial learning rate of 1e-4.

All T1-weighted MRI scans were pre-processed using HCP Prefreesurfer pipeline [3] and registered to a 1 mm^3 MNI space [12,13] to estimate blood pressure as described in Fig. 2. In addition, bias-corrected T1 MRI images were registered to the native T2* space. For the T2*-weighted MRI scans, N4 bias field correction [24] was performed, followed by skull-stripping using SynthStrip [9]. Finally, the co-registered T1- and T2*-weighted MRI images were used as multi-modal inputs.

Table 1. Quantitative results of cerebral microbleed segmentation. The best performance is highlighted in **bold**, and the second-best is underlined.

Method	Venue	In-house Dataset				VALDO2021			
		DSC↑	F_1↑	SEN↑	FP_{avg}↓	DSC↑	F_1↑	SEN↑	FP_{avg}↓
UNETR++ [22]	IEEE TMI '24	44.02	61.17	60.47	0.990	37.48	42.63	42.05	0.711
nnU-Net [10]	Nat. Methods '21	51.68	67.54	66.56	0.423	36.56	44.21	43.09	0.395
nnU-Net ResEncL [11]	MICCAI '24	47.69	66.71	64.93	0.385	41.08	47.40	49.12	0.447
CMB-UNETR++ [16]	BIBM '24	47.28	65.34	62.23	0.750	41.40	47.54	46.33	0.684
AG-nnUNet [15]	MICCAI '24	47.02	65.97	66.14	0.433	39.82	44.08	45.34	0.395
AG-nnUNet ResEncL [15]	MICCAI '24	45.85	**68.43**	67.09	0.385	44.66	48.16	46.88	**0.184**
CLIP-driven-UNet [17,18]	MedIA '24	44.05	50.36	44.46	**0.221**	34.15	40.79	39.04	0.368
UniSeg [26]	MICCAI '23	48.85	63.40	69.82	1.163	40.09	46.02	56.52	1.421
BP-nnUNet (Ours)	MICCAI '25	**57.48**	67.95	**70.97**	0.269	**49.01**	**56.98**	**57.18**	0.316

3.3 Evaluation Metrics

In addition to the Dice score (DSC), we used F_1 score, sensitivity (SEN), and the average number of false positives per subject (FP_{avg}) as the evaluation metrics. To assess whether the predicted CMB lesions were correct, we measured the distance between the centroid of each predicted lesion and its nearest ground truth by applying a distance threshold of 5 mm, following the CMB detection criteria of the MICCAI VALDO 2021 challenge [23].

3.4 Quantitative and Qualitative Results

Table 1 summarizes the performance of our proposed framework and competing methods. We compared two task-specific models, CMB-UNETR++ [16] and AG-nnUNet [15], with their respective backbones, UNETR++ [22] and nnU-Net [10]. In addition, we included nnU-Net with a residual encoder (i.e., nnU-Net ResEncL [11]) and two prompt-driven universal models, CLIP-driven-UNet [17, 18] and UniSeg [26] as the SOTA baselines. All baseline methods were trained for binary CMB segmentation, as they were not originally designed for multi-label segmentation. By contrast, our BP-nnUNet was trained to predict two classes: lobar and deep MBs. For a fair comparison with the baselines, the predictions of BP-nnUNet for lobar and deep MBs were converted into binary format.

Our BP-nnUNet achieved either the best or second-best performance across all metrics. On the in-house dataset, BP-nnUNet achieved the best Dice score and sensitivity, while achieving a close second in F_1 score and FP_{avg}. On the VALDO 2021 dataset, BP-nnUNet outperformed the second-best baselines by a significant margin in the Dice and F_1 scores. This emphasizes the robustness and generalizability of our framework to unseen datasets. Figure 4 shows the superiority of BP-nnUNet in minimizing both false positives and false negatives, compared with the most competitive baselines (AG-nnUNet ResEncL [15] and

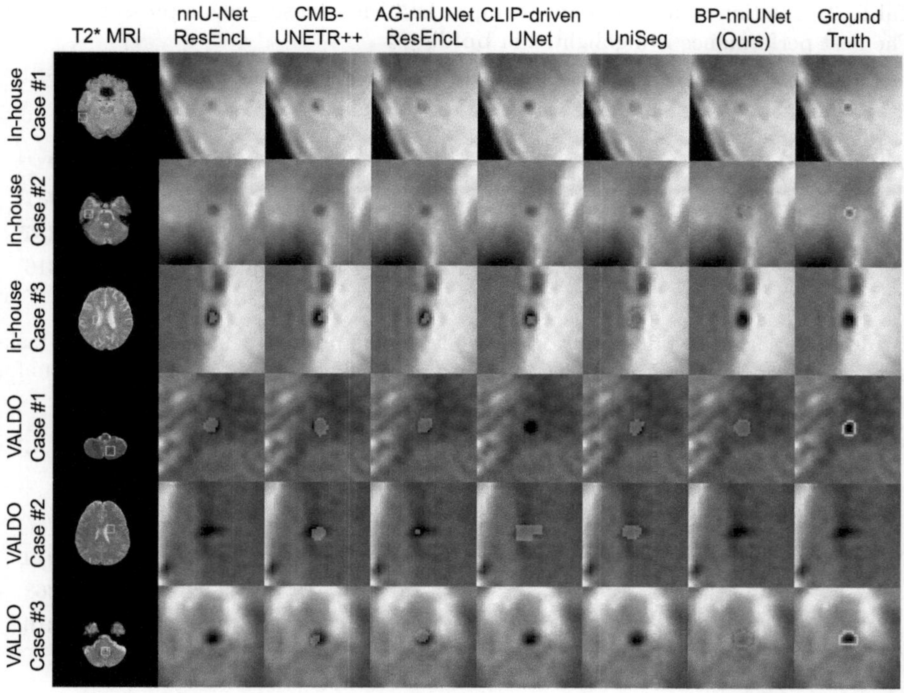

Fig. 4. Qualitative results of cerebral microbleed segmentation. Predictions are given in red while ground truth is given in yellow. (Color figure online)

UniSeg [26]). These findings highlight the importance of designing input prompts that range from task-level to subject-level prompts.

3.5 Ablation Analysis

We conducted ablation studies to evaluate the effectiveness of the BP data on nnU-Net ResEncL [11]. We considered three scenarios: *no prompt* (nnU-Net ResEncL baseline), *medical text prompt* for both lobar and deep MBs, and our *joint prompt* of medical text and BP data. Table 2 shows that our *joint prompt* consistently outperformed other scenarios in all metrics except FP_{avg}. Even the *medical text prompt* scenario outperformed all baselines in the Dice and F_1 scores, particularly on the unseen datasets, demonstrating the superiority of our CMB subtyping approach in designing input prompts. Furthermore, incorporating case-level features, such as BP, significantly enhances the overall performance of our BP-nnUNet, at the minor cost of only a slight increase in false positives on unseen datasets. Hence, BP plays a crucial role in boosting the CMB segmentation performance.

We also compared our anatomically-aware joint prompt fusion module with other fusion methods, as shown in Table 3. Specifically, *cross-attention* refers to

Table 2. Ablation results on input prompts. BP denotes using blood pressure prompt. The best performance is highlighted in **bold**.

Prompt		In-house Dataset				VALDO2021			
Text	BP	DSC↑	F_1↑	SEN↑	FP_{avg}↓	DSC↑	F_1↑	SEN↑	FP_{avg}↓
		47.69	66.71	64.93	0.385	41.08	47.40	49.12	0.447
✓		46.55	67.55	67.05	0.442	45.41	55.13	53.29	**0.237**
✓	✓	**57.48**	**67.95**	**70.97**	**0.269**	**49.01**	**56.98**	**57.18**	0.316

Table 3. Ablation results on fusion methods. Single prompt denotes a common task-level prompt for cerebral microbleeds. Joint prompt denotes subtype-level prompts for both lobar and deep microbleeds. The best performance is highlighted in **bold**.

	In-house Dataset				VALDO2021			
Fusion Method	DSC↑	F_1↑	SEN↑	FP_{avg}↓	DSC↑	F_1↑	SEN↑	FP_{avg}↓
Cross-attention	44.52	66.90	65.06	0.337	45.80	51.07	49.84	**0.211**
Single Prompt Fusion	49.40	65.87	63.83	0.327	41.51	42.22	42.65	0.368
Joint Prompt Fusion	**57.48**	**67.95**	**70.97**	**0.269**	**49.01**	**56.98**	**57.18**	0.316

the application of cross-attention between imaging features and joint prompt embeddings. *Single prompt fusion* refers to the use of a medical text prompt for a single class of CMB, whereas *joint prompt fusion* incorporates subtype-level prompts for both lobar and deep MBs. Table 3 shows the superiority of *joint prompt fusion* on both the in-house and external datasets. Our ablation study highlights the importance of CMB subtyping, which enables prompt fusion in a joint fashion. By contrast, adopting a single task-level prompt was particularly unsuccessful on unseen datasets. This aligns with the poor performance of the CLIP-driven-UNet [17,18], which was also trained with a task-level prompt derived from a pre-trained CLIP model. Therefore, applying subtype-level prompts is essential for enhancing the CMB segmentation performance.

4 Conclusion

In this paper, we propose a joint prompt of medical text and BP data to fully leverage the clinical characteristics of CMB subtypes. By enriching medical text prompts from a task level to a subtype level and leveraging pathological details through BP prompt, our framework achieved promising performance compared with SOTA CMB segmentation methods and universal prompt-driven models. Ablation studies revealed the effectiveness of our three design components: CMB subtyping, incorporation of BP, and joint prompt fusion, all of which significantly enhanced the performance on unseen datasets. Specifically, employing medical text prompts based on CMB subtypes outperformed the baseline methods on unseen datasets. Leveraging BP data effectively exploited subject-level features, significantly boosting the CMB segmentation performance. Lastly, our joint prompt fusion effectively utilized both BP and subtype-level prompts, delivering the best performance via BP-nnUNet.

Acknowledgments. We greatly thank the organizers and participants of the MICCAI VALDO 2021 challenge for the use of the public dataset. This study was supported by the National Research Foundation of Korea (NRF) grant funded by the Korea government (MSIT) (RS-2025-00518533), K-Brain Project of the National Research Foundation (NRF), funded by the Korean government (MSIT) (RS-2023-00265393), National Research Foundation (RS-2024-00408040), AI Graduate School Support Program (Sungkyunkwan University) (RS-2019-II190421), ICT Creative Consilience program (IITP-2025-RS-2020-II201821), and the Artificial Intelligence Innovation Hub program (RS-2021-II212068).

Disclosure of Interests. The authors have no competing interests to declare that are relevant to the content of this article.

References

1. Alfaro-Almagro, F., et al.: Image processing and quality control for the first 10,000 brain imaging datasets from UK biobank. Neuroimage **166**, 400–424 (2018)
2. Chen, P., et al.: Multilayer meta-matching: translating phenotypic prediction models from multiple datasets to small data. Imaging Neurosci. **2**, 1–22 (2024)
3. Glasser, M.F., et al.: The minimal preprocessing pipelines for the human connectome project. Neuroimage **80**, 105–124 (2013)
4. Greenberg, S.M., et al.: Microbleeds versus macrobleeds: evidence for distinct entities. Stroke **40**(7), 2382–2386 (2009)
5. Greenberg, S.M., et al.: Cerebral microbleeds: a guide to detection and interpretation. Lancet Neurol. **8**(2), 165–174 (2009)
6. Gregoire, S., et al.: The microbleed anatomical rating scale (mars) reliability of a tool to map brain microbleeds. Neurology **73**(21), 1759–1766 (2009)
7. He, K., Chen, X., Xie, S., Li, Y., Dollár, P., Girshick, R.: Masked autoencoders are scalable vision learners. In: 2022 IEEE/CVF Conference on Computer Vision and Pattern Recognition (CVPR), pp. 15979–15988 (2022)
8. He, T., et al.: Meta-matching as a simple framework to translate phenotypic predictive models from big to small data. Nat. Neurosci. **25**(6), 795–804 (2022)
9. Hoopes, A., Mora, J.S., Dalca, A.V., Fischl, B., Hoffmann, M.: Synthstrip: skull-stripping for any brain image. Neuroimage **260**, 119474 (2022)
10. Isensee, F., Jaeger, P.F., Kohl, S.A., Petersen, J., Maier-Hein, K.H.: nnU-Net: a self-configuring method for deep learning-based biomedical image segmentation. Nat. Methods **18**(2), 203–211 (2021)
11. Isensee, F., et al.: nnU-Net revisited: a call for rigorous validation in 3D medical image segmentation. In: 27th International Conference on Medical Image Computing and Computer-Assisted Intervention (MICCAI), pp. 488–498 (2024). https://doi.org/10.1007/978-3-031-72114-4_47

12. Jenkinson, M., Bannister, P., Brady, M., Smith, S.: Improved optimization for the robust and accurate linear registration and motion correction of brain images. Neuroimage **17**(2), 825–841 (2002)
13. Jenkinson, M., Smith, S.: A global optimisation method for robust affine registration of brain images. Med. Image Anal. **5**(2), 143–156 (2001)
14. Kwon, J., Kim, J., Park, H.: Leveraging segmentation-guided spatial feature embedding for overall survival prediction in glioblastoma with multimodal magnetic resonance imaging. Comput. Methods Programs Biomed. **255**, 108338 (2024)
15. Kwon, J., Seo, S.W., Park, H.: Anatomically-guided segmentation of cerebral microbleeds in t1-weighted and t2*-weighted MRI. In: 27th International Conference on Medical Image Computing and Computer-Assisted Intervention (MICCAI), pp. 24–33 (2024). https://doi.org/10.1007/978-3-031-72069-7_3
16. Kwon, J., Seo, S.W., Park, H.: Enhancing cerebral microbleed segmentation with pretrained UNETR++. In: 2024 IEEE International Conference on Bioinformatics and Biomedicine (BIBM), pp. 3372–3377. IEEE (2024). https://doi.org/10.1109/BIBM62325.2024.10822393
17. Liu, J., et al.: Clip-driven universal model for organ segmentation and tumor detection. In: 2023 IEEE/CVF International Conference on Computer Vision (ICCV), pp. 21095–21107 (2023)
18. Liu, J., et al.: Universal and extensible language-vision models for organ segmentation and tumor detection from abdominal computed tomography. Med. Image Anal. **97**, 103226 (2024)
19. Ma, J., et al.: Loss odyssey in medical image segmentation. Med. Image Anal. **71**, 102035 (2021)
20. Peng, H., Gong, W., Beckmann, C.F., Vedaldi, A., Smith, S.M.: Accurate brain age prediction with lightweight deep neural networks. Med. Image Anal. **68**, 101871 (2021)
21. Potter, G.M., Sarah Keir, M.: Improving inter-rater agreement about brain microbleeds: development of the brain observer microbleed. Stroke **35**, 1831–1835 (2004)
22. Shaker, A.M., Maaz, M., Rasheed, H., Khan, S., Yang, M.H., Khan, F.S.: Unetr++: delving into efficient and accurate 3d medical image segmentation. IEEE Trans. Med. Imaging **43**(9), 3377–3390 (2024)
23. Sudre, C.H., et al.: Where is VALDO? VAscular lesions detection and segmentation challenge at MICCAI 2021. Med. Image Anal. **91**, 103029 (2024)
24. Tustison, N.J., et al.: N4ITK: improved N3 bias correction. IEEE Trans. Med. Imaging **29**(6), 1310–1320 (2010)
25. Wulan, N., et al.: Translating phenotypic prediction models from big to small anatomical MRI data using meta-matching. Imaging Neurosci. **2**, 1–21 (2024)
26. Ye, Y., Xie, Y., Zhang, J., Chen, Z., Xia, Y.: Uniseg: a prompt-driven universal segmentation model as well as a strong representation learner. In: 26th International Conference on Medical Image Computing and Computer-Assisted Intervention (MICCAI), pp. 508–518 (2023). https://doi.org/10.1007/978-3-031-43898-1_49
27. Zhang, S., et al.: A multimodal biomedical foundation model trained from fifteen million image–text pairs. NEJM AI **2**(1) (2024)

C^2MAOT: Cross-modal Complementary Masked Autoencoder with Optimal Transport for Cancer Segmentation in PET-CT Images

Jiaju Huang[1], Shaobin Chen[1], Xinglong Liang[2], Xiao Yang[3], Zhuoneng Zhang[1], Yue Sun[1], Ying Wang[3], and Tao Tan[1](✉)

[1] Faculty of Applied Sciences, Macao Polytechnic University, Macao, China
taotan@mpu.edu.mo
[2] Netherlands Cancer Institute, Amsterdam, Netherlands
[3] The Fifth Affiliated Hospital, Sun Yat-sen University, Zhuhai, China

Abstract. Accurate cancer segmentation in PET-CT images is crucial for oncology, yet remains challenging due to lesion diversity, data scarcity, and modality heterogeneity. Existing methods often struggle to effectively fuse cross-modal information and leverage self-supervised learning for improved representation. In this paper, we introduce C^2MAOT, a Cross-modal Complementary Masked Autoencoder with Optimal Transport framework for PET-CT cancer segmentation. Our method employs a novel modality-complementary masking strategy during pre-training to explicitly encourage cross-modal learning between PET and CT encoders. Furthermore, we integrate an optimal transport loss to guide the alignment of feature distributions across modalities, facilitating robust multi-modal fusion. Experimental results on two datasets demonstrate that C^2MAOT outperforms existing state-of-the-art methods, achieving significant improvements in segmentation accuracy across five cancer types. These results establish our proposed method as an effective approach for tumor segmentation in PET-CT imaging. Our code is available at https://github.com/hjj194/c2maot.

Keywords: PET-CT Segmentation · Cross-modal Fusion · Self-supervised Learning

1 Introduction

Positron Emission Tomography-Computed Tomography (PET-CT) is a key imaging modality in oncology, combining the metabolic insights provided by PET with the anatomical details from CT [1]. PET images, using fluorodeoxyglucose (FDG) as a radiotracer, highlights areas of high glucose uptake, which are often indicative of tumors, while CT images provide high-resolution images crucial for precise lesion localization and visualizing organ boundaries [25]. Therefore, accurate tumor segmentation in PET-CT images is essential for enhancing tumor

J. C. Gee et al. (Eds.): MICCAI 2025, LNCS 15960, pp. 87–97, 2026.
https://doi.org/10.1007/978-3-032-04927-8_9

detection, assessing disease staging, and evaluating treatment responses. This capability plays a critical role in clinical applications such as radiation therapy planning, surgical guidance, and ongoing disease monitoring [5,11,32].

Recent advancements in deep learning have shown promising results in automatic tumor segmentation in medical imaging. Models based on convolutional neural networks (CNNs) have demonstrated significant capabilities in feature extraction and segmentation tasks [10,17,19,20,33], while transformer-based architectures have shown exceptional performance in capturing long-range dependencies and contextual information across image regions [4,6,14,28]. Despite these advances, tumor segmentation in PET-CT still faces three major challenges. (1) Data scarcity: The limited availability of labeled PET-CT data and the high cost of manual annotation hinder model training. (2) Lesion diversity: Tumors vary in size, shape, and location, making accurate segmentation difficult. This is particularly problematic for metastatic tumors that can appear in multiple, spatially distributed locations. (3) Modality heterogeneity: PET and CT provide complementary but distinct information, and existing methods struggle to effectively align feature distributions between these modalities and maintain consistent feature representations.

Current approaches have attempted to address these challenges but with limited success. To tackle data scarcity, self-supervised learning (SSL) has gained significant attention for its ability to leverage unlabeled data in medical imaging [18]. Approaches such as contrastive learning (e.g., SimCLR [7], MoCo [16]), masked image modeling (e.g., MAE [15], SimMIM [31]), and self-distillation methods (e.g., DINO [3], BYOL [13]) have shown promise in pre-training models on large datasets without requiring manual annotation. However, most SSL methods focus on individual modalities and often fail to capture the complementary information provided by multiple imaging modalities, such as PET and CT. For the lesion diversity challenge, augmentation techniques and multi-scale architectures have been proposed [19], but these methods often fail to generalize across different tumor types and locations. To address modality heterogeneity, early fusion methods relied on simple strategies like concatenation or weighted averaging of features from each modality [2]. While these approaches provide some benefit, they often struggle to fully exploit the complementary nature of the modalities due to differences in their resolutions, intensities, and noise characteristics [21]. More sophisticated approaches, such as attention mechanism-based models, have been developed to enable more nuanced interactions between modalities [23]. Cross-modal attention mechanisms effectively capture contextual dependencies by dynamically weighting modality-specific features. Despite these efforts, challenges remain in effectively aligning feature distributions between modalities and maintaining consistency in feature representation across different modalities.

In this paper, we propose a novel pre-training segmentation framework, C^{2}MAOT, which integrates cross-modal complementary masking and optimal transport (OT)-guided multi-modal fusion. Our method addresses three key challenges in multi-modal tumor segmentation. First, we tackle data scarcity

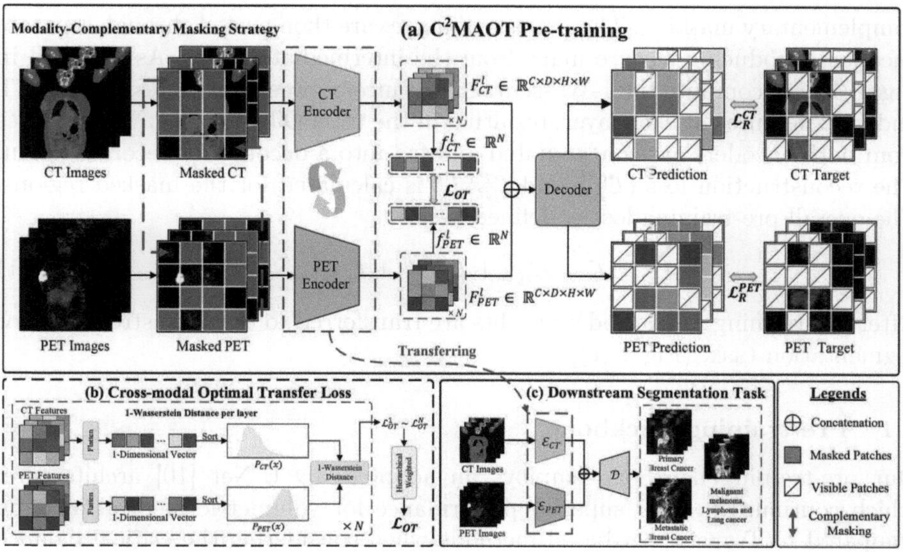

Fig. 1. Overview of the proposed C²MAOT framework. It consists of two main components: pre-training and fine-tuning for downstream segmentation tasks.

through a self-supervised pre-training approach that effectively leverages unlabeled PET-CT data. Second, our method addresses lesion diversity via a spatially-aware complementary masking strategy specifically designed for the distributed nature of tumors. By forcing the model to reconstruct complementary regions across modalities, it enhances cross-modal learning within the PET-CT domain, improving the representation of diverse tumor patterns present in the training distribution. Third, we overcome modality heterogeneity by introducing an OT loss to guide the alignment of feature distributions between PET and CT, enabling more effective multi-modal fusion and ensuring consistency in feature representations across modalities.

Our contributions are three-fold: (1) We introduce a novel cross-modal complementary masking strategy for self-supervised pre-training, which explicitly facilitates information exchange between PET and CT encoders while improving the model's ability to handle spatially diverse tumor presentations. (2) An optimal transport loss guides the alignment of feature distributions across modalities, leading to a more effective fusion of PET and CT features. (3) Extensive evaluation of C²MAOT on two datasets demonstrates that our method significantly outperforms SOTA methods, achieving substantial improvements in segmentation accuracy across five tumor types.

2 Methodology

The framework of our proposed C²MAOT is shown in Fig. 1. During the pre-training phase (Fig. 1 (a)), PET and CT images are first masked using modality-complementary masking. These masked images are then passed through separate encoders, producing feature maps from the intermediate layers. As depicted in Fig. 1 (b), we compute the 1-Wasserstein distance between the corresponding CT and PET features at each layer, resulting in the total OT loss $\mathcal{L}_{OT}$. The features from both encoders are concatenated and fed into a decoder for reconstruction. The reconstruction loss ($\mathcal{L}_R^{CT}$ and $\mathcal{L}_R^{PET}$) is calculated for the masked regions. The overall pre-training loss is defined as:

$$\mathcal{L} = \mathcal{L}_R^{CT} + \mathcal{L}_R^{PET} + \mathcal{L}_{OT}. \tag{1}$$

After pre-training, the encoder weights are transferred to the downstream tumor segmentation task. (Fig. 1 (c)).

2.1 Pre-training Backbone

Our pre-training backbone employs an adapted 3D U-Net [10] architecture, which continues to offer superior performance for volumetric medical imaging compared to Transformer-based methods when considering the critical balance between model effectiveness, parameter efficiency, and computational demands. Departing from conventional multimodal approaches like nnU-Net [20] that channel-concatenate modalities for single-encoder processing, our architecture implements dual independent encoders specifically optimized for PET-CT imaging. This allows the model to explicitly extract modality-specific features in the encoder stage, enhancing the representation of each modality's unique characteristics. The decoder receives skip connections and bottleneck features from both encoders, which are concatenated along the channel dimension to effectively combine the modality-specific information.

2.2 Modality-Complementary Masked Autoencoder

Inspired by He et al. [15], our pre-training employs a mask-and-reconstruction strategy. While vanilla MAE randomly masks patches in a single modality, PET and CT images—acquired simultaneously for the same patient—share similar anatomical and semantic structures. Therefore, we propose a modality-complementary masking strategy by randomly masking 50% of the 3D patches ($8 \times 8 \times 8$) in a complementary manner: for corresponding regions, if a PET patch is masked, the CT patch is kept intact, and vice versa. This design strategically compels the network to learn and integrate complementary information from the paired modality to complete the masked modality's missing data during reconstruction, effectively fostering deep cross-modal representation learning. Consistent with MAE, the reconstruction loss is computed solely over the masked regions using the standard L2 (MSE) loss formulation. The reconstruction losses for the PET and CT are defined as follows:

$$\mathcal{L}_{\mathrm{R}}^{PET} = \frac{1}{N}\sum_{i=1}^{N} \|\mathbf{P}_i^{PET} - \hat{\mathbf{P}}_i^{PET}\|_2^2, \quad \mathcal{L}_{\mathrm{R}}^{CT} = \frac{1}{N}\sum_{i=1}^{N} \|\mathbf{P}_i^{CT} - \hat{\mathbf{P}}_i^{CT}\|_2^2, \tag{2}$$

where $\mathbf{P}_i^{PET}$ and $\hat{\mathbf{P}}_i^{PET}$ represent the original and reconstructed voxel intensities for the i-th voxel of the PET image, respectively, and similarly for CT. These separate losses allow the model to optimize each modality independently, ensuring better reconstruction accuracy for both PET and CT images during the training process.

2.3 Optimal Transport Loss for Cross-Modal Feature Alignment

In order to effectively fuse features from PET and CT images while mitigating modality discrepancies, we introduce an Optimal Transport (OT) loss. OT is a method for measuring the difference between two modality feature distributions, with the goal of minimizing the transportation cost between them [26]. PET and CT images often exhibit differences in intensity distributions, resolution, and noise characteristics, making direct comparison or fusion of their features challenging. The OT loss helps to align the features from both modalities, facilitating better fusion while preserving each modality's unique characteristics. The core idea behind the OT distance is that, given two distributions, we aim to find an optimal mapping that minimizes the transportation cost between the two distributions. The cost is determined by calculating the distance between corresponding features. To achieve this, we use the 1-Wasserstein distance [27], which provides an effective way to measure the minimal transportation cost between two distributions. In our implementation, the features from PET and CT are extracted using two separate encoders, resulting in feature maps denoted as $\mathbf{F}_{PET}$ and $\mathbf{F}_{CT}$, respectively. The 1-Wasserstein distance is defined as:

$$W_1(\mathbf{F}_{PET}, \mathbf{F}_{CT}) = \inf_{\gamma \in \Gamma(\mathbf{F}_{PET}, \mathbf{F}_{CT})} \mathbb{E}_{(\mathbf{x},\mathbf{y})\sim\gamma}[\,|\mathbf{x} - \mathbf{y}\|], \tag{3}$$

here, $\mathbb{E}$ represents the expectation, which refers to the average distance between the feature pairs $(\mathbf{x}, \mathbf{y})$. inf is the infimum, representing the minimum transportation cost across all possible transport plans. In simpler terms, this equation means that we seek to find a mapping γ that minimizes the cost of transferring features from the PET distribution to the CT distribution. To prioritize deeper features in the alignment process, we introduce a layer-specific weighting factor w_l, which increases exponentially with depth. Deeper layers tend to capture more abstract, modality-invariant semantic information that is critical for downstream tasks like tumor segmentation. In practice, we set the weights to follow an exponential growth, i.e., $w_l = \alpha^l$, where $\alpha > 1$ is a constant that controls the rate of growth. The final OT loss $\mathcal{L}_{OT}$ is computed as the weighted sum of 1-Wasserstein distances across all layers:

$$\mathcal{L}_{OT} = \sum_{l=1}^{N} w_l \cdot W_1(\mathbf{F}_{PET}^{(l)}, \mathbf{F}_{CT}^{(l)}), \tag{4}$$

where N is the number of layers in the encoder, $\mathbf{F}_{PET}^{(l)}$ and $\mathbf{F}_{CT}^{(l)}$ represent the feature maps at the l-th layer for PET and CT, respectively, and w_l is the weight associated with the l-th layer.

3 Experiments and Results

3.1 Datasets and Evaluation Metric

Datasets. The study utilized PET-CT data from two sources: the AutoPET III Dataset [12] and an In-house Breast Dataset. The AutoPET III Dataset provided 1,014 cases (501 tumor-positive, 513 control) covering lung cancer, lymphoma, and melanoma. The In-house Breast Dataset contributed 393 cases (287 tumor-positive, 106 control) of primary and metastatic breast cancer (BC). We allocated 50% of the AutoPET III Dataset for pre-training, with the remaining data used alongside the In-house Breast Dataset for downstream segmentation evaluation.

Evaluation Metric. Model performance was assessed using four metrics: Dice Similarity Coefficient (DSC), Intersection over Union (IoU), recall, and precision. DSC and IoU measure spatial overlap between predicted and ground truth segmentations, while recall and precision evaluate detection accuracy and reliability respectively.

3.2 Implementation Details

Experiments were conducted using Python 3.11, PyTorch 2.4.1, and Ubuntu 22.04 on dual NVIDIA 4090 GPUs (24GB memory each), with a patch size of $128 \times 128 \times 128$. Pre-training employed the Adam optimizer (initial learning rate: 1×10^{-4}, batch size: 2) for 500 epochs, including a 10-epoch warm-up period. Learning rate decay followed a polynomial strategy:

$$\text{lr} = \text{initial lr} \times \left(1 - \frac{\text{epoch}}{\text{max epoch}}\right)^{0.9}. \tag{5}$$

For both pre-training and tumor segmentation fine-tuning, we applied comprehensive data augmentation including geometric transformations, intensity modifications, and random noise. Fine-tuning maintained the initial learning rate and decay strategy, with training combining cross-entropy and dice loss over 1000 epochs with 250 iterations each.

3.3 Quantitative and Qualitative Results

Comparison with Existing SOTA Methods. To evaluate the effectiveness of our method, we compared it with advanced segmentation methods including nnUNet [20], UNETR [28], 3DUX-Net [22], and U-Mamba [24], as well as SSL methods such as MoCov3 [8], MAE3D [9], Swin-MM [29], and VoCo [30] that utilize 3D UNet as the backbone. As shown in Table 1, our method achieved the highest or near-highest scores across all metrics including DSC, IoU, recall, and precision. Quantitatively, the average DSC surpasses the second-best method by 1.15% and the average IoU improves by 1.04%. Notably, with the exception of MoCov3 (which may be attributed to its requirement for large batch sizes to obtain sufficient negative samples), self-supervised pretraining generally enhanced the performance of baseline models, demonstrating the potential of self-supervised methods in improving segmentation performance. The qualitative results in Fig. 2 further illustrate that our method generates segmentation results closely aligned with ground truth, with fewer instances of over-segmentation and under-segmentation, particularly for dispersed lesions such as metastatic breast cancer, lymphoma, and melanoma.

Table 1. Comparison for tumor segmentation. The evaluation metrics include DSC, IoU, Recall, and Precision in (%). Best results are shown in **bold**, while second-best results are <u>underlined</u>.

Category	Metrics	From Scratch				With General SSL		With Medical SSL		
		nnUNet	UNETR	3DUX-Net	U-Mamba	MoCov3	MAE3D	Swin-MM	VoCo	Proposed
Primary Breast Cancer	DSC	83.32	79.13	78.88	82.80	82.17	84.43	85.25	<u>87.02</u>	**88.18**
	IoU	71.41	65.47	65.13	70.65	69.74	73.28	74.46	<u>77.49</u>	**78.86**
	Recall	83.64	78.79	79.36	**90.27**	82.04	85.44	87.52	88.27	<u>89.39</u>
	Precision	84.28	82.85	81.60	83.48	81.89	85.23	86.62	**87.84**	<u>87.37</u>
Metastatic Breast Cancer	DSC	51.56	54.38	55.49	44.50	49.72	55.85	<u>57.65</u>	57.13	**59.47**
	IoU	34.73	37.34	38.40	28.62	33.08	38.74	<u>40.50</u>	40.27	**42.32**
	Recall	55.35	62.56	59.64	49.89	52.41	57.83	64.72	<u>71.46</u>	**72.43**
	Precision	54.25	54.74	58.28	50.38	51.28	56.27	63.39	<u>72.11</u>	**73.66**
AutoPET III Lesions	DSC	50.33	46.58	46.59	34.85	50.45	53.48	55.76	**57.40**	<u>57.34</u>
	IoU	33.63	30.36	30.37	21.10	33.73	36.50	38.66	**40.50**	<u>40.19</u>
	Recall	38.85	37.09	36.68	34.15	37.72	40.53	39.48	<u>41.29</u>	**42.50**
	Precision	45.24	43.18	44.10	37.72	43.63	<u>46.85</u>	**47.32**	46.43	46.22
Average	DSC	61.74	60.03	60.32	54.05	60.78	64.59	66.22	<u>67.18</u>	**68.33**
	IoU	46.59	44.39	44.63	40.12	45.52	49.51	51.21	<u>52.75</u>	**53.79**
	Recall	59.28	59.48	58.56	58.10	57.39	61.27	63.91	<u>67.01</u>	**68.11**
	Precision	61.26	60.26	61.33	57.19	58.93	62.78	65.78	<u>68.79</u>	**69.08**

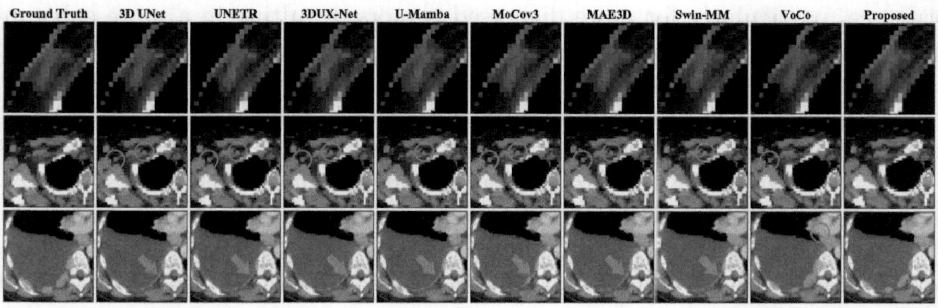

Fig. 2. Visualization of tumor segmentation. Red: primary breast cancer; Green: metastatic breast cancer; Orange: lung cancer, lymphoma and melanoma. Cyan: circles and arrows indicate over- and under-segmentation errors, respectively. (Color figure online)

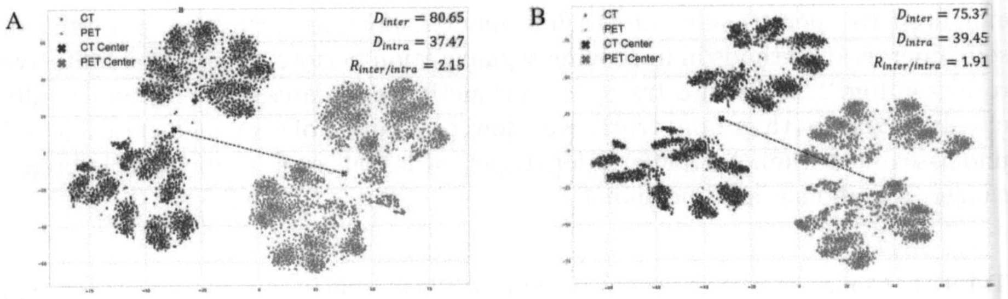

Fig. 3. T-SNE visualization of the CT and PET feature spaces. (A) Without OT loss, (B) With OT loss. Blue represents CT features, yellow represents PET features. (Color figure online)

Table 2. Ablation study results on the average DSC (%).

Experimental Setting			Category			
Baseline	OT-Loss	MC-Mask	Primary-BC	Metastatic-BC	AutoPET	AVG
✓			85.12	53.42	52.08	63.54
✓	✓		86.03	55.14	53.78	64.98
✓		✓	85.74	57.13	55.26	66.04
✓	✓	✓	**88.18**	**59.47**	**57.34**	**68.33**

Ablation Study. Table 2 demonstrates the contribution of key components to the model's performance. Starting with the baseline model (trained from scratch), we first introduce the OT loss, which leads to improvements in the DSC across all tumor types, with an average DSC increase of 1.44%. Incorporating the Modality-Complementary Mask (MC-Mask) further enhances performance, particularly for more dispersed tumors, resulting in a more substantial average DSC increase of 2.5%. The effect of OT loss can be visualized in Fig. 3, which illustrates its impact on feature distribution alignment. As shown, OT loss reduces the inter-modality distance while increasing intra-modality dispersion, decreasing the inter-to-intra ratio from 2.15 to 1.91, which indicates a more optimal feature space organization where different modalities are better aligned while preserving richer feature representations within each modality. This demonstrates that OT loss effectively promotes feature space consolidation while preserving modality-specific characteristics, facilitating more robust cross-modal fusion.

4 Conclusion

In this study, we proposed C^{2}MAOT, a novel pre-training segmentation framework specifically designed to address the challenges of cancer segmentation in PET-CT images. Our method leveraged a modality-complementary masking strategy within a masked autoencoder architecture to promote explicit cross-modal interaction and learning between PET and CT encoders. By incorporating an OT loss, our proposed method effectively aligned feature distributions from both modalities, leading to enhanced and robust multi-modal feature fusion. Extensive experimental results on two datasets, covering a range of cancer types, conclusively demonstrated that our method achieved state-of-the-art segmentation performance, surpassing existing supervised and self-supervised approaches. The significant improvements observed underscored the efficacy of our complementary masking and OT guided fusion in capturing and integrating the complementary information inherent in PET-CT images. This work established C^{2}MAOT as a promising approach for advancing accurate and reliable cancer segmentation in multi-modal medical imaging.

Acknowledgements. This work was supported by the Science and Technology Development Fund of Macao (0105/2022/A).

Disclosure of Interests. The authors declare no competing interests.

References

1. Aide, N., Lasnon, C., et al.: Advances in PET/CT technology: an update. In: Seminars in Nuclear Medicine. vol. 52, pp. 286–301. Elsevier (2022)
2. Azam, M.A., et al.: A review on multimodal medical image fusion: compendious analysis of medical modalities, multimodal databases, fusion techniques and quality metrics. Comput. Biol. Med. **144**, 105253 (2022)
3. Caron, M., et al.: Emerging properties in self-supervised vision transformers. In: Proceedings of the IEEE/CVF International Conference on Computer Vision, pp. 9650–9660 (2021)
4. Chen, J., et al.: TransUnet: rethinking the U-Net architecture design for medical image segmentation through the lens of transformers. Med. Image Anal. (2024)
5. Chen, S., et al.: Fit-net: feature interaction transformer network for pathologic myopia diagnosis. IEEE Trans. Med. Imaging **42**(9), 2524–2538 (2023)
6. Chen, S., et al.: Multi-risk factors joint prediction model for risk prediction of retinopathy of prematurity. EPMA J. **15**(2), 261–274 (2024)
7. Chen, T., Kornblith, S., Norouzi, M., Hinton, G.: A simple framework for contrastive learning of visual representations. In: International Conference on Machine Learning, pp. 1597–1607. PMLR (2020)
8. Chen, X., Xie, S., He, K.: An empirical study of training self-supervised vision transformers. In: Proceedings of the IEEE/CVF International Conference on Computer Vision, pp. 9640–9649 (2021)
9. Chen, Z., Agarwal, D., Aggarwal, K., Safta, W., et al.: Masked image modeling advances 3D medical image analysis. In: Proceedings of the IEEE/CVF Winter Conference on Applications of Computer Vision, pp. 1970–1980 (2023)

10. Çiçek, Ö., Abdulkadir, A., Lienkamp, S.S., Brox, T., Ronneberger, O.: 3D U-Net: learning dense volumetric segmentation from sparse annotation. In: Ourselin, S., Joskowicz, L., Sabuncu, M.R., Unal, G., Wells, W. (eds.) MICCAI 2016. LNCS, vol. 9901, pp. 424–432. Springer, Cham (2016). https://doi.org/10.1007/978-3-319-46723-8_49
11. Gao, Y., Tan, T., et al.: Multi-modal longitudinal representation learning for predicting neoadjuvant therapy response in breast cancer treatment. IEEE J. Biomed. Health Inf. (2025)
12. Gatidis, S., et al.: A whole-body FDG-PET/CT dataset with manually annotated tumor lesions. Sci. Data **9**(1), 601 (2022)
13. Grill, J.B., Strub, F., et al.: Bootstrap your own latent-a new approach to self-supervised learning. In: Advances in Neural Information Processing Systems, vol. 33, pp. 21271–21284 (2020)
14. Hatamizadeh, A., et al.: Unetr: transformers for 3D medical image segmentation. In: Proceedings of the IEEE/CVF Winter Conference on Applications of Computer Vision, pp. 574–584 (2022)
15. He, K., Chen, X., Xie, S., Li, Y., Dollár, P., Girshick, R.: Masked autoencoders are scalable vision learners. In: Proceedings of the IEEE/CVF Conference on Computer Vision and Pattern Recognition, pp. 16000–16009 (2022)
16. He, K., Fan, H., Wu, Y., Xie, S., Girshick, R.: Momentum contrast for unsupervised visual representation learning. In: Proceedings of the IEEE/CVF Conference on Computer Vision and Pattern Recognition, pp. 9729–9738 (2020)
17. Huang, J., Chen, S., Liang, X., Sun, Y., Hu, M., Tan, T.: All-in-one multi-organ segmentation in 3D CT images via self-supervised and cross-dataset learning. In: 2025 IEEE 22nd International Symposium on Biomedical Imaging (ISBI), pp. 1–5. IEEE (2025)
18. Huang, S.C., Pareek, A., Jensen, M., Lungren, M.P., Yeung, S., Chaudhari, A.S.: Self-supervised learning for medical image classification: a systematic review and implementation guidelines. NPJ Digit. Med. **6**(1), 74 (2023)
19. Huang, Z., et al.: Stu-net: scalable and transferable medical image segmentation models empowered by large-scale supervised pre-training. arXiv preprint arXiv:2304.06716 (2023)
20. Isensee, F., Jaeger, P.F., Kohl, S.A., Petersen, J., Maier-Hein, K.H.: nnU-Net: a self-configuring method for deep learning-based biomedical image segmentation. Nat. Methods **18**(2), 203–211 (2021)
21. Khan, S.U., Khan, M.A., Azhar, M., Khan, F., Lee, Y., Javed, M.: Multimodal medical image fusion towards future research: a review. J. King Saud Univ. Comput. Inf. Sci. **35**(8), 101733 (2023)
22. Lee, H.H., Bao, S., Huo, Y., Landman, B.A.: 3D UX-Net: a large kernel volumetric convnet modernizing hierarchical transformer for medical image segmentation. arXiv preprint arXiv:2209.15076 (2022)
23. Lu, J., Chen, J., Cai, L., Jiang, S., Zhang, Y.: H2ASeg: hierarchical adaptive interaction and weighting network for tumor segmentation in PET/CT images. In: International Conference on Medical Image Computing and Computer-Assisted Intervention, pp. 316–327. Springer (2024). https://doi.org/10.1007/978-3-031-72111-3_30
24. Ma, J., Li, F., Wang, B.: U-mamba: enhancing long-range dependency for biomedical image segmentation. arXiv preprint arXiv:2401.04722 (2024)
25. Ming, Y., Wu, N., et al.: Progress and future trends in PET/CT and PET/MRI molecular imaging approaches for breast cancer. Front. Oncol. **10**, 1301 (2020)

26. Montesuma, E.F., Mboula, F.M.N., Souloumiac, A.: Recent advances in optimal transport for machine learning. IEEE Trans. Pattern Anal. Mach. Intell. (2024)
27. Panaretos, V.M., Zemel, Y.: Statistical aspects of wasserstein distances. Ann. Rev. Stat. Appl. **6**(1), 405–431 (2019)
28. Tang, Y., Yang, D., et al.: Self-supervised pre-training of swin transformers for 3D medical image analysis. In: Proceedings of the IEEE/CVF Conference on Computer Vision and Pattern Recognition, pp. 20730–20740 (2022)
29. Wang, Y., Li, Z., et al.: SwinMM: masked multi-view with swin transformers for 3d medical image segmentation. In: International Conference on Medical Image Computing and Computer-assisted Intervention, pp. 486–496. Springer (2023). https://doi.org/10.1007/978-3-031-43898-1_47
30. Wu, L., Zhuang, J., Chen, H.: Voco: a simple-yet-effective volume contrastive learning framework for 3D medical image analysis. In: Proceedings of the IEEE/CVF Conference on Computer Vision and Pattern Recognition, pp. 22873–22882 (2024)
31. Xie, Z., et al.: Simmim: a simple framework for masked image modeling. In: Proceedings of the IEEE/CVF Conference on Computer Vision and Pattern Recognition, pp. 9653–9663 (2022)
32. Zhang, T., Tan, T., et al.: Predicting breast cancer types on and beyond molecular level in a multi-modal fashion. NPJ Breast Cancer **9**(1), 16 (2023)
33. Zhang, Z., Han, L., Sun, Y., Tan, T., et al.: UniMRISegnet: universal 3D network for various organs and cancers segmentation on multi-sequence MRI. IEEE J. Biomed. Health Inf. (2024)

CATVis: Context-Aware Thought Visualization

Tariq Mehmood[1(✉)], Hamza Ahmad[2], Muhammad Haroon Shakeel[3], and Murtaza Taj[1]

[1] Lahore University of Management Sciences, Lahore, Pakistan
{23030009,murtaza.taj}@lums.edu.pk
[2] Forman Christian College, University, Lahore, Pakistan
241548068@formanite.fccollege.edu.pk
[3] Arbisoft, Lahore, Pakistan
haroon.shakeel@arbisoft.com
http://cvlab.lums.edu.pk

Abstract. EEG-based brain-computer interfaces (BCIs) have shown promise in various applications, such as motor imagery and cognitive state monitoring. However, decoding visual representations from EEG signals remains a significant challenge due to their complex and noisy nature. We thus propose a novel 5-stage framework for decoding visual representations from EEG signals: (1) an EEG encoder for concept classification, (2) cross-modal alignment of EEG and text embeddings in CLIP feature space, (3) caption refinement via re-ranking, (4) weighted interpolation of concept and caption embeddings for richer semantics, and (5) image generation using a pre-trained Stable Diffusion model. We enable context-aware EEG-to-image generation through cross-modal alignment and re-ranking. Experimental results demonstrate that our method generates high-quality images aligned with visual stimuli, outperforming SOTA approaches by 27.08% in Classification Accuracy, 15.21% in Generation Accuracy and reducing Fréchet Inception Distance by 36.61%, indicating superior semantic alignment and image quality. The code is available at https://github.com/CVLABLUMS/CATVis.

Keywords: Brain-Computer Interfaces (BCIs) · EEG Decoding · Visual Reconstruction · Cross-Modal Alignment

1 Introduction

Brain-computer interfaces (BCIs) decode neural signals to enable communication, particularly for individuals with severe neuromotor impairments. Electroencephalography (EEG), a non-invasive and affordable technique, is widely used in BCIs due to its high temporal resolution, despite challenges such as

T. Mehmood and H. Ahmad—These authors contributed equally to this work and are co-corresponding authors.

J. C. Gee et al. (Eds.): MICCAI 2025, LNCS 15960, pp. 98–108, 2026.
https://doi.org/10.1007/978-3-032-04927-8_10

low spatial resolution [26], low signal-to-noise ratio [3], and artifact contamination [12]. Decoding visual representations from EEG remains difficult, as traditional feature-based approaches struggled with limited datasets, while recent deep learning advances have improved reconstruction fidelity [8,25].

Existing approaches like ThoughtViz [24] and Brain2Image [11] primarily focus on visualizing the conceptual information captured by neural signals, often overlooking the contextual elements, such as color, setting, or interaction. By neglecting these finer details, current methods provide only a coarse approximation of the visual stimulus. To address this gap, we propose a framework that not only improves the concept recognition but also captures the finer details by incorporating textual descriptions as an intermediary representation.

Our framework employs the EEG–image pairs dataset from [14,22], in which each EEG sample is paired with a corresponding visual stimulus. We begin by generating a caption for each stimulus to obtain a ground-truth description that serves as the basis for alignment in the CLIP [16] feature space. Next, our encoder classifies the EEG signals into conceptual categories, and we interpolate the concept and caption embeddings to form a semantically rich conditioning vector for Stable Diffusion (SD), enabling the generation of high-fidelity images.

Our end-to-end approach leverages state-of-the-art techniques in natural language processing and computer vision to create a complete pipeline from neural activity to visual output. Our contributions can be summarized as follows:

- Unlike prior work [11,24], which focused on conceptual decoding (class labels), we enable context-aware (descriptive information) EEG-to-image generation through cross-modal alignment and re-ranking, boosting generation accuracy by 15.21% over baselines.
- Unlike resource-intensive self-supervised methods [2,7], we achieve a 27.08% higher classification accuracy using a supervised EEG Conformer [21], demonstrating that robust EEG representations can be learned without extensive pretraining.
- We modulate the emphasis on concept vs. context. This increases the semantic fidelity of reconstructed images, aligning them more closely with the true visual content.

2 Related Work

Decoding visual information from brain activity and reconstructing images has gained significant attention in recent research. Initial work in this domain leveraged the discriminative power of GANs to directly generate realistic images from brain activity [11,13,20,24]. Brain2Image [11] used an LSTM to classify EEG into 40 ImageNet classes, conditioning VAE and GAN generation with discriminative features. ThoughtViz [24] built on this with 1D/2D-CNNs to classify EEG into object classes, driving GAN synthesis. Yet, both [11,24] rely on class-discriminative features, limiting their capacity to capture contextual details beyond basic conceptual decoding, thus reducing their utility for detailed visual reconstruction.

Diffusion models, inspired by non-equilibrium thermodynamics, have emerged as an effective generative modeling approach [17]. DreamDiffusion [2] fine-tunes Stable Diffusion [17] by conditioning on robust EEG features obtained through a Masked Autoencoder approach. These features are further aligned with CLIP image embeddings to improve cross-modal consistency, enabling high-quality image generation. However, DreamDiffusion relies on computationally intensive self-supervised pretraining, which limits its scalability and accessibility, particularly for resource-constrained settings.

In contrast, BrainVis [7] removes the need for external data by training a robust EEG encoder with Latent Masked Modeling, leveraging both time-domain and frequency-domain features while incorporating reconstruction and classification objectives. These features are aligned with interpolated semantic embeddings (coarse-grained labels and fine-grained captions). The aligned embeddings are used as conditioning for cascaded diffusion models, resulting in better semantic accuracy of generated images. However, BrainVis' complex cascaded diffusion architecture increases computational overhead.

3 Methodology

Our current methodology comprises five main components: (1) Conformer-Based EEG Encoder for concept classification; (2) Cross-modal alignment in CLIP feature space between EEG and caption embeddings; (3) Caption refinement through re-ranking of retrieved embeddings from CLIP feature space; (4) Weighted interpolation of concept and caption for semantically rich embedding; (5) Image generation using pretrained Stable Diffusion (Fig. 1).

3.1 EEG Encoder for Concept Classification

Unlike self-supervised approaches [2,7] that rely on self-supervised reconstruction objectives, we adopt a supervised EEG Conformer [21] encoder, trained from scratch on our dataset. The architecture leverages spatio-temporal convolutions and a stacked Transformer encoder to capture EEG signals' local and long-term dependencies, with a fully connected classification head to predict class labels.

3.2 Cross-Modal Alignment

Our cross-modal alignment framework learns to project EEG signals into CLIP's joint embedding space through contrastive learning. We adapt the EEG Conformer architecture by replacing its final classification layer with a linear projection to 768 dimensions followed by L2-normalization. The caption embeddings are obtained via CLIP's pretrained ViT-L/14 text encoder. We optimize the following symmetric contrastive loss:

$$\mathcal{L} = \frac{1}{2}(\mathcal{L}_{\text{EEG}\rightarrow\text{Text}} + \mathcal{L}_{\text{Text}\rightarrow\text{EEG}}) \tag{1}$$

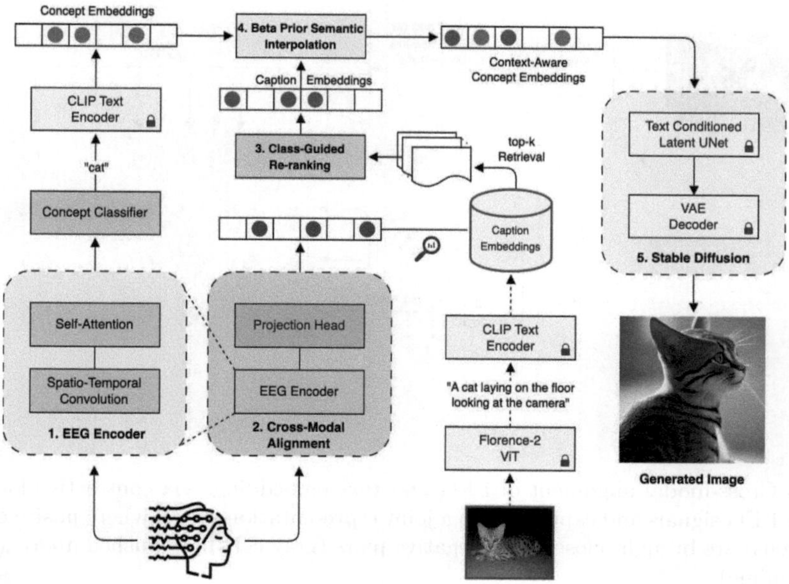

Fig. 1. Overview of our proposed five-stage framework for reconstructing images from EEG signals. The numbered modules correspond to the key steps in the methodology.

To be more explicit, we have two directional terms. For the EEG→Text direction:

$$\mathcal{L}_{\text{EEG}\rightarrow\text{Text}} = -\sum_{i=1}^{B} \log\left(\frac{\exp\left(\mathbf{e}_i^{\text{EEG}} \cdot \mathbf{e}_i^{\text{text}} / \tau\right)}{\sum_{j=1}^{B} \exp\left(\mathbf{e}_i^{\text{EEG}} \cdot \mathbf{e}_j^{\text{text}} / \tau\right)}\right), \tag{2}$$

where $\mathbf{e}_i^{\text{EEG}}$ is the normalized embedding for the EEG sample i, $\mathbf{e}_j^{\text{text}}$ is the embedding for the text sample j, B is the mini-batch size, and τ is the temperature parameter. Similarly, for the Text→EEG direction:

$$\mathcal{L}_{\text{Text}\rightarrow\text{EEG}} = -\sum_{i=1}^{B} \log\left(\frac{\exp\left(\mathbf{e}_i^{\text{text}} \cdot \mathbf{e}_i^{\text{EEG}} / \tau\right)}{\sum_{j=1}^{B} \exp\left(\mathbf{e}_i^{\text{text}} \cdot \mathbf{e}_j^{\text{EEG}} / \tau\right)}\right). \tag{3}$$

In this setup, each EEG embedding $\mathbf{e}_i^{\text{EEG}}$ is "pulled closer" to its corresponding text embedding $\mathbf{e}_i^{\text{text}}$ and pushed away from non-matching captions in the batch. Figure 2 demonstrates this alignment process.

3.3 Caption Refinement

Through cross-modal alignment, we align EEG representations with CLIP embeddings, which contain fine-grained semantic information. However, because EEG signals reflect only the subject's perception rather than every detail of the captioned image, they often fail to capture finer contextual cues. As a result, the neural response typically centers on the main concept and some contextual elements, leading to alignment inconsistencies when EEG embeddings are compared with CLIP text embeddings containing richer semantic information.

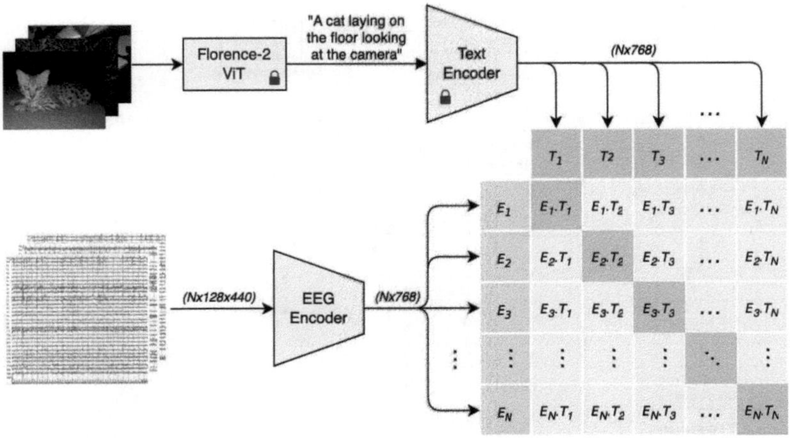

Fig. 2. Cross-modal alignment of EEG and text embeddings via contrastive loss. We project EEG signals and captions into a joint representation space, where positive pairs (blue cells) are brought closer, and negative pairs (gray cells) are pushed apart. (Color figure online)

To address this limitation, we employ a two-stage retrieval process:

1. **Initial Retrieval.** Given an EEG embedding $\mathbf{e}^{\text{EEG}} \in \mathcal{Q}$ (the *query*) and a collection of text embeddings $\mathcal{X} = \{\mathbf{e}_1^{\text{text}}, \ldots, \mathbf{e}_n^{\text{text}}\}$ of n caption embeddings belonging to the entire corpus, we retrieve the top-k most similar captions in the CLIP feature space as:

$$\arg \overset{(k)}{\max_{x \in \mathcal{X}}} \cos\big(\mathbf{e}^{\text{EEG}}, \mathbf{e}^{\text{text}}\big), \tag{4}$$

   This operation returns a subset of $\mathcal{X}$ that best match the neural response.
2. **Class-Guided Re-ranking.** These top-k candidates are then re-ranked according to their similarity to the predicted CLIP-encoded class label obtained from our concept classifier, ensuring that the final caption aligns with the primary object most likely perceived by the subject. This re-ranking step yields 6.99% higher Generation Accuracy (GA), 6.30% higher Inception Score (IS), and 3.43% lower Fréchet Inception Distance (FID) versus without re-ranking (see Table 2).

3.4 Beta Prior Semantic Interpolation

To balance object-focused conceptual and fine-grained descriptive information, we interpolate the predicted concept embedding from our EEG classifier with the re-ranked caption embedding. Let $\mathbf{e}^{\text{class}} \in \mathbb{R}^{768}$ be the CLIP-encoded class label predicted by the EEG Conformer, and $\mathbf{e}^{\text{text}} \in \mathbb{R}^{768}$ the top-ranked CLIP caption embedding. We sample $\lambda \sim \text{Beta}(\alpha, \beta)$ and compute:

$$\mathbf{z} = \lambda \mathbf{e}^{\text{class}} + (1 - \lambda)\mathbf{e}^{\text{text}}, \tag{5}$$

where $\mathbf{z}$ conditions the diffusion process.

3.5 Stable Diffusion

We use a pre-trained Stable Diffusion model [17] to transform the EEG-derived conditioning vector $\mathbf{z}$ into photorealistic images. Specifically, we condition the denoising process on $\mathbf{z}$ by integrating it into the cross-attention mechanism of Stable Diffusion's UNet, guiding the reverse diffusion process to reconstruct the image latent. Finally, the VAE decoder converts the latent representation into the pixel space, yielding the generated image. By leveraging Stable Diffusion's generative capabilities with EEG-driven embeddings, our approach aims to generate images that capture both the primary object class and finer contextual details inferred from neural activity.

4 Experiments

4.1 Dataset

We use the corrected EEG-ImageNet dataset from [14], which addresses preprocessing flaws in the original release [22], such as the inclusion of a DC component that introduced EEG drift and artificially boosted performance. This dataset has been the subject of methodological discussion: [1] highlights potential experimental confounds, including block-level stimulus leakage and classifier bias. In response, [15] provide a detailed rebuttal, citing cognitive neuroscience literature in support of block designs and contesting the validity of the critique's single-subject, low-frequency evaluation setup. While the EEG-ImageNet dataset stands as valid and citable, newer alternatives such as [9] also offer a valuable direction for future work.

The EEG-ImageNet dataset [14] features EEG signals from 6 subjects exposed to 2000 ImageNet images (40 concepts, 50 images each) via Rapid Serial Visual Presentation, recorded with a 128-electrode headset. Each trial yields $12,000$ EEG-image pairs, with signals (128 channels, 440 timepoints) preprocessed to retain the $20-460$ ms interval and band-pass filtered at $55-95$ Hz for cognitive relevance [14]. We adopt the provided train/validation/test splits. Ground-truth captions are generated using Microsoft Florence-2 Large [27] to pair EEG signals with image descriptions for cross-modal alignment.

4.2 Implementation Details

We implement our framework in PyTorch (Python 3.10), using the EEG Conformer [21] as the encoder backbone, modified to output 768-dimensional embeddings for CLIP alignment. The temporal convolution uses $k = 40$ kernels of size $(1, 25)$, while the spatial convolution applies the same number of kernels with a size of $(ch, 1)$, where $ch = 128$. The encoder is trained for 120 epochs with Adam ($\eta = 0.0001$) on cross-entropy loss. The cross-modal alignment module is trained separately for 100 epochs, minimizing InfoNCE loss ($\tau = 0.07$) with random sampling for mini-batches. Interpolation coefficients are sampled from a Beta distribution ($\alpha = 10, \beta = 10$). Image generation uses pre-trained Stable Diffusion v1-5 [17], conditioned via CLIP ViT-L/14 embeddings.

4.3 Evaluation Metrics

EEG Classification Accuracy (CA). CA measures the EEG encoder's ability to extract discriminative latent features from EEG signals, with higher values indicating better class prediction accuracy.

***N*-way Top-*K* Classification Accuracy of Generation (GA).** Following [2,4,7], GA evaluates semantic correctness of generated images using an N-way top-K classification task. We use a pre-trained ImageNet1K classifier [5] to compare ground-truth and generated images, selecting N-1 random classes plus the ground-truth class, and checking if the top-K prediction matches the ground truth over 50 trials. We set $N = 50$ and $K = 1$, consistent with prior evaluations.

Inception Score (IS). IS [18] assesses image quality and diversity using a pre-trained Inception-v3 classifier [23]. It measures classifier confidence and class distribution diversity, where higher scores reflect better quality and variety.

Fréchet Inception Distance (FID). FID [10] evaluates generation quality by computing the feature distribution distance between ground-truth and generated images using Inception-v3 [23]. Lower FID values indicate better alignment.

5 Results and Discussion

Our framework demonstrates significant advancements in decoding visual representations from EEG signals, as evidenced by both quantitative metrics and qualitative outputs. Table 1 shows that our supervised EEG Conformer achieves a Top-1 CA of 61.09%, surpassing BrainVis [7] by 27.08%, highlighting the encoder's ability to extract robust, discriminative features without resource-intensive pretraining. For image generation, our method achieves a GA of 56.78% for Subject 4 (Table 4) and 52.65% on average (Table 3), outperforming DreamDiffusion by 23.91% (Subject 4) and BrainVis by 15.21% (average), alongside an IS of 37.49 and a FID of 80.29, reflecting a 36.61% reduction in FID compared to BrainVis. These results underscore improved semantic alignment and image quality, as visually confirmed in Fig. 3, where generated images closely resemble their ground-truth stimuli in both conceptual and contextual details. Collectively, these findings validate the efficacy of our five-stage approach.

5.1 Ablation Studies

We ablated our EEG Conformer and interpolation strategy to assess their roles. Table 5 shows removing the Transformer reduces Top-1 CA by 5.76%, as it captures global temporal dependencies across EEG channels, vital for long-range neural patterns. Excluding the convolutional block drops CA by 45.77%, since it extracts local spatio-temporal features essential for stimulus-specific responses. Table 6 tests interpolation weights: a concept-heavy weight (0.75) yields the highest GA (61.48%) by emphasizing class-specific accuracy from the EEG classifier,

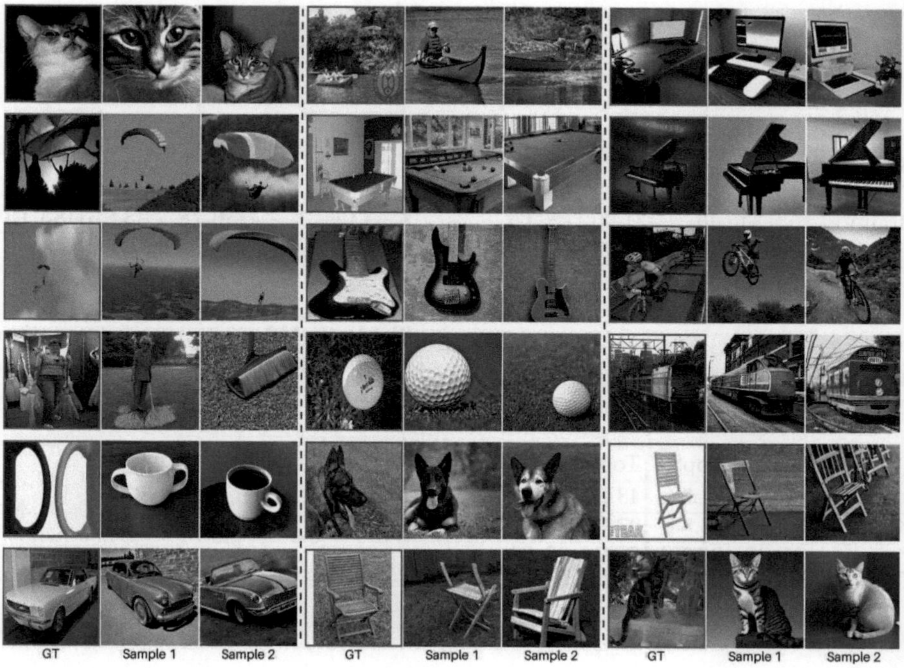

Fig. 3. Images generated from EEG signals using our framework. Red boxes show visual stimuli, followed by two generated samples. First 15 images (i.e., the first 5 rows) are the same visual stimuli used in Fig. 6 of [7] to allow qualitative comparison. (Color figure online)

Table 1. Comparison of EEG classification accuracy (%) with existing methods.

Method	Top1↑	Top3 ↑	Top5↑	F1 ↑
Brain2Image	16.48	37.90	55.24	0.16
KD-STFT [6]	41.20	75.30	87.82	0.40
BrainVis	47.66	79.23	91.20	0.43
CATVis	**61.09**	**93.84**	**98.15**	**0.61**

Table 2. Effect of caption re-ranking on generation performance.

Method	GA (%↑)	IS ↑	FID ↓
w/o Re-ranking	53.07	32.46	86.07
w/ Re-ranking	**56.78**	**34.51**	**83.12**

while a caption-heavy weight (0.75) lowers GA (53.93%) due to potential noise in descriptive details. The Beta(10,10) scheme balances both, achieving the best IS (35.79) by enhancing diversity via captions. These results validate the synergy of encoder components and the interpolation's role in balancing conceptual and contextual fidelity for Stable Diffusion conditioning.

Table 3. Comparison with image reconstruction baselines on all six subjects.

Method	GA (%↑)	IS ↑	FID ↓
Brain2Image	–	5.07	–
ESG-ADA [19]	-	10.82	174.13
BrainVis	45.46	31.00	126.66
CATVis	**52.65**	**37.50**	**80.30**

Table 4. Comparison with image reconstruction baselines on Subject 4.

Method	GA (%↑)	IS ↑	FID ↓
DreamDiffusion	45.80	–	–
BrainVis	49.27	31.53	121.02
CATVis	**56.78**	**34.51**	**83.13**

Table 5. Classification accuracy (%) of EEG Encoder w/wo CNN and w/wo Transformer Encoder (TE)

CNN	TE	Top1 ↑	Top3 ↑	Top5 ↑	F1 ↑
–	✓	15.32	34.50	48.27	0.16
✓	–	55.33	90.84	97.70	0.55
✓	✓	**61.09**	**93.84**	**98.15**	**0.61**

Table 6. Ablation on concept–caption interpolation with weights ω_{concept}, ω_{caption}

ω_{concept}	ω_{caption}	GA (%↑)	IS ↑	FID ↓
0.75	0.25	**61.48**	34.34	**82.33**
0.25	0.75	53.93	30.50	83.30
0.5	0.5	56.79	34.50	86.66
Beta(10,10)		56.86	**35.79**	84.13

6 Conclusion

Our five-stage framework advances EEG-based visual reconstruction, achieving a 27.08% higher CA and 15.21% improved GA over state-of-the-art methods, with a 36.61% FID reduction. By integrating supervised EEG encoding, CLIP alignment, caption refinement, and interpolation, we capture both conceptual and contextual details, outperforming prior approaches without extensive pre-training. This work offers a scalable solution for BCI applications, with future potential in refining cross-modal alignment and exploring real-time visualization.

Disclosure of Interests. The authors have no competing interests to declare that are relevant to the content of this article.

References

1. Ahmed, H., Wilbur, R.B., Bharadwaj, H.M., Siskind, J.M.: Confounds in the data–comments on "decoding brain representations by multimodal learning of neural activity and visual features." IEEE Trans. Pattern Anal. Mach. Intell. **44**(12), 9217–9220 (2021)
2. Bai, Y., Wang, X., Cao, Y.p., Ge, Y., Yuan, C., Shan, Y.: DreamDiffusion: generating high-quality images from brain EEG signals. arXiv preprint arXiv:2306.16934 (2023)
3. Chaddad, A., Wu, Y., Kateb, R., Bouridane, A.: Electroencephalography signal processing: a comprehensive review and analysis of methods and techniques. Sensors **23**(14), 6434 (2023)

4. Chen, Z., Qing, J., Xiang, T., Yue, W.L., Zhou, J.H.: Seeing beyond the brain: conditional diffusion model with sparse masked modeling for vision decoding. In: IEEE International Conference on Computer Vision and Pattern Recognition, pp. 22710–22720. IEEE (2023)
5. Dosovitskiy, A., et al.: An image is worth 16 ×16 words: transformers for image recognition at scale. In: International Conference on Learning Representations (2021)
6. Ferrante, M., Boccato, T., Bargione, S., Toschi, N.: Decoding EEG signals of visual brain representations with a CLIP based knowledge distillation. In: ICLR Workshop on Learning from Time Series For Health (2024)
7. Fu, H., Wang, H., Chin, J.J., Shen, Z.: BrainVis: exploring the bridge between brain and visual signals via image reconstruction. In: IEEE International Conference on Acoustics, Speech and Signal Processing, pp. 1–5. IEEE (2025)
8. Gong, S., Xing, K., Cichocki, A., Li, J.: Deep learning in EEG: advance of the last ten-year critical period. IEEE Trans. Cogn. Dev. Syst. **14**(2), 348–365 (2022)
9. Guenther, S., Kosmyna, N., Maes, P.: Image classification and reconstruction from low-density EEG. Sci. Rep. **14**(1), 16436 (2024)
10. Heusel, M., Ramsauer, H., Unterthiner, T., Nessler, B., Hochreiter, S.: GANs trained by a two time-scale update rule converge to a local Nash equilibrium. In: Advances in Neural Information Processing Systems, vol. 30 (2017)
11. Kavasidis, I., Palazzo, S., Spampinato, C., Giordano, D., Shah, M.: Brain2Image: converting brain signals into images. In: ACM International Conference on Multimedia, pp. 1809–1817. ACM (2017)
12. Mannan, M.M.N., Kamran, M.A., Jeong, M.Y.: Identification and removal of physiological artifacts from electroencephalogram signals: a review. IEEE Access **6**, 30630–30652 (2018)
13. Mishra, R., Sharma, K., Jha, R.R., Bhavsar, A.: NeuroGAN: image reconstruction from EEG signals via an attention-based GAN. Neural Comput. Appl. **35**(12), 9181–9192 (2023)
14. Palazzo, S., Spampinato, C., Kavasidis, I., Giordano, D., Schmidt, J., Shah, M.: Decoding brain representations by multimodal learning of neural activity and visual features. IEEE Trans. Pattern Anal. Mach. Intell. **43**(11), 3833–3849 (2020)
15. Palazzo, S., Spampinato, C., Kavasidis, I., Giordano, D., Schmidt, J., Shah, M.: Rebuttal to "Comments on 'Decoding brain representations by multimodal learning of neural activity and visual features"'. IEEE Trans. Pattern Anal. Mach. Intell. **46**(12), 11540–11542 (2024)
16. Radford, A., et al.: Learning transferable visual models from natural language supervision. In: International Conference on Machine Learning, pp. 8748–8763. PMLR (2021)
17. Rombach, R., Blattmann, A., Lorenz, D., Esser, P., Ommer, B.: High-resolution image synthesis with latent diffusion models. In: IEEE International Conference on Computer Vision and Pattern Recognition, pp. 10684–10695. IEEE (2022)
18. Salimans, T., Goodfellow, I., Zaremba, W., Cheung, V., Radford, A., Chen, X.: Improved techniques for training GANs. In: Advances in Neural Information Processing Systems, vol. 29 (2016)
19. Singh, P., Dalal, D., Vashishtha, G., Miyapuram, K., Raman, S.: Learning robust deep visual representations from EEG brain recordings. In: IEEE Winter Conference on Applications of Computer Vision, pp. 7553–7562. IEEE (2024)
20. Singh, P., Pandey, P., Miyapuram, K., Raman, S.: EEG2IMAGE: image reconstruction from EEG brain signals. In: IEEE International Conference on Acoustics, Speech and Signal Processing, pp. 1–5. IEEE (2023)

21. Song, Y., Zheng, Q., Liu, B., Gao, X.: EEG conformer: convolutional transformer for EEG decoding and visualization. IEEE Trans. Neural Syst. Rehabil. Eng. **31**, 710–719 (2023)
22. Spampinato, C., Palazzo, S., Kavasidis, I., Giordano, D., Souly, N., Shah, M.: Deep learning human mind for automated visual classification. In: IEEE International Conference on Computer Vision and Pattern Recognition, pp. 6809–6817. IEEE (2017)
23. Szegedy, C., Vanhoucke, V., Ioffe, S., Shlens, J., Wojna, Z.: Rethinking the inception architecture for computer vision. In: IEEE International Conference on Computer Vision and Pattern Recognition. IEEE (2016)
24. Tirupattur, P., Rawat, Y.S., Spampinato, C., Shah, M.: ThoughtViz: visualizing human thoughts using generative adversarial network. In: ACM International Conference on Multimedia, pp. 950–958. ACM (2018)
25. Vallabhaneni, R.B., et al.: Deep learning algorithms in EEG signal decoding application: a review. IEEE Access **9**, 125778–125794 (2021)
26. Värbu, K., Muhammad, N., Muhammad, Y.: Past, present, and future of EEG-based BCI applications. Sensors **22**(9), 3331 (2022)
27. Xiao, B., et al.: Florence-2: advancing a unified representation for a variety of vision tasks. In: IEEE International Conference on Computer Vision and Pattern Recognition, pp. 4818–4829. IEEE (2024)

CBrain: Cross-Modal Learning for Brain Vigilance Detection in Resting-State fMRI

Chang Li[1(✉)], Yamin Li[1], Haatef Pourmotabbed[1], Shengchao Zhang[2], Jorge A. Salas[1], Sarah E. Goodale[1,3], Roza G. Bayrak[1], and Catie Chang[1]

[1] Vanderbilt University, Nashville, TN, USA
chang.li@vanderbilt.edu
[2] Rhode Island Hospital (Brown University Health), Providence, RI, USA
[3] Vanderbilt University Medical Center, Nashville, TN, USA

Abstract. Detecting human vigilance states (e.g., natural shifts between alertness and drowsiness) from functional magnetic resonance imaging (fMRI) data can provide novel insight into the whole-brain patterns underlying these critical states. Moreover, as a person's vigilance levels are closely tied to their behavior and brain activity, vigilance state can strongly influence the results of fMRI studies. Therefore, the ability to annotate fMRI scans with vigilance information can also enable clearer and more robust results in fMRI research. However, well-established vigilance indicators are derived from other modalities such as behavioral responses, electroencephalography (EEG), and pupillometry, which are not typically available in fMRI data collection. While previous works indicate the promise of distinguishing vigilance states from fMRI alone, EEG data can provide reliable vigilance indicators that complement and augment fMRI domain information. Here, we propose **CBrain**: **C**ross-modal learning for **Brain** vigilance detection in resting-state fMRI. Our model transfers EEG vigilance information into an fMRI latent space in training, and predicts human vigilance states using only fMRI data in testing, addressing the need for external vigilance indicators. Experimental results demonstrate CBrain's ability to predict vigilance states across different individuals at a granularity of 10-fMRI-frames with an **81.07%** $mF1$ score on a test set of unseen subjects. Additionally, our generalization experiments highlight the model's potential to estimate vigilance in an unseen task and in resting-state fMRI scans collected with a different scanner at a different site. Source code: https://github.com/neurdylab/CBrain.

Keywords: fMRI · EEG · brain patterns · cross-modal learning

1 Introduction

Functional magnetic resonance imaging (fMRI) data provides whole-brain blood-oxygen-level-dependent (BOLD) signals at millimeter-scale resolution. During fMRI scans, subjects often experience natural shifts between alertness and

J. C. Gee et al. (Eds.): MICCAI 2025, LNCS 15960, pp. 109–119, 2026.
https://doi.org/10.1007/978-3-032-04927-8_11

drowsiness, also known as fluctuations in *vigilance*. Deriving brain-wide patterns associated with vigilance provides a window into brain activity associated with this critical brain state [3,13]. Further, vigilance state fluctuations strongly influence subjects' behavior and brain activity [13]. As a result, modeling the effects of vigilance on fMRI data, and providing accurate vigilance state annotations, can facilitate fMRI research and enable a more robust interpretation results across a broad range of studies.

Well-established vigilance indicators can be derived from measures such as EEG and pupillometry. However, collecting these additional data requires specialized hardware, additional setup time, and expertise in acquiring and denoising these signals. Resting-state fMRI scans also lack behavioral measures, which can also convey vigilance information. The lack of ground-truth vigilance indicators in most fMRI datasets motivates the development of methods to infer vigilance fluctuations directly from fMRI data. Previous work has demonstrated the feasibility of detecting vigilance from fMRI [8,23,31]. However, these methods are either limited in temporal resolution, predicting vigilance states from functional connectivity windows of ≥ 40 s [23,26] or they have high temporal resolution but are limited in their ability to quantify corresponding states and state transitions across scans (as opposed to relative variations within scans) [8], or are unsupervised methods that need additional information to label the discovered states [31]. fMRI foundation models [1,2,6,29] are demonstrating promise in applications such as fMRI signal reconstruction and disease modeling. However, to our knowledge, such models have not yet been leveraged for predicting vigilance states from fMRI.

EEG is highly complementary to fMRI: fMRI has high spatial resolution but lower temporal resolution, and measures a blood-oxygen response; while EEG has lower spatial resolution but high temporal resolution, and measures neuro-electrical activity. Moreover, EEG data is rich with vigilance-related information [19]. Combining fMRI and EEG may thus provide improved spatiotemporal features, and allows for incorporating EEG vigilance information into latent representations of fMRI data. We hypothesize that leveraging EEG to train an fMRI vigilance-detection model may increase a model's ability to discern vigilance states. Currently, fMRI vigilance labeling strategies are not yet established as ground truth. By training fMRI models to predict EEG-derived labels, we address an open challenge: enabling vigilance detection from fMRI alone.

We propose **CBrain**: **C**ross-modal learning for **Brain** vigilance detection, a cross-modal architecture that leverages EEG to enhance fMRI-based vigilance detection. We train our model on simultaneous EEG-fMRI data, transferring vigilance-related cross-modal EEG knowledge into the fMRI latent space. During testing, CBrain uses only 10-frame fMRI data patches as input and achieves an **81.07%** $mF1$ score in predicting vigilance states for unseen test-set subjects, addressing the challenge of distinguishing brain states in fMRI without relying on other data modalities. Our experiments on another EEG-fMRI dataset, acquired on a different scanner, demonstrate our model's ability to generalize to both resting-state and task scans. These results show that CBrain has the potential to enrich existing fMRI scans, including those in large public datasets that lack vigilance measures [17,24], with new information about vigilance state.

2 Method

Our model is trained on 10-frame fMRI data patches and corresponding EEG data patches. We perform fMRI intra-modal learning, fMRI-EEG cross-modal learning, and prediction head training in one stage. In testing, our model predicts vigilance states based only on the fMRI data input, as shown in Fig. 1.

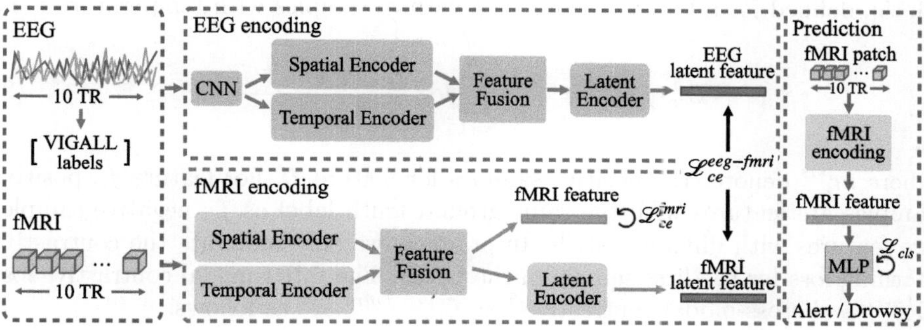

Fig. 1. CBrain's pipeline. We train our model on paired 10-frame fMRI data and EEG data segments. In testing, our model takes only fMRI data as input.

Obtaining Patch-Wise Vigilance Ground-Truth: In a simultaneous EEG-fMRI dataset, for each fMRI scan X with T frames, we calculate frame-wise vigilance score $V_{1..T}$ by applying Vigilance Algorithm Leipzig [19] on its paired EEG data Y, and convert the integer values to range (−1, 0, 1) following [21]. We sum frame-wise labels for every 10-frame fMRI patch and assign the ground truth as alert if the sum exceeds −5, and drowsy otherwise.

Cross-Modal Contrastive Learning: For each 10-frame fMRI data segment x, consisting of data from 64 regional and 2 global fMRI time courses (described in Sect. 3.1), we first extract its spatial features f_x^s and temporal features f_x^t by performing attention on the spatial and temporal axes using spatial and temporal transformer-based [25] encoders respectively. Then we fuse f_x^s and f_x^t with a given feature fusion ratio to obtain fMRI-domain spatial-temporal features f_x by:

$$f_x = ratio * f_x^s + (1 - ratio) * f_x^t \tag{1}$$

For the fMRI segment's corresponding 26-channel EEG data patch y, we apply a 1D CNN downsampling layer, then harvest EEG spatial features f_y^s and temporal features f_y^t from EEG spatial and temporal transformer encoders in the same manner as in fMRI domain. We fuse these EEG spatial and temporal features using the same feature ratio as fMRI features. We map fMRI features f_x and EEG features f_y to a common latent space using our latent encoding modules, each consisting of two blocks of a linear layer followed by one ReLU activation, obtaining f_x^{latent} and f_y^{latent}. Then, we apply contrastive learning on the fMRI features and cross-modal latent space, allowing the model to capture intrinsic vigilance-state differences and to bridge EEG brain-state-related features into the fMRI domain. In the fMRI domain, we perform contrastive learning on fMRI fused features f_x to increase the fMRI features' discriminability. For cross-modal learning, we apply contrastive loss on the fMRI latent features f_x^{latent} and corresponding EEG latent features f_y^{latent}, bringing EEG brain-state information into a shared latent space with fMRI features. We use contrastive learning loss [4,15], defined for a batch of features f with m samples with temperature τ as:

$$\mathcal{L}_{ce}(f_{1..m}) = -\frac{1}{m}\sum_{i=1}^{m}\log\frac{\sum_{k=1}^{m_i^{pos}} e^{f_i^\top f_k/\tau}}{\sum_{j=1}^{m} e^{f_i^\top f_j/\tau}}, \tag{2}$$

where m_i^{pos} denotes the positive samples for feature f_i. For feature f_i, positive samples are features with the same ground truth label as f_i; negative samples are features with different labels. In practice, we only integrate the contrastive learning loss for the first sample in the batch. The fMRI-modal contrastive loss $\mathcal{L}_{ce}^{fmri}$ and cross-modal contrastive loss $\mathcal{L}_{ce}^{eeg-fmri}$ are as follows:

$$\mathcal{L}_{ce}^{fmri} = \mathcal{L}_{ce}(f_{x_{1..m}}) \tag{3}$$

$$\mathcal{L}_{ce}^{eeg-fmri} = \mathcal{L}_{ce}(f_{x_{1..m}}^{latent}) + \mathcal{L}_{ce}(f_{y_{1..m}}^{latent}) + \mathcal{L}_{ce}(f_{x_{1..m}}^{latent}, f_{y_{1..m}}^{latent}) \tag{4}$$

Vigilance State Prediction: Given a 10-frame segment of fMRI data from an unseen subject, we first extract its spatial-temporal features without mapping it to the latent space, and feed it to a classification MLP head with three hidden layers, each comprising a linear layer, batch normalization, and LeakyReLU activation. The output layer is a linear layer without normalization. The MLP head is trained using cross-entropy loss $\mathcal{L}_{cls}$. The overall training loss can be formulated as:

$$\mathcal{L} = \mathcal{L}_{cls} + \mathcal{L}_{ce}^{fmri} + \mathcal{L}_{ce}^{eeg-fmri} \tag{5}$$

Model Training: We train our model in an end-to-end manner, with the feature fusion ratio as 0.5, contrastive learning temperature τ as 0.1, the weights of $\mathcal{L}_{cls}$, $\mathcal{L}_{ce}^{fmri}$, $\mathcal{L}_{ce}^{eeg-fmri}$ as 0.5, 0.1, and 0.1. We train our model for 50 epochs using AdamW optimizer, a learning rate of 7e-4, a weight decay of 0.1, and a batch size of 32 on a single NVIDIA RTX 6000 GPU. Training on the full dataset takes less than one hour. We adopt a linear warm up from 1e-6 for 20 epochs. We develop our code based on the frameworks in [15,18] to perform cross-modal contrastive learning for vigilance detection. CBrain's theoretical complexity is as follows: encoder [25]: $O(n^2 * d)$, MLP: $O(m * d_{1..n})$, contrastive learning [4]: $O(m^2)$ (n: sequence length, d: feature dim, m: batch size).

3 Experiments

3.1 Training and External Validation Datasets

For model training, we used a simultaneous EEG-fMRI dataset comprising 29 resting-state scans from 22 healthy subjects. Subjects provided written informed consent, and protocols were approved by the Institutional Review Board. Functional MRI data was acquired using a multi-echo EPI sequence (3T scanner, TR = 2100 ms). 32-channel EEG data was collected at 5 KHz, synchronized to the scanner's 10 MHz clock. Briefly, for fMRI data preprocessing, slice-timing correction, motion coregistration, multi-echo ICA denoising, alignment to MNI152 space, and 3 mm spatial smoothing were performed. Using the Dictionary of Functional Modes atlas [5], 64 regions of interest (ROIs) were extracted. Additionally, 1st through 4th-order polynomial trends and 6 motion parameters were regressed out of the data. The global signal with and without motion parameters were calculated and added to the input as two additional channels. We lagged the fMRI data by 2 TRs (4.2 sec) with respect to the EEG data to account for hemodynamic response delay. EEG data was corrected for MRI and ballistocardiogram artifacts and downsampled to 250 Hz. We use 26 channels in training(EMG/ECG channels excluded). We obtain fMRI (and paired EEG) data segments with a sliding time window of 10 fMRI frames and a step size of 5 fMRI frames. Please refer to [12] for the dataset and preprocessing details. The training set includes 80% of the subjects (17 subjects, 23 scans), and the test set comprises 20% (5 subjects, 6 scans), with no subject overlap between training and testing splits. Noisy EEG segments were identified according to [7] and were excluded from training and testing. Another EEG-fMRI dataset, acquired at a different site and with different fMRI paradigms, was used for external validation (with no data used in training) [8]. This dataset comprises scans from 14 healthy subjects (3T, TR = 2100 ms), who collectively underwent 16 eye-closed-rest scans and 12 scans collected during the delivery of intermittent auditory tones (also with eyes closed). We preprocessed this dataset the same way as the dataset described above.

3.2 Main Results

The main results, and comparisons with baselines trained with fMRI data only, are shown in Table 1. Our baselines include: fMRI classification models with strong performance in [20], timeseries models tailored for brain imaging tasks in [28], an fMRI foundation model [2] and the hierarchical SVM with high accuracy in a sleep staging task [23]. **CBrain** achieves a macro mean-F1 ($mF1$) score of 81.07% in the testing set, surpassing all baselines. The comparison between full-CBrain with fMRI-only CBrain and fMRI-only baselines shows that incorporating EEG data boosts CBrain's performance (3.22%), supporting our key contribution that EEG knowledge enhances fMRI-based brain state detection. This demonstrates the power of integrating complementary modalities and the potential of cross-modal supervision as an emerging paradigm in this field.

Table 1. CBrain's performance on testing set and eye-closed-rest external validation scans, compared to baselines, with (subject-wise mean ± std) for best performing models. $F1_d$: $F1_{drowsy}$, $F1_a$: $F1_{alert}$. **Bold with underline:** best performance. Underline: second-best performance. Random guess: output random predictions on brain states.

Methods	Testing Set			External Validation (ecr)		
	$F1_d$	$F1_a$	$mF1$	$F1_d$	$F1_a$	$mF1$
random guess	55.71	42.20	48.95(52.2 ± 9.7)	50.32	46.31	48.31(57.0 ± 9.6)
BrainLM [2]	64.82	37.25	51.04	59.72	37.13	48.43
BolT [1]	73.63	51.52	62.58	65.10	61.39	63.25
Medformer [28]	78.40	50.88	64.64	72.95	56.41	64.68
Nonformer [14]	76.86	55.40	66.13	76.40	69.94	73.17
BrainNetCNN [10]	78.98	58.77	68.87	72.69	66.70	69.69
SVM [23]	82.40	58.18	70.29	79.61	70.18	74.89
meanLSTM [20]	81.05	65.51	73.28(68.5 ± 16.1)	78.52	77.13	77.82(77.2 ± 15.0)
attnMLP [20]	85.39	70.53	77.96(70.4 ± 17.2)	79.10	75.85	77.48(78.2 ± 14.3)
meanMLP [20]	84.59	69.57	77.08(70.2 ± 17.0)	83.41	81.20	**82.31**(81.0 ± 13.3)
CBrain (fMRI-only)	84.50	71.21	77.85(71.2 ± 15.0)	79.54	77.35	78.44(77.3 ± 17.2)
CBrain	86.20	75.95	**81.07**(75.5 ± 12.2)	79.63	78.01	78.82(78.6 ± 15.7)

CBrain also yields consistently high classification $mF1$ in the eye-closed-rest (ecr) external validation data. CBrain trained with fMRI data alone attains a competitive performance of 77.85% $mF1$ in the testing set and 78.44% $mF1$ in the ecr external validation set, only inferior to the full CBrain model and the attention MLP [20], demonstrating the effectiveness of our intra-model contrastive learning in differentiating vigilance states. We acknowledge that meanMLP [20], a very recent model, outperforms CBrain in the external validation set; the reason for this difference merits further investigation. We speculate that the slight drop in CBrain's performance on the external validation data might be caused by differences in dataset statistics, and future work might examine potential improvements through 1) data normalization strategies, and 2) encoding demographic and hardware information to handle inter-subject and inter-site variability.

3.3 Ablation Studies and Qualitative Analysis

Comparison with EEG Foundation Models: In Table 2, we compare our EEG encoder with state-of-the-art EEG foundation models that benefit from pretraining techniques to extract high-quality EEG features [9,11,27,30] compared in [27]. We finetune the released checkpoints on our training set to extract EEG features and use 1D CNN for downsampling before applying our latent encoder.

Our original model has the best $mF1$ score, demonstrating that by fusing spatial-temporal features, we can extract high-quality EEG features that better assist fMRI models in improving the fMRI latent space's discriminability. Leveraging pre-trained EEG features leads to high performance, as utilizing BIOT [30], BENDR [11], and LaBraM [9] reached an $mF1$ that exceeded all fMRI-only models in the testing set. This reveals that EEG knowledge can enhance the fMRI model's understanding of brain states.

Training Loss: CBrain performs best with all loss components as shown in Table 3. These results further demonstrate that complementary EEG knowledge can improve the fMRI model's ability to differentiate brain patterns.

Generalization: CBrain has a strong performance on the unseen external validation dataset collected at a different site and scanner, suggesting its generalizability to unseen data (Table 4). A comparison between CBrain trained on fMRI data alone versus EEG-fMRI data shows that integrating EEG improves its robustness on external datasets beyond resting-state scans.

Qualitative Analysis: Figure 2 shows the predictions of the full CBrain model, together with ground-truth vigilance states and the corresponding EEG spectrograms. In Fig. 3, we visualize the fMRI feature space for the testing set and the external validation eye-closed-rest scans. UMAP [16] scatter plots (first column) show that our fMRI encoders can successfully extract fMRI features with strong brain-state discriminability. We also visualize the average alert and drowsy fMRI features in the test and external validation (ecr) sets and project them onto the brain surface map. In line with prior work [31], the drowsy patterns show greater sensorimotor and visual activity. These results demonstrate that CBrain can generate consistent fMRI embeddings for different brain states.

Table 2. Comparison of EEG encoders on the testing set (with subject-wise mean ± std). **Bold:** best performance.

Method	$F1_{drowsy}$	$F1_{alert}$	$mF1$	mAC
CBrain w/ EEGPT [27]	83.89	70.18	77.03(72.6 ± 13.6)	79.08
CBrain w/ BIOT [30]	86.08	72.60	79.34(71.9 ± 15.9)	81.54
CBrain w/ BENDR [11]	85.99	75.42	80.71(75.6 ± 14.4)	82.15
CBrain w/ LaBraM [9]	85.61	75.98	80.79(75.3 ± 14.0)	82.00
CBrain	86.20	75.95	**81.07**(75.5 ± 12.2)	**82.46**

Table 3. Ablations on training loss with training weights. **Bold:** best performance.

Method	$\mathcal{L}_{cls}$	$\mathcal{L}_{ce}^{fmri}$	$\mathcal{L}_{ce}^{eeg-fmri}$	$F1_{drowsy}$	$F1_{alert}$	$mF1$	mAC
	0.5	–	–	81.86	67.10	74.48	76.62
CBrain	0.5	0.1	–	83.21	69.96	76.59	78.46
	0.5	0.1	0.1	86.20	75.95	**81.07**	**82.46**

Table 4. Generalization experiments on external validation dataset in eye-closed-rest and eye-closed-task scans with (subject-wise mean ± std). $F1_d$: $F1_{drowsy}$, $F1_a$: $F1_{alert}$. **Bold:** best performance.

Methods	Eye-closed-rest			Eye-closed-task		
	$F1_d$	$F1_a$	$mF1$	$F1_d$	$F1_a$	$mF1$
CBrain (fMRI only)	79.54	77.35	78.44(77.3 ± 17.2)	76.09	48.77	62.43(64.2 ± 21.5)
CBrain	79.63	78.01	**78.82**(78.6 ± 15.7)	74.84	52.27	63.56(63.2 ± 22.3)

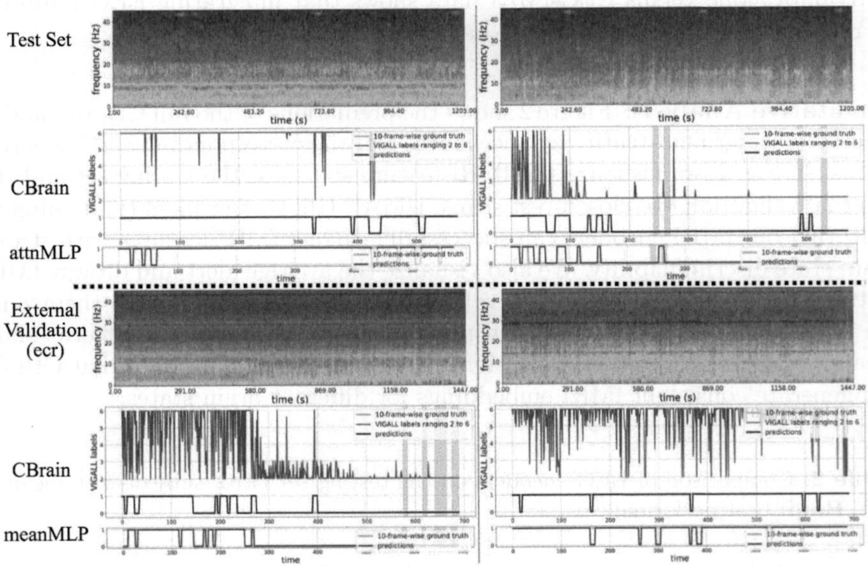

Fig. 2. Visualization of CBrain's predictions on the testing set and external validation (eye-closed-rest) scans, along with the best-performing baselines (testing set: attnMLP [20], ecr scans: meanMLP [20]). Top plots of each panel: EEG spectrograms [22]. Bottom plots of each panel: model predictions (blue), baseline predictions (purple), the 10-frame ground truth (orange), the original frame-wise integer Vigilance Algorithm Leipzig (VIGALL) labels (green), noisy EEG segments (gray bars). Higher labels indicate higher alertness levels [19]. The overlap between predicted and ground truth labels reflects strong performance.

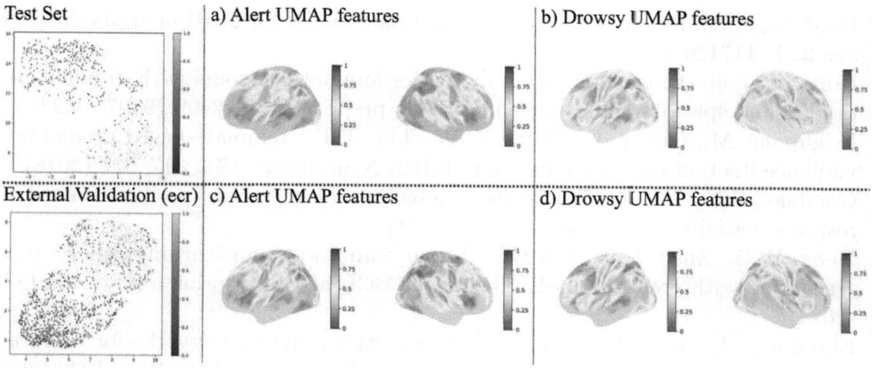

Fig. 3. Qualitative analysis of CBrain's extracted fMRI features. Each row illustrates the UMAP [16] embedding of fMRI features within the dataset, and the mapping of average alert and drowsy fMRI features on the brain surface in both hemispheres.

4 Conclusion

We propose **CBrain**: **C**ross-modal learning for **Brain** vigilance detection from resting-state fMRI, which transfers cross-modal vigilance knowledge from EEG to the fMRI domain. Our model accurately predicts vigilance state from fMRI data in a 10-fMRI-frame granularity. This approach can provide vigilance-state annotations to fMRI scans, including public databases, in the common scenario where dedicated vigilance measures (such as EEG and eye behavior) are not simultaneously recorded. Our model's robustness in generalization tasks supports the idea that cross-modal EEG information can enhance the discriminability of brain state patterns within single-modal fMRI data.

Acknowledgments. This work was supported by NIH grants R01 NS112252, F99 AG079810, and T32 EB021937. The external dataset was acquired in the Advanced MRI Section of the NINDS, NIH.

Disclosure of Interests. The authors have no competing interests.

References

1. Bedel, H.A., Sivgin, I., Dalmaz, O., Dar, S.U., Çukur, T.: Bolt: fused window transformers for fMRI time series analysis. Med. Image Anal. **88**, 102841 (2023)
2. Caro, J.O., et al.: BrainLM: a foundation model for brain activity recordings. bioRxiv (2023)
3. Chang, C., et al.: Tracking brain arousal fluctuations with fMRI. Proc. Natl. Acad. Sci. **113**(16), 4518–4523 (2016)
4. Chen, T., Kornblith, S., Norouzi, M., Hinton, G.: A simple framework for contrastive learning of visual representations. In: International Conference on Machine Learning, pp. 1597–1607. PMLR (2020)

5. Dadi, K., et al.: Fine-grain atlases of functional modes for fMRI analysis. Neuroimage **221**, 117126 (2020)
6. Dong, Z., e al.: Brain-jepa: brain dynamics foundation model with gradient positioning and spatiotemporal masking. arXiv preprint arXiv:2409.19407 (2024)
7. Falahpour, M., Chang, C., Wong, C.W., Liu, T.T.: Template-based prediction of vigilance fluctuations in resting-state fMRI. Neuroimage **174**, 317–327 (2018)
8. Goodale, S.E., et al.: fMRI-based detection of alertness predicts behavioral response variability. Elife **10**, e62376 (2021)
9. Jiang, W.B., Zhao, L.M., Lu, B.L.: Large brain model for learning generic representations with tremendous EEG data in BCI. arXiv preprint arXiv:2405.18765 (2024)
10. Kawahara, J., et al.: BrainNetCNN: convolutional neural networks for brain networks; towards predicting neurodevelopment. Neuroimage **146**, 1038–1049 (2017)
11. Kostas, D., Aroca-Ouellette, S., Rudzicz, F.: Bendr: using transformers and a contrastive self-supervised learning task to learn from massive amounts of EEG data. Front. Hum. Neurosci. **15**, 653659 (2021)
12. Li, Y., et al.: Neurobolt: resting-state EEG-to-fMRI synthesis with multi-dimensional feature mapping. arXiv preprint arXiv:2410.05341 (2024)
13. Liu, T.T., Falahpour, M.: Vigilance effects in resting-state fMRI. Front. Neurosci. **14**, 321 (2020)
14. Liu, Y., Wu, H., Wang, J., Long, M.: Non-stationary transformers: exploring the stationarity in time series forecasting. In: Advances in Neural Information Processing Systems, vol. 35, pp. 9881–9893 (2022)
15. Lu, Y., Xu, C., Wei, X., Xie, X., Tomizuka, M., Keutzer, K., Zhang, S.: Open-vocabulary point-cloud object detection without 3D annotation. In: Proceedings of the IEEE/CVF Conference on Computer Vision and Pattern Recognition, pp. 1190–1199 (2023)
16. McInnes, L., Healy, J., Melville, J.: Umap: uniform manifold approximation and projection for dimension reduction. arXiv preprint arXiv:1802.03426 (2018)
17. Miller, K.L., et al.: Multimodal population brain imaging in the UK biobank prospective epidemiological study. Nat. Neurosci. **19**(11), 1523–1536 (2016)
18. Misra, I., Girdhar, R., Joulin, A.: An end-to-end transformer model for 3D object detection. In: Proceedings of the IEEE/CVF International Conference on Computer Vision, pp. 2906–2917 (2021)
19. Olbrich, S., Fischer, M.M., Sander, C., Hegerl, U., Wirtz, H., Bosse-Henck, A.: Objective markers for sleep propensity: comparison between the multiple sleep latency test and the vigilance algorithm leipzig. J. Sleep Res. **24**(4), 450–457 (2015)
20. Popov, P., Mahmood, U., Fu, Z., Yang, C., Calhoun, V., Plis, S.: A simple but tough-to-beat baseline for fMRI time-series classification. Neuroimage **303**, 120909 (2024)
21. Pourmotabbed, H., et al.: Multimodal state-dependent connectivity analysis of arousal and autonomic centers in the brainstem and basal forebrain. bioRxiv (2024)
22. Prerau, M.J., Brown, R.E., Bianchi, M.T., Ellenbogen, J.M., Purdon, P.L.: Sleep neurophysiological dynamics through the lens of multitaper spectral analysis. Physiology **32**(1), 60–92 (2017)
23. Tagliazucchi, E., von Wegner, F., Morzelewski, A., Borisov, S., Jahnke, K., Laufs, H.: Automatic sleep staging using fMRI functional connectivity data. Neuroimage **63**(1), 63–72 (2012)
24. Van Essen, D.C., et al.: The WU-MINN human connectome project: an overview. Neuroimage **80**, 62–79 (2013)

25. Vaswani, A.: Attention is all you need. In: Advances in Neural Information Processing Systems (2017)
26. Wang, C., Ong, J.L., Patanaik, A., Zhou, J., Chee, M.W.: Spontaneous eyelid closures link vigilance fluctuation with fMRI dynamic connectivity states. Proc. Natl. Acad. Sci. **113**(34), 9653–9658 (2016)
27. Wang, G., Liu, W., He, Y., Xu, C., Ma, L., Li, H.: EEGPT: pretrained transformer for universal and reliable representation of EEG signals. In: The Thirty-eighth Annual Conference on Neural Information Processing Systems (2024)
28. Wang, Y., Huang, N., Li, T., Yan, Y., Zhang, X.: Medformer: a multi-granularity patching transformer for medical time-series classification. arXiv preprint arXiv:2405.19363 (2024)
29. Wei, Z., Dan, T., Ding, J., Wu, G.: Neuropath: a neural pathway transformer for joining the dots of human connectomes. arXiv preprint arXiv:2409.17510 (2024)
30. Yang, C., Westover, M., Sun, J.: Biot: biosignal transformer for cross-data learning in the wild. In: Advances in Neural Information Processing Systems, vol. 36 (2024)
31. Zhang, S., Goodale, S.E., Gold, B.P., Morgan, V.L., Englot, D.J., Chang, C.: Vigilance associates with the low-dimensional structure of fMRI data. Neuroimage **267**, 119818 (2023)

CENet: Context Enhancement Network for Medical Image Segmentation

Afshin Bozorgpour[1], Sina Ghorbani Kolahi[2], Reza Azad[1], Ilker Hacihaliloglu[3], and Dorit Merhof[1,4(✉)]

[1] Faculty of Informatics and Data Science, University of Regensburg, Regensburg, Germany
dorit.merhof@ur.de
[2] Department of Industrial and Systems Engineering, Tarbiat Modares University, Tehran, Iran
[3] Department of Radiology and Department of Medicine, University of British Columbia, Vancouver, BC, Canada
[4] Fraunhofer Institute for Digital Medicine MEVIS, Bremen, Germany

Abstract. Medical image segmentation, particularly in multi-domain scenarios, requires precise preservation of anatomical structures across diverse representations. While deep learning has advanced this field, existing models often struggle with accurate boundary representation, variability in organ morphology, and information loss during downsampling, limiting their accuracy and robustness. To address these challenges, we propose the Context Enhancement Network (CENet), a novel segmentation framework featuring two key innovations. First, the Dual Selective Enhancement Block (DSEB) integrated into skip connections enhances boundary details and improves the detection of smaller organs in a context-aware manner. Second, the Context Feature Attention Module (CFAM) in the decoder employs a multi-scale design to maintain spatial integrity, reduce feature redundancy, and mitigate overly enhanced representations. Extensive evaluations on both radiology and dermoscopic datasets demonstrate that CENet outperforms state-of-the-art (SOTA) methods in multi-organ segmentation and boundary detail preservation, offering a robust and accurate solution for complex medical image analysis tasks. The source code is publicly available at https://github.com/xmindflow/cenet.

Keywords: Medical Image Segmentation · Feature Enhancement · Multi-scale Representation

1 Introduction

Medical image segmentation, driven by advancements in deep learning and computer vision, is a critical tool for extracting semantically meaningful information from raw medical datasets. It enables the precise pixel-wise delineation of anatomical structures, organs, and lesions, which are often characterized by

J. C. Gee et al. (Eds.): MICCAI 2025, LNCS 15960, pp. 120–129, 2026.
https://doi.org/10.1007/978-3-032-04927-8_12

diverse shapes, appearances, and pathological conditions [1]. This capability is essential for clinical applications and computer-aided diagnosis systems. Among the most prominent approaches for segmentation are Fully Convolutional Neural Networks (FCNs), particularly the U-Net architecture [19] and its variants. These models leverage an encoder-decoder structure to capture multiscale representations: the encoder extracts contextual information, while the decoder upsamples compressed features to produce precise, localized predictions. Skip connections further enhance this process by preserving fine-grained spatial details [17], such as boundaries, and enabling the decoder to reconstruct predictions more accurately using high-quality, contextualized features from the encoder.

Despite their strengths, Convolutional Neural Networks (CNNs) inherently struggle to model global contextual relationships due to the limited receptive field of convolutional kernels. This limitation often leads to suboptimal performance in multiscale segmentation tasks involving complex structures [21]. To address this, various strategies have been proposed, including deformable convolutions [2], dilated convolutions [8], spatial pyramid pooling [6], and the integration of attention mechanisms into high-level semantic feature maps [22]. More recently, the Vision Transformer (ViT) [7] has emerged as a promising alternative, utilizing self-attention mechanisms to effectively model long-range dependencies and achieve SOTA performance. However, while ViTs excel at capturing global context, they often underperform in modeling local representations and context. Additionally, their quadratic computational complexity makes them inefficient for large-scale applications [11].

Despite advances in CNNs and Transformer networks, their hierarchical structures, reliance on downsampling, and self-attention mechanisms often compromise boundary details and fine-grained semantic representation, limiting their ability to capture multiscale features essential for complex organ and lesion morphologies in medical images. Although hybrid CNN-Transformer architectures [5,9,12] and localized self-attention mechanisms [4,10] have been explored, their focus on global representations and fixed receptive fields restricts accurate segmentation of deformable structures across scales. Many approaches, such as [26], focus on body features over edge information, which is crucial for accurate segmentation and detail reconstruction. While studies like [13,16] separately integrate fine-grained features (e.g., boundaries), the absence of proper control mechanisms often results in noise, decoder-stage degradation, and inefficient network learning with strong inductive bias.

To address these challenges, we propose the Context Enhancement Network (CENet), a novel framework for medical image segmentation. ❶ The Dual Selective Enhancement Block (DSEB) refines fine-grained features by leveraging coarse guidance from the previous decoder, amplifying salient regions while filtering irrelevant information to maintain contextual balance. ❷ The decoder includes the Context Feature Attention Module (CFAM), which uses depthwise dilated convolutions and a context-aware gating mechanism for multiscale feature representation while addressing over-enhancement through adaptive rectification. ❸ Evaluations on radiology (*Synapse*, *ACDC*) and dermoscopic datasets (*PH*^{2}, *HAM10000*) show that CENet outperforms SOTA methods in precise and context-aware segmentation.

2 Method

In this study, we propose a novel Context Enhancement Network (CENet) designed to enhance feature representation and improve segmentation accuracy by leveraging hierarchical feature extraction and refined contextual information to effectively capture global contextual dependencies and local spatial details. The network is systematically structured into three key components: a Pyramid Vision Transformer V2 (PvT-V2) [26] backbone for multi-scale feature extraction, DSEB to contextually enrich skip connections, and CFAM for multi-scale representation and semantic feature refinement. The overall architecture of the proposed method is illustrated in Fig. 1.

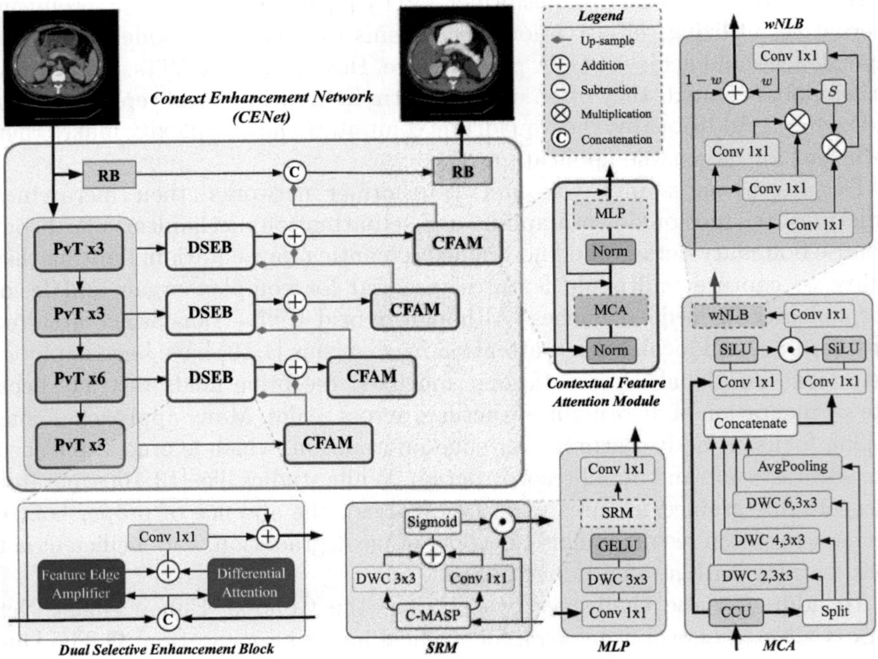

Fig. 1. The Contextual Enhancement Network (CENet) uses a pretrained encoder to generate multi-resolution features, processed by the DSEB as skip connections. The decoder refines these features via the CFAM, which includes a CCU, MCA, wNLB, and MLP.

2.1 Dual Selective Enhancement Block (DSEB)

The DSEB improves CENet's skip connections for precise, boundary-sensitive medical image segmentation. After concatenating encoder features with upsampled CFAM outputs (using EUC Block [18]) as decoder signals, the input is processed through two stages: the Feature Edge Amplifier (FEA) and Differential Attention (DiffAtt) [28]. The FEA enhances skip connection feature maps by refining edges and spatial details crucial for boundary detection and the localization of small structures. It begins by performing multi-scale downsampling $\mathcal{D}(F, s)$ and upsampling $U(F, s)$ on the input feature map F, defined as:

$$F_{u1} = U(\mathcal{D}(F, s_1), s_0), \quad F_{u2} = U(\mathcal{D}(F, s_2), s_0), \tag{1}$$

where $s_1 = 0.75$, $s_2 = 0.5$, and $s_0 = 1.0$ are the scales. The difference between these upsampled features isolates refined edge details, computed as $F_{\text{edge}} = |F_{u1} - F_{u2}|$. These features are added back to the original feature map F, weighted by $\lambda \in \mathbb{R}^d$, resulting in the enhanced feature map:

$$\tilde{F} = F + \lambda F_{\text{edge}}. \tag{2}$$

The DiffAtt, inspired by NLP, reduces attention noise and selectively enhances meaningful context from the concatenated input features. Query and key vectors are split into two groups, and separate attention maps are computed. By subtracting these maps, issues such as imbalanced token importance in visual contexts (e.g., boundary regions or small structures) are addressed, while redundant attention is removed. The refined attention weights are combined with the value vector, reducing irrelevant context and focusing on critical structures, including fine details like edges. By combining FEA and DiffAtt, DSEB improves feature detail, reinforces localized boundaries while highlighting salient context, surpassing irrelevant regions, and improves segmentation of intricate structures, achieving better overall accuracy.

2.2 Contextual Feature Attention Module (CFAM)

The CFAM, a Transformer-based decoder, improves hierarchical feature processing in CENet while avoiding over-enhanced representations. It refines features through four interconnected components for coherent information flow. First, the **Channel Calibration Unit (CCU)** processes the input $F \in \mathbb{R}^{H \times W \times C}$ to recalibrate channel-wise features, enhancing their capacity to capture diverse characteristics for precise segmentation. The CCU employs Global Multi-Aspect Pooling (G-MASP), combining average pooling ($\mathcal{P}_{\text{avg}}(F)$), max pooling ($\mathcal{P}_{\max}(F)$), and standard deviation pooling ($\mathcal{P}_{\text{std}}(F)$) to generate a global descriptor $\mathbf{g} \in \mathbb{R}^{3C}$. This descriptor drives a channel-wise attention mechanism that reduces the channel capacity to C and adaptively reweights feature maps, prioritizing significant channels for improved diversity and representation. The CCU's formulation is given as:

$$\begin{aligned} \mathbf{g} &= [\mathcal{P}_{\text{avg}}(F); \mathcal{P}_{\max}(F); \mathcal{P}_{\text{std}}(F)], \\ \mathbf{s} &= \sigma\left(f_{1\times1}\left(\text{GELU}\left(f_{1\times1}(\mathbf{g})\right)\right)\right), \quad F' = F \odot \mathbf{s}, \end{aligned} \tag{3}$$

where $\mathbf{s} \in \mathbb{R}^C$ is weights, $F' \in \mathbb{R}^{H\times W\times C}$ is the reweighted output, $\odot$ is point-wise multiplication, $f_{1\times1}$ is point-wise convolution, $[;]$ is concatenation, and σ is sigmoid.

Following the CCU, CFAM utilizes the **Multi-scale Contextual Aggregator (MCA)** to refine F' by capturing spatial context across multiple scales. The input $F' \in \mathbb{R}^{C\times H\times W}$ is split into four parts: $F'_i \in \mathbb{R}^{C_i\times H\times W}$, where F'_1, F'_2, F'_3 share equal channel dimensions ($C_1 = C_2 = C_3$), and F'_4 contains the less than 10 percent of all the channels, ensuring $C_1 + C_2 + C_3 + C_4 = C$. Most channels are allocated to the first three branches for the convolution operations, while the fourth branch handles global patterns via average pooling. The splits F'_1, F'_2, F'_3 are processed with parallel dilated depth-wise convolutions (f_{dk}) using dilation rates (e.g., 3, 5, and 8), respectively, while F'_4 undergoes average pooling. The outputs are concatenated and refined through a $f_{1\times1}$ convolution with SiLU activation. A context-aware gating mechanism then adjusts feature importance, suppressing redundancy and emphasizing salient features, resulting in robust multi-scale representations for improved segmentation. The MCA is expressed as:

$$\begin{aligned} F_{\text{cat}} &= \left[f_{d_3}(F'_1), f_{d_5}(F'_2), f_{d_8}(F'_3), \mathcal{C}_{avg}(F'_4)\right], \\ F_{\text{MCA}} &= \left(\text{SiLU}(f_{1\times1}(F_{\text{cat}})) \odot \text{SiLU}(f_{1\times1}(F'))\right) + F, \end{aligned} \tag{4}$$

where $\mathcal{C}_{avg}$ represents channel-wise average pooling, f_{d_k} denotes the depth-wise convolution operator with a kernel size of $k \times k$ and dilation rate $d_k \in \{6, 8, 12\}$.

To adjust feature enhancement based on the accumulation of the noise from overly enhanced representations that come from previous layers and corresponding DSEB blocks after feature aggregation, the **weighted Non-local Block (wNLB)** prevents noise accumulation by modelling long-range spatial dependencies and adaptively denoising features while preserving critical details. It uses a self-attention mechanism with a learnable weighting parameter as a specific instance of non-local (NL) operations [27]. The final stage of the CFAM uses an enhanced MLP block, equipped with **Spatial Calibration Module (SRM)**. SRM applies Channel-wise Multi-Axis Spatial Pooling (C-MASP) using parallel pooling ($\mathcal{C}_{avg}, \mathcal{C}_{max}, \mathcal{C}_{std}$) to create a spatial descriptor $G \in \mathbb{R}^{3\times H\times W}$. This descriptor is processed by parallel point-wise convolution that captures pixel-wise relations, and depth-wise convolution (f_k^{dw}) captures neighborhood interactions. The combined output (S) recalibrates the feature map via element-wise multiplication, then passes through a linear layer in the enhanced MLP for rich feature representations. Formally:

$$\begin{aligned} S = \sigma(f_{1\times1}(G) + f_k^{dw}(G)), \quad G = \left[\mathcal{C}_{avg}(F_{\text{MCA}}), \mathcal{C}_{max}(F_{\text{MCA}}), \mathcal{C}_{std}(F_{\text{MCA}})\right], \\ F_{\text{recal}} = F_{\text{MCA}} \odot S. \end{aligned} \tag{5}$$

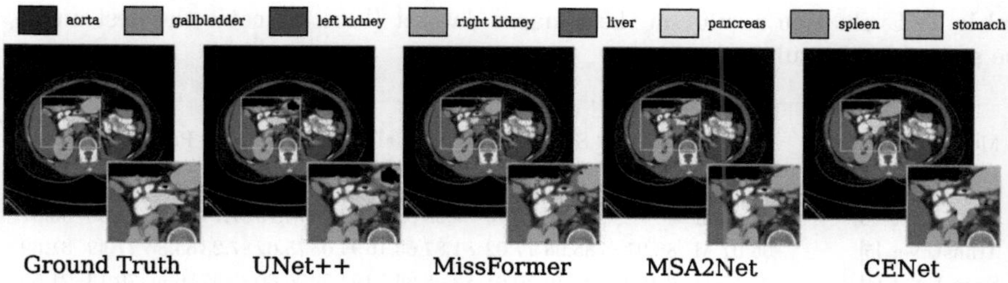

Fig. 2. Visual comparison of the proposed method versus others on the Synapse dataset.

3 Experiments and Results

3.1 Datasets and Implementation Details

The performance of CENet was evaluated on four datasets. The model, developed in PyTorch and trained on an NVIDIA A100 GPU (80GB), used the ImageNet-pretrained PVTv2-b2 encoder [26] at a 224 × 224 input resolution. For the *Synapse* dataset (30 CT scans), 18 scans were used for training and 12 for validation [14], following TransUNet's protocol [5]. Training involved 250 epochs, a batch size of 16, and an SGD optimizer with a 0.05 learning rate. On the *ACDC* [3] dataset (100 cardiac MRI scans), the split was 70 training, 10 validation, and 20 testing cases, with 150 epochs, a batch size of 12, and an Adam optimizer (learning rate 0.0001). For skin lesion segmentation, the *PH2* dataset (80 samples) and *HAM10000* (10,015 images) were trained for 100 epochs with a batch size of 16 and an Adam optimizer (learning rate 0.0001), using preprocessing/augmentation from [1]. The model also integrates BDoU Loss [23].

3.2 Results

Radiology: The performance of CENet on radiological datasets was evaluated, with Table 1 presenting the quantitative results on the Synapse dataset using DSC and HD metrics. Our approach significantly outperforms existing CNN-based methods. CENet also shows enhanced learning capabilities compared to Hybrid models, with improvements of 0.29% over MSA^{2}Net, respectively. These results underscore CENet's ability in segmenting various organs. Figure 2 provides a visual representation of CENet's performance in segmenting various organs, demonstrating CENet's accuracy in multi-scale segmentation of the kidneys, pancreas, and stomach. Furthermore, Table 2 emphasizes the effectiveness of our method against SOTA approaches on the ACDC dataset for cardiac segmentation in MRI images. CENet achieves the highest average DICE score of 92.18%. Moreover, CENet excels in all three organ segmentation tasks, regardless of morphological variations.

Table 1. Evaluation results on the Synapse dataset (blue indicates the best and red the second best results).

Methods	Params	FLOPs	Spl.	RKid.	LKid.	Gal.	Liv.	Sto.	Aor.	Pan.	Average	
											DSC↑	HD95↓
R50 U-Net [5]	30.42 M	-	85.87	78.19	80.60	63.66	93.74	74.16	87.74	56.90	74.68	36.87
TransUNet [5]	96.07 M	88.91 G	85.08	77.02	81.87	63.16	94.08	75.62	87.23	55.86	77.49	31.69
Swin-UNet [4]	27.17 M	6.16 G	90.66	79.61	83.28	66.53	94.29	76.60	85.47	56.58	79.13	21.55
HiFormer-B [9]	25.51 M	8.05 G	90.99	79.77	85.23	65.23	94.61	81.08	86.21	59.52	80.39	14.70
VM-UNet [20]	44.27 M	6.52 G	89.51	82.76	86.16	69.41	94.17	81.40	86.40	58.80	81.08	19.21
PVT-EMCAD-B2 [18]	26.76 M	5.60 G	92.17	84.10	88.08	68.87	95.26	83.92	88.14	68.51	83.63	15.68
MSA²Net [12]	112.77 M	15.56 G	92.69	84.24	88.30	74.35	95.59	84.03	89.47	69.30	84.75	13.29
2D D-LKA Net [2]	101.64 M	19.92 G	91.22	84.92	88.38	73.79	94.88	84.94	88.34	67.71	84.27	20.04
CENet (Ours)	33.39 M	12.76 G	94.58	86.33	91.12	68.10	96.19	83.89	89.56	70.54	85.04	8.84

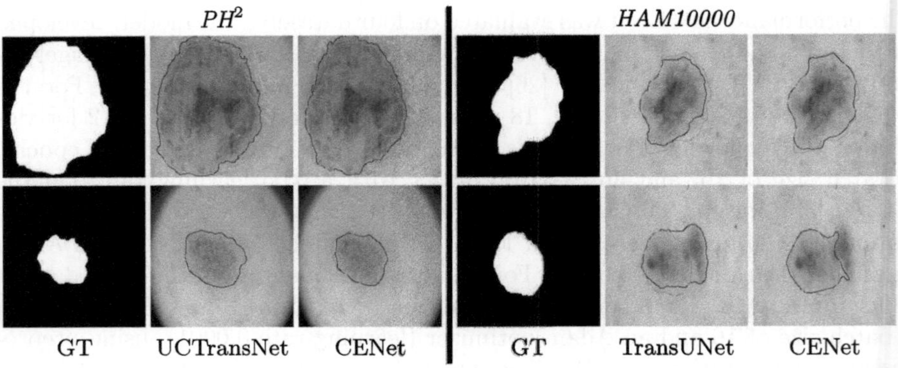

Fig. 3. Qualitative comparison of CENet and previous methods across skin benchmarks.

Dermoscopy: Table 2 evaluates our proposed network on two skin lesion segmentation datasets using DSC and ACC metrics. CENet outperforms CNN-based, Transformer-based, and hybrid methods, demonstrating superior performance and generalization. It surpasses Swin-Unet [15], U-Net [19], and UCTransNet [24], achieving DSC score improvements of 0.58% and 1.45% on PH^2 and HAM10000 datasets, respectively. Moreover, Fig. 3 showcases CENet's superiority in capturing intricate structures and producing precise boundaries through effective boundary integration.

Ablation Study. To evaluate CENet's components, ablation studies were conducted to assess their efficiency, performance, and placement within the DSEB and CFAM modules (Table 3). Results show that optimal performance occurs when the wNLB is positioned at the end of the MCA and CCU, while the DiffAttn and FEA operate in parallel within the DSEB, outperforming alternative

Table 2. Evaluation results on the skin benchmarks (*PH^2* and *HAM10000*) and *ACDC* dataset.

Methods	PH^2 DSC	PH^2 ACC	HAM10000 DSC	HAM10000 ACC
U-Net [19]	89.36	92.33	91.67	95.67
TransUNet [5]	88.40	92.00	93.53	96.49
Swin-Unet [4]	94.49	96.78	92.63	96.16
DeepLabv3+ [6]	92.02	95.03	92.51	96.07
Att-UNet [17]	90.03	92.76	92.68	96.10
UCTransNet [24]	90.93	94.08	93.46	96.84
MissFormer [10]	85.50	90.50	92.11	96.21
CENet	95.04	97.19	94.71	97.04

Methods	aDSC	RV	MYO	LV
R50+UNet [5]	87.55	87.10	80.63	94.92
R50+AttnUNet [5]	86.75	87.58	79.20	93.47
TransUNet[5]	89.71	88.86	84.53	95.73
ViT+CUP [5]	81.45	81.46	70.71	92.18
Swin-UNet [4]	90.00	88.55	85.62	95.83
R50+ViT+CUP [5]	87.57	86.07	81.88	94.75
MT-UNet [25]	90.43	86.64	89.04	95.62
MISSFormer [10]	90.86	89.55	88.04	94.99
CENet	92.18	90.90	89.63	95.99

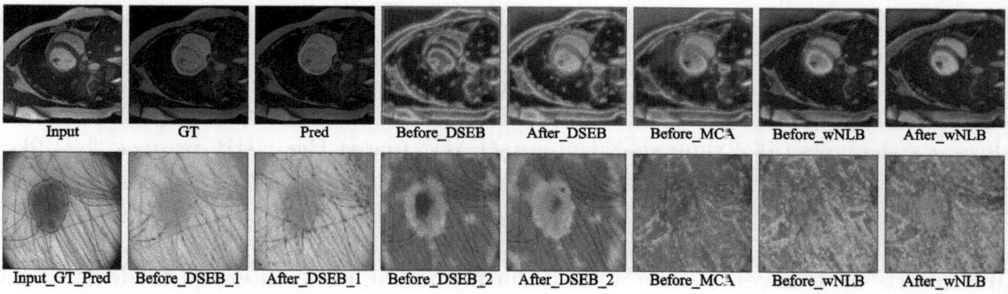

Fig. 4. Feature visualization in CENet: first row shows an *ACDC* sample, second row a *PH^2* example

configurations. Feature analyses (Fig. 4) on the *ACDC* and *PH^2* datasets further demonstrate that in the first row, attention maps before applying the DSEB are diffuse, failing to focus on key regions. After applying the DSEB, attention becomes more focused, and the wNLB refines MCA multiscale recalibrated fea-

Table 3. Effect of different components of CENet on PH^2 dataset. #FLOPs are reported in (G), and P indicates the Parameters in Millions (M). All results are averaged over three runs. The best results are shown in bold.

Components				FLOPs (G)	Params (M)	Performance (DICE) PH^2
DSEB		CFAM				
FEA	DiffAtt	wNLB	CCU			
No	No	No	No	7.53	29.86	94.08
Yes	No	Yes	No	9.22	31.41	94.27
Yes	Yes	No	No	11.16	31.83	94.42
Yes	Yes	Yes	No	12.84	33.37	94.79
Yes	Yes	Yes	Yes	12.84	33.39	**95.04**

ture maps by suppressing irrelevant details and emphasizing salient regions. In the second row, the DSEB enhances feature focus at different CENet levels, while the wNLB acts as a denoising mechanism, suppressing over-attended details (e.g., hairs) and directing attention to critical regions.

4 Conclusion

The proposed CENet improves context-aware medical image segmentation with a DSEB for boundary-sensitive and context amplification and a CFAM for multiscale representation and reduces feature redundancy. Evaluations demonstrate CENet's superior accuracy and boundary preservation over SOTA methods, providing a robust solution for complex segmentation tasks.

Acknowledgments. The authors gratefully acknowledge the computational and data resources provided by the Leibniz Supercomputing Centre. Also, the authors gratefully acknowledge the scientific support and HPC resources provided by the Erlangen National High-Performance Computing Center (NHR@FAU) of the Friedrich-Alexander-Universität Erlangen-Nürnberg (FAU) under the NHR project "b213da." NHR funding is provided by federal and Bavarian state authorities. NHR@FAU hardware is partially funded by the German Research Foundation (DFG) âĂŞ 440719683.

Disclosure of Interests. The authors have no competing interests to declare that are relevant to the content of this article.

References

1. Azad, R., et al.: Medical image segmentation review: the success of U-Net. IEEE Trans. Pattern Anal. Mach. Intell. (2024)
2. Azad, R., et al.: Beyond self-attention: deformable large kernel attention for medical image segmentation (2023)
3. Bernard, O., et al.: Deep learning techniques for automatic MRI cardiac multi-structures segmentation and diagnosis: is the problem solved? IEEE Trans. Med. Imaging **37**(11), 2514–2525 (2018)
4. Cao, H., et al.: Swin-UNet: UNet-like pure transformer for medical image segmentation. In: Proceedings of the European Conference on Computer Vision Workshops(ECCVW) (2022)
5. Chen, J., et al.: TransUNet: transformers make strong encoders for medical image segmentation. arXiv preprint arXiv:2102.04306 (2021)
6. Chen, L.C., Zhu, Y., Papandreou, G., Schroff, F., Adam, H.: Encoder-decoder with atrous separable convolution for semantic image segmentation. In: Proceedings of the European conference on computer vision (ECCV), pp. 801–818 (2018)
7. Dosovitskiy, A., et al.: An image is worth 16 × 16 words: transformers for image recognition at scale. arXiv preprint arXiv:2010.11929 (2020)
8. Guo, M.H., Lu, C.Z., Liu, Z.N., Cheng, M.M., Hu, S.M.: Visual attention network. Comput. Visual Media **9**(4), 733–752 (2023)
9. Heidari, M., et al.: Hiformer: hierarchical multi-scale representations using transformers for medical image segmentation. In: Proceedings of the IEEE/CVF Winter Conference on Applications of Computer Vision, pp. 6202–6212 (2023)

10. Huang, X., Deng, Z., Li, D., Yuan, X., Fu, Y.: Missformer: an effective transformer for 2D medical image segmentation. IEEE Trans. Med. Imaging (2022). https://doi.org/10.1109/TMI.2022.3230943
11. Jamil, S., Jalil Piran, M., Kwon, O.J.: A comprehensive survey of transformers for computer vision. Drones **7**(5), 287 (2023)
12. Kolahi, S.G., et al.: Msa^{2}net: multi-scale adaptive attention-guided network for medical image segmentation. arXiv preprint arXiv:2407.21640 (2024)
13. Kuang, H., Wang, Y., Liang, Y., Liu, J., Wang, J.: Bea-Net: body and edge aware network with multi-scale short-term concatenation for medical image segmentation. IEEE J. Biomed. Health Inform. **27**(10), 4828–4839 (2023). https://doi.org/10.1109/JBHI.2023.3304662
14. Landman, B., Xu, Z., Igelsias, J., Styner, M., Langerak, T., Klein, A.: Miccai multi-atlas labeling beyond the cranial vault–workshop and challenge. In: Proceedings of the MICCAI Multi-Atlas Labeling Beyond Cranial Vault—Workshop Challenge, vol. 5, p. 12 (2015)
15. Liu, Z., et al.: Swin transformer: Hierarchical vision transformer using shifted windows. In: Proceedings of the IEEE/CVF International Conference on Computer Vision, pp. 10012–10022 (2021)
16. Ma, Q., Mao, K., Wang, G., Xu, L., Zhao, Y.: LCAUnet: a skin lesion segmentation network with enhanced edge and body fusion. arXiv preprint arXiv:2305.00837 (2023)
17. Oktay, O., et al.: Attention U-Net: learning where to look for the pancreas. arXiv preprint arXiv:1804.03999 (2018)
18. Rahman, M.M., Munir, M., Marculescu, R.: EMCAD: efficient multi-scale convolutional attention decoding for medical image segmentation. In: Proceedings of the IEEE/CVF Conference on Computer Vision and Pattern Recognition, pp. 11769–11779 (2024)
19. Ronneberger, O., Fischer, P., Brox, T.: U-net: convolutional networks for biomedical image segmentation. In: Medical Image Computing and Computer-Assisted Intervention–MICCAI 2015: 18th International Conference, Munich, Germany, October 5–9, 2015, Proceedings, Part III 18, pp. 234–241. Springer (2015)
20. Ruan, J., Xiang, S.: VM-Unet: vision mamba UNet for medical image segmentation. arXiv preprint arXiv:2402.02491 (2024)
21. Shamshad, F., et al.: Transformers in medical imaging: a survey. Med. Image Anal. **88**, 102802 (2023)
22. Sohn, K., et al.: Visual prompt tuning for generative transfer learning (2022)
23. Sun, F., Luo, Z., Li, S.: Boundary difference over union loss for medical image segmentation (2023)
24. Wang, H., Cao, P., Wang, J., Zaiane, O.R.: UCTransNet: rethinking the skip connections in u-net from a channel-wise perspective with transformer. In: Proceedings of the AAAI Conference on Artificial Intelligence, vol. 36, pp. 2441–2449 (2022)
25. Wang, H., et al.: Mixed transformer u-net for medical image segmentation. In: ICASSP 2022-2022 IEEE International Conference on Acoustics, Speech and Signal Processing (ICASSP), pp. 2390–2394. IEEE (2022)
26. Wang, W., et al.: Pvt v2: improved baselines with pyramid vision transformer. Comput. Visual Media **8**(3), 415–424 (2022)
27. Wang, X., Girshick, R., Gupta, A., He, K.: Non-local neural networks. In: Proceedings of the IEEE Conference on Computer Vision and Pattern Recognition, pp. 7794–7803 (2018)
28. Ye, T., et al.: Differential transformer. arXiv preprint arXiv:2410.05258 (2024)

Co-Seg: Mutual Prompt-Guided Collaborative Learning for Tissue and Nuclei Segmentation

Qing Xu[1,2], Wenting Duan[1], and Zhen Chen[3](✉)

[1] University of Lincoln, Brayford Pool, UK
[2] University of Nottingham, Nottinghamshire, UK
[3] Yale University, New Haven, CT 06510, USA
zchen.francis@gmail.com

Abstract. Histopathology image analysis is critical yet challenged by the demand of segmenting tissue regions and nuclei instances for tumor microenvironment and cellular morphology analysis. Existing studies focused on tissue semantic segmentation or nuclei instance segmentation separately, but ignored the inherent relationship between these two tasks, resulting in insufficient histopathology understanding. To address this issue, we propose a Co-Seg framework for collaborative tissue and nuclei segmentation. Specifically, we introduce a novel co-segmentation paradigm, allowing tissue and nuclei segmentation tasks to mutually enhance each other. To this end, we first devise a region-aware prompt encoder (RP-Encoder) to provide high-quality semantic and instance region prompts as prior constraints. Moreover, we design a mutual prompt mask decoder (MP-Decoder) that leverages cross-guidance to strengthen the contextual consistency of both tasks, collaboratively computing semantic and instance segmentation masks. Extensive experiments on the PUMA dataset demonstrate that the proposed Co-Seg surpasses state-of-the-arts in the semantic, instance and panoptic segmentation of tumor tissues and nuclei instances. The source code is available at https://github.com/xq141839/Co-Seg

Keywords: Collaborative Learning · Tissue Segmentation · Nuclei Segmentation · Mutual Prompt

1 Introduction

Medical image segmentation plays a crucial role in clinical applications and has received extensive attention in the research [5,7,10,29]. Particularly, histopathology image analysis is challenged by the delineation of tissue regions and further separation of individual nuclei within the regions to assess tissue subtypes and tumor grading [36]. These requirements have led to two significant segmentation tasks in the field, *i.e.*, tissue semantic segmentation and nuclei instance segmentation.

J. C. Gee et al. (Eds.): MICCAI 2025, LNCS 15960, pp. 130–140, 2026.
https://doi.org/10.1007/978-3-032-04927-8_13

For medical semantic segmentation, existing methods [24,26,34,35] rely on hierarchical encoding and decoding to enhance multi-scale awareness for precise mask generation. In particular, ViT-based [1] and Mamba-based [17] architectures model long-range dependencies to gain global contexts of target regions. For instance segmentation, a series of works [2,6,10,30] adopted different distance proxy maps to improve the boundary understanding of instances. Recent SAM-based methods [3–5,16,19,21,29,33] have revealed impressive medical semantic and instance segmentation performance by manually providing corresponding task prompts.

Despite the advancements, existing methods [1,8,18,31,32] usually focus on the optimization of tissue semantic or nuclei instance segmentation tasks, which can only be supervised by the isolated knowledge. In fact, both tasks are highly correlated as they aim to achieve the adequate perception and understanding of histopathology images. For example, accurately identifying nuclei provides valuable cues for understanding the underlying tissue structures, while tissue segmentation can aid in localizing nuclei. Therefore, this strong interdependence motivates us to develop a collaborative approach that integrates tissue and nuclei segmentation to advance state-of-the-art histopathology image analysis.

To overcome this bottleneck, we propose **Co-Seg**, a collaborative tissue and nuclei segmentation framework that allows semantic and instance segmentation tasks to mutually enhance each other. As illustrated in Fig. 1(c), the Co-Seg, based on a co-segmentation paradigm, improves segmentation mask quality by capturing contextual dependencies between the two tasks. Specifically, we first devise the region-aware prompt encoder (RP-Encoder) to extract high-quality semantic and instance prompts from the target region of both tasks, guiding segmentation decoding. We further introduce a mutual prompt mask decoder (MP-Decoder), which jointly computes semantic and instance maps by leveraging cross-guidance. This collaborative learning approach achieves contextual consistency, reducing prediction errors in both tasks. Experimental results on melanoma tissue and nuclei segmentation tasks demonstrate that our Co-Seg achieves remarkable performance over state-of-the-art segmentation methods.

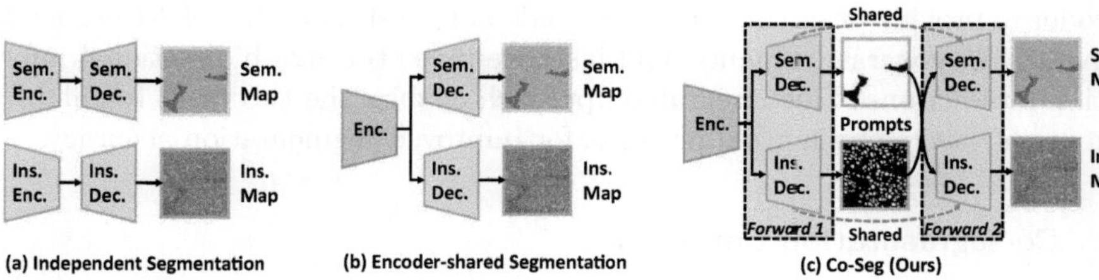

Fig. 1. Comparison of our Co-Seg and existing works in tissue and nuclei segmentation. (a) Two independent networks for tissue and nuclei segmentation. (b) A shared image encoder but separated task decoders for tissue and nuclei segmentation. (c) Our Co-Seg leverages mutual prompts for collaborative tissue and nuclei segmentation.

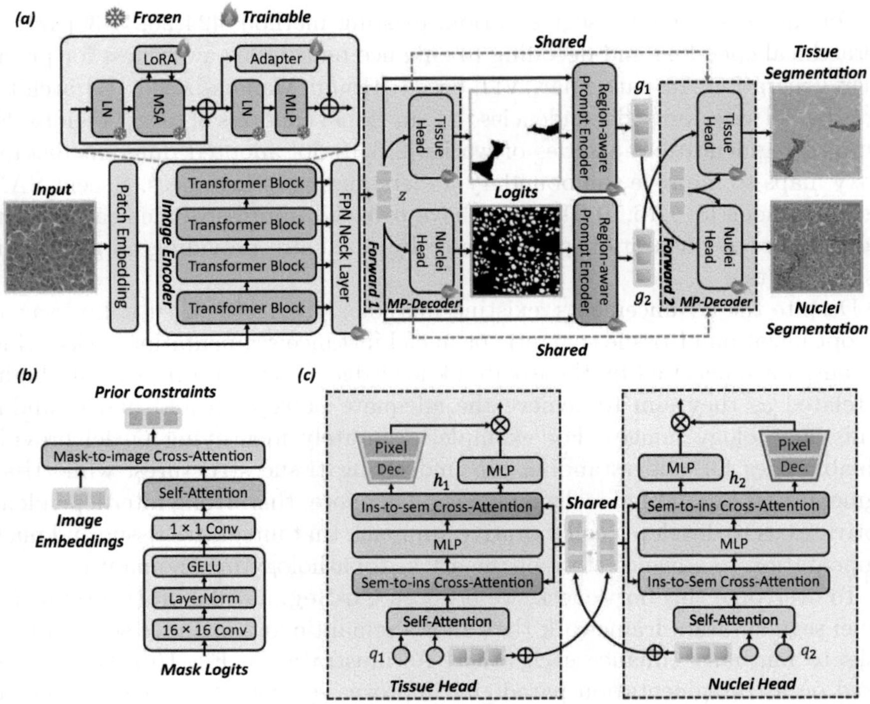

Fig. 2. (a) The overview of the proposed Co-Seg framework for collaborative tissue and nuclei segmentation, consisting of (b) RP-Encoder and (c) MP-Decoder. Co-Seg fully exploits complementary information by leveraging mutual prompts.

2 Methodology

As elaborated in Fig. 2, we devise the Co-Seg framework with the co-segmentation paradigm, achieving the mutual optimization between tissue and nuclei segmentation in medical imaging. To accomplish this, we design the RP-Encoder to provide prior constraints to guide both tasks, and the MP-Decoder to cooperatively generate semantic and instance maps through bidirectional information interactions. This integrated approach enables the two tasks to enhance each other, leveraging mutual prompts for improved segmentation accuracy.

2.1 Co-segmentation Paradigm

Existing tissue [9,13,22] and nuclei [3,20,23] segmentation methods decouple the parameter space of semantic and instance segmentation, disrupting their inter-dependencies, degrading the performance. To address this issue, we propose the co-segmentation paradigm that leverages closed-loop bidirectional interaction to realize dual-task collaborative optimization. Specifically, our goal is to train a model $f_\Theta : x \rightarrow \{y_1, y_2\}$, where $\Theta = \{\theta_1, \theta_2\}$ represents the parameters for semantic segmentation θ_1 and instance segmentation θ_2, with y_1 and y_2 as the corresponding segmentation masks. To model the interdependence between the two tasks, we define their joint prediction of y_1 and y_2 as:

$$p(y_1, y_2|x, \theta_1, \theta_2) = p(y_1|x, \theta_1) \cdot p(y_2|y_1, x, \theta_2) = p(y_2|x, \theta_2) \cdot p(y_1|y_2, x, \theta_1). \quad (1)$$

Note that Eq. (1) reveals the mutual dependency between semantic and instance segmentation through intertwined conditional probabilities. It states that $p(y_1, y_2|x, \theta_1, \theta_2)$, the joint probability of obtaining both segmentations given the image and model parameters, can be decomposed in two symmetrical ways:

- $p(y_1|x, \theta_1) \cdot p(y_2|y_1, x, \theta_2)$ suggests that once the semantic segmentation y_1 is known, it directly influences the prediction of the instance segmentation y_2.
- Similarly, $p(y_2|x, \theta_2) \cdot p(y_1|y_2, x, \theta_1)$ implies that knowing the instance segmentation y_2 affects the outcome of the semantic segmentation y_1.

Both expressions are mathematically equivalent because they describe the same joint probability from different perspectives, highlighting that each segmentation task provides crucial context that enhances the accuracy of the other. This reciprocal relationship underscores the need for a collaborative approach, where both segmentation tasks are optimized together rather than in isolation, leveraging the full context available from each task to improve overall segmentation results. Translating this relationship to the gradient calculations for θ_1 and θ_2, we have:

$$\nabla_{\theta_i} \ell_i = -\nabla_{\theta_i} \mathbb{E}\left[\log p(y_i|x, \theta_i)\right] - \nabla_{\theta_i} \mathbb{E}\left[\log p(y_j|y_i, x, \theta_j)\right] + \nabla_{\theta_i} I(y_1; y_2|x), \quad (2)$$

for $i, j \in \{1, 2\}, i \neq j$ where $\nabla_\theta I(y_1; y_2|x) \approx \mathbb{E}\left[\nabla_\theta \log \frac{p(y_1, y_2|x)}{p(y_1|x)p(y_2|x)}\right]$. This term captures the mutual information between y_1 and y_2, ensuring that both segmentation tasks benefit from shared feature learning. The first term of Eq. (2) represents the main segmentation loss gradient, while the second term incorporates the implicit gradient of task interdependencies, modeling the complementarity between θ_1 and θ_2. This leads to the following optimization update rule:

$$\theta_1 \leftarrow \theta_1 - \eta \nabla_{\theta_1} \ell_1, \quad \theta_2 \leftarrow \theta_2 - \eta \nabla_{\theta_2} \ell_2, \quad (3)$$

where η is the learning rate. In this way, the co-segmentation paradigm breaks the barrier of isolating gradient flows from each other in multi-task learning, enabling our Co-Seg framework to achieve the bidirectional optimization of tissue and nuclei segmentation.

2.2 Region-Aware Prompt Encoder

The proposed co-segmentation paradigm requires establishing implicit interdependencies between tissue and nuclei segmentation tasks. To achieve this, we introduce the RP-Encoder that leverages mask prompts to capture task-specific target regions as prior constraints, as presented in Fig. 2(b). Specifically, we first employ several convolutional blocks to extract features m_i from mask logits μ_i. After that, the RP-Encoder performs self-attention, followed by cross-attention with shared image embeddings z to generate a set of dense prompts g_i, as follows:

$$m_i = \text{Conv}_2(\text{GELU}(\text{LN}(\text{Conv}_1(\mu_i)))), \tag{4}$$

$$g_i = \text{CrossAttention}(\text{SelfAttention}(m_i), z, z), \tag{5}$$

where Conv_1 denotes a 16×16 convolutions for downsampling, Conv_2 is a 1×1 convolution for channel expansion, $\text{LN}(\cdot)$ is LayerNorm. These operations encode the structural context of mask logits. On this basis, we apply RP-Encoder to semantic and instance mask logits, effectively obtaining region-aware prompts: g_1 and g_2 as prior constraints of Co-Seg for tissue and nuclei segmentation tasks.

2.3 Mutual Prompt Mask Decoder

Following the co-segmentation paradigm, we propose the MP-Decoder to enforce bidirectional interactions between tissue and nuclei segmentation, ensuring mutual refinement rather than treating them as isolated tasks, as illustrated in Fig. 2(c). Specifically, the MP-Decoder contains tissue and nuclei heads that leverage two sets of query embeddings: q_1 and q_2 to save the decoding information of both tasks. To leverage complementary effects between tasks, we first perform self-attention on each query, followed by cross-attention with the other task's prompts, as follows:

$$q_i' = \text{CrossAttention}(\text{SelfAttention}(q_i), z \oplus g_j, z \oplus g_j), \tag{6}$$

where $\oplus$ represents the element-wise addition. This is followed by applying MLP to ensure the refined query embeddings are expressive. The MP-Decoder further applies the reverse cross-attention to generate task-specific image embeddings h_1 and h_2, as follows:

$$h_i = \text{CrossAttention}(z \oplus g_j, q_i', q_i'). \tag{7}$$

On this basis, nuclei context prompts can enhance tissue segmentation consistency, while tissue boundary feedback provides prior constraints for nuclei discrimination. Following the standard SAM [15], each segmentation head adopts a pixel decoder to upsample the refined task-specific image embedding. Finally, our MP-Decoder generates tissue and nuclei segmentation predictions by performing the dot product between their upscaled image embeddings and corresponding task queries. Overall, the proposed MP-Decoder utilizes the bidirectional interactions of tissue and nuclei segmentation information to mutually improve the quality of segmentation masks of our Co-Seg framework.

Table 1. Comparison with state-of-the-arts on tissue semantic segmentation.

Methods	Dice	mIoU	HD
nnUnet [14]	90.90	84.48	236.03
Swin-Umamba [17]	89.67	82.75	265.43
EMCAD [24]	90.96	84.61	306.36
SAC [21]	91.19	85.01	295.57
H-SAM [5]	91.47	85.47	246.16
Co-Seg	**92.51**	**87.18**	**206.72**

Table 2. Comparison with state-of-the-arts on nuclei instance segmentation

Methods	F1	Prec.	Rec.	AJI
HoverNet [6]	75.53	70.53	82.80	64.40
CellPose [30]	75.30	71.12	81.79	65.15
CPPNet [2]	74.96	69.58	83.06	64.63
CellViT [10]	76.04	70.21	83.51	66.13
PromptNucSeg [29]	76.47	71.81	83.02	66.71
Co-Seg	**79.70**	**77.05**	**83.61**	**69.14**

2.4 Optimization Pipeline

To construct our Co-Seg framework, we first adopt Hirea ViT [27] as the shared image encoder for tissue and nuclei segmentation, ensuring the consistency of feature learning. In particular, we load the pre-trained SAM2-L [25] to initialize corresponding parameters and freeze these weights to preserve pre-trained knowledge. Additionally, we insert LoRA [12] and Adapter [11] into attention and FFN layers to achieve parameter-efficient fine-tuning from natural to histopathology domains. Based on the co-segmentation paradigm, Co-Seg includes two decoding forwards, where the tissue and nuclei heads of the MP-Decoder share parameters in these two stages. Specifically, the MP-Decoder leverages the image embedding generated from the image encoder to calculate binary segmentation masks y'_1 and y'_2 of both tasks without any prompts. Then, we deliver them to the RP-Encoder to provide tissue and nuclei prompts. In the second forward, they will be directly transferred back to the MP-Decoder as prior constraints. Finally, the MP-Decoder utilizes cross-guidance to co-generate tissue and nuclei segmentation masks. The training of Co-Seg consists of two parts: (1) the optimization of tissue g_1 and nuclei g_2 prior constraints and (2) the joint optimization of tissue and nuclei segmentation. The overall loss is formularized as:

$$\mathcal{L}_{\text{CoSeg}} = \underbrace{\lambda_1 \mathcal{L}_{\text{sem}}^{g_1}(\hat{y}'_1, y'_1) + \lambda_2 \mathcal{L}_{\text{ins}}^{g_2}(\hat{y}'_2, y'_2)}_{\text{Prior Constraint Loss}} + \underbrace{\mathcal{L}_{\text{sem}}(\hat{y}_1, y_1) + \mathcal{L}_{\text{ins}}(\hat{y}_2, y_2)}_{\text{Segmentation Loss}}, \tag{8}$$

where λ_1 and λ_2 are factors to adjust the contribution of each term. By optimizing $\mathcal{L}_{\text{CoSeg}}$, our Co-Seg achieves accurate tissue and nuclei segmentation with superior performance across different histopathology images.

3 Experiments

3.1 Experimental Setup

To validate the effectiveness of the proposed Co-Seg, we adopt the melanoma-specific histopathology dataset: PUMA [28] for tissue semantic and nuclei instance segmentation tasks. It contains 206 histopathology images of 1024×1024 resolution and adopts a common split of training, validation, and test sets as 7:1:2. We perform all experiments on a NVIDIA A5000 GPU using PyTorch.

For fair comparisons, we implement all tissue and nuclei segmentation methods with the same training settings and configurations. We utilize the pre-trained SAM ViT-H [15] structure as the image encoder of medical SAM baselines. We apply the optimizer using Adam with an initial learning rate of 1×10^{-4} and use the exponential decay strategy to adjust the learning rate with a factor of 0.98. The batch size and epochs are set to 16 and 300. We leverage the combination of cross-entropy loss and Dice loss to supervise the tissue segmentation. For nuclei segmentation, we follow the standard combination loss of Focal loss, Dice loss, MSE loss and MSGE loss [6,10]. The loss coefficient λ_1 and λ_2 are set to 2 and 1. Our Co-Seg introduces additional 9.1% learnable parameters to the baseline (in Table 4) during fine-tuning.

Table 3. Comparison with state-of-the-arts on histopathology panoptic segmentation.

Methods	PQ_{tissue}	DQ_{tissue}	SQ_{tissue}	PQ_{nuclei}	DQ_{nuclei}	SQ_{nuclei}
nnUnet [14]	57.34	68.27	70.13	58.60	72.83	80.24
HoverNet [6]	58.95	67.25	69.92	60.91	74.96	81.04
CellViT [10]	60.55	71.32	69.17	62.91	76.59	81.95
H-SAM [5]	61.62	71.17	70.81	62.68	76.39	81.86
PromptNucSeg [29]	61.58	70.84	70.92	63.37	77.41	81.79
Co-Seg	**63.09**	**72.82**	**70.99**	**66.11**	**79.70**	**82.80**

Table 4. Ablation study of Co-Seg on PUMA tissue and nuclei Segmentation.

P	*D*	*C*	Tissue Semantic Seg.			Nuclei Instance Seg.			
			Dice	mIoU	HD	F1	Prec.	Rec.	AJI
			90.98	84.71	287.19	76.35	73.18	82.27	66.59
✓			91.23	85.18	261.37	77.01	73.25	83.01	67.08
✓	✓		91.76	86.13	246.56	78.19	74.91	82.54	67.93
✓	✓	✓	**92.51**	**87.18**	**206.72**	**79.70**	**77.05**	**83.61**	**69.14**

3.2 Comparison with State-of-the-Art Methods

Tissue Semantic Segmentation. We first evaluate the performance of all models in the tissue semantic segmentation. As illustrated in Table 1, we observe that classical semantic segmentation methods are inferior to medical SAMs, *e.g.*, H-SAM [5] surpasses EMCAD [24] with a 0.51% Dice increase. Remarkably, our Co-Seg achieves the best performance with a P-value < 0.005, Dice of 92.51%, and the lowest HD of 206.72, indicating precise tissue boundary localization.

Nuclei Instance Segmentation. The comparison of nuclei instance segmentation is shown in Table 2. Our Co-Seg framework reveals the overwhelming performance of four metrics. In particular, Co-Seg achieves a significant advantage over the second-best PromptNucSeg [29], *e.g.*, a P-value < 0.001, a F1-score increase of 3.23%, and an AJI increase of 2.43%.

Histopathology Panoptic Segmentation. We further comprehensively evaluate the performance of our Co-Seg with the panoptic segmentation protocol. As presented in Table 3, our Co-Seg benefits from the novel co-segmentation paradigm, achieving the best performance of all six metrics, with a remarkable PQ increase of 1.51% and 2.74% in tissue and nuclei segmentation tasks. The qualitative comparison is visualized in Fig. 3.

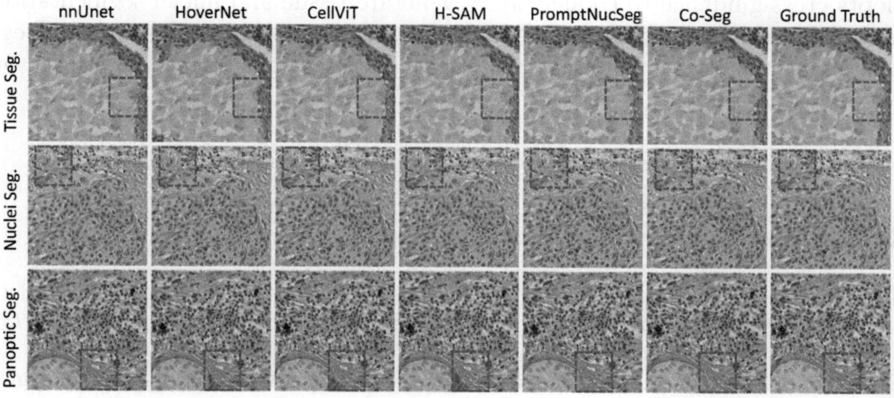

Fig. 3. Qualitative comparison of three evaluation protocols on tissue semantic and nuclei instance segmentation. Benefiting from the mutual reinforcement of semantic and instance segmentation tasks, our Co-Seg can delineate precise tissue regions and segment accurate nuclei, containing fewer false positives.

3.3 Ablation Study

To investigate the effectiveness of our proposed co-segmentation paradigm C, RP-Encoder P and MP-Decoder D, we further conduct the comprehensive ablation study on the tissue semantic and nuclei instance segmentation of the PUMA dataset, as illustrated in Table 4. By removing the tailored modules from Co-Seg, we construct two independent Hiera ViT-based semantic and instance segmentation networks as the ablation baseline. By introducing the RP-Encoder, the performance achieves Dice and F1-score increases of 0.25% and 0.66% on tissue semantic and nuclei instance segmentation, respectively. Moreover, we investigate the effect of combined RP-Encoder and MP-Decoder, resulting in superior performance, with the Dice of 91.76% and the F1-score of 78.19%. Finally, we establish the framework using the proposed co-segmentation paradigm. The result proves that this design can significantly prompt the semantic and instance segmentation capabilities. In this way, these ablation experiments demonstrate the effectiveness of the RP-Encoder, MP-Decoder and co-segmentation paradigm in our Co-Seg framework.

4 Conclusion

In this work, we identify the potential relationship between tissue and nuclei segmentation in histopathology and propose a novel co-segmentation paradigm to establish the Co-Seg framework for promoting nuclei and tissue segmentation mutually. It comprises two modules: the RP-Encoder aims to provide task-specific region prompts by perceiving the target area, and the MP-Decoder adopts cross-guidance to cooperatively generate tissue and nuclei segmentation masks. Extensive experiments on the melanoma tissue and nuclei instance segmentation dataset demonstrate that Co-Seg outperforms existing methods by remarkable margins.

Disclosure of Interests. The authors declare no competing interests.

References

1. Chen, J., et al.: Transunet: rethinking the u-net architecture design for medical image segmentation through the lens of transformers. Med. Image Anal. **97**, 103280 (2024)
2. Chen, S., Ding, C., Liu, M., Cheng, J., Tao, D.: Cpp-net: context-aware polygon proposal network for nucleus segmentation. IEEE Trans. Image Process. **32**, 980–994 (2023)
3. Chen, Z., Xu, Q., Liu, X., Yuan, Y.: Un-sam: domain-adaptive self-prompt segmentation for universal nuclei images. Med. Image Anal. p. 103607 (2025)
4. Chen, Z., et al.: Asi-seg: audio-driven surgical instrument segmentation with surgeon intention understanding. In: IROS, pp. 13773–13779. IEEE (2024)
5. Cheng, Z., Wei, Q., Zhu, H., Wang, Y., Qu, L., Shao, W., Zhou, Y.: Unleashing the potential of sam for medical adaptation via hierarchical decoding. In: CVPR, pp. 3511–3522 (2024)
6. Graham, S., et al.: Hover-net: simultaneous segmentation and classification of nuclei in multi-tissue histology images. Med. Image Anal. **58**, 101563 (2019)
7. Guo, X., Chen, Z., Yuan, Y.: Complementary network with adaptive receptive fields for melanoma segmentation. In: ISBI, pp. 2010–2013. IEEE (2020)
8. Hao, Z., Quan, H., Lu, Y.: Emf-former: An efficient and memory-friendly transformer for medical image segmentation. In: MICCAI, pp. 231–241. Springer (2024)
9. He, A., Wang, K., Li, T., Du, C., Xia, S., Fu, H.: H2former: an efficient hierarchical hybrid transformer for medical image segmentation. IEEE Trans. Med. Imaging **42**(9), 2763–2775 (2023)
10. Hörst, F., et al.: Cellvit: vision transformers for precise cell segmentation and classification. Med. Image Anal. **94**, 103143 (2024)
11. Houlsby, N., Giurgiu, A., Jastrzebski, S., Morrone, B., De Laroussilhe, Q., Gesmundo, A., Attariyan, M., Gelly, S.: Parameter-efficient transfer learning for nlp. In: ICML. pp. 2790–2799. PMLR (2019)

12. Hu, E.J., et al.: LoRA: low-rank adaptation of large language models. In: ICLR (2022)
13. Ibtehaz, N., Kihara, D.: Acc-unet: a completely convolutional unet model for the 2020s. In: MICCAI, pp. 692–702. Springer (2023)
14. Isensee, F., Jaeger, P.F., Kohl, S.A., Petersen, J., Maier-Hein, K.H.: nnu-net: a self-configuring method for deep learning-based biomedical image segmentation. Nat. Methods **18**(2), 203–211 (2021)
15. Kirillov, A., et al.: Segment anything. In: ICCV, pp. 4015–4026, October 2023
16. Li, C., et al.: Flaws can be applause: Unleashing potential of segmenting ambiguous objects in sam. In: NeurIPS, vol. 37, pp. 45578–45599 (2024)
17. Liu, J., et al.: Swin-umamba: mamba-based unet with imagenet-based pretraining. In: MICCAI, pp. 615–625. Springer (2024)
18. Liu, Y., Zhu, H., Liu, M., Yu, H., Chen, Z., Gao, J.: Rolling-unet: revitalizing mlp's ability to efficiently extract long-distance dependencies for medical image segmentation. In: AAAI, pp. 3819–3827 (2024)
19. Lou, Z., et al.: Nusegdg: integration of heterogeneous space and gaussian kernel for domain-generalized nuclei segmentation. Knowledge-Based Systems, p. 113641 (2025)
20. Meng, Z., Dong, J., Zhang, B., Li, S., Wu, R., Su, F., Wang, G., Guo, L., Zhao, Z.: Nusea: Nuclei segmentation with ellipse annotations. IEEE J. Biomed, Health Inform (2024)
21. Na, S., Guo, Y., Jiang, F., Ma, H., Huang, J.: Segment any cell: a sam-based auto-prompting fine-tuning framework for nuclei segmentation. arXiv preprint arXiv:2401.13220 (2024)
22. Nam, J.H., Syazwany, N.S., Kim, S.J., Lee, S.C.: Modality-agnostic domain generalizable medical image segmentation by multi-frequency in multi-scale attention. In: CVPR, pp. 11480–11491 (2024)
23. Nam, S., Jeong, J., Luna, M., Chikontwe, P., Park, S.H.: Pronet: point refinement using shape-guided offset map for nuclei instance segmentation. In: MICCAI, pp. 528–538. Springer (2023)
24. Rahman, M.M., Munir, M., Marculescu, R.: Emcad: Efficient multi-scale convolutional attention decoding for medical image segmentation. In: CVPR, pp. 11769–11779 (2024)
25. Ravi, N., et al.: SAM 2: segment anything in images and videos. In: ICLR (2025)
26. Ronneberger, O., Fischer, P., Brox, T.: U-net: Convolutional networks for biomedical image segmentation. In: Medical image computing and computer-assisted intervention–MICCAI 2015: 18th international conference, Munich, Germany, October 5-9, 2015, proceedings, part III 18. pp. 234–241. Springer (2015)
27. Ryali, C., et al.: Hiera: A hierarchical vision transformer without the bells-and-whistles. In: ICML, pp. 29441–29454. PMLR (2023)
28. Schuiveling, M., et al.: A novel dataset for nuclei and tissue segmentation in melanoma with baseline nuclei segmentation and tissue segmentation benchmarks. GigaScience (2025)
29. Shui, Z., et al.: Unleashing the power of prompt-driven nucleus instance segmentation. In: ECCV, pp. 288–304. Springer (2024)
30. Stringer, C., Wang, T., Michaelos, M., Pachitariu, M.: Cellpose: a generalist algorithm for cellular segmentation. Nat. Methods **18**(1), 100–106 (2021)
31. Valanarasu, J.M.J., Patel, V.M.: Unext: Mlp-based rapid medical image segmentation network. In: MICCAI, pp. 23–33. Springer (2022)

32. Xing, Z., Ye, T., Yang, Y., Liu, G., Zhu, L.: Segmamba: Long-range sequential modeling mamba for 3d medical image segmentation. In: MICCAI. pp. 578–588. Springer (2024)
33. Xu, Q., et al.: Esp-medsam: efficient self-prompting sam for universal domain-generalized medical image segmentation. arXiv preprint arXiv:2407.14153 (2024)
34. Xu, Q., et al.: Hrmedseg: unlocking high-resolution medical image segmentation via memory-efficient attention modeling. arXiv preprint arXiv:2504.06205 (2025)
35. Xu, Q., Ma, Z., Na, H., Duan, W.: Dcsau-net: a deeper and more compact split-attention u-net for medical image segmentation. Comput. Biol. Med. **154**, 106626 (2023)
36. Zhao, T., et al.: A foundation model for joint segmentation, detection and recognition of biomedical objects across nine modalities. Nature methods, pp. 1–11 (2024)

Cycle Context Verification for In-Context Medical Image Segmentation

Shishuai Hu[1,2], Zehui Liao[1,2], Liangli Zhen[2], Huazhu Fu[2(✉)], and Yong Xia[1,3,4(✉)]

[1] National Engineering Laboratory for Integrated Aero-Space-Ground-Ocean Big Data Application Technology, School of Computer Science and Engineering, Northwestern Polytechnical University, Xi'an 710072, China

[2] Institute of High Performance Computing, Agency for Science, Technology and Research, Singapore 138632, Singapore

hzfu@ieee.org

[3] Ningbo Institute of Northwestern Polytechnical University, Ningbo 315048, China

[4] Research & Development Institute of Northwestern Polytechnical University in Shenzhen, Shenzhen 518057, China

yxia@nwpu.edu.cn

Abstract. In-context learning (ICL) is emerging as a promising technique for achieving universal medical image segmentation, where a variety of objects of interest across imaging modalities can be segmented using a single model. Nevertheless, its performance is highly sensitive to the alignment between the query image and in-context image-mask pairs. In a clinical scenario, the scarcity of annotated medical images makes it challenging to select optimal in-context pairs, and fine-tuning foundation ICL models on contextual data is infeasible due to computational costs and the risk of catastrophic forgetting. To address this challenge, we propose Cycle Context Verification (CCV), a novel framework that enhances ICL-based medical image segmentation by enabling self-verification of predictions and accordingly enhancing contextual alignment. Specifically, CCV employs a cyclic pipeline in which the model initially generates a segmentation mask for the query image. Subsequently, the roles of the query and an in-context pair are swapped, allowing the model to validate its prediction by predicting the mask of the original in-context image. The accuracy of this secondary prediction serves as an implicit measure of the initial query segmentation. A query-specific prompt is introduced to alter the query image and updated to improve the measure, thereby enhancing the alignment between the query and in-context pairs. We evaluated CCV on seven medical image segmentation datasets using two ICL foundation models, demonstrating its superiority over existing methods. Our results highlight CCV's ability to enhance ICL-based segmentation, making it a robust solution for universal medical image segmentation. The code will be available at https://github.com/ShishuaiHu/CCV.

S. Hu and Z. Liao—Contributed equally.

J. C. Gee et al. (Eds.): MICCAI 2025, LNCS 15960, pp. 141–151, 2026.
https://doi.org/10.1007/978-3-032-04927-8_14

Keywords: Medical image segmentation · In-context learning · Test-time optimization

1 Introduction

Universal medical image segmentation has garnered significant attention due to the increasing need for medical professionals to segment and analyze various objects of interest across diverse imaging modalities, rather than focusing on a single organ or lesion [4,5,9,10,12–14,29,32]. In-context learning (ICL) [7], which enables a single model to segment varied targets through contextual guidance [4], has shown promise in addressing this challenge.

Initially developed for large language models (LLMs), ICL leverages contextual information to improve predictions [7] and has been successfully adapted to both natural [3,25,26] and medical image segmentation [4]. In a typical ICL framework for image segmentation, a test image (query) is processed using in-context image-mask pairs that exemplify the target structure. When trained on diverse, large-scale datasets, these models exhibit universal segmentation capabilities [3,4,25,26]. Despite their potential, the performance of these models is highly sensitive to the alignment between the query image and the in-context pairs [31]. Consequently, many research efforts have been devoted to selecting optimal in-context pairs for each test query [8,21,22,27,28]. However, the ideal context for a given query may not exist within the available in-context pairs. Moreover, in medical image segmentation, particularly when dealing with underrepresented targets, the availability of suitable in-context pairs is often limited [6], making their selection impractical. Furthermore, fine-tuning the ICL model on these in-context pairs is not feasible due to computational costs and the risk of catastrophic forgetting [11]. In this paper, we suggest an alternative approach: enhancing the context rather than selecting optimal pairs or fine-tuning the ICL model to improve segmentation performance.

Current methods for visual context enhancement in ICL are limited. One relevant approach is InMeMo [30], which introduces task-specific prompts added to the borders of the context image to guide the ICL model during segmentation. However, InMeMo requires a substantial amount of labeled data to train these task-specific prompts, and the data must correspond to the same task as the test query image. This is particularly impractical in medical image segmentation due to the scarcity of suitable in-context pairs [6]. To overcome this limitation, we propose learning a query-specific prompt to enhance the context for each query image and optimize it with the limited available in-context pairs during testing. However, defining an optimization objective at test time to fully exploit labeled in-context pairs remains a challenge.

In the field of LLMs, test time scaling is gaining increasing attention [15,20]. This approach incorporates constraints, such as budget forcing [15], to prompt the model to verify its outputs and correct reasoning errors. Inspired by the concept of answer verification, we propose a cycle context verification mechanism (see Fig. 1), which allows the ICL model to double-check its predicted mask and optimize the query-specific prompt to improve segmentation. Specifically, given a test query image and an associated in-context pair, the ICL model generates an initial segmentation mask for the query. We then swap the roles of the query and in-context pair: the query image and its predicted mask become the new context pair, while the original in-context image becomes the new query image. By taking the swapped query and context pair as input, the ICL model can double-check its original prediction for the query image and generate a mask prediction for the in-context image. The accuracy of the original query prediction indicates the alignment between the query image and the context pair, which also impacts the accuracy of the in-context image prediction. In other words, the accuracy of the in-context prediction implies the accuracy of the original query prediction. Since the in-context mask is available, the accuracy of the in-context prediction can be directly evaluated. This enables optimization of the query-specific prompt, thus improving the accuracy of the in-context prediction and, in turn, enhancing the query segmentation.

In this paper, we introduce Cycle Context Verification (CCV) for in-context medical image segmentation. CCV employs a cyclic pipeline during testing that allows the ICL model to verify its query prediction. A learnable, query-specific prompt is incorporated into the query image to improve the alignment between the query and the in-context pairs. This prompt is iteratively updated to improve the accuracy of the in-context prediction, leading to better query segmentation. We evaluate the proposed method against other techniques using two ICL backbones on seven medical image segmentation datasets, demonstrating its superior performance over existing methods. The contributions of this work are three-fold:

1. We propose a novel, plug-and-play cycle context verification pipeline that enables the ICL model to double-check its query predictions at inference time.
2. Unlike traditional prompt learning, which tunes a unified prompt across the entire dataset, we introduce a learnable, query-specific prompt that is optimized to improve the alignment between each query image and its in-context pairs, thereby enhancing segmentation performance.
3. Our experiments on seven medical image segmentation datasets, using two ICL foundation models, show that the proposed method outperforms existing techniques.

2 Method

2.1 Problem Formulation and Method Overview

Let $\mathcal{M}$ denote a pretrained ICL model for medical image segmentation, and let $S = \{(x_n, y_n)\}_{n=1}^{N}$ represent a set of in-context image-mask pairs, where N is the number of available pairs. S can be the limited available pairs or context pairs selected through popular context selection methods [8,18,31]. Given a test query image x_t, the objective is to enhance the alignment between S and x_t

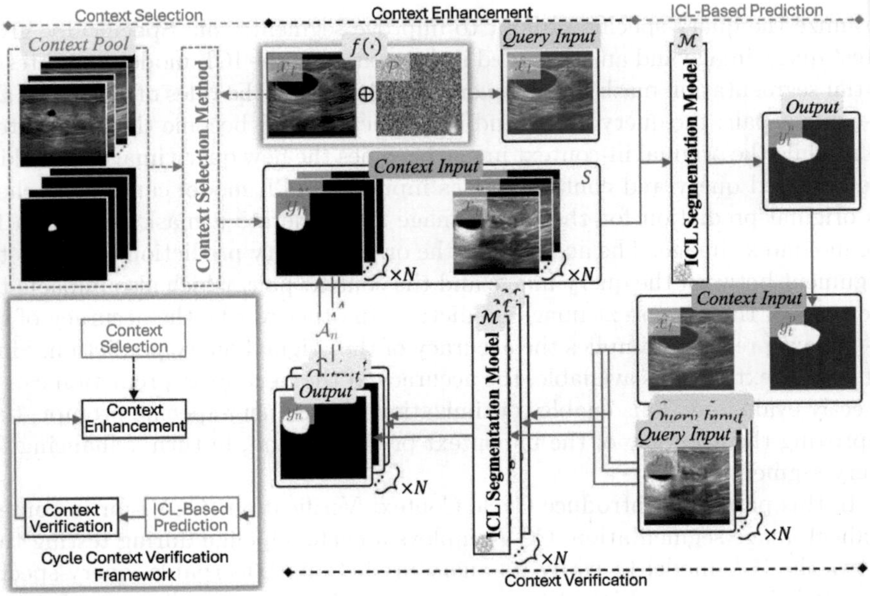

Fig. 1. Diagram of the proposed CCV framework. The symbol $\oplus$ denotes the addition of the query-specific prompt to the query image. The red arrows represent the cycle data flow. (Color figure online)

to improve the predicted segmentation mask $\hat{y_t^p} = \mathcal{M}(f(x_t), S)$, while keeping model $\mathcal{M}$ frozen. Here, $f(\cdot)$ represents a function that modifies x_t.

The proposed CCV framework introduces a Cycle ICL pipeline, which enables the model to double-check its query prediction y_t^p iteratively, a proxy accuracy measurement $\mathcal{A}$ to evaluate the accuracy of y_t^p, and a learnable query-specific prompt $\mathcal{P}$ to refine x_t towards generating more accurate y_t^p. In the Cycle ICL pipeline, a segmentation mask y_t^p is firstly predicted by $\mathcal{M}$ as usual. The model then swaps the roles of the query and in-context pair, treating the query and its predicted mask as a new in-context pair while designating the original in-context image as the new query. This process yields a secondary prediction y_n^p, whose accuracy, measured using $\mathcal{A}$, serves as an indirect estimate of the reliability of y_t^p. The query-specific prompt $\mathcal{P}$ is iteratively optimized to maximize the accuracy measurement $\mathcal{A}$. Once converged, the enhanced inputs $f(x_t)$ and S are fed to $\mathcal{M}$ to yield the final prediction $\hat{y_t^p}$. An overview of the framework is illustrated in Fig. 1. We now delve into its details.

2.2 Cycle ICL Pipeline for Context Verification

Given a set of in-context pairs $S = \{(x_n, y_n)\}_{n=1}^N$, a pretrained ICL model $\mathcal{M}$, and a query image x_t, the initial query prediction y_t^p is generated as:

$$y_t^p = \mathcal{M}(x_t, S). \tag{1}$$

To enable self-verification, the model's input configuration is modified. The predicted mask y_t^p is paired with the original query image x_t to create a new in-context pair $\{(x_t, y_t^p)\}$, while the original in-context image x_n is designated as the new query. The model then generates a secondary prediction y_n^p:

$$y_n^p = \mathcal{M}(x_n, \{(x_t, y_t^p)\}). \tag{2}$$

Since the ground truth mask y_n for x_n is available, the accuracy of y_n^p can be quantitatively evaluated using the Dice Similarity Coefficient (DSC) $\mathcal{A}_n = \mathcal{D}(y_n^p, y_n)$. The overall accuracy $\mathcal{A}$ across all in-context pairs is then computed as:

$$\mathcal{A} = \frac{\sum_{n=1}^{N} \mathcal{A}_n}{N}. \tag{3}$$

The segmentation quality of $\mathcal{M}$ is contingent on the context provided. If the initial query prediction y_t^p deviates substantially from the ground truth, incorporating it into the context may degrade the accuracy of y_n^p. Thus, $\mathcal{A}$ serves as an implicit measure of the reliability of y_t^p.

2.3 Query-Specific Prompt for Context Enhancement

To improve segmentation accuracy, the objective is to maximize $\mathcal{A}$. However, directly optimizing $\mathcal{M}$ may potentially lead to model collapse, where the model becomes overfitted to the verification step without improving y_t^p. To circumvent this, we introduce a query-specific prompt $\mathcal{P}$, which is learnable and spatially aligned with the query image x_t. By adding $\mathcal{P}$ to x_t, the correspondence between the query and in-context images is expected to be enhanced, thereby yielding improved segmentation outcomes. The transformation of the query image can be expressed as:

$$\hat{x_t} = f(x_t) = x_t + \mathcal{P}. \tag{4}$$

Using the modified query $\hat{x_t}$, we recompute $\hat{y_t^p}$, $\hat{y_n^p}$, $\hat{\mathcal{A}}_n$, and $\hat{\mathcal{A}}$. The loss function is defined as:

$$\mathcal{L} = 1 - \hat{\mathcal{A}}. \tag{5}$$

Minimizing $\mathcal{L}$ encourages the model to refine $\mathcal{P}$ iteratively. A higher $\hat{\mathcal{A}}$ indicates an improved performance of the predicted masks $\{\hat{y_n^p}\}_{n=1}^N$, thereby improving the reliability of y_t^p. To maintain computational efficiency, an accuracy threshold $\mathcal{T}$ is established. If $\hat{\mathcal{A}} > \mathcal{T}$, prompt optimization is halted. Additionally, a maximum iteration limit $\mathcal{E}$ is imposed to prevent excessive optimization. Once $\mathcal{P}$ is optimized, the final segmentation mask $\hat{y_t^p}$ is generated by feeding $\hat{x_t}$ and S into $\mathcal{M}$.

3 Experiments and Analysis

3.1 Experimental Setup

ICL Backbones. The proposed method is evaluated using UniverSeg [4] and SegGPT [26] as ICL backbones. UniverSeg is an ICL backbone specifically

Table 1. The DSC (%) of the competing methods and those equipped with CCV.

$\mathcal{M}$	Methods	BUS	CQE	GLS	NPP	OCT	ROC	ROD	USC	Average
UniverSeg [4]	RS	47.49	87.21	60.50	28.84	34.97	57.72	77.83	80.79	59.42
	RS+CCV	50.99	87.79	60.71	33.52	38.63	60.26	79.65	81.92	$61.68_{\uparrow 2.26}$
	VPR [31]	54.13	90.68	60.06	32.34	44.72	61.25	78.76	83.91	63.23
	VPR+CCV	57.28	91.11	62.82	36.38	47.73	63.39	80.27	85.22	$65.53_{\uparrow 2.30}$
	DualSC [8]	53.64	88.60	62.29	34.47	46.34	62.10	79.30	84.07	63.85
	DualSC+CCV	57.58	91.11	62.91	36.84	48.03	63.66	80.44	85.11	$65.71_{\uparrow 1.86}$
	InMeMo [30]	48.01	86.73	61.98	28.56	34.53	54.41	79.33	81.27	59.35
	InMeMo+CCV	51.67	87.34	62.15	31.53	37.34	58.01	80.11	82.37	$61.32_{\uparrow 1.97}$
SegGPT [26]	RS	42.05	91.00	76.66	59.89	35.85	67.28	91.38	80.88	68.12
	RS+CCV	46.26	91.79	80.76	64.96	37.67	71.12	92.46	84.13	$71.14_{\uparrow 3.02}$
	VPR [31]	55.92	93.44	85.09	68.29	58.26	68.06	91.91	82.64	75.45
	VPR+CCV	60.71	93.85	86.46	72.04	59.32	72.13	92.59	85.54	$77.83_{\uparrow 2.38}$
	DualSC [8]	57.43	93.55	85.94	68.97	53.54	70.92	92.00	83.23	75.70
	DualSC+CCV	60.79	93.88	86.73	71.58	60.64	72.19	92.62	85.38	$77.98_{\uparrow 2.28}$
	InMeMo [30]	59.68	95.37	81.85	63.50	46.42	78.75	93.22	82.83	75.20
	InMeMo+CCV	61.77	95.57	83.09	66.26	46.85	82.18	94.03	86.20	$76.99_{\uparrow 1.79}$

designed for medical image segmentation. SegGPT, on the other hand, is an ICL backbone primarily trained on large-scale natural image segmentation datasets.

Datasets and Evaluation Metrics. We selected seven medical image segmentation datasets encompassing eight tasks across seven medical imaging modalities[1] to evaluate the proposed method. These include the BreastUS (BUS, 647 images) [2], COVID-QU-Ex (CQE, 5826 images) [23], GlaS (GLS, 165 images) [19], NeoPolyp (NPP, 1000 images) [16], OCT-CME (OCT, 1460 images) [1], REFUGE (ROC and ROD, 1200 images) [17], and UWaterlooSkinCancer (USC, 206 images) [24] datasets. Each dataset is partitioned into proportions of 3:1:1, designated for test samples, validation samples, and candidate in-context samples, respectively. Notably, these datasets are not included in the training sets of either UniverSeg or SegGPT. The DSC is used as the evaluation metric to assess the segmentation performance.

Implementation Details. The proposed method is implemented using the PyTorch framework and tested on a single NVIDIA 3090 GPU. For the UniverSeg model, the context size is set to eight. All input images are resized to 128 × 128, converted to gray-scale images, and normalized to the intensity range of [0, 1]. For the SegGPT backbone, the context size is set to one. All images are

[1] These datasets can be accessed through https://huggingface.co/datasets/microsoft/BiomedParseData.

resized to 448 × 448, converted to three-channel color images, and normalized by subtracting the mean and dividing by the standard deviation. The query-specific prompt $\mathcal{P}$ is zero-initialized. The threshold $\mathcal{T}$ and the iteration limitation $\mathcal{E}$ are empirically set to 0.9 and 20 respectively.

3.2 Experimental Results and Analysis

We employ three in-context pair selection methods as baselines to select in-context pairs from the candidate in-context dataset for each test query sample, including: Random Selection (RS), which involves the random selection of in-context pairs from the candidate in-context dataset, VPR [31], which fine-tunes a pretrained CLIP image encoder [18] to extract and rank features, subsequently selecting the top-K in-context pairs, and DualSC [8], which incorporates mask similarity into the ranking process upon VPR. We also incorporate InMeMo [30], a context enhancement method that prompts the model regarding the specific segmentation task to be performed, as a baseline method. The proposed method is designed to be complementary to these existing approaches. Therefore, we integrate the proposed method with these baselines and report the performance.

Comparison using UniverSeg as Backbone. The Dice scores attained by other competing methods, as well as those equipped with the proposed method, are presented in the first part of Table 1. The results indicate that our method improves the overall segmentation performance of RS, VPR, and DualSC, emphasizing the necessity of enhancing context during inference. Furthermore, RS equipped with our method outperforms InMeMo, affirming that query-specific prompts can be superior to task-specific prompts. Our approach also enhances the performance of InMeMo, indicating that the query-specific prompt design is complementary to task-specific prompt.

Comparison using SegGPT as Backbone. The Dice scores for our model and competing methods using SegGPT as backbone are presented in the second part of Table 1. Although using fewer in-context pairs, SegGPT still outperforms UniverSeg, highlighting the strong generalization capabilities of SegGPT. Even though, the proposed method enhances the overall segmentation performance across all competing approaches, further demonstrating its advantages.

Table 2. Performance of removing cycle ICL pipeline and prompt.

Method		Average
Cycle ICL Pipeline	Prompt Optimiz.	
×	×	58.34
✓	×	57.24
×	✓	58.57
✓	✓	60.65

Table 3. Performance of four variants of prompt optimization.

Method				Average
Border Prompt	Image Prompt	Update First	Update Second	
×	✓	✓	×	59.76
×	✓	×	✓	59.37
×	✓	✓	✓	60.65
✓	×	✓	✓	58.82

3.3 Ablation Analysis

To evaluate the effectiveness of the proposed method in a limited context scenario, we select a small number of in-context pairs, *i.e.*, eight for UniverSeg, from the candidate in-context dataset for each task, utilizing them as context to guide the segmentation and evaluate the performance of the ICL model and the proposed method. The effectiveness of the cycle ICL pipeline and the query-specific prompt optimization is also evaluated in the limited context scenario via removing these two components respectively. UniverSeg is used as the ICL backbone for the ablation experiments. The results are shown in Table 2. Removing prompt optimization represents employing solely the cycle ICL pipeline and updating the entire ICL model with a small learning rate 1e-4 for 20 iterations. Removing Cycle ICL pipeline represents employing only the query-specific prompt and minimizing the entropy of $\hat{y_t^p}$ to update $\mathcal{P}$. The performance drop of removing prompt optimization can be attributed to the model collapse when updating the entire ICL model. The results reveal that both the cyclical design and the query-specific prompt-based optimization are essential for the functioning of the proposed method.

Analysis of Query-Specific Prompt. We compare four configurations in the limited context scenario using the cycle ICL pipeline: use image prompt (CCV), use border prompt [30] instead of image prompt, update the image prompt only in the first stage when generating $\hat{y_t^p}$, and update the image prompt only in the second stage when generating $\hat{y_n^p}$. The results is shown in Table 3. It indicates that the performance of using border prompt is inferior to that of using image prompt. Notably, freezing the image prompt during the generation of either $\hat{y_t^p}$ or $\hat{y_n^p}$ is found to be less effective than updating the image prompt in these two stages. This observation underscores the necessity for the updating of $\mathcal{P}$ to enhance the alignment between the query image and in-context pairs throughout these stages. We visualized examples of the learned query-specific prompt $\mathcal{P}$ in Fig. 2. The predictions generated by the ICL model for x_t and $\hat{x}_t$ are also illustrated. Additionally, the ground truth segmentation maps are

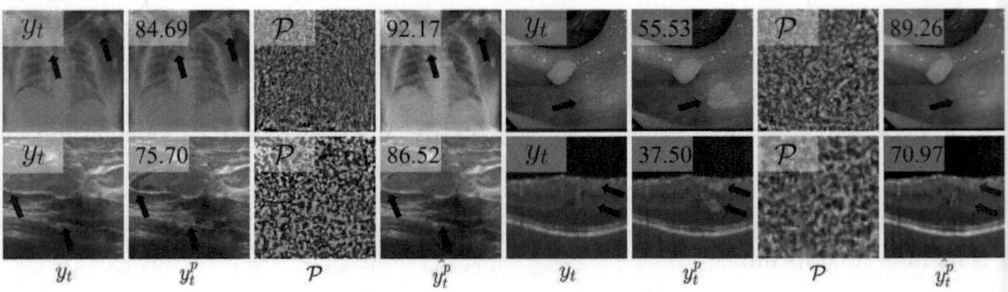

Fig. 2. Visualization of four learned query-specific prompt $\mathcal{P}$, corresponding predictions $\hat{y_t^p}$ and original predictions y_t^p. Together with ground truth y_t. The dice scores of y_t^p and $\hat{y_t^p}$ is also displayed for comparison. Note that $\mathcal{P}$ has been rescaled to [0, 255] for a better visualization. Best viewed in color.

provided for reference. The visualization demonstrates that the learned $\mathcal{P}$ can effectively alter x_t towards generating markedly improved segmentation results, thereby underscoring the validity of the proposed methodology.

4 Conclusion

In this paper, we introduced the Cycle Context Verification (CCV) framework to enhance in-context medical image segmentation, particularly under limited in-context data. By implementing a cycle ICL pipeline, the ICL model is capable of double-checking its predictions. The integration and optimization of a query-specific prompt enhance the alignment between the test query image and the associated in-context pairs. Experimental results demonstrate that CCV not only improves existing context selection and enhancement methods, but also elevates segmentation performance in challenging limited context scenarios. Furthermore, ablation studies elucidate the design of the cycle ICL pipeline and query-specific prompt-based optimization. These findings highlight the potential of CCV to advance in-context learning based universal medical image segmentation.

Acknowledgments. This work was supported in part by the National Natural Science Foundation of China under Grants 62171377 and 92470101, in part by the "Pioneer" and "Leading Goose" R&D Program of Zhejiang, China, under Grant 2025C01201(SD2), in part by the Ningbo Clinical Research Center for Medical Imaging under Grant 2021L003 (Open Project 2022LYKFZD06), in part by the Shenzhen Science and Technology Program under Grant JCYJ20220530161616036, and in part by the Innovation Foundation for Doctor Dissertation of Northwestern Polytechnical University under Grants CX2023016 and CX2022056.

Disclosure of Interests. The authors have no competing interests to declare that are relevant to the content of this article.

References

1. Ahmed, Z., Panhwar, S.Q., Baqai, A., Umrani, F.A., Ahmed, M., Khan, A.: Deep learning based automated detection of intraretinal cystoid fluid. Int. J. Imaging Syst. Technol. **32**(3), 902–917 (2022)
2. Al-Dhabyani, W., Gomaa, M., Khaled, H., Fahmy, A.: Dataset of breast ultrasound images. Data Brief **28**, 104863 (2020)
3. Bar, A., Gandelsman, Y., Darrell, T., Globerson, A., Efros, A.: Visual prompting via image inpainting. Adv. Neural. Inf. Process. Syst. **35**, 25005–25017 (2022)
4. Butoi, V.I., Ortiz, J.J.G., Ma, T., Sabuncu, M.R., Guttag, J., Dalca, A.V.: Universeg: Universal medical image segmentation. In: Proceedings of the IEEE/CVF International Conference on Computer Vision, pp. 21438–21451 (2023)
5. Chen, C., Miao, J., Wu, D., Zhong, A., Yan, Z., Kim, S., Hu, J., Liu, Z., Sun, L., Li, X., et al.: Ma-sam: Modality-agnostic sam adaptation for 3d medical image segmentation. Med. Image Anal. **98**, 103310 (2024)

6. Cheng, Z., Wang, S., Xin, T., Zhou, T., Zhang, H., Shao, L.: Few-shot medical image segmentation via generating multiple representative descriptors. IEEE Trans. Med. Imaging (2024)
7. Dong, Q., et al.: A survey on in-context learning. In: Proceedings of the 2024 Conference on Empirical Methods in Natural Language Processing, pp. 1107–1128 (2024)
8. Gao, J., Lao, Q., Kang, Q., Liu, P., Du, C., Li, K., Zhang, L.: Boosting your context by dual similarity checkup for in-context learning medical image segmentation. IEEE Trans. Med. Imaging (2024)
9. Huang, Y., et al.: Segment anything model for medical images? Med. Image Anal. **92**, 103061 (2024)
10. Kirillov, A., et al.: Segment anything. In: Proceedings of the IEEE/CVF International Conference on Computer Vision, pp. 4015–4026 (2023)
11. Lin, Y., et al.: Speciality vs generality: an empirical study on catastrophic forgetting in fine-tuning foundation models. arXiv preprint arXiv:2309.06256 (2023)
12. Ma, J., He, Y., Li, F., Han, L., You, C., Wang, B.: Segment anything in medical images. Nat. Commun. **15**(1), 654 (2024)
13. Mazurowski, M.A., Dong, H., Gu, H., Yang, J., Konz, N., Zhang, Y.: Segment anything model for medical image analysis: an experimental study. Med. Image Anal. **89**, 102918 (2023)
14. Mei, J., Zhou, T., Huang, K., Zhang, Y., Zhou, Y., Wu, Y., Fu, H.: A survey on deep learning for polyp segmentation: techniques, challenges and future trends. Visual Intell. **3**(1), 1 (2025)
15. Muennighoff, N., et al.: s1: Simple test-time scaling. arXiv preprint arXiv:2501.19393 (2025)
16. Ngoc Lan, P., et al.: Neounet: towards accurate colon polyp segmentation and neoplasm detection. In: Advances in Visual Computing: 16th International Symposium, ISVC 2021, virtual event, October 4-6, 2021, proceedings, part II, pp. 15–28. Springer (2021)
17. Orlando, J.I., et al.: Refuge challenge: a unified framework for evaluating automated methods for glaucoma assessment from fundus photographs. Med. Image Anal. **59**, 101570 (2020)
18. Radford, A., et al.: Learning transferable visual models from natural language supervision. In: International Conference on Machine Learning, pp. 8748–8763. PMLR (2021)
19. Sirinukunwattana, K., Snead, D.R., Rajpoot, N.M.: A stochastic polygons model for glandular structures in colon histology images. IEEE Trans. Med. Imaging **34**(11), 2366–2378 (2015)
20. Snell, C., Lee, J., Xu, K., Kumar, A.: Scaling llm test-time compute optimally can be more effective than scaling model parameters. arXiv preprint arXiv:2408.03314 (2024)
21. Sun, Y., Chen, Q., Wang, J., Wang, J., Li, Z.: Exploring effective factors for improving visual in-context learning. arXiv preprint arXiv:2304.04748 (2023)
22. Suo, W., Lai, L., Sun, M., Zhang, H., Wang, P., Zhang, Y.: Rethinking and improving visual prompt selection for in-context learning segmentation. In: European Conference on Computer Vision, pp. 18–35. Springer (2024)
23. Tahir, A.M., et al.: Covid-qu-ex dataset. Kaggle (2021). https://doi.org/10.34740/kaggle/dsv/3122958
24. Venugopal, V., Joseph, J., Das, M.V., Nath, M.K.: Dtp-net: a convolutional neural network model to predict threshold for localizing the lesions on dermatological macro-images. Comput. Biol. Med. **148**, 105852 (2022)

25. Wang, X., Wang, W., Cao, Y., Shen, C., Huang, T.: Images speak in images: a generalist painter for in-context visual learning. In: Proceedings of the IEEE/CVF Conference on Computer Vision and Pattern Recognition, pp. 6830–6839 (2023)
26. Wang, X., Zhang, X., Cao, Y., Wang, W., Shen, C., Huang, T.: Seggpt: towards segmenting everything in context. In: Proceedings of the IEEE/CVF International Conference on Computer Vision, pp. 1130–1140 (2023)
27. Wu, C., Restrepo, D., Shuai, Z., Liu, Z., Shen, L.: Efficient in-context medical segmentation with meta-driven visual prompt selection. In: International Conference on Medical Image Computing and Computer-Assisted Intervention, pp. 255–265. Springer (2024)
28. Xu, C., Liu, C., Wang, Y., Yao, Y., Fu, Y.: Towards global optimal visual in-context learning prompt selection. arXiv preprint arXiv:2405.15279 (2024)
29. Zeng, Q., Xie, Y., Lu, Z., Xia, Y.: A human-in-the-loop method for pulmonary nodule detection in ct scans. Visual Intell. **2**(1), 19 (2024)
30. Zhang, J., Wang, B., Li, L., Nakashima, Y., Nagahara, H.: Instruct me more! random prompting for visual in-context learning. In: Proceedings of the IEEE/CVF Winter Conference on Applications of Computer Vision, pp. 2597–2606 (2024)
31. Zhang, Y., Zhou, K., Liu, Z.: What makes good examples for visual in-context learning? Adv. Neural. Inf. Process. Syst. **36**, 17773–17794 (2023)
32. Zhao, T., et al.: A foundation model for joint segmentation, detection and recognition of biomedical objects across nine modalities. Nature Methods, pp. 1–11 (2024)

DGMIR: Dual-Guided Multimodal Medical Image Registration Based on Multi-view Augmentation and On-Site Modality Removal

Gao Le[1,2], Yucheng Shu[1,2,3(✉)], Lihong Qiao[1,2,3], Lijian Yang[1,2], Bin Xiao[1,2], Weisheng Li[1,2], and Xinbo Gao[1,2]

[1] Chongqing University of Posts and Telecommunications, Chongqing 400065, China
d240201011@stu.cqupt.edu.cn,
{shuyc,qiaolh,xiaobin,liws,gaoxb}@cqupt.edu.cn, eeejyang@gmail.com
[2] Chongqing Key Laboratory of Image Cognition, Chongqing 400065, China
[3] Chongqing Key Laboratory of Precision Diagnosis and Treatment for Kidney Diseases, Chongqing 400065, China

Abstract. Multi-modal medical image registration integrates complementary information from various modalities to deliver comprehensive visual insights for disease diagnosis, treatment planning, surgical navigation, etc. However, current methods often suffer from artifacts, computational overhead, or insufficient handling of modality-specific interference. Moreover, they still rely on specialized modules, such as generative trans-modal units, additional encoders, or handcrafted modality-invariant operators, without fully exploiting the inherent potential of registration features. To address these drawbacks in multimodal medical image registration, we propose a novel registration framework. First, a plug-and-play architecture is proposed to directly process multi-scale heterogeneous features, with active guidance only during deformation field generation stage. Second, we introduce a multi-view feature reorganization module that dynamically optimizes feature distributions via adaptive relation computation and global calibration. Finally, an in-network modality removal module is introduced to leverage multi-scale adaptive convolutions to explicitly eliminate modality-specific interference. Extensive experiments on the BraTS2018 and Learn2Reg2021 datasets confirm that our proposed method achieves state-of-the-art performance on multiple multimodal medical image registration metrics. (https://github.com/St-Antonio/DGMIR).

Keywords: multimodal medical image registration · feature disentanglement · feature removal

1 Introduction

Deformable image registration (DIR) is a fundamental technique in medical image processing. It aims to establish spatial alignment between two or more

J. C. Gee et al. (Eds.): MICCAI 2025, LNCS 15960, pp. 152–162, 2026.
https://doi.org/10.1007/978-3-032-04927-8_15

images, facilitating a wide range of clinical downstream workflows, such as disease diagnosis, treatment planning, surgical navigation, etc. Recent advances in deep learning have driven significant progress in DIR, leading to notable improvements in both computational speed and registration accuracy [2–4].

However, advancements in imaging technology have led to widespread use of multimodal medical images like CT, MRI, etc. Their complementary diagnostic value makes multimodal image registration a critical clinical requirement. Nevertheless, due to variations in acquisition protocols, the same or similar anatomical structures often exhibit significant visual differences across imaging modalities. While existing registration frameworks are still limited by this semantic-appearance mismatch phenomenon, recent research has increasingly focused on multimodal medical image registration methodologies.

Among these researches, image translation-based strategies are widely employed [5–8]. By mapping images from different modalities to a unified modality via generative network, multimodal registration can be simplified to a monomodal task, notably reducing the influence of visual heterogeneity. However, these methods often introduce artifacts and require additional computational resources [1]. Additionally, some studies [9,10] have introduced cross-modal-interaction networks to enhance feature correlation between multimodal images. However, such methods merely integrate generic modules, such as attention mechanisms, into the registration task, failing to address the fundamental challenge of modality-specific confounding factors in cross-modal feature alignment.

Recently, some methods attempt to distill task-specific representations essential for multimodal registration [11–13]. For instance, Qin et al. [11] introduced a feature space decomposition approach, separating the shared shape space from the modality-specific appearance space. Wang et al. [12] developed distinct encoders to separately extract general and structural features, further optimizing structural representation via a self-similarity module. Mok et al. [13] proposed to leverage neighborhood self-similarity and anatomy-based contrastive learning to achieve highly discriminative, contrast-invariant representations.

In sum, while existing methods demonstrate competent multimodal registration performance, they predominantly rely on additional network components, such as generative trans-modal units, specialized encoding branches, or handcrafted modality-invariant feature engineering strategies, without fully exploiting the potential inherent in the registration features themselves. Therefore, in this paper, we argue that the scaling-law should not be taken for granted: before stacking various fancy network modules, it is imperative to first focus on generalized feature learning frameworks and conduct an in-depth analysis to explore their upper limits in multimodal registration tasks. Therefore, in this study, we propose a novel multimodal DIR model, namely DGMIR, to address the aforementioned issues. Our main contributions are as follows:

- We propose a flexible multimodal registration framework that directly integrates multi-scale heterogeneous features in a plug-and-play manner. By implementing active guidance only during the deformation field decoding and generation phases, our approach iteratively achieves accurate multimodal registration.

- We introduce a multi-view feature reorganization guidance module that dynamically modulates feature distributions through adaptive relation computation and global calibration factor, thereby enhancing the discriminative power of multimodal registration features.
- We propose a modality on-site removal guidance module, which leverages adaptive mean convolutions across multiple scales within the registration backbone, to explicitly learn and eliminate modality information for the continual optimization of the deformation field.

Extensive experiments on two public multimodal datasets, BraTS2018 [20] and Learn2Reg2021 [21], demonstrate our method's superiority over state-of-the-art approaches, achieving significantly improvements in multimodal registration.

2 Methodology

2.1 Overall Structure

The overall structure of the network is shown in Fig. 1. Given a pair of fixed and moving image, I_F and I_M, the model outputs a deformation field ϕ: $U_\theta(I_F, I_M) = \phi$, where U_θ denotes the registration network.

The encoder adopts a four-stage architecture, where each stage contains just one convolutional layer followed by a ReLU activation function. The number of output channels in each stage is $\{16, 32, 32, 64\}$, with the first stage preserving the original resolution and the subsequent three stages successively reducing the feature map size by half. The encoder produces two sets of feature maps: $\{F_1, F_2, F_3, F_4\}$ and $\{M_1, M_2, M_3, M_4\}$. The decoder adopts a coarse-to-fine manner across four stages, each comprising a sequential combination of a multi-view feature reorganization guidance module and a modality on-site removal guidance module.

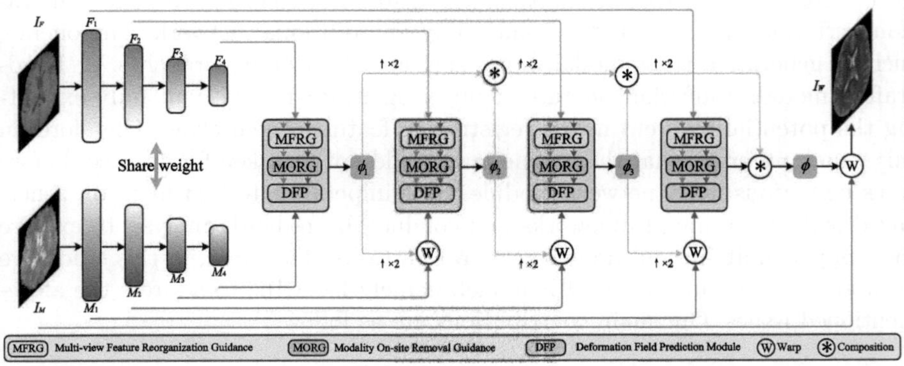

Fig. 1. The Overall structure of the proposed network.

2.2 Multi-view Feature Reorganization Guidance (MFRG)

Through preliminary experiments, we discovered that there are notable disparities between multi-modal image features due to different imaging protocols, and these disparities are predominantly manifested as gaps in the statistical distribution within the channel dimension [22]. In particular, the mean and maximum are two key statistics within each channel, which can offer different views to guide feature modulation. Therefore, we propose a multi-view guided feature enhancement mechanism, which integrates aforementioned statistics indicators to evaluate the importance of channel with different distribution. By combining these two distribution indicators, we achieve a more comprehensive characterization of feature distribution. Moreover, we incorporate a learnable global calibration factor to guide holistic feature expression by adaptively fusing these complementary indicators. Finally, we quantify inter-modal feature relationships and selectively strengthen the feature representation. The specific structure is shown in Fig. 2.

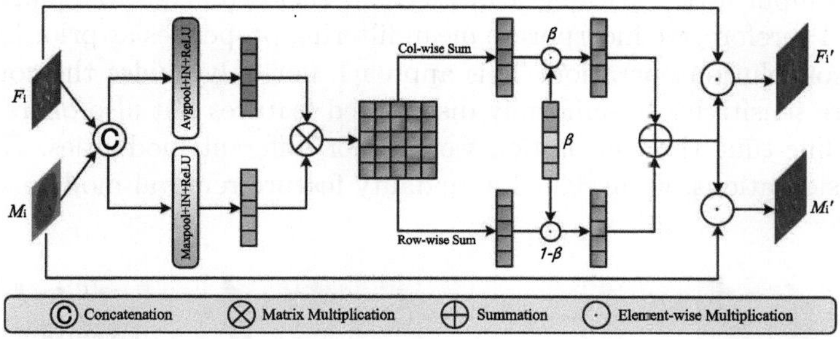

Fig. 2. Multi-view Feature Reorganization Guidance module.

For fixed and moving image features $F_i, M_i \in \mathbb{R}^{c \times h \times w \times d}$, where i indicates the encoder stage, $h \times w \times d$ specifies the spatial dimensions of the feature map, and c denotes channel number. We begin by concatenating F_i and M_i, then independently applying max pooling and average pooling layers to the combined features. Subsequently, we use two $1 \times 1 \times 1$ convolution layers to both shrink and recover the feature descriptors based on the pooled outputs, thereby eliminating redundant information while retaining salient features. After deriving the feature global expressions, we calculate inter-modal cross-correlation and intra-modal self-correlation through matrix multiplication. Inspired by [23,24], we utilize row-wise and column-wise summations to capture correlations between the channels and quantify feature distribution's importance relative to all others, then integrate these importance scores with a calibration factor β. Ultimately, feature expressions are refined through element-wise multiplication of the adaptive enhancement factor and the original features. By introducing β, the network not only adaptively modulates feature expressiveness but also suppresses extraneous representations.

2.3 Modality On-Site Removal Guidance (MORG)

In our multimodal registration experiments, we found the modality feature exhibits smaller gradient variations and a relatively uniform distribution. We also decomposed the image into high and low-frequency components and reconstructing it only with the high-frequency part, we found that the modality-independent features can be effectively preserved. Based on these observations, we assume that uniformly distributed modality features may interfere the inter-modal correspondence to accurately predicting the deformation field. However, current methods do not explicitly exclude interference from these modality-specific features. Therefore, we propose a guided coding method that explicitly decodes modality information and progressively eliminates modality-specific features at multiple scales during the decoding process. To the best of our knowledge, this marks the first attempt to explicitly perform modality removal within the feature decoding and flow generation process.

In digital image processing, mean filtering attenuates high-frequency components while preserving low-frequency information. It utilizes a fixed-size sliding window, computing the arithmetic mean of the enclosed pixels to replace the center pixel. Therefore, we incorporate mean filtering properties as prior knowledge into the convolution operation. This approach not only guides the convolution to be more sensitivity to uniformly distributed features but also offers the flexibility to fine-tune the convolution weights for different modalities. Guided by these considerations, we designed a modality feature removal module.

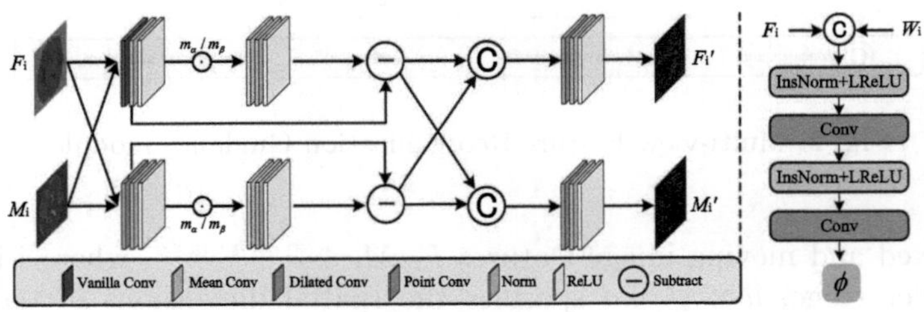

Fig. 3. Left: Modality On-site Removal Guidance module. Right: Deformation Field Prediction module.

Next, we provide the detailed implementation of our method as illustrated in Fig. 3. Initially, we apply both a 3×3×3 vanilla convolution and a dilated convolution to the input features, respectively. These two distinct convolution techniques facilitate feature extraction across varying receptive field ranges. Next, the vanilla convolution outputs undergo a 3×3×3 mean convolution, whereas the dilated convolution outputs undergo a 5×5×5 mean convolution. Specifically, the weights of these mean convolutions are initialized using a uniform distribution. Subsequently, we subtract the mean convolution output from the vanilla convolution output, producing de-modality features at the current scale.

Finally, the subtracted outputs are concatenated and fed into a pointwise convolution layer, yielding the module's final output. Where m_α and m_β are learnable parameters act as modality-specific parameters to dynamically adjust feature representations for different modalities by Hadamard product with features.

Eventually, we adopt a coarse-to-fine strategy for predicting the deformation field. In the initial decoding stage, the network inputs are the original encoder features F_1 and M_1. In later stages, the features become F_i and $M_i \circ \phi_{i-1}(W_i)$, where ϕ_{i-1} is obtained through $2\times$ upsampling of the previous stage's sub-deformation field, and the operator $\circ$ is applied through the Spatial Transformation Network [14]. Lastly, the deformation field at each stage is derived by merging the output of the deformation field prediction module (DFP) with the sub-deformation field from the preceding stage. Implementation specifics of the DFP appear in Fig. 3, and its key formula is presented below:

$$\phi_i = \begin{cases} \phi_i^{'}, \phi_i^{'} = \text{DFP}(F_i, M_i) i = 1 \\ (\phi_{i-1} \circ \phi_i^{'}) + \phi_i^{'}, \phi_i^{'} = \text{DFP}(F_i, M_i \circ \phi_{i-1}) i = 2, 3, 4 \end{cases} \tag{1}$$

3 Experiments

Datasets and Experimental Details: In this study, we employ two public datasets: BraTS2018 and Learn2Reg2021. BraTS2018 comes from the MICCAI Brain Tumor Segmentation (BraTS) 2018 Challenge, consisting of 285 imaging volumes across four MRI modalities (T1, T1ce, T2, FLAIR), each accompanied by expert segmentation labels. All scans measure 240×240×155 at 1 mm isotropic resolution. We validate our proposed model using T1 and T2 modalities. Since the original dataset is pre-aligned, we introduce random misalignment by applying elastic transformations and Gaussian smoothing to T2 images, followed by cropping them to 160×192×128. Finally, we randomly split the dataset into training, validation, and test sets, with 185, 35, and 65 examples, respectively. Learn2Reg2021 (L2R2021) is an abdominal multimodal dataset comprising CT and MRI scans from the same patients, each accompanied by expert segmentation labels. All scans measure 192×160×192 at 1 mm isotropic resolution and have been rigidly pre-aligned. The dataset is then divided into training, validation, and test subsets, comprising 5, 1, and 2 scans respectively.

We implement our network in PyTorch and train it with the Adam optimizer with a learning rate of 1e-4, using a batch size of 1 for 500 epochs. The result for each method is reported on the same PC with 3.2GHz CPU and RTX 4090 GPU. For evaluation, we use Dice score (DCS), 95% Hausdorff distance (HD95), and the percentage of negative values of the Jacobian determinant ($\%|J_\phi| \leq 0$) to assess registration performance.

In this study, we use $\mathcal{L}_{\text{MIND}} = \frac{1}{N}\sum_{x\in\Omega}\sum_{n\in\mathcal{N}} \exp(-||D_F(x,n) - D_W(x,n)||^2)$ [19] as similarity loss function, where $D(x,n) = \left(\frac{\|I(x)-I(x+n)\|^2}{V(x)}\right)$, $I(x)$ represents the intensity value of the image at position x, $(x+n)$ denotes a point within the neighborhood, $V(x)$ is the local variance at position x. When segmentation labels are available, we also introduce a weakly supervised loss

Table 1. The quantitative results of different registration methods on two datasets.

	BraTS2018			Learn2Reg2021		
	DSC (%)	HD95	$\%\lvert J_\phi\rvert \leq 0$	DSC (%)	HD95	$\%\lvert J_\phi\rvert \leq 0$
VoxelMorph [3]	75.5	2.546	0.07e-4	59.9	24.77	7.31e-3
TransMorph [2]	77.1	2.334	0.84e-4	60.9	24.42	1.07e-2
CorrMLP [18]	77.2	2.379	0.07e-4	63.6	25.37	2.19e-4
GroupMorph [16]	76.8	2.267	0.85e-4	60.0	24.09	4.56e-3
TransMatch [17]	76.8	2.256	0.06e-4	61.8	**22.69**	6.13e-3
ModeT [15]	74.4	2.602	0.66e-4	63.9	27.27	3.77e-4
Ours	**79.4**	**1.931**	1.52e-4	**65.5**	23.67	9.08e-4

function $\mathcal{L}_{\text{Dice}} = -\frac{1}{C}\sum_{i=1}^{C}\frac{2|S_{Fi}\cap S_{Wi}|}{|S_{Fi}|+|S_{Wi}|}$, where S_F and S_W represent the segmentation labels of the fixed and warped images, respectively, and C denotes the number of categories in the segmentation labels. In addition to similarity-based loss, we incorporate a regularization term $\mathcal{L}_{\text{reg}} = \sum \|\nabla\varphi\|^2$ to ensure the continuity and smoothness of the deformation field. The final loss function is $\mathcal{L} = \mathcal{L}_{\text{MIND}} + \lambda_1\mathcal{L}_{\text{Dice}} + \lambda_2\mathcal{L}_{\text{reg}}$, where λ_1 and λ_2 are the trade-off parameters between different terms, and in this experiment, they are set to 1 and 0.5, respectively.

We compare our model with a series of state-of-the-art deep learning methods, including VoxelMorph [3], TransMorph [2], GroupMorph [16], TransMatch [17], ModeT [15], and CorrMLP [18]. All comparison methods use the same loss function aforementioned, and all hyperparameter settings are kept unchanged with the original code.

Quantitative and Qualitative Analysis: Table 1 presents the quantitative comparison results of various methods on two datasets. In both datasets, the proposed DGMIR demonstrates a substantial advantage across all primary evaluation metrics. In terms of the DCS metric, our method surpasses the second-best approach by 2.2% and 1.6% on BraTS2018 and L2R2021, respectively. Regarding the HD95 metric, our method outperforms all competitors on BraTS2018 but trails the top-performing method by 0.98 on L2R2021. We also visualized the registration results of all methods on BraTS2018, as shown on the left panel of Fig. 4. To confirm the statistical significance of these results, we conducted two-sided Wilcoxon signed-rank tests on the DSC and HD95 metrics, comparing our method against the second-best approach in each dataset. The obtained p-values ($p \ll 0.05$) consistently reject the null hypothesis, confirming the statistical significance of our method's superior performance. In addition, the same order of magnitude or lower $\%|J_\phi| \leq 0$ results in the table show that DGMIR maintains a good anatomical topology while maintaining excellent registration accuracy.

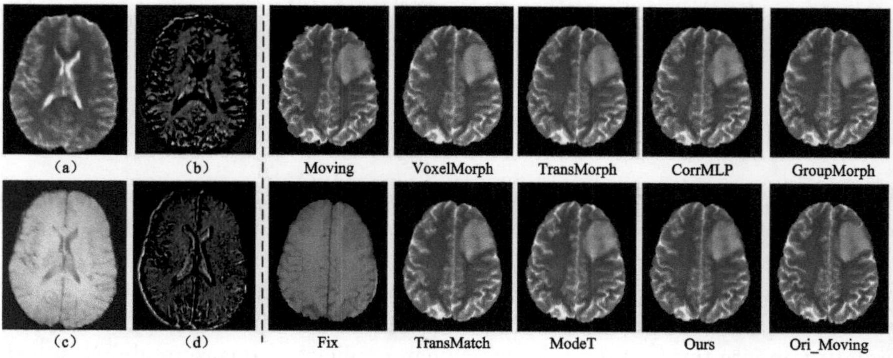

Fig. 4. Left: Visualization of the output of the MORG, where (a) and (c) denote moving and fixed feature from difference modalities (T1 and T2), and (b) and (d) denote the feature after modality removal. Right: Visualization of the registration results, where Ori_Moving denotes the original volume without random elastic transformation.

Table 2. The results of ablation experiment.

Methods	DSC(%)	HD95	$\%\lvert J_\phi \rvert \leq 0$
Baseline	69.5	2.936	-
Baseline+MFRG	78.3	2.155	2.35e-4
Baseline+MORG	78.4	1.999	1.53e-4
Baseline+MORG(*no_grad*)	77.6	2.132	1.53e-4
Baseline+MORG(*ord_conv*)	77.6	2.171	1.18e-4
Baseline+MFRG(*lncc*)	76.5	2.237	1.84e-3
Baseline+MFRG+MORG(*lncc*)	78.9	2.130	1.47e-4
Baseline+MFRG+MORG(Ours)	79.4	1.931	1.52e-4

To assess the effectiveness of the proposed modules, we conducted systematic ablation experiments on the BraTS2018 dataset, as presented in Table 2. The baseline model includes only the deformation field prediction module in the decoder. Experimental findings reveal that adding MFRG and MORG individually boosts DSC by 8.8% and 8.9%, respectively, confirming each component's independent effectiveness. Furthermore, we examine the efficacy of mean convolution in MORG. In Table 2, where *no_grad* signifies that the mean convolution weights are not learned via backpropagation, *ord_conv* denotes using the standard convolution, *lncc* denotes LNCC loss function. We also visualize the MORG output on the right panel of Fig. 4, where the features exhibit a high degree of uniformity after modality removal yet preserve the underlying structural information. By integrating the tabular data with the visualization results, one can see that mean convolution effectively isolates and extracts modality-specific features. Notably, employing the image intensity-based LNCC loss function yields results comparable to MIND. This is chiefly because, after the modality removal step, the feature space predominantly contains structural features resembling a unimodal distribution, allowing unimodal loss to perform comparably well.

4 Conclusion

We introduce a flexible multimodal medical image registration network that achieves better registration results by actively guiding the prediction of deformation fields at the decoding stage. A multi-view feature reorganization guidance module integrates the global representation of feature distributions under different views and selectively enhances their representation. In addition, we employ a modality on-site removal module, which can explicitly capture modality-specific features and progressively eliminate them during decoding. The experimental results confirm the superior performance of the proposed method.

Acknowledgements. This work was supported by the National Natural Science Foundation of China (Nos. 62331008, 62276040), the Natural Science Foundation of Chongqing (Nos. CSTB2023NSCQ-LZX0047, CSTB2024TIAD-KPX0040), National Key Research and Development Program (2024YFB4710100), the Technology Innovation and Application Development Key Projects of Chongqing (CSTB2025TIAD-KPX0009).

Disclosure of Interests. The authors have no competing interests to declare that are relevant to the content of this article.

References

1. Kong, L., Lian, C., Huang, D., Hu, Y., Zhou, Q.: Breaking the dilemma of medical image-to-image translation. Adv. Neural. Inf. Process. Syst. **34**, 1964–1978 (2021)
2. Chen, J., Frey, E.C., He, Y., Segars, W.P., Li, Y., Du, Y.: Transmorph: transformer for unsupervised medical image registration. Med. Image Anal. **82**, 102615 (2022)
3. Balakrishnan, G., Zhao, A., Sabuncu, M.R., Guttag, J., Dalca, A.V.: VoxelMorph: a learning framework for deformable medical image registration. IEEE Trans. Med. Imaging **38**(8), 1788–1800 (2019)
4. Shu, Y., Wang, H., Xiao, B., Bi, X., Li, W.: Medical image registration based on uncoupled learning and accumulative enhancement. In: International Conference on Medical Image Computing and Computer-Assisted Intervention, pp. 3–13. Springer, Cham (2021)
5. Chen, Z. K., Wei, J., Li, R.: Unsupervised Multi-Modal Medical Image Registration via Discriminator-Free Image-to-Image Translation. In: IJCAI. IJCAI Proceedings (2022)
6. Liu, Y., Wang, W., Li, Y., Lai, H., Huang, S., Yang, X.: Geometry-consistent adversarial registration model for unsupervised multi-modal medical image registration. IEEE J. Biomed. Health Inform. **27**(7), 3455–3466 (2023)

7. Wei, D., et al.: SLIR: synthesis, localization, inpainting, and registration for image-guided thermal ablation of liver tumors. Med. Image Anal. **65**, 101763 (2020)
8. Huang, Z., Chen, B.: Unsupervised Multi-Modal Medical Image Registration via query-selected attention and decoupled Contrastive Learning. In 2024 IEEE International Conference on Multimedia and Expo (ICME), pp. 1–6. IEEE (2024)
9. Song, X., et al.: Cross-modal attention for multi-modal image registration. Med. Image Anal. **82**, 102612 (2022)
10. Zhang, J., Qing, C., Li, Y., Wang, Y.: BCSwinReg: a cross-modal attention network for CBCT-to-CT multimodal image registration. Comput. Biol. Med. **171**, 107990 (2024)
11. Qin, C., Shi, B., Liao, R., Mansi, T., Rueckert, D., Kamen, A.: Unsupervised deformable registration for multi-modal images via disentangled representations. In: International Conference on Information Processing in Medical Imaging, pp. 249-261. Springer, Cham (2019)
12. Wang, Z., Wang, H., Ni, D., Xu, M., Wang, Y.: Encoding matching criteria for cross-domain deformable image registration. Med. Phys. (2024)
13. Mok, T.C., et al.: Modality-Agnostic Structural Image Representation Learning for Deformable Multi-Modality Medical Image Registration. In: Proceedings of the IEEE/CVF Conference on Computer Vision and Pattern Recognition, pp. 11215–11225 (2024)
14. Jaderberg, M., Simonyan, K., Zisserman, A., et al.: Spatial transformer networks. Advances in Neural Information Processing Systems, pp. 2017–2025 (2015)
15. Wang, H., Ni, D., Wang, Y.: ModeT: Learning deformable image registration via motion decomposition transformer. In International Conference on Medical Image Computing and Computer-Assisted Intervention, pp. 740–749. Springer, Cham (2023)
16. Tan, Z., Zhang, L., Lv, Y., Ma, Y., Lu, H.: GroupMorph: medical image registration via grouping network with contextual fusion. IEEE Trans. Med. Imaging (2024)
17. Chen, Z., Zheng, Y., Gee, J.C.: Transmatch: a transformer-based multilevel dual-stream feature matching network for unsupervised deformable image registration. IEEE Trans. Med. Imaging **43**(1), 15–27 (2023)
18. Meng, M., Feng, D., Bi, L., Kim, J.: Correlation-aware coarse-to-fine MLPs for deformable medical image registration. In Proceedings of the IEEE/CVF Conference on Computer Vision and Pattern Recognition, pp. 9645–9654 (2024)
19. Heinrich, M.P., Jenkinson, M., Bhushan, M., Matin, T., Gleeson, F.V., Brady, M., Schnabel, J.A.: MIND: modality independent neighbourhood descriptor for multi-modal deformable registration. Med. Image Anal. **16**(7), 1423–1435 (2012)
20. Gollub, R.L., et al.: The MCIC collection: a shared repository of multi-modal, multi-site brain image data from a clinical investigation of schizophrenia. Neuroinformatics **11**, 367–388 (2013)
21. Hering, A., et al.: Learn2Reg: comprehensive multi-task medical image registration challenge, dataset and evaluation in the era of deep learning. IEEE Trans. Med. Imaging, **42**(3), 697-712. (2022)
22. Hu, S., Liao, Z., Xia, Y. Devil is in channels: contrastive single domain generalization for medical image segmentation. In International Conference on Medical Image Computing and Computer-Assisted Intervention, pp. 14–23, October 2023
23. Gao, Z., Xie, J., Wang, Q., Li, P. Global second-order pooling convolutional networks. In Proceedings of the IEEE/CVF Conference on Computer Vision and Pattern Recognition, pp. 3024–3033 (2019)

24. Sun, H., et al.: Unsupervised bidirectional contrastive reconstruction and adaptive fine-grained channel attention networks for image dehazing. Neural Netw. **176**, 106314 (2024)

Diffusion-Based Virtual Staining from Polarimetric Mueller Matrix Imaging

Xiaoyu Zheng[1], Jing Wen[2], Jiaxin Zhuang[1], Yao Du[1], Jing Cong[3], Limei Guo[4], Lin Luo[2(✉)], Chao He[5(✉)], and Hao Chen[1]

[1] Department of Computer Science and Engineering, Hong Kong University of Science and Technology, Hong Kong, China
[2] College of Engineering, Peking University, Beijing, China
luol@pku.edu.cn
[3] Beijing Institute of Collaborative Innovation, Beijing, China
[4] Department of Pathology, School of Basic Medical Science, Peking University Health Science Center, Peking University Third Hospital, Beijing, China
[5] Department of Engineering Science, University of Oxford, Oxford X1 3PJ, UK
chao.he@eng.ox.ac.uk

Abstract. Polarization, as a new optical imaging tool, has been explored to assist in the diagnosis of pathology. Moreover, converting the polarimetric Mueller Matrix (MM) to standardized stained images becomes a promising approach to help pathologists interpret the results. However, existing methods for polarization-based virtual staining are still in the early stage, and the diffusion-based model, which has shown great potential in enhancing the fidelity of the generated images, has not been studied yet. In this paper, a Regulated Bridge Diffusion Model (RBDM) for polarization-based virtual staining is proposed. RBDM utilizes the bidirectional bridge diffusion process to learn the mapping from polarization images to other modalities such as H&E and fluorescence. And to demonstrate the effectiveness of our model, we conduct the experiment on our manually collected dataset, which consists of 18,000 paired polarization, fluorescence and H&E images, due to the unavailability of the public dataset. The experiment results show that our model greatly outperforms other benchmark methods. Our data and code are available at https://github.com/xiaoyu-z/RBDM/

Keywords: Virtual Staining · Polarization · Image-to-image Translation

1 Introduction

Polarization imaging technology enables non-invasive acquisition of birefringence characteristics and surface microstructure information at sub-resolution scales of tissue structures by analyzing the modulation properties of biological tissues on the polarization state of incident light waves (such as Stokes vectors, Mueller matrices, and other parameters) [1]. This unique imaging mechanism

J. C. Gee et al. (Eds.): MICCAI 2025, LNCS 15960, pp. 163–173, 2026.
https://doi.org/10.1007/978-3-032-04927-8_16

imparts high sensitivity in detecting pathological features such as collagen fiber alignment and cell membrane integrity, demonstrating significant potential for clinical diagnosis applications in recent years, including tumor boundary delineation and tissue fibrosis evaluation [2,3]. However, due to the human visual system's lack of direct perception of polarization states, the rich physical information contained in polarization images needs to be converted into grayscale or pseudo-color images through complex post-processing algorithms for visualization. This secondary encoding process results in significant differences between the imaging results and the cognitive paradigms of pathologists, severely limiting the interpretability and acceptability of polarization imaging technology in clinical diagnosis [4].

In traditional pathological practice, fluorescence imaging and Hematoxylin and Eosin (H&E) staining techniques have become the gold standard due to their clear biological relevance. Fluorescence labeling enables functional imaging at the molecular level through specific antibody-antigen reactions [5], while H&E staining provides intuitive histomorphological information through differential staining of cell nuclei and cytoplasm [6]. The emergence of virtual staining technology has offered novel solutions to improve the efficiency of computational pathology. This technology establishes a non-linear mapping relationship between unstained tissue images and target staining modalities through deep learning models, achieving significant advances in the field of computational pathology in recent years [7]. Mainstream methods primarily rely on Generative Adversarial Networks (GANs) and their variants, realizing cross-modal image translation via Cycle-Consistency constraints [8–10]. Recent research has further explored the application of Diffusion Models in virtual staining, leveraging their progressive denoising mechanism to enhance the morphological fidelity of staining boundaries [11]. However, diffusion methods for polarization data has not been investigated in previous works.

In this study, we integrate the Mueller matrix full-polarization imaging system with a diffusion-based virtual staining framework to establish an image-to-image translation model from raw polarization features to standard Hematoxylin and Eosin (H&E) as well as fluorescence images, namely the Regulated Bridge Diffusion Model (RBDM). By bridging polarization data to diagnostically familiar stains, our method unlocks a staining-free, low-cost, high-fidelity paradigm for precision pathology. The contributions are summarized in the following aspects:

(1) First Public Multi-modality Polarization Pathology Dataset: We construct and release a benchmark dataset that contains 18,000 strictly registered image patches with polarization, H&E, and fluorescence modalities.

(2) First Diffusion Model from Polarization to Other Modalities: We design a diffusion model to achieve stable mapping from 16-channel Mueller matrix imaging to the stained image domain via a regulated diffusion process.

(3) State-of-the-art Performance Validation: Our model significantly outperforms other benchmark methods on our dataset in both polarization-to-fluorescence and polarization-to-H&E tasks.

2 Related Work

2.1 Image-to-Image Translation

Image-to-image Translation (I2I) is an important research direction in the field of computer vision, which aims to transform images between different domains while maintaining the semantic consistency of the content. Recent advances in deep learning, particularly GANs and diffusion models, have significantly propelled progress in this field. Isola et al. proposed Pix2Pix, a Conditional-GAN (cGAN) framework for paired image translation, achieving success in tasks such as semantic segmentation and colorization [12]. Wang et al. extended this with Pix2PixHD, improving resolution and detail through a multi-scale architecture [13]. To realize structural fidelity, PyramidPix2pix [14] introduced a pyramid network architecture that progressively refines translations on multiple scales, effectively preserving hierarchical features in high-resolution outputs.

Despite their success, GAN-based methods often suffer from training instability and mode collapse, limiting their ability to generate high-quality complex images. Recent advances in diffusion models have addressed these limitations. Li et al. introduced BBDM [15], modeling I2I as a stochastic Brownian bridge process to directly learn domain transformations without iterative refinement. Xia et al. developed DiffI2I [16], which introduces a compact I2I prior extraction network and dynamic I2I transformer to guide denoising, enabling fewer iterations and lighter models for high-quality translation.

2.2 Polarization Imaging in Virtual Staining

Polarization imaging has emerged as a powerful tool in biomedical imaging due to its sensitivity to tissue micro-structural and anisotropic properties [17]. Recent advancements in polarization imaging have shown its potential in virtual staining [18]. Si et al. [19] proposed a deep learning-based approach using a cGAN structure to translate MM images into bright-field microscopy equivalents, addressing low throughput issues while preserving structural insights. Fan et al. [20] further simplified data acquisition by introducing an unsupervised CycleGAN framework trained on unpaired MM and H&E-stained images, eliminating the need for spatial registration.

These studies highlight the potential of polarization imaging in virtual staining, providing new auxiliary indicators for pathological diagnosis, and improving the value of multi-modality data through deep learning. But in those studies, there is a lack of model comparisons and Principal Component Analysis (PCA) is relied on for dimension reduction, which discards critical MM information.

3 Multi-modality Polarization Pathology Dataset

We present the Multi-modality Polarization Pathology Dataset (MPPD), the first breast cancer pathology dataset integrating Mueller matrix polarimetric imaging, H&E staining, and fluorescence staining. The MPPD comprises 7 breast cancer patient samples, consisting of 18,000 pairs of 600×600 patches with precisely spatially aligned samples. This multi-modality framework enables the comprehensive characterization of pathological samples in structural, molecular, and physical domains.

3.1 Data Collection

To achieve high-efficiency Mueller matrix imaging, we developed an adaptive full-vector imaging system that integrates hardware control, image acquisition, and data analysis modules. The acquisition process begins with the use of specialized slide scanners to obtain Whole-Slide Images (WSIs) of H&E staining and fluorescence staining. Subsequently, the samples are placed on a stage for polarimetric imaging. After autofocus determines the optimal focal plane, the system scans the sample region-by-region with a preset step size, recording the Mueller matrix data for each region. Given that samples are large and a single shot cannot cover the entire slice, an ITKMontage [21] module is employed for multi-modality image registration and stitching. To address issues such as size consistency and missing data in multi-modality image stitching, the system automatically generates configuration files to record tile position information. It also utilizes boundary detection and virtual tile generation techniques. Additionally, grayscale uniformity correction and brightness normalization are applied. Ultimately, this process results in the formation of complete panoramic WSI images under a 4× objective lens.

3.2 Data Processing

We designed a multi-stage registration pipeline to achieve precise spatial alignment for pixel-level correlation across modalities. Initial pre-processing standardizes the image scales and removes extraneous blank regions to mitigate feature interference. The polarization-derived gray modality is designated as the reference framework due to its consistent correspondence with other modalities. Feature matching is performed using the SuperPoint algorithm [22], a deep learning-based detector that identifies cross-modal keypoints for robust correspondence

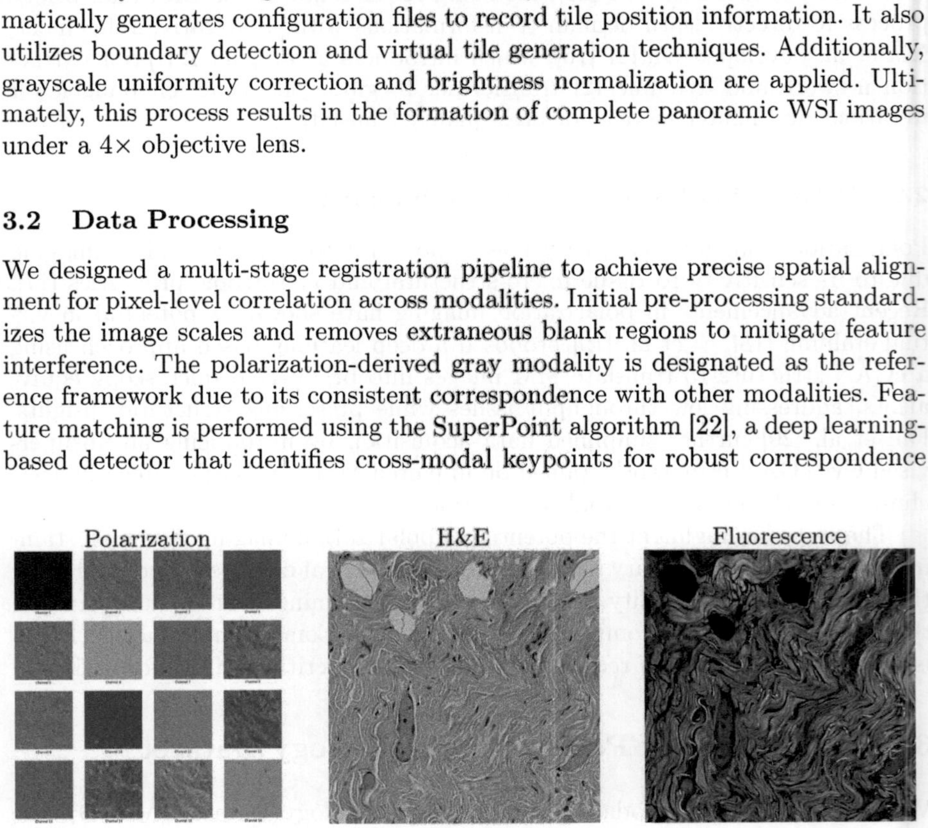

Fig. 1. Polarization-H&E-Fluorescence Dataset Visualization

estimation. Rigid registration optimizes global alignment via affine transformations. Subsequent non-rigid registration refines local deformations using B-spline modeling [23] combined with curvature regularization [24] to prevent artifacts. After registration, all modalities are resampled to 2304×1296 pixels and aligned with the polarization field-of-view (FOV). Then we applied adaptive thresholding to exclude tissue-free or artifact-dominated regions. The validated FOVs are partitioned into 600×600 pixel patches using a sliding window approach, preserving multi-modality alignment through strict spatial correspondence. This process produced 18,000 high-quality patches for downstream analysis (Fig. 1).

4 Methods

We extend the Brownian Bridge Diffusion Models (BBDM) [15] to the polarization domain with additional regulations. Figure 2 shows the general structure of our Regulated Bridge Diffusion Model (RBDM).

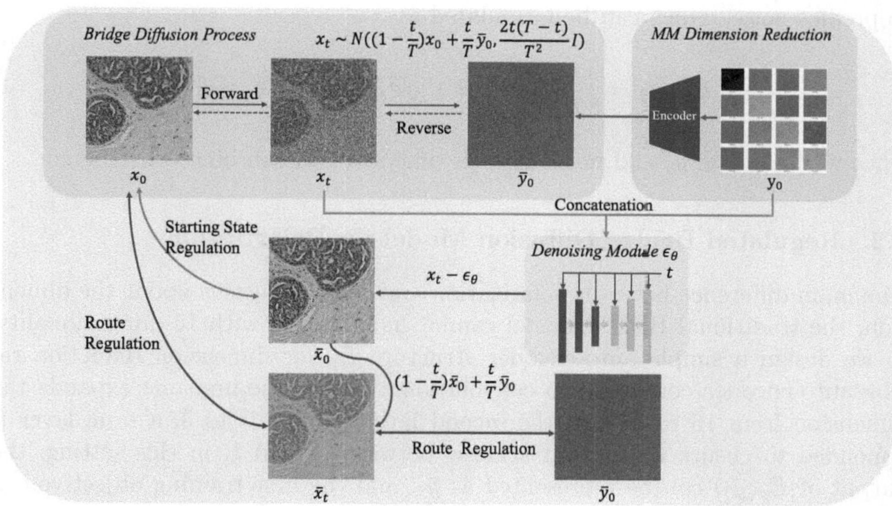

Fig. 2. Overall Framework of Regulated Bridge Diffusion Model

4.1 Bridge Diffusion Model

Given an image x_0 from domain A and image y_0 from domain B, the Bridge Diffusion Model aims to perform successful image-to-image translation from y_0 to x_0. During training, translation will go through the forward and reverse processes. In the forward process, the state distribution at each time step t can be formulated as

$$q(x_t|x_0, y_0) = N(x_t; (1-\frac{t}{T})x_0 + \frac{t}{T}y_0, \sigma_t I) \tag{1}$$

Here T is the total number of time steps, σ_t stands for the variance, which is $\frac{2t(T-t)}{T^2}$, and it is easy to observe that at staring state and ending state, the σ_t is 0 and the mean value equals to x_0 and y_0 respectively, which makes it a fixed mapping process. And the intermediate state x_t can therefore be represented as

$$x_t = (1 - \frac{t}{T})x_0 + \frac{t}{T}y_0 + \sqrt{\sigma_t}\epsilon \tag{2}$$

where $\epsilon \sim N(0, I)$.

The reverse process starts with y_0 and aims to predict x_{t-1} given x_t. It can be formulated as

$$p_\theta(x_{t-1}|x_t, y_0) = N(x_{t-1}; \mu_\theta(x_t, t), \tilde{\sigma}_t I) \tag{3}$$

where $\tilde{\sigma}_t$ is the variance of the noise and $\mu_\theta(x_t, t)$ is the predicted mean value of the noise.

The training objective of the bridge diffusion model is to optimize the Evidence Lower Bound (ELBO) and is proven equivalent to training a network ϵ_θ to predict noise, which can be formulated as

$$L_{BDM} = \left\| \frac{t}{T}(y_0 - x_0) + \sqrt{\sigma_t}\epsilon - \epsilon_\theta(x_t, t, y_0) \right\| \tag{4}$$

Once we know the y_0 and noise at each time step, x_0 can be reconstructed.

4.2 Regulated Bridge Diffusion Model for Polarization

One main difference between polarization and other images is about the dimension, the traditional BBDM model cannot handle data with 16-dimensionality, so we design a simple auto-encoder structure E_p for dimension reduction ra. The auto-encoder contains two convolutional layers, the first one expands the dimension from 16 to 64 and the second layer reduces it to 3, a tanh layer is appended to ensure the output scale is between -1 and 1, in this setting, the output of $E_p(y_0)$ can be represented as $\bar{y}_0$, and the first training objective can be written as

$$L_1 = \left\| \frac{t}{T}(\bar{y}_0 - x_0) + \sqrt{\sigma_t}\epsilon - \epsilon_\theta(x_t, t, y_0) \right\| \tag{5}$$

However, since the auto-encoder is trainable, intermediate states generated by the bridge diffusion process may lack consistency during training, leading to suboptimal results. Therefore, we design two other regulation modules to enforce stability across diffusion timesteps to enhance performance, denoted as Starting State Regulation (SSR) and Route Regulation (RR).

Starting State Regulation Given any time step t, x_0 can actually be estimated by the following formula during training

$$\begin{aligned}\bar{x}_0 &= x_t - \epsilon_\theta(x_t, t, y_0) \\ &= (1 - \frac{t}{T})x_0 + \frac{t}{T}\bar{y}_0 + \sqrt{\sigma_t}\epsilon - \epsilon_\theta(x_t, t, y_0)\end{aligned} \tag{6}$$

To ensure the estimated quality, we apply the perceptual loss [25] to narrow the difference between feature maps of x_0 and $\bar{x}_0$ using the pre-trained VGG [26] network. The starting state regulation loss can be expressed as

$$L_2 = \|F(x_0) - F(\bar{x}_0)\| \tag{7}$$

where F stands for feature extraction network.

Route Regulation The route regulation is according to the attribute of the Bridge Diffusion Model. When the time step t is close to T, the intermediate image $\bar{x}_t$ generated should be more similar to y_0 than x_0, and when the time step t is close to 0, the image $\bar{x}_t$ should be more like x_0 without considering noise. MS-SSIM loss [27] is applied to measure the multi-scale structural similarity between the ground truth and the generated image, and the higher the value of the MS-SSIM is, the more similar the two images are. So the route regulation loss is then represented as

$$L_3 = max(0, \frac{T - 2t}{T}[M(\bar{x}_t, \bar{y}_0) - M(\bar{x}_t, x_0)]) \tag{8}$$

where M means MS-SSIM value, and in order to train the whole network, we use $\bar{x}_0$ to generate new $\bar{x}_t$. The equation is

$$\bar{x}_t = (1 - \frac{t}{T})\bar{x}_0 + \frac{t}{T}\bar{y}_0 \tag{9}$$

Thus the overall training objective for our model becomes

$$L = L_1 + L_2 + L_3 \tag{10}$$

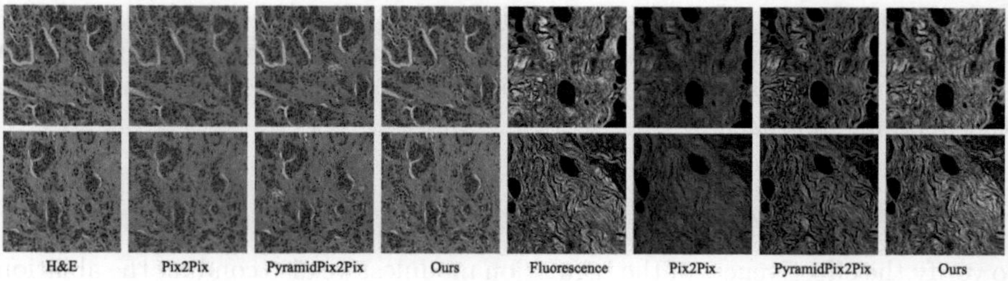

Fig. 3. Visualization Results of Our Model and Other Baselines

Table 1. Model Performance from Polarization to H&E and Fluorescence

Task	Model	PSNR↑	SSIM↑	FID↓	LPIPS↓
	Pix2Pix	15.42±3.50	0.19±0.10	215.93	0.60
Polar->H&E	Pyramid Pix2pix	19.68±2.68	0.41±0.13	62.02	0.38
	Ours	**20.93±3.25**	**0.53±0.18**	**19.67**	**0.26**
	Pix2Pix	16.13±1.28	0.26±0.05	116.73	0.62
Polar-> Fluo	Pyramid Pix2pix	18.09±2.69	0.32±0.13	65.79	0.52
	Ours	**18.52±2.32**	**0.44±0.12**	**43.03**	**0.39**

5 Experiments

5.1 Implementation Detail

We use our original dataset for performance evaluation. 70% patches are selected for training, and the rest are reserved for testing. All patches are resized to 256 × 256 and normalized to $[-1, 1]$. For training, the total number of epochs is set to 40, and the Adam optimizer with a learning rate of $1e^{-4}$ is used. The batch size is 2 and the total number of time steps is set to 500 with a sample step of 100.

5.2 Result

To make a fair comparison, we select multiple evaluation metrics and benchmark methods. The widely used metrics including Structural Similarity Index Measure (SSIM) [28], Peak Signal-to-Noise Ratio (PSNR) [29], Fréchet Inception Distance (FID) [30] and Learned Perceptual Image Patch Similarity (LPIPS) [31] are calculated to measure the image quality and similarity. The baseline methods are GAN-based model for paired data, including Pix2Pix [12] and Pyramid-Pix2pix [14]. The results of polarization-to-H&E and polarization-to-fluorescence are shown in Table 1. The task of transforming polarization to fluorescence is more difficult as the fluorescence is more complex. The best results are highlighted in bold. And it is obvious that our model significantly outperforms other methods in terms of all the evaluation indicators. We also picked four patches for the visualization of the image generation outcomes, the results are shown in Fig. 3, the results from our model are more accurate and detailed compared to other baseline approaches.

5.3 Ablation Study

To verify the effectiveness of the regulation modules, we also conduct the ablation study by adding the Starting State Regulation (SSR) and the Route Regulation (RR). The result is shown in Table 2, which indicates that the diffusion-based model achieves better performance compared to previous baselines, and both SSR and RR help improve the performance of the model.

Table 2. Ablation Result of Starting State Regulation and Route Regulation

Module		Polar->H&E				Polar-> Fluo			
SSR	RR	PSNR↑	SSIM ↑	FID↓	LPIPS↓	PSNR↑	SSIM↑	FID↓	LPIPS↓
×	×	20.66±3.37	0.51±0.19	21.21	0.27	18.13±2.44	0.42±0.13	55.89	0.42
✓	×	20.71±3.31	0.52±0.20	20.80	0.26	18.34±2.41	0.43±0.13	49.12	0.40
✓	✓	20.93±3.25	0.53±0.18	19.67	0.26	18.52±2.32	0.44±0.12	43.03	0.39

6 Conclusion

In this paper, we extend the bridge diffusion model to the polarization domain and propose first public multi-modality polarization dataset named MPPD and a novel diffusion-based model named RBDM, which realizes the mapping from polarimetric Mueller matrix to standardized stained domain. We also conduct experiments on our MPPD datasets and the RBDM model achieves the state-of-the-art performance, which indicates a promising direction for polarization-based virtual staining in the future.

Acknowledgments. We would like to thank the support of St John's College, the University of Oxford, and the Royal Society University Research Fellowship (URF\R1\241734) (C.H.). And this work was supported by the National Natural Science Foundation of China (No. 62202403).

Disclosure of Interests. The authors have no competing interests to declare that are relevant to the content of this article.

References

1. He, C., He, H., Chang, J., Chen, B., Ma, H., Booth, M.J.: Polarisation optics for biomedical and clinical applications: a review. Light: Sci. Appl. **10**(1), 194 (2021)
2. Layden, D., Ghosh, N., Vitkin, A.: Quantitative polarimetry for tissue characterization and diagnosis. Advanced Biophotonics: Tissue Optical Sectioning **5**, 73–108 (2013)
3. Yao, Y., Wan, J., Zhang, F., Dong, Y., Chen, L., Ma, H.: Correlation of image textures of a polarization feature parameter and the microstructures of liver fibrosis tissues. J. Innov. Opt. Health Sci. **16**(05), 2241004 (2023)
4. Gao, W.: Parameterizing depolarizing mueller matrices of anisotropic media–a review. Waves in Random and Complex Media, pp. 1–37 (2023)
5. Bancroft, J.D., Gamble, M.: Theory and Practice of Histological Techniques, 7th ed. Elsevier (2019)
6. Deng, L., Chen, C., Yu, W., Shao, C., Shen, Z., Wang, Y., He, C., Li, H., Liu, Z., He, H., et al.: Influence of hematoxylin and eosin staining on linear birefringence measurement of fibrous tissue structures in polarization microscopy. J. Biomed. Opt. **28**(10), 102909–102909 (2023)
7. Bai, B., Yang, X., Li, Y., Zhang, Y., Pillar, N., Ozcan, A.: Deep learning-enabled virtual histological staining of biological samples. Light: Sci. Appl. **12**(1), 57 (2023)

8. Zhu, J.Y., Park, T., Isola, P., Efros, A.A.: Unpaired image-to-image translation using cycle-consistent adversarial networks. In: Proceedings of the IEEE International Conference on Computer Vision. pp. 2223–2232 (2017)
9. Ma, J., Chen, H.: Efficient supervised pretraining of swin-transformer for virtual staining of microscopy images. IEEE Trans. Med. Imaging **43**(4), 1388–1399 (2023)
10. Zhang, W., Yang, D., Che, H., Ran, A.R., Cheung, C.Y., Chen, H.: Unpaired optical coherence tomography angiography image super-resolution via frequency-aware inverse-consistency gan. IEEE J. Biomed. Health Inf. (2024)
11. Shen, Y., Ke, J.: Staindiff: Transfer stain styles of histology images with denoising diffusion probabilistic models and self-ensemble. In: International Conference on Medical Image Computing and Computer-Assisted Intervention, pp. 549–559. Springer (2023)
12. Isola, P., Zhu, J.Y., Zhou, T., Efros, A.A.: Image-to-image translation with conditional adversarial networks. In: Proceedings of the IEEE Conference on Computer Vision and Pattern Recognition, pp. 1125–1134 (2017)
13. Wang, T.C., Liu, M.Y., Zhu, J.Y., Tao, A., Kautz, J., Catanzaro, B.: Highresolution image synthesis and semantic manipulation with conditional gans. In: Proceedings of the IEEE conference on Computer Vision and Pattern Recognition, pp. 8798–8807 (2018)
14. Liu, S., Zhu, C., Xu, F., Jia, X., Shi, Z., Jin, M.: Bci: Breast cancer immunohistochemical image generation through pyramid pix2pix. In: Proceedings of the IEEE/CVF conference on Computer Vision and Pattern Recognition, pp. 1815–1824 (2022)
15. Li, B., Xue, K., Liu, B., Lai, Y.K.: Bbdm: Image-to-image translation with brownian bridge diffusion models. In: Proceedings of the IEEE/CVF Conference on Computer Vision and Pattern Recognition, pp. 1952–1961 (2023)
16. Xia, B., et al.: Diffi2i: efficient diffusion model for image-to-image translation. IEEE Trans. Pattern Anal. Mach. Intell. (2024)
17. Guan, C., Zeng, N., He, H.: Review of polarization-based technology for biomedical applications. J. Innov. Optic. Health Sci. **18**(02), 2430002 (2025)
18. Wang, W., Lim, L.G., Srivastava, S., Bok-Yan So, J., Shabbir, A., Liu, Q.: Investigation on the potential of mueller matrix imaging for digital staining. J. Biophotonics **9**(4), 364–375 (2016)
19. Si, L., Li, N., Huang, T., Du, S., Dong, Y., Yao, Y., Ma, H.: Computational image translation from mueller matrix polarimetry to bright-field microscopy. J. Biophotonics **15**(3), e202100242 (2022)
20. Fan, J., Zeng, N., He, H., He, C., Liu, S., Ma, H.: Generating bright-field images of stained tissue slices from mueller matrix polarimetric images with cyclegan using unpaired dataset. J. Innov. Optic. Health Sci. **18**(02), 2343003 (2025)
21. Zukić, D., Jackson, M., Dimiduk, D., Donegan, S., Groeber, M., McCormick, M.: Itkmontage: a software module for image stitching. Integrating Mater. Manuf. Innov. **10**, 115–124 (2021)
22. DeTone, D., Malisiewicz, T., Rabinovich, A.: Superpoint: Self-supervised interest point detection and description. In: Proceedings of the IEEE Conference on Computer Vision and Pattern Recognition Workshops, pp. 224–236 (2018)
23. Unser, M., Aldroubi, A., Eden, M.: B-spline signal processing. i. theory. IEEE Trans. Signal Process. **41**(2), 821–833 (1993)
24. Myronenko, A., Song, X., Carreira-Perpinan, M.: Non-rigid point set registration: coherent point drift. Advances in Neural Information Processing Systems **19** (2006)

25. Johnson, J., Alahi, A., Fei-Fei, L.: Perceptual Losses for Real-Time Style Transfer and Super-Resolution. In: Leibe, B., Matas, J., Sebe, N., Welling, M. (eds.) ECCV 2016. LNCS, vol. 9906, pp. 694–711. Springer, Cham (2016). https://doi.org/10.1007/978-3-319-46475-6_43
26. Simonyan, K., Zisserman, A.: Very deep convolutional networks for large-scale image recognition. arXiv preprint arXiv:1409.1556 (2014)
27. Wang, Z., Simoncelli, E.P., Bovik, A.C.: Multiscale structural similarity for image quality assessment. In: The Thrity-Seventh Asilomar Conference on Signals, Systems & Computers, 2003, vol. 2, pp. 1398–1402. IEEE (2003)
28. Wang, Z., Bovik, A.C., Sheikh, H.R., Simoncelli, E.P.: Image quality assessment: from error visibility to structural similarity. IEEE Trans. Image Process. **13**(4), 600–612 (2004)
29. Hore, A., Ziou, D.: Image quality metrics: Psnr vs. ssim. In: 20th International Conference on Pattern Recognition, pp. 2366–2369. IEEE (2010)
30. Heusel, M., Ramsauer, H., Unterthiner, T., Nessler, B., Hochreiter, S.: Gans trained by a two time-scale update rule converge to a local nash equilibrium. Advances in Neural Information Processing Systems **30** (2017)
31. Zhang, R., Isola, P., Efros, A.A., Shechtman, E., Wang, O.: The unreasonable effectiveness of deep features as a perceptual metric. In: Proceedings of the IEEE Conference on Computer Vision and Pattern Recognition, pp. 586–595 (2018)

Dual Correlation-Aware Mamba for Microvascular Obstruction Identification in Non-contrast Cine Cardiac Magnetic Resonance

Yige Yan[1], Jun Cheng[2(✉)], Xulei Yang[2], Shuang Leng[3,4], Ru San Tan[3,4], Liang Zhong[3,4(✉)], and Jagath C. Rajapakse[1]

[1] College of Computing and Data Science, Nanyang Technological University, Singapore, Singapore
asjagath@ntu.edu.sg

[2] Institute for Infocomm Research, Agency for Science, Technology and Research (A*STAR), 1 Fusionopolis Way, Connexis, 138632 Singapore, Singapore
cheng_jun@i2r.a-star.edu.sg

[3] National Heart Centre Singapore, 5 Hospital Drive, 169609 Singapore, Singapore

[4] Duke-NUS Medical School, National University of Singapore, Singapore, Singapore
gmszl@nus.edu.sg

Abstract. Microvascular obstruction (MVO) is a key prognostic factor in acute myocardial infarction, with affected patients experiencing higher mortality rates. Currently, late gadolinium enhancement cardiac magnetic resonance (CMR) is the gold standard for MVO identification. However, it is unsuitable for patients with renal impairment, who make up 20% of all patients. Recent studies have demonstrated the feasibility of using non-contrast cine CMR to identify MVO. Despite this, existing methods struggle to effectively learn crucial motion features, as they implicitly model motion dynamics while overlooking regional wall motion abnormalities, which are important for MVO identification. To this end, we introduce a Dual Correlation-aware Mamba, which includes an Adjacent Frame Correlation (AFC) module and a Diastolic Frame Correlation (DFC) module to address these limitations. The AFC module calculates the correlations through adjacent frames to explicitly model the motion dynamics. The DFC module learns correlations between the diastolic frame and others. Leveraging the diastolic frame as a reference, this module highlights regional abnormalities and guides motion learning. Experimental results demonstrate that our method outperforms competing methods, potentially providing a non-contrast tool for MVO identification. The code is available at https://github.com/code-koukai/Dual-Correlation-Mamba.

Keywords: Cardiac Magnetic Resonance · Microvascular Obstruction · Contrast-free Technology · Motion Features

---

Y. Yan—The work was done during an internship at I^{2}R, A*STAR.

J. C. Gee et al. (Eds.): MICCAI 2025, LNCS 15960, pp. 174–184, 2026.
https://doi.org/10.1007/978-3-032-04927-8_17

1 Introduction

Acute myocardial infarction remains a leading cause of death worldwide [3]. Microvascular obstruction (MVO), a frequent complication, is strongly linked to increased mortality and serves as a critical prognostic indicator [10]. Clinically, identifying the presence of MVO is essential for initial risk stratification and guiding treatment strategies [21]. Late gadolinium enhancement (LGE) cardiac magnetic resonance (CMR) is the gold standard for MVO identification [1,19]. However, it is unsuitable for patients with renal impairment, who make up 20% of cases [20]. Therefore, there is an increasing need for non-contrast MVO identification. Cine CMR, a non-contrast dynamic imaging technique, has emerged as a promising alternative [5,7,16]. It captures motion throughout the cardiac cycle, enabling the detection of regional wall motion abnormalities (RWMA) and myocardial dysfunction associated with MVO [6]. These findings highlight the potential of cine CMR for non-contrast MVO identification [8,28,29].

In cine CMR, MVO is not visible on a single frame due to the absence of contrast agents, making it necessary to analyze motion across multiple frames for its identification. Some studies have leveraged motion analysis in cine CMR to assess myocardial function and pathology [11,14,17,23,24,26]. However, the inherent complexity and small size of MVO motion complicate the analysis, making identification more challenging. Yan *et al.* [27] demonstrate the feasibility of MVO identification in cine CMR by extracting spatiotemporal features. However, this method implicitly models motion dynamics, failing to sufficiently capture the motion feature. Furthermore, they overlook the RWMA induced by MVO, which is crucial for MVO identification in cine CMR.

In this paper, we propose a Dual Correlation-aware Mamba (DCM), which includes an Adjacent Frame Correlation (AFC) module and a Diastolic Frame Correlation (DFC) module to address these issues. Our contributions lie in explicitly modeling motion dynamics while leveraging the diastolic frame as a reference to guide motion learning. Specifically, to explicitly model the motion dynamics, the AFC module computes correlations between adjacent frames to estimate pixel displacements. The features are enhanced by displacement estimation and aggregated by a bidirectional temporal Mamba [9]. Inspired by the important role of RWMA in MVO diagnosis [15,25], the DFC module designates the diastolic frame as a reference, highlighting RWMA variations throughout the cardiac cycle. It learns correlations between the diastolic frame and the other frames and aggregates the motion information through a bidirectional spatial Mamba. Experiments demonstrate that our proposed method enhances motion feature extraction, leading to improved non-contrast MVO identification. The key contributions of our work are as follows:

1. We introduce Dual Correlation-aware Mamba, a novel method for MVO identification that explicitly models motion dynamics and guides motion learning to enhance motion feature extraction in non-contrast cine CMR;
2. Our proposed method includes the Adjacent Frame Correlation module to explicitly model the motion by computing correlations between adjacent

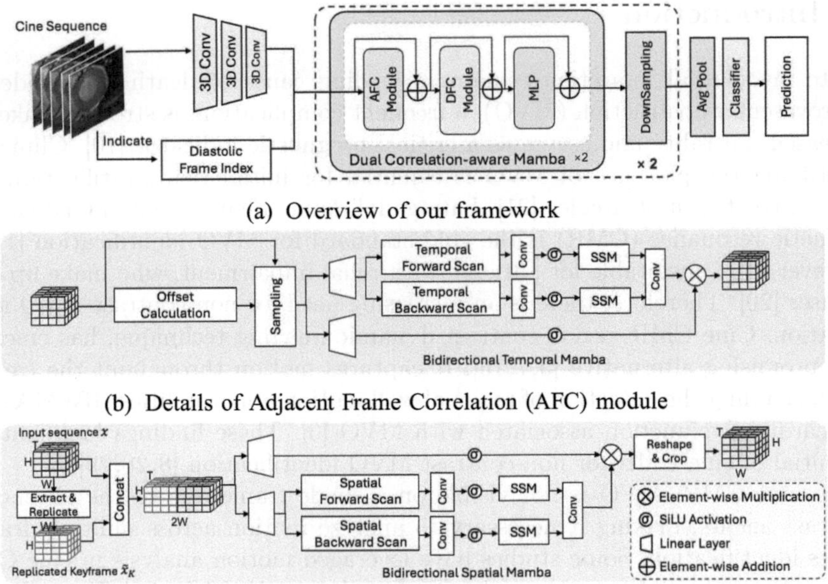

(a) Overview of our framework

(b) Details of Adjacent Frame Correlation (AFC) module

(c) Details of Diastolic Frame Correlation (DFC) module

Fig. 1. The illustration of our method includes: (a) the overview of our framework, (b) the details of the Adjacent Frame Correlation module, and (c) the details of the Diastolic Frame Correlation module.

frames, and the Diastolic Frame Correlation module to guide motion learning by learning correlations between the diastolic frame and the other frames;
3. Experiment results demonstrate that our method outperforms competing methods, effectively enhancing motion feature extraction and offering a novel non-contrast solution for MVO identification.

2 Method

We propose a novel framework for non-contrast MVO identification that computes the correlations between adjacent frames and learns the correlations between the diastolic frame and others to enhance motion feature extraction using non-contrast cine CMR. The model, illustrated in Fig. 1a, extracts features from a cine sequence using three 3D CNN layers. These features are refined using Dual Correlation-aware Mamba (DCM) blocks, each comprising an Adjacent Frame Correlation (AFC) module and a Diastolic Frame Correlation (DFC) module, to enhance motion extraction. The DFC module leverages the diastolic frame index to locate the diastolic frame. Two DCM blocks and a downsampling layer form a stage, with features processed through two sequential stages for progressive learning. Finally, the features undergo average pooling and a classifier layer to generate the prediction for the presence of MVO.

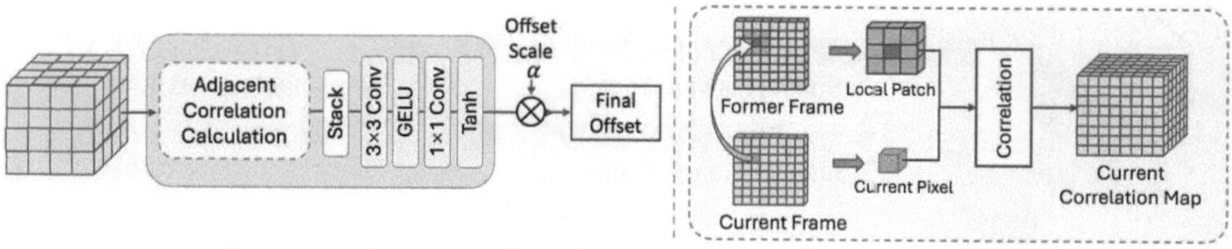

(a) Offset Calculation module (b) Adjacent Correlation Calculation

Fig. 2. Details of (a) Offset Calculation module and (b) Adjacent Correlation Calculation. Adjacent Correlation Calculation uniformly processes adjacent frame pairs along the temporal dimension and stacks the results.

2.1 Dual Correlation-Aware Mamba

MVO identification in cine CMR is challenging due to the induced complex motion and its small size. Existing methods implicitly model motion, leading to inadequate motion extraction. They also overlooked the crucial RWMA associated with MVO. We propose a Dual Correlation-aware Mamba, which includes an AFC module for explicit motion modeling and a DFC module to guide the model's motion learning on motion abnormalities to address these limitations.

The AFC module predicts displacement offsets between adjacent frames, explicitly modeling motion representation. The processed features are then passed to the DFC module for enhanced motion learning. The DFC module utilizes the diastolic frame as a global reference for the other frames, guiding the model's motion learning. Finally, the fused features are fed into an MLP, which performs nonlinear transformation to further refine the overall feature representation. A residual connection is applied after each module.

Adjacent Frame Correlation Module. Existing methods implicitly model motion in cine CMR while lacking dedicated mechanisms to focus on motion learning, resulting in insufficient motion extraction. However, motion extraction plays a crucial role in MVO identification. In light of this, we propose the AFC module, which explicitly models cardiac motion by computing correlations between adjacent frames, enhancing the model's motion extraction capabilities.

Figure 1billustrates the details of the AFC module. Given an input sequence $X \in \mathbb{R}^{B\times C\times T\times H\times W}$ (where B is the batch size, C the channels, T the time steps, and H and W the height and width), we first feed it into the Offset Calculation module, which is depicted in Fig. 2a. Specifically, we extract adjacent frame pairs X_i and $X_{i+1} \in \mathbb{R}^{B\times C\times H\times W}$ for each frame index $i \in \{1,\ldots,T-1\}$. For each pair, we compute local adjacent correlations within a symmetric window $W = \{(\Delta h, \Delta w) \mid \Delta h, \Delta w \in [-r, r]\}$, as illustrated in Fig. 2b:

$$\mathrm{Corr}_i(b,h,w,\Delta h,\Delta w) = \frac{\langle X_{i+1}(b,\cdot,h,w),\ X_i(b,\cdot,h-\Delta h,w+\Delta w)\rangle}{\sqrt{C}}, \quad (1)$$

where r is the window radius and $\langle\cdot,\cdot\rangle$ denotes the inner product. The resulting Corr_i of shape $[B,H,W,2r+1,2r+1]$ is reshaped into $\hat{\mathrm{Corr}}_i \in \mathbb{R}^{B\times(2r+1)^2\times H\times W}$

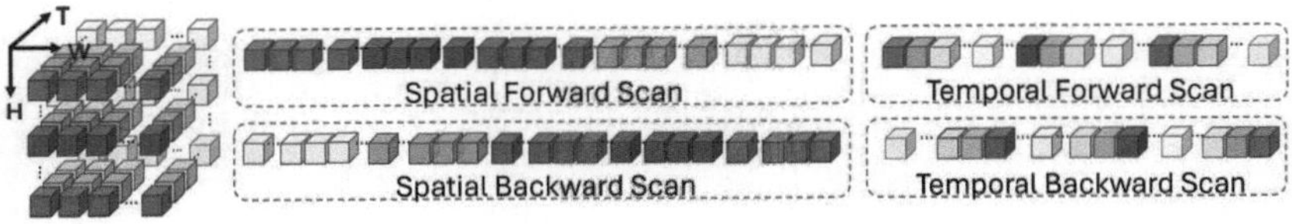

Fig. 3. Illustration of the bidirectional spatial and temporal scans. The arrangements of differently colored blocks illustrate the scanning orders.

and fed into a network $f(\cdot)$, which consists of two convolutional layers with GELU activation, to predict offsets:

$$dh, dw = \alpha \cdot \tanh\Big(f\big(\hat{\text{Corr}}_i\big)\Big), \tag{2}$$

where α is a offset scale factor and $dh, dw \in \mathbb{R}^{B\times H\times W}$ are predicted offsets. Here, we set $\alpha = 3$. Using bilinear interpolation, the features are resampled based on the dh and dw, yielding the final aligned feature representation:

$$\hat{X}_i(b,\cdot,h,w) = \mathcal{I}\big(X_i,\ h + dh(b,h,w), w + dw(b,h,w)\big), \tag{3}$$

where $\mathcal{I}(\cdot)$ denotes the interpolation function. Next, we employ bidirectional temporal Mamba for aggregation. We stack $\hat{X}_i$ temporally and perform temporal forward and backward scans, as illustrated in Fig. 3, forming two sequences. These sequences are then processed using selective state space models (SSMs) [9] and fused to obtain the final representation. This approach explicitly models myocardial motion, enhancing motion modeling effectiveness.

Diastolic Frame Correlation Module Existing methods overlook the RWMA caused by MVO, missing its value for MVO identification. We propose the DFC module, as depicted in Fig. 1c

Given an input sequence $x \in \mathbb{R}^{B\times C\times T\times H\times W}$, we extract the diastolic frame as the keyframe x_K and replicate it along the temporal dimension to obtain $\tilde{x}_K$. Then $\tilde{x}_K$ is concatenated with x along the width dimension, forming a joint representation $x_{\text{joint}} \in \mathbb{R}^{B\times C\times T\times H\times (2W)}$. Next, x_{joint} undergoes spatial forward and backward scans to generate bidirectional sequences, as shown in Fig. 3. These sequences are processed by SSMs to establish global correspondences between each frame and the keyframe. Finally, the features are reshaped to their original form, with the last W elements along the width dimension extracted as the final feature. In this way, we can build a diastole-focused correlation comparison, which highlights how cardiac motion evolves relative to the diastolic frame.

2.2 Loss Function

To train our model for MVO identification, we use the cross-entropy (CE) loss for classification with coarse-grained mask regularization [27] and mean square error (MSE) loss. The total loss is defined as:

Table 1. Performance Comparison with Competing Methods

Model	AUC	Specificity	Accuracy	Recall	F1-score	Params
VST [12]	0.5878	0.5714	0.5413	0.5000	0.4786	27.58M
VIVIT [2]	0.5571	0.5974	0.5413	0.4642	0.4602	41.98M
TimesFormer [4]	0.5797	0.5389	0.5226	0.5000	0.4686	76.54M
TAM [13]	0.6372	0.6818	0.6241	0.5446	0.5495	24.80M
P3D [18]	0.6321	0.6627	0.6385	0.5909	0.5253	43.72M
C3D [22]	0.6140	0.5779	0.5677	0.5536	0.5188	63.32M
CGMR [27]	0.7231	0.7337	0.6842	0.6161	0.6216	43.72M
Our Method	**0.7687**	**0.7532**	**0.7218**	**0.6786**	**0.6726**	39.91M

$$L_{total} = L_{\mathrm{CE}} + \lambda \cdot L_{\mathrm{MSE}}, \tag{4}$$

where L_{CE} is the CE loss, L_{MSE} is the MSE loss, and λ is a weighting factor balancing the two loss components. Based on the work [27], we set $\lambda = 0.05$.

3 Experimental Results

3.1 Dataset and Metrics

The dataset includes 816 cases from acute myocardial infarction patients, acquired with a 3T Siemens scanner from local hospitals. Each case contains a cine sequence and an LGE-derived MVO annotation. The dataset was randomly split into training (550 cases, 221 MVO-positive) and test (266 cases, 112 MVO-positive) sets. We evaluated our model's performance using Area Under Curve (AUC), Specificity, Accuracy, Recall, and F1-score.

3.2 Implementation Details

We trained the model for 10K iterations with a batch size of 8, using AdamW optimizer with an initial learning rate of 1E-5. The learning rate was scheduled using a linear warm-up for the first 1,500 iterations, followed by a polynomial decay policy with power 1.0 until iteration 10K. All models are trained on an Nvidia A100 GPU. A fixed random seed was set. Data augmentation was used, including flipping, blurring, and noise addition.

3.3 Performance Comparison with Competing Methods

Since non-contrast MVO identification is still in the early stages with limited research, we compared our model against competing methods, including CGMR [27] and other computer vision approaches, as summarized in Table 1. Results show our approach effectively identifies the presence of MVO, achieving an

Table 2. Ablation Study of the Proposed Modules

AFC	DFC	AUC	Specificity	Accuracy	Recall	F1-score	Params
✓	✓	**0.7687**	**0.7532**	**0.7218**	**0.6786**	**0.6726**	39.91M
	✓	0.7439	0.7273	0.7030	0.6696	0.6550	38.54M
✓		0.7378	0.7273	0.7068	0.6786	0.6609	38.95M
		0.7312	0.7142	0.6992	0.6786	0.6556	37.58M

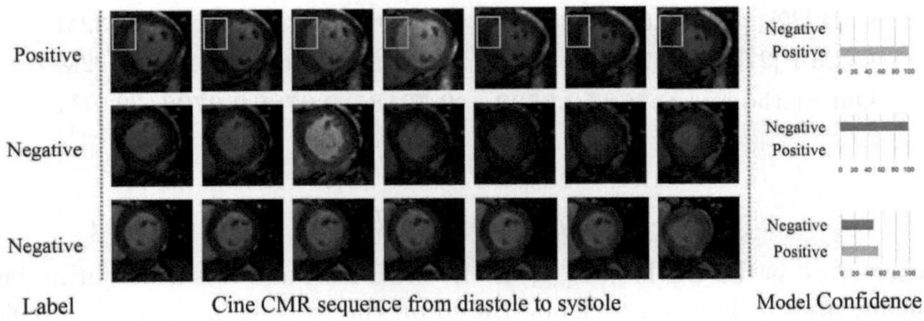

Fig. 4. Examples for good and bad cases. The first two examples show successful cases, while the last displays a failed case. The cine sequences highlight diastole-to-systole dynamics, with the leftmost frames in diastole and the rightmost frames in systole. Yellow boxes indicate the locations of MVOs. (Color figure online)

AUC of 0.7687 and surpassing competing methods. Figure 4 illustrates that our model can accurately identify MVO in most cases. However, individual variability causes some non-MVO cases to exhibit motion patterns similar to those induced by MVO, making identification challenging.

3.4 Ablation Studies

Ablation Study of the Proposed Modules. We evaluated the impact of our proposed modules on overall performance using four configurations: (1) the full framework, (2) without the AFC module, (3) without the DFC module, and (4) without both. In these experiments, the AFC module was replaced with a bidirectional temporal Mamba, while the DFC module was substituted with a bidirectional spatial Mamba. As shown in Table 2, both modules contribute to the performance. Replacing Mamba with the self-attention mechanism resulted in an AUC of 0.7150, demonstrating the effectiveness of our design.

Ablation Study of the Adjacent Frame Correlation Module. Table 3 presents the impact of AFC settings on performance. The results indicate that both an optimal radius and learnable offsets improve performance. Additionally, we visualized the AFC module's motion modeling, as shown in Fig. 5.

Table 3. Ablation Study of the Adjacent Frame Correlation Module

Settings	AUC	Specificity	Accuracy	Recall	F1-score
Radius=[1,1]	0.7451	0.7208	0.6955	0.6607	0.6463
Radius=[2,1]	**0.7687**	**0.7532**	**0.7218**	**0.6786**	**0.6726**
Radius=[3,1]	0.7583	0.7467	0.7180	0.6786	0.6691
Fixed Offset	0.7227	0.6753	0.6729	0.6696	0.6329

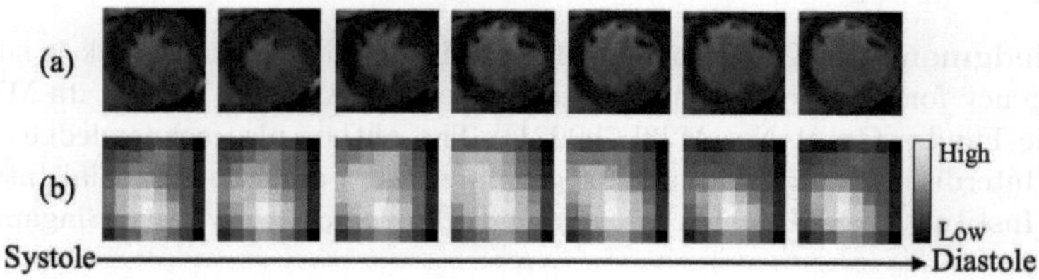

Fig. 5. Visualization of the magnitude of (dh, dw) in Offset Calculation module. (a) shows the cine CMR from the systole to the diastole, and (b) presents the corresponding heatmap. The heatmap effectively captures myocardial motion degrees from systole to diastole, which are initially low, then increase over time, and eventually decrease.

Keyframe Selection Ablation. We investigated the effect of keyframe selection in the DFC module. For clarity, the entire cardiac cycle was linearly scaled to the interval [0, 1], with 0 marking the start of the cycle and 1 representing the end-diastolic (or near end-diastolic) phase. Keyframes chosen at 1/6, 1/2, and 2/3 were compared with using the frame at 1. Figure 6a demonstrates the importance of selecting the end-diastolic or near end-diastolic frame as the keyframe.

Number of Stages Ablation. We investigated the impact of the number of stages on our framework. Figure 6b shows that selecting two stages achieves the best performance.

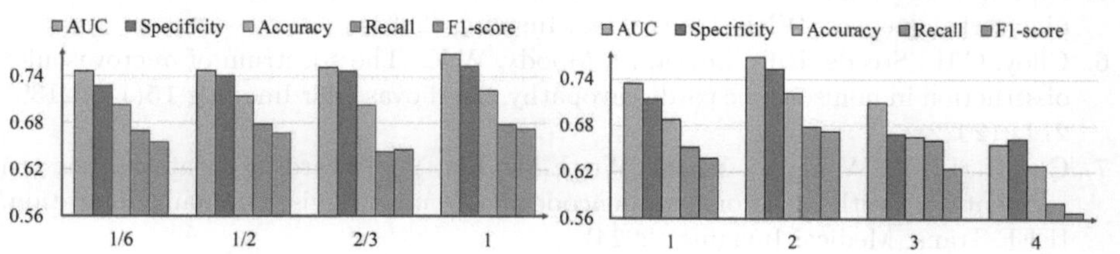

(a) Keyframe Selection Ablation (b) Number of Stages Ablation

Fig. 6. Visualization of performance comparison with (a) different keyframe selection and (b) different number of stages.

4 Conclusion

This paper introduces a Dual Correlation-aware Mamba for MVO identification in non-contrast cine CMR, including an Adjacent Frame Correlation module to explicitly model the dynamics and a Diastolic Frame Correlation module to guide motion learning. Results indicate our method enhances motion feature extraction and outperforms competing methods. Future work will refine MVO delineation and distinguish it from other cardiac lesions causing myocardial abnormalities.

Acknowledgments and Disclosure of Funding. This research work is supported by the Agency for Science, Technology and Research (A*STAR) under its MTC Programmatic Funds (Grant No. M23L7b0021). The author also acknowledges support from the Interdisciplinary Graduate Programme (IGP) and the Artificial Intelligence Research Institute @ NTU (AI-X), Nanyang Technological University, Singapore.

Disclosure of Interests. The authors have no competing interests to declare that are relevant to the content of this article.

References

1. Amyar, A., et al.: Gadolinium-free cardiac mri myocardial scar detection by 4d convolution factorization. In: International Conference on Medical Image Computing and Computer-Assisted Intervention, pp. 639–648. Springer (2023). https://doi.org/10.1007/978-3-031-43895-0_60
2. Arnab, A., Dehghani, M., Heigold, G., Sun, C., Lučić, M., Schmid, C.: Vivit: a video vision transformer. In: Proceedings of the IEEE/CVF International Conference on Computer Vision, pp. 6836–6846 (2021)
3. Barnes, M., Heywood, A.E., Mahimbo, A., Rahman, B., Newall, A.T., Macintyre, C.R.: Acute myocardial infarction and influenza: a meta-analysis of case-control studies. Heart **101**(21), 1738–1747 (2015)
4. Bertasius, G., Wang, H., Torresani, L.: Is space-time attention all you need for video understanding? In: Proceedings of the 38th International Conference on Machine Learning, vol. 2, p. 4 (2021)
5. Bi, N., et al.: Segmorph: concurrent motion estimation and segmentation for cardiac mri sequences. IEEE Trans. Med. Imaging (2024)
6. Choy, C.H., Steeds, R.P., Leyva, F., Moody, W.E.: The spectrum of microvascular obstruction in nonischemic cardiomyopathy. Cardiovascular Imaging **15**(12), 2139–2144 (2022)
7. Cui, H., Li, Y., Wang, Y., Xu, D., Wu, L.M., Xia, Y.: Towards accurate cardiac mri segmentation with variational autoencoder-based unsupervised domain adaptation. IEEE Trans. Medical Imaging (2024)
8. Gräni, C., Stark, A.W., Fischer, K., et al.: Diagnostic performance of cardiac magnetic resonance segmental myocardial strain for detecting microvascular obstruction and late gadolinium enhancement in patients presenting after a st-elevation myocardial infarction. Front. Cardiovascular Med. **9**, 909204 (2022)
9. Gu, A., Dao, T.: Mamba: Linear-time sequence modeling with selective state spaces. arXiv preprint arXiv:2312.00752 (2023)

10. Hamirani, Y.S., Wong, A., Kramer, C., Salerno, M.: Effect of microvascular obstruction and intramyocardial hemorrhage by cmr on lv remodeling and outcomes after myocardial infarction: a systematic review and meta-analysis. JACC Cardiovasc. Imaging **7**(9), 940–52 (2014)
11. Liu, X., et al.: Tagged-to-cine mri sequence synthesis via light spatial-temporal transformer. In: International Conference on Medical Image Computing and Computer-Assisted Intervention, pp. 701–711. Springer (2024). https://doi.org/10.1007/978-3-031-72104-5_6
12. Liu, Z., et al.: Video swin transformer. In: Proceedings of the IEEE/CVF Conference on Computer Vision and Pattern Recognition, pp. 3202–3211 (2022)
13. Liu, Z., Wang, L., Wu, W., Qian, C., Lu, T.: Tam: temporal adaptive module for video recognition. In: Proceedings of the IEEE/CVF International Conference on Computer Vision, pp. 13708–13718 (2021)
14. Meng, Q., Bai, W., Liu, T., O'regan, D.P., Rueckert, D.: Mesh-based 3d motion tracking in cardiac mri using deep learning. In: International Conference on Medical Image Computing and Computer-Assisted Intervention. pp. 248–258. Springer (2022). https://doi.org/10.1007/978-3-031-16446-0_24
15. Nijveldt, R., Beek, A.M., Hirsch, A., et al.: Functional recovery after acute myocardial infarction: comparison between angiography, electrocardiography, and cardiovascular magnetic resonance measures of microvascular injury. J. Am. Coll. Cardiol. **52**(3), 181–189 (2008)
16. Qi, R., Li, X., Xu, L., Zhang, J., Zhang, Y., Xu, C.: Cardiac physiology knowledge-driven diffusion model for contrast-free synthesis myocardial infarction enhancement. In: International Conference on Medical Image Computing and Computer-Assisted Intervention, pp. 200–210. Springer (2024). https://doi.org/10.1007/978-3-031-72378-0_19
17. Qiu, J., et al.: Myops-net: myocardial pathology segmentation with flexible combination of multi-sequence cmr images. Med. Image Anal. **84**, 102694 (2023)
18. Qiu, Z., Yao, T., Mei, T.: Learning spatio-temporal representation with pseudo-3d residual networks. In: Proceedings of the IEEE/CVF International Conference on Computer Vision, pp. 5533–5541 (2017)
19. Reimer, K.A., Lowe, J.E., Rasmussen, M.M., Jennings, R.B.: The wavefront phenomenon of ischemic cell death. 1. myocardial infarct size vs duration of coronary occlusion in dogs. Circulation **56**(5), 786–794 (1977)
20. Shroff, G.R., Frederick, P.D., Herzog, C.A.: Renal failure and acute myocardial infarction: clinical characteristics in patients with advanced chronic kidney disease, on dialysis, and without chronic kidney disease. Am. Heart J. **163**(3), 399–406 (2012)
21. Symons, R., Pontone, G., Schwitter, J., et al.: Long-term incremental prognostic value of cardiovascular magnetic resonance after st-segment elevation myocardial infarction: a study of the collaborative registry on cmr in stemi. JACC: Cardiovascular Imaging **11**(6), 813–825 (2018)
22. Tran, D., Bourdev, L., Fergus, R., Torresani, L., Paluri, M.: Learning spatiotemporal features with 3d convolutional networks. In: Proceedings of the IEEE/CVF International Conference on Computer Vision, pp. 4489–4497 (2015)
23. Tripathi, P.C., et al.: Tensor-based multimodal learning for prediction of pulmonary arterial wedge pressure from cardiac mri. In: International Conference on Medical Image Computing and Computer-Assisted Intervention, pp. 206–215. Springer (2023). https://doi.org/10.1007/978-3-031-43990-2_20

24. Vimalesvaran, K., et al.: Detecting aortic valve pathology from the 3-chamber cine cardiac mri view. In: International Conference on Medical Image Computing and Computer-Assisted Intervention, pp. 571–580. Springer (2022). https://doi.org/10.1007/978-3-031-16431-6_54
25. Wu, K.C.: Cmr of microvascular obstruction and hemorrhage in myocardial infarction. J. Cardiovasc. Magn. Reson. **14**(1), 72 (2012)
26. Xu, C., Xu, L., Ohorodnyk, P., Roth, M., Chen, B., Li, S.: Contrast agent-free synthesis and segmentation of ischemic heart disease images using progressive sequential causal gans. Med. Image Anal. **62**, 101668 (2020)
27. Yan, Y., et al.: Coarse-grained mask regularization for microvascular obstruction identification from non-contrast cardiac magnetic resonance. In: International Conference on Medical Image Computing and Computer-Assisted Intervention, pp. 231–241. Springer (2024). https://doi.org/10.1007/978-3-031-72378-0_22
28. Zhang, Q., Burrage, M.K., Shanmuganathan, M., et al.: Artificial intelligence for contrast-free mri: scar assessment in myocardial infarction using deep learning-based virtual native enhancement. Circulation **146**(20), 1492–1503 (2022)
29. Zhu, Y., Cheng, J., Cui, Z.X., Ren, J., Wang, C., Liang, D.: Sre-cnn: a spatiotemporal rotation-equivariant cnn for cardiac cine mr imaging. In: International Conference on Medical Image Computing and Computer-Assisted Intervention, pp. 679–689. Springer (2024). https://doi.org/10.1007/978-3-031-72104-5_65

EdgeANet: A Transformer-based Edge Representation Learning Network for Canine X-ray Verification

In-Gyu Lee[1], Jun-Young Oh[1], Hyewon Choi[2], Tae-Eui Kam[3], Namsoon Lee[2], Sang-Hwan Hyun[2], Euijong Lee[1], and Ji-Hoon Jeong[1(✉)]

[1] Department of Computer Science, Chungbuk National University, Cheongju, Republic of Korea
{ingyu.lee,jy.oh,kongjjagae,jh.jeong}@chungbuk.ac.kr

[2] College of Veterinary Medicine, Chungbuk National University, Cheongju, Republic of Korea
{hyewon0203,ultravet,shhyun}@chungbuk.ac.kr

[3] Department of Artificial Intelligence, Korea University, Seoul, Republic of Korea
kamte@korea.ac.kr

Abstract. Artificial intelligence (AI) has shown great potential in medical imaging, yet its adoption in veterinary medicine remains limited due to data scarcity and anatomical complexity. This study introduces a novel transformer-based edge representation learning network for verifying rotated vertebral bodies in canine thoracic X-ray images. The proposed method integrates a localization module to identify the spinous process, a transformer encoder for global feature extraction using a self-attention mechanism, and an edge encoder to enhance feature extraction of fine-grained details, improving classification performance. Experimental results demonstrate that our method achieves superior accuracy, precision, and recall, outperforming state-of-the-art (SOTA) methods with a classification accuracy of 0.7838. Furthermore, the ablation study confirms that including the proposed encoders significantly impacts performance, demonstrating their effectiveness in improving classification accuracy. These findings highlight the importance of multi-scale feature extraction in veterinary imaging and suggest that EdgeANet can be a valuable tool for AI-assisted X-ray verification in veterinary and human medical applications.

Keywords: Veterinary AI · Edge detection · X-ray verification · Deep learning · Medical imaging

1 Introduction

Artificial intelligence (AI) is revolutionizing data analysis, enabling decision-making and predictive capabilities [1,2]. In medicine, AI supports disease diagnosis, treatment planning, and prognosis prediction [3,4]. Integrating AI into

J. C. Gee et al. (Eds.): MICCAI 2025, LNCS 15960, pp. 185–195, 2026.
https://doi.org/10.1007/978-3-032-04927-8_18

veterinary imaging also has the potential to standardize diagnosis, reduce human error, and enhance clinical efficiency [5,6]. As pets are increasingly seen as family members, there is a growing need for disease prevention, early diagnosis, and effective treatments [7]. However, the adoption of AI in veterinary medicine, similar to human healthcare, remains slow due to challenges such as the need for high-quality labeled datasets, the complexity of anatomical and behavioral variations, and a shortage of specialized AI practitioners familiar with veterinary applications [8,9]. Despite these challenges, AI holds great potential to enhance diagnostic accuracy, reduce workload for veterinarians, and improve overall pet healthcare [10].

In veterinary medicine, imaging is essential since animal patients cannot communicate symptoms [11]. While computed tomography and magnetic resonance imaging provide high-resolution insights, they require anesthesia, making them costly and risky [12]. Consequently, radiography is more widely used due to its affordability and accessibility, emphasizing the need to validate X-ray image quality [13]. Due to its diagnostic importance, AI-driven X-ray analysis is an active research area in veterinary medicine. Li et al. [14] developed a regressive vision transformer model for canine cardiomegaly classification using vertebral heart scale prediction, achieving state-of-the-art (SOTA) performance and improving interpretability for clinicians. Banzato et al. [13] conducted the first study assessing canine thoracic X-ray quality, highlighting the impact of technical errors like improper positioning on diagnostic reliability. AI-based methods have also been explored for diagnosing conditions such as stifle joint diseases and hip dysplasia [5,15,16]. Given the critical role of X-rays in veterinary diagnostics, ensuring proper image acquisition is essential.

Medical imaging in veterinary medicine relies on structural analysis, where fine-grained details such as bone contours, fractures, and subtle misalignments are crucial for diagnosis [17,18]. Traditional deep learning models, especially convolutional neural networks (CNN), primarily focus on learning high-level features, often neglecting small-scale structural variations [19]. However, in radiographic images, the presence of minor rotations, deformations, or anomalies in vertebral structures can critically impact the accuracy of diagnosis [20]. In this study, edge representation learning is critical in highlighting these structural changes. Enhancing edge-based features enables a model to better distinguish between normal and abnormal cases, improving classification accuracy.

Our contributions can be summarized as follows:

- We propose EdgeANet, a novel transformer-based edge representation learning method for identifying small-scale errors, such as rotated spinous processes, in thoracic X-ray images. The proposed method determines whether X-ray images are accurately captured by detecting small objects that indicate common imaging errors, such as spinal rotation.
- We develop a module that enhances model performance in detecting minor errors and improving X-ray verification and medical imaging applications. This method has the potential to advance veterinary diagnostics and extend to human medical imaging in the future.

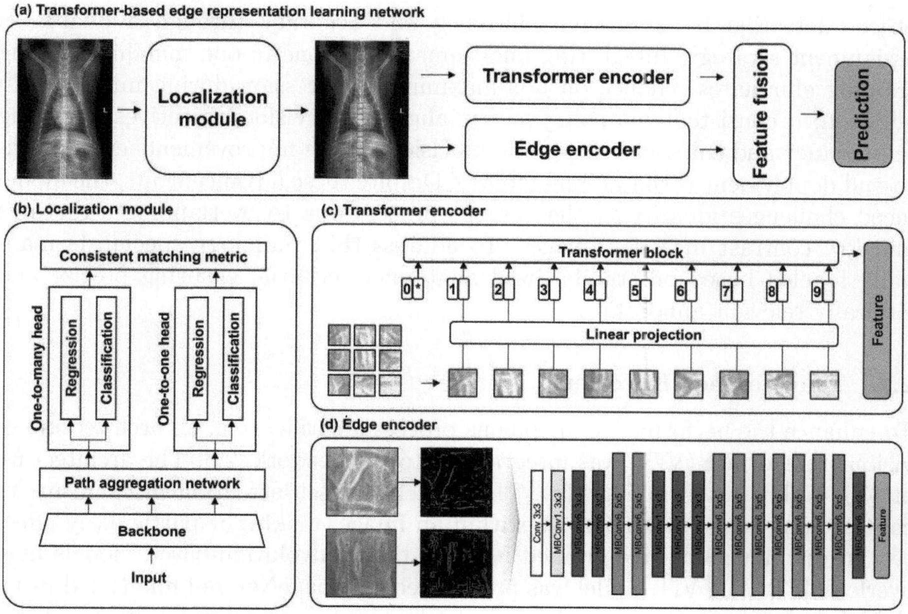

Fig. 1. Overall flow of the proposed network. The proposed network takes a raw image as input and identifies the spinous process through the localization module shown in (b). Subsequently, global and fine-grained features are extracted through the transformer encoder in (c) and the edge encoder in (d). These features are then fused to produce the final classification result.

2 Methods

Figure 1. illustrates our proposed transformer-based edge representation learning network for verifying thoracic X-ray images by detecting rotated spinous processes. The network is composed of three components. The first component, the localization module, identifies the spinous process within the vertebrae. The second component, the transformer encoder, extracts global contextual features, while the third component, the edge encoder, captures fine-grained edge details to enhance structural analysis. Thoracic dorsoventral and ventrodorsal X-ray images are first processed through the localization module, which detects the spinous process. The identified region then extracts features through the transformer encoder and edge encoder, enabling the model to learn global and edge representations. These extracted features are fused using summation to classify the spinous process as normal or abnormal.

2.1 Localization Module

In this study, the architecture of YOLOv10 [21] was employed for image localization, as illustrated in Fig. 1(b). The localization module enhances real-time object detection by improving efficiency and accuracy through a dual-label assignment strategy. Integrating one-to-many and one-to-one matching during training eliminates reliance on non-maximum suppression during inference. In addition, a consistent matching metric aligns supervision objectives, reducing redundancy and enhancing detection precision. These improvements enable end-to-end deployment with minimal latency. Despite these advancements, the model faced challenges identifying the spinous process due to overlapping structures and low contrast in X-ray images. To address this, radiology specialists manually labeled based on standardized anatomical criteria, ensuring precise and clinically relevant annotations.

2.2 Transformer Encoder

To enhance the performance of spinous process identification, an architecture of vision transformer (ViT) was integrated into the network [22]. The architecture of ViT is illustrated in Fig. 1(c). The model uses self-attention mechanisms to capture long-range dependencies within an image, making it particularly effective in distinguishing fine-grained features that convolution-based models may overlook [23]. The ViT model was pre-trained on ImageNet and fine-tuned using the labeled dataset. The training process involved augmentations such as rotation, flipping, and contrast adjustments to simulate variations encountered in clinical settings. This method improved the model's ability to discern subtle differences in the spinous process region, even in challenging cases. Self-attention computation includes a relative position bias $B \in \mathbb{R}^{M^2 \times M^2}$ for each head. The self-attention mechanism of the transformer block is formulated as follows:

$$\text{Attention}(Q, K, V) = \text{SoftMax}\left(\frac{QK^T}{\sqrt{d}} + B\right)V \tag{1}$$

where $Q, K, V \in \mathbb{R}^{M^2 \times d}$ are the query, key, and value matrices, d is the dimension of the query and key, and M^2 is the number of patches in the window.

2.3 Edge Encoder

An edge feature extraction method was incorporated to refine the detection and classification process. The architecture of the edge encoder is illustrated in Fig. 1(d). Canny edge detection was applied during preprocessing to highlight critical structures while suppressing background noise [24]. It was selected for its robust Gaussian noise filtering and effective non-maximum suppression with tunable thresholds, which enabled reliable edge enhancement on X-ray images and contributed to performance gains. The lower and upper thresholds for Canny edge detection were set to fixed intensity values of 40 and 75 respectively based on the 0 to 255 grayscale pixel range. The extracted edge features were fused with the transformer encoder outputs to construct a hybrid feature representation. This integration improved the model's ability to identify spinous processes, particularly under low contrast or anatomical overlap. An example output of the Canny edge detection is shown in Fig. 2.

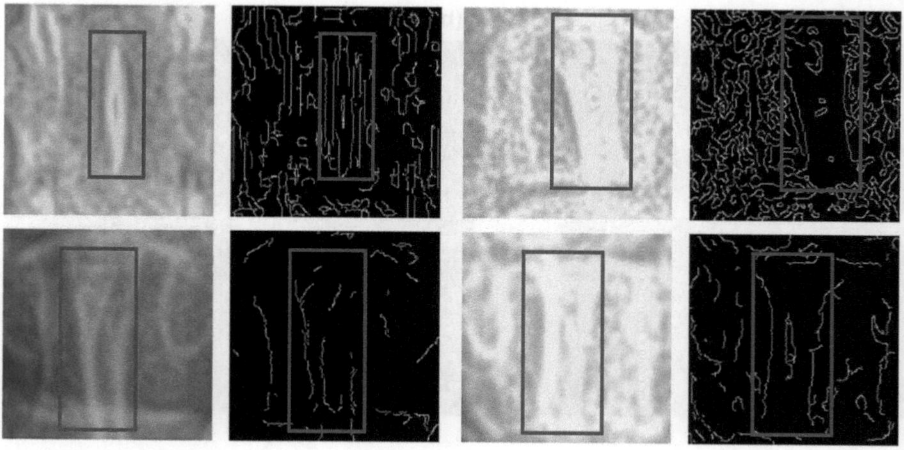

Fig. 2. Example of Canny edge detection. This method can make the networks capture local features of the spinous process.

3 Experiments

3.1 Dataset and Experimental Setup

Dataset. The abnormal and normal spinous processes are depicted in Fig. 3. Radiographic images were obtained using appropriate exposure settings based on abdominal thickness measurements, 200 to 300 mA, 60 to 80 kVp, and a focal film distance of 100 cm. Thorax radiographic images were taken using a digital radiographic system (Toshiba RotanodeTM, Tokyo, Japan). These images were processed with a radiographic system (BLADE v1., Median International Co., Anyang, Republic of Korea).

X-ray images taken for disease diagnosis were filtered and used to select the study subjects. Among the captured X-ray images, those in which the vertebral bodies were rotated were collected and used as abnormal data. In contrast, non-rotated images were collected as the control group and used as normal data. To mitigate annotator bias and ensure objectivity in the absence of external datasets for validation, labeling was conducted through rigorous multi-cross validation by eight veterinarians, each with 1 to 5 years of clinical experience and 1 to 2 years of specialization in veterinary radiology. The labeling criteria were based on the Textbook of Veterinary Diagnostic Radiology by Donald Thrall, one of the most authoritative references in the field [25]. To ensure consistency, only cases confirmed by at least six of the eight veterinarians were included in the dataset. The dataset included samples from various species, comprising 90 abnormal and 100 normal thoracic X-ray images. Although the sample size is limited, this reflects the practical realities of veterinary clinical imaging.

Experimental Setup. Our experiments were conducted on Ryzen 7 7800X3D CPU and four RTX 4090 GPUs. The number of epochs for all methods was set to

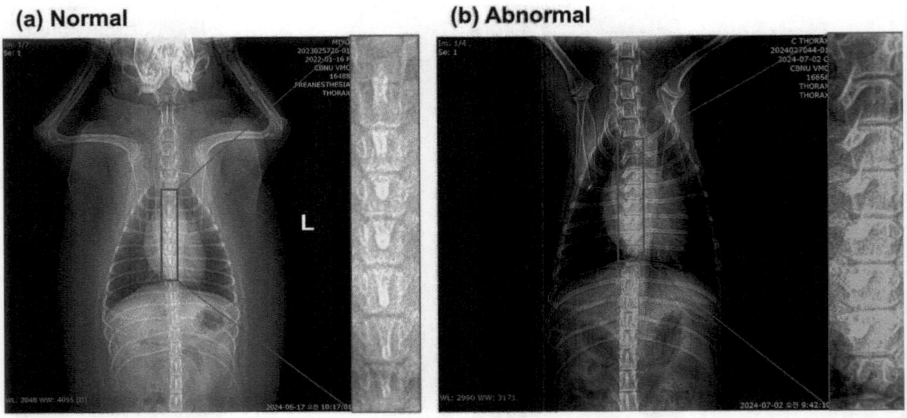

Fig. 3. Example of thoracic X-ray images. (a) presents a normal thoracic X-ray image without spinal rotation, while (b) presents an abnormal thoracic X-ray image showing rotated spinous processes.

100, with the Adam optimizer and a learning rate of 2e-4. The dataset was first split into a 4:1 ratio for the train and test sets. Then, 4-fold cross-validation was applied to the train set for model evaluation. Quantitative metrics, including classification accuracy, precision, recall, F1-score, and confusion matrix, were used to assess the model's performance [26].

3.2 Comparison with SOTA Methods

The study compared the performance of various classification models, including transformer-based architectures such as Swin Transformer [27], as well as CNN-based models like ConvNeXt [28], ResNet [29], DenseNet [30], and EfficientNet [31]. These models achieve SOTA performance or demonstrate performance close to it on benchmark datasets such as ImageNet for image classification tasks. Comparing our model with these architectures allows us to evaluate its efficiency and effectiveness (Table 1).

The left side of the table presents the highest performance achieved during 4-fold cross-validation. In contrast, the right side shows the test results of the best-performing model from each fold, with the average and standard deviation calculated accordingly. The experimental results demonstrate that EdgeANet significantly outperforms other state-of-the-art methods in validation and test performance across multiple evaluation metrics. It consistently achieves the highest accuracy across multiple folds, with values reaching up to 0.8684 in fold 3, which is on par or superior to the best-performing models in individual folds.

EdgeANet achieves the highest average test accuracy of 0.7838, clearly outperforming all competing models including EfficientNet. In addition, it registers the highest precision of 0.6962, recall of 0.8623, and F1-score of 0.7813, demonstrating superior accuracy and a balance between precision and recall. Unlike

Table 1. Quantitative evaluation metrics of 4-fold cross-validation and test performance. We highlight the best performance in bold.

Model	Best validation performance				Average test performance			
	Fold 1	Fold 2	Fold 3	Fold 4	Accuracy	Precision	Recall	F1-score
Swin Transformer [27]	0.7632	0.7895	0.7632	0.7368	0.6411 (± 0.0336)	0.4183 (± 0.0221)	0.6232 (± 0.0140)	0.5691 (± 0.0055)
ResNet [29]	0.7368	0.7895	0.8421	0.8158	0.6619 (± 0.0107)	0.5747 (± 0.0808)	0.6911 (± 0.0308)	0.6405 (± 0.0367)
DenseNet [30]	**0.8158**	**0.8684**	0.8421	**0.8421**	0.6773 (± 0.0141)	0.5639 (± 0.0190)	0.7311 (± 0.0207)	0.6722 (± 0.0140)
EfficientNet [31]	0.7895	**0.8684**	0.8421	**0.8421**	0.6847 (± 0.0303)	0.5967 (± 0.0703)	0.7077 (± 0.0276)	0.6722 (± 0.0483)
ConvNeXt [28]	**0.8158**	**0.8684**	0.7368	0.7632	0.7114 (± 0.0466)	0.5845 (± 0.0690)	0.7748 (± 0.0506)	0.7078 (± 0.0478)
EdgeANet (Ours)	**0.8158**	0.8158	**0.8684**	**0.8421**	**0.7838** (± 0.0310)	**0.6962** (± 0.0276)	**0.8623** (± 0.0258)	**0.7813** (± 0.0271)

ConvNeXt or DenseNet, which achieve competitive results in specific folds but fail to maintain consistency across test metrics, EdgeANet ensures stable and superior performance across all evaluation criteria.

EdgeANet's high recall score of 0.8623 suggests that it captures more relevant instances than other models, making it particularly useful in applications where missing key classifications could have significant consequences. These results establish EdgeANet as the leading model, highlighting its robustness and practical applicability in real-world scenarios that demand high accuracy and generalization.

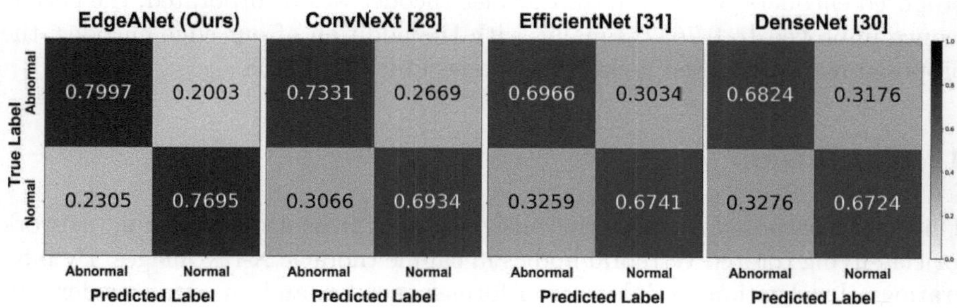

Fig. 4. Confusion matrix of the four best-performing models.

Our experimental results are visualized through the confusion matrix presented in Fig. 4, which includes only the top four models with the highest test accuracy. This study aims to accurately distinguish between normal and abnormal data by effectively extracting features from small structures, such as the spinous process, in X-ray images for verification. EdgeANet demonstrates outstanding performance in this task, having achieved a TP of over 0.7997, outperforming EfficientNet, DenseNet, and ConvNeXt, which achieved a TP in the range of 0.6824 to 0.7331. Regarding the normal class, the proposed method shows performance with an TN of 0.7695, while others showed lower reliability, leading to a higher misclassification rate for normal cases.

The model uses a transformer encoder to extract global features and an edge encoder to capture local structural details, allowing for a more comprehensive representation of the X-ray images. By combining these feature extraction techniques, the model enhances classification accuracy, particularly in detecting subtle abnormalities that conventional methods may overlook.

Table 2. Ablation study of adapted encoders.

Model	Transformer encoder	Edge encoder	Accuracy
Baseline	✗	✗	0.7368
w/o Edge encoder	✓	✗	0.7632
EdgeANet (Ours)	✓	✓	**0.8158**

3.3 Ablation Study

In Table 2, we present an ablation study on different components of our network. The results demonstrate that our methods enhance classification accuracy by extracting deep features through edge-based representation learning. The baseline model achieved a classification accuracy of 0.7368 without any of the proposed encoders. When the transformer encoder was incorporated, the performance improved to 0.7632. Finally, with the addition of our edge encoder, the proposed model achieved a classification accuracy of 0.8158.

4 Conclusion

This study proposed a transformer-based edge representation learning network for classifying rotated vertebral bodies in canine thoracic X-ray images. By integrating a localization module, a transformer encoder, and an edge encoder, our model effectively captures global and fine-grained features critical for accurate classification. Comparative analysis against widely used deep learning models demonstrated that our method achieves higher classification accuracy and improved recall rates, reinforcing its robustness for veterinary applications. The findings highlight the importance of multi-aspect feature extraction in enhancing the precision of diagnostic imaging models, particularly in cases where traditional CNN-based architectures struggle with subtle anatomical variations. Although the dataset size is limited due to the practical constraints of veterinary clinical imaging, our results provide a foundation for future study. We plan to expand the dataset using generative augmentation techniques such as generative adversarial networks to improve generalizability, adapt the network for other anatomical regions, and explore real-world deployment in veterinary clinics. This study underscores the potential of AI-assisted imaging analysis in advancing veterinary diagnostics and ensuring higher standards of pet healthcare.

Acknowledgments. This work was partly supported by Innovative Human Resource Development for Local Intellectualization program through the Institute of Information & Communications Technology Planning & Evaluation (IITP) grant funded by the Korea government (MSIT) (IITP-2025-RS-2020-II201462, 20%), grant funded by the Institute of Information & Communications Technology Planning & Evaluation (IITP) grant funded by the Korea government (MSIT) (RS-2021-II212068, Artificial Intelligence Innovation Hub), Korea Institute for Advancement of Technology (KIAT) grant funded by the Korea government (MOTIE) (No. P0025399), grant funded by the Korea Institute of Planning and Evaluation for Technology in Food, Agriculture and Forestry (IPET) through the Agriculture and Food Convergence Technologies Program for Research Manpower Development Program funded by the Ministry of Agriculture, Food and Rural Affairs (MAFRA) (RS-2024-00398561), and grant funded by the National Research Foundation of Korea (NRF) grant funded by the Korea government(MSIT) (RS-2024-00456480).

Disclosure of Interests. The authors have no competing interests to declare that are relevant to the content of this article.

References

1. Davenport, T., Kalakota, R.: The potential for artificial intelligence in healthcare. Future Healthc. J. **6**(2), 94–98 (2019)
2. Hou, J., Xu, J., Chen, H.: Concept-attention whitening for interpretable skin lesion diagnosis. In: International Conference on Medical Image Computing and Computer-Assisted Intervention, pp. 113–123. Springer (2024)
3. Topol, E.J.: High-performance medicine: the convergence of human and artificial intelligence. Nat. Med. **25**(1), 44–56 (2019)
4. Biswas, K., et al.: Adaptive smooth activation function for improved organ segmentation and disease diagnosis. In: International Conference on Medical Image Computing and Computer-Assisted Intervention, pp. 65–74. Springer (2024)
5. Hennessey, E., DiFazio, M., Hennessey, R., Cassel, N.: Artificial intelligence in veterinary diagnostic imaging: a literature review. Vet. Radiol. Ultrasound **63**, 851–870 (2022)
6. Bouhali, O., Bensmail, H., Sheharyar, A., David, F., Johnson, J.P.: A review of radiomics and artificial intelligence and their application in veterinary diagnostic imaging. Vet. Sci. **9**(11), 620 (2022)
7. Ly, L.H., Gordon, E., Protopopova, A.: Exploring the relationship between human social deprivation and animal surrender to shelters in British Columbia, Canada. Front. Vet. Sci. **8**, 656597 (2021)
8. Benjamens, S., Dhunnoo, P., Meskó, B.: The state of artificial intelligence-based FDA-approved medical devices and algorithms: an online database. NPJ Digit. Med. **3**(1), 118 (2020)

9. Yuan, Z., Fang, Z., Huang, Z., Wu, F., Yao, Y.F., Li, Y.: Adapting pre-trained generative model to medical image for data augmentation. In: International Conference on Medical Image Computing and Computer-Assisted Intervention, pp. 79–89. Springer (2024)
10. Appleby, R.B., Basran, P.S.: Artificial intelligence in veterinary medicine. J. Am. Vet. Med. Assoc. **260**(8), 819–824 (2022)
11. Burti, S., Banzato, T., Coghlan, S., Wodziniski, M., Bendazzoli, M., Zotti, A.: Artificial intelligence in veterinary diagnostic imaging: perspectives and limitations. Res. Vet. Sci. 105317 (2024)
12. Portier, K., Ida, K.K.: Anesthetic risk and complications in veterinary medicine (2020)
13. Banzato, T., Wodzinski, M., Burti, S., Vettore, E., Muller, H., Zotti, A.: An AI-based algorithm for the automatic evaluation of image quality in canine thoracic radiographs. Sci. Rep. **13**(1), 17024 (2023)
14. Li, J., Zhang, Y.: Regressive vision transformer for dog cardiomegaly assessment. Sci. Rep. **14**(1), 1539 (2024)
15. Shim, H., et al.: Deep learning-based diagnosis of stifle joint diseases in dogs. Vet. Radiol. Ultrasound **64**(1), 113–122 (2023)
16. McEvoy, F.J., et al.: Deep transfer learning can be used for the detection of hip joints in pelvis radiographs and the classification of their hip dysplasia status. Vet. Radiol. Ultrasound **62**(4), 387–393 (2021)
17. Steffey, M.A., Garcia, T.C., Gorney, A., Stover, S.M., Zwingenberger, A.L.: Computed tomography-derived structural analysis for the likelihood of pathologic fracture in canine antebrachial osteosarcoma. Vet. Radiol. Ultrasound (2024)
18. Soltero-Rivera, M.M., Nguyen, R., Goldschmidt, S.L., Hatcher, D.C., Arzi, B.: Diagnostic yield of dental radiography and digital tomosynthesis for the identification of anatomic structures in cats. Front. Vet. Sci. **11**, 1408807 (2024)
19. Azulay, A., Weiss, Y.: Why do deep convolutional networks generalize so poorly to small image transformations? J. Mach. Learn. Res. **20**(184), 1–25 (2019)
20. Miao, K.H., et al.: Radiological diagnosis and advances in imaging of vertebral compression fractures. J. Imaging **10**(10), 244 (2024)
21. Wang, A., et al.: YOLOv10: Real-time end-to-end object detection. arXiv preprint arXiv:2405.14458 (2024)
22. Dosovitskiy, A.: An image is worth 16x16 words: Transformers for image recognition at scale. arXiv preprint arXiv:2010.11929 (2020)
23. Mehta, S., Rastegari, M.: MobileViT: Light-weight, general-purpose, and mobile-friendly vision transformer. arXiv preprint arXiv:2110.02178 (2021)
24. Canny, J.: A computational approach to edge detection. IEEE Trans. Pattern Anal. Mach. Intell. **6**, 679–698 (1986)
25. Thrall, D.E.: Textbook of Veterinary Diagnostic Radiology. Elsevier Health Sciences, Amsterdam (2012)
26. Goutte, C., Gaussier, E.: A probabilistic interpretation of precision, recall and f-score, with implication for evaluation. In: European Conference on Information Retrieval, pp. 345–359. Springer (2005)
27. Liu, Z., et al.: Swin transformer: Hierarchical vision transformer using shifted windows. In: IEEE/CVF International Conference on Computer Vision, pp. 10012–10022 (2021)
28. Liu, Z., Mao, H., Wu, C.Y., Feichtenhofer, C., Darrell, T., Xie, S.: A ConvNet for the 2020s. In: IEEE/CVF Conference on Computer Vision and Pattern Recognition, pp. 11976–11986 (2022)

29. He, K., Zhang, X., Ren, S., Sun, J.: Deep residual learning for image recognition. In: IEEE Conference on Computer Vision and Pattern Recognition, pp. 770–778 (2016)
30. Huang, G., Liu, Z., Van Der Maaten, L., Weinberger, K.Q.: Densely connected convolutional networks. In: IEEE Conference on Computer Vision and Pattern Recognition, pp. 4700–4708 (2017)
31. Tan, M., Le, Q.: EfficientNet: Rethinking model scaling for convolutional neural networks. In: International Conference on Machine Learning, pp. 6105–6114. PMLR (2019)

EEG-DINO: Learning EEG Foundation Models via Hierarchical Self-distillation

Xujia Wang[1,2], Xuhui Liu[1,2], Xi Liu[1], Qian Si[2], Zhaoliang Xu[2], Yang Li[2(✉)], and Xiantong Zhen[1(✉)]

[1] Central Research Institue, United Imaging Healthcare, Co., Ltd., Beijing, China
xiantong.zhen@cri-united-imaging.com
[2] Beihang University, Beijing, China
liyang@buaa.edu.cn

Abstract. Electroencephalography (EEG) provides a non-invasive window into the brain's electrical activity, playing an essential role in various brain-computer interface (BCI) and healthcare applications. In this paper, we propose EEG-DINO, a novel foundation model for EEG encoding based on a hierarchical self-distillation framework. By multi-view semantic alignment, the model is able to extract multi-level semantic features from EEG data, which captures a wide range of semantic information, increasing the robustness against noise and variances inherent in complex EEG signals. Moreover, acknowledging the unique heterogeneous spatial-temporal dependencies in EEG signals, we design a channel-aware sampling mechanism and a decoupled positional embedding scheme. They independently address spatial and temporal dimensions, enabling the model to capture the intricate structural characteristics of EEG signals. We train EEG-DINO on a large-scale EEG corpus spanning over 9000 h, which consistently achieves state-of-the-art performance on multiple downstream tasks (The pre-trained weights and code for fine-tuning are available at: https://huggingface.co/eegdino/EEG-DINO). These results demonstrate the great effectiveness of our self-distillation framework for EEG encoding.

Keywords: Electroencephalography (EEG) · Self-supervised learning · Foundation models · Self-Distillation · EEG pre-training

1 Introduction

Electroencephalography (EEG) records electrical activity on the scalp, providing high temporal resolution, making it ideal for real-time brain activity monitoring [15]. EEG is extensively used in the diagnosis and monitoring of neurological disorders, including epilepsy [19], sleep disorders [22], neurodegenerative diseases [3] and neuroprosthetics [1,2]. Effective encoding of EEG signals has emerged as a vital step in effectively performing complex and diverse tasks, as it enables the extraction of meaningful information from raw data, thereby enhancing the accuracy and reliability [17].

J. C. Gee et al. (Eds.): MICCAI 2025, LNCS 15960, pp. 196–205, 2026.
https://doi.org/10.1007/978-3-032-04927-8_19

Early EEG analysis methods primarily employ deep learning (DL) with various neural networks [9,16,18] to learn signal features. However, they typically rely on supervised learning tailored to specific tasks or datasets, leading to challenges in generalization, especially due to limited data, high noise levels, stable nature over short time intervals, and significant variations in EEG signal formats. Recently, building on advancements of self-supervised learning (SSL) in various fields [6,12], many works [8,20,21] propose EEG foundation models pre-trained on large-scale unlabeled EEG data, demonstrating the potential of SSL in learning meaningful spatiotemporal features that generalize across subjects and recording sessions. Nonetheless, these methods usually adopt reconstructive objectives, which are ill-suited to handle the issues of the low signal-to-noise ratio and the stable nature of EEG signals over short time intervals. As a result, they tend to focus on noise reconstruction, limiting the ability of models to capture discriminative neurophysiological patterns and meaningful temporal variations.

In this paper, we propose EEG-DINO, the first distillation-based foundation model for EEG encoding. EEG-DINO adopts a hierarchical self-distillation framework within a shared network architecture, enabling knowledge distillation through self-supervised learning across multiple augmented views. This framework leverages multi-view learning to achieve semantic alignment, enhancing robustness and facilitating the extraction of multi-level semantic features from complex, noisy EEG signals. Moreover, we introduce a channel-aware sampling mechanism and a decoupled positional embedding scheme. They exploit the unique structural characteristics of EEG signals by handling spatial and temporal dimensions separately, enabling effective modeling of the heterogeneous spatial-temporal dependencies inherent in EEG data. We note that our work focuses on EEG foundation model learning based on the self-distillation principles and thereby is fundamentally different from early EEG distillation works [4,5,14], which distill a pre-trained model to a small model for supervised learning on specific tasks. We extensively evaluate EEG-DINO on multiple downstream tasks, where it demonstrates superior performance compared to counterpart methods, setting new state-of-the-arts, highlighting the promise of our distillation-driven self-supervised learning for EEG encoding.

2 Method

This section outlines the hierarchical self-distillation framework with a teacher-student architecture to pre-train EEG-DINO, as shown in Fig. 1. Given the raw EEG signals represented as $X \in \mathbb{R}^{C \times T}$, where C denotes the number of channels and T is the number of timestamps, we first devise a channel-aware sampling mechanism to obtain multi-view inputs, allowing the model to capture multi-level semantic information.

Subsequently, we utilize the time-frequency embedding (TFE) [20] to project the raw signal X to a set of EEG tokens $\mathcal{E}$. Notably, an EEG token is defined as a 1-second temporal segment across all channels. Meanwhile, we introduce the decoupled positional embedding (DPE) scheme to encode spatial and temporal

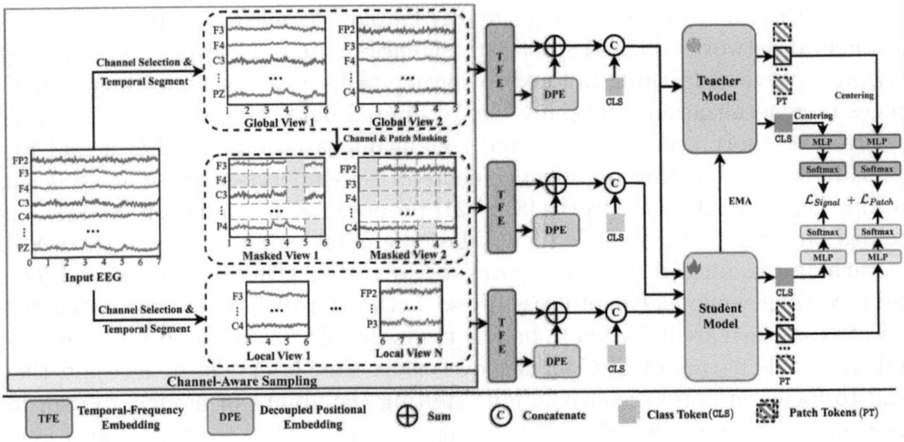

Fig. 1. Overview of the hierarchical self-distillation framework to pre-train EEG-DINO.

positional dependencies separately, which is then integrated to $\mathcal{E}$ to form the input EEG tokens of the transformer for pre-training. In this context, the model learns a set of features that capture temporal and spatial patterns in EEG data. As a result, the pre-trained EEG-DINO can be adapted to conduct specific downstream tasks.

2.1 Tailored Sampling and Positional Embedding for EEG Signals

Channel-Aware Sampling. EEG inherently exhibits spatially distributed patterns across sensor channels, where each channel corresponds to localized neural activity in distinct brain regions. To exploit this spatial-temporal characteristic, we design a channel-aware sampling mechanism specifically for EEG to create three view types, forming the hierarchy for self-distillation across views. As shown in Fig. 1, unlike conventional image cropping that extracts contiguous spatial regions, our mechanism uniquely operates on EEG signals by adaptively sampling subsets of channels and temporal segments to construct diverse perspectives. Global views retain a moderate proportion of sparse channels and continuous temporal windows to preserve broad spatial-temporal patterns, local views employ aggressive spatiotemporal reduction to focus on localized dynamics and masked views integrate channel-wise and temporal-patch masking on global views to simulate incomplete observations. This pipeline creates 12 diverse perspectives (2 global, 2 masked, 8 local) per sample to encourages the model to learn robust representations resilient to channel variations, following the perspective configuration in DINO.

Decoupled Positional Embedding. After tokenizing multi-views signals using TFE [20], the DPE performs spatial channel encoding and dynamic temporal encoding separately. The former applies a learnable projection that maps the one-hot channel vector into an embedding $\mathcal{P}_c$. The later dynamically performs channel-wise 1D convolutions along the temporal axis to produce the temporal embedding $\mathcal{P}_t$. This design decouples channel and temporal embeddings, which enables robust generalization across structural characteristics. The final patch embedding combines both components through summation:

$$\text{Embed}(X) = \mathcal{P}_c + \mathcal{P}_t + \mathcal{E} \tag{1}$$

2.2 Pre-training via Hierarchical Self-distillation

Teacher-Student Architecture. Our framework consists of a teacher model and a student model, instantiating DINO-v2 principles for EEG representation learning. The input sequence for this architecture is constructed by concatenating the class token (CLS) with patch embeddings. As shown in Fig. 1, the teacher processes only global views to maintain stable target representations. The student operates on all view types to produce robust feature representations. To ensure stable training, the teacher parameters are updated through exponential moving average (EMA) [7] of the student model parameters. This momentum update mechanism prevents abrupt changes in the target representations while allowing gradual refinement.

Hierarchical Self-distillation. We propose a hierarchical self-distillation framework, where knowledge distillation is performed across views at different levels, which forces the foundation model to learn robust representations.

Signal-Level Distillation. The distillation at this level applies to CLS of teacher from global views and CLS of student from global, local and masked views. Following DINO v2 [12], as shown in Fig. 1, CLS is passed into an MLP layer followed by a softmax operation to produce probability vector. For the teacher model, we apply centering to the logits (pre-softmax outputs) by subtracting a momentum-updated mean vector to obtain the probability vector. The cross-entropy loss is then computed: $\mathcal{L}_* = -\sum p_t \log p_s$ where $*$ denotes any one of the views, p_t and p_t are the probability vectors from teacher and student models, respectively. We can calculate weighted losses from global, local, and masked views at the signal-level distillation:

$$\mathcal{L}_{Signal} = \mathcal{L}_{Global} + \mathcal{L}_{Local} + \mathcal{L}_{Masked} \tag{2}$$

where $\mathcal{L}_{Global}$ is the cross-entropy loss between teacher and student models for global views; similarly, $\mathcal{L}_{Local}$ and $\mathcal{L}_{Masked}$ are losses for local and masked views, respectively.

Patch-Level Distillation. This distillation applies to the patch tokens (PT) of the teacher model from the global views and the patch tokens PT of the student model from masked views [23]. As shown in Fig. 1, the respective probability vectors p_t and p_s from the teacher and student models are obtained in a similar way to compute the loss: $\mathcal{L}_{Patch} = -\sum_i p_t \log p_s$ over all masked patch tokens indexed by i at patch-level distillation.

Total Loss. The final total loss for optimization is computed as the average of the two weighted loss components:

$$\mathcal{L} = \mathcal{L}_{Signal} + \mathcal{L}_{Patch} \tag{3}$$

The student model obtained from the pre-training stage will be used as the foundation model to be adapted to downstream tasks by either full-parameter fine-tuning or linear probing.

3 Experiments and Results

3.1 Dataset and Setup

Pre-training Dataset. We perform pre-training on the Temple University Hospital EEG (TUEG) dataset [11], one of the largest publicly available EEG datasets. This dataset contains over 30,000 clinical EEG recordings from more than 16,000 patients. Following CBraMod [20], the raw EEG data undergoes several preprocessing steps: we select 19 common channels from the international 10-20 system (e.g., FP1, FP2, F3, F4, etc.), resample all signals to 200 Hz and split them into 30-second epochs. Finally, a total of 1,109,545 EEG samples, over 9000 h in duration, are retained for pre-training.

Downstream Datasets. To demonstrate the generalization of our model, we systematically evaluate our model on several different types of downstream datasets:

TUEV [11] is a curated subset of the TUEG database annotated for six typical events. Following protocols from BIOT [21], EEG signals recorded via 19 standardized bipolar electrode pairs (10-20 system) were downsampled to 200 Hz, and segmented into 112,491 5-second epochs. Training subjects were split into an 80% training set and a 20% validation set.

TUAB [11] represents a curated selection from the TUEG database and have been annotated as normal or abnormal. We have used the same preprocessing method as TUEV. Finally, all EEG signals are resampled to 200 Hz and divided into 409,455 10-second 19-channel samples.

SEED-V [10] is a multimodal emotion recognition dataset featuring five emotions. It uses EEG recordings from 16 participants (open-source version) across three sessions with 15 trials each. EEG signals, originally recorded at 1000 Hz using 62 electrodes and downsampled to 200 Hz, were segmented into 117,744 one-second segments. Each session's trials are evenly divided (5:5:5) into three subsets for balanced experimental validation, following CBraMod's setup [20].

Environments and Settings. The experiments are implemented by Python 3.11.11, Torch 2.5.1+cu124, on eight H800 GPUs. All the models are optimized on training set, selected from the validation set and evaluated on the test set. We obtain five sets of results with different random seeds and report the mean and standard deviation values. We devise three different configurations: EEG-DINO-(S)mall, EEG-DINO-(M)edium , and EEG-DINO-(L)arge as shown in Table 1.

Table 1. Hyperparameters for EEG-DINO pre-training

Model	Layers	Hidden Size	MLP Size	Params
EEG-DINO-S	12	200	512	4.6M
EEG-DINO-M	16	512	1024	33M
EEG-DINO-L	24	1024	2048	201M

Baselines and Metrics. The baselines of foundation models are from BIOT [21], LaBraM-(B)ase [8], CBraMod [20] and the non-foundation models baselines are from CNN-(T)ransformer [13] and ST-(T)ransformer [16]. To make head-to-head comparisons, we fine-tune BIOT, LaBraM-B and CBraMod based on their public code, pre-trained weights and public parameter settings under the same dataset seeds as ours. We employ Balanced Accuracy (BA), AUC-PR and AUROC as evaluation metrics for binary classification. And for multi-class classification, we employ BA, Cohen's Kappa and Weighted F1 Score. For model optimization and selection, we designate AUROC as the monitoring metric for binary classification tasks and Cohen's Kappa for multi-class classification tasks.

3.2 Results

We performed experiments comparing linear probing and full-parameter fine-tuning, where full-parameter fine-tuning updates all model weights including the backbone and the added classification head, and linear probing updates classification head only, keeping backbone parameters fixed.

Linear Probing. Tables 2 and 3 compare EEG-DINO-S/M/L with state-of-the-art self-supervised EEG models across three benchmark datasets using a linear probing protocol. EEG-DINO-S significantly outperforms counterpart methods on multiple datasets (TUEV, SEED-V, TUAB) in terms of all metrics, while using fewer or similar parameters. Our EEG-DINO performs exceptionally well even without full-parameter fine-tuning, which demonstrate that EEG-DINO as a foundational model has a strong encoding capability. Furthermore, increasing model capacity (from Small to Medium and Large) consistently enhances EEG representation learning, suggesting that scaling data and parameters leads to more generalized pattern extraction for diverse downstream EEG tasks.

Table 2. The results of linear probing on TUEV and SEED-V

Dataset	Methods	Params	BA(%)	Cohen's Kappa	Weighted F1
TUEV	BIOT [21]	3.2M	33.27 ± 2.56	0.3835 ± 0.0554	0.6792 ± 0.0288
	LaBraM-B [8]	5.8M	34.61 ± 2.25	0.3968 ± 0.0329	0.6974 ± 0.0161
	CBraMod [20]	4.0M	32.46 ± 2.72	0.3884 ± 0.1824	0.6889 ± 0.0625
	EEG-DINO-S	4.6M	54.82 ± 1.06	0.5673 ± 0.0023	0.7861 ± 0.0024
	EEG-DINO-M	33M	58.80 ± 0.68	0.6180 ± 0.0145	0.8111 ± 0.0066
	EEG-DINO-L	201M	$\mathbf{60.54} \pm 0.53$	$\mathbf{0.6419} \pm 0.0122$	$\mathbf{0.8214} \pm 0.0045$
SEED-V	BIOT [21]	3.2M	24.61 ± 2.87	0.0798 ± 0.0361	0.2489 ± 0.0257
	LaBraM-B [8]	5.8M	25.21 ± 2.67	0.0854 ± 0.0342	0.2543 ± 0.0265
	CBraMod [20]	4.0M	25.36 ± 2.57	0.0842 ± 0.0384	0.2568 ± 0.0275
	EEG-DINO-S	4.6M	29.81 ± 0.35	0.1273 ± 0.0052	0.3035 ± 0.0063
	EEG-DINO-M	33M	33.65 ± 0.56	0.1707 ± 0.0047	0.3426 ± 0.0052
	EEG-DINO-L	201M	$\mathbf{35.79} \pm 0.33$	$\mathbf{0.1984} \pm 0.0029$	$\mathbf{0.3652} \pm 0.0041$

Table 3. The reults of linear probing on TUAB

Methods	Params	BA(%)	AUC-PR	AUROC
BIOT [21]	3.2M	73.08 ± 0.29	0.7849 ± 0.0036	0.8013 ± 0.0058
LaBraM-B [8]	5.8M	74.57 ± 0.14	0.8081 ± 0.0009	0.8115 ± 0.0014
CBraMod [20]	4.0M	67.85 ± 1.33	0.7721 ± 0.0259	0.7826 ± 0.0452
EEG-DINO-S	4.6M	78.41 ± 0.08	0.8666 ± 0.0006	0.8706 ± 0.0004
EEG-DINO-M	33M	79.15 ± 0.11	0.8680 ± 0.0008	0.8763 ± 0.0010
EEG-DINO-L	201M	$\mathbf{79.63} \pm 0.07$	$\mathbf{0.8701} \pm 0.0014$	$\mathbf{0.8814} \pm 0.0007$

Fine-Tuning. As shown in Tables 4 and 5, our experiments demonstrate that EEG-DINO-S outperforms the baselines in all metrics on the TUEV and the SEED-V datasets, showing enhanced inter-subject agreement and better generalization. In binary detection on the TUAB dataset, our model performs similarly to the baselines while utilizing fewer parameters, highlighting its efficiency in maintaining high performance with reduced complexity. Scaling to EEG-DINO-M/L further boosts performance. Additionally, we use a randomly initialized model (EEG-DINO-S*) for fine-tuning as a contrast to the pre-trained version. It underperforms compared to the pre-trained versions, which further verifies the effectiveness of our pre-training approach. While it still exceeds supervised CNN-Transformer baselines, underscoring the architecture's effectiveness.

Ablation Study. To further evaluate the effectiveness of our proposed techniques, we conduct three ablation studies on pre-training: w/o decoupled positional embedding (DPE), random masking strategy and full cropping strategy. The results are presented in Fig. 2, which demonstrates that all three techniques are

Table 4. The results of fine-tuning on TUEV and SEED-V

Dataset	Methods	Params	BA(%)	Cohen's Kappa	Weighted F1
TUEV	CNN-T [13]	3.2M	40.87 ± 1.61	0.3815 ± 0.0134	0.6854 ± 0.0293
	ST-T [16]	3.5M	39.84 ± 2.28	0.3765 ± 0.0306	0.6823 ± 0.0190
	BIOT [21]	3.2M	52.81 ± 2.25	0.5273 ± 0.0249	0.7492 ± 0.0082
	LaBraM-B [8]	5.8M	64.09 ± 0.65	0.6637 ± 0.0093	0.8312 ± 0.0052
	CBraMod [20]	4.0M	59.42 ± 1.32	0.5818 ± 0.0149	0.7817 ± 0.0201
	EEG-DINO-S*	4.6M	50.20 ± 1.74	0.4620 ± 0.3960	0.7307 ± 0.2072
	EEG-DINO-S	4.6M	65.16 ± 0.64	0.6654 ± 0.0082	0.8356 ± 0.0046
	EEG-DINO-M	33M	66.11 ± 0.95	0.6739 ± 0.0123	0.8357 ± 0.0054
	EEG-DINO-L	201M	$\mathbf{66.79} \pm 0.57$	$\mathbf{0.6809} \pm 0.0094$	$\mathbf{0.8398} \pm 0.0049$
SEED-V	CNN-T [13]	3.2M	36.78 ± 0.78	0.2072 ± 0.0183	0.3642 ± 0.0088
	ST-T [16]	3.5M	30.52 ± 0.72	0.1083 ± 0.0121	0.2833 ± 0.0105
	BIOT [21]	3.2M	38.37 ± 1.87	0.2261 ± 0.0262	0.3856 ± 0.0203
	LaBraM-B [8]	5.8M	39.76 ± 1.38	0.2386 ± 0.0209	0.3974 ± 0.0111
	CBraMod [20]	4.0M	38.99 ± 0.25	0.2414 ± 0.0013	0.3977 ± 0.0058
	EEG-DINO-S*	4.6M	38.83 ± 0.33	0.2399 ± 0.0046	0.3963 ± 0.0036
	EEG-DINO-S	4.6M	40.63 ± 0.45	0.2564 ± 0.0067	0.4092 ± 0.0060
	EEG-DINO-M	33M	41.38 ± 0.32	0.2727 ± 0.0055	0.4234 ± 0.0052
	EEG-DINO-L	201M	$\mathbf{41.77} \pm 0.40$	$\mathbf{0.2801} \pm 0.0051$	$\mathbf{0.4315} \pm 0.0042$

*Random initialization before fine-tuning.

Table 5. The reults of fine-tuning on TUAB

Methods	Params	BA(%)	AUC-PR	AUROC
CNN-T [13]	3.2M	77.77 ± 0.22	0.8433 ± 0.0039	0.8461 ± 0.0013
ST-T [16]	3.5M	79.66 ± 0.23	0.8521 ± 0.0026	0.8707 ± 0.0019
BIOT [21]	3.2M	79.59 ± 0.57	0.8792 ± 0.0023	0.8815 ± 0.0043
LaBraM-B [8]	5.8M	81.40 ± 0.19	0.8965 ± 0.0016	0.9022 ± 0.0009
CBraMod [20]	4.0M	80.91 ± 0.32	0.8906 ± 0.0018	0.8831 ± 0.0030
EEG-DINO-S*	4.6M	79.17 ± 0.16	0.8662 ± 0.0017	0.8708 ± 0.0021
EEG-DINO-S	4.6M	81.37 ± 0.36	0.8906 ± 0.0021	0.8981 ± 0.0016
EEG-DINO-M	33M	81.55 ± 0.32	0.8963 ± 0.0011	0.9018 ± 0.0017
EEG-DINO-L	201M	$\mathbf{82.07} \pm 0.24$	$\mathbf{0.9012} \pm 0.0015$	$\mathbf{0.9100} \pm 0.0009$

*Random initialization before fine-tuning.

effective across the three datasets used. This indicates that these techniques are well-suited for EEG signals and can exploit the unique structural characteristics, enhancing our model's performance and robustness.

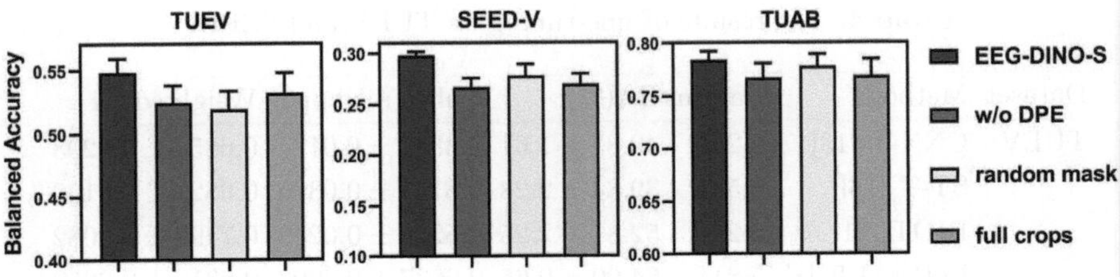

Fig. 2. The results of ablation study on TUEV, SEED-V and TUAB (linear probing)

4 Conclusion

This paper introduces EEG-DINO, a novel distillation-driven self-supervised learning framework for EEG signal encoding. By integrating hierarchical self-distillation with multi-view semantic alignment, channel-aware sampling, and decoupled positional embedding, our approach effectively captures discriminative spatiotemporal EEG signal patterns. Extensive evaluations demonstrate superior performance of EEG-DINO over counterparts. Our work underscores the potential of self-supervised knowledge distillation for robust EEG representation learning, paving the way for adaptive, scalable neurotechnology solutions.

Acknowledgments. This study was partially funded by the National Natural Science Foundation of China (Grant No. 62176068, 623B2011, 62325301 and U24B20186).

Disclosure of Interests. The authors have no competing interests to declare that are relevant to the content of this article.

References

1. Al-Saegh, A., Dawwd, S.A., Abdul-Jabbar, J.M.: Deep learning for motor imagery EEG-based classification: a review. Biomed. Signal Process. Control **63**, 102172 (2021)
2. Altaheri, H., et al.: Deep learning techniques for classification of electroencephalogram (EEG) motor imagery (mi) signals: a review. Neural Comput. Appl. **35**(20), 14681–14722 (2023)
3. Babiloni, C., et al.: Measures of resting state EEG rhythms for clinical trials in Alzheimer's disease: recommendations of an expert panel. Alzheimer's Dementia **17**(9), 1528–1553 (2021)
4. Fan, C., et al.: Dgsd: Dynamical graph self-distillation for EEG-based auditory spatial attention detection. Neural Netw. **179**, 106580 (2024)
5. Ferrante, M., Boccato, T., Bargione, S., Toschi, N.: Decoding visual brain representations from electroencephalography through knowledge distillation and latent diffusion models. Comput. Biol. Med. **178**, 108701 (2024)
6. He, K., Chen, X., Xie, S., Li, Y., Dollár, P., Girshick, R.: Masked autoencoders are scalable vision learners. In: Proceedings of the IEEE/CVF Conference on Computer Vision and Pattern Recognition, pp. 16000–16009 (2022)

7. He, K., Fan, H., Wu, Y., Xie, S., Girshick, R.: Momentum contrast for unsupervised visual representation learning. In: Proceedings of the IEEE/CVF Conference on Computer Vision and Pattern Recognition, pp. 9729–9738 (2020)
8. Jiang, W.B., Zhao, L.M., Lu, B.L.: Large brain model for learning generic representations with tremendous EEG data in bci (2024). arXiv preprint arXiv:2405.18765
9. Lawhern, V.J., Solon, A.J., Waytowich, N.R., Gordon, S.M., Hung, C.P., Lance, B.J.: Eegnet: a compact convolutional neural network for EEG-based brain-computer interfaces. J. Neural Eng. **15**(5), 056013 (2018)
10. Liu, W., Qiu, J.L., Zheng, W.L., Lu, B.L.: Comparing recognition performance and robustness of multimodal deep learning models for multimodal emotion recognition. IEEE Trans. Cogn. Dev. Syst. **14**(2), 715–729 (2021)
11. Obeid, I., Picone, J.: The temple university hospital EEG data corpus. Front. Neurosci. **10**, 196 (2016)
12. Oquab, M., et al.: Dinov2: Learning robust visual features without supervision (2023). arXiv preprint arXiv:2304.07193
13. Peh, W.Y., Yao, Y., Dauwels, J.: Transformer convolutional neural networks for automated artifact detection in scalp EEG. In: 2022 44th Annual International Conference of the IEEE Engineering in Medicine and Biology Society (EMBC), pp. 3599–3602. IEEE (2022)
14. Peng, R., et al.: Multi-branch mutual-distillation transformer for EEG-based seizure subtype classification. IEEE Trans. Neural Syst. Rehabil. Eng. **32**, 831–839 (2024)
15. Ranjan, R., Sahana, B.C., Bhandari, A.K.: Deep learning models for diagnosis of schizophrenia using EEG signals: emerging trends, challenges, and prospects. Arch. Comput. Methods Eng. **31**(4), 2345–2384 (2024)
16. Song, Y., Jia, X., Yang, L., Xie, L.: Transformer-based spatial-temporal feature learning for EEG decoding (2021). arXiv preprint arXiv:2106.11170
17. Sun, J., et al.: Adaptive spatiotemporal encoding network for cognitive assessment using resting state EEG. NPJ Digit. Med. **7**(1), 375 (2024)
18. Supakar, R., Satvaya, P., Chakrabarti, P.: A deep learning based model using RNN-LSTM for the detection of schizophrenia from EEG data. Comput. Biol. Med. **151**, 106225 (2022)
19. Tasci, I., et al.: Epilepsy detection in 121 patient populations using hypercube pattern from EEG signals. Inf Fusion **96**, 252–268 (2023)
20. Wang, J., et al.: Cbramod: A criss-cross brain foundation model for EEG decoding (2024). arXiv preprint arXiv:2412.07236
21. Yang, C., Westover, M., Sun, J.: Biot: Biosignal transformer for cross-data learning in the wild. Adv. Neural. Inf. Process. Syst. **36**, 78240–78260 (2023)
22. Zhao, W., et al.: EEG spectral analysis in insomnia disorder: a systematic review and meta-analysis. Sleep Med. Rev. **59**, 101457 (2021)
23. Zhou, J., et al.: ibot: Image BERT pre-training with online tokenizer (2021). arXiv preprint arXiv:2111.07832

Frequency-Enhanced Multi-granularity Context Network for Efficient Vertebrae Segmentation

Jian Shi[1], Tianqi You[1], Pingping Zhang[2(✉)], Hongli Zhang[3], Rui Xu[1], and Haojie Li[4]

[1] School of Software Technology and DUT-RU International School of Information Science and Engineering, Dalian University of Technology, Dalian, China
[2] School of Future Technology, School of Artificial Intelligence, Dalian University of Technology, Dalian, China
zhpp@dlut.edu.cn
[3] The Fifth Affiliated Hospital of Zhengzhou University, Zhengzhou, China
[4] College of Computer Science and Engineering, Shandong University of Science and Technology, Qingdao, China

Abstract. Automated and accurate segmentation of individual vertebra in 3D CT and MRI images is essential for various clinical applications. Due to the limitations of current imaging techniques and the complexity of spinal structures, existing methods still struggle with reducing the impact of image blurring and distinguishing similar vertebrae. To alleviate these issues, we introduce a Frequency-enhanced Multi-granularity Context Network (FMC-Net) to improve the accuracy of vertebrae segmentation. Specifically, we first apply wavelet transform for lossless downsampling to reduce the feature distortion in blurred images. The decomposed high and low-frequency components are then processed separately. For the high-frequency components, we apply a High-frequency Feature Refinement (HFR) to amplify the prominence of key features and filter out noises, restoring fine-grained details in blurred images. For the low-frequency components, we use a Multi-granularity State Space Model (MG-SSM) to aggregate feature representations with different receptive fields, extracting spatially-varying contexts while capturing long-range dependencies with linear complexity. The utilization of multi-granularity contexts is essential for distinguishing similar vertebrae and improving segmentation accuracy. Extensive experiments demonstrate that our method outperforms state-of-the-art approaches on both CT and MRI vertebrae segmentation datasets. The source code is publicly available at https://github.com/anaanaa/FMCNet.

Keywords: Vertebrae Segmentation · Wavelet Transform · Frequency Feature · Multi-granularity · State Space Model

J. C. Gee et al. (Eds.): MICCAI 2025, LNCS 15960, pp. 206–216, 2026.
https://doi.org/10.1007/978-3-032-04927-8_20

1 Introduction

The spine is vital to the body, supporting the organ function and serving as the core of the neural pathways. Any injury to the spine can impair neural conduction and lead to associated disorders [1]. Automatic segmentation and identification of individual vertebra in the spine are crucial for the treatment of orthopedic diseases, as they play a key role in preoperative diagnosis, treatment planning, surgical guidance, and postoperative assessment [2–4]. Although significant efforts have been made to achieve highly accurate vertebrae segmentation, the performance still remains suboptimal due to the limitations of current imaging techniques and the complexity of spinal structures [5]. As illustrated in Fig. 1, there are two main challenges in this task: (1) The intrinsic properties of different 3D medical imaging modalities, such as high noise levels and low contrast in CT scans, and non-isotropic spatial resolution in MRI images, lead to image blurring and distortion. Examples of these phenomena in CT and MRI images are illustrated in Fig. 1 (a) and Fig. 1 (b), respectively. (2) The high similarity in shape and structure between adjacent vertebrae presents a significant challenge for accurate segmentation. Figure 1 (c) demonstrates the impact of homogeneous structures on vertebrae segmentation, showing that partial voxel predictions for adjacent vertebrae are incorrectly assigned the same label.

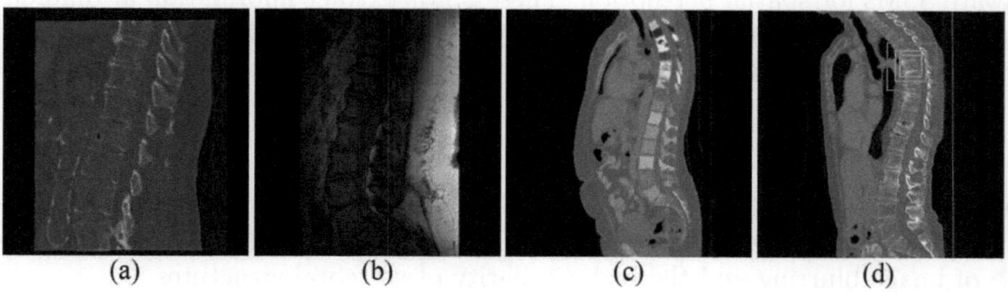

Fig. 1. Challenging examples in vertebrae segmentation. (a) and (b) present examples of image blurring caused by current imaging techniques in CT and MRI modalities, respectively. (c) presents examples of the challenges in distinguishing similar vertebrae. (d) highlights the importance of multi-granularity contextual information contained in the features of different receptive fields for distinguishing similar vertebrae.

Existing vertebrae segmentation methods typically employ an encoder-decoder structure and have achieved commendable results [6,7]. However, previous methods often overlook the critical impact of downsampling techniques. The max-pooling and bilinear interpolation are commonly utilized, frequently resulting in a substantial loss of fine-grained details during the downsampling process. This loss is especially harmful for blurred images, as retaining fine-grained details is essential for accurate segmentation. Besides, vertebrae segmentation requires sufficient contextual information to distinguish vertebrae. Existing Convolutional Neural Networks (CNNs) rely on local convolutional kernels to extract features, which limits their ability to extract global and geometric features. Although Transformer-based methods [8,9] excel in capturing long-range dependencies, their self-attention mechanism has a high computational complexity. Recently, the State Space Model (SSM) [10,11] could capture the long-range dependencies with linear complexity, providing a promising solution to the limitations of CNNs and Transformers. While SSM lacks consideration of spatially-varying contexts, this limitation reduces its sensitivity to multi-scale contextual information in similar vertebrae, impairing its ability to distinguish them.

To mitigate the aforementioned issues, we introduce a Frequency-enhanced Multi-granularity Context Network (FMC-Net) for accurate vertebrae segmentation. To begin with, we replace traditional downsampling methods with wavelet transform to reduce the loss of detailed information in blurred images. The wavelet transform achieves downsampling by decomposing the features into high and low-frequency components, which are then processed separately in the subsequent steps. High-frequency components contain a wealth of image details, textures, and edge information, but they also include noises. Therefore, we employ a High-frequency Feature Refinement (HFR) to amplify the prominence of key features while filtering out noises, effectively restoring details and textures in blurred images. Low-frequency components contain the most image content, we propose a Multi-granularity State Space Model (MG-SSM) to comprehensively extract low-frequency information. Specifically, MG-SSM contains granularity-aware SSMs for spatial perception. These SSMs extract information at different granularities to distinguish similar vertebrae. We evaluate the proposed method on both CT and MRI vertebrae segmentation datasets. Extensive experimental results demonstrate that our approach achieves state-of-the-art performance.

Our main contributions can be summarized as follows:

- We propose FMC-Net that leverages wavelet transform to avoid information loss and enhances high and low-frequency features to address the challenges of image blurring and the high similarity of vertebral structures.
- For the high-frequency components, we introduce the HFR to restore fine-grained details. For the low-frequency components, we employ the MG-SSM to capture spatially-varying long-range dependencies.
- Experiments on VERSE2019 and LUMBAR datasets demonstrate the superior performance of FMC-Net compared to other state-of-the-art approaches.

2 Method

The Overall Architecture: As shown in Fig. 2, given a 3D input volume $I \in \mathbb{R}^{D\times H\times W}$, D, H and W represent the depth, height, and width of the volume, respectively. FMC-Net aims to achieve accurate vertebrae segmentation by employing a wavelet-based encoder-decoder structure integrated with HFR and MG-SSM. Firstly, in the encoder stage, wavelet transform enables lossless downsampling, separating the features into high-frequency and low-frequency components. The HFR and MG-SSM are respectively applied to these components, yielding enhanced feature representations. These enhanced components are fused as the final features at each stage, effectively capturing both semantic and detailed information. In the decoder stage, Wavelet Transform Upsampling (WTU) employs wavelet transform-based feature fusion during upsampling, enabling the decoder to effectively leverage the information from the encoder.

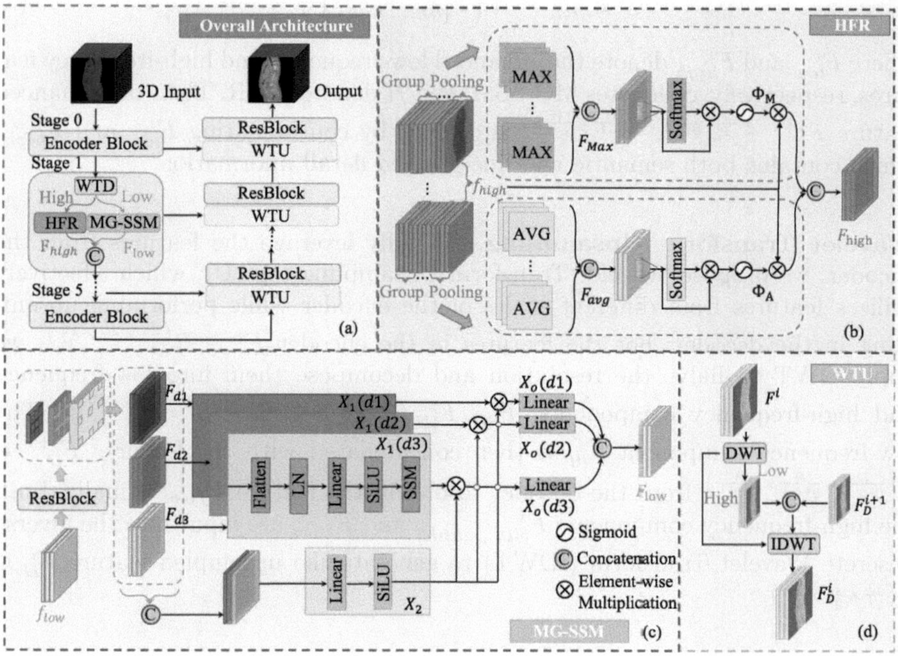

Fig. 2. (a) is the overall architecture of the proposed FMC-Net. (b) and (c) are HFR and MG-SSM, which enhance the high-frequency and low-frequency components, respectively. (d) is WTU, which utilizes features from different scales of the encoder while performing upsampling in the decoder.

2.1 Wavelet Transform-Based Sampling

Wavelet Transform Downsampling. Inspired by Guo *et al.* [12], we implement Wavelet Transform Downsampling (WTD) using the Discrete Wavelet Transform (DWT) [13] to alleviate the irreversible distortion and information loss caused by downsampling. For the feature $F^i \in \mathbb{R}^{\left(\frac{D}{2^i}\times\frac{H}{2^i}\times\frac{W}{2^i}\right)}$ of the i-th ($i \in (0, 1, \ldots, 5)$) stage encoder, we employ DWT to decompose the feature F^i into eight sub-wavelet bands, i.e., low-frequency component F^i_{lll}, and high-frequency components $F^i_{\{llh,\ldots,hhh\}}$ in horizontal, vertical, and depth dimensions (which are x-axis, y-axis and z-axis):

$$\{F^i_{lll}, F^i_{llh}, F^i_{lhl}, F^i_{lhh}, F^i_{hll}, F^i_{hlh}, F^i_{hhl}, F^i_{hhh}, \} = DWT(F^i), \tag{1}$$

where $F^i_{\{lll,\ldots,hhh\}} \in \mathbb{R}^{\left(\frac{D}{2^{i+1}} \times \frac{H}{2^{i+1}} \times \frac{W}{2^{i+1}}\right)}$. As low-frequency components contain the majority of the essential information, while high-frequency components capture intricate details, textures, and edge information, we focus on enhancing both of these components:

$$F^i_{low} = \mathcal{S}\left(F^i_{lll}\right), \tag{2}$$

$$F^i_{high} = \mathcal{H}\left(F^i_{\{llh,\ldots,hhh\}}\right), \tag{3}$$

where F^i_{low} and F^i_{high} denote the enhanced low-frequency and high-frequency features, respectively. $\mathcal{S}$ denotes MG-SSM, and $\mathcal{H}$ denotes HFR. Then the enhanced feature $F^{i+1} \in \mathbb{R}^{\left(\frac{D}{2^{i+1}} \times \frac{H}{2^{i+1}} \times \frac{W}{2^{i+1}}\right)}$ is derived by concatenating F^i_{low} and F^i_{high}, which contains both semantic and fine-grained detail information.

Wavelet Transform Upsampling. To fully leverage the features from the encoder, we propose Wavelet Transform Upsampling (WTU), which effectively utilizes features from different scales of the encoder while performing upsampling in the decoder. For the features in the encoder $F^i \in \mathbb{R}^{\left(\frac{D}{2^i} \times \frac{H}{2^i} \times \frac{W}{2^i}\right)}$, we apply DWT to halve the resolution and decompose them into low-frequency and high-frequency components, F^i_{lll}, $F^i_{\{llh,\ldots,hhh\}} \in \mathbb{R}^{\left(\frac{D}{2^{i+1}} \times \frac{H}{2^{i+1}} \times \frac{W}{2^{i+1}}\right)}$. The low-frequency component F^i_{lll} is then concatenated with the features $F^{i+1}_D \in \mathbb{R}^{\left(\frac{D}{2^{i+1}} \times \frac{H}{2^{i+1}} \times \frac{W}{2^{i+1}}\right)}$ from the decoder to obtain the feature F_{fuse}. Finally, both the high-frequency components $F^i_{\{llh,\ldots,hhh\}}$ and F_{fuse} are input into the inverse Discrete Wavelet Transform (IDWT) to generate the upsampled feature $F^i_D \in \mathbb{R}^{\left(\frac{D}{2^i} \times \frac{H}{2^i} \times \frac{W}{2^i}\right)}$.

2.2 High-Frequency Feature Refinement

High-frequency components capture detailed textures, but also contain noises. We employ HFR to effectively leverage high-frequency information. The high-frequency components f_{high} (comprising $F_{llh}, \ldots, F_{hhh}$) are processed through two distinct paths: one focuses on amplifying the prominence of key features, and the other filters out noises, thereby facilitating the restoration of fine-grained details and textures in blurred images. Both paths employ spatial attention mechanisms to dynamically modulate different frequency components during training. In the amplification path, the spatial attention mechanism Φ_M can be formulated as:

$$F_{\text{Max}} = \mathcal{F}^{7C \to C}_{3\times3}([\text{GroupMax}(F_{llh}), \ldots, \text{GroupMax}(F_{hhh})]), \tag{4}$$

$$\Phi_M = \sigma(\text{Softmax}(F_{\text{Max}}) \otimes F_{\text{Max}}), \tag{5}$$

where σ denotes the Sigmoid function, $[\cdot]$ is the concatenation operation, and GroupMax is the group max-pooling operation. Each high-frequency component is divided into 2^i groups for pooling, and i denotes the number of encoder stages. After pooling, each component undergoes a dimension transformation that reduces the channel dimension from 2^i to C. $\mathcal{F}_{3\times3}^{7C\rightarrow C}$ is a 3×3 convolutional layer that transforms the dimension of the concatenated high-frequency feature from $7C$ to C. The concatenation operation aggregates the amplified features of each high-frequency component after max-pooling operations, while the Softmax operation computes the contribution of each component to the spatial attention map. Meanwhile, the denoising path follows the same structure as the amplification path, except that the average pooling is used instead of max-pooling. Similarly, a spatial attention map Φ_A is generated, which smooths the features of each component and effectively filters out noises. The final enhanced high-frequency features can be obtained by the following formula:

$$F_{high} = [\Phi_M \otimes F_{\{llh,\dots,hhh\}}, \Phi_A \otimes F_{\{llh,\dots,hhh\}}]. \tag{6}$$

Here, the spatial attention map Φ_M is multiplied with each high-frequency component to emphasize the key features, while Φ_A performs a similar operation to filter out noises from each component. Finally, the amplified and denoised components are concatenated to obtain the enhanced high-frequency component.

2.3 Multi-granularity State Space Model

For low-frequency component f_{low}, we propose MG-SSM to effectively utilize the rich information for distinguishing similar vertebrae. Firstly, we employ a residual block to extract local features. Then we use three parallel 3D depth-wise convolutional layers with different dilation rates to capture deep features with different scales of receptive fields. Subsequently, these features (F_{d1}, F_{d2}, F_{d3}) are respectively fed into three Visual State Space Modules (VSSMs) to learn multi-scale contexts. This process can be defined as following:

$$\begin{aligned} F_{dj} &= \text{DWConv}_{dj}(\text{ResBlock}(F_{lll})), j \in (1,2,3), \\ X_1(dj) &= \text{SSM}\left(\phi\left(\text{Linear}\left(\text{LN}(F_{dj})\right)\right)\right), \end{aligned} \tag{7}$$

where dj represents the j-th dilation rate, ϕ denotes the Sigmoid Linear Unit (SiLU) activation function [14], and $X_1(dj)$ is the first branch of each VSSM. Next, F_{d1}, F_{d2}, and F_{d3} are concatenated to generate multi-scale features while simultaneously capturing the inter-scale relationships. And the concatenated feature is then fed into the second branch of VSSM:

$$X_2 = \phi(\text{Conv}([F_{d1}, F_{d2}, F_{d3}]), \tag{8}$$

The features from both branches are combined to produce the output:

$$X_o(dj) = \text{Linear}(X_1(dj) \otimes X_2), \tag{9}$$

where X_2 is shared among three VSSMs. Finally, each VSSM output is passed through a linear layer, followed by concatenation to generate the final enhanced low-frequency component. MG-SSM addresses the limitations of a single SSM in capturing spatially-varying features and provides multi-granularity contextual information for distinguishing similar vertebrae.

3 Experiments

Datasets and Metrics. We evaluate the effectiveness of our approach on two public vertebrae segmentation datasets. The VERSE2019 dataset [15] consists of 160 CT scans with 26 segmentation classes, covering vertebrae from the cervical to the lumbar regions. The LUMBAR dataset [16] contains 156 MRI scans from multiple scanner models and vendors, including T1-weighted, contrast-enhanced T1-weighted, and T2-weighted sequences. It provides segmentation labels for lumbar vertebrae and intervertebral discs. We use only the lumbar vertebrae labels and conduct experiments on T1-weighted images due to their superior quality. Following the previous works [15,21], we use two standard metrics to evaluate segmentation performance: Dice Similarity Coefficient (DSC) and 95th percentile of Hausdorff Distance (HD95).

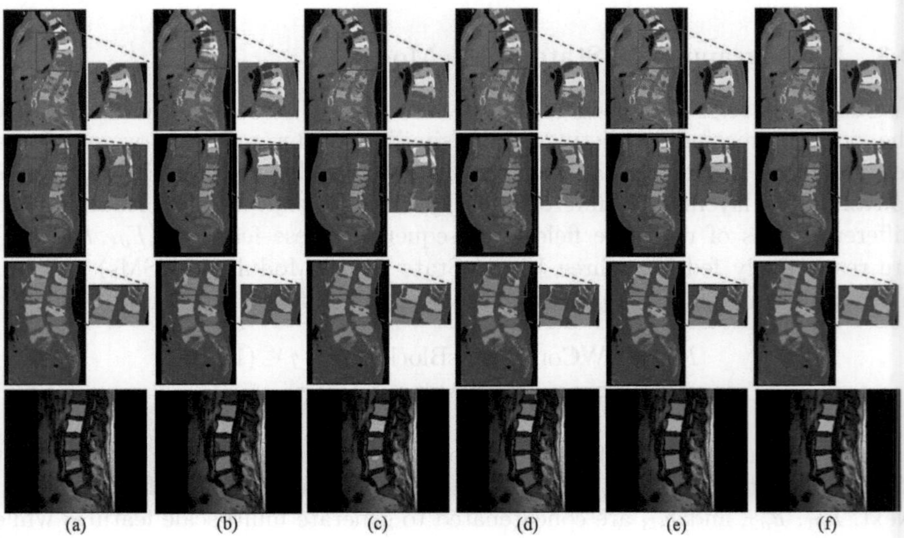

Fig. 3. Qualitative results with different methods. The top three rows display segmentation results from the VERSE2019 dataset, covering cervical, thoracic, and lumbar regions with various regions and views, whereas the bottom row shows results from the LUMBAR dataset. (a) nn-UNet, (b) UMamba, (c) MambaClnix, (d) Segmamba, (e) FMC-Net (Ours), (f) Ground Truth.

Table 1. Comparison of different methods on VERSE2019 and LUMBAR datasets. Cer., Tho., Lum. denote cervical, thoracic, and lumbar vertebrae, respectively.

Methods	DSC (%)					HD95 (mm) ↓				
	VERSE2019				LUMBAR	VERSE2019				LUMBAR
	Cer.	Tho.	Lum.	Mean	Lum.	Cer.	Tho.	Lum.	Mean	Lum.
nn-UNet	77.89	74.77	70.95	74.73	61.76	**1.21**	5.12	**3.94**	3.74	39.38
CoTR	74.56	67.15	64.57	68.61	72.55	7.93	10.15	11.39	9.83	31.87
Swin-UNetR	79.30	56.90	60.90	64.73	47.75	2.12	17.15	9.07	11.00	97.28
UMamba	83.54	79.34	71.01	78.28	72.47	1.60	**3.13**	4.75	3.09	27.59
SegMamba	73.90	74.63	65.43	72.22	78.17	5.70	4.86	11.70	6.73	17.53
MambaClinix	83.94	71.26	67.93	74.01	55.79	1.91	7.80	5.54	5.61	29.09
Ours	**85.46**	**80.74**	**71.83**	**79.92**	**79.13**	1.36	3.32	4.47	**3.04**	**17.13**

Table 2. Ablation experiments with different components on LUMBAR dataset.

Baseline	DWT-Sample	HFR	MG-SSM	DSC%	HD95
✓				71.09	42.77
✓	✓			71.75	22.12
✓	✓	✓		75.30	24.52
✓	✓		✓	76.56	18.67
✓			✓	75.94	16.72
✓	✓	✓	✓	79.13	17.13

Implementation Details. The proposed method is implemented in PyTorch and runs on an NVIDIA-A800-SXM4-80GB GPU. It is integrated within the nnU-Net framework for automatic data processing and augmentation, and employs a combined loss of weighted cross-entropy and Dice losses at each decoder stage. The model is trained on the VERSE2019 dataset with a patch size of (128, 160, 96) for 1000 epochs and on the LUMBAR dataset with a patch size of (32, 256, 160) for 600 epochs, both using a learning rate of 0.01.

Comparison with State-of-the-Art Methods. We compare the proposed model with three categories of methods to evaluate its effectiveness. The first category includes CNN-based methods, such as nn-UNet [17]. The second category includes Transformer-based approaches, comprising Swin-UNetR [18] and CoTR [19]. The third category consists of Mamba-based methods, including UMamba [20], SegMamba [21] and MambaClinix [22]. The evaluation results on the VERSE2019 and LUMBAR datasets are presented in Table 1. The VERSE2019 dataset is very challenging due to the inclusion of cervical, thoracic, and lumbar regions with varying fields of view. Although the performance improvements of our method over existing approaches may be modest in certain regions, it still achieves a new state-of-the-art. Compared with the best-performing Mamba-based method, SegMamba, our approach improves the mean DSC per class by 1.64%. On the LUMBAR dataset, which contains only lumbar vertebrae, our method demonstrates a more substantial performance gain, achieving 79.13% in DSC and 17.13 in HD95. Figure 3 shows representative visualizations from both datasets for qualitative comparison. The visualizations demonstrate that our method has a significant advantage in distinguishing similar vertebrae, showcasing its robustness and effectiveness in the vertebrae segmentation task.

Ablation Study. To evaluate the effectiveness of each component within the proposed model, we conduct ablation studies on the LUMBAR dataset. Table 2 illustrates the contribution of each component to the performance. Specifically, we employ a standard U-shaped model with additive residual blocks as the baseline. Subsequently, we incrementally integrate the DWT-Sample, HFR, and MG-SSM to evaluate the effectiveness of each component. Here, DWT-Sample refers to the combination of WTD and WTU, which are utilized for downsampling and upsampling during the encoder and decoder stages, respectively. The results demonstrate that each component significantly enhances the overall performance, confirming the effectiveness of each individual component.

4 Conclusion

In this paper, we propose a novel vertebrae segmentation network, FMC-Net, which effectively addresses the challenges of image blurring and the difficulty in distinguishing similar vertebrae. We take advantage of the wavelet transform to prevent the information loss, decomposing the features into low-frequency and high-frequency components for feature enhancement. HFR refines the high-frequency components, successfully restoring fine-grained details and alleviating image blurring. MG-SSM extracts spatially-varying contexts and long-range dependencies, which enable the full utilization of the rich information contained in the low-frequency components. Extensive experiments demonstrate that FMC-Net outperforms state-of-the-art approaches on both CT and MRI vertebrae segmentation datasets.

Acknowledgments. This study was supported by the Fundamental Research Funds for the Central Universities with No. DUT24YG135 and DUT23YG232.

Disclosure of Interests. All authors declare that they have no conflicts of interest.

References

1. Wang, H., Song, Q., Yin, R., Ma, R.: B-spine: learning B-spline curve representation for robust and interpretable spinal curvature estimation. In: Proceedings of the AAAI Conference on Artificial Intelligence, vol. 38, No. 6, pp. 5381-5389 (2024)

2. Pang, S., et al.: SpineParseNet: spine parsing for volumetric MR image by a two-stage segmentation framework with semantic image representation. IEEE Trans. Med. Imaging **40**(1), 262–273 (2020)
3. Lessmann, N., Van Ginneken, B., De Jong, P.A., Išgum, I.: Iterative fully convolutional neural networks for automatic vertebra segmentation and identification. Med. Image Anal. **53**, 142–155 (2019)
4. Wu, H., Zhang, J., Fang, Y., Liu, Z., Wang, N., Cui, Z., Shen, D.: Multi-view vertebra localization and identification from ct images. In: Medical Image Computing and Computer Assisted Intervention, pp. 136-145 (2023)
5. Tao, R., Liu, W., Zheng, G.: Spine-transformers: vertebra labeling and segmentation in arbitrary field-of-view spine CTs via 3D transformers. Med. Image Anal. **75**, 102258 (2022)
6. Shi, W., et al.: Attention gate based dual-pathway network for vertebra segmentation of X-ray spine images. IEEE J. Biomed. Health Inform. **26**(8), 3976–3987 (2022)
7. Masuzawa, N., Kitamura, Y., Nakamura, K., Iizuka, S., Simo-Serra, E.: Automatic segmentation, localization, and identification of vertebrae in 3D CT images using cascaded convolutional neural networks. In: Medical Image Computing and Computer Assisted Intervention, pp. 681-690 (2020)
8. You, X., Gu, Y., Liu, Y., Lu, S., Tang, X., Yang, J.: EG-Trans3DUNet: a single-staged transformer-based model for accurate vertebrae segmentation from spinal CT images. In: International Symposium on Biomedical Imaging, pp. 1-5 (2022)
9. You, X., Gu, Y., Liu, Y., Lu, S., Tang, X., Yang, J.: VerteFormer: a single-staged Transformer network for vertebrae segmentation from CT images with arbitrary field of views. Med. Phys. **50**(10), 6296–6318 (2023)
10. Smith, J. T., Warrington, A., Linderman, S. W.: Simplified state space layers for sequence modeling. In: International Conference on Learning Representations (2022)
11. Gu, A., Dao, T.: Mamba: linear-time sequence modeling with selective state spaces. In: First Conference on Language Modeling (2024)
12. Xu, G., Liao, W., Zhang, X., Li, C., He, X., Wu, X.: Haar wavelet downsampling: a simple but effective downsampling module for semantic segmentation. Pattern Recogn. **143**, 109819 (2023)
13. Li, Q., Shen, L., Guo, S., Lai, Z.: Wavelet integrated CNNs for noise-robust image classification. In: Proceedings of the IEEE/CVF Conference on Computer Vision and Pattern Recognition, pp. 7245-7254 (2020)
14. Elfwing, S., Uchibe, E., Doya, K.: Sigmoid-weighted linear units for neural network function approximation in reinforcement learning. Neural Netw. **107**, 3–11 (2018)
15. Sekuboyina, A., et al.: VerSe: a vertebrae labelling and segmentation benchmark for multi-detector CT images. Med. Image Anal. **73**, 102166 (2021)
16. Khalil, Y.A., et al.: Multi-scanner and multi-modal lumbar vertebral body and intervertebral disc segmentation database. Scientific Data **9**(1), 97 (2022)
17. Isensee, F., Jaeger, P.F., Kohl, S.A., Petersen, J., Maier-Hein, K.H.: nnU-Net: a self-configuring method for deep learning-based biomedical image segmentation. Nat. Methods **18**(2), 203–211 (2021)
18. Tang, Y., et al.: Self-supervised pre-training of swin transformers for 3d medical image analysis. In: Proceedings of the IEEE/CVF Conference on Computer Vision and Pattern Recognition, pp. 20730-20740 (2022)
19. Xie, Y., Zhang, J., Shen, C., Xia, Y.: Cotr: efficiently bridging cnn and transformer for 3d medical image segmentation. In: Medical Image Computing and Computer Assisted Intervention, pp. 171-180 (2021)

20. Ma, J., Li, F., Wang, B.: U-mamba: Enhancing long-range dependency for biomedical image segmentation. arXiv preprint arXiv:2401.04722 (2024)
21. Xing, Z., Ye, T., Yang, Y., Liu, G., Zhu, L.: Segmamba: long-range sequential modeling mamba for 3d medical image segmentation. In: Medical Image Computing and Computer-Assisted Intervention, pp. 578-588 (2024)
22. Bian, C., et al.: Mambaclinix: hierarchical gated convolution and mamba-based u-net for enhanced 3d medical image segmentation. arXiv preprint arXiv:2409.12533 (2024)

Global and Local Contrastive Learning for Joint Representations from Cardiac MRI and ECG

Alexander Selivanov[1(✉)], Philip Müller[1,2], Özgün Turgut[1,2], Nil Stolt-Ansó[1,4], and Daniel Rueckert[1,2,3,4]

[1] Chair for AI in Healthcare and Medicine, Technical University of Munich (TUM) and TUM University Hospital, Munich, Germany
alexander.selivanov@tum.de
[2] School of Medicine, Klinikum rechts der Isar,TUM, Munich, Germany
[3] Department of Computing, Imperial College London, London, UK
[4] Munich Center for Machine Learning (MCML), Munich, Germany

Abstract. An electrocardiogram (ECG) is a widely used, cost-effective tool for detecting electrical abnormalities in the heart. However, it cannot directly measure functional parameters, such as ventricular volumes and ejection fraction, which are crucial for assessing cardiac function. Cardiac magnetic resonance (CMR) is the gold standard for these measurements, providing detailed structural and functional insights, but is expensive and less accessible. To bridge this gap, we propose PTACL (**P**atient and **T**emporal **A**lignment **C**ontrastive **L**earning), a multimodal contrastive learning framework that enhances ECG representations by integrating spatio-temporal information from CMR. PTACL uses global patient-level contrastive loss and local temporal-level contrastive loss. The global loss aligns patient-level representations by pulling ECG and CMR embeddings from the same patient closer together, while pushing apart embeddings from different patients. Local loss enforces fine-grained temporal alignment within each patient by contrasting encoded ECG segments with corresponding encoded CMR frames. This approach enriches ECG representations with diagnostic information beyond electrical activity and transfers more insights between modalities than global alignment alone, all without introducing new learnable weights. We evaluate PTACL on paired ECG-CMR data from 27,951 subjects in the UK Biobank. Compared to baseline approaches, PTACL achieves better performance in two clinically relevant tasks: (1) retrieving patients with similar cardiac phenotypes and (2) predicting CMR-derived cardiac function parameters, such as ventricular volumes and ejection fraction. Our results highlight the potential of PTACL to enhance non-invasive cardiac diagnostics using ECG. The code is available at: https://github.com/alsalivan/ecgcmr.

Keywords: Contrastive Learning · Time-Alignment · ECG · MRI

J. C. Gee et al. (Eds.): MICCAI 2025, LNCS 15960, pp. 217–227, 2026.
https://doi.org/10.1007/978-3-032-04927-8_21

1 Introduction

Cardiovascular diseases (CVD) remain a leading cause of mortality worldwide [28]. Electrocardiograms (ECGs) are a non-invasive and cost-effective tool for detecting electrical heart abnormalities (e.g., arrhythmias [1]). However, ECGs can't directly assess structural and functional cardiac properties like ventricular volumes, myocardial mass, or ejection fraction. In contrast, cardiac magnetic resonance (CMR) is the gold standard for these assessments [11], but it is expensive, time-consuming, and requires specialized expertise, limiting its accessibility.

To integrate information from both modalities, researchers have applied contrastive learning to align ECG and CMR embeddings globally [5,18,25], pulling same-patient representations closer while pushing apart different patients. However, these approaches treat ECG and CMR as holistic representations, overlooking finer temporal relationships. Since CMR captures dynamic sequences across the cardiac cycle while ECG records continuous electrical activity, global alignment alone lacks the temporal precision needed to effectively integrate both modalities.

To address this limitation, we introduce PTACL (**P**atient and **T**emporal **A**lignment **C**ontrastive **L**earning), a contrastive learning framework that combines global and local alignment. The global loss aligns ECG and CMR embeddings at the patient level, while the local loss enforces fine-grained temporal alignment by contrasting ECG-derived segments—encoded tokens from a single heartbeat representation—with their corresponding CMR frame representations at the same cardiac phase (Fig. 1). Notably, the local alignment is completely parameter-free, introducing no additional learnable parameters. Our method is fully self-supervised and is trained on paired ECG-CMR data from 27,951 subjects in the UK Biobank [21]. We evaluate PTACL on patient retrieval and phenotype regression, assessing how well ECG embeddings capture CMR-derived cardiac function. Our results demonstrate that local contrastive alignment improves ECG representations, enhancing their diagnostic utility for assessing cardiac structure and function.

2 Related Works

Self-supervised learning has advanced both single modality and multimodal representation learning. For single modality tasks, masked autoencoders (MAE) [7] have proven highly effective in learning meaningful representations from unlabeled data. They have been widely applied to signals such as ECG [4,16,19,26,30,32], speech, and audio [8], as well as medical imaging, where 3D MAEs [6,24] learn spatial-temporal features from volumetric scans [15,22,33].

In multimodal learning, most contrastive approaches are inspired by CLIP [17], which aligns per-patient representations from different modalities in a shared embedding space, pulling similar samples together while pushing apart unrelated ones. This framework has been applied in medical domains, such as ECG-CMR integration [5,18,25] and X-ray–text association [23,27,34], enabling

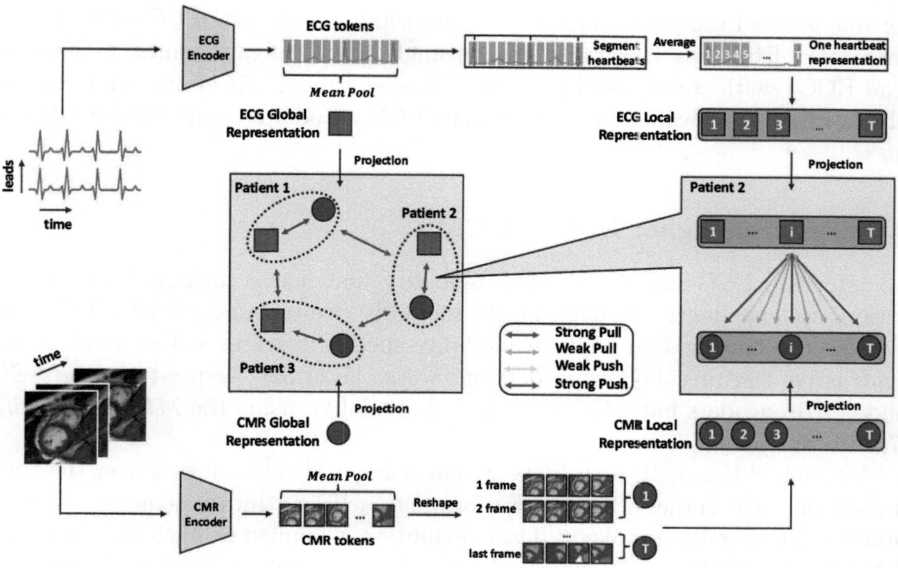

Fig. 1. Overview of PTACL. We train a multimodal model on pairs of ECG and CMR using a combination of a global contrastive loss (left square) and a local contrastive loss (right square). The global loss aligns patient-level ECG and CMR embeddings by pulling representations from the same patient closer together while pushing apart those from different patients. The local loss enforces fine-grained temporal alignment by contrasting ECG-derived segments—encoded from a single-heartbeat representation—with their corresponding CMR frames. In the figure, representations from 1 to T start at the end-diastole (ED) phase. Overlays in the tokens (CMR and ECG) are for visualization purposes only.

zero-shot classification and improved diagnostic insights. However, these methods primarily focus on global alignment, treating each modality as a single representation and potentially overlooking fine-grained local interactions.

To address this, recent work has started exploring local contrastive learning for medical applications. For example, GLORIA [9] applies a global-local learning scheme to image-text pairs, while Seibold et al. [20] refine CLIP by contrasting local image features with radiology reports, and LoVT [14] aligns image subregions with text embeddings. These methods mirror broader trends in vision-language research [12,29,31].

3 Methods

Our framework consists of three phases: single modality pre-training (SMP) (Sect. 3.1), multimodal contrastive learning (Sect. 3.2), and inference. First, we pre-train separate ECG and CMR encoders using MAE [7] on unlabeled data. Then, we jointly train both encoders with substantial augmentations, applying a global contrastive loss for patient-level alignment and a local contrastive loss for fine-grained temporal alignment. The local loss contrasts ECG-derived time segments—fixed length segments from a single heartbeat representation of multi-lead ECG—with corresponding encoded frames from CMR at the same cardiac phase. Finally, we evaluate the pre-trained ECG encoder on multiple downstream tasks.

3.1 Pre-training for ECG and CMR

To integrate ECG and CMR, each modality first learns meaningful representations independently through single modality pre-training (SMP). This step enables the encoders to capture modality-specific features before multimodal contrastive learning. Following previous works [8,24,32], we pre-train the ECG and CMR encoders independently on each modality, using the MAE framework [7] in both cases.

We embed the multi-lead ECG signals using a 1D convolution over the time dimension with kernel size p_t, transforming each lead into a sequence of non-overlapping time-based tokens. The convolution is applied on each lead independently, with weights shared across all leads. To learn temporal and lead structure, we add learnable positional embeddings before randomly masking a portion of tokens.

For CMR, we treat the 2D+T short-axis sequence as a multi-frame input and apply a 3D convolution with kernel size $[t, p, p]$, where t defines the temporal window and p controls the spatial patch size. This generates a sequence of spatio-temporal tokens, each representing a localized region of the heart across time. To learn spatial and temporal structure, we add learnable positional embeddings before masking.

In both cases, a transformer encoder processes only the visible tokens, while a lightweight transformer decoder reconstructs the masked tokens. The Mean Squared Error (MSE) loss is computed only on the masked tokens, comparing the reconstructed output to the original uncorrupted input.

3.2 Multimodal Contrastive Pre-training

To learn robust cross-modal representations, we employ two contrastive losses: a global loss for patient-level alignment and a local loss for fine-grained temporal alignment. The global loss ensures that embeddings from the same patient are similar while pushing apart embeddings from different patients. The local loss further refines alignment by matching ECG-derived temporal embeddings with corresponding CMR frames. ECG local embeddings consist of T segments, each spanning p_t time steps from a single heartbeat representation of multi-lead ECG, while CMR embeddings represent encoded frames from the cardiac sequence.

Global Contrastive Loss: For a batch B, let $z_{\mathcal{E}}, z_{\mathcal{C}}$ be the ECG and CMR embeddings, obtained via average pooling over token representations followed by separate projection layers. We enforce cross-modal alignment using the InfoNCE loss, following [17]. The loss is defined symmetrically for ECG → CMR and CMR→ECG directions. The ECG→CMR contrastive loss is:

$$\mathcal{L}_{\text{ECG}\to\text{CMR}} = \frac{1}{B}\sum_{i=1}^{B} -\log \frac{\exp\left(\cos\left(z_{\mathcal{E}}^{(i)}, z_{\mathcal{C}}^{(i)}\right)/\tau\right)}{\sum_{j=1}^{B}\exp\left(\cos\left(z_{\mathcal{E}}^{(i)}, z_{\mathcal{C}}^{(j)}\right)/\tau\right)}, \tag{1}$$

where τ is the temperature parameter, $z_{\mathcal{E}}$ and $z_{\mathcal{C}}$ are the projected embeddings, and $\cos(\cdot,\cdot)$ denotes cosine similarity. The final global contrastive loss $\mathcal{L}_{\text{global}}$ is a weighted combination of both directions.

Local Contrastive Loss: CMR acquisition is guided by ECG gating, where imaging data is continuously recorded over multiple heartbeats while simultaneously capturing the ECG signal. The detected R peaks are later used to align and reconstruct a single representative heartbeat cycle. This physiological alignment motivates our local contrastive loss formulation.

Local Embedding Formation: For CMR, each frame is divided into spatial tokens representing $p \times p$ image patches and temporal tokens capturing sequences of t frames. Local embeddings are obtained by averaging spatial tokens within each temporal token, producing $z_{\mathcal{C}}^{(i,t)}$, where t indexes the CMR frames. For ECG, each token represents a segment of length p_t from a single lead. To standardize heartbeat representations across patients, we extract tokens between consecutive R-peaks for each detected heartbeat. These variable-length sequences are interpolated to a fixed length of T tokens. Finally, we compute the mean across multiple heartbeats to obtain the ECG local embeddings for patient i, denoted as $z_{\mathcal{E}}^{(i,k)}$, where k indexes the ECG time segments.

Temporal Alignment and Local Loss: Following [10], we enforce temporal alignment between $z_{\mathcal{E}}^{(i,k)}$ (ECG) and $z_{\mathcal{C}}^{(i,t)}$ (CMR) local embeddings, using symmetrical alignment matrix $\mathcal{P}_{k,t}$:

$$\mathcal{P}_{k,t} \propto \begin{cases} \delta_{k,t} & \sigma = 0 \quad \text{(hard alignment)} \\ \exp\left(-\frac{1}{2}\left(\frac{\text{dist}_{k,t}}{\sigma}\right)^2\right) & \sigma > 0 \quad \text{(soft alignment)} \end{cases} \tag{2}$$

Here $\sum_{t=1}^{T} \mathcal{P}_{k,t} = 1$ and $\delta_{k,t}$ ensures exact matches for $\sigma = 0$, while for $\sigma > 0$, alignment is softened using a Gaussian weighting based on the normalized temporal distance $\text{dist}_{k,t}$ (Fig. 2).

The local contrastive loss is defined symmetrically for ECG → CMR and CMR → ECG directions. The local per-patient, per-segment loss for ECG → CMR:

$$\ell_{\text{ECG}\to\text{CMR}}^{(i,k)} = -\sum_{t=1}^{T} \mathcal{P}_{k,t} \log\left[\frac{\exp\left(\cos(z_{\mathcal{E}}^{(i,k)}, z_{\mathcal{C}}^{(i,t)})/\tau\right)}{\sum_{t'=1}^{T}\exp\left(\cos(z_{\mathcal{E}}^{(i,k)}, z_{\mathcal{C}}^{(i,t')})/\tau\right)}\right]. \tag{3}$$

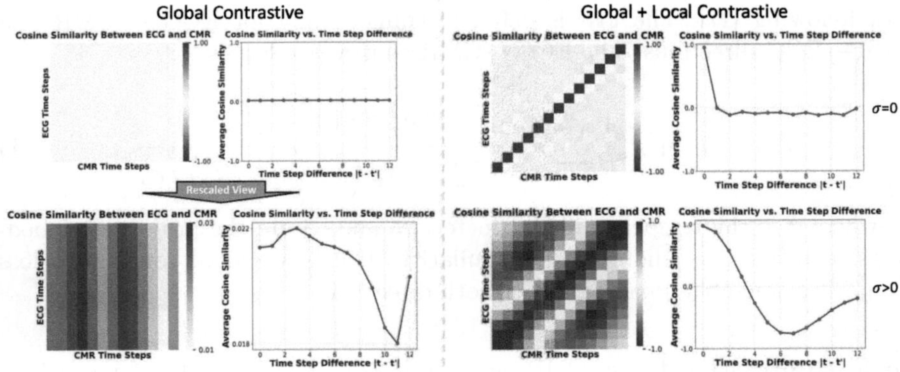

Fig. 2. Impact of local contrastive loss on temporal alignment between time steps from both modalities computed during inference. The left side, using only global contrastive learning, shows no clear temporal correspondence between modalities. The right side demonstrates the effect of adding local contrastive loss (PTACL), leading to improved alignment. For $\sigma = 0$, the model enforces strict one-to-one time step matching, while for $\sigma > 0$, Gaussian weighting enables smoother alignment. Our method achieves temporal alignment between ECG and CMR time steps, which is absent in the global-only approach.

The total local contrastive loss for ECG $\rightarrow$ CMR is obtained by averaging $\ell^{(i,k)}_{\text{ECG}\rightarrow\text{CMR}}$ over all ECG segments k and patients in the batch. The final local contrastive loss $\mathcal{L}_{\text{local}}$ is a weighted combination of both directions.

Total Loss: The final objective combines both global and local contrastive losses to ensure alignment at both patient and temporal levels. The total loss is:

$$L = \mathcal{L}_{\text{global}} + \beta \cdot \mathcal{L}_{\text{local}} \tag{4}$$

where β controls the relative contribution of the local contrastive loss.

4 Experiments

We evaluate our model using paired 12-lead resting ECG and short-axis CMR data from the UK Biobank [21], with 27,951 patients for training and 5,991 for test. CMR sequences contain 50 frames, with the middle short-axis slice cropped around the cardiac center (84 × 84) and min-max normalized. ECGs are 10-second, 12-lead recordings sampled 500 Hz. ECGs are preprocessed via an 8-level wavelet transform, following [13,32], for baseline wander and noise removal and followed by z-score normalization. CMR augmentations include random resized cropping, time sampling, rotation, flipping, color jitter, Gaussian blurring, and noise, while ECG augmentations include Fourier surrogate transformations, random cropping (2500 points), jittering, and rescaling. We evaluate 10 CMR-derived cardiac phenotypes [2]: LV/RV end-diastolic and end-systolic volumes (EDV, ESV), stroke volume (SV), ejection fraction (EF), LV mass (M), and LV cardiac output (CO).

4.1 Retrieval of Similar Patients

To assess the clinical utility of our learned ECG embeddings, we test their ability to retrieve patients with similar CMR-derived phenotypes. For each query ECG, we rank all patients in the CMR embedding database using cosine similarity. A match is counted if the candidate's phenotype is within $\pm 0.5\,\sigma$ of the query's value, where σ is the standard deviation of that phenotype across test data. We compare two embedding strategies: a baseline model (global) and PTACL (global+local). While our global model already demonstrates strong retrieval performance, incorporating a local loss further improves precision and ranking. This enhancement is clinically significant as it enables more accurate patient stratification. Table 1 summarizes Precision@k (P@k), mean rank (MnR), and median rank (MdR) for key heart phenotypes. For reference, values in parentheses indicate performance when retrieving k random patients, reported only for $k = 1$ and $k = 10$ to save space.

Table 1. Retrieval results for CMR-derived heart phenotypes using ECG embeddings. We compare baseline (global $\triangle$) and PTACL (global+local $\square$) models, reporting performance as $\triangle/\square$ (random), with random retrieval in parentheses. A match is defined as a candidate whose phenotype differs from the query by no more than $\pm 0.5\,\sigma$. The inclusion of local loss in PTACL consistently improves patient retrieval with higher precision and lower rank scores across all phenotypes, demonstrating its potential to enhance clinical patient stratification.

	P@1↑	P@3↑	P@10↑	P@15↑	MnR↓	MdR↓
EDV_{LV}	.492 / **.510** (.291)	.462 / **.463**	.426 / **.427** (.287)	**.416** / .415	3.1 / **2.9**	2.0 / **1.0**
ESV_{LV}	.501 / **.524** (.303)	.469 / **.480**	.438 / **.442** (.284)	.429 / **.432**	3.0 / 3.0	1.0 / 1.0
SV_{LV}	.458 / **.472** (.294)	.417 / **.423**	.387 / **.388** (.296)	**.380** / .377	3.4 / **3.2**	2.0 / 2.0
EF_{LV}	.431 / **.436** (.278)	.392 / **.396**	.355 / **.359** (.282)	.349 / **.351**	3.9 / **3.8**	2.0 / 2.0
CO_{LV}	.439 / **.453** (.295)	.407 / **.417**	.377 / **.378** (.284)	.368 / .368	3.9 / **3.6**	2.0 / 2.0
M_{LV}	.547 / **.561** (.293)	.516 / **.522**	.479 / **.486** (.294)	.471 / **.474**	2.6 / 2.6	1.0 / 1.0
EDV_{RV}	.500 / **.516** (.280)	.467 / **.477**	.433 / **.441** (.287)	.426 / **.430**	3.0 / **2.8**	2.0 / **1.0**
ESV_{RV}	.518 / **.531** (.290)	.485 / **.494**	.455 / **.460** (.281)	.446 / **.451**	2.8 / **2.6**	1.0 / 1.0
SV_{RV}	.458 / **.473** (.285)	.420 / **.428**	.386 / **.390** (.289)	.380 / **.382**	3.8 / **3.7**	2.0 / 2.0
EF_{RV}	.433 / **.443** (.288)	.397 / **.405**	.370 / **.373** (.290)	.364 / **.365**	4.3 / **4.2**	2.0 / 2.0

4.2 Regression of Cardiac Phenotypes

We evaluate the regression performance of different pre-training strategies on various cardiac phenotypes using R^2 scores. Table 2 presents the results for models trained on ECG, CMR, or paired ECG-CMR data. CMR single modality pre-training (SMP) reveals the upper bound of imaging information, while the lower performance of ECG SMP highlights the challenges of using ECG alone. Multimodal learning aims to reduce this gap. Among self-supervised approaches, MAE [7] consistently outperforms SimCLR [3] on both modalities, although fully supervised single modality training remains a robust baseline. Our Global model serves as a strong baseline, already outperforming previous works. Adding a local contrastive loss (PTACL) further improves performance across all cardiac phenotypes. Importantly, PTACL achieves these gains without introducing additional learnable parameters, making it a computationally efficient enhancement.

Table 2. Regression performance (R^2) for different pre-training methods. Models are evaluated with either linear probing (L) or full fine-tuning (F) using attention pooling (A). Our Global model outperforms previous multimodal approaches. Adding a local contrastive loss (PTACL) further improves results across all phenotypes without introducing additional learnable parameters. SMP stands for single modality pre-training.

Method	EDV_{LV}	ESV_{LV}	SV_{LV}	EF_{LV}	CO_{LV}	M_{LV}	EDV_{RV}	ESV_{RV}	SV_{RV}	EF_{RV}
CMR $R^2 \uparrow$										
Random Init^{L}	.015	.017	.009	.008	.003	.021	.018	.020	.009	.007
SimCLR^{L} [3]	.572	.612	.466	.513	.418	.628	.618	.681	.463	.520
MAE^{L} [7]	.809	.772	.678	.489	.571	.824	.799	.784	.657	.492
Supervised^{F}	.814	.789	.665	.470	.536	.823	.801	.808	.623	.473
ECG $R^2 \uparrow$										
Random Init^{*,L}	-4.8	-3.8	-7.	-38.8	-7.6	18.3	.9	3.6	-9.6	.9
SimCLR^{L} [3]	.278	.261	.196	.085	.156	.343	.308	.317	.205	.130
MAE^{L} [7]	.439	.425	.332	.218	.252	.524	.466	.481	.333	.254
Supervised^{F}	.458	.443	.333	.198	.252	.516	.481	.491	.340	.237
Multimodal (ECG+CMR) $R^2 \uparrow$										
ECCL^{1,F} [5]	.372	.348	.270	.176	.212	.449	.397	.397	.281	.203
CMAE^{2,L} [18]	.451	.380	.316	.103	.281	.536	.490	.445	.320	.129
MMCL^{3,F,A} [25]	.498	.497	.360	.245	-	.597	.527	.534	.375	.248
PTACL^{1,4,L}	.359	.334	.256	.114	.187	.406	.381	.386	.253	.149
PTACL^{1,5,6,L}	.509	.496	.383	.252	.286	.614	.537	.553	.387	.295
Global^{6,L}	.507	.499	.377	.258	.281	.612	.534	.553	.380	.297
PTACL^{1,6,L}	**.514**	**.500**	**.386**	**.255**	**.289**	**.616**	**.540**	**.557**	**.389**	**.299**
PTACL^{1,6,F,A}	**.544**	**.537**	**.408**	**.283**	**.310**	**.645**	**.568**	**.582**	**.408**	**.315**

* $\times 10^{-5}$, ^{1} Single mid SAX slice, ^{2} LAX + reconstruction loss, ^{3} Three SAX slices, ^{4} No SMP, ^{5} Local loss changed to maximising cosine similarity for matching time steps (no negatives), ^{6} With SMP, std ~0.001–0.002, ^{L} Linear probing, ^{F} Full fine-tuning, ^{A} Attention pooling.

4.3 Implementation Details

Our approach employs transformer-based architectures for both modalities. The CMR model has 4 layers, 8 attention heads, a patch [2, 12, 12], and a 90% mask ratio (13M parameters). The ECG model has 6 layers, 8 heads, a patch 50, and a 75% mask ratio (19M parameters). Both use a 1-layer, 4-head decoder. During multimodal pre-training, the first two layers are frozen, leaving 6.3M and 12.6M trainable parameters for CMR and ECG, respectively. The contrastive loss follows the $\mathcal{L}_{out}^{sup}$ variant from [10]. We use $T = 13$, $\beta = 1.0$. Regression is performed with $\sigma = 0.1$ and global/local temperatures of 0.1/1.0, retrieval uses $\sigma = 0.0$ with temperatures of 0.07/0.07. Linear probing is performed with ordinary least squares regression. Results are averaged over five random seeds.

5 Conclusion and Discussion

We introduced PTACL, a contrastive learning framework that enhances ECG representations through patient-level and time step alignment with CMR. Unlike prior methods focused solely on global alignment, PTACL captures finer temporal correspondences, leading to better retrieval and phenotype regression, all without adding extra learnable parameters. Our approach relies on only a single middle short-axis slice from the CMR, demonstrating that minimal imaging data can still significantly enrich ECG. However, PTACL relies on paired ECG-CMR data, and our evaluation is limited to a single cohort. Future work could explore generalization to external datasets, extend PTACL to other modalities (e.g., ultrasound), or reduce reliance on paired multimodal data.

Acknowledgments. This research has been conducted using the UK Biobank Resource under Application Number 87802. This work is funded by the European Research Council (ERC) project Deep4MI (884622). A. S. is funded via the EVUK program ("Next-generation AI for Integrated Diagnostics") of the Free State of Bavaria.

Disclosure of Interests. The authors have no competing interests to declare that are relevant to the content of this article.

References

1. Ansari, Y., Mourad, O., Qaraqe, K., Serpedin, E.: Deep learning for ECG arrhythmia detection and classification: An overview of progress for period 2017–2023. Front. Physiol. **14**, 1246746 (2023)
2. Bai, W., Suzuki, H., Huang, J., Francis, C., Wang, S., Tarroni, G., et al.: A population-based phenome-wide association study of cardiac and aortic structure and function. Nat. Med. **26**(10), 1654–1662 (2020)
3. Chen, T., Kornblith, S., Norouzi, M., Hinton, G.: A simple framework for contrastive learning of visual representations. In: ICML, pp. 1597–1607. PMLR (2020)
4. Chien, H.Y.S., Goh, H., Sandino, C.M., Cheng, J.Y.: MAEEG: Masked autoencoder for EEG representation learning. In: NeurIPS Workshop (2022). https://arxiv.org/abs/2211.02625

5. Ding, Z., Hu, Y., Li, Z., Zhang, H., Wu, F., Xiang, Y., et al.: Cross-modality cardiac insight transfer: A contrastive learning approach to enrich ECG with CMR features. In: Linguraru, M.G., et al. (eds.) Medical Image Computing and Computer Assisted Intervention. MICCAI 2024. LNCS, vol. 15003, pp. 109–119. Springer, Cham (2024). https://doi.org/10.1007/978-3-031-72384-1_11
6. Feichtenhofer, C., Fan, H., Li, Y., He, K.: Masked autoencoders as spatiotemporal learners. In: Koyejo, S., et al. (eds.) Advances in NeurIPS, vol. 35, pp. 35946–35958 (2022)
7. He, K., Chen, X., Xie, S., Li, Y., Dollár, P., Girshick, R.: Masked autoencoders are scalable vision learners. In: CVPR, pp. 16000–16009 (2022)
8. Huang, P.Y., Xu, H., Li, J., Baevski, A., Auli, M., Galuba, W., et al.: Masked autoencoders that listen. In: NeurIPS, vol. 35, 28708–28720 (2022)
9. Huang, S.C., Shen, L., Lungren, M.P., Yeung, S.: GLoRIA: A multimodal global-local representation learning framework for label-efficient medical image recognition. In: ICCV, pp. 3942–3951 (2021)
10. Khosla, P., Teterwak, P., Wang, C., Sarna, A., Tian, Y., Isola, P., et al.: Supervised contrastive learning. In: Larochelle, H., et al. (eds.) Advances in NeurIPS. vol. 33, pp. 18661–18673 (2020)
11. Leiner, T., Bogaert, J., Friedrich, M.G., Mohiaddin, R., et al.: SCMR position paper (2020) on clinical indications for cardiovascular magnetic resonance. J. Cardiovasc. Magn. Reson. **22**(1), 76 (2020)
12. Ma, Y., Xu, G., Sun, X., Yan, M., Zhang, J., Ji, R.: X-CLIP: End-to-end multi-grained contrastive learning for video-text retrieval. In: ACM MM, pp. 638–647. ACM (2022)
13. Martis, R.J., Acharya, U.R., Min, L.C.: ECG beat classification using PCA, LDA, ICA and discrete wavelet transform. Biomed. Signal Process. Control **8**(5), 437–448 (2013)
14. Müller, P., Kaissis, G., Zou, C., Rueckert, D.: Joint learning of localized representations from medical images and reports. In: Avidan, S., Brostow, G., Cissé, M., Farinella, G.M., Hassner, T. (eds.) Computer Vision. ECCV 2022. LNCS, vol. 13686. Springer, Cham (2022). https://doi.org/10.1007/978-3-031-19809-0_39
15. Munk, A., Ambsdorf, J., Llambias, S.N., Nielsen, M.: AMAES: Augmented masked autoencoder pretraining on public brain MRI data for 3D-native segmentation. arXiv preprint arXiv:2408.00640 (2024)
16. Na, Y., Park, M., Tae, Y., Joo, S.: Guiding masked representation learning to capture spatio-temporal relationship of electrocardiogram. In: ICLR (2024)
17. Radford, A., Kim, J.W., Hallacy, C., Ramesh, A., Goh, G., Agarwal, S., et al.: Learning transferable visual models from natural language supervision. In: ICML, pp. 8748–8763. PMLR (2021)
18. Radhakrishnan, A., Friedman, S.F., Khurshid, S., Ng, K., Batra, P., Lubitz, S.A., et al.: Cross-modal autoencoder framework learns holistic representations of cardiovascular state. Nat. Commun. **14**(1), 2436 (2023)
19. Sawano, S., Kodera, S., Takeuchi, H., Sukeda, I., Katsushika, S., et al.: Masked autoencoder-based self-supervised learning for electrocardiograms to detect left ventricular systolic dysfunction. In: NeurIPS Workshop (2022)
20. Seibold, C., Reiß, S., Sarfraz, M.S., Stiefelhagen, R., Kleesiek, J.: Breaking with fixed set pathology recognition through report-guided contrastive training. In: Wang, L., Dou, Q., Fletcher, P.T., Speidel, S., Li, S. (eds.) Medical Image Computing and Computer Assisted Intervention. MICCAI 2022. LNCS, vol. 13435, pp. 690–700. Springer, Cham (2022). https://doi.org/10.1007/978-3-031-16443-9_66

21. Sudlow, C., Gallacher, J., Allen, N., Beral, V., Burton, P., Danesh, J., et al.: UK Biobank: An open access resource for identifying the causes of a wide range of complex diseases of middle and old age. PLoS Med. **12**(3), e1001779 (2015)
22. Tang, Y., Yang, D., Li, W., Roth, H.R., Landman, B., et al.: Self-supervised pre-training of Swin transformers for 3D medical image analysis. In: CVPR, pp. 20730–20740 (2022)
23. Tiu, E., Talius, E., Patel, P., Langlotz, C.P., Ng, A.Y., Rajpurkar, P.: Expert-level detection of pathologies from unannotated chest x-ray images via self-supervised learning. Nat. Biomed. Eng. **6**(12), 1399–1406 (2022)
24. Tong, Z., Song, Y., Wang, J., Wang, L.: VideoMAE: Masked autoencoders are data-efficient learners for self-supervised video pre-training. Adv. NeurIPS **35**, 10078–10093 (2022)
25. Turgut, Ö., Müller, P., Hager, P., Shit, S., Starck, S., Menten, M.J., et al.: Unlocking the diagnostic potential of electrocardiograms through information transfer from cardiac magnetic resonance imaging. Med. Image Anal. **101**, 103451 (2025)
26. Wang, G., Wang, Q., Iyer, G., Nag, A., John, D.: Unsupervised pre-training using masked autoencoders for ECG analysis. In: BioCAS, pp. 1–5 (2023)
27. Wang, Z., Wu, Z., Agarwal, D., Sun, J.: MedCLIP: Contrastive learning from unpaired medical images and text. In: Goldberg, Y., et al. (eds.) EMNLP, p. 3876. ACL (2022)
28. World Health Organization: WHO Mortality Database (2025). https://www.who.int/data/data-collection-tools/who-mortality-database. Accessed 31 Jan 2025
29. Yang, J., Bisk, Y., Gao, J.: TACo: Token-aware cascade contrastive learning for video-text alignment. In: ICCV, pp. 11562–11572 (2021)
30. Yang, S., Lian, C., Zeng, Z., Xu, B., Su, Y., et al.: Masked self-supervised ECG representation learning via multiview information bottleneck. Neural Comput. Appl. **36**(14), 7625–7637 (2024)
31. Yao, L., Huang, R., Hou, L., Lu, G., Niu, M., Xu, H., et al.: FILIP: Fine-grained interactive language-image pre-training. In: ICLR (2022)
32. Zhang, H., Liu, W., Shi, J., Chang, S., Wang, H., He, J., et al.: MaeFE: Masked autoencoders family of electrocardiogram for self-supervised pretraining and transfer learning. IEEE Trans. Instrum. Meas. **72**, 1–15 (2023)
33. Zhang, Y., Chen, C., Shit, S., Starck, S., Rueckert, D., Pan, J.: Whole heart 3D+T representation learning through sparse 2D cardiac MR images. In: Linguraru, M.G., et al. (eds.) Medical Image Computing and Computer Assisted Intervention. MICCAI 2024. LNCS, vol. 15001, pp. 359–369. Springer, Cham (2024). https://doi.org/10.1007/978-3-031-72378-0_34
34. Zhang, Y., Jiang, H., Miura, Y., Manning, C.D., Langlotz, C.P., et al.: Contrastive learning of medical visual representations from paired images and text. In: Lipton, Z., et al. (eds.) MLHC, vol. 182, pp. 2–25. PMLR (2022)

HAGE: Hierarchical Alignment Gene-Enhanced Pathology Representation Learning with Spatial Transcriptomics

Thao M. Dang, Haiqing Li, Yuzhi Guo, Hehuan Ma, Feng Jiang, Yuwei Miao, Qifeng Zhou, Jean Gao, and Junzhou Huang(✉)

Department of Computer Science and Engineering, The University of Texas at Arlington, Arlington, TX 76019, USA
jzhuang@uta.edu

Abstract. Histopathology images capture tissue morphology, while spatial transcriptomics (ST) provides spatially resolved gene expression, offering complementary molecular insights. However, acquiring ST data is costly and time-consuming, limiting its practical use. To address this, we propose HAGE (**H**ierarchical **A**lignment **G**ene-**E**nhanced), a framework that enhances pathology representation learning by predicting gene expression directly from histological images and integrating molecular context into the pathology model. HAGE leverages gene-type embeddings, which encode relationships among genes, guiding the model in learning biologically meaningful expression patterns. To further improve alignment between histology and gene expression, we introduce a hierarchical clustering strategy that groups image patches based on molecular and visual similarity, capturing both local and global dependencies. HAGE consistently outperforms existing methods across six datasets. In particular, on the HER2+ breast cancer cohort, it significantly improves the Pearson correlation coefficient by 8.0% and achieves substantial reductions in mean squared error and mean absolute error by 18.1% and 38.0%, respectively. Beyond gene expression prediction, HAGE improves downstream tasks, such as patch-level cancer classification and whole-slide image diagnostics, demonstrating its broader applicability. To the best of our knowledge, HAGE is the first framework to integrate gene co-expression as prior knowledge into a pathology image encoder via a cross-attention mechanism, enabling more biologically informed and accurate pathology representations. https://github.com/uta-smile/gene_expression.

Keywords: spatial transcriptomics · pathology images · multimodal contrastive learning · hierarchical alignment

1 Introduction

Histopathology images offer rich morphological insights into tissue architecture and cellular features, while spatial transcriptomics (ST) provides location-specific gene expression data that can illuminate disease progression and patient

J. C. Gee et al. (Eds.): MICCAI 2025, LNCS 15960, pp. 228–238, 2026.
https://doi.org/10.1007/978-3-032-04927-8_22

outcomes. However, acquiring ST data is costly and resource-intensive, whereas H&E-stained histology images are routinely generated in clinical settings [2]. This disparity motivates the development of methods to infer gene expression directly from histopathology. By leveraging paired pathology images and spatial expression spots, deep learning systems can extract visual features predictive of various molecular biomarkers, enabling more efficient molecular profiling without the need for expensive ST sequencing.

In typical ST data, each tissue spot corresponds to a patch in a whole slide image (WSI), providing expression measurements for thousands of genes. Deep learning has been explored to map histopathology to transcriptomic data. For instance, ST-Net [1] employs a CNN-based encoder for direct prediction, while HisToGene [2] and Hist2ST [3] enhance ST-Net by incorporating spatial relationships through Vision Transformers and graph-based models. THItoGene [4] further utilizes graph neural networks and dynamic convolutional networks. However, these methods rely on predefined spatial assumptions, which may not generalize well to heterogeneous tissues such as cancer, where gene expression patterns are highly variable [5]. Another research direction explores contrastive learning to align histology and gene expression representations in a shared latent space, such as BLEEP [5] and mclSTExp [6]. While these methods improve generalization, they often face data limitations when trained from scratch. Moreover, most existing methods treat gene targets as independent outputs, focusing solely on image-expression alignment and neglecting critical gene-gene relationships. Although these methods represent progress, a more biologically grounded framework is needed to fully integrate histology and transcriptomics.

Motivated by these challenges, we adopt the UNI [8] foundation model and introduce a hierarchical clustering strategy to efficiently align patches sharing morphological and molecular similarities. The use of foundation models for extracting image representations has been widely adopted in histology [9]. To move beyond purely visual features, we further strengthen our approach by integrating different modalities, as fusion has shown significant potential across a range of tasks and domains in bioinformatics [10–15]. Since co-expression correlations indicate functional associations, as genes in the same biological process are often co-regulated [16], we incorporate gene embeddings from Gene2Vec [17], trained on co-expression data, into the image encoder through cross-attention, capturing co-regulation patterns that better align histological features with gene expression. This enriched biological context helps capture "invisible" signals where microenvironments induce significant gene expression variability.

Our contributions are threefold: 1) We propose a novel patch-level framework that integrates gene co-expression into the pathology image encoder, leveraging gene co-activation patterns as biologically grounded prior knowledge to guide representation learning. This enables the encoder to attend to visual regions with awareness of gene-gene coordination, improving biological interpretability. 2) We introduce a hierarchical clustering strategy that aligns image-expression pairs across local and global levels, enhancing efficiency and predictive accuracy. 3) Comprehensive evaluations on six datasets across gene expression prediction, patch-level cancer classification, and WSI-level cancer classification demonstrate state-of-the-art performance and more biologically grounded representations.

2 Methodology

2.1 Preliminaries

Our learning pipeline is illustrated in Fig. 1. We denote the set of gene embeddings as $\mathbf{G} = \{\mathbf{g}_1, \mathbf{g}_2, \dots, \mathbf{g}_{|G|}\}$ with $\mathbf{G} \in \mathbb{R}^{|G| \times d_g}$, retrieved from the Gene2Vec [17] library. Given histopathology image patch with features $\mathbf{x} \in \mathbb{R}^{d_i}$ extracted by a frozen UNI model, our Gene-informed Image Encoder (GE), $f_{GE}(\cdot)$, fuses the image features with $\mathbf{G}$ via cross-attention to yield gene-guided image embeddings: $\mathbf{z}^i \leftarrow f_{GE}(\mathbf{x}, \mathbf{G}) \in \mathbb{R}^d$. In parallel, the raw gene expression vector $\mathbf{y} \in \mathbb{R}^{d_e}$ is normalized and fed into a three-layer MLP Expression Encoder, $f_{EE}(\cdot)$, to produce expression embeddings: $\mathbf{z}^e \leftarrow f_{EE}(\mathbf{y}) \in \mathbb{R}^d$.

As stated in Sect. 1, our goal is to infer gene expression from images. Therefore, improving the alignment between modalities is crucial for accuracy. To this end, we align the paired embeddings $\{\mathbf{z}_i^i, \mathbf{z}_i^e\}_{i=1}^N$ in the local view. Specifically, we enforce cross-modality consistency using the CyCLIP [18] loss, denoted as $\mathcal{L}_{Cy}$, and ensure robust alignment using the well-known CLIP [19] loss. Due to space limitations, we only provide the formulation for the CyCLIP loss, and refer the reader to [19] for details on the CLIP loss. The CyCLIP loss is computed as:

$$\mathcal{L}_{Cy} = \frac{1}{N} \sum_{j=1}^{N} \sum_{k=1}^{N} \left(\langle \mathbf{z}_j^i, \mathbf{z}_k^e \rangle - \langle \mathbf{z}_k^i, \mathbf{z}_j^e \rangle \right)^2 + \frac{1}{N} \sum_{j=1}^{N} \sum_{k=1}^{N} \left(\langle \mathbf{z}_j^i, \mathbf{z}_k^i \rangle - \langle \mathbf{z}_j^e, \mathbf{z}_k^e \rangle \right)^2, \quad (1)$$

with $\langle \cdot, \cdot \rangle$ represents the inner product.

In the global view, we apply k-means to raw expression $\mathbf{y}$ to form k_1 clusters. Within each cluster, we further cluster the corresponding $\mathbf{x}$ into k_2 subclusters. Let $\{\mathbf{c}_i^e, \mathbf{c}_i^i\}_{i=1}^{k_1 \times k_2}$ denote the centroids in the expression and image spaces, respectively; these centroid pairs are aligned using the CLIP loss [19], $\mathcal{L}_{CLGlobal}$.

A three-layer MLP Expression Predictor, $f_{Pred}(\cdot)$, then maps the gene-guided image embeddings to predicted expression vectors: $\hat{\mathbf{y}} \leftarrow f_{Pred}(\mathbf{z}^i)$. We train $f_{Pred}(\cdot)$ with a Pearson Correlation Coefficient loss:

$$\mathcal{L}_{PCC} = 1 - \frac{1}{|G|} \sum_{j=1}^{|G|} \frac{\sum_{i=1}^{N} \left(\hat{y}_{ij} - \bar{\hat{y}}_j \right) \left(y_{ij} - \bar{y}_j \right)}{\sqrt{\sum_{i=1}^{N} \left(\hat{y}_{ij} - \bar{\hat{y}}_j \right)^2 + \epsilon} \sqrt{\sum_{i=1}^{N} \left(y_{ij} - \bar{y}_j \right)^2 + \epsilon}}, \quad (2)$$

where $\bar{\hat{y}}_j$ and $\bar{y}_j$ are the mean of predicted expression and target expression for gene j across all N samples, respectively. While ϵ is a small constant for numerical stability. The total loss is defined as:

$$\mathcal{L} = \lambda_1 \mathcal{L}_{Cy} + \lambda_2 \mathcal{L}_{CLLocal} + \lambda_3 \mathcal{L}_{CLGlobal} + \lambda_4 \mathcal{L}_{PCC}. \quad (3)$$

At inference, given a patch features $\mathbf{x}$, we compute: $\hat{\mathbf{y}} = f_{Pred}(f_{GE}(\mathbf{x}, \mathbf{G}))$, thereby predicting the gene expression profile for each patch.

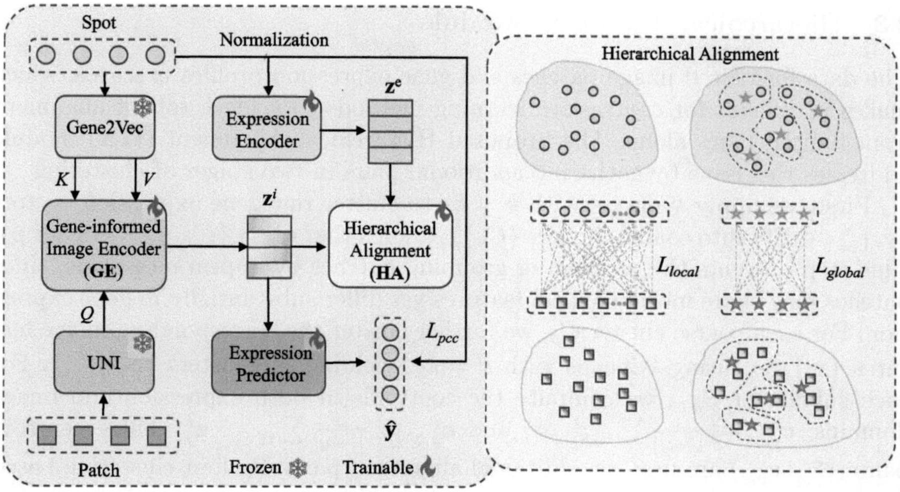

Fig. 1. HAGE predicts gene expression from pathology images by aligning expression profiles and image patches in a shared embedding space. 1) Given a list of interesting gene names, the corresponding gene embeddings retrieved from Gene2Vec are fused with image features from a frozen foundation model via GE. 2) Expression features are encoded through an MLP-based Expression Encoder. 3) HA aligns gene-guided image and expression embeddings at both local and global levels using contrastive learning. 4) The Expression Predictor predicts gene expression from gene-informed image features. At inference, given a WSI, the model outputs the predicted gene expression, while GE remains applicable to other downstream tasks.

2.2 Gene-Informed Image Encoder

We design a gene-informed image encoder (**GE**) that merges patch-level image embeddings with gene embeddings derived from Gene2Vec. Traditional image encoders that lack explicit gene information merely regress expression values, without capturing gene identity or correlations. By integrating Gene2Vec embeddings, our encoder learns both individual gene features and geneâĂŞgene relationships, allowing it to align visual features with biological context.

We project the image features $\{\mathbf{x}_i\}_{i=1}^{N} \in \mathbb{R}^{d_i}$ and the gene embeddings $\{\mathbf{g}_j\}_{j=1}^{|G|} \in \mathbb{R}^{d_g}$ into a common space of dimension d using learnable projection matrices. Treating the visual features as queries and the gene features as keys and values, we employ a cross-attention mechanism [20] to fuse the two modalities. Standard residual connections and a two-layer MLP block are then applied to refine the fused representations. The overall transformation is given by: $\mathbf{z}_i^i \leftarrow f(\mathbf{x}_i) = f_{MLP}\Big(f_{Attn}(W_x\mathbf{x}_i, \{W_g\mathbf{g}_j\}_{j=1}^{|G|}) + W_x\mathbf{x}_i\Big)$, yielding gene-informed image embeddings that are enriched with biologically grounded information. Notably, the trained GE module is reusable in other downstream tasks beyond gene expression prediction, enabling broader applicability in biomedical applications (see Sect. 3.2 for details).

2.3 Hierarchical Alignment Module

The data for paired image patches and gene expression profiles is scarce, which makes it difficult for contrastive learning methods to achieve robust alignment from limited pairs alone. The proposed Hierarchical Alignment (**HA**) module addresses this issue by gathering additional pairs in two stages of clustering.

First, we apply k-means with $k = k_1$ to cluster raw gene expression vectors $\{\mathbf{y}_i\}_{i=1}^{N} \in \mathbb{R}^{d_e}$ into coarse clusters $\{C_p\}_{p=1}^{k_1}$, such that $C_p = \{i : \mathbf{y}_i \sim \text{cluster } p\}$. This step prevents the problem of grouping patches by appearance alone, since patches may share morphological features yet differ substantially in gene expression. For each coarse cluster C_p, we further cluster the corresponding image features $\{\mathbf{x}_i\}_{i \in C_p}$ using k-means with $k = k_2$, yielding subclusters $\{S_{p,q}\}_{q=1}^{k_2}$. For each subcluster $S_{p,q}$, we compute the centroids in both expression and image domains: $\mathbf{c}_{p,q}^{e} = \frac{1}{|S_{p,q}|} \sum_{j \in S_{p,q}} \mathbf{z}_j^{e}$ and $\mathbf{c}_{p,q}^{i} = \frac{1}{|S_{p,q}|} \sum_{j \in S_{p,q}} \mathbf{z}_j^{i}$. These centroid pairs $(\mathbf{c}_{p,q}^{e}, \mathbf{c}_{p,q}^{i})$ are used as additional alignment pairs. By first clustering based on gene expression, and then on morphology, this two-tier approach provides a global view of the data and adds more training pairs.

3 Experiments

3.1 Implementation Details

Datasets. For the gene expression prediction task, following the setting of related works [3,4,6], we select 32 breast samples and 12 skin samples from HER2+ [21] and cSCC [22] datasets, respectively. To assess the quality of the learned embeddings, we evaluate patch-level classification on PCAM [23] and SkinCancer [24] datasets. For WSI-level diagnosis, we use TCGA-BRCA [25] for cancer subtype classification (e.g., IDC and ILC) and the SLN-Breast [26] for positive or negative carcinoma prediction. Totally, six tasks are conducted.

Gene List Selection. We apply a standard gene selection pipeline as described in [6]. We compile a list of the top 1,000 highly variable genes for each WSI and intersect these lists to identify common genes across all WSIs. We then refine the gene list by intersecting it with the Gene2Vec dictionary, yielding a final set of 771 genes for the breast dataset and 168 genes for the skin dataset.

Baselines and Evaluation Metrics. For the gene expression prediction task, we compare our approach against six SOTA methods: ST-Net [1], HisToGene [2], Hist2ST [3], THItoGene [4], BLEEP [5], and mclSTExp [6]. To ensure fairness, we adopt the leave-one-out protocol used by previous studies [1,6] and perform five experimental runs with different training and validation splits for each left-out subject. Following related works [3–6], performance is evaluated using Mean Squared Error (MSE), Mean Absolute Error (MAE), and Pearson Correlation Coefficient (PCC) metrics. In particular, we compute the PCC for all genes in the gene list as well as for the top 50 highly expressed genes (HEG).

Other Downstream Tasks. To ensure a fair comparison, we do not evaluate against the six baselines mentioned earlier. Instead, we report the performance of our gene-informed image encoder and UNIv1 [8]. Notably, none of the datasets used in our study overlap with the testing datasets, eliminating potential data leakage. We employ 5-fold cross-validation and assess performance using Accuracy, AUC, and Recall, reporting the mean and standard deviation.

3.2 Results

Gene Expression Prediction. After leaving one subject out, we split the remaining data into training and validation sets at a 4:1 ratio and run five trials per subject. Training lasts 15 epochs for HER2+ and 100 for cSCC, with early stopping after 5 epochs. Both breast and skin cancer experiments use same hyperparameters: $\lambda_1 = \lambda_4 = 10$, $\lambda_2 = \lambda_3 = 1$, $k_1 = 30$, $k_2 = 2$, and $d = 1024$. CLIP loss temperature is fixed at $\tau = 0.07$. We optimize with Adam (learning rate 10^{-4}, weight decay 10^{-5}) and StepLR (step size 10, decay factor $\gamma = 0.5$).

Table 1. Gene expression prediction performance of various methods.

Dataset	HER2+			
Methods	PCC (All) ↑	PCC (HEG) ↑	MSE ↓	MAE ↓
ST-Net [1]	0.0561 ± 0.017	0.0134 ± 0.013	0.5312 ± 0.008	0.6306 ± 0.011
HisToGene [2]	0.0842 ± 0.015	0.0711 ± 0.014	0.5202 ± 0.014	0.6422 ± 0.005
Hist2ST [3]	0.1443 ± 0.013	0.1849 ± 0.015	0.5135 ± 0.009	0.6087 ± 0.013
THItoGene [4]	0.1726 ± 0.018	0.2809 ± 0.013	0.5012 ± 0.011	0.5956 ± 0.009
BLEEP [5]	0.1873 ± 0.005	0.2909 ± 0.016	0.6015 ± 0.016	0.5824 ± 0.004
mclSTExp [6]	0.2304 ± 0.011	0.3866 ± 0.021	0.5897 ± 0.013	0.5813 ± 0.008
HAGE (ours)	**0.2489 ± 0.001**	**0.4458 ± 0.003**	**0.4830 ± 0.005**	**0.3606 ± 0.002**
Dataset	cSCC			
Methods	PCC (All) ↑	PCC (HEG) ↑	MSE ↓	MAE ↓
ST-Net [1]	0.0012 ± 0.022	0.0018 ± 0.015	0.6806 ± 0.006	0.6404 ± 0.003
HisToGene [2]	0.0771 ± 0.024	0.0919 ± 0.012	0.6805 ± 0.012	0.6234 ± 0.007
Hist2ST [3]	0.1838 ± 0.011	0.2175 ± 0.016	0.6748 ± 0.017	0.6107 ± 0.006
THItoGene [4]	0.2373 ± 0.009	0.2719 ± 0.012	0.6546 ± 0.006	0.6012 ± 0.019
BLEEP [5]	0.2449 ± 0.017	0.3122 ± 0.027	0.5163 ± 0.007	0.5399 ± 0.015
mclSTExp [6]	0.3235 ± 0.019	0.4261 ± 0.016	0.4302 ± 0.005	0.5208 ± 0.009
HAGE (ours)	**0.3397 ± 0.006**	**0.4607 ± 0.007**	**0.4248 ± 0.002**	**0.3296 ± 0.002**

Table 1 summarizes HAGE's performance against six SOTA methods on HER2+ and cSCC datasets. On HER2+, HAGE achieves a PCC of 0.2489 for all genes and 0.4458 for the top 50 HEG, improving by 8.0% and 15.3%, respectively, over the second-best method. It also attains an MSE of 0.4830 and MAE of 0.3606, reducing errors by 18.1% and 38.0% relative to mclSTExp.

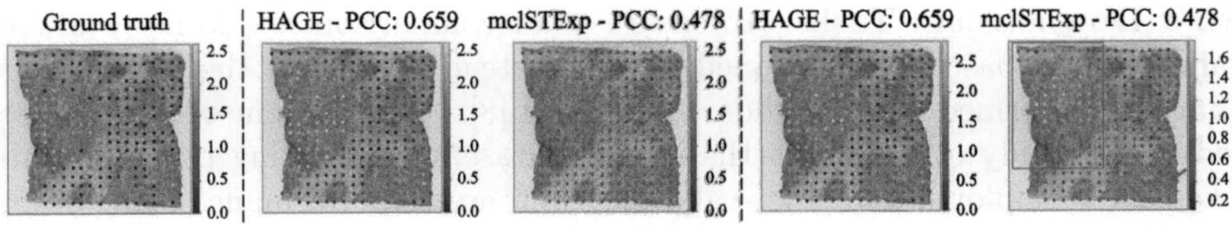

Fig. 2. Visualization of ITGB6 gene expression predictions. (Left) Ground truth for slide D2 in the HER2+ dataset, (middle) fixed scale, and (right) variable scale.

On cSCC, HAGE achieves a PCC of 0.3397 (all genes) and 0.4607 (HEG), improving by roughly 5.0% and 8.1% over mclSTExp. While MSE improvement is modest, MAE decreases significantly by 36.7%. These results confirm HAGE's consistent advantage in expression prediction across cancer types.

Figure 2 visualizes ITGB6 [27], a key diagnostic biomarker for breast cancer, particularly in the HER2+ dataset. High ITGB6 expression implies tumor presence; if the model can effectively localize such regions, it demonstrates potential for clinical interpretation. Our method achieves a PCC of 0.659, significantly outperforming the SOTA method, mclSTExp (0.478). In the variable scale setting, both HAGE and mclSTExp capture the overall trend, but the latter fails to accurately reflect expression levels at certain spots. In the fixed scale setting, which compares predictions to absolute ground truth values, mclSTExp struggles to identify tumor regions and distinguish signals from connective and invasive tumor tissues, whereas HAGE provides a clearer, more precise prediction.

Tumor Classification. *Patch-level.* We evaluate our learned representations using a linear probing protocol for patch-level tumor classification. The classifier is a single linear layer that projects the input embeddings onto the target classes. Both the breast (e.g., PCAM) and skin (e.g., SkinCancer) linear probing tasks are trained for 10 epochs using cross-entropy loss. We report the average performance of our five models and the UNIv1 model.

Slide-level. SLN-Breast followed the preprocessing steps and implementation settings described in [28], while TCGA-BRCA was preprocessed using the default settings of CLAM [29]. We use the pretrained gene-guided encoder with the best performance in PCAM classification as the image encoder and adopt ABMIL [30] to aggregate patch-level features into slide-level embeddings. Despite

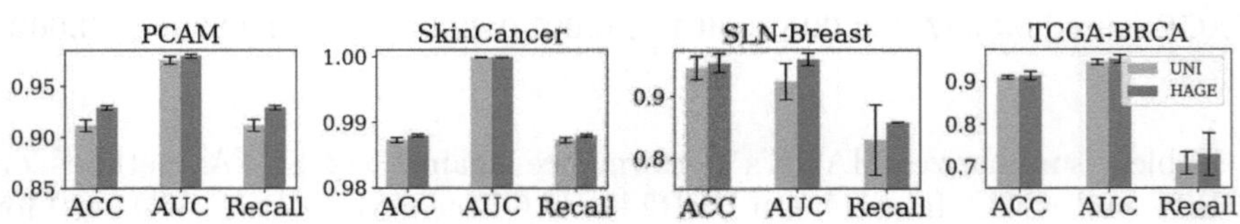

Fig. 3. Comparison of the gene-informed image encoder in HAGE and the UNIv1 image encoder on patch-level and slide-level classification across breast and skin datasets.

Table 2. Impact of each HAGE component on gene prediction on the HER2+ dataset.

Architecture	PCC (A) ↑	PCC (H) ↑	MSE ↓	MAE ↓
HAGE w/o both HA and GE (I)	0.2330	0.4131	0.5081	0.3547
HAGE w/o Gene-informed Encoder (II)	0.2338	0.4164	0.5076	0.3543
HAGE w/o Hierarchical Alignment (III)	0.2474	0.4410	0.4842	0.3613
HAGE with Single-layer cluster align. (IV)	0.2461	0.4390	0.4812	0.3603

using standard mechanisms and lightweight modeling choices, the integration of gene co-expression enhances downstream performance. Figure 3 shows that our gene-informed encoders consistently outperform UNIv1 in both patch-level and slide-level classification tasks on breast and skin datasets.

3.3 Ablation Study

Impact of Each Component. Table 2 ablates two key components of HAGE: the gene-informed encoder (GE) and hierarchical alignment (HA). Removing both (I) significantly reduces performance, confirming that morphological features alone are insufficient for gene expression prediction. Retaining HA without GE (II), or GE without HA (III), improves performance over the baseline but still falls short of the full HAGE. While clustering reduces MSE and MAE for both HA and its variant, only HA consistently improves all metrics. Replacing the two-level hierarchical alignment with a single-layer cluster alignment (IV) degrades performance, suggesting that cluster formation impacts results more than simply adjusting their number (see Table 3 for details). These results emphasize the necessity of integrating gene-level knowledge into the image encoder and the effectiveness of hierarchical alignment in capturing molecular patterns.

Impact of Cluster. We conduct a grid search over k_1 and k_2 to evaluate the robustness of the HA component. As shown in Table 3, performance remains relatively stable across a practical range of cluster settings (e.g., $k_1 \in \{10, 20, 30, 40\}$, $k_2 \in \{2, 3\}$), indicating that our method is not overly sensitive to moderate adjustments in these hyperparameters. While clustering facilitates

Table 3. Grid search on the impact of cluster settings on the HER2+ dataset.

	$k_2 = 2$				$k_2 = 3$			
Metrics	PCC (A)	PCC (H)	MSE	MAE	PCC (A)	PCC (H)	MSE	MAE
$k_1 = 10$	0.2510	0.4499	0.4856	0.3606	0.2480	0.4433	0.4832	0.3611
$k_1 = 20$	0.2480	0.4422	0.4817	0.3605	0.2485	0.4447	0.4862	0.3567
$k_1 = 30$	0.2503	0.4490	0.4823	0.3604	0.2463	0.4405	0.4833	0.3612
$k_1 = 40$	0.2493	0.4468	0.4841	0.3603	0.2501	0.4504	0.4826	0.3606

the alignment of patches with similar molecular characteristics, robust performance across variations in cluster size is desirable, as large result changes under different configurations may suggest noise or biologically implausible groupings. In practice, we set $k_1 = 30$ and $k_2 = 2$ to balance performance and efficiency.

The primary factor influencing performance is how clusters are formed. We test an alternative approach that relies solely on gene expression, creating a single-layer clustering by setting $k_1 = 60$ and removing the subcluster step. Although this approach results in the same total number of centroids as our hierarchical structure, it yields lower performance (Table 2).

4 Conclusion

We introduce HAGE (Hierarchical Alignment Gene-Enhanced), a framework that integrates gene features and hierarchical clustering to infer spatial transcriptomic signals from histopathology images. By incorporating gene-specific information into the image encoder via cross-attention, HAGE learns representations that are both visually and biologically informative. Experiments on six datasets demonstrate its effectiveness, showing stronger correlations, lower error metrics, and adaptability across diverse tissues and clinical contexts. Overall, HAGE offers a robust, gene-aware pathology representation that enhances both expression inference and downstream task performance.

Acknowledgments. This work was partially supported by US National Science Foundation IIS-2412195, CCF-2400785, the Cancer Prevention and Research Institute of Texas (CPRIT) award (RP230363), the National Institutes of Health (NIH) R01 award (1R01AI190103-01) and Microsoft Accelerate Foundation Models Research (2024).

Disclosure of Interests. The authors have no competing interests to declare that are relevant to the content of this article.

References

1. He, B., et al.: Integrating spatial gene expression and breast tumour morphology via deep learning. Nat. Biomed. Eng. **4**, 827–834 (2020). https://doi.org/10.1038/s41551-020-0578-x
2. Pang, M., Kenong, S., Li, M.: Leveraging information in spatial transcriptomics to predict super-resolution gene expression from histology images in tumors. bioRxiv 2021.11.28.470212. https://doi.org/10.1101/2021.11.28.470212
3. Zeng, Y., et al.: Spatial transcriptomics prediction from histology jointly through transformer and graph neural networks. Brief. Bioinform. **23**, 1–11 (2022). https://doi.org/10.1093/bib/bbac297
4. Jia, Y., Liu, J., Chen, L., Zhao, T., Wang, Y.: THItoGene: a deep learning method for predicting spatial transcriptomics from histological images. Brief. Bioinform. **25**, 1–10 (2024). https://doi.org/10.1093/bib/bbad464
5. Xie, R., et al.: Spatially resolved gene expression prediction from histology images via bi-modal contrastive learning. Adv. Neural. Inf. Process. Syst. **36**, 70626–70637 (2023)

6. Min, W., Shi, Z., Zhang, J., Wan, J., Wang, C.: Multimodal contrastive learning for spatial gene expression prediction using histology images. Brief. Bioinform. **25**, bbae551 (2024). https://doi.org/10.1093/bib/bbae551
7. Yang, Y., Hossain, M.Z., Stone, E.A., Rahman, S.: exemplar guided deep neural network for spatial transcriptomics analysis of gene expression prediction. In: Proceedings of the Winter Conference on Applications of Computer Vision (2023). https://doi.org/10.1109/WACV56688.2023.00501
8. Chen, R.J., Ding, T., Lu, M.Y., et al.: Towards a general-purpose foundation model for computational pathology. Nat. Med. **30**, 850–862 (2024). https://doi.org/10.1038/s41591-024-02857-3
9. Dang, T.M., et al.: MFMF: multiple foundation model fusion networks for whole slide image classification. In Proceedings of the 15th ACM International Conference on Bioinformatics, Computational Biology and Health Informatics (BCB 2024) (2024)
10. Li, H., Wang, C., Zhao, G., He, Z., Wang, Y., Sun, Z.: Sclera-transfuse: fusing Swin transformer and CNN for accurate sclera segmentation. In: 2023 IEEE International Joint Conference on Biometrics (IJCB) (2023). https://doi.org/10.1109/IJCB57857.2023.10448814
11. Wang, C., Li, H., Zhang, Y., Zhao, G., Wang, Y., Sun, Z.: Sclera-TransFuse: fusing vision transformer and CNN for accurate sclera segmentation and recognition. IEEE Trans. Biomet. Behav. Ident. Sci. **6**(4), 575–590 (2024). https://doi.org/10.1109/TBIOM.2024.3415484
12. Dang, T.M., et al.: AVET: a novel transform function to improve cancellable biometrics security. IEEE Trans. Inf. Forensics Secur. **18**, 758–772 (2023). https://doi.org/10.1109/TIFS.2022.3230212
13. Jiang, F., et al.: GTE a graph learning framework for prediction of T-cell receptors and epitopes binding specificity. Brief. Bioinform. **25**, bbae343 (2024). https://doi.org/10.1093/bib/bbae343
14. Wang, C., Li, H., Ma, W., et al.: MetaScleraSeg: an effective meta-learning framework for generalized sclera segmentation. Neural Comput. Appl. **35**, 21797–21826 (2023). https://doi.org/10.1007/s00521-023-08937-8
15. Jiang, F., et al.: AlphaEpi: enhancing B cell epitope prediction with AlphaFold 3. In: Proceedings of the 15th ACM International Conference on Bioinformatics, Computational Biology and Health Informatics (BCB 2024). https://doi.org/10.1145/3698587.3701389
16. Jiang, F., et al.: AlphaEpi: enhancing B cell epitope prediction with AlphaFold 3. In: Proceedings of the 15th ACM International Conference on Bioinformatics, Computational Biology and Health Informatics (BCB 2024). https://doi.org/10.1145/3698587.3701389
17. Du, J., Jia, P., Dai, Y., et al.: Gene2vec: distributed representation of genes based on co-expression. BMC Genom. **20**, 82 (2019). https://doi.org/10.1186/s12864-018-5370-x
18. Goel, S., Bansal, H., Bhatia, S., Rossi, R.A., Vinay, V., Grover, A.: CYCLIP: cyclic contrastive language-image pretraining. In: Advances in Neural Information Processing Systems (2022)
19. Radford, A., et al.: Learning transferable visual models from natural language supervision. In: Proceedings of the International Conference on Machine Learning (2021)
20. Vaswani, A., et al.: Attention is all you need. In: Proceedings of the 31st International Conference on Neural Information Processing Systems (2017)

21. Andersson, A., et al.: Lundeberg: spatial deconvolution of HER2-positive breast cancer delineates tumor-associated cell type interactions. Nat. Commun. **12**, 6012 (2021). https://doi.org/10.1038/s41467-021-26271-2
22. Ji, A.L., et al.: Multimodal analysis of composition and spatial architecture in human squamous cell carcinoma. Cell **182**, 497–514 (2020). https://doi.org/10.1016/j.cell.2020.05.039
23. Veeling, B.S., Linmans, J., Winkens, J., Cohen, T., Welling, M.: Rotation Equivariant CNNs for Digital Pathology. arXiv:1806.03962 (2018)
24. Kriegsmann, K., et al.: Deep learning for the detection of anatomical tissue structures and neoplasms of the skin on scanned histopathological tissue sections [Data set] (2023). https://doi.org/10.11588/DATA/7QCR8S heiDATA, V1
25. Lingle, W., et al.: The Cancer Genome Atlas Breast Invasive Carcinoma Collection (TCGA-BRCA) (Version 3) [Data set]. The Cancer Imaging Archive (2016). https://doi.org/10.7937/K9/TCIA.2016.AB2NAZRP
26. Campanella, G., Hanna, M.G., Brogi, E., Fuchs, T.J.: Breast Metastases to Axillary Lymph Nodes [Data set]. The Cancer Imaging Archive (2019). https://doi.org/10.7937/tcia.2019.3xbn2jcc
27. Desai, K., Nair, M.G., Prabhu, J.S., et al.: High expression of integrin β6 in association with the Rho-Rac pathway identifies a poor prognostic subgroup within HER2 amplified breast cancers. Cancer Med. **5**(8), 2000–2011 (2016). https://doi.org/10.1002/cam4.756
28. Dang, T.M., et al.: Huang: abnormality-aware multimodal learning for WSI classification. Front. Med. **12**, 1546452 (2025). https://doi.org/10.3389/fmed.2025.1546452
29. Lu, M.Y., Williamson, D.F.K., Chen, T.Y., et al.: Data-efficient and weakly supervised computational pathology on whole-slide images. Nat. Biomed. Eng. **5**, 555–570 (2021). https://doi.org/10.1038/s41551-020-00682-w
30. Ilse, M., Tomczak, J.M., Welling, M.: Attention-based deep multiple instance learning. In: Proceedings of the International Conference on Machine Learning (2018)

Hierarchical Self-supervised Adversarial Training for Robust Vision Models in Histopathology

Hashmat Shadab Malik[1(✉)], Shahina Kunhimon[1], Muzammal Naseer[2], Fahad Shahbaz Khan[1,3], and Salman Khan[1,4]

[1] Mohamed Bin Zayed University of Artificial Intelligence (MBZUAI), Abu Dhabi, UAE
{hashmat.malik,shahina.kunhimon,fahad.khan,salman.khan}@mbzuai.ac.ae
[2] Khalifa University, Abu Dhabi, UAE
muhammadmuzammal.naseer@ku.ac.ae
[3] Linköping University, Linköping, Sweden
[4] Australian National University, Canberra, Australia

Abstract. Adversarial attacks pose significant challenges for vision models in critical fields like healthcare, where reliability is essential. Although adversarial training has been well studied in natural images, its application to biomedical and microscopy data remains limited. Existing self-supervised adversarial training methods overlook the hierarchical structure of histopathology images, where patient-slide-patch relationships provide valuable discriminative signals. To address this, we propose *Hierarchical Self-Supervised Adversarial Training* (`HSAT`), which exploits these properties to craft adversarial examples using multi-level contrastive learning and integrate it into adversarial training for enhanced robustness. We evaluate `HSAT` on multiclass histopathology dataset OpenSRH and the results show that `HSAT` outperforms existing methods from both biomedical and natural image domains. `HSAT` enhances robustness, achieving an average gain of 54.31% in the white-box setting and reducing performance drops to 3–4% in the black-box setting, compared to 25–30% for the baseline. These results set a new benchmark for adversarial training in this domain, paving the way for more robust models. Code and models are available at https://github.com/HashmatShadab/HSAT.

Keywords: Adversarial Robustness · Self-Supervised Learning

1 Introduction

Computer vision has made significant strides in recent years, improving performance across various tasks. However, deep learning models remain vulnerable to adversarial attacks, where subtle, imperceptible perturbations to images cause

J. C. Gee et al. (Eds.): MICCAI 2025, LNCS 15960, pp. 239–249, 2026.
https://doi.org/10.1007/978-3-032-04927-8_23

incorrect model responses [3,9,14,24,29,30]. These attacks are broadly classified into *white-box attacks*, where the attacker has full access to the model, and *black-box attacks*, where access is limited, and adversarial examples are crafted using query-based [2,4,27] or transfer-based methods [10,26,28]. This vulnerability poses significant risks, especially in healthcare, where the reliability of vision models is crucial for accurate medical diagnoses and treatment planning.

Previous works in natural image domains has explored adversarial training strategies, including supervised [24,31,33] and self-supervised learning (SSL) methods [11,19,23], to enhance model robustness. However, there has been limited focus on assessing the robustness of vision models in the medical domain, where reliable performance is crucial. While SSL methods are gaining traction in medical domains, especially in histopathology [5,7,17], self-supervised adversarial training for robust representation learning of such images remains unexplored. In this work, we bridge this gap by investigating self-supervised adversarial training for learning robust visual features of histopathology images.

In self-supervised adversarial training, contrastive learning is preferred over masking-based reconstruction due to its discriminative objective, which aligns with adversarial robustness goals and effectively enhances vision model robustness [11,19,23]. Existing methods primarily focus on instance- or patch-level discriminative learning, overlooking the hierarchical structure present in biomedical data. Gigapixel slides in medical imaging are typically tiled into smaller patches, forming a patient-slide-patch hierarchy where all samples from a single patient correspond to a shared diagnosis [17]. This multi-level correlation provides a strong, underutilized discriminative signal that can be harnessed to improve adversarial training. Inspired by this, we propose *Hierarchical Self-Supervised Adversarial Training* (`HSAT`)—a novel approach that harnesses the hierarchical correlations between patches, slides, and patients to enhance adversarial training. `HSAT` first generates adversarial examples in the *maximization step* by utilizing a hierarchical contrastive objective, resulting in the generation of strong adversarial examples that confuse the model. This is followed by a *minimization step*, to reduce the loss induced by these adversarial perturbations. By jointly optimizing these multi-level objectives, `HSAT` strengthens the model's adversarial robustness, resulting in superior performance compared to current approaches [11,19,23].

We benchmark `HSAT` on the multiclass `OpenSRH` [16] microscopic cancer dataset, comparing it with state-of-the-art non-adversarial self-supervised methods and instance-level adversarial training approaches. `HSAT` significantly enhances adversarial robustness, providing stronger protection against both white-box and black-box attacks. We summarize our key contributions as follows:

❶ We propose a hierarchy-wise adversarial perturbation method that leverages the correlation between patient, slide, and patch levels in biomedical microscopy images to craft adversarial examples.

❷ Our proposed framework, `HSAT`, mitigates adversarial vulnerabilities by minimizing confusion from hierarchical adversarial perturbations, promoting robust representation learning across patient, slide, and patch levels.

❸ HSAT achieves state-of-the-art adversarial performance on the OpenSRH dataset, outperforming current self-supervised adversarial training approaches adapted from the natural image domain.

2 Related Work

Self-supervised Learning in Histopathology: Advancements in microscopy have enabled the digitization of tissue imaging, resulting in large-scale digital datasets, much of which remain unannotated due to the expertise required for labeling. Therefore, SSL has become a key method for pretraining models in histopathology, with techniques such as contrastive learning [8], reconstruction from partial signals [1], and knowledge distillation [6]. Notably, contrastive learning, which leverages instance-level identity by contrasting different views of the same image, has proven effective in learning rich representations. Recent efforts in histopathology have utilized contrastive learning to design hierarchical contrastive objectives for patch representations [17]. However, these models are vulnerable to adversarial attacks, and to our knowledge, no work has explored adversarial self-supervised training to improve robustness.

Adversarial Robustness and Self-supervised Learning: Adversarial training strategies have been extensively explored in the self-supervised learning domain for natural image data to improve model robustness [11,19,23]. Several instance-level adversarial contrastive learning methods [10,11,23] leverage contrastive loss to generate adversarial examples, strengthening model resilience. However, these methods predominantly focus on instance- or patch-level adversarial training, often neglecting the hierarchical relations inherent in more complex data, such as histopathology images. As a result, directly adapting these approaches to histopathology data may lead to suboptimal robustness. Existing adversarial methods in histopathology typically address attacks and defenses at the patch or instance level [12,13,18,20,25], overlooking the multi-scale relationships within the data. To bridge this gap, we propose a novel self-supervised adversarial training framework that explicitly captures the multi-level correlation in histopathology images, offering a more effective and robust training paradigm.

3 Methodology

In this section, we introduce our approach, HSAT—a hierarchical self-supervised adversarial training framework designed for training vision models on histopathology images. The core idea behind HSAT is that adversarial examples can be generated hierarchically (see Fig. 1), not only by *pushing* an image away from its augmented versions (*Patch Positives*) but also from positive paired images at the slide and patient level (*Slide and Patient Positives*). This process forms the maximization step. To counter these perturbations, the model is trained to *pull* the adversarial counterparts of positive pairs closer in the feature space. Ultimately, HSAT strengthens model robustness across all hierarchical levels. To the best of our knowledge, we are the first to investigate self-supervised adversarial training in histopathology images and integrate hierarchy-based multiple discriminative tasks for generating adversarial examples for training.

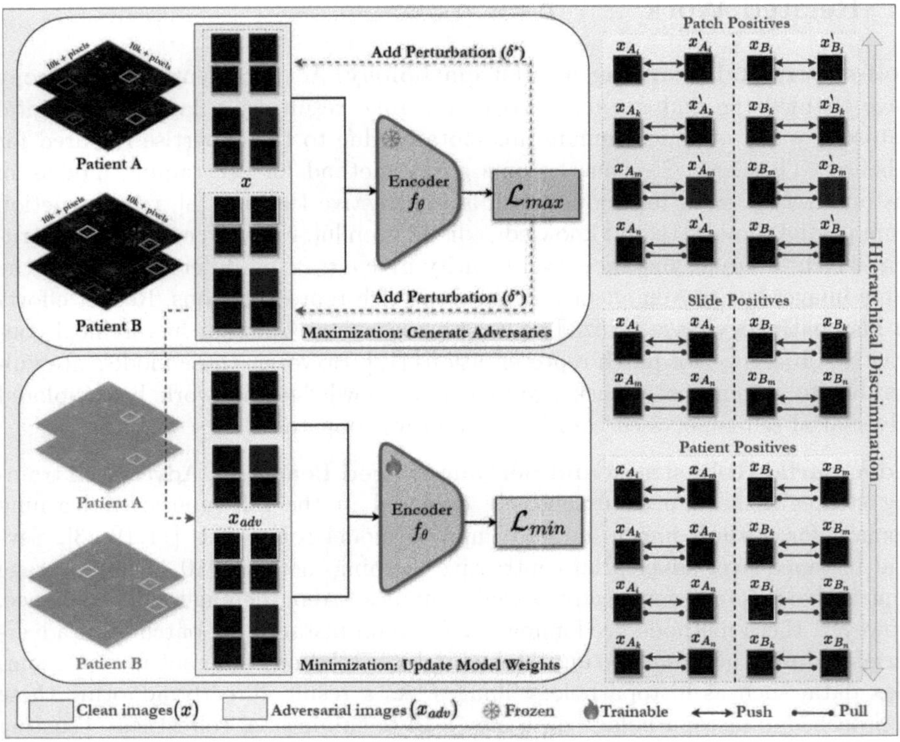

Fig. 1. **HSAT Framework**: Adversarial examples are generated by maximizing the distance between images and their corresponding positive pairs across patch, slide, and patient *(maximization)*. The model is then updated by minimizing the hierarchical loss on adversarial images *(minimization)*, enforcing robust feature learning at all levels.

3.1 Hierarchical Self-Supervised Adversarial Training

In histopathology, vision tasks follow a three-tier data hierarchy: gigapixel Whole Slide Images (WSIs or slides) are divided into sub-sampled patches, with patches linked to slides and slides to patients. Recent work [17] leverages this structure using self-supervised contrastive objectives with positive pairs sampled across all the levels. Given a batch of n patients, each with n_s slides, n_p patches per slide, and n_a augmentations per patch, the encoder f_θ extracts features from the patches, which are then projected into latent representations z. These representations are used to compute the hierarchical contrastive loss, with positive pairs defined at three levels: patch (same patch augmentations), slide (patches from the same slide), and patient (patches from the same patient). Therefore hierarchical contrastive loss on a set of images $\mathcal{I}$ is defined as follows:

$$\mathcal{L}_{con}^{l} = -\sum_{i \in \mathcal{I}} \frac{1}{|\mathcal{H}_l(i)|} \sum_{h \in \mathcal{H}_l(i)} \log \left(\frac{\exp\left((\mathbf{z}_i \cdot \mathbf{z}_h)/\tau\right)}{\sum_{p \in \mathcal{P}(i)} \exp\left((\mathbf{z}_i \cdot \mathbf{z}_p)/\tau\right)} \right) \tag{1}$$

Here, $\mathbf{z}_i$ is the feature representation of patch i, $\mathcal{H}_l(i)$ is the set of positive pairs at hierarchical level l (patch, slide, or patient), while $\mathcal{P}(i)$ denotes all the other patches in $\mathcal{I}$, except patch i. $(\mathbf{z}_i \cdot \mathbf{z}_h)$ measures cosine similarity, between the feature vectors, while τ is the temperature parameter. However, as shown in Sect. 4, both [17] and instance-based self-supervised adversarial training from the natural domain [19] are vulnerable adversarial attacks resulting in significant performance drop (see Table 1). To achieve optimal robustness, we propose a simple yet effective approach; adversarially training a self-supervised model using adversarial examples crafted through *hierarchy-wise* attacks. We first describe the attack, followed by our training strategy to enhance the model's robustness.

Hierarchy-Wise Adversarial Attack: Given an input image x, we craft an adversarial image x_{adv} by maximizing the hierarchical contrastive loss across all levels; patch, slide, and patient. The adversarial perturbation δ^* is obtained by solving the optimization problem $\delta^* = \arg\max_{\|\delta\|_\infty \leq \epsilon} \mathcal{L}_{\max}(x + \delta, \mathcal{I})$ where:

$$\mathcal{L}_{\max}(x + \delta, \mathcal{I}) = \sum_{l} \mathcal{L}_{con}^{l}(f_\theta(x + \delta), f_\theta(\mathcal{I})), \quad \forall l \in \{\text{patch, slide, patient}\} \tag{2}$$

The perturbation δ is constrained by the ℓ_∞ norm to remain within a perturbation budget ϵ. The resulting adversarial image $x_{\text{adv}} = x + \delta$ is designed to confuse the model not only with its corresponding augmented patch x' but also with the positive paired patches at the slide and patient levels. We use Projected Gradient Descent (PGD) to optimize the perturbation δ iteratively, ensuring it stays within the defined ℓ_∞ bound. This multi-task maximization objective provides significantly more diversity and discriminative signal compared to current instance-level adversarial training methods [11,19,23], enabling the model to learn better robust representations.

Hierarchial Self-supervised Adversarial Training (HSAT): We propose HSAT, a min-max optimization framework to learn robust representations via self-supervised contrastive learning. The inner maximization step, $\mathcal{L}_{\max}$ generates adversarial perturbations by crafting a heirarchy-wise attack. The outer minimization step, $\mathcal{L}_{\min}$, updates the encoder f_θ to minimize the heirarchial contrastive loss, encouraging robust feature representations across patch, slide, and patient levels. The overall objective is given by:

$$\min_{\theta} \mathcal{L}_{\min}(\theta) = \mathbb{E}_{x \sim \mathcal{D}} \sum_{l} \mathcal{L}_{con}^{l}(f_\theta(x_{adv}), f_\theta(\mathcal{I})),\ \forall l \in \{\text{patch, slide, patient}\} \tag{3}$$

By jointly optimizing across patch, slide, and patient levels, HSAT captures richer, more resilient representations, reducing the risk of overfitting to specific adversarial patterns as well as improving its generalization.

4 Experiments

Dataset: OpenSRH [16], an optical microscopy dataset of brain tumor biopsies and resections, was used for training and validation. It comprises 303 patients across 7 classes. Data is split into 243 patients for training and 60 for validation. Whole slide images (WSIs) are patched into 300 × 300 patches, yielding 229K training and 56K validation patches.

Implementation Details: We conduct experiments using ResNet-50 [15] and WideResNet-50 [32], each coupled with an MLP layer to project embeddings into 128-dimensional latent space, as outlined in [17]. For both non-adversarial hierarchical training [17], and our HSAT framework, models are trained for 40,000 epochs with an effective batch size $(n \cdot n_a \cdot n_s \cdot n_p)$ of 512. We employ the AdamW optimizer [22] with an initial learning rate of 0.001 for the first 10% of iterations, followed by a cosine decay scheduler. During training, we utilize weak (horizontal and vertical flips) and strong (affine transformations, color jittering, and random erasing) augmentations. Specifically for HSAT, we implement a perturbation warmup strategy for the first 5,000 epochs and dynamic data augmentations [23] to stabilize the training. Adversarial examples in the maximization step are generated by optimizing $\mathcal{L}_{max}$ using (PGD) attack for 5 steps at $\epsilon = \frac{8}{255}$. We evaluate three HSAT variants: (1) HSAT-Patient, capturing the full patient-slide-patch hierarchy $(n_a = n_s = n_p = 2)$; (2) HSAT-Slide, incorporating slide-level information $(n_a = n_s = 2, n_p = 1)$; and (3) HSAT-Patch, using only patch-level information, aligning with current instance-level adversarial training methods $(n_a = 2, n_s = n_p = 1)$. By keeping the effective batch size constant, we ensure that the baseline adversarial training approach, HSAT-Patch, operates under an equivalent computational training budget. Our results focus mainly on HSAT-Patient, denoted as HSAT, using the full term only for comparisons.

Evaluation: We use k-Nearest Neighbors (kNN) for quantitative evaluation, freezing the pretrained visual backbone to compute embeddings for the train and validation splits on OpenSRH. A kNN classifier predicts class labels by matching test patches to the nearest training embeddings. We report accuracy (Acc) and mean class accuracy (MCA) at patch, slide, and patient levels, following [17]. For adversarial evaluation, we generate adversarial examples across various perturbation budgets using iterative attacks (10 steps) that minimize cosine similarity between clean and adversarial feature representations.

Robustness Against White-Box Attacks: In this section, we examine the robustness of vision models in white-box settings, using iterative attacks, such as PGD [24], BIM [21] and MIFGSM [10]. Adversarial examples are crafted on the OpenSRH validation set with a perturbation budget of $\epsilon = \{\frac{4}{255}, \frac{8}{255}\}$ and 10 attack iterations. Models trained with our HSAT framework are compared

Table 1. Our proposed HSAT results in significant gain in robustness against different *white box* attacks at perturbation budgets of $\epsilon = \frac{4}{255}$ and $\epsilon = \frac{8}{255}$ *(higher is better).*

Training	Model	Clean		PGD-4		BIM-4		MIFGSM-4		PGD-8		BIM-8		MIFGSM-8	
		Acc	MCA	Acc	MCA	Acc	MCA	Acc	MCA	Acc	MCA	Acc	MCA	Acc	MCA
Patch Classification															
Training from Scratch															
[17]	ResNet-50	83.10	82.29	10.25	9.84	10.27	9.85	11.15	10.34	5.52	5.01	5.57	5.02	5.97	5.31
HSAT-Patch	ResNet-50	44.84	39.63	44.67	39.65	44.67	39.65	44.80	39.80	44.23	39.76	44.22	39.75	44.26	39.88
HSAT	ResNet-50	66.43	65.61	64.07	61.36	64.11	61.40	65.11	62.40	57.18	53.48	57.10	53.38	58.52	54.76
Training from ImageNet weights															
[17]	ResNet-50	80.07	77.78	23.61	22.11	23.32	21.84	21.76	21.03	13.07	12.80	13.06	12.75	12.53	12.71
[17]	WideResNet-50	79.24	76.46	32.28	26.53	32.24	26.51	28.95	23.86	20.37	17.21	20.90	17.25	19.24	15.97
HSAT-Patch	ResNet-50	50.80	44.36	51.08	44.95	51.06	44.94	51.10	44.96	50.29	45.22	50.29	45.22	50.41	45.31
HSAT	ResNet-50	66.48	66.33	64.07	61.36	64.11	61.40	65.11	62.40	56.97	55.00	56.98	55.04	57.92	56.06
HSAT	WideResNet-50	68.35	67.76	60.51	59.27	60.54	59.32	61.44	60.19	53.13	51.26	53.12	51.26	54.44	52.73
Slide Classification															
Training from Scratch															
[17]	ResNet-50	89.11	88.24	9.73	8.93	9.69	8.89	12.45	11.27	3.11	2.66	4.28	3.57	4.67	3.87
HSAT-Patch	ResNet-50	58.37	53.30	57.20	52.18	57.09	52.15	57.98	52.75	56.81	52.11	56.81	52.11	57.98	53.22
HSAT	ResNet-50	76.26	75.09	77.04	73.58	77.00	73.53	77.43	73.87	75.10	69.74	75.00	69.57	74.71	69.69
Training from ImageNet weights															
[17]	ResNet-50	85.60	83.86	27.63	26.11	27.24	25.70	25.29	24.86	10.51	10.61	10.12	10.06	10.51	10.86
[17]	WideResNet-50	85.60	83.62	45.91	38.74	45.91	38.74	40.08	34.21	21.40	18.13	20.62	17.45	19.46	16.27
HSAT-Patch	ResNet-50	60.31	54.69	61.87	56.08	61.82	56.03	61.48	55.80	60.74	55.75	60.71	55.73	60.70	55.73
HSAT	ResNet-50	77.43	76.87	75.51	73.65	75.49	73.63	75.88	73.21	71.21	67.91	71.60	68.20	71.20	67.89
HSAT	WideResNet-50	80.93	79.25	77.82	74.42	77.82	74.42	76.65	73.56	71.21	66.50	70.43	65.66	70.82	66.19
Patient Classification															
Training from Scratch															
[17]	ResNet-50	90.00	91.08	10.00	8.73	10.00	8.73	13.33	11.59	3.33	2.86	3.33	3.86	5.00	4.29
HSAT-Patch	ResNet-50	55.00	51.62	55.00	51.68	54.89	51.60	54.75	50.96	56.72	54.83	56.67	54.63	56.61	54.62
HSAT	ResNet-50	76.67	73.03	78.33	79.00	78.33	79.00	76.67	76.15	76.67	77.29	76.67	73.03	75.00	71.73
Training from ImageNet weights															
[17]	ResNet-50	86.67	88.20	31.67	28.95	28.33	26.35	25.00	23.49	11.67	10.32	11.60	10.26	11.59	10.28
[17]	WideResNet-50	85.00	85.34	41.67	35.41	41.67	35.41	36.67	31.67	23.33	20.19	21.67	18.76	18.33	16.03
HSAT-Patch	ResNet-50	58.33	55.77	60.28	58.63	60.00	58.60	60.04	58.53	60.13	58.59	60.00	58.33	60.02	58.21
HSAT	ResNet-50	78.33	78.72	75.00	74.85	75.00	74.85	75.00	74.85	70.00	67.84	71.67	69.42	71.67	69.42
HSAT	WideResNet-50	80.00	80.17	76.67	73.03	78.33	75.89	75.00	73.29	66.67	64.24	66.67	64.24	68.33	65.67

against non-adversarial hierarchical training methods for histopathology images [17] and instance-level adversarial training methods proposed for natural images ($\mathcal{L}_{max} = \mathcal{L}_{min} = \mathcal{L}_{patch}$), referred to as HSAT-Patch. As reported in Table 1, our approach consistently outperforms both baselines. Under the PGD attack at $\epsilon = \frac{8}{255}$ using a ResNet-50 backbone, we achieve gains in adversarial robustness of 43.90%, 60.70%, and 58.33% compared to [17], and 6.68%, 10.51%, and 10% on patch, slide, and patient classification, respectively compared to instance-level adversarial training (HSAT-Patch). Furthermore, we observe a drop in clean performance compared to [17] which is an expected trade-off between adversarial and clean accuracy. However, our method still surpasses current instance-level adversarial training methods on clean performance, with improvements of 15.68%, 17.12%, and 20% in patch, slide, and patient classification, respectively.

Robustness Against Black-Box Attacks: In this section, we evaluate and compare the robustness of different models against transfer-based *black-box* attacks. These attacks exploit the ***transferability*** property of adversarial examples i.e., adversarial examples crafted on a surrogate model transfer to unknown target models. To measure transferability, all vision models are used interchangeably as both surrogate and target models. In Table 2, performance drop on adversarial examples (crafted using `PGD` at $\epsilon = \frac{8}{255}$) relative to clean samples is reported using `Acc-D` and `MCA-D`. We observe that target vision models trained using our `HSAT` framework are significantly more robust against transferability of adversarial examples crafted across different surrogate models. On average, `HSAT` trained target models show a performance drop of around $3 - 4\%$, while target models trained using [17] show a performance drop of around $25 - 30\%$.

Table 2. Performance against transfer-based *black-box* attacks *(lower is better)*: `HSAT`-trained models show strong robustness, while baseline models remain highly vulnerable.

Target →	Type	ResNet-50 ([17])		ResNet-50 (HSAT)		WResNet-50 ([17])		WResNet-50 (HSAT)		ResNet-101 ([17])	
Surrogate ↓		Acc-D	MCA-D	Acc-D	MCA-D	Acc-D	MCA-D	Acc-D	MCA-D	Acc-D	MCA-D
ResNet-50 ([17])	Patch	66.81	64.81	1.42	3.08	35.64	37.99	2.48	4.32	39.49	41.98
	Slide	74.71	72.94	1.55	3.56	32.68	37.34	2.72	5.26	37.35	39.60
	Patient	75.00	77.88	0	1.27	31.67	38.63	3.33	5.58	36.66	43.33
ResNet-50 (HSAT)	Patch	20.48	25.67	9.54	11.38	20.62	25.76	4.81	7.43	23.69	29.68
	Slide	15.95	22.24	6.22	8.96	18.28	24.93	5.83	9.41	12.45	18.87
	Patient	16.67	25.46	8.33	10.88	21.67	28.50	6.67	9.74	15.00	23.17
WResNet-50 ([17])	Patch	32.21	36.60	1.56	3.32	58.51	59.37	2.39	4.35	35.72	40.37
	Slide	18.67	26.83	1.94	4.20	64.20	65.73	3.50	6.84	24.51	30.75
	Patient	15.00	24.42	0	1.27	61.67	65.15	1.67	4.28	26.66	35.05
WResNet-50 (HSAT)	Patch	20.09	24.73	3.63	5.27	20.20	24.99	15.34	16.62	23.92	29.40
	Slide	12.45	18.65	3.11	5.81	15.56	22.53	10.50	13.35	12.45	18.22
	Patient	16.67	25.17	5.00	8.29	16.67	24.19	13.33	15.93	13.33	20.45
ResNet-101 ([17])	Patch	54.06	53.54	2.59	4.41	52.46	51.95	3.53	5.50	70.48	68.36
	Slide	45.91	47.43	4.28	6.76	45.91	48.13	4.67	7.19	77.82	74.36
	Patient	43.34	48.63	5.00	7.15	41.67	46.09	3.33	5.71	78.33	78.19
Average	Patch	31.71	35.13	**2.33**	**4.02**	32.23	35.18	**3.30**	**5.40**	30.68	35.36
	Slide	23.24	28.79	**2.72**	**5.08**	28.11	33.23	**4.18**	**7.17**	21.69	26.86
	Patient	22.92	30.92	**2.50**	**4.49**	27.92	34.35	**3.75**	**6.33**	22.91	30.50

Ablations: The ablation studies for *Hierarchical Adversarial Training* (`HSAT`) are presented in Tables 3 and 4. Table 3 examines the effect of increasing hierarchical discrimination levels during adversarial training, progressing from patch-level (`HSAT-Patch`) to slide-level (`HSAT-Slide`) and patient-level (`HSAT-Patient`). The results show a consistent improvement in adversarial robustness as the level of discrimination increases, highlighting the benefits of leveraging hierarchical structures in biomedical data. Table 4 investigates the impact of different loss terms used to generate adversarial examples in the maximization step of HSAT. Starting with patch-level loss $\mathcal{L}_{con}^{patch}$, we observe a performance boost when slide-level loss $\mathcal{L}_{con}^{slide}$ is added, with further gains upon incorporating patient-level loss $\mathcal{L}_{con}^{patient}$. These findings underscore the effectiveness of multi-level adversarial training, demonstrating that aligning robust representations across hierarchical levels enhances model resilience against adversarial attacks.

Table 3. Comparing adversarial training at different levels of discrimination.

Training	Patch		Slide		Patient	
	Acc	MCA	Acc	MCA	Acc	MCA
HSAT-Patch	50.29	45.22	60.70	55.73	60.00	58.92
HSAT-Slide	55.07	50.20	69.32	64.69	68.33	66.29
HSAT-Patient	56.97	55.00	71.21	67.91	70.00	67.84

Table 4. Comparing effect of using different types of loss in the maximization step.

$\mathcal{L}_{max}$	Patch		Slide		Patient	
	Acc	MCA	Acc	MCA	Acc	MCA
$\mathcal{L}_{con}^{patch}$	50.19	47.52	63.59	63.27	62.00	63.55
$\mathcal{L}_{con}^{patch} + \mathcal{L}_{con}^{slide}$	49.19	46.70	65.76	66.09	63.33	66.67
$\mathcal{L}_{con}^{p} + \mathcal{L}_{con}^{s} + \mathcal{L}_{con}^{pt}$	56.97	55.00	71.21	67.91	70.00	67.84

5 Conclusion

We introduce *Hierarchical Self-Supervised Adversarial Training* (HSAT), a novel approach designed for improving adversarial robustness in biomedical image analysis. HSAT integrates multi-level discriminative learning into adversarial training, capturing the hierarchical structure within biomedical data. Unlike traditional self-supervised adversarial training methods, HSAT forces models to learn resilient features across both fine-grained and coarse-grained levels. Our experiments on OpenSRH benchmark demonstrate that HSAT significantly enhances resistance to adversarial attacks, outperforming existing methods.

Disclosure of Interests. The authors have no competing interests to declare that are relevant to the content of this article.

References

1. Boyd, J., Liashuha, M., Deutsch, E., Paragios, N., Christodoulidis, S., Vakalopoulou, M.: Self-supervised representation learning using visual field expansion on digital pathology. In: Proceedings of the IEEE/CVF International Conference on Computer Vision, pp. 639–647 (2021)
2. Brendel, W., Rauber, J., Bethge, M.: Decision-based adversarial attacks: reliable attacks against black-box machine learning models. In: International Conference on Learning Representations (ICLR) (2018)
3. Carlini, N., Wagner, D.: Towards evaluating the robustness of neural networks. In: 2017 IEEE symposium on security and privacy (SP), pp. 39–57. IEEE (2017)
4. Chen, P.Y., Zhang, H., Sharma, Y., Yi, J., Hsieh, C.J.: Zoo: zeroth order optimization based black-box attacks to deep neural networks without training substitute models. In: Proceedings of the 10th ACM Workshop on Artificial Intelligence and Security (2017)
5. Chen, R.J., et al.: Towards a general-purpose foundation model for computational pathology. Nat. Med. **30**(3), 850–862 (2024)
6. Chen, R.J., Krishnan, R.G.: Self-supervised vision transformers learn visual concepts in histopathology. arXiv preprint arXiv:2203.00585 (2022)
7. Ciga, O., Xu, T., Martel, A.L.: Self supervised contrastive learning for digital histopathology. Mach. Learn. App. **7**, 100198 (2022). https://doi.org/10.1016/j.mlwa.2021.100198, https://www.sciencedirect.com/science/article/pii/S2666827021000992
8. Ciga, O., Xu, T., Martel, A.L.: Self supervised contrastive learning for digital histopathology. Mach. Learn. App. **7**, 100198 (2022)
9. Croce, F., Hein, M.: Reliable evaluation of adversarial robustness with an ensemble of diverse parameter-free attacks. In: International Conference on Machine Learning, pp. 2206–2216. PMLR (2020)
10. Dong, Y., et al.: Boosting adversarial attacks with momentum. In: Proceedings of the IEEE/CVF Conference on Computer Vision and Pattern Recognition (CVPR) (2018)
11. Fan, L., Liu, S., Chen, P.Y., Zhang, G., Gan, C.: When does contrastive learning preserve adversarial robustness from pretraining to finetuning? Adv. Neural. Inf. Process. Syst. **34**, 21480–21492 (2021)
12. Foote, A., Asif, A., Azam, A., Marshall-Cox, T., Rajpoot, N., Minhas, F.: Now you see it, now you don't: adversarial vulnerabilities in computational pathology. arXiv preprint arXiv:2106.08153 (2021)
13. Ghaffari Laleh, N., et al.: Adversarial attacks and adversarial robustness in computational pathology. Nat. Commun. **13**(1), 5711 (2022)
14. Goodfellow, I.J., Shlens, J., Szegedy, C.: Explaining and harnessing adversarial examples. arXiv preprint arXiv:1412.6572 (2014)
15. He, K., Zhang, X., Ren, S., Sun, J.: Deep residual learning for image recognition. corr abs/1512.03385 (2015)
16. Jiang, C., et al.: OpenSRH: optimizing brain tumor surgery using intraoperative stimulated Raman histology. Adv. Neural. Inf. Process. Syst. **35**, 28502–28516 (2022)
17. Jiang, C., et al.: Hierarchical discriminative learning improves visual representations of biomedical microscopy. In: Proceedings of the IEEE/CVF Conference on Computer Vision and Pattern Recognition, pp. 19798–19808 (2023)

18. Kanca, E., Gulsoy, T., Ayas, S., Kablan, E.B.: Generating robust adversarial images in medical image classification using GANs. In: 2024 Innovations in Intelligent Systems and Applications Conference (ASYU), pp. 1–5 (2024). https://doi.org/10.1109/ASYU62119.2024.10757024
19. Kim, M., Tack, J., Hwang, S.J.: Adversarial self-supervised contrastive learning. Adv. Neural. Inf. Process. Syst. **33**, 2983–2994 (2020)
20. Kumar, D., Sharma, A., Narayan, A.: Attacking CNNs in histopathology with snap: Sporadic and naturalistic adversarial patches (student abstract). Proc. AAAI Conf. Artif. Intell. 38(2121), 23550–23551 (2024). https://doi.org/10.1609/aaai.v38i21.30468, https://ojs.aaai.org/index.php/AAAI/article/view/30468
21. Kurakin, A., Goodfellow, I.J., Bengio, S.: Adversarial examples in the physical world. In: Artificial Intelligence Safety and Security, pp. 99–112. Chapman and Hall/CRC (2018)
22. Loshchilov, I., Hutter, F.: Decoupled weight decay regularization. arXiv preprint arXiv:1711.05101 (2017)
23. Luo, R., Wang, Y., Wang, Y.: Rethinking the effect of data augmentation in adversarial contrastive learning. arXiv preprint arXiv:2303.01289 (2023)
24. Madry, A., Makelov, A., Schmidt, L., Tsipras, D., Vladu, A.: Towards deep learning models resistant to adversarial attacks. arXiv preprint arXiv:1706.06083 (2017)
25. Maliamanis, T.V., Apostolidis, K.D., Papakostas, G.A.: How resilient are deep learning models in medical image analysis? the case of the moment-based adversarial attack (MB-ADA). Biomedicines **10**(10), 2545 (2022)
26. Malik, H.S., Kunhimon, S.K., Naseer, M., Khan, S., Khan, F.S.: Adversarial pixel restoration as a pretext task for transferable perturbations. arXiv preprint arXiv:2207.08803 (2022)
27. Narodytska, N., Kasiviswanathan, S.: Simple black-box adversarial attacks on deep neural networks. In: Computer Vision and Pattern Recognition (CVPR) Workshop (2017)
28. Naseer, M.M., Khan, S.H., Khan, M.H., Shahbaz Khan, F., Porikli, F.: Cross-domain transferability of adversarial perturbations. In: Advances in Neural Information Processing Systems (NeurIPS) (2019)
29. Su, J., Vargas, D.V., Sakurai, K.: One pixel attack for fooling deep neural networks. IEEE Trans. Evol. Comput. **23**(5), 828–841 (2019)
30. Szegedy, C., et al.: Intriguing properties of neural networks. arXiv preprint arXiv:1312.6199 (2013)
31. Wong, E., Rice, L., Kolter, J.Z.: Fast is better than free: revisiting adversarial training. arXiv preprint arXiv:2001.03994 (2020)
32. Zagoruyko, S., Komodakis, N.: Wide residual networks. arXiv preprint arXiv:1605.07146 (2016)
33. Zhang, H., Yu, Y., Jiao, J., Xing, E., El Ghaoui, L., Jordan, M.: Theoretically principled trade-off between robustness and accuracy. In: International Conference on Machine Learning, pp. 7472–7482. PMLR (2019)

High-Precision Mixed Feature Fusion Network Using Hypergraph Computation for Cervical Abnormal Cell Detection

Jincheng Li[1], Danyang Dong[1], Menglin Zheng[1], Jingbo Zhang[1], Yueqin Hang[1], Lichi Zhang[2], and Lili Zhao[1(✉)]

[1] School of Artificial Intelligence and Computer Science, Nantong University, Nantong 226019, China
ylzh@ntu.edu.cn

[2] School of Biomedical Engineering, Shanghai Jiao Tong University, Shanghai 200030, China

Abstract. Automatic detection of abnormal cervical cells from Thin-prep Cytologic Test (TCT) images is a critical component in the development of intelligent computer-aided diagnostic systems. However, existing algorithms typically fail to effectively model the correlations of visual features, while these spatial correlation features actually contain critical diagnostic information. Furthermore, no detection algorithm has the ability to integrate inter-correlation features of cells with intra-discriminative features of cells, lacking a fusion strategy for the end-to-end detection model. In this work, we propose a hypergraph-based cell detection network that effectively fuses different types of features, combining spatial correlation features and deep discriminative features. Specifically, we use a Multi-level Fusion Sub-network (MLF-SNet) to enhance feature extraction capabilities. Then we introduce a Cross-level Feature Fusion Strategy with Hypergraph Computation module (CLFFS-HC), to integrate mixed features. Finally, we conducted experiments on three publicly available datasets, and the results demonstrate that our method significantly improves the performance of cervical abnormal cell detection. The code is publicly available at https://github.com/ddddoreen/HyperMF2-Cell-Detection.

Keywords: Cervical abnormal cell detection · Hypergraph computation · Hypergraph neural networks · Feature fusion

1 Introduction

Cervical cancer is ranked as the fourth most common cancer globally [1]. If diagnosed early, it can be effectively treated and cured [2]. Thin-prep Cytologic Test (TCT) is the most widely utilized and effective screening method for detecting cervical abnormal and precancerous lesions [3]. Despite its widespread clinical application and significant contribution to reducing cervical cancer mortality,

J. C. Gee et al. (Eds.): MICCAI 2025, LNCS 15960, pp. 250–259, 2026.
https://doi.org/10.1007/978-3-032-04927-8_24

the analysis of Whole Slide Images (WSI), which often contain billions of pixels, remains a time-consuming and error-prone task for pathologists [4]. Therefore, there is an urgent need for automated cervical cell analysis methods to assist cytologists in evaluating cervical pathology images more efficiently, accurately, and objectively.

Recently, various deep learning-based detection methods have been applied in TCT screening to identify cervical abnormal cells. Specifically, Shi et al. [5] developed a new classification approach for cervical cells using Graph Convolutional Networks (GCN), which aims to enhance classification accuracy by exploring the latent relationships within cervical cell images. Fei et al. [6] achieved robust detection of cervical abnormal cells through a distillation method with local-scale consistency refinement, thereby assisting pathologists in making more accurate diagnoses. Additionally, Chai et al. [7] introduced an innovative dual-path discriminative detection network (DPD-Net), which effectively extracts and distinguishes features of normal and abnormal cells. These studies highlight the potential of deep learning in improving the accuracy of abnormal cervical cell detection.

Although the aforementioned methods have made significant progress in TCT screening, several issues remain to be addressed: 1) In clinical practice, pathologists typically assess target cells by comparing them with their surrounding cells, particularly when the cells exhibit ambiguous classifications. However, existing detection algorithms generally lack effective modeling of the visual feature correlations between cells, overlooking the important diagnostic information contained in spatial correlation features. 2) Cellular lesions are often accompanied by morphological abnormalities in the nucleus, such as enlargement, irregular membranes and unclear boundaries. Considering both the inter-cellular correlations and the intra-discriminative features of the cells could substantially enhance detection performance. However, no existing detection algorithm effectively combines spatial inter-correlation features with intra-discriminative features of cells due to the lack of a comprehensive fusion strategy.

To address these issues, we propose a novel mixed feature fusion network using hypergraph computation for cervical abnormal cell detection. We adopt the classic one-stage detection framework [8] and appropriately integrate the hypergraph computation module at multiple levels of an end-to-end detection network [9]. Specifically, we use a Multi-level Fusion Sub-network (MLF-SNet) integrating three distinct types of convolution to enrich the diversity of feature representations, thereby enhancing the feature extraction ability. Then we introduce a Cross-level Feature Fusion Strategy with Hypergraph Computation module (CLFFS-HC). This strategy combines inter-correlation features of cells at the individual level with intra-discriminative features of cells at the population level, unifying them into an end-to-end framework to seamlessly integrate mixed features. Experimental results on three publicly available datasets demonstrate the superiority of our method in cervical abnormal cell detection.

2 Method

The proposed framework, illustrated in Fig. 1, consists of three progressive computational phases to enhance cervical abnormal cell detection. **First, in the WSI preprocessing stage**, we implement a sliding window strategy to extract 640×640 image patches from Whole Slide Images. This dimension addresses the inherent challenges of high-resolution pathological image processing while maintaining cellular-level details. **Second, for hierarchical feature extraction**, we develop a Multi-level Fusion Sub-network (MLF-SNet) to overcome the limitations of conventional one-stage detectors. Compared to standard YOLO architectures with limited multi-level feature integration capabilities, our MLF-SNet employs parallel convolution pathways to enrich the diversity of feature representations. **Third, in the feature enhancement phase**, we introduce a novel hypergraph computation approach [10] through a Cross-Level Feature Fusion Strategy (CLFFS-HC). This module constructs dynamic hyperedges connecting cross-level feature representations from different network layers, combining inter-correlation features of cells at the individual level with intra-discriminative features of cells at the population level.

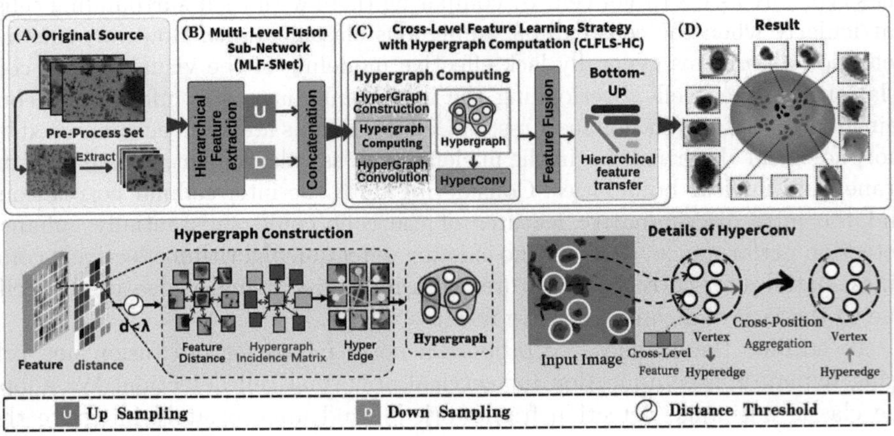

Fig. 1. The overview of our proposed framework, (A): Data Preprocessing, (B): MLF-SNet performs feature extraction on the preprocessed images, (C): CLFFS-HC achieves complex interactions of different types and levels of mixed features across multiple levels, resulting in high-precision detection results. The implementation of Hypergraph Construction and Hypergraph Convolution is explained in detail.

2.1 Multi-level Fusion Sub-Network (MLF-SNet)

To enhance the capture of spatial correlations between cellular structures, we introduce hypergraph computation into our design. A hypergraph $\mathcal{G} = (\mathcal{V}, \mathcal{E})$

consists of vertices $\mathcal{V}$ and hyperedges $\mathcal{E}$. We extract five feature layers from the backbone to construct the vertices $\mathcal{V}$. To fuse multi-level information, inspired by multi-level convolution [11], we design a multi-level fusion sub-network during feature extraction process. This subnetwork includes a 1×1 convolution for channel feature recalibration, deformable convolution (DCN) [12] for effective spatial correlation information, and retains the C3K2 convolutional module to achieve hierarchical feature preservation. This approach allows the model to adapt its feature extraction and fusion techniques based on varying inputs, facilitating a comprehensive modeling of multi-level features. The detailed implementation is illustrated in Fig. 2.

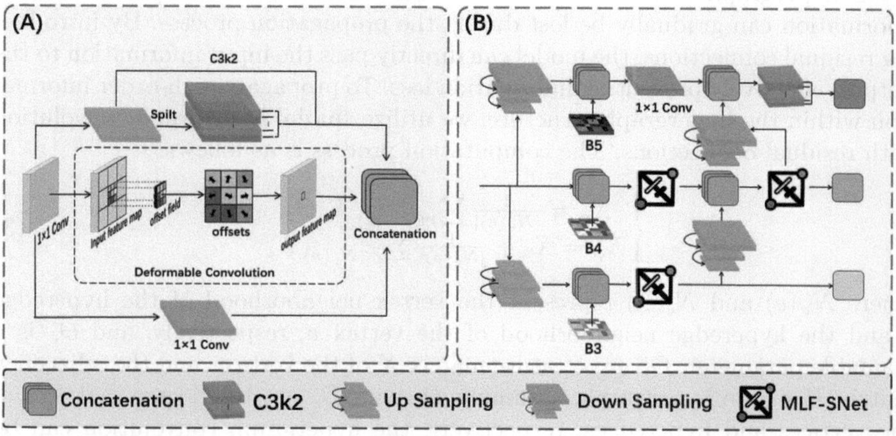

Fig. 2. (A): Architecture of MLF-SNet, featuring channel recalibration (1×1 conv), spatial adaptation (DCN), and hierarchical feature retention (C3K2 module); (B): Architecture of the fusion strategy and the Bottom-Up Pathway.

As illustrated in Fig. 2(A), MLF-SNet processes raw cell images through iterative layer fusion, producing enhanced multi-scale features. A final 1×1 convolution compresses these features into an optimized output. This combination enables adaptive feature extraction while maintaining critical spatial relationships between cells. The process works synergistically with CLFFS-HC to improve detection accuracy.

2.2 Cross-Level Feature Fusion Strategy with Hypergraph Computation Module (CLFFS-HC)

To address inadequate integration of inter-correlation and intra-discriminative cellular features, we design the Cross-level Feature Fusion Strategy with Hypergraph Computation module (CLFFS-HC). As described in Sect. 2.1, we construct the set of hypergraph vertices $\mathcal{V}$. To model the neighborhood relationships within the hypergraph module, we construct the set of hyperedges $\mathcal{E}$ through a distance threshold λ. Specifically, for each feature point x_u, we find all feature points that are less than λ distance away and form a hyperedge with them and x_v. A hyperedge e can be represented as:

$$e = \{u \mid ||x_u - x_v||_2 < \lambda, u \in V\}, \tag{1}$$

The Euclidean Norm $||\cdot||_2$ represents all such hyper-edges forming the hyperedge set E. The incidence matrix H of the hypergraph $\mathcal{G} = (\mathcal{V}, \mathcal{E})$ is defined as:

$$H_{ve} = \begin{cases} 1, \text{ if } v \in \mathcal{E} \\ 0, \text{ if } v \notin \mathcal{E} \end{cases} \tag{2}$$

In hypergraph convolutional networks, as the number of layers increases, information can gradually be lost during the propagation process. By introducing residual connections, the model can directly pass the input information to the output, effectively preventing information loss. To propagate high-order information within the hypergraph structure, we utilize spatial hypergraph convolution with residual connections. The computation process is as follows:

$$\begin{cases} X_e = \frac{1}{|N_v(e)|} \sum_{v \in N_v(e)} X_v \Theta_e, \\ X_v' = X_v + \frac{1}{|N_e(v)|} \sum_{e \in N_e(v)} X_e \end{cases} \tag{3}$$

where $N_{\mathcal{V}}(e)$ and $N_e(\mathcal{V})$ represent the vertex neighborhood of the hyperedge e and the hyperedge neighborhood of the vertex v, respectively, and Θ_e is a trainable parameter. Given a feature matrix X of the vertices and the adjacency matrix H of the hypergraph, assuming $D_{\mathcal{V}}$ and D_e are the degree matrices of the vertices and hyperedges, respectively, the hypergraph convolution can be expressed as:

$$HyperConv(X, H) = X + D_v^{-1} H D_e^{-1} H^T X \Theta \tag{4}$$

$D_{\mathcal{V}}^{-1} H D_e^{-1}$ calculates the normalized adjacency matrix, $D_{\mathcal{V}}^{-1} H D_e^{-1} H^T X$ represents the aggregation of vertex features through hyperedges to capture higher-order relationships between vertices. Θ is a learnable parameter matrix used for transforming the aggregated features, enhancing the model's expressiveness. A residual connection of X is also introduced to preserve the original feature information and prevent information loss.

The feature maps of the five layers extracted by the backbone are set as B_1, B_2, B_3, B_4, B_5, respectively. These features are fused through channel concatenation to obtain a mixed feature X_m that contains information from all layers. This X_m is then used as input for hypergraph convolution, resulting in a high-order feature representation X_{hyper}:

$$X_{hyper} = HyperConv(X_m, H) \tag{5}$$

The hypergraph convolution output X_{hyper} contains high-order correlation information across levels. To optimize feature integration and ensure channel alignment, we implement a fusion strategy incorporating the Bottom-Up Pathway [13]. As shown in Fig. 2(B), X_{hyper} undergoes progressive combination with the backbone features B_3, B_4, B_5 to generate multi-level outputs N_1, N_2, N_3. Unlike the original model limited to adjacent-layer fusion, our method can directly fuse the feature information extracted from the backbone. This design significantly mitigates connectivity disparities between features at different network depths. Deep layer features inherently lose spatial resolution due to increased abstraction. Our Bottom-Up Pathway addresses this limitation by strategically transferring high-resolution details from shallow layers to deeper ones, preserving critical structural information.

In summary, CLFFS-HC overcomes the limitations of traditional grid structures, enhancing the comprehensiveness and granularity of feature representations. By combining inter-correlation features of cells at the individual level with intra-discriminative features of cells at the population level, this strategy successfully integrates mixed features.

3 Experimental Results

3.1 Dataset and Experimental Setup

Dataset: To evaluate the effectiveness of the proposed model, we conducted extensive experiments on public datasets that cover different staining and imaging conditions. The data used for testing include: (A) Alibaba Tianchi competition [14], (B) Comparison Detector [15], and (C) CRIC [16]. Dataset A consists of slide images of cervical smears labeled by professional doctors, with each slide image obtained using a 20x digital scanner. Dataset B comprises 7,410 cervical cytology images with 11 types of abnormal cell lesions. Note that, for comparison with the SOTA model [17], we only selected four types of abnormal cells for the experiment, namely ASC-US, ASC-H, LSIL, and HSIL cells. There are 400 pap smear cervical images in the dataset C in total. The nuclei of epithelial cells are addressed in six classes according to the Bethesda system, including Negative, ASC-US, ASC-H, LSIL, HSIL, and SCC.

Experimental Details: Our implementation is based on MMdetection [18], with all parameters optimized using Adam [19]. The model is implemented in PyTorch on a single Nvidia Tesla P100 GPU. Performance is evaluated using two metrics: COCO-style Average Precision(AP) and Average Recall(AR). We calculate the AP over multiple IoU thresholds from 0.5 to 0.95 with a step size of 0.05, and separately evaluate the AP at an IoU threshold of 0.5 (denoted as AP.5).

3.2 Evaluation of Cervical Abnormal Cell Detection

Comparative Analysis: We compare the performance of our proposed method with existing methods for cervical lesion detection and representative methods for object detection. To ensure fair comparison, we employed consistent preprocessing methods across all experiments for each dataset. Specifically, the sliding window method is used to crop dataset A into 640 × 640 images, no preprocessing is applied to dataset B, and the images of dataset C are augmented using imgaug. We will release the processed images in our publicly available code repository. The results are shown in Table 1. Our method achieves superior performance across all evaluation metrics, especially with a significant improvement in the detection performance of AP.5. These results demonstrate the effectiveness of incorporating hypergraph computation into cervical lesion detection.

Table 1. Comparison of different methods on three datasets

Method	dataset A [14]			dataset B [15]			dataset C [16]		
	AP	AP.5	AR	AP	AP.5	AR	AP	AP.5	AR
Faster R-CNN [20]	50.5	74.1	66.4	18.3	36.6	33.2	33.6	58.7	52.5
RetinaNet [21]	50.6	80.7	68.2	14.9	29.3	33.8	32.2	55.3	49.5
Sparse R-CNN [22]	69.2	86.5	76.2	18.0	35.0	44.3	30.6	53.8	52.8
Cascade R-CNN [23]	44.4	69.5	62.5	18.4	34.7	33.0	35.9	59.0	54.7
YOLOX [24]	28.5	55.2	60.6	17.1	33.7	28.1	29.1	48.3	44.2
RT-DETR [25]	72.2	88.9	80.1	21.3	36.9	40.2	32.9	55.9	50.4
SOTA for A [25]	72.2	88.9	80.1						
SOTA for B [17]				24.6	44.7	46.6			
SOTA for C [7]							35.2	61.9	55.3
ours	79.2	97.5	91.4	29.8	51.4	54.4	44.6	77.0	72.4

Ablation Study: Ablation studies were conducted on three public datasets to evaluate the contribution of each component. As shown in Table 2, our method demonstrates consistent improvements across various metrics compared to the baseline YOLO11n model. In particular, the significant improvement in recall indicates that our model has effectively addressed the issue of false negatives by incorporating the idea of hypergraph computation.

Table 2. Results of the Ablation study

Method	dataset A [14]			dataset B [15]			dataset C [16]		
	AP	AP.5	AR	AP	AP.5	AR	AP	AP.5	AR
baseline [9]	74.0	86.5	78.4	21.7	43.0	47.8	32.6	56.7	50.7
+MLF-SNet	79.6	90.2	84.2	23.9	45.2	50.1	36.1	61.0	55.4
+CLFFS-HC	82.4	91.4	82.7	25.4	46.7	52.1	38.9	69.3	63.2
+MLF-SNet and CLFFS-HC	79.2	97.5	91.4	29.8	51.4	54.4	44.6	77.0	72.4

Visualization Map: To visually demonstrate clinical relevance, we present comparative detection maps of representative samples in Fig. 3. Our method successfully identifies subtle lesion patterns frequently missed by conventional approaches, while maintaining superior specificity against complex backgrounds. This dual capability of reducing both false negatives and false positives significantly enhances diagnostic reliability, particularly in challenging cases with ambiguous cellular morphology. The visual evidence aligns with quantitative metrics, substantiating our method's potential as a clinically viable decision-support tool.

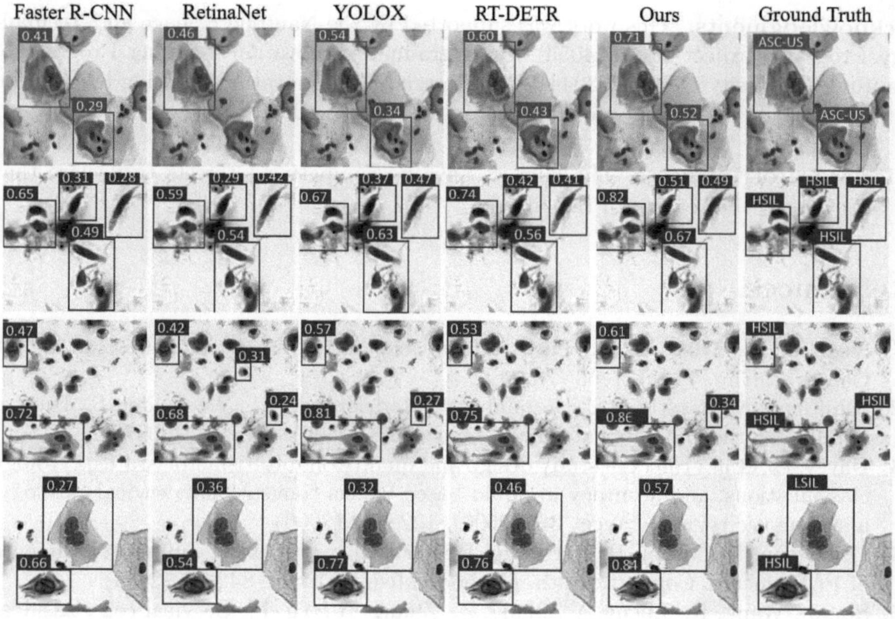

Fig. 3. Comparative Experiment Visualization Map.

4 Conclusion

In this paper, we have proposed a novel hypergraph-based network for cervical abnormal cell detection that effectively addresses two critical limitations in existing methods: the inadequate modeling of spatial correlations and the insufficient integration of inter-cellular and intra-cellular features. Our main contributions include: (1) the development of a Multi-level Fusion Sub-network (MLF-SNet) that enriches feature representation through diverse convolution operations, and (2) the introduction of a Cross-level Feature Fusion Strategy with Hypergraph

Computation (CLFFS-HC) that effectively combines inter-correlation and intra-discriminative cellular features. Extensive experiments on three publicly available datasets demonstrate the superior performance of our proposed method compared to existing approaches. The ablation studies further validate the effectiveness of each component in our framework, particularly highlighting the significant improvement in detection accuracy through hypergraph computation. This work provides promising directions for future research in automated cervical cell analysis. Future work could focus on extending the framework to handle more diverse pathological conditions and optimizing the computational efficiency for real-time clinical applications.

Acknowledgments. This work was supported by the Nantong Science and Technology Program Project (JC2024055), Undergraduate Innovation Training Program of Nantong University (X202510304412), Undergraduate Innovation Training Program of School of Artificial Intelligence and Computer Science (202504).

Disclosure of Interests. The authors have no competing interests to declare that are relevant to the content of this article.

References

1. Siegel, R.L., Miller, K.D., Fuchs, H.E., Jemal, A.: Cancer statistics, 2021. CA Cancer J. Clin. **71**(1), 7–33 (2021)
2. Schiffman, M., Castle, P.E., Jeronimo, J., Rodriguez, A.C., Wacholder, S.: Human papillomavirus and cervical cancer. Lancet **370**(9590), 890–907 (2007)
3. Davey, E., et al.: Effect of study design and quality on unsatisfactory rates, cytology classifications, and accuracy in liquid-based versus conventional cervical cytology: a systematic review. Lancet **367**(9505), 122–132 (2006)
4. Bengtsson, E., Malm, P.: Screening for cervical cancer using automated analysis of PAP-smears. Comput. Math. Methods Med. **2014**(1), 842037 (2014)
5. Shi, J., Wang, R., Zheng, Y., Jiang, Z., Zhang, H., Yu, L.: Cervical cell classification with graph convolutional network. Comput. Methods Programs Biomed. **198**, 105807 (2021)
6. Fei, M., et al.: Robust cervical abnormal cell detection via distillation from local-scale consistency refinement. In: Greenspan, H., et al. (eds.) MICCAI 2023. LNCS, vol. 14225, pp. 652–661. Springer, Cham (2023). https://doi.org/10.1007/978-3-031-43987-2_63
7. Chai, S., et al.: DPD-Net: dual-path proposal discriminative network for abnormal cell detection in cervical cytology images. Biomed. Signal Process. Control **89**, 105887 (2024)
8. Redmon, J., Divvala, S., Girshick, R., Farhadi, A.: You only look once: unified, real-time object detection. In: Proceedings of the IEEE Conference on Computer Vision and Pattern Recognition, pp. 779–788 (2016)
9. Khanam, R., Hussain, M.: YOLOv11: an overview of the key architectural enhancements. arXiv preprint arXiv:2410.17725 (2024)
10. Feng, Y., You, H., Zhang, Z., Ji, R., Gao, Y.: Hypergraph neural networks. In: Proceedings of the AAAI Conference on Artificial Intelligence, vol. 33, pp. 3558–3565 (2019)

11. Szegedy, C., et al.: Going deeper with convolutions. In: Proceedings of the IEEE Conference on Computer Vision and Pattern Recognition, pp. 1–9 (2015)
12. Dai, J., et al.: Deformable convolutional networks. In: Proceedings of the IEEE International Conference on Computer Vision, pp. 764–773 (2017)
13. Lin, T.Y., Dollár, P., Girshick, R., He, K., Hariharan, B., Belongie, S.: Feature pyramid networks for object detection. In: Proceedings of the IEEE Conference on Computer Vision and Pattern Recognition, pp. 2117–2125 (2017)
14. Tianchi Competition: Dataset (2019). https://tianchi.aliyun.com/competition/. Accessed Nov 2019
15. Liang, Y., Tang, Z., Yan, M., Chen, J., Liu, Q., Xiang, Y.: Comparison detector for cervical cell/clumps detection in the limited data scenario. Neurocomputing **437**, 195–205 (2021)
16. Rezende, M.T., et al.: CRIC searchable image database as a public platform for conventional PAP smear cytology data. Sci. Data **8**(1), 151 (2021)
17. Fei, M., et al.: Distillation of multi-class cervical lesion cell detection via synthesis-aided pre-training and patch-level feature alignment. Neural Netw. **178**, 106405 (2024)
18. Chen, K., et al.: MMDetection: open MMLab detection toolbox and benchmark. arXiv preprint arXiv:1906.07155 (2019)
19. Kingma, D.P., Ba, J.: Adam: a method for stochastic optimization. arXiv preprint arXiv:1412.6980 (2014)
20. Ren, S., He, K., Girshick, R., Sun, J.: Faster R-CNN: towards real-time object detection with region proposal networks. In: Advances in Neural Information Processing Systems, vol. 28 (2015)
21. Lin, T.Y., Goyal, P., Girshick, R., He, K., Dollár, P.: Focal loss for dense object detection. In: Proceedings of the IEEE International Conference on Computer Vision, pp. 2980–2988 (2017)
22. Sun, P., et al.: Sparse R-CNN: end-to-end object detection with learnable proposals. In: Proceedings of the IEEE/CVF Conference on Computer Vision and Pattern Recognition, pp. 14454–14463 (2021)
23. Cai, Z., Vasconcelos, N.: Cascade R-CNN: delving into high quality object detection. In: Proceedings of the IEEE Conference on Computer Vision and Pattern Recognition, pp. 6154–6162 (2018)
24. Ge, Z., Liu, S., Wang, F., Li, Z., Sun, J.: YOLOX: exceeding YOLO series in 2021. arXiv preprint arXiv:2107.08430 (2021)
25. Zhao, Y., et al.: DETRs beat YOLOs on real-time object detection. In: Proceedings of the IEEE/CVF Conference on Computer Vision and Pattern Recognition, pp. 16965–16974 (2024)

Hybrid-View Attention Network for Clinically Significant Prostate Cancer Classification in Transrectal Ultrasound

Zetian Feng[1], Juan Fu[2], Xuebin Zou[2], Hongsheng Ye[2], Hong Wu[1], Jianhua Zhou[2(✉)], and Yi Wang[1(✉)]

[1] Smart Medical Imaging, Learning and Engineering (SMILE) Lab, Medical UltraSound Image Computing (MUSIC) Lab, School of Biomedical Engineering, Shenzhen University Medical School, Shenzhen University, Shenzhen, China
onewang@szu.edu.cn

[2] The Department of Ultrasound, State Key Laboratory of Oncology in South China, Guangdong Provincial Clinical Research Center for Cancer, Sun Yat-sen University Cancer Center, Guangzhou, China
zhoujh@sysucc.org.cn

Abstract. Prostate cancer (PCa) is a leading cause of cancer-related mortality in men, and accurate identification of clinically significant PCa (csPCa) is critical for timely intervention. Transrectal ultrasound (TRUS) is widely used for prostate biopsy; however, its low contrast and anisotropic spatial resolution pose diagnostic challenges. To address these limitations, we propose a novel hybrid-view attention (HVA) network for csPCa classification in 3D TRUS that leverages complementary information from transverse and sagittal views. Our approach integrates a CNN-transformer hybrid architecture, where convolutional layers extract fine-grained local features and transformer-based HVA models global dependencies. Specifically, the HVA comprises intra-view attention to refine features within a single view and cross-view attention to incorporate complementary information across views. Furthermore, a hybrid-view adaptive fusion module dynamically aggregates features along both channel and spatial dimensions, enhancing the overall representation. Experiments are conducted on an in-house dataset containing 590 subjects who underwent prostate biopsy. Comparative and ablation results prove the efficacy of our method. *The code is available at* https://github.com/mock1ngbrd/HVAN.

Keywords: Clinically significant prostate cancer · Transrectal ultrasound · Attention mechanism · Computer-aided diagnosis

1 Introduction

Prostate cancer (PCa) is one of the most common malignancies worldwide and the second leading cause of cancer-related deaths in men [12]. Prostate biopsy

Z. Feng and J. Fu—Contribute equally to this work.

J. C. Gee et al. (Eds.): MICCAI 2025, LNCS 15960, pp. 260–269, 2026.
https://doi.org/10.1007/978-3-032-04927-8_25

is the golden standard for diagnosing PCa. Based on pathological features of prostate tissue, the diagnosis is classified into clinically significant PCa (csPCa), clinically insignificant PCa (cisPCa), and benign prostate hyperplasia [8]. Identifying csPCa is crucial for timely intervention, as it has a poor prognosis and requires immediate treatment [9].

Multiparametric magnetic resonance imaging (mp-MRI) is recommended for prostate biopsy, offering detailed localization of PCa [9]. However, its limited availability and complicated operation hinder its widespread adoption. Transrectal ultrasound (TRUS), with its low cost, ease of use, and real-time imaging capabilities, is commonly used for biopsy guidance [2]. Yet, the low contrast of TRUS poses challenges in accurately identifying and diagnosing PCa [1], leading to unnecessary biopsies and increased surgical risks [7]. Thus, methods to enhance accurate identification of csPCa in TRUS are critical.

Recently, deep learning approaches have been proposed for csPCa identification in TRUS. Sun *et al.* [14] use a 3D convolutional neural network (CNN) with prostate mask guidance for csPCa classification in TRUS. Wu *et al.* [17] propose a multi-modality fusion network for csPCa classification, leveraging information from B-mode and shear wave elastography. Later, Wu *et al.* [18] extend their work by introducing few shot segmentation task to enhance the capability of classification encoder. These methods succeed in identifying csPCa via additional guidance of prostate mask or elastic information. However, they all focus on the transverse view, neglecting the complementary information in the sagittal view, which can help confirm any suspicious lesions [4]. Therefore, it highlights the need for incorporating transverse-sagittal-view information to improve csPCa identification. Combining information from different views has been explored in other medical imaging contexts [5,10,13], but most approaches are limited to 2D space thus cannot be directly employed to analyze TRUS scan videos from both transverse and sagittal views.

In this study, we propose a hybrid-view attention (HVA) network for the accurate csPCa classification in 3D TRUS. The proposed HVA not only refines the features within each view, but also leverages cross-view information to compensate for information loss induced by the large slice thickness. Additionally, we introduce a fusion module that adaptively aggregates features across both channel and spatial dimensions. Comparative and ablation experiments on a large TRUS dataset with biopsy-proven PCa validate the efficacy of our method. Our main contributions are summarized as follows:

- To the best of our knowledge, this is the first study for the classification of csPCa by simultaneously analyzing both transverse and sagittal TRUS scans. To enable effective hybrid-view feature learning, we design a novel architecture that combines CNNs for local feature extraction and transformers for global feature modeling.
- We introduce a hybrid-view attention, comprising intra-view attention (IVA) and cross-view attention (CVA). The IVA focuses on capturing fine-grained features within a single view, while the CVA integrates complementary information from orthogonal views.

- To effectively aggregate hybrid-view information, a dynamic feature fusion module is proposed to adaptively re-weight features along both channel and spatial dimensions.

2 Method

Figure 1(a) illustrates the architecture of our hybrid-view attention network, which employs a dual-stream encoder to extract and aggregate features from different views for classification. The network leverages a CNN-transformer hybrid design, where each encoder is organized into four stages. At each stage, two convolutional blocks first capture local features, followed by a HVA module that models global contextual information. The HVA module comprises two specialized attention modules: the IVA module refines features within a single view in the imaging plane, while the CVA module leverages complementary information from an orthogonal view to further enhance feature quality. Finally, a hybrid-view adaptive fusion (HVAF) module dynamically aggregates the refined features across both channel and spatial dimensions, yielding a comprehensive representation for accurate classification.

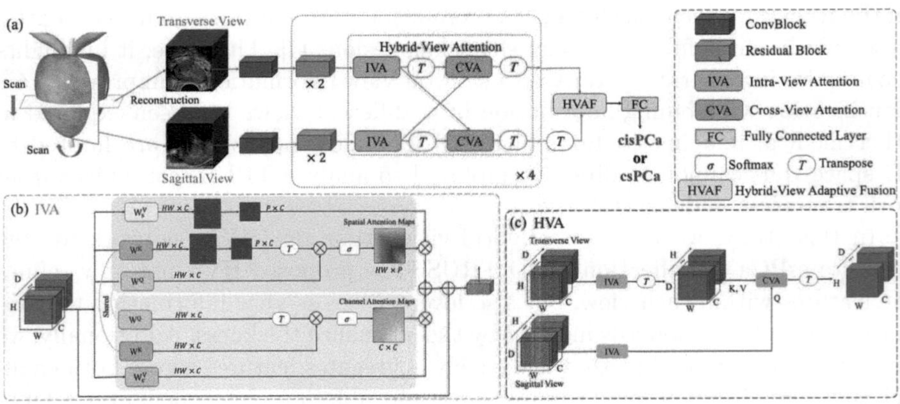

Fig. 1. Overview of the proposed network: (a) network architecture, (b) intra-view attention module, and (c) the detailed structure in the hybrid-view attention module.

2.1 CNN-Transformer Network for Hybrid-View Learning

The proposed network employs a dual-stream encoder to capture view-specific features, facilitating intra- and inter-view feature learning. Each stream consists of a stem layer followed by four down-sampling stages, each integrating a CNN-transformer hybrid design.

To establish local-to-global modeling, each stage comprises two residual convolutional blocks [3], IVA and CVA modules, successively. The two attention modules leverage attention mechanisms in transformer [15] for long-range dependency modeling. Specifically, the residual blocks extract fine-grained spatial details and local structures, while the IVA module enhances features within a single view by capturing long-range dependencies. The following CVA module further refines features by incorporating complementary spatial information from the orthogonal view. Together, these two modules form the HVA module, modeling global features across different TRUS scan views.

By combining CNN-based local feature extraction with attention-driven global modeling, the proposed network effectively captures both fine-grained details and long-range dependencies, leveraging complementary features from different views for the csPCa classification task.

2.2 Hybrid-View Attention

To model global dependencies across TRUS scans from different views, we introduce a hybrid-view attention transformer module, which integrates both intra-view and cross-view attention modules. As shown in Fig. 1, the IVA module refines features within a single view, while the CVA module leverages high-quality spatial information from an orthogonal view to enhance feature representation.

Intra-view Attention. To fully leverage the high in-plane spatial resolution of TRUS images, we propose IVA module to capture feature representations within the imaging plane of a specific view, capturing fine-grained spatial dependencies crucial for the subsequent cross-view learning.

Given transverse and sagittal feature maps $F_t, F_s \in \mathbb{R}^{B \times C \times H \times W \times D}$, where B is the batch size, C is the number of channels, and $H \times W \times D$ are spatial dimensions, note that the transverse plane corresponds to $H \times W$, while the sagittal plane corresponds to $W \times D$. To capture fine-grained information within the imaging view, the imaging axis is merged into the batch dimension, resulting in transverse features $\tilde{F}_t \in \mathbb{R}^{BD \times C \times H \times W}$ and sagittal features $\tilde{F}_s \in \mathbb{R}^{BH \times C \times W \times D}$, on which IVA is applied. Inspired by advanced attention mechanisms for contextual spatial-channel feature aggregation, the IVA module is built upon an efficient paired-attention block [11]. Each block consists of spatial and channel attention modules, designed to jointly capture spatial relationships and channel-wise dependencies.

Specifically, the IVA module employs a shared queries-keys mechanism to improve computational efficiency while preserving representational richness. Taking the transverse view as an example, the input features $\tilde{F}_t$ are processed in parallel by the spatial and channel attention modules, as shown in Fig. 1(b). A shared set of query and key embeddings is learned, while separate value embeddings are used for spatial and channel attention. Given $\tilde{F}_t \in \mathbb{R}^{BD \times C \times H \times W}$, we first project $\tilde{F}_t$ into query Q_{shared}, key K_{shared}, spatial value $V_{spatial}$, and channel value $V_{channel}$ using four distinct linear layers:

$$Q_{shared} = W^Q \tilde{F}_t, \quad K_{shared} = W^K \tilde{F}_t, \quad V_{spatial} = W_s^V \tilde{F}_t, \quad V_{channel} = W_c^V \tilde{F}_t.$$

These projections have dimensions $HW \times C$. To reduce the computational burden of spatial attention, K_{shared} and $V_{spatial}$ are projected into a lower-dimensional space using learnable matrices, from $HW \times C$ to $P \times C$, where $P \ll HW$. The spatial attention is then computed as:

$$F_{spatial} = softmax(\frac{Q_{shared} K_{proj}^T}{\gamma_s}) \cdot V_{proj}, \tag{1}$$

where Q_{shared}, K_{proj}, and V_{proj} denote shared queries, projected shared keys, projected spatial values, and γ_s is a learnable scaling parameter. The channel attention captures inter-dependencies between feature channels by applying a matrix multiplication operation in the channel dimension between the channel values and channel attention maps. The channel attention is defined as follows:

$$F_{channel} = V_{channel} \cdot softmax(\frac{Q_{shared}^T K_{shared}}{\gamma_c}), \tag{2}$$

where Q_{shared}, K_{shared}, and $V_{channel}$ denote shared queries, shared keys, channel values, and γ_c is a learnable scaling parameter. Finally, the outputs of the spatial and channel attention modules are summed, together with the $\tilde{F}_t$. The aggregated features are reshaped back to $\mathbb{R}^{B \times C \times H \times W \times D}$, and further refined using a residual block to enhance representations.

Cross-View Attention. The slice thickness results in lower spatial resolution along the scanning direction comparing to the in-plane space. While interpolation can address this imbalance, it does not provide any information gain and may introduce additional errors. To address this issue, we propose a CVA module, which incorporates information from the orthogonal view to enhance feature representation.

The CVA module operates similarly to the IVA module, but instead of refining features in a single view, it applies attention across the orthogonal views. Features from one view serve as queries to refine the features of the other view, capturing high-quality complementary orthogonal information to enrich the feature maps and improve performance in cross-view learning. Specifically, the CVA module utilizes parallel spatial and channel attention modules. Taking transverse view as an example (see Fig. 1(c)), CVA is applied on the sagittal plane, using sagittal view features as high-quality queries, and vice versa. Given feature maps $F_t, F_s \in \mathbb{R}^{B \times C \times H \times W \times D}$, the process begins by reshaping transverse and sagittal view features into 2D sagittal planar features, denoted as $\tilde{F}_{ts} \in \mathbb{R}^{BH \times C \times W \times D}$ and $\tilde{F}_s \in \mathbb{R}^{BH \times C \times W \times D}$, respectively. K_{shared}, $V_{spatial}$ and $V_{channel}$ are projected from $\tilde{F}_{ts}$, and Q_{shared} is projected from $\tilde{F}_s$. Finally, we apply the same computations as in IVA to refine the features, following Eqs. 1 and 2. After computing the HVA, the feature maps are reshaped back to $\mathbb{R}^{B \times C \times H \times W \times D}$.

2.3 Hybrid-View Adaptive Fusion

We propose the HVAF module to enhance the feature aggregation from different views by dynamically re-weighting features across both channel and spatial dimensions. As illustrated in Fig. 2, the HVAF module first concatenates the feature maps of both views, then applies channel attention to emphasize discriminative channels, followed by spatial attention to refine location-specific details within each view. The final refined features are concatenated to produce the fusion output.

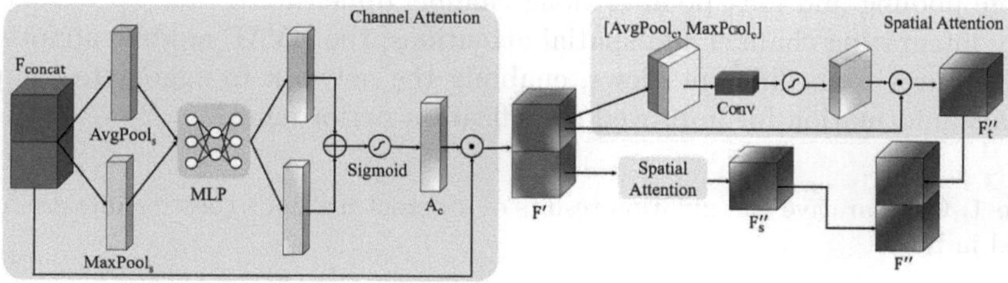

Fig. 2. The structure of the proposed hybrid-view adaptive fusion module.

As shown in Fig. 2, the channel attention mechanism in HVAF captures informative features across channels using both global average pooling and max pooling. A gating mechanism based on multi-layer perceptron (MLP) adaptively reweights feature maps. Given feature maps $F_t, F_s \in \mathbb{R}^{C\times H\times W\times D}$, the channel attention process is formulated as follows:

$$F_{concat} = Concat(F_t, F_s), \tag{3}$$

$$F' = A_c(F_{concat}) \odot F_{concat}, \tag{4}$$

where $Concat$ is the concatenation operation along the channel dimension, $\odot$ denotes element-wise multiplication, and $A_c \in \mathbb{R}^{C\times 1\times 1\times 1}$ is the channel attention map calculated as:

$$A_c = \sigma(MLP(AvgPool_s(F_{concat})) + MLP(MaxPool_s(F_{concat}))), \tag{5}$$

where σ is sigmoid function, MLP is a shared multilayer perceptron, $AvgPool_s$ and $MaxPool_s$ are global average pooling and max pooling along spatial dimensions. Note that the attention maps are broadcasted accordingly when multiplying, that is to say the channel attention map is broadcasted across the spatial dimensions and vice versa.

The spatial attention mechanism in HVAF is applied to each view independently. It highlights salient spatial regions by combining both average and max pooling across channels, enabling the model to focus on important anatomical semantic information. The features of different views are separated from F':

$$F'_t = F'[: C], F'_s = F'[C :], \tag{6}$$

then the spatial attention is computed as follows:

$$F_t'' = A_s(F_t') \odot F_t', F_s'' = A_s(F_s') \odot F_s', \tag{7}$$

$$F'' = Concat(F_t'', F_s''), \tag{8}$$

where $F'' \in \mathbb{R}^{C \times H \times W \times D}$ is the final fused features, $A_s \in \mathbb{R}^{1 \times H \times W \times D}$ is the spatial attention map calculated as:

$$A_s = \sigma(Conv(Concat(AvgPool_c(F), MaxPool_c(F)))), \tag{9}$$

where $Conv$ is a $3 \times 3 \times 3$ convolution, $AvgPool_c$ and $MaxPool_c$ are global average pooling and max pooling along channel dimensions.

By integrating channel and spatial attentions, the HVAF module adaptively fuses features from different views, enabling the network to aggregate complementary information for improved classification performance.

Table 1. Comparative and ablation results of different methods (best results are highlighted in bold).

TV	SV	IVA	CVA	HVAF	AUC	F1-score	Accuracy	Sensitivity	Specificity
✓					0.7562	0.8000	0.7171	0.8269	0.4792
✓		✓			0.7808	0.823	0.7566	0.8269	0.6042
	✓				0.7015	0.7729	0.6908	0.7692	0.5208
	✓	✓			0.7600	0.8246	0.7368	0.9038	0.3750
✓	✓				0.7622	0.8241	0.7500	0.8558	0.5208
✓	✓	✓			0.8149	0.8235	0.7237	**0.9423**	0.2500
✓	✓	✓	✓		0.8249	0.8042	0.7566	0.7308	**0.8125**
✓	✓	✓	✓	✓	**0.8534**	**0.8611**	**0.8026**	0.8942	0.6042
DMVFN [13]					0.7622	0.7795	0.7171	0.7308	0.6875
AADNN [10]					0.7893	0.7897	0.7303	0.7404	0.7083
MVMT [5]					0.7843	0.8295	0.7566	0.8653	0.5208
MMFN [17]					0.7812	0.8225	0.7303	0.9135	0.3333

3 Experiments

Dataset. The experiments were carried out on an in-house 3D TRUS dataset collected from the Cancer Center of Sun Yat-Sen University. The study was conducted retrospectively and therefore receiving a waiver of approval from the institutional review board. The dataset comprises 590 pairs of TRUS volumes scanned from transverse and sagittal views, collected from 590 patients who underwent TRUS-guided transperineal biopsy followed by whole-gland prostatectomy. It includes 401 csPCa and 189 cisPCa cases. We randomly split the dataset into training set (438 pairs, 297 csPCa) and testing set (152 pairs, 104 csPCa). All volumes were resized to $128 \times 128 \times 128$.

Implementation Details. The method was implemented in PyTorch on a NVIDIA Tesla V100 GPU with 32G memory. We trained the network using Adam optimizer with an initial learning rate of 10^{-4} for 100 epochs. Focal loss [6] was employed. The following quantitative metrics were used to evaluate the classification performance [16]: area under the ROC curve (AUC), F1-score (F1), accuracy, sensitivity, and specificity.

Results. We first assessed the contribution of each designed component through ablation experiments. Firstly, we evaluated single-view performance using only one view (TV or SV) and one stream of encoder. Secondly, we analyzed the impact of the IVA module in both single-view and multi-view settings. Thirdly, we examined the effect of the CVA module. Finally, we accessed the HVAF module by replacing it with simple feature concatenation. Table 1 shows the ablation results. Figure 3(a) presents the ROC curves. It can be observed that simply concatenating features from both views improved AUC compared to single-view model. This highlights the complementary nature of multi-view images. The addition of the IVA module enhanced the model performance in both single-view and multi-view settings, while the CVA module further improved AUC and accuracy of the model. Notably, the HVAF module led to substantial gains in AUC, F1-score, and accuracy.

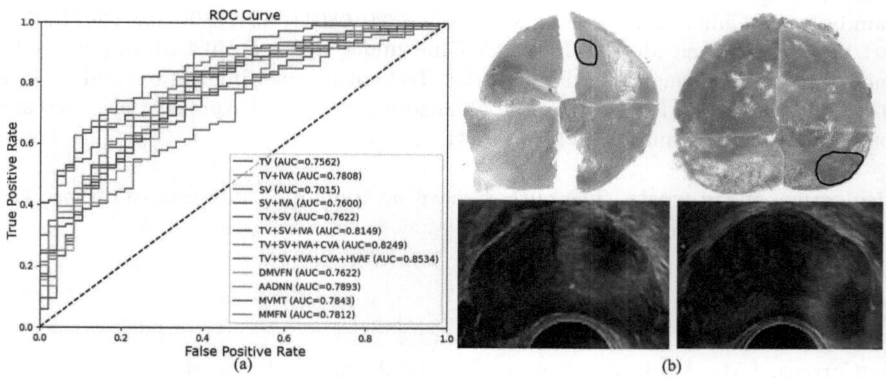

Fig. 3. (a) The receiver operating characteristic (ROC) curves of different methods on the testing set. (b) The class activation mapping (CAM) images generated using our network, and the pathological images with annotated csPCa locations.

We further compared our method with cutting-edge multi-view fusion approaches [5,10,13]. Since these approaches are applied in other 2D medical imaging contexts, we adapted them to 3D by replacing 2D convolution and other 2D operations with their 3D counterparts. Additionally, we compared a related multi-modal fusion model for csPCa classification [17]. As shown in Table 1, our method consistently outperformed these approaches in terms of AUC, F1-score and accuracy, underscoring its efficacy in identifying csPCa in 3D TRUS through our hybrid-view learning method.

Figure 3(b) visualizes two examples of 2D slices with class activation mapping (CAM) overlaid. They are extracted from two csPCa TRUS volumes, respectively. Figure 3(b) also shows their corresponding pathological images with annotated csPCa locations. It can be observed a favorable agreement of the suggested csPCa regions.

4 Conclusion

We propose a hybrid-view attention network for csPCa classification in 3D TRUS. The hybrid-view attention comprises intra-view attention to refine features within a single view and cross-view attention to leverage complementary information across orthogonal views. In addition, a hybrid-view adaptive fusion module dynamically aggregates features along both channel and spatial dimensions. Extensive experiments on an in-house TRUS dataset demonstrate that our approach effectively outperforms single-view methods and state-of-the-art multi-view techniques, underscoring the capability of our hybrid-view learning in improving prostate cancer diagnosis. Future work is to validate our method on a large multi-center dataset to evaluate its generalization performance.

Acknowledgments. This work was supported in part by the National Natural Science Foundation of China under Grants 81971631, 8320108011, 62471306 and 62071305, in part by the Shenzhen Medical Research Fund under Grant D2402010, in part by the Guangdong-Hong Kong Joint Funding for Technology and Innovation under Grant 2023A0505010021, and in part by the Guangdong Basic and Applied Basic Research Foundation under Grant 2022A1515011241.

Disclosure of Interests. The authors have no competing interests to declare that are relevant to the content of this article.

References

1. Correas, J.M.: Advanced ultrasound in the diagnosis of prostate cancer. World J. Urol. **39**, 661–676 (2021)
2. Grey, A.D., et al.: Multiparametric ultrasound versus multiparametric MRI to diagnose prostate cancer (CADMUS): a prospective, multicentre, paired-cohort, confirmatory study. Lancet Oncol. **23**(3), 428–438 (2022)
3. He, K., Zhang, X., Ren, S., Sun, J.: Deep residual learning for image recognition. In: Proceedings of the IEEE Conference on Computer Vision and Pattern Recognition (CVPR), pp. 770–778 (2016)
4. Hendrikx, A., Safarik, L., Hammerer, P.: TRUS and biopsy: practical aspects. Eur. Urol. **41**(6), I–X (2002)
5. Huang, H., et al.: Personalized diagnostic tool for thyroid cancer classification using multi-view ultrasound. In: Wang, L., Dou, Q., Fletcher, P.T., Speidel, S., Li, S. (eds.) MICCAI 2022. LNCS, vol. 13433, pp. 665–674. Springer, Cham (2022). https://doi.org/10.1007/978-3-031-16437-8_64

6. Lin, T.Y., Goyal, P., Girshick, R., He, K., Dollár, P.: Focal loss for dense object detection. In: Proceedings of the IEEE International Conference on Computer Vision, pp. 2980–2988 (2017)
7. Loeb, S., et al.: Systematic review of complications of prostate biopsy. Eur. Urol. **64**(6), 876–892 (2013)
8. Matoso, A., Epstein, J.I.: Defining clinically significant prostate cancer on the basis of pathological findings. Histopathology **74**(1), 135–145 (2019)
9. Mottet, N., et al.: EAU-EANM-ESTRO-ESUR-SIOG guidelines on prostate cancer—2020 update. Part 1: screening, diagnosis, and local treatment with curative intent. Eur. Urol. **79**(2), 243–262 (2021)
10. Pi, Y., Zhao, Z., Xiang, Y., Li, Y., Cai, H., Yi, Z.: Automated diagnosis of bone metastasis based on multi-view bone scans using attention-augmented deep neural networks. Med. Image Anal. **65**, 101784 (2020)
11. Shaker, A.M., Maaz, M., Rasheed, H., Khan, S., Yang, M.H., Khan, F.S.: UNETR++: delving into efficient and accurate 3D medical image segmentation. IEEE Trans. Med. Imaging **43**(9), 3377–3390 (2024)
12. Siegel, R.L., Kratzer, T.B., Giaquinto, A.N., Sung, H., Jemal, A.: Cancer statistics, 2025. CA Cancer J. Clin. (2025)
13. Song, D., Zhang, Z., Li, W., Yuan, L., Zhang, W.: Judgment of benign and early malignant colorectal tumors from ultrasound images with deep multi-view fusion. Comput. Methods Programs Biomed. **215**, 106634 (2022)
14. Sun, Y.K., et al.: Three-dimensional convolutional neural network model to identify clinically significant prostate cancer in transrectal ultrasound videos: a prospective, multi-institutional, diagnostic study. EClinicalMedicine **60** (2023)
15. Vaswani, A., et al.: Attention is all You need. In: Advances in Neural Information Processing Systems, vol. 30 (2017)
16. Wang, H., et al.: A narrative review of image processing techniques related to prostate ultrasound. Ultrasound Med. Biol. **51**(2), 189–209 (2025)
17. Wu, H., et al.: Multi-modality transrectal ultrasound video classification for identification of clinically significant prostate cancer. In: 2024 IEEE International Symposium on Biomedical Imaging (ISBI), pp. 1–5. IEEE (2024)
18. Wu, H., et al.: Towards multi-modality fusion and prototype-based feature refinement for clinically significant prostate cancer classification in transrectal ultrasound. In: Linguraru, M.G., et al. (eds.) MICCAI 2024. LNCS, vol. 15005, pp. 724–733. Springer, Cham (2024). https://doi.org/10.1007/978-3-031-72086-4_68

IKAN: Interactive KAN with Modulation Fusion for Medical Image Segmentation

Sihan Liu[1,2], Tonghua Wan[1,2], Yuxin Cai[1,2], Shengcai Chen[3], Bo Hu[3], Yan Wan[3], and Wu Qiu[1,2](✉)

[1] College of Life Science and Technology, Huazhong University of Science and Technology, Wuhan, Hubei, China
wuqiu@hust.edu.cn

[2] Advanced Biomedical Imaging Facility, Wuhan, Hubei, China

[3] Union Hospital, Tongji Medical College, Huazhong University of Science and Technology, Wuhan, Hubei, China

Abstract. Interactive segmentation in medical imaging remains challenged by progressive loss of crucial interaction cues (click responsiveness, boundary fidelity) in deep networks. To address this limitation, we propose Interactive Kolmogorov-Arnold Network with adaptive modulation (*IKAN*), a unified framework that synergistically preserves interaction signals through spline-activated basis functions while enabling iterative anatomical refinement. The architecture achieves enhanced diagnostic fidelity by integrating three core components: hierarchical multi-scale feature extraction through Hierarchical Inception and Channel Attention Module (HICAM), dual-branch adaptive probability modulation for backbone/side-feature fusion, and click density-guided prediction sharpening. By dynamically correlating user-provided clicks with multi-modal data patterns, our method resolves ambiguous boundaries in complex clinical scenarios. Evaluated across OCT, BUSI, and AISD datasets, our method demonstrates enhanced segmentation accuracy in complex clinical scenarios, outperforming state-of-the-art approaches through systematic preservation and amplification of diagnostic interaction cues. The code is available at https://github.com/Umbrellaliu/iKAN.

Keywords: Medical Image Interactive Segmentation · Probability Modulation · KAN · Carotid Artery OCT

1 Introduction

Interactive segmentation plays a crucial role in medical image analysis, especially in clinical settings where accurate delineation of anatomical structures is essential [18]. Unlike fully automated segmentation, which relies only on image data, interactive segmentation combines human cues (e.g., clicks or scribbles) and coarse segmentation from the previous iteration [3,15,19]. This dual guidance

S. Liu and T. Wan—Contributed equally.

J. C. Gee et al. (Eds.): MICCAI 2025, LNCS 15960, pp. 270–279, 2026.
https://doi.org/10.1007/978-3-032-04927-8_26

improves segmentation accuracy and allows for iterative refinement based on expert input, increasing clinical reliability.

However, current interactive segmentation methods have limitations. Many approaches simply add or concatenate the image, click map, and previous probability map, assuming each modality contributes equally [14,15]. This naive fusion ignores the distinct roles of each input: user cues provide localized, high-priority signals, while previous segmentation results offer useful context. As the network deepens, these cues can weaken, leading to suboptimal performance. Additionally, the iterative nature of interactive segmentation, where new inputs refine previous results, is often not fully realized. Many methods fail to properly merge new user inputs with historical probability maps, resulting in a refinement process that doesn't fully capture the contributions of the cues [12]. This can undermine the precision and robustness needed for clinical applications.

Large models tailored for medical tasks, such as SAMMed2D [4] and MedSAM [13], have been recently introduced. While these models have achieved notable progress in certain areas, they may not fully adapt to domain-specific knowledge and features, leading to decreased performance in specialized tasks.

This study presents a novel interactive segmentation framework with three key advancements: 1) a KAN framework with fusion modules that leverages multi-source inputs and iterative refinements; 2) a hierarchical multi-scale feature extraction module with channel attention to prioritize informative cues; 3) a dual-branch fusion strategy with adaptive probability modulation, combining backbone and side-branch features while refining pixel values near user clicks for precise guidance.

2 Methodology

KAN's mathematically transparent function composition and enhanced nonlinear modeling capabilities provide inherent interpretability for capturing complex anatomical patterns in medical imaging. Building on UKAN [11], we develop an enhanced backbone through three architectural innovations: wavelet convolutions integrated into KAN blocks enable multi-frequency feature learning while preserving interpretability; a Multi-Scale Cross-Attention (MSCA) module [2] bridges CNN blocks and KAN layers via spatial-channel feature recalibration; and adaptive fusion strategies prioritize error-prone regions through coordinated preprocessing and modulation. The overall architecture and interactive loop of IKAN is depicted in Fig. 1.

The framework operates through three sequential phases (Fig. 1): 1) Hierarchical Preprocessing: Input tensor $X = [X_{\text{image}}, X_{\text{coord}}, X_{\text{prev}}]$ undergoes multi-scale feature extraction via HICAM's channel-attentive spatial pyramid. 2) Modulated Feature Extraction: The enhanced UKAN backbone processes features while a parallel ResNet branch derives modulation signals from current clicks and previous probability maps. 3) Adaptive Decoding: Features fused by the attentional feature fusion (AFF) [5] in bottleneck layers are decoded through residual connections preserving spatial fidelity, culminating in 1×1 convolution and softmax classification.

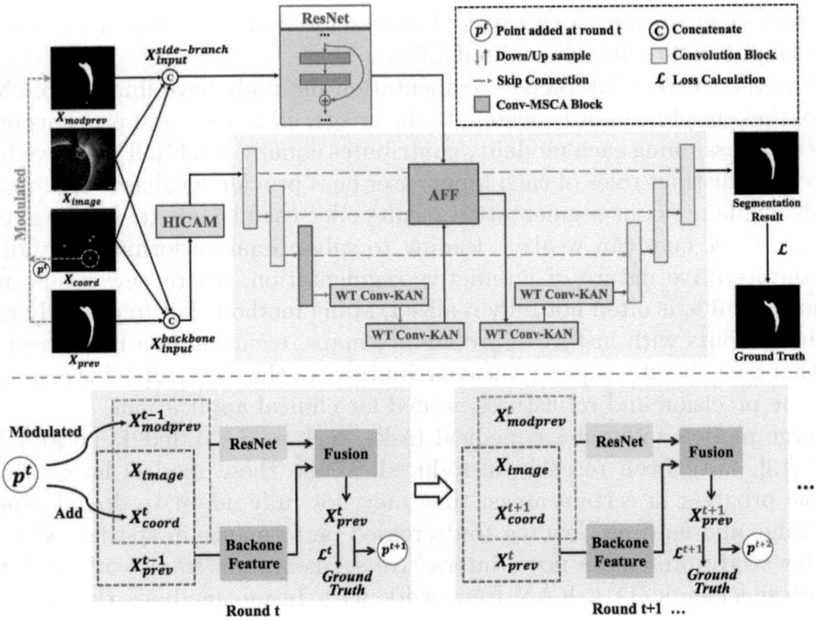

Fig. 1. Network architecture and Interactive Loop of IKAN

The IKAN framework establishes an iterative refinement loop through sequential interaction rounds denoted by t. In each round t, the system modulates the previous probability map X_{prev}^{t-1} with newly added user clicks p^t to generate enhanced guidance X_{modprev}^{t-1}, computes the current probability map X_{prev}^{t} through network processing and propagates X_{prev}^{t} as the initial probability map for round $t+1$.

The framework's closed-loop mechanism drives progressive refinement of probability maps by integrating HICAM's hierarchical context aggregation, MSCA-driven CNN/KAN synergy, and AFF-guided fusion of UKAN features with ResNet-modulated click adjustments—collectively enabling precise segmentation through iterative updates.

Our loss function combines a weighted SoftIoU [8] with iteration-specific scaling iterloss [16] to adaptively balance training objectives, where T denotes the total iterations and W_t follows $W_{t+1} > W_t$ to progressively prioritize fine-grained predictions. Each iteration loss is defined as:

$$\mathcal{L}_t = 1 - \frac{\sum(\hat{y}_t \odot y \odot w_{\text{pixel}})}{\sum(\max(\hat{y}_t, y) \odot w_{\text{pixel}})}, \quad \mathcal{L}_{\text{Total}} = \sum_{t=1}^{T} W_t \mathcal{L}_t \tag{1}$$

with $\hat{y}_t \in [0, 1]$ as the predicted probability tensor at iteration t, $y \in \{-1, 0, 1\}$ as the ground truth (where -1 indicates ignored regions), and $w_{\text{pixel}} = \mathbb{I}(y \neq -1)$ masking valid pixels.

2.1 Hierarchical Inception and Channel Attention Module (HICAM)

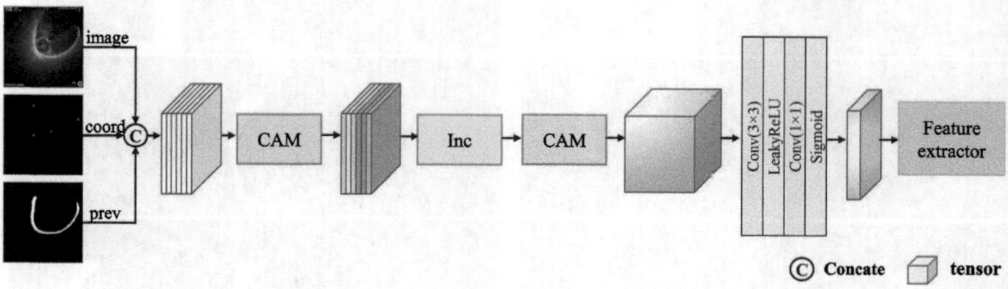

Fig. 2. Structure of HICAM

The Hierarchical Inception and Channel Attention Module (HICAM) processes multi-source inputs through a cascaded architecture to enhance interactive segmentation performance. As illustrated in Fig. 2, the module operates in three sequential stages.

The input sources, including the raw image X_{image}, user interaction coordinates X_{coord}, and prior probability map X_{prev}, are concatenated into tensor $\mathbf{X}$, which then undergoes channel-wise attention modulation (CAM) via squeeze-and-excitation operations adaptively reweighted to emphasize the importance of each input type [7]. The processed features are then processed through a multi-scale inception structure with varying convolutional kernel sizes [17], capturing spatial patterns efficiently. Finally, a second CAM layer refines the multi-scale features obtained from the inception module, enabling the model to better focus on deeper, more informative cues. The complete transformation can be formally expressed as:

$$\mathbf{Z} = \text{CAM}_{\text{final}}(\text{Inc}(\text{CAM}_{\text{init}}(\mathbf{X}))) \tag{2}$$

HICAM enhances feature focus while filtering out irrelevant information, improving segmentation accuracy. Its integration of inception-based multi-scale extraction and adaptive channel weighting makes it a critical component in interactive segmentation networks.

2.2 Enhanced Probability Map Modulation

We improve the modulation mechanism of Lee et al. [10] by addressing two limitations: (1) ineffective correction for large initial prediction errors and (2) over-expanded modulation radius. Our enhanced approach introduces three components:

1) Adaptive Modulation Radius: The radius R is constrained within $[R_{\min}, R_{\max}]$:

$$R = \min\left(R_{\max}, \max\left(R_{\min}, \frac{1}{2}\min_{v \in O} \|u - v\|\right)\right) \tag{3}$$

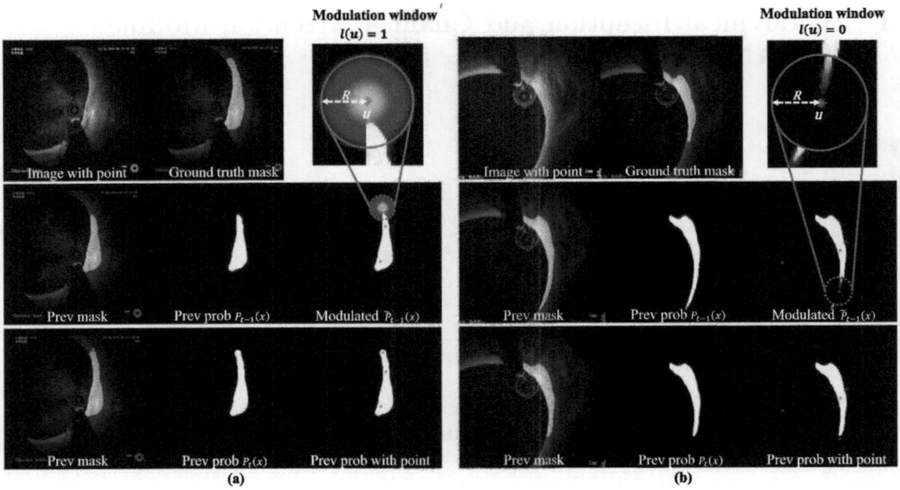

Fig. 3. Visualization of modulation effects for (a) positive clicks (enhancement) and (b) negative clicks (suppression). The color gradient indicates probability values.

where u represents the current click position, $O = \{v_1, v_2, \dots\}$ is the set of all previous clicks of the type opposite to u. This ensures balanced spatial context by considering distances to all relevant opposite-type clicks.

2) Distance-based Modulation Mask: A linear mask $M(x)$ with intensity α:

$$M(x) = \begin{cases} \alpha \left(1 - \frac{\|x-u\|}{R}\right), & \|x-u\| \leq R \\ 0, & \text{otherwise} \end{cases} \tag{4}$$

where x represents the spatial coordinates of an arbitrary pixel in the image.

3) Refined Probability Update: The modulated probability becomes:

$$\tilde{P}_{t-1}(x) = \begin{cases} P_{t-1}(x)^{1/\gamma} + M(x), & l(u) = 1 \wedge \|x-u\| \leq R \\ P_{t-1}(x)^{\gamma} - M(x), & l(u) = 0 \wedge \|x-u\| \leq R \\ P_{t-1}(x), & \text{otherwise} \end{cases} \tag{5}$$

where $P_{t-1}(x)$ is the previous prediction probability at location x, $l(u) \in \{0, 1\}$ is the click label (1 for positive/foreground click, 0 for negative/background click), $\gamma > 1$ is the adaptive gamma factor that sharpens probabilities for positive clicks and smooths them for negative clicks.

As shown in Fig. 3, this design enforces $P(x) = 1$ at positive click centers and $P(x) = 0$ at negative click centers, with smooth distance-based transitions. The adaptive radius prevents excessive modulation scope while maintaining spatial relationships to existing clicks.

The enhanced modulation mechanism offers three primary advancements compared to the original framework: 1) rigorous boundary enforcement at user-specified click locations through precise probability assignment (1.0 for positive clicks, 0.0 for negative clicks), eliminating regional ambiguity; 2) improved probability transition smoothness via linear distance masking, outperforming gamma correction approaches; 3) adaptive radius constraints that prevent over-modulation in sparse interaction scenarios, effectively preserving spatial relationships between local refinements and global context. These optimizations collectively strengthen iterative segmentation robustness, especially when addressing substantial initial prediction inaccuracies or spatially distributed user inputs.

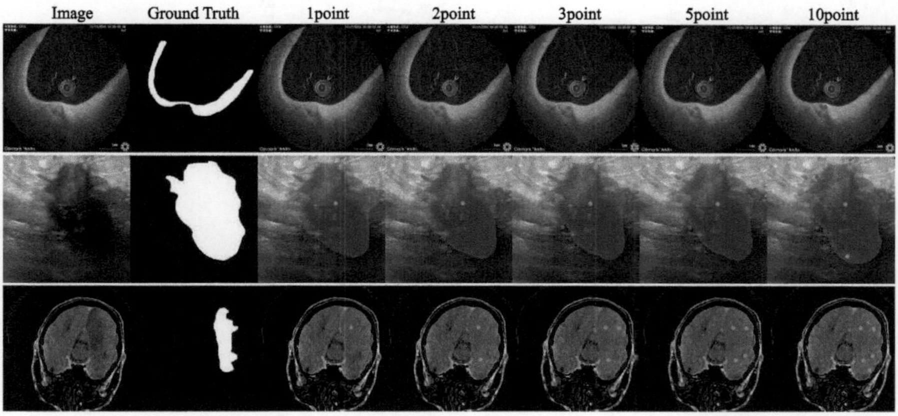

Fig. 4. Results of IKAN, from top to bottom are carotid artery OCT, BUSI, AISD.

2.3 Dual-Branch Interactive Fusion

User-provided positive/negative clicks iteratively guide prediction refinement in interactive segmentation. Our dual-branch architecture (Fig. 1) synergizes global context and localized correction. The backbone branch processes concatenated inputs $X = [X_{\text{image}}, X_{\text{coord}}, X_{\text{prev}}]$ to maintain multi-scale anatomical context. Simultaneously, the ResNet side branch [6] enhances spatial modulation of error-prone regions using $X = [X_{\text{image}}, X_{\text{coord}}, X_{\text{modprev}}]$ where X_{modprev} encodes previous predictions adjusted by current user clicks.

The extracted features F_{backbone} and F_{side} are fused via attentional feature fusion(AFF) [5], where $\mathcal{A}$ generates spatial attention weights α through channel-wise averaging and convolution operations:

$$F_{\text{fused}} = \alpha \cdot F_{\text{backbone}} + (1-\alpha) \cdot F_{\text{side}}, \quad \alpha = \sigma(\mathcal{A}([F_{\text{backbone}}, F_{\text{side}}])) \tag{6}$$

Our dual-branch fusion architecture synergizes global context preservation with user-guided refinement to enhance interactive segmentation accuracy. The backbone branch processes the original image, coordinate maps, and prior predictions to capture multi-scale contextual features, while the spatially modulated side branch selectively enhances regions marked by user annotations. This spatial modulation explicitly prioritizes error-prone regions, thereby improving segmentation precision. An attention-based fusion mechanism dynamically balances these complementary streams—preserving anatomical coherence through global context while incorporating real-time user corrections—ensuring robust performance across diverse interaction patterns. The modular architecture facilitates flexible customization of interaction handling and feature integration strategies.

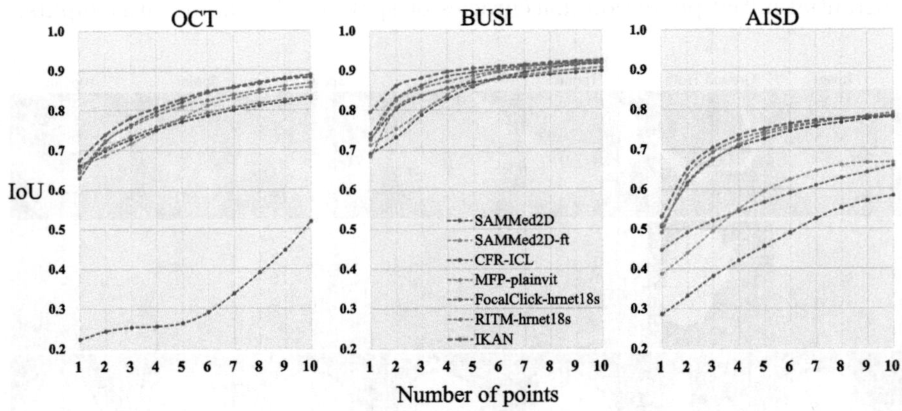

Fig. 5. IoU improvement with increasing number of points.

3 Experiments and Results

Datasets and Pre-processing: The following image sets including our own dataset and publicly available datasets are used to evaluate the performance of our approaches: (1) Private carotid artery OCT dataset (1,142 training and 453 test samples), which consists of OCT image pullbacks from 36 patients with fibrous cap lesions. (2) BUSI breast ultrasound dataset (452 training and 75 test) which collected in 2018 covering female patients aged 25 to 75 years [1]. (3) AISD acute ischemic stroke dataset (2,945 training and 841 test) [9] which contains 397 Non-Contrast-enhanced CT (NCCT) scans of acute ischemic stroke.

Hyper-Parameters Settings: The data augmentation techniques employed during training include random resizing, horizontal flipping, padding, random cropping, and adjustments to brightness and contrast. We trained on an NVIDIA RTX 4080 GPU with 16 GB of memory. We use the Adam optimizer with an initial learning rate of 5×10^{-4}, $\beta_1 = 0.9$, $\beta_2 = 0.999$, $\epsilon = 1 \times 10^{-8}$ and MultiStepLR scheduler reducing rates by 10× at epochs over 220 with early stopping employed to prevent overfitting. The batch size is set to 4, and the input images are resized to 256 × 256. Extensive experiments on the validation set determined the optimal iteration loss weight $W_t \in \{1, 2, 3\}$ and the modulated radius $[R_{\min}, R_{\max}]$ is [6, 25].

Evaluation Metrics: We adopt rigorous clinical standards using Number of Clicks (NoC) at three thresholds: NoC@80, NoC@85, and NoC@90. Our click-based method is grounded on NoBRS [15], which optimizes image segmentation results through user interactions. Simulated clicks are generated until reaching target IoU or max 20 clicks.

Results: We conducted a comparative evaluation of our approach with several state-of-the-art methods, including RITM-hrnet18s [15], FocalClick-hrnet18s [3], MFP-plainvit [10], CFR-ICL [16], SAMMed2D [4] and SAMMed2D-ft which is obtained by fine-tuning the SAMMed2D on the same datasets. These evaluations were performed across both publicly available datasets (BUSI and AISD) and our own carotid artery OCT dataset. As shown in Table 1, IKAN achieves higher accuracy than all baseline methods. The framework requires minimal user interactions to attain competitive segmentation quality (Fig. 5). The ablation studies confirm the critical contributions of both HICAM and adaptive mask fusion modules outperforming the versions with individual components removed with all modules enabled. These components collectively enhance segmentation precision in complex medical imaging scenarios, as evidenced by the performance improvements in Table 2 and visual results in Fig. 4.

Table 1. Comparison of segmentation performance

Method	OCT			BUSI			AISD		
NoC@	80	85	90	80	85	90	80	85	90
RITM-hrnet18s [15]	4.73	7.54	12.95	2.41	3.41	6.45	9.70	13.46	17.68
FocalClick-hrnet18s [3]	3.98	6.16	**11.25**	1.92	3.13	6.01	13.4	15.99	18.59
MFP-plainvit [10]	5.59	8.32	13.85	2.52	3.87	7.55	9.36	13.50	17.66
CFR-ICL [16]	5.90	8.62	14.38	2.60	3.61	7.09	9.92	13.93	17.84
SAMMed2D-ft [4]	5.82	9.81	16.55	2.89	3.77	6.75	14.71	17.33	19.27
SAMMed2D [4]	14.99	17.6	19.75	3.41	5.07	7.84	16.11	17.92	19.28
IKAN^{Proposed}	**3.56**	**5.98**	12.35	**1.57**	**2.29**	**4.59**	**9.24**	**13.26**	**17.41**

Table 2. Ablation Study Results

KAN	HICAM	Fusion	Modulation	AFF	NoC@80	NoC@85	NoC@90
✓	×	×	×	×	4.00	6.40	12.39
✓	✓	×	×	×	3.80	6.14	**11.9**
✓	✓	✓	×	×	3.83	6.25	12.41
✓	✓	✓	✓	×	3.79	6.14	12.51
×	✓	✓	✓	✓	4.01	6.43	12.56
✓	✓	✓	✓	✓	**3.56**	**5.98**	12.35

4 Discussion and Conclusion

This study introduces IKAN (Interactive KAN framework), a clinically viable medical image segmentation system built upon an enhanced UKAN backbone. Designed to address complex anatomical segmentation challenges, IKAN achieves high precision and robustness across diverse medical imaging scenarios. The framework integrates multiple innovative components, such as HICAM and an adaptive dual-branch fusion mechanism with dynamic modulation, enabling effective adaptation to various imaging modalities and clinical tasks. Particularly effective in demanding applications like carotid artery segmentation in OCT imaging and demonstrates superior performance in resource-constrained environments where high-quality annotated data are limited. Quantitative evaluations across three benchmark datasets reveal that IKAN enhances input utilization efficiency compared to conventional approaches while maintaining critical interactive cues throughout network processing.

In conclusion, IKAN provides an efficient solution for interactive segmentation, improving performance in clinically challenging, low-annotation scenarios.

Acknowledgments. The authors are grateful to the High-Performance Computing platform of Huazhong University of Science and Technology and the Supercomputing Platform of Hubei Medical Devices Quality Supervision and Test Institute.

Disclosure of Interests. No competing interests to declare.

References

1. Al-Dhabyani, W., Gomaa, M., Khaled, H., Fahmy, A.: Dataset of breast ultrasound images. Data Brief **28**, 104863 (2020)
2. Chen, J., Wan, L., Zhu, J., Xu, G., Deng, M.: Multi-scale spatial and channel-wise attention for improving object detection in remote sensing imagery. IEEE Geosci. Remote Sens. Lett. **17**(4), 681–685 (2019)
3. Chen, X., Zhao, Z., Zhang, Y., Duan, M., Qi, D., Zhao, H.: FocalClick: towards practical interactive image segmentation. In: Proceedings of the IEEE/CVF Conference on Computer Vision and Pattern Recognition, pp. 1300–1309 (2022)
4. Cheng, J., et al.: SAM-Med2d. arXiv preprint arXiv:2308.16184 (2023)
5. Dai, Y., Gieseke, F., Oehmcke, S., Wu, Y., Barnard, K.: Attentional feature fusion. In: Proceedings of the IEEE/CVF Winter Conference on Applications of Computer Vision, pp. 3560–3569 (2021)
6. He, K., Zhang, X., Ren, S., Sun, J.: Deep residual learning for image recognition. In: Proceedings of the IEEE Conference on Computer Vision and Pattern Recognition, pp. 770–778 (2016)
7. Hu, J., Shen, L., Sun, G.: Squeeze-and-excitation networks. In: Proceedings of the IEEE Conference on Computer Vision and Pattern Recognition, pp. 7132–7141 (2018)
8. Huang, Y., Tang, Z., Chen, D., Su, K., Chen, C.: Batching soft IoU for training semantic segmentation networks. IEEE Signal Process. Lett. **27**, 66–70 (2019)

9. Liang, K., et al.: Symmetry-enhanced attention network for acute ischemic infarct segmentation with non-contrast CT images. In: de Bruijne, M., et al. (eds.) MICCAI 2021. LNCS, vol. 12907, pp. 432–441. Springer, Cham (2021). https://doi.org/10.1007/978-3-030-87234-2_41
10. Lee, C., Lee, S.H., Kim, C.S.: MFP: making full use of probability maps for interactive image segmentation. In: 2024 IEEE/CVF Conference on Computer Vision and Pattern Recognition (CVPR), pp. 4051–4059. IEEE (2024)
11. Li, C., et al.: U-KAN makes strong backbone for medical image segmentation and generation. arXiv preprint arXiv:2406.02918 (2024)
12. Lin, Z., Zhang, Z., Chen, L.Z., Cheng, M.M., Lu, S.P.: Interactive image segmentation with first click attention. In: Proceedings of the IEEE/CVF Conference on Computer Vision and Pattern Recognition, pp. 13339–13348 (2020)
13. Ma, J., He, Y., Li, F., Han, L., You, C., Wang, B.: Segment anything in medical images. Nat. Commun. **15**(1), 654 (2024)
14. Sofiiuk, K., Petrov, I., Barinova, O., Konushin, A.: F-BRS: rethinking backpropagating refinement for interactive segmentation. In: Proceedings of the IEEE/CVF Conference on Computer Vision and Pattern Recognition (CVPR), June 2020
15. Sofiiuk, K., Petrov, I.A., Konushin, A.: Reviving iterative training with mask guidance for interactive segmentation. In: 2022 IEEE International Conference on Image Processing (ICIP), pp. 3141–3145. IEEE (2022)
16. Sun, S., Xian, M., Xu, F., Capriotti, L., Yao, T.: CFR-ICL: cascade-forward refinement with iterative click loss for interactive image segmentation. In: Proceedings of the AAAI Conference on Artificial Intelligence, vol. 38, pp. 5017–5024 (2024)
17. Szegedy, C., et al.: Going deeper with convolutions. In: Proceedings of the IEEE Conference on Computer Vision and Pattern Recognition, pp. 1–9 (2015)
18. Wang, G., et al.: DeepiGeoS: a deep interactive geodesic framework for medical image segmentation. IEEE Trans. Pattern Anal. Mach. Intell. **41**(7), 1559–1572 (2018)
19. Xu, N., Price, B., Cohen, S., Yang, J., Huang, T.S.: Deep interactive object selection. In: Proceedings of the IEEE Conference on Computer Vision and Pattern Recognition, pp. 373–381 (2016)

Knowledge Bridges the Intent Gap: Contextual Fusion in Medical Fine-Grained Segmentation

Hengyuan Zhang[1,2], Peng Qiao[1,2(✉)], Wenyu Li[1,2], Yan Jia[1,2], and Yong Dou[1,2(✉)]

[1] College of Computer Science and Technology, National University of Defense Technology, Changsha, China
{zhanghengyuan,pengqiao,wenyu18,jia.yan20,yongdou}@nudt.edu.cn
[2] National Key Laboratory of Parallel and Distributed Computing, National University of Defense Technology, Changsha, China

Abstract. Segment Anything Model (SAM) has been widely used in common medical image segmentation for its great zero-shot generalization by providing points or box as prompt. However, we find that SAM and its variants do not cope well with complex fine-grained segmentation tasks such as kidney anatomical structure segmentation due to the discrepancy between the model's interpretation of the task and the actual intent conveyed by the prompts. This paper introduces a new approach called Knowledge SAM (KSAM). By providing a pair of example image and corresponding fine-grained segmentation mask as the knowledge prompt, model can utilize the contextual information to better understand the meaning of the unseen fine-grained segmentation task. To accommodate knowledge prompts, we design two modules specifically designed for knowledge prompt feature fusion. KSAM outperforms the SAM models based on different prompts across both our proposed kidney anatomical structure dataset and REFUGE. Notably, our approach demonstrates competitive performance while offering better extensibility on new tasks compared with prompt-free methods.

Keywords: SAM · fine-grained segmentation · knowledge prompt

1 Introduction

Segment Anything Model (SAM) has gained widespread adoption as a visual foundation model for various segmentation tasks, including abdominal organ segmentation [2,22] in medical images and instance segmentation [21]. However, SAM's prompt methods including point and box prompt are designed for regular targets in natural images, which make SAM not suited for medical segmentation tasks with complex target shapes. Specifically, in fine-grained segmentation

H. Zhang and P. Qiao—Contributed equally to this research.

J. C. Gee et al. (Eds.): MICCAI 2025, LNCS 15960, pp. 280–290, 2026.
https://doi.org/10.1007/978-3-032-04927-8_27

tasks such as kidney anatomical structure segmentation [4,8,9,16,20,25], simple prompts make it difficult for the model to fully comprehend the task and accurately identify the target objects for segmentation. Additionally, labeled datasets for such fine-grained segmentation tasks are scarce. Consequently, developing effective strategies to fine-tune SAM and design better prompt methods that allow the model to fully understand the segmentation task remain a challenging yet worthwhile research problem. Vanilla SAM usually have the problems: **1)** SAM inherently relies on high quality sparse prompts, which are difficult to generate automatically in practical applications. Moreover, even minor variations in the prompts can significantly influence the model's performance. **2)** Fine-tuning typically requires a substantial amount of data.

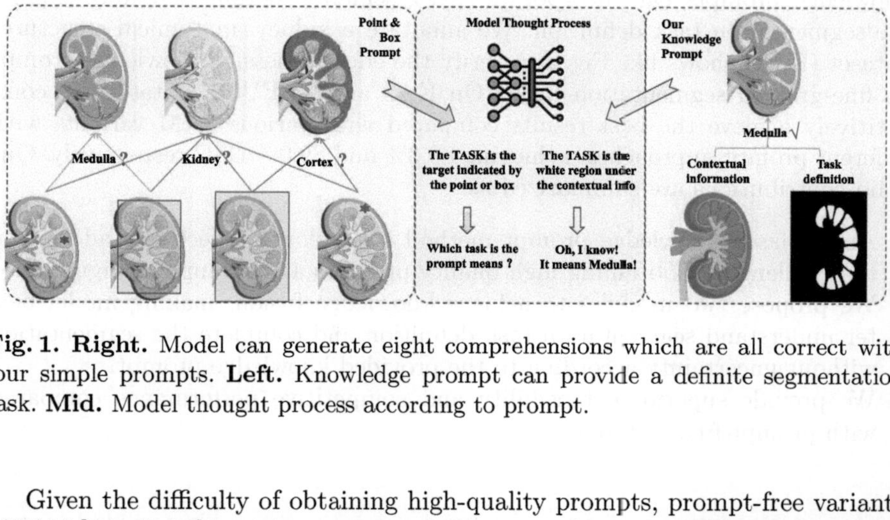

Fig. 1. Right. Model can generate eight comprehensions which are all correct with four simple prompts. **Left.** Knowledge prompt can provide a definite segmentation task. **Mid.** Model thought process according to prompt.

Given the difficulty of obtaining high-quality prompts, prompt-free variants of SAM [6,7,17,27] have been proposed. However, these approaches rigidly solidify the segmentation capabilities of model to predefined tasks. Other works such as [10,15,26,28], prompt the model by providing a similar image with an existing mask. These methods select the most related point by calculating similarity pixel-by-pixel with the target image, allowing the model to segment without sparse prompts. However, the accuracy of similarity-based point selection can be limited, potentially leading to significant false prompting issues [24].

To solve Problem. 2), some efficient fine-tuning methods [5,18,22,27] reduce the number of training parameters. Nevertheless, the demand for a large amount of training data still exists. For some complex tasks, it is difficult to obtain enough data, which can lead to overfitting. At the same time, we find that point and box prompt can only provide position information, which may not be well-suited for the segmentation of medical images. Target organs usually have similar characteristics with surrounding tissues in medical images. SAM is highly sensitive to object edges, tending to segment the entire organ when given a point prompt, even if the origin goal is to segment a specific structure within the organ. For example, in fine-grained kidney structure segmentation, the model's comprehension of the task can vary significantly when using point prompts or box prompts, as shown in Fig. 1. This happens when the region indicated by point or box has multiple semantic labels.

To address the comprehension ambiguity problem, we propose a novel prompt method called knowledge prompt, which leverages richer contextual information to enhance the model's understanding of the segmentation task. A knowledge prompt is an image pair comprising a template and its corresponding mask, where the mask specifies the target region need to be segmented. The template provides rich contextual semantics about the target's surroundings as prior knowledge. Guided by the mask, KSAM accurately segments the target region by maintaining contextual awareness of adjacent anatomical structures. Compared with point or box prompts which only provide positional information, knowledge prompts can incorporate richer semantic details and fully express the segmentation task definition. We annotate a kidney anatomical structures dataset (KAS) shows like Fig. 2 to verify the effectiveness of knowledge prompt on fine-grained segmentation tasks. On KAS and REFUGE dataset, we competitively achieve the best results compared with various SAM variants with different prompt approaches, achieving 89.3% and 89.9% DSC respectively. Our main contributions are summarized as:

- An unbiased knowledge prompt method is employed, effectively addressing the challenges of obtaining high quality prompts and prompt ambiguity.
- We propose two modules to achieve knowledge fusion, making model better understand segmentation task definition and complete the segmentation without uncertainty according to the provided knowledge prompt.
- We provide superior extensibility and competitive performance compared with prompt-free methods.

2 Method

2.1 Knowledge-Based Prompt

Due to the specificity and complexity of medical images, it is necessary to provide richer prompt information than simple points or boxes to accomplish fine-grained segmentation tasks. As shown in Fig. 3, we demonstrate the predictions using point prompts on kidney anatomical structure segmentation tasks. The soft prediction results show that the model produces completely opposite segmentation outcomes. The model inaccurately predicts lower confidence in regions that should have high-confidence features.

Therefore, we propose knowledge prompt (KP) to solve the problem in fine-grained segmentation task. Such approach contains enough medical prior knowledge to guide the model. The pipline of KSAM and related modules are shown in Fig. 4. Given the image to be segmented $X \in \mathbb{R}^{3\times H\times W}$, the knowledge prompt $KP = \{KP_{instance}, KP_{task}\}$. Where $KP_{instance} \in \mathbb{R}^{3\times H\times W}$, $KP_{task} \in \mathbb{R}^{H\times W}$, model $F(\cdot)$, with output $Pred = F(X, KP_{instance}, KP_{task})$, $Pred \in \{0,1\}^{H\times W}$,

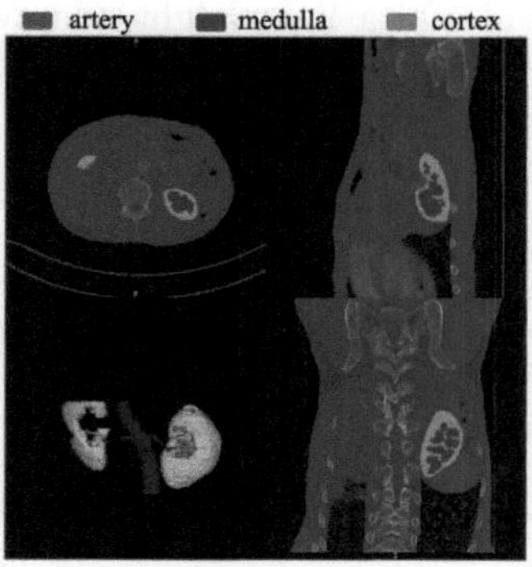

Fig. 2. Overview of the KAS, including planes in the Axial, Sagittal and Coronal and a 3D reconstruction view.

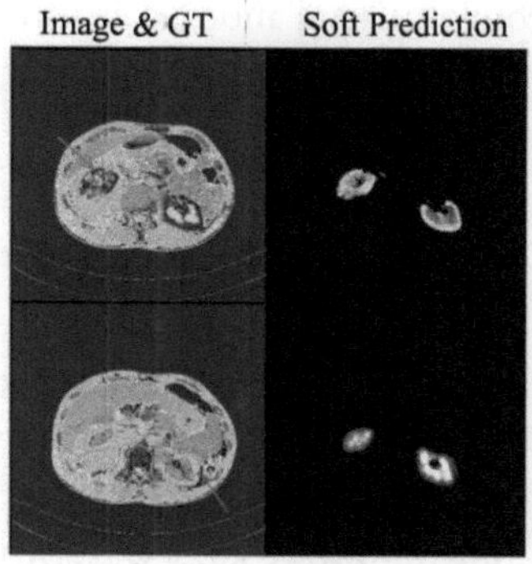

Fig. 3. Using point prompt, soft predictions from MedSA [22]. Points pointed by arrows indicate the target object.

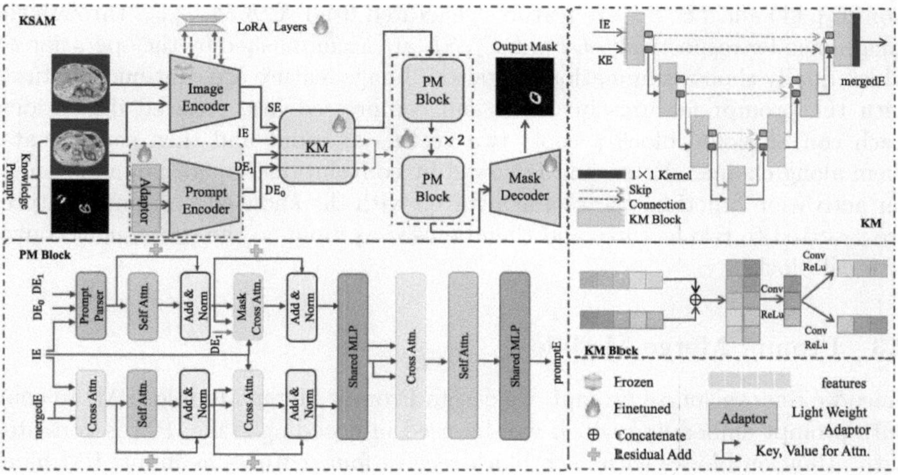

Fig. 4. The overview of KSAM and detailed structures of proposed modules.

each value of the prediction mask corresponds to whether the pixel matches the segmentation task under knowledge prompt KP.

We fuse input image with the knowledge prompt through the proposed Knowledge Merge module to enhance the origin input features. $KP_{instance}$ can be seen as the contextual of the segmentation task that KP_{task} indicates, which containing plenty of information about the target object. The enhanced feature *mergedE* with contextual information of the target object can help decoder to achieve better segmentation. Prompt Merge module compromising four blocks learns prompt embedding corresponding to KP which represents the defination of the segmentation task. Finally, *mergedE* together with the fused prompt embedding *promptE* are decoded by Mask Decoder to obtain the final segmentation mask. Since vanilla SAM do not support such prompt, we try to reuse the convolutional module in prompt encoder and add a lightweight adapter to support RGB images.

$$DE_0, DE_1 = PrEnc(PrAdaptor(KP_{instance}, KP_{task})) \tag{1}$$

$$Output = MaskDec(mergedE, promptE) \tag{2}$$

where *PrAdaptor* and *PrEnc* denote lightweight adapter and the original prompt encoder of the SAM respectively. $MaskDec$ represents the mask decoder of vanilla SAM and *Output* is the prediction.

2.2 Knowledge Merge Module

We use Knowledge Merge module(KM) to enhance the original input The fusion module adopts an architecture similar to VNet [14] shown in Fig. 4. The module is consist of convolutional blocks and handles two branches. DE_0, DE_1 come from Eq. (1) and IE, SE are features generated from $X, KP_{Instance}$ through the image encoder respectively. DE_0, DE_1, SE are accomplished by the operation $\odot$, which finally choose summation operation. Image feature are continuously fused with the prompt features in depth and compressed into reduced dimensions. Each convolutional block accepts two vectors as input and then concatenates them along channel dimension. Followed by convolutional operation and nonlinear activation function, the vector involved with the knowledge prompt features are reverted to two features and then be used as input to the subsequent convolutional block.

2.3 Prompt Merge Module

Since we remove sparse prompt, we design Prompt Merge Module (PM) to compute prompt embedding which will be used in decoding stage. PM is consisted with attention blocks with each block receive four features as input. PM block have two branches and have symmetrical structure. We reference the Prompt Parser structure [23] instead of cross attention in prompt branch for the fusion of images, knowledge prompts and task definition.

Besides that, we use a learnable mask attention in [6] to enhance the prompt features. The mask further helps the model to retain and enhance the features relevant to the segmentation task. Finally, the two features are transformed by a shared MLP to obtain the prompt embedding.

$$X_{maskAtt} = \text{sigmoid}(DE_1) \otimes Att \tag{3}$$

$\otimes$ denotes the pixel-by-pixel multiplication of the matrix and *Att* represents attention similarity. By using the sigmoid function, task features DE_1 is mapped to $(0, 1)$. The feature of the background will be mapped to a small non-zero value. After multiplication, both the target foreground features can be effectively preserved according to the mask, while the background part of the features are added to increase the robustness.

2.4 Decoder Change and Loss Function

The decoder maintains the structure of the vanilla SAM and decodes the feature vectors *mergedE* and *promptE* to get the final segmentation result. The loss we used is the weighted sum of the Binary Cross Entropy loss and the Dice loss.

3 Experiments

3.1 Dataset

KAS. The contrast-enhanced CT images come from the partner hospitals and have obtained authorization. We annotate the areas of artery and more complex structures including cortex and medulla using 3D Slicer under the guidance of experienced medical specialists. Given the difficulty and rarity of fine-grained segmentation tasks, the amount of available training data is limited. To further increase task complexity, we use CT images from a patient with severe damage to the left kidney. These images are significantly different from the normal kidney form. This choice allows us to assess whether the model can learn the essential features of the organ's structure, rather than relying on fixed morphological features. In the end, we collected a total of 75 samples.

REFUGE is a fundus dataset containing 400 images available for training. Since a pair of fundus image consists of optic cup as well as disc part, where the cup is surrounded by the disc, it is difficult to distinguish between the cup and the disc in a clear sense when using traditional point prompt or box prompt. The processing settings for the data follow [22].

3.2 Implementation Details

The implementation of the experiments is based on Pytorch and all models except OnePrompt are trained on a single NVIDIA V100 GPU. The LoRA fine-tuning method used is referenced to [22] with the rank of LoRA set to 4. We use a ViT-B backbone to perform all experiments, and the input images as well as the knowledge prompt have the same resolution of 512×512. The optimizer used is Adam with the setting of $\beta_1 = 0.9, \beta_2 = 0.999$. The warmup mechanism is added while training. The maximum epoch and learning rate are set to 300 and 0.0001. The metric for evaluation is Dice Similarity Coefficient (DSC) [11].

3.3 Results

As shown in Table 1, we compare KSAM with several prevalent variants, encompassing point-based, box-based, prompt-free, instance-based, and traditional methods. The outputs of each methods is shown in Fig. 5.

On KAS Dataset, the use of point and box prompt is usually more effective in segmenting artery due to the fact that artery is an independent structure with a clearer outline structure, and is less prone to the problem of unclear

Table 1. Results of different Prompt models. Evaluated on KAS Dataset and REFUGE by DSC. The 1st, 2nd and 3rd-best performances are highlighted by red, orange and yellow respectively. († represents model is fully finetuned.)

Types	Methods	KAS				REFUGE		
		Artery	Medulla	Cortex	Avg↑	Disc	Cup	Avg↑
Point/Box Prompt	MedSA [22]	0.795	0.580	0.620	0.665	0.807	0.644	0.726
	MedSAM [12](bbox 20pixel †)	0.929	0.815	0.878	0.874	0.951	0.818	0.884
	MedSAM [12](bbox 30pixel †)	0.923	0.656	0.856	0.812	0.951	0.462	0.707
	MedSAM2 [13](bbox 20pixel †)	0.927	0.879	0.692	0.833	0.958	0.832	0.895
	MedSAM2 [13](bbox 30pixel †)	0.926	0.873	0.771	0.857	0.956	0.636	0.796
	SAM-Med2D [5](one point)	0.865	0.791	0.702	0.786	0.774	0.660	0.717
	SAM-Med2D [5](bbox 20pixel)	0.904	0.805	0.718	0.809	0.886	0.681	0.784
prompt-free Methods	HSAM [6]	0.921	0.891	0.878	0.897	0.721	0.701	0.711
	AutoSAM [7]	0.930	0.876	0.880	0.895	0.608	0.663	0.636
	SAMed [27]	0.914	0.874	0.875	0.888	0.586	0.538	0.562
Traditional Methods	PANet [19]	0.086	0.348	0.295	0.243	0.340	0.535	0.438
	TransUnet [3]	0.888	0.865	0.845	0.866	0.540	0.486	0.513
	SwinUnet [1]	0.829	0.867	0.822	0.840	0.732	0.726	0.729
Instance prompt	PerSAM [28]	0.071	0.079	0.169	0.106	0.399	0.316	0.358
	OnePrompt [23](SegLab)	0.131	0.442	0.488	0.354	0.800	0.420	0.610
	ours	0.922	0.882	0.881	0.893	0.931	0.866	0.899

task designation. However, it performs bad on medulla and cortex, which have complex structures, and is unable to correctly understand the structure that the cue is trying to decompose. Although MedSAM shows a similar performance reaching an 87.8% DSC, it adopts a fully finetuned training approach while pretraining on more than 1 million medical data images yet still lower than our method. For instance prompt methods, mainstream methods usually obtain point prompts according to similarity and reuse the original prompt encoder. Obviously there is no guarantee that the most related point is always right. On REFUGE Dataset, we outperform all the other methods reaching an 89.9% DSC for the special structure of organs.

All the other methods perform poorly in the cup category. Although MedSAM as well as MedSAM2 using fully finetune achieve close results, when the prompt provided are disturbed, the performance is severely degraded up to a performance gap of almost 18%.

Although the prompt-free method HSAM achieves 0.4% higher performance than ours, HSAM is unscalable when facing an unseen task, as its ability is fixed during the training stage.

We first fine-tune KSAM on REFUGE and then refine the weights obtained from REFUGE KAS. The same parameter settings are used in both stages, and no additional fine-tuning tricks are employed. We evaluate the performance on the REFUGE and KAS separately every 10 epochs using the weights refine-tuned on KAS. The results obtained are shown as Fig. 6.

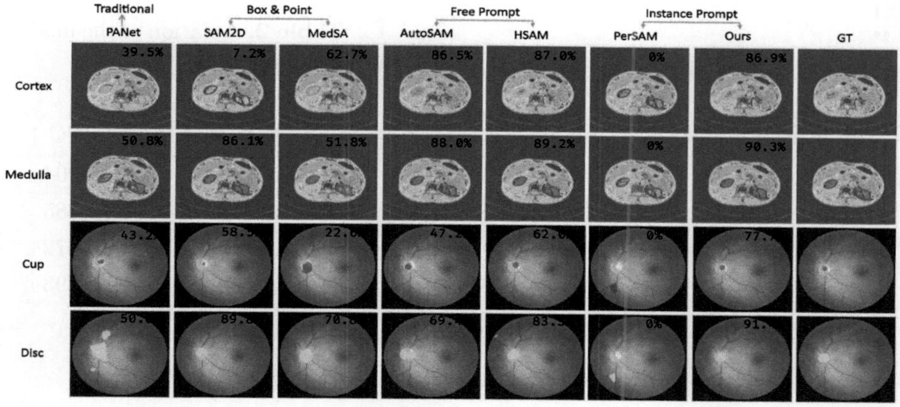

Fig. 5. Comparison of output results. Each column represents segmentation results for different tasks on KAS and REFUGE.

We achieve an average DSC of 67.15% at 25 epoch which has already surpassed MedSA that need 300 epochs. When fine-tuned by 15 epochs, we still achieve 73.9% DSC on the REFUGE, which outperform all baselines except for MedSAM and SAM-Med2D which using fully finetune and pretrained on millions of medical images. For Traditional Methods as well as Free Prompt Methods, they have no capability in this experimental setup. The tasks have already been fixed at the beginning, so they will completely forget what have learned before.

3.4 Ablation Study

As shown in Table 2, we conduct ablation experiments on KM and PM. Due to KP serving as a essential structure which can not be simply eliminated, it is used in all experiments. The direct use of KP results in poor performance because vanilla SAM does not support this prompt method, and no additional components are adapted. Adding PM improves performance by 8.1%, enhancing prompt understanding. Adding KM further boosts performance by 36.8%, significantly enhancing the encoded features through knowledge prompt integration. The best result is achieved when all modules are active, enabling the model to incorporate knowledge features and discriminate between prompt tasks. It should be noted that experiment 2 is similar with OnePrompt when just use PM and it is logical to have similar performance (38.5% vs. 35.4%).

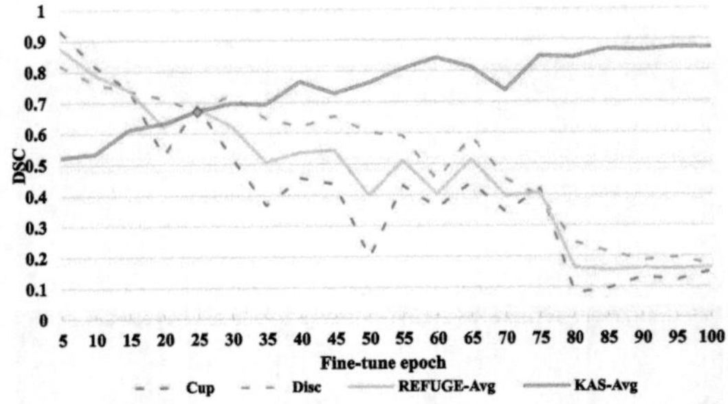

Fig. 6. Scalability test on datasets.

Table 2. Ablation of the modules proposed.

ID	KP	KM	PM	DSC ↑
1	✓	-	-	0.304
2	✓	-	✓	0.385
3	✓	✓	-	0.672
4	✓	✓	✓	0.893

4 Conclusion

We use knowledge prompts to address the issue of misinterpretation in complex fine-grained tasks. KSAM achieves higher performance with partial fine-tuning, even surpassing some models that are fully fine-tuned. The performance of scalability far exceeds the prompt free approach while maintaining the same level of segmentation capability. These results demonstrate that our method offers a novel approach to help visual prompt models better understand what to segment.

Acknowledgments. Thanks to everyone for their efforts in this work.

Disclosure of Interests. The authors have no competing interests to declare that are relevant to the content of this article.

References

1. Cao, H., et al.: Swin-Unet: Unet-like pure transformer for medical image segmentation. In: Karlinsky, L., Michaeli, T., Nishino, K. (eds.) ECCV 2022. LNCS, vol. 13803, pp. 205–218. Springer, Cham (2022). https://doi.org/10.1007/978-3-031-25066-8_9
2. Chen, F., Tang, J., Wang, P., Wang, T., Li, S., Deng, T.: DEAP-3DSAM: decoder enhanced and auto prompt SAM for 3D medical image segmentation. In: 2024 IEEE International Conference on Bioinformatics and Biomedicine (BIBM), pp. 1852–1859 (2024). https://doi.org/10.1109/BIBM62325.2024.10822764
3. Chen, J., et al.: TransUNet: transformers make strong encoders for medical image segmentation. arXiv preprint arXiv:2102.04306 (2021)
4. Chen, Z., et al.: Comprehensive 3D analysis of the renal system and stones: segmenting and registering non-contrast and contrast computed tomography images. Inf. Syst. Front., 1–15 (2024)
5. Cheng, J., et al.: SAM-MED2D. arXiv preprint arXiv:2308.16184 (2023)

6. Cheng, Z., et al.: Unleashing the potential of SAM for medical adaptation via hierarchical decoding. In: Proceedings of the IEEE/CVF Conference on Computer Vision and Pattern Recognition, pp. 3511–3522 (2024)
7. Hu, X., Xu, X., Shi, Y.: How to efficiently adapt large segmentation model (SAM) to medical images. arXiv preprint arXiv:2306.13731 (2023)
8. Jin, C., Shi, F., Xiang, D., Zhang, L., Chen, X.: Fast segmentation of kidney components using random forests and ferns. Med. Phys. **44**(12), 6353–6363 (2017)
9. Korfiatis, P., et al.: Automated segmentation of kidney cortex and medulla in CT images: a multisite evaluation study. J. Am. Soc. Nephrol. **33**(2), 420–430 (2022)
10. Liu, Y., Zhu, M., Li, H., Chen, H., Wang, X., Shen, C.: Matcher: segment anything with one shot using all-purpose feature matching. arXiv preprint arXiv:2305.13310 (2023)
11. Luo, X., et al.: Word: a large scale dataset, benchmark and clinical applicable study for abdominal organ segmentation from CT image. Med. Image Anal. **82**, 102642 (2022)
12. Ma, J., He, Y., Li, F., Han, L., You, C., Wang, B.: Segment anything in medical images. Nat. Commun. **15**(1), 654 (2024)
13. Ma, J., et al.: Segment anything in medical images and videos: benchmark and deployment. arXiv preprint arXiv:2408.03322 (2024)
14. Milletari, F., Navab, N., Ahmadi, S.A.: V-Net: fully convolutional neural networks for volumetric medical image segmentation. In: 2016 Fourth International Conference on 3D Vision (3DV), pp. 565–571. IEEE (2016)
15. Tang, L., Jiang, P.T., Xiao, H., Li, B.: Towards training-free open-world segmentation via image prompt foundation models. Int. J. Comput. Vis., 1–15 (2024)
16. Tang, Y., et al.: Renal cortex, medulla and pelvicaliceal system segmentation on arterial phase CT images with random patch-based networks. In: Medical Imaging 2021: Image Processing, vol. 11596, pp. 379–386. SPIE (2021)
17. Tianrun, C., et al.: Sam fails to segment anything? SAM-adapter: adapting SAM in underperformed scenes: camouflage shadow and more. arXiv preprint arXiv:2304.09148 (2023)
18. Wang, H., et al.: SAM-MED3D: towards general-purpose segmentation models for volumetric medical images. arXiv preprint (2023)
19. Wang, K., Liew, J.H., Zou, Y., Zhou, D., Feng, J.: PANet: few-shot image semantic segmentation with prototype alignment. In: Proceedings of the IEEE/CVF International Conference on Computer Vision, pp. 9197–9206 (2019)
20. Wang, S., et al.: CPNet: cycle prototype network for weakly-supervised 3D renal compartments segmentation on CT images. In: de Bruijne, M., et al. (eds.) MICCAI 2021. LNCS, vol. 12902, pp. 592–602. Springer, Cham (2021). https://doi.org/10.1007/978-3-030-87196-3_55
21. Wang, X., et al.: SAM-NeRF: NeRF-based 3D instance segmentation with segment anything model. In: Wand, M., Malinovská, K., Schmidhuber, J., Tetko, I.V. (eds.) ICANN 2024. LNCS, vol. 15017, pp. 434–448. Springer, Cham (2024). https://doi.org/10.1007/978-3-031-72335-3_30
22. Wu, J., et al.: Medical SAM adapter: adapting segment anything model for medical image segmentation. arXiv preprint arXiv:2304.12620 (2023)
23. Wu, J., Xu, M.: One-prompt to segment all medical images. In: Proceedings of the IEEE/CVF Conference on Computer Vision and Pattern Recognition, pp. 11302–11312 (2024)
24. Xu, Z., Chen, Q.: NubbleDrop: a simple way to improve matching strategy for prompted one-shot segmentation. arXiv preprint arXiv:2405.11476 (2024)

25. Yang, X., Le Minh, H., Cheng, K.T.T., Sung, K.H., Liu, W.: Renal compartment segmentation in dce-mri images. Med. Image Anal. **32**, 269–280 (2016)
26. Yin, F., Li, J., Wei, Y., Zhang, W., Xu, C.: SAMS: one-shot learning for the segment anything model using similar images. In: 2024 International Joint Conference on Neural Networks (IJCNN), pp. 1–8. IEEE (2024)
27. Zhang, K., Liu, D.: Customized segment anything model for medical image segmentation. arXiv preprint arXiv:2304.13785 (2023)
28. Zhang, R., et al.: Personalize segment anything model with one shot. arXiv preprint arXiv:2305.03048 (2023)

Learning Concept-Driven Logical Rules for Interpretable and Generalizable Medical Image Classification

Yibo Gao[1], Hangqi Zhou[1], Zheyao Gao[2], Bomin Wang[1], Shangqi Gao[3], Sihan Wang[1], and Xiahai Zhuang[1](✉)

[1] Fudan University, Shanghai, China
ybgao22@m.fudan.edu.cn, zxh@fudan.edu.cn
[2] The Chinese University of Hong Kong, Hong Kong, China
[3] University of Cambridge, Cambridge, UK
https://zmiclab.github.io

Abstract. The pursuit of decision safety in clinical applications highlights the potential of transparent methods in medical imaging. While concept-based models offer local concept explanations (instance-level), they often neglect the global decision logic (dataset-level). Moreover, these models often suffer from *concept leakage*, where unintended information within soft concept representations undermines both interpretability and generalizability. To address these limitations, we propose **Concept Rule Learner** (CRL), a novel framework to learn Boolean logical rules from binary visual concepts. CRL employs logical layers to capture concept correlations and extract clinically meaningful rules, thereby providing both local and global interpretability. The results from two tasks demonstrate that CRL achieves competitive performance with existing interpretable methods while improving generalizability to out-of-distribution data. The code of our work is available at https://github.com/obiyoag/crl

Keywords: Explainable-AI · Concept Learning · Rule-based Model

1 Introduction

Deep learning models, especially those operating as black boxes, have shown great promise in medical imaging applications [11,13,14]. However, the high standards of trust and accountability required in healthcare have spurred growing interest in transparent models [20,24], where "*explainability*" and "*logic*" have been emphasized as two aspects by the FDA principles [23]. "*Logic*" refers to the decision rules underlying model predictions, similar to clinical practice, where medical professionals assess symptoms and make decisions based on established clinical guidelines or rules. While recent research increasing focuses on "*explainability*" via concept explanations, less attention has been paid to "*logic*" rules for medical imaging applications.

J. C. Gee et al. (Eds.): MICCAI 2025, LNCS 15960, pp. 291–300, 2026.
https://doi.org/10.1007/978-3-032-04927-8_28

Concept Bottleneck Models (CBMs) [12] are a prevalent framework for providing concept explanations. In CBMs, a concept predictor generates explanable concepts, which are then used by a label predictor to make final predictions. Based on CBMs, Concept Embedding Models (CEMs) [7] enhance predictive performance by employing high-dimensional concept embeddings. However, these models only provide explicit local explanations by focusing on individual predictions, detailing how predicted concepts influence each decision [17,27]. *They offer insights for particular instances but may not fully capture the overall decision logic across the entire medical dataset.* For more comprehensive interpretability, integrating both local concept explanations and global logical rules is essential [21,28] for transparent medical decision-making. A recent work, named Deep Concept Reasoner (DCR) [2], explores to extract syntactic logical rules from concept embeddings. However, because DCR relies on fuzzy logic and concept embeddings, its decision-making process is less transparent.

Moreover, the above concept-based models suffer from *concept leakage* [15], where the label predictors inadvertently exploit unintended image information from soft concepts (*i.e.*, probabilities or embeddings). *The leakage compromises both interpretability and generalizability* [16]. For interpretability, predictions are influenced not only by the intended concepts but also by image information encoded within the soft concept representations. The concept predictor no longer needs to faithfully predict the concepts for accurate label predictions. Regarding generalizability, reliance on leaked information may lead to overfitting, rendering the models less robust to distribution shifts. Hard CBMs, which use binary concepts (*i.e.*, 0 and 1) for the label predictor, reduce leakage as these binary values inherently carry less extraneous information. However, they often exhibit poor performance as they predict concepts independently, neglecting the correlations between concepts.

To address the above limitations and achieve transparent medical decision-making, we propose **Concept Rule Learner** (CRL), a framework to learn Boolean logical rules from medical data. Inspired by prior works on neuro-symbolic learning [6,25,26], CRL extends logical layers to capture correlations between binary visual concepts, mitigating the issue of concept leakage. Each logical layer comprises a conjunction layer and a disjunction layer, which perform `AND` and `OR` operations, respectively. The connections across these layers generate flexible decision rules, and the final prediction for an input image is derived by aggregating the contributions of triggered rules through linear weights. By making decisions based on domain-invariant logical rules, CRL not only incorporates global interpretability, but also improves generalizability to unseen domains. Our main contributions can be summarized as follows:

- We propose CRL, a novel framework that learns Boolean logical rules from binary visual concepts to model concept correlations, while mitigating the issue of concept leakage.
- CRL not only delivers concept explanations for individual predictions but also extracts decision rules for the entire datasets, thereby unifying local and global interpretations.

- We evaluate the effectiveness of the proposed method on two medical image classification tasks. The experimental results demonstrate that our approach could extract meaningful concept logical rules and exhibits superior generalizability to the unseen dataset.

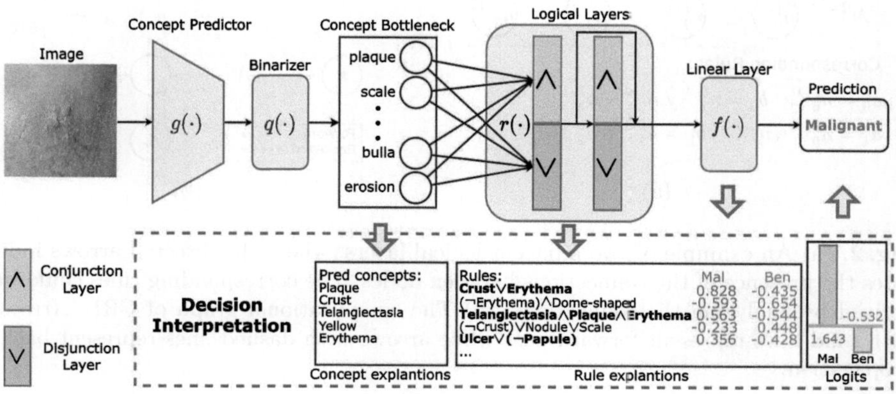

Fig. 1. The architecture of Concept Rule Learner (CRL). CRL is a composition of four functions ($f \circ r \circ q \circ g$), where $g(\cdot)$ represents the concept predictor, $r(\cdot)$ denotes a stack of logical layers, $q(\cdot)$ means discretization and $f(\cdot)$ corresponds to a linear layer. The decision interpretation is presented in the dashed box, illustrating how concepts and rules contribute to the final prediction.

2 Method

2.1 Model Architecture

Given an image $\mathbf{x} \in \mathbb{R}^n$ annotated with a task label $y \in \{1, \cdots, L\}$ and K concept labels $\mathbf{c} \in \{0,1\}^K$, our objective is to develop a **Concept Rule Learner** (CRL) capable of performing medical image classification through concept-driven logical rules. As illustrated in Fig. 1, the CRL framework is **a quadruple of functions** $(\boldsymbol{g}, \boldsymbol{q}, \boldsymbol{r}, \boldsymbol{f})$, whose composition $(f \circ r \circ q \circ g)$ predicts the label based on the derived concepts and rules. **The first function** $\boldsymbol{g} : \mathbb{R}^n \to [0,1]^K$, referred to as the concept predictor, learns a mapping from the input image $\mathbf{x}$ to a set of concept activations $\hat{\mathbf{c}} = g(\mathbf{x}) \in [0,1]^K$, where $\hat{c}_i$ represents the probability supporting the presence of the i-th concept. **The second function** $\boldsymbol{q} : [0,1]^K \to \{0,1\}^K$ serves as a binarizer that converts the concept activations into binary values. **The third function** $\boldsymbol{r} : \{0,1\}^K \to \{0,1\}^R$, consists of a series of logical layers, mapping the activated concepts to a set of rule activations $\mathbf{r} = r(q(\hat{\mathbf{c}})) \in \{0,1\}^R$, where R denotes the total number of extracted logical rules. A rule activation $r_i = 1$ indicates that the i-th rule is activated for the input image $\mathbf{x}$ and $r_i = 0$ indicates otherwise. **The fourth function** $\boldsymbol{f} : \{0,1\}^R \to \mathbb{R}^L$, represents the final linear layer, which learns the mapping from the activated logical rules to the task logits $\hat{\mathbf{y}} = f(\hat{\mathbf{r}}) \in \mathbb{R}^L$.

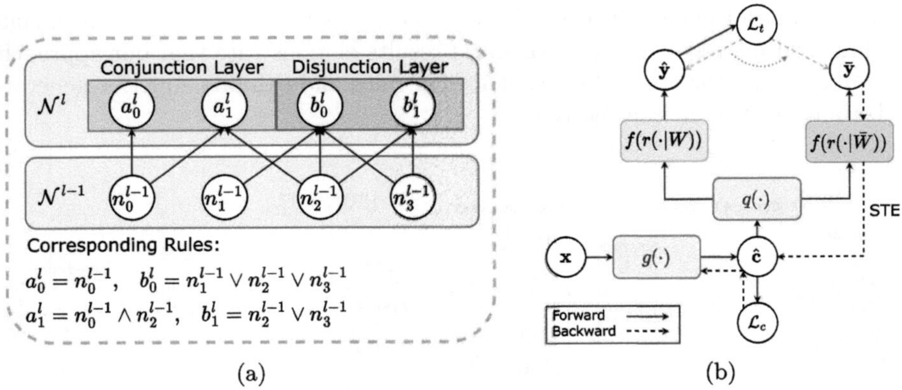

Fig. 2. (a) An example of two adjacent logical layers, where the directed arrows indicates the presence of the connections between nodes. The corresponding rules could be derived by analyzing the connections. (b) The computational graph of CRL. Arrows with solid lines represent forward pass while arrows with dashed lines represent back-propagation.

2.2 Logical Operation Modelling

To learn logical rules from concept activations, the function $r(\cdot)$ adopts a series of logical layers to model the concept-based rules with Boolean logic. Let $\mathcal{N}^l$ denote the l-th logical layer, where n_j^l represents the j-th node within the layer. The output of layer $\mathcal{N}^l$ is a vector containing the values of all nodes, denoted as $\mathbf{n}^l$. As shown in Fig. 2a, each logical layer $\mathcal{N}^l$ consists of a conjunction layer and a disjunction layer, denoted by $\mathcal{A}^l$ and $\mathcal{B}^l$, respectively. The i-th node in the conjunction layer, denoted as a_i^l, represents the conjunction (AND operation) of nodes from the preceding layer connected to it. Conversely, the i-th node in the disjunction layer, denoted as b_i^l, encapsulates the disjunction (OR operation) of its connected predecessors. Formally, the nodes in the conjunction and disjunction layers are defined as:

$$a_i^{(l)} = \bigwedge_{W_{i,j}^{(l,0)}=1} n_j^{l-1}, \qquad b_i^l = \bigvee_{W_{i,j}^{(l,1)}=1} n_j^{l-1}, \tag{1}$$

where $W^{(l,0)}$ denotes the adjacency matrix of the conjunction layer $\mathcal{A}^l$ and the previous layer $\mathcal{N}^{l-1}$, with $W_{i,j}^{(l,0)} \in \{0,1\}$. Specifically, $W_{i,j}^{(l,0)} = 1$ indicates the presence of an edge connecting a_i^l to n_j^{l-1}, while $W_{i,j}^{(l,0)} = 0$ indicates the absence. Similarly, $W^{(l,1)}$ is the adjacency matrix of the disjunction layer $\mathcal{B}^l$ and $\mathcal{N}^{l-1}$. These adjacency matrices are treated as the weight matrices for the logical layers, analogous to weight matrices in neural networks. The output of the l-th layer is the concatenation of the outputs of the conjunction and disjunction layers, *i.e.*, $\mathbf{n}^l = \mathbf{a}^l \oplus \mathbf{b}^l$, where $\mathbf{a}^l$ and $\mathbf{b}^l$ are the outputs of $\mathcal{A}^l$ and $\mathcal{B}^l$ respectively. Figure 2a illustrates an example of two adjacent logical layers, where the directed arrows indicates the presence of the connections between nodes. By examining the weights $W^{(l,0)}$ and $W^{(l,1)}$, we could derive the corresponding rules in both conjunctive and disjunctive normal forms.

2.3 Training Paradigm

While the logical layers are capable of expressing Boolean operations, their non-differentiable structure makes CRL challenging to optimize. To address this issue, we introduce continuous logical layers in training. These layers are differentiable and share the same parameters as the discrete counterparts, enabling end-to-end optimization while preserving the interpretability of the original design.

Let $\bar{W}^{(l,0)}, \bar{W}^{(l,1)} \in [0,1]$ denote the continuous weight matrices of conjunction and disjunction layers. To make Eq. (1) differentiable, we leverage the logical activation functions introduced by [19]:

$$Conj(\mathbf{n}, \bar{W}_i) = P(\prod_{j=1}^{N} F_c(n_j, \bar{W}_{i,j})), \quad Disj(\mathbf{n}, \bar{W}_i) = 1 - P(\prod_{j=1}^{N} F_d(n_j, \bar{W}_{i,j})),$$

where N is the number of nodes, $F_c(n,w) = 1 - w(1-n)$ and $F_d(n,w) = 1 - n \cdot w$. If $\mathbf{n}$ and W_i are both binary vectors, then $Conj(\mathbf{n}, W_i) = \bigwedge_{W_{i,j}=1} \mathbf{n}_j$ and $Disj(\mathbf{n}, W_i) = \bigvee_{W_{i,j}=1} \mathbf{n}_j$. $P(x) = 1/(1 - \log x)$ is a projection function to prevent gradients vanishing. After using continuous weights and logical activation functions, the nodes in continuous logical layers are defined as follows:

$$\bar{a}_i^l = Conj(\bar{\mathbf{n}}^{l-1}, \bar{W}_i^{(l,0)}), \qquad \bar{b}_i^l = Disj(\bar{\mathbf{n}}^{l-1}, \bar{W}_i^{(l,1)}).$$

By employing continuous logical layers, CRL facilitates end-to-end training. The computational graph of CRL is illustrated in Fig. 2b. As shown, the concept predictor generates concept activations $\hat{\mathbf{c}} = g(\mathbf{x})$ from input images. The binarized concept activations are passed to both the discrete and continuous logical layers to generate the task predictions given by $\hat{\mathbf{y}} = f(r(q(\hat{\mathbf{c}}))|W)$ and $\bar{\mathbf{y}} = f(r(q(\hat{\mathbf{c}}))|\bar{W})$. The objective function is defined as:

$$\mathcal{L} = \mathcal{L}_t(\hat{\mathbf{y}}, y) + \mathcal{L}_c(\hat{\mathbf{c}}, \mathbf{c}) + \lambda ||\bar{W}||_2, \tag{2}$$

where $\mathcal{L}_t$ is the cross-entropy loss, $\mathcal{L}_c$ represents the mean cross-entropy loss across all training concepts and λ is the regularization hyperparameter that controls the complexity of the logical layers. For backpropagation, as indicated by the dotted arrow in Fig. 2b, the gradients $\partial\mathcal{L}_t/\partial\hat{\mathbf{y}}$ are grafted onto the backward pass of $f(r(\cdot|\bar{W}))$. To address the non-differentiability of the binarizer, we utilize a Straight-Through Estimator (STE) [10], which approximates the gradients for the discretization.

2.4 Decision Interpretation

As illustrated in the dashed box of Fig. 1, CRL provides concept explanations for individual predictions as well as global logical rules for the entire dataset, thereby integrating both local and global interpretability. After training, the weights of logical layers $r(\cdot)$ are analyzed to summarize R logical rules, where R depends on the number of nodes and the hyperparameter λ in Eq. (2). During inference, the concept predictor $g(\cdot)$ first extracts concepts from images and matches them against the learned rules. Each rule is associated with a set of class-specific weights from the linear layer $f(\cdot)$. When a rule is triggered by a match (bold in Fig. 1), its corresponding weights are added to the overall class logits $\hat{\mathbf{y}}$. The final prediction is determined by the cumulative logits, which aggregate the contribution of all matched rules.

3 Experiments

3.1 Experimental Setup

Tasks and Datasets: To evaluate the proposed method across different medical scenarios, we assess two tasks: skin disease diagnosis and white blood cell (WBC) classification. **Skin disease diagnosis:** We employ the Fitzpatrick17k (`F17k`) dataset [9] and the Diverse Dermatology Images (`DDI`) dataset [4], incorporating concept annotations from the SkinCon dataset [5]. The SkinCon dataset comprises 48 concepts annotated by board-certified dermatologists. Following the approach in [18], we focus our training and testing to images labeled as *Benign* (Ben) or *Malignant* (Mal). **WBC classification:** We utilize the `PBC` dataset [1], along with the concept annotations from the WBCAtt dataset [22]. The WBCAtt dataset contains 24 morphological attributes, and the classification includes five distinct classes.

Implementation Details: We employ ResNet-34 pretrained on the ImageNet dataset, as the backbone network. The balance parameter between concept and task loss is set to 1. We use the AdamW optimizer with an initial learning rate of 5×10^{-5} and a weight decay of 0.01. The learning rate decays to zero following a cosine scheduler. Models are trained for 300 epochs with a batch size of 64. CRL is implemented with two logical layers, each comprising 256 nodes. The hyperparameter λ set to 5×10^{-6} to control the complexity of the rules. For evaluation, we report the average accuracy (ACC) and F1 score (F1) for both concept prediction and diagnosis tasks. For skin disease diagnosis, we perform 5-fold cross-validation. For WBC classification, we adopt the original data split from [22], conducting experiments with three different random seeds.

3.2 Results

Model Utility Analysis: To showcase the classification utility of the proposed method, we compare CRL with other concept-based methods on F17k and PBC datasets. The comparative methods include CBM [12], align-CBM [18], CEM [7], evi-CEM [8] and DCR [2]. Among these, align-CBM integrates clinical knowledge to prioritize the most relevant, while evi-CEM employs evidential learning to model concept uncertainty. Hard-CBM, which serves as a baseline, only accepts

Table 1. Performance comparison on skin disease diagnosis and WBC classification tasks. **Bold** text indicates the best results, while underlined text denotes the second-best results. The symbol $\star$ denotes methods employ binary concept values, while $\dagger$ indicates methods with global interpretability.

Dataset	Method	Concept Metric		Diagnosis Metric	
		ACC(%)	F1(%)	ACC(%)	F1(%)
F17k	CBM [12]	$91.41_{\pm 0.19}$	$\mathbf{59.44}_{\pm 0.97}$	$\underline{76.37}_{\pm 3.45}$	$\underline{76.30}_{\pm 3.45}$
	align-CBM [18]	$89.77_{\pm 0.52}$	$58.98_{\pm 1.41}$	$75.93_{\pm 2.39}$	$75.83_{\pm 2.41}$
	CEM [7]	$91.85_{\pm 0.55}$	$\underline{59.11}_{\pm 2.06}$	$76.26_{\pm 2.59}$	$76.17_{\pm 2.52}$
	evi-CEM [8]	$\underline{92.04}_{\pm 0.80}$	$58.65_{\pm 1.71}$	$\mathbf{76.47}_{\pm 1.69}$	$\mathbf{76.39}_{\pm 1.66}$
	hard-CBM*	$86.36_{\pm 1.09}$	$49.57_{\pm 0.48}$	$73.51_{\pm 2.94}$	$73.43_{\pm 2.92}$
	DCR† [2]	$91.03_{\pm 0.80}$	$49.40_{\pm 1.00}$	$75.05_{\pm 2.12}$	$74.97_{\pm 2.12}$
	CRL*†	$\mathbf{92.80}_{\pm 0.47}$	$52.39_{\pm 0.28}$	$75.95_{\pm 3.09}$	$75.90_{\pm 3.08}$
PBC	CBM [12]	$\underline{95.21}_{\pm 0.10}$	$\mathbf{91.65}_{\pm 0.15}$	$98.93_{\pm 0.14}$	$98.44_{\pm 0.22}$
	align-CBM [18]	$95.01_{\pm 0.23}$	$89.95_{\pm 0.59}$	$99.14_{\pm 0.05}$	$\underline{99.25}_{\pm 0.52}$
	CEM [8]	$94.98_{\pm 0.34}$	$\underline{90.86}_{\pm 0.69}$	$\underline{99.43}_{\pm 0.12}$	$99.23_{\pm 0.18}$
	evi-CEM [8]	$94.44_{\pm 0.91}$	$89.18_{\pm 2.09}$	$\mathbf{99.57}_{\pm 0.03}$	$\mathbf{99.42}_{\pm 0.03}$
	hard-CBM*	$64.53_{\pm 2.42}$	$56.24_{\pm 2.74}$	$98.22_{\pm 0.48}$	$97.52_{\pm 0.73}$
	DCR† [2]	$92.79_{\pm 0.43}$	$82.92_{\pm 0.64}$	$98.93_{\pm 0.32}$	$98.47_{\pm 0.49}$
	CRL*†	$\mathbf{95.32}_{\pm 0.26}$	$90.51_{\pm 0.73}$	$98.67_{\pm 0.25}$	$98.03_{\pm 0.42}$

binary concepts without logical layers. The comparison results are presented in Table 1. From the results, we observe that CRL achieves comparable predictive performance with other CBM variants, though it exhibits a performance trade-off compared to methods based on concept embeddings. Notably, when comparing CRL and hard-CBM, both of which utilize binary concepts, CRL significantly outperforms hard-CBM. This improvement can be attributed to the logical layers in CRL, which capture the concept correlations with logical operations.

Model Generalizability Analysis: To evaluate generalizability, we assess the out-of-domain (OOD) performance of the models on the unseen `DDI` dataset, using models trained on the source `F17k` dataset. As reported in Table 2, we can observe that concept models relying on soft concepts (probabilities or embeddings) exhibit relatively large performance drops, whereas methods employing binary concept values experience smaller declines. This suggests that binary concepts could help mitigate concept leakage. Notably, although both CRL and hard-CBM utilize binary concepts, CRL achieves a diagnostic ACC of 73.46, outperforming the second-best method by 9.21 while exhibiting the smallest performance drop. This improvement can be attributed to the domain-invariant logical rules extracted by logical layers, which enhance robustness to distribution shifts.

Table 2. Performance comparison on the unseen DDI dataset. **Bold** text indicates the best results, while <u>underlined</u> text denotes the second-best results. The symbol ⋆ denotes methods employ binary concept values, while † indicates methods with global interpretability.

Method	OOD Performance↑				Performance Drop↓			
	Concept Metric		Diagnosis Metric		Concept Metric		Diagnosis Metric	
	ACC(%)	F1(%)	ACC(%)	F1(%)	ACC(%)	F1(%)	ACC(%)	F1(%)
CBM [12]	$90.59_{\pm 0.27}$	<u>$53.09_{\pm 0.44}$</u>	$62.36_{\pm 4.05}$	$57.49_{\pm 2.02}$	0.82	6.35	14.01	18.81
align-CBM [18]	$89.34_{\pm 0.25}$	$51.47_{\pm 0.22}$	$63.13_{\pm 4.01}$	<u>$59.66_{\pm 2.18}$</u>	0.33	7.51	12.80	<u>16.17</u>
CEM [7]	<u>$91.66_{\pm 0.26}$</u>	**$53.21_{\pm 0.57}$**	$62.39_{\pm 2.59}$	$57.42_{\pm 2.01}$	<u>0.19</u>	5.90	13.87	18.75
evi-CEM [8]	$91.35_{\pm 0.20}$	$52.17_{\pm 0.64}$	<u>$64.25_{\pm 2.04}$</u>	$58.82_{\pm 1.62}$	0.69	6.48	12.12	17.48
hard-CBM⋆	$84.63_{\pm 1.20}$	$48.48_{\pm 0.28}$	$63.77_{\pm 2.35}$	$56.96_{\pm 1.52}$	1.73	<u>1.09</u>	<u>9.74</u>	16.47
DCR† [2]	$90.97_{\pm 0.49}$	$47.89_{\pm 0.72}$	$62.23_{\pm 4.09}$	$57.50_{\pm 2.35}$	**0.06**	1.51	12.82	17.47
CRL⋆†	**$92.14_{\pm 0.20}$**	$51.78_{\pm 0.19}$	**$73.46_{\pm 2.36}$**	**$63.42_{\pm 1.27}$**	0.66	**0.61**	**2.49**	**12.48**

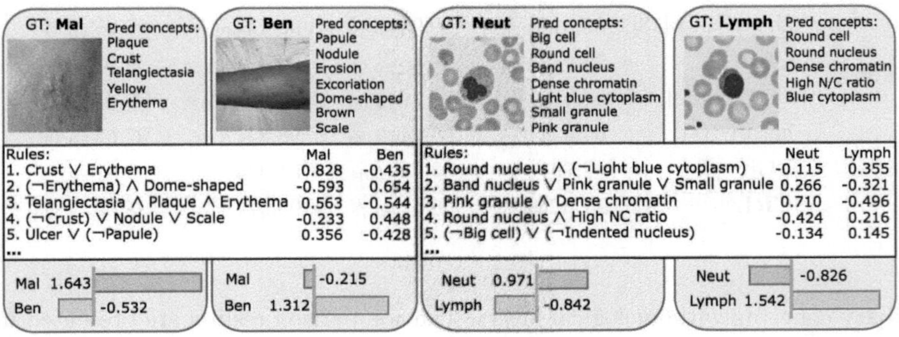

Fig. 3. An illustration of concept logical rules for both skin disease diagnosis task (left) and WBC classification (right). Note: We only present the rule weights and logits of *Neutrophil* (Neut) and *Lymphocyte* (Lymph) for WBC classification.

Model Interpretability Analysis: To illustrate that CRL could generate meaningful logical rules, we present the rules obtained from both skin disease diagnosis and WBC classification tasks, as illustrated in Fig. 3. For skin disease diagnosis, we observe that concepts `Telangiectasia`, `Ulcer` and `Crust` are associated with *Malignant*, aligning with established clinical knowledge [3]. In case of WBC classification, the rules indicate that *Lymphocytes* are characterized by a `High NC ratio` and a `Round nucleus`, while *Neutrophils* typically exhibit `Pink granules` and `Dense chromatin`, consistent with clinical observations [22]. The case studies demonstrate that CRL can effectively extract concept-based logical rules that are clinically meaningful, offering both local concept explanations and global rule explanations for the entire medical dataset.

4 Conclusion

This paper introduces CRL, a framework for interpretable medical image classification that mitigates concept leakage and unifies local and global interpretability. By employing binary concepts and learnable logical layers, CRL effectively models concept correlations and extracts clinically meaningful decision rules. Experiments on two medical image classification tasks demonstrate that CRL achieves competitive performance and exhibits superior generalizability to unseen data.

Acknowledgments. This work was funded by the National Natural Science Foundation of China (grant No. 62372115) and the Shanghai Municipal Education Commission-Artificial Intelligence Initiative to Promote Research Paradigm Reform and Empower Disciplinary Advancement Plan (grant no. 24KXZNA13).

Disclosure of Interests. The authors have no competing interests to declare that are relevant to the content of this article.

References

1. Acevedo, A., Merino, A., Alférez, S., Ángel Molina, Boldú, L., Rodellar, J.: A dataset of microscopic peripheral blood cell images for development of automatic recognition systems. Data Brief **30**, 105474 (2020)
2. Barbiero, P., et al.: Interpretable neural-symbolic concept reasoning. In: International Conference on Machine Learning (2023)
3. Bolognia, J., Schaffer, J., Cerroni, L.: Dermatology. Elsevier (2017)
4. Daneshjou, R., et al.: Disparities in dermatology AI performance on a diverse, curated clinical image set. Sci. Adv. **8**(31) (2022)
5. Daneshjou, R., Yuksekgonul, M., Cai, Z.R., Novoa, R.A., Zou, J.: SkinCon: a skin disease dataset densely annotated by domain experts for fine-grained debugging and analysis. In: Neural Information Processing Systems (2022)
6. Duan, X., Wang, X., Zhao, P., Shen, G., Zhu, W.: DeepLogic: joint learning of neural perception and logical reasoning. IEEE Trans. Pattern Anal. Mach. Intell. **45**(4), 4321–4334 (2023)
7. Espinosa Zarlenga, M., et al.: Concept embedding models: beyond the accuracy-explainability trade-off. In: Advances in Neural Information Processing Systems, vol. 35, pp. 21400–21413 (2022)
8. Gao, Y., Gao, Z., Gao, X., Liu, Y., Wang, B., Zhuang, X.: Evidential concept embedding models: towards reliable concept explanations for skin disease diagnosis. In: International Conference on Medical Image Computing and Computer Assisted Intervention (2024)
9. Groh, M., et al.: Evaluating deep neural networks trained on clinical images in dermatology with the Fitzpatrick 17k dataset. In: Proceedings of the IEEE/CVF Conference on Computer Vision and Pattern Recognition, pp. 1820–1828 (2021)
10. Hubara, I., Courbariaux, M., Soudry, D., El-Yaniv, R., Bengio, Y.: Binarized neural networks. In: Advances in Neural Information Processing Systems, vol. 29 (2016)

11. Isensee, F., Jaeger, P.F., Kohl, S.A.A., Petersen, J., Maier-Hein, K.H.: nnU-Net: a self-configuring method for deep learning-based biomedical image segmentation. Nat. Methods **18**(2), 203–211 (2021)
12. Koh, P.W., et al.: Concept Bottleneck Models. In: Proceedings of the 37th International Conference on Machine Learning, pp. 5338–5348 (2020)
13. Li, C., et al.: LLaVA-Med: training a large language-and-vision assistant for biomedicine in one day. arXiv preprint arXiv:2306.00890 (2023)
14. Ma, J., He, Y., Li, F., Han, L., You, C., Wang, B.: Segment anything in medical images. Nat. Commun. **15**, 654 (2024)
15. Mahinpei, A., Clark, J., Lage, I., Doshi-Velez, F., Pan, W.: Promises and pitfalls of black-box concept learning models. arXiv preprint arXiv:2106.13314 (2021)
16. Margeloiu, A., Ashman, M., Bhatt, U., Chen, Y., Jamnik, M., Weller, A.: Do concept bottleneck models learn as intended? In: International Conference on Learning Representations workshop (2021)
17. Oikarinen, T., Das, S., Nguyen, L.M., Weng, T.W.: Label-Free Concept Bottleneck Models. In: International Conference on Learning Representations (2023)
18. Pang, W., Ke, X., Tsutsui, S., Wen, B.: Integrating clinical knowledge into concept bottleneck models. In: International Conference on Medical Image Computing and Computer-Assisted Intervention (MICCAI) (2024)
19. Payani, A., Fekri, F.: Learning algorithms via neural logic networks. arXiv preprint arXiv:1904.01554 (2019)
20. Rudin, C.: Stop explaining black box machine learning models for high stakes decisions and use interpretable models instead. Nat. Mach. Intell. **1**(5), 206–215 (2019)
21. Schrouff, J., et al.: Best of both worlds: local and global explanations with human-understandable concepts. arXiv preprint arXiv:2106.08641 (2021)
22. Tsutsui, S., Pang, W., Wen, B.: WBCAtt: a white blood cell dataset annotated with detailed morphological attributes. In: Advances in Neural Information Processing Systems (NeurIPS) (2023)
23. U.S. Food and Drug Administration: Transparency for machine learning-enabled medical devices: Guiding principles (2024)
24. van der Velden, B.H.M., Kuijf, H.J., Gilhuijs, K.G.A., Viergever, M.A.: Explainable artificial intelligence (XAI) in deep learning-based medical image analysis. Med. Image Anal. **79**, 102470 (2022)
25. Wang, Z., Zhang, W., Liu, N., Wang, J.: Learning interpretable rules for scalable data representation and classification. IEEE Trans. Pattern Anal. Mach. Intell. **46**(02), 1121–1133 (2024)
26. Yang, G., Song, L.: Learn to explain efficiently via neural logic inductive learning. In: International Conference on Learning Representations (2023)
27. Yang, Y., Panagopoulou, A., Zhou, S., Jin, D., Callison-Burch, C., Yatskar, M.: Language in a Bottle: Language Model Guided Concept Bottlenecks for Interpretable Image Classification. In: Proceedings of the IEEE/CVF Conference on Computer Vision and Pattern Recognition, pp. 19187–19197 (2023)
28. Zhang, Y., Tiňo, P., Leonardis, A., Tang, K.: A survey on neural network interpretability. IEEE Trans. Emerg. Top. Comput. Intell. **5**(5), 726–742 (2021)

Location-Aware Parameter Fine-Tuning for Multimodal Image Segmentation

Sicong Gao, Maurice Pagnucco, and Yang Song(✉)

The University of New South Wales, Sydney, Australia
z5306254@ad.unsw.edu.au

Abstract. Accurate segmentation of lung infection regions is critical for early diagnosis and quantitative assessment of disease severity. However, existing segmentation methods largely depend on high-quality, manually annotated data. Although some approaches have attempted to alleviate the reliance on detailed annotations by leveraging radiology reports, their complex model architectures often hinder practical training and widespread clinical deployment. With the advent of large-scale pre-trained foundation models, efficient and lightweight segmentation frameworks have become feasible. In this work, we propose a novel segmentation framework that utilizes CLIP to generate multimodal high-quality prompts, including coarse mask, point, and text prompts, which are subsequently fed into the Segment Anything Model 2 (SAM2) to produce the final segmentation results. To fully exploit the informative content of medical reports, we introduce a localization loss that extracts positional cues from the text to guide the model in localizing potential lesion regions. Experiments on the CT dataset MosMedData+ and the X-ray dataset QaTa-COV19 demonstrate that our method achieves state-of-the-art performance while requiring only minimal parameter fine-tuning. These results highlight the effectiveness and clinical potential for pulmonary infection segmentation.

Keywords: Medical Image Segmentation · Vision-language Model · Transfer Learning

1 Introduction

Lung diseases pose a significant global health challenge, with radiological imaging such as X-ray and Computed Tomography (CT) playing a crucial role in early detection, especially during outbreaks like COVID-19 [4,8,19]. While convolutional neural network (CNN) [1,10] and Transformer [3,6] based segmentation methods have advanced the field, their dependence on extensive pixel-level annotations limits scalability. Reducing annotation needs without sacrificing segmentation accuracy remains a critical challenge [11,13,15,20,23].

To tackle this issue, several recent studies have begun exploring textual information to guide segmentation, and existing multimodal methods for medical

J. C. Gee et al. (Eds.): MICCAI 2025, LNCS 15960, pp. 301–311, 2026.
https://doi.org/10.1007/978-3-032-04927-8_29

imaging predominantly mimic multimodal approaches from the natural image domain [24]. For example, Li et al. [14] proposed LViT, the first method to integrate CNN and Transformer architectures for text-guided segmentation. Similarly, GuideDecoder [26] performs multimodal fusion at the decoder stage, and MMI-UNet [2] applies multimodal fusion in the encoder stage. TGANet [21] proposes a text-guided attention mechanism for polyp segmentation. MTPTN [9] introduces progressive text-based prior prompts to generate multimodal features for nuclei segmentation.

On the other hand, existing methods have not considered that medical reports differ from the textual descriptions associated with natural images. They usually contain various localization terms. As illustrated by the blue text in Fig. 1, this information concisely describes the location of the lesion area. However, current existing approaches heavily rely on attention mechanisms to align the text with the visual modality and extract additional information from the text. Such reliance fails to explicitly capture effective medical cues (e.g., localization information), making it challenging to guide the visual modality in lesion localization accurately. To address this limitation, we propose a location-aware method that employs a global positional feature classification (GPFC) loss to guide the visual modality in accurately localizing lesion regions. By using lesion localization classification as a loss function, we validate whether the model has successfully learned localization cues from the text. Additionally, incorporating such classification at multiple stages further ensures that the text-image encoders can extract effective global features, thereby generating high-quality prompts.

In addition, large-scale pre-trained models have shifted segmentation toward efficient fine-tuning. Foundation models like CLIP [17] and the recent release of SAM2 [18] have further advanced this direction. Although SAM2, an extension of SAM [12] with a larger training dataset and faster performance, offers promising results, it still requires high-quality prompts to avoid segmentation errors [22]. Therefore, it remains challenging to effectively adapt SAM2 to downstream tasks. In this work, we leverage CLIP to generate the prompts required by SAM2 while freezing the text and image encoder and fine-tuning only the bridging module during encoding. To fully exploit CLIP's multimodal capabilities, we introduce a Query-guided Attention Fusion (QGAF) module, which differs from previous approaches that rely on simple attention fusion. The QGAF module acts as a bridge between CLIP's text and image encoders, using learnable query tokens to extract lesion location information from text, enabling CLIP to generate high-quality prompts that meet SAM2's requirements. Our method reduces the burden of pixel-level annotation by leveraging available medical report text.

In summary: (1) We propose a text-guided, location-aware segmentation framework that uses global positional feature classification (GPFC) to precisely localize lesions from positional cues in medical reports. (2) We design a Query-guided Attention Fusion (QGAF) module for fine-tuning CLIP to generate three effective prompts (coarse mask, point, and text) for SAM2. (3) Experimental results on the MosMedData+ [16] and QaTa-COV19 [7] datasets demonstrate our model achieves state-of-the-art performance with minimal parameter fine-tuning.

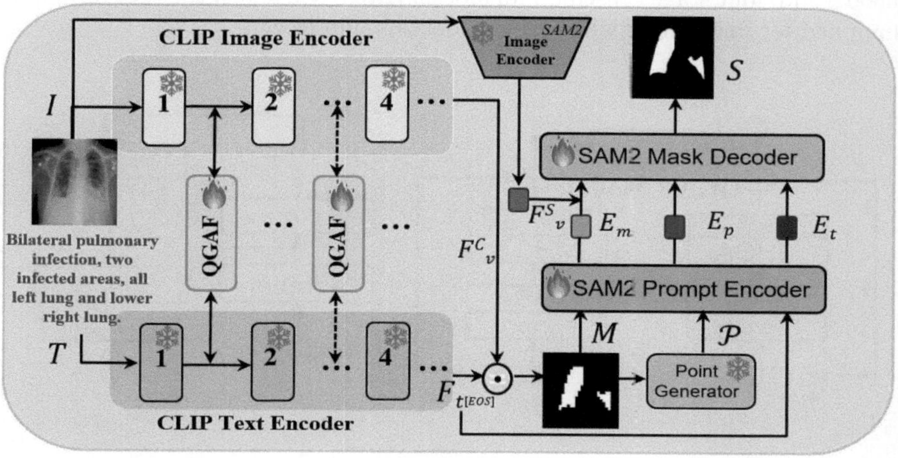

Fig. 1. An overview of our proposed framework, highlighting that only minimal parameter fine-tuning is required. The details of Query-guided Attention Fusion (QGAF) module are shown in Fig. 2.

2 Methodology

Figure 1 shows an overview of our proposed method. Given an image I and a medical report T, we employ CLIP as the encoder and use the QGAF module to fuse multimodal information for extracting image features ${F^C}_v$ (Sect. 2.1). After that, we generate the coarse mask prompt, point prompt, and text prompt required by SAM2. The SAM2 prompt encoder then processes these prompts to obtain the corresponding encodings E_m, E_p, and E_t (Sect. 2.2). The SAM2 mask decoder subsequently produces the final segmentation mask S (Sect. 2.3). The whole process is trained using a combination of location-aware loss and segmentation loss (Sect. 2.4). All encoders remain frozen during training.

2.1 Text and Image Encoder

Given an image $I \in \mathbb{R}^{H \times W \times 3}$, we extract image features ${F^C}_v \in \mathbb{R}^{N \times C_i}$ using the CXR-CLIP image encoder E_I^C. The corresponding medical report T is processed by the CXR-CLIP text encoder E_T^C to obtain text features $F_t \in \mathbb{R}^{L \times C_t}$. CXR-CLIP [25] is a pretrained model on relevant medical datasets based on CLIP. We utilize the MedSAM2 image encoder E_I^S to extract image features $F_v^S \in \mathbb{R}^{N \times C_i}$ for final segmentation. MedSAM2 [28] is a pretrained model by fine-tuning SAM2 on several medical datasets. Here, C_i and C_t denote the dimensions of the extracted image and text features respectively, N denotes the number of image patches, L signifies the length (i.e., the number of tokens) of the text description. Throughout the process, we partition the CXR-CLIP image-text encoder into four stages based on its blocks. After each stage, the QGAF module is applied for fine-tuning while all encoders remain frozen.

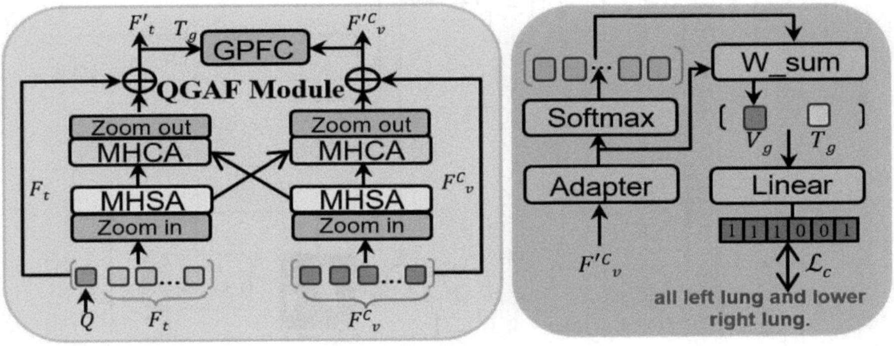

Fig. 2. Overview of the proposed module: the left side shows the Query-guided Attention Fusion (QGAF) module, while the right side presents the Global Positional Feature Classification (GPFC) component in QGAF.

Query-Guided Attention Fusion Module. As shown in Fig. 2, during the image encoding stage, we fuse textual information using the QGAF bridging module to learn localization knowledge for lesion regions. Unlike previous methods that rely on simple attention mechanisms, we employ a learnable query token $Q \in \mathbb{R}^{T \times C_t}$ (T is the token length) and concatenate it with the text feature F_t. To save memory and improve efficiency, we apply a zoom-in operation to reduce the feature dimension to $C_f = 64$, then perform multimodal fusion through self-attention and cross-attention mechanisms, and finally restore the original dimension with a zoom-out operation. The process is formulated as follows:

$$F'_t = F_t + MHCA(MHSA([Q, F_t]), MHSA({F^C}_v)), \tag{1}$$

$$F'^C{}_v = {F^C}_v + MHCA(MHSA({F^C}_v), MHSA([Q, F_t])). \tag{2}$$

where, $[\cdot]$ denotes concatenation. For simplicity, the zoom-in and zoom-out operations are not shown in the equations; they are implemented via a feed-forward network (FFN) that reduces and then restores the feature dimension; MHCA is the multihead cross-attn. We describe the GPFC component in Sect. 2.4.

2.2 Prompt Generation

Coarse Mask Prompt. After four encoder stages, we obtain the final image feature ${F^C}_v$ and text feature F_t. We use the end-of-sequence token $F_{t[EOS]} \in$

$\mathbb{R}^{1\times C_t}$ as the global text feature. Next, we reshape ${F^C}_v$ into $h_i \times w_i \times C_i$ and perform element-wise multiplication with the global text feature to generate the pseudo mask $M \in \mathbb{R}^{h_i\times w_i\times 1}$: $M = F_{t[EOS]} \odot {F^C}_v$. This approach, based on contrastive learning, computes the similarity score between the global text feature and each image pixel feature. A higher similarity score indicates a greater likelihood that the pixel belongs to a lesion region.

Point Prompt. Our point prompts are generated from the coarse mask. From the coarse mask M, we derive probability values for each of the $h_i \times w_i$ points indicating the likelihood of belonging to a lesion area. We then select the Top-k points as point prompts. First, candidate points with a probability above a threshold ($\theta > 0.9$) are identified. The first point is chosen as the candidate with the highest likelihood and is added to the set of selected points $\mathcal{P}$. Subsequent points are determined based on the criterion $Dis_{max}(p, \mathcal{P})$, where point p is selected if it is the farthest from the already selected points. This approach ensures that the chosen points are not only highly probable but also well distributed, covering various lesion regions rather than clustering in a single area.

Text Prompt. We use the end-of-sequence text token $F_{t[EOS]} \in \mathbb{R}^{1\times C_t}$ from the final text encoder layer as the global information. Then a simple feed-forward network (FFN) process is performed to generate the text prompt.

2.3 Mask Decoder

The three types of prompts are encoded by the MedSAM2 prompt encoder to obtain the coarse mask E_m, point prompt E_p, and text prompt E_t. The SAM2 image encoder extracts the image features F_v^S, which, combined with E_m and the other prompts, are fed into the MedSAM2 mask decoder to yield the final segmentation S. The overall process is expressed as:

$$S = SAM2_{Dec}(F_v^S + E_m, [E_p, E_t]). \tag{3}$$

2.4 Loss Design

Global Positional Feature Classification Component. The proposed GPFC component (Fig. 2, right) classifies global positions by partitioning the lung into six lesion regions (upper, middle, and lower lobes for both lungs). To ensure multimodal consistency, the module leverages both the global image feature V_g and global text feature T_g. Firstly, an adapter (comprising two linear layers with activation) is used to refine the image feature. A Softmax function then assigns regional weights, which are aggregated via a weighted sum to produce V_g. Simultaneously, T_g is extracted from the EOS token of the text representation V_t. The global image feature V_g and text feature T_g are concatenated and passed through a classification layer to yield the prediction. Since the GPFC is performed with each encoder block. So, for encoder block i, we have $\ell_{c_i} = \text{BCE}(\text{Linear}([V_g, T_g]), \mathcal{Y})$, where $\mathcal{Y}$ denotes the ground-truth label and

BCEWithLogitsLoss is adopted for backpropagation. The overall classification loss is defined as $\ell_c = \sum_{i=1}^{4} \ell_{c_i}$, since we have 4 encoder blocks.

Segmentation Loss. We employ a loss function that combines Dice and Cross-Entropy terms on both coarse mask M and final segmentation S to evaluate segmentation performance. Let U be the total number of pixels and V be the number of classes. For each pixel u and class v (positive or negative), $\hat{y}_{u,v}$ denotes the predicted probability that pixel u belongs to class v, and $y_{u,v}$ is the corresponding ground-truth label. $\ell_{\text{Dice}} = 1 - \frac{1}{UV} \sum_{u=1}^{U} \sum_{v=1}^{V} \frac{2|\hat{y}_{u,v} \cap y_{u,v}|}{|\hat{y}_{u,v}|+|y_{u,v}|}$, $\ell_{\text{CE}} = -\frac{1}{U} \sum_{u=1}^{U} \sum_{v=1}^{V} [\, y_{u,v} \log(\hat{y}_{u,v})]$. The overall segmentation loss ℓ_s is formed by equally weighting these two terms: $\ell_s = 0.5\,\ell_{\text{Dice}} + 0.5\,\ell_{\text{CE}}$. Therefore, our proposed loss ℓ can be formed as:

$$\ell = \lambda_1 \ell_s(M) + \lambda_2 \ell_s(S) + \lambda_3 \ell_c. \tag{4}$$

3 Experiments

Datasets. We evaluated our method on two publicly available COVID-19 medical image datasets: QaTa-COV19 [7] and MosMedData+ [16]. QaTa-COV19, compiled jointly by Qatar University and Tampere University, comprises 9258 chest X-ray images of COVID-19 cases, partitioned into training (5716 samples), validation (1429 samples), and testing (2113 samples) sets. Each image is accompanied by a lesion mask that delineates the infected regions. MosMedData+ is a large-scale lung CT dataset containing 2729 CT scan slices from multiple sources, with 2183 training, 273 validation, and 273 testing samples—all annotated with ground-truth infection masks. Moreover, we use the medical reports generated in [14] to enable multimodal learning.

Implementation Details. We trained our model using the PyTorch framework with a batch size of 5 and an initial learning rate of 3×10^{-4}. The model was optimized using AdamW with a weight decay of 0.001, and a cosine annealing learning rate schedule was applied with a warm-up period of 1000 iterations. The training ran for 70 epochs. We set the length of the learnable query tokens T to 10 and selected $Top3$ point prompts. The loss hyperparameters λ_1, λ_2, and λ_3 were set to 0.3, 0.6, and 0.1, respectively. All experiments were conducted on a system equipped with an RTX-6000 48 GB GPU and 128 GB of RAM.

Quantitative Results. We conducted comparative experiments on five unimodal methods and five multimodal methods, among which LViT, GuideDecoder, and MMI-UNet were proposed explicitly for these two datasets. As shown in Table 1, our method achieves consistently the best performance on both datasets. Notably, our model requires only 16.9M trainable parameters, demonstrating its efficiency and ability to transfer pretrained large-scale foundation models, thereby saving training resources. Furthermore, when compared with standalone CXR-CLIP and MedSAM2, our approach, despite using fewer parameters, yields superior performance, proving the feasibility of fine-tuning.

Additionally, as shown in Fig. 3, our proposed method enhances the localization of the lesion area, with segmentation results closely aligned with the actual lesions. This validates the effectiveness of our location-aware approach. In contrast, CXR-CLIP and MedSAM2 tends to misclassify normal regions as lesions, which further underscores the efficiency of our prompt generation strategy. Overall, our method effectively overcomes the limitations of both CLIP and SAM2.

Table 1. Performance comparison of various segmentation models on MosMedData+ and QaTa-COV19 datasets.

Method	Text	Param(M)	MosMedData+		QaTa-COV19	
			Dice	mIoU	Dice	mIoU
U-Net++ [27]	✗	74.5	0.7175	0.5839	0.7962	0.7025
nnUNet [10]	✗	19.1	0.7259	0.6036	0.8042	0.7081
TransUNet [5]	✗	105	0.7124	0.5844	0.7863	0.6913
SwinUNet [3]	✗	82.3	0.6329	0.5019	0.7807	0.6834
MedSAM2 [28]	✗	38.9	0.5427	0.4109	0.7536	0.6428
CXR-CLIP [25]	✓	136.6	0.7628	0.6327	0.8325	0.7409
LAVT [24]	✓	118.6	0.7329	0.6041	0.7928	0.6989
LViT [14]	✓	29.7	0.7457	0.6133	0.8366	0.7511
GuideDecoder [26]	✓	44	0.7775	0.6360	0.8978	0.8145
MMI-UNet [2]	✓	56.2	<u>0.7842</u>	<u>0.6450</u>	<u>0.9088</u>	<u>0.8328</u>
Ours	✓	16.9	**0.7981**	**0.6572**	**0.9206**	**0.8489**

Table 2. Ablation study on the QGAF module and GPFC loss.

Ablation	MosMedData+		QaTa-COV19	
	Dice	mIoU	Dice	mIoU
Baseline (w/o QGAF)	0.6948	0.5791	0.8036	0.7301
QGAF (w/o GPFC)	0.7627	0.6478	0.9012	0.8227
QGAF (Ours)	**0.7981**	**0.6572**	**0.9206**	**0.8489**

Ablation Studies. We conducted experiments to validate our components. Firstly, an ablation study on the QGAF module (Table 2) demonstrates its ability to learn multimodal information for segmentation tasks. We compared a baseline without the QGAF module and evaluated two variants—with and without the GPFC loss. The results indicate that our QGAF module effectively captures cross-modal information, while the GPFC loss further enhances lesion localization by integrating positional data extracted from radiology reports. Specifically,

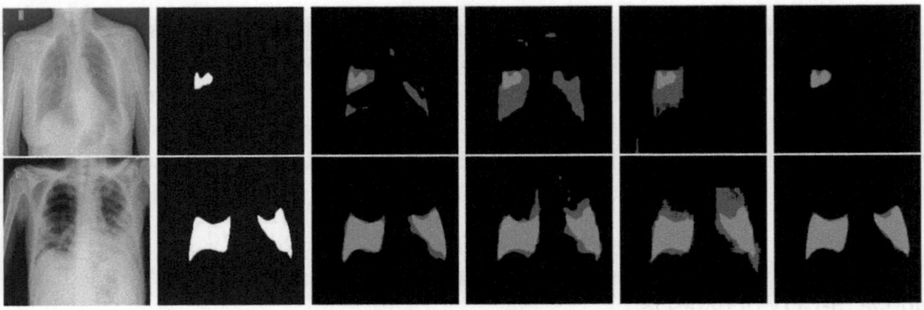

Fig. 3. Visualization of segmentation outcomes on the QaTa-COV19 dataset. In the images, green represents true positives, red denotes false negatives, and blue signifies false positives. (Color figure online)

on MosMedData+, our method improves the Dice score by 10.33% and mIoU by 7.81%, and on QaTa-COV19, the Dice score increases by 11.70% and mIoU by 11.88%. Furthermore, Table 3 shows the QGAF module achieves optimal performance when using the length 10 learnable query token, which effectively leverages text information. Table 4 reveals that using the Top-3 point prompts generally results in superior segmentation performance compared to the Top-4, despite the latter showing a slight increase of 0.38% in mIoU on MosMedData+. Table 5 highlights MedSAM2's dependence on high-quality prompts, as their absence significantly degrades performance. Our analysis of three prompts reveals that the coarse mask prompt has the most significant impact on segmentation results—improving the Dice score by 23.20% and mIoU by 23.45% on MosMedData+, and by 18.65% and 21.03% on QaTa-COV19. This is followed by point and text prompts. Overall, the results demonstrate the critical importance of high-quality prompts for effectively transferring SAM2 to downstream tasks.

Table 3. Ablation study on the length (T) of learnable query token Q.

T	MosMedData+		QaTa-COV19	
	Dice	mIoU	Dice	mIoU
5	0.7742	0.6273	0.8847	0.7911
10	**0.7981**	**0.6572**	**0.9206**	**0.8489**
15	0.7829	0.6501	0.9182	0.8339
20	0.7641	0.6176	0.8922	0.8001

Table 4. Ablation Study on Top-k point prompts selection for MedSAM2

Top-k	MosMedData+		QaTa-COV19	
	Dice	mIoU	Dice	mIoU
1	0.7650	0.6220	0.8620	0.7940
2	0.7820	0.6400	0.9170	0.8481
3	**0.7981**	0.6572	**0.9206**	**0.8489**
4	0.7935	**0.6610**	0.9071	0.8215

Table 5. Ablation study on three different prompts.

Coarse Mask	Points	Text	MosMedData+		QaTa-COV19	
			Dice	mIoU	Dice	mIoU
✗	✗	✗	0.4816	0.3527	0.6914	0.5829
✓	✗	✗	0.7136	0.5872	0.8779	0.7932
✓	✓	✗	0.7731	0.6392	0.8930	0.8154
✓	✓	✓	**0.7981**	**0.6572**	**0.9206**	**0.8489**

4 Conclusions

In this study, we propose a location-aware segmentation framework that leverages CLIP to generate three high-quality prompts for SAM2. By fine-tuning the QGAF bridging module with minimal training, our framework achieves high accuracy. In addition, the GPFC localization loss, derived from medical reports, enables our method to precisely localize lesion regions. This work offers a new approach for future lightweight multimodal medical segmentation.

Disclosure of Interests. The authors have no competing interests to declare that are relevant to the content of this article.

References

1. Azad, R., et al.: Medical image segmentation review: the success of U-Net. IEEE Trans. Pattern Anal. Mach. Intell. (2024)
2. Bui, P.N., Le, D.T., Choo, H.: Visual-textual matching attention for lesion segmentation in chest images. In: Linguraru, M.G., et al. (eds.) MICCAI 2024. LNCS, vol. 15009, pp. 702–711. Springer, Cham (2024). https://doi.org/10.1007/978-3-031-72114-4_67
3. Cao, H., et al.: Swin-Unet: Unet-like pure transformer for medical image segmentation. In: Karlinsky, L., Michaeli, T., Nishino, K. (eds.) ECCV 2022. LNCS, vol. 13803, pp. 205–218. Springer, Cham (2022). https://doi.org/10.1007/978-3-031-25066-8_9
4. Chaunzwa, T.L., et al.: Deep learning classification of lung cancer histology using CT images. Sci. Rep. **11**(1), 1–12 (2021)
5. Chen, J., et al.: TransUnet: transformers make strong encoders for medical image segmentation. arXiv preprint arXiv:2102.04306 (2021)
6. Chen, J., et al.: TransUNet: rethinking the U-Net architecture design for medical image segmentation through the lens of transformers. Med. Image Anal. **97**, 103280 (2024)
7. Degerli, A., Kiranyaz, S., Chowdhury, M.E., Gabbouj, M.: OSEGNet: operational segmentation network for COIVD-19 detection using chest X-ray images. In: 2022 IEEE International Conference on Image Processing (ICIP), pp. 2306–2310. IEEE (2022)
8. Fan, D.P., et al.: Inf-Net: automatic COVID-19 lung infection segmentation from CT images. IEEE Trans. Med. Imaging **39**(8), 2626–2637 (2020)

9. Han, X., Chen, Q., Xie, Z., Li, X., Yang, H.: Multiscale progressive text prompt network for medical image segmentation. Comput. Graph. **116**, 262–274 (2023)
10. Isensee, F., Jaeger, P.F., Kohl, S.A., Petersen, J., Maier-Hein, K.H.: nnU-Net: a self-configuring method for deep learning-based biomedical image segmentation. Nat. Methods **18**(2), 203–211 (2021)
11. Kim, K., et al.: LLM-Guided multi-modal multiple instance learning for 5-year overall survival prediction of lung cancer. In: Linguraru, M.G., et al. (eds.) MICCAI 2024. LNCS, vol. 15003, pp. 239–249. Springer, Cham (2024). https://doi.org/10.1007/978-3-031-72384-1_23
12. Kirillov, A., et al.: Segment anything. In: Proceedings of the IEEE/CVF International Conference on Computer Vision, pp. 4015–4026 (2023)
13. Lei, Y., Li, Z., Shen, Y., Zhang, J., Shan, H.: CLIP-lung: textual knowledge-guided lung nodule malignancy prediction. In: Greenspan, H., et al. (eds.) MICCAI 2023. LNCS, vol. 14226, pp. 403–412. Springer, Cham (2023). https://doi.org/10.1007/978-3-031-43990-2_38
14. Li, Z., et al.: LViT: language meets vision transformer in medical image segmentation. IEEE Trans. Med. Imaging (2023)
15. Luo, Y., et al.: CARL: cross-aligned representation learning for multi-view lung cancer histology classification. In: Greenspan, H., et al. (eds.) MICCAI 2023. LNCS, vol. 14224, pp. 358–367. Springer, Cham (2023). https://doi.org/10.1007/978-3-031-43904-9_35
16. Morozov, S.P., et al.: MosMedData: chest CT scans with COVID-19 related findings dataset. arXiv preprint arXiv:2005.06465 (2020)
17. Radford, A., et al.: Learning transferable visual models from natural language supervision. In: International Conference on Machine Learning, pp. 8748–8763. PMLR (2021)
18. Ravi, N., et al.: SAM 2: segment anything in images and videos. arXiv preprint arXiv:2408.00714 (2024)
19. Singh, T., Mishra, S., Kalra, R., Satakshi, Kumar, M., Kim, T.: COVID-19 severity detection using chest X-ray segmentation and deep learning. Sci. Rep. **14**(1), 19846 (2024)
20. Su, H., Lei, H., Guoliang, C., Lei, B.: Cross-graph interaction and diffusion probability models for lung nodule segmentation. In: Linguraru, M.G., et al. (eds.) MICCAI 2024. LNCS, vol. 15001, pp. 482–492. Springer, Cham (2024). https://doi.org/10.1007/978-3-031-72378-0_45
21. Tomar, N.K., Jha, D., Bagci, U., Ali, S.: TGANet: text-guided attention for improved polyp segmentation. In: Wang, L., Dou, Q., Fletcher, P.T., Speidel, S., Li, S. (eds.) MICCAI 2022. LNCS, vol. 13433, pp. 151–160. Springer, Cham (2022). https://doi.org/10.1007/978-3-031-16437-8_15
22. Xiong, X., et al.: SAM2-UNet: segment anything 2 makes strong encoder for natural and medical image segmentation. arXiv preprint arXiv:2408.08870 (2024)
23. Yang, H., Shen, L., Zhang, M., Wang, Q.: Uncertainty-guided lung nodule segmentation with feature-aware attention. In: Wang, L., Dou, Q., Fletcher, P.T., Speidel, S., Li, S. (eds.) MICCAI 2022. LNCS, vol. 13435, pp. 44–54. Springer, Cham (2022). https://doi.org/10.1007/978-3-031-16443-9_5
24. Yang, Z., Wang, J., Tang, Y., Chen, K., Zhao, H., Torr, P.H.: LAVT: language-aware vision transformer for referring image segmentation. In: Proceedings of the IEEE/CVF Conference on Computer Vision and Pattern Recognition, pp. 18155–18165 (2022)

25. You, K., et al.: CXR-CLIP: toward large scale chest X-ray language-image pre-training. In: Greenspan, H., et al. (eds.) MICCAI 2023. LNCS, vol. 14221, pp. 101–111. Springer, Cham (2023). https://doi.org/10.1007/978-3-031-43895-0_10
26. Zhong, Y., Xu, M., Liang, K., Chen, K., Wu, M.: Ariadne's thread: using text prompts to improve segmentation of infected areas from chest X-ray images. In: Greenspan, H., et al. (eds.) MICCAI 2023. LNCS, vol. 14223, pp. 724–733. Springer, Cham (2023). https://doi.org/10.1007/978-3-031-43901-8_69
27. Zhou, Z., Siddiquee, M.M.R., Tajbakhsh, N., Liang, J.: UNet++: redesigning skip connections to exploit multiscale features in image segmentation. IEEE Trans. Med. Imaging **39**(6), 1856–1867 (2019)
28. Zhu, J., Qi, Y., Wu, J.: Medical SAM 2: segment medical images as video via segment anything model 2. arXiv preprint arXiv:2408.00874 (2024)

Lymph Node Metastasis Classification with Prototype-Guided Multiple Instance Aggregation and Heterogeneous Feature Fusion

Haoshen Li[1,2], Tashan Ai[3(✉)], Yirui Wang[2], Zhanghexuan Ji[2], Qinji Yu[2,4], Le Lu[2], Bin Dong[5,6], Li Zhang[1], Xianghua Ye[7], Kuaile Zhao[3(✉)], and Dakai Jin[2]

[1] Center for Data Science, Peking University, Beijing, China
[2] DAMO Academy, Alibaba Group, Washington, D.C., USA
[3] Fudan University Shanghai Cancer Center, Shanghai, China
18111230048@fudan.edu.cn, kuaile_z@sina.com
[4] Shanghai Jiao Tong University, Shanghai, China
[5] Beijing International Center for Mathematical Research and the New Cornerstone Science Laboratory, Peking University, Beijing, China
[6] Center for Machine Learning Research, Peking University, Beijing, China
[7] The First Affiliated Hospital, Zhejiang University, Hangzhou, China

Abstract. Lymph node metastasis diagnosis in computed tomography (CT) scans is an essential yet very challenging task for esophageal cancer staging and treatment planning. Recent advances in deep learning have markedly improved the performance in lymph node (LN) metastasis classification. However, these methods often focus more on the averaged features of all CT slices containing a 3D LN instance, lacking effective fusion of key slice-wise features, which is important in the LN metastasis analysis by physicians. In addition, existing deep learning models are trained using CT scans in an end-to-end fashion, thus lacking the explicit incorporation of clinically relevant meta-imaging features (i.e., morphological and radiomic features). Meta-imaging features play a crucial role in LN assessment and may not be effectively captured by direct end-to-end deep learning models. To address these issues, we formulate the 3D LN metastasis classification as a multiple instance learning (MIL) problem by extracting and fusing slice-level features (instance) into a comprehensive bag representation. Building on this, we propose a two-streamed MIL framework with a prototype-guided aggregation method that effectively captures LN characteristics at both local and global scales. Furthermore, a multi-scale multi-source fusion module is introduced to integrate the heterogeneous meta-imaging features with deep learning features, enhancing the comprehensive representation of LN. Five-fold cross-validation on a cohort of 284 esophageal cancer patients with 809 pathology-confirmed LN instances demonstrate the superiority of our methods compared to the state-of-the-art approaches with +2.66% in AUROC and +4.81% in sensitivity improvements.

Keywords: Lymph node metastasis · Multiple instance learning · Heterogeneous Feature Fusion

J. C. Gee et al. (Eds.): MICCAI 2025, LNCS 15960, pp. 312–321, 2026.
https://doi.org/10.1007/978-3-032-04927-8_30

1 Introduction

Esophageal cancer (EC) ranks as the sixth leading cause of cancer-related deaths globally, representing 1 out of every 20 cancer deaths [19]. Lymph node (LN) metastasis serves as one of the critical prognostic indicators in EC. Accurate identification of preoperative LN metastasis is crucial for guiding the treatment decisions (whether to proceed with surgery or opt for neoadjuvant therapy) and formulating treatment strategies (surgical resection area and radiotherapy clinical target volume [CTV]) [1]. Therefore, LN metastasis assessment holds significant clinical importance in EC diagnosis and management.

Assessing LN metastatic status in CT is challenging even for experienced physicians. It depends on various factors, such as global characteristics (*e.g.*, size, shape), localized imaging characteristics (*e.g.*, intensity inhomogeneity, textures), etc. All of these characteristics contribute to the LN status, but none serve as a standalone predictive factor. For example, the existing criteria, such as RECIST [16], morphology [15] and texture characteristics [2], show limited performance in previous studies [3,5,10,12,20] (*e.g.*, sensitivity ranged from 39.7% to 67.2% with specificity around 80.0%). Thus, proposing an effective computer-aided diagnosis (CAD) solution for this task is highly desirable.

Previous work have used deep learning for LN abnormality diagnosis. Roth *et al.* develops a two-stage convolutional neural network (CNN) network with a 2.5D universal image decomposition representation and random aggregation to detect and classify enlarged LNs [13,14]. Lee *et al.* attempts various CNNs architectures to diagnose cervical LN metastasis in CT using 202 patients with thyroid cancer [9]. Kann *et al.* train a dual 3D network, which jointly learns the global and local features of LNs, to classify metastatic LNs and extranodal extension (ENE) using CT scans of 270 patients [8]. Li *et al.* proposes a semi-supervised framework to better handle unlabeled LNs, which achieves the state-of-the-art classification performance [11]. Although these methods demonstrate their effectiveness, they still face two major limitations. First, previous methods focus more on the averaged features of all CT slices containing a 3D LN instance, lacking effective fusion of key slice-wise features. For example, the leading 2.5D method [11] simply merges multiple sets of slice features by concatenation, making it difficult to identify key slice information when handling with LNs in thinner CT scan that can occupy more than 10 to 20 CT axial slices. For metastatic LNs, its malignant characteristics are more likely to be evident in partial slices than in all slices. Therefore, it is important to integrate and capture the key slice features that most correlate to the metastatic status. Second, previous deep learning-based methods also lack explicit consideration of clinic-relevant meta-imaging features, such as long, short axis diameters, heterogeneity intensity, central necrosis, etc. These meta-imaging features may not be effectively captured by the direct end-to-end deep learning model.

To address these limitations, we formulate the 3D LN metastasis classification as a multiple instance learning (MIL) problem by extracting and combining slice-level features (instance) into a comprehensive bag representation. Building on this, we propose a two-streamed MIL framework with a prototype-

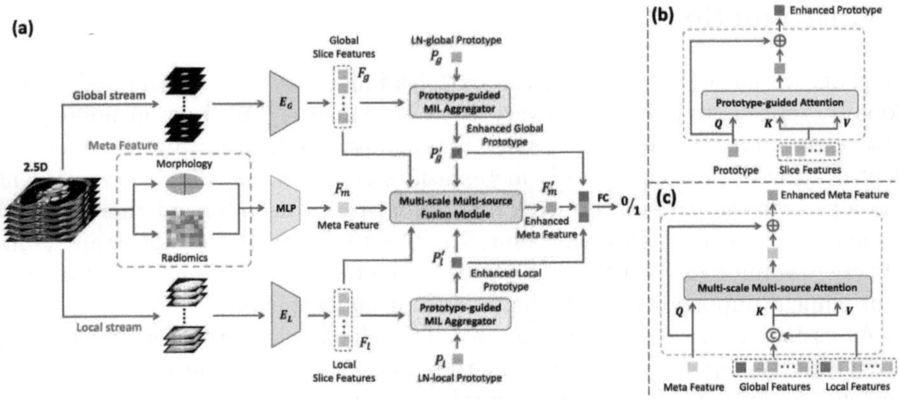

Fig. 1. (a) Illustrates the pipeline. A LN CT patch is transposed into multiple 3-slice images with original and zoom-in size, as the inputs for the global and local stream, respectively. In addition, extracted morphology and radiomics features from the LN patch are introduced as the meta feature. The prototype-guided MIL aggregator is introduced to fuse slice features into the improved LN prototype, as shown in (b). The Multi-scale Multi-source Fusion Module is introduced to enhance the meta feature by multi-scale image features, as shown in (c).

guided aggregation method that captures LN characteristics at both local and global scales. Furthermore, a multi-scale multi-source fusion module is introduced to integrate the heterogeneous meta-imaging features (such as morphological and radiomics features) with deep learning features, enhancing the comprehensive representation of LN. We collected and curated 284 EC patients with 809 pathology-confirmed LN instances. In five-fold cross-validation, our proposed method exhibits superior performance, significantly exceeding the state-of-the-art LN classification approaches and other MIL-based methods and achieving the highest AUROC score of 84.68% (+2.66% improvement) in this challenging task.

2 Method

An overview of our method is shown in Fig. 1. We develop a two-stream MIL framework with slice-wise prototype-guided MIL aggregation module to enhance the key LN features in both streams, as detailed in Sect. 2.1. To integrate meta-imaging features with deep learning features, a multi-scale multi-source fusion module is introduced that enhances meta-features by computing cross-attention with multi-scale LN imaging features, as explained in Sect. 2.2.

2.1 Two-Streamed Prototype-Guided Multiple Instance Aggregation

Metastatic LN imaging characteristics may only be present in a few slices of an LN 3D CT volume. Inspired by this observation, we naturally formulate the 3D

LN classification problem as an MIL task by extracting and fusing slice-level features (MIL instance) into a comprehensive bag representation for volume-level LN metastasis classification. Our two-streamed slice-wise MIL framework is shown in Fig. 1.

While global characteristics (*e.g.*, size, shape) can indicate LN metastasis, the use of localized features (*e.g.*, texture, inhomogeneity) can facilitate more accurate and robust LN metastasis modeling. Motivated by this, instead of extracting imaging features solely from the original CT input, we take a two-stream approach to simultaneously encode both the original and resized input that spatially zoomed in or out to fill the input slice, which implicitly promotes learning localized fine-scale features. This two-stream encoding method not only provides complementary features for subsequent multi-scale feature fusion but acts as a mutual regularization to avoid the network overfitting to a single-scale input. Previous methods either directly extract LN features from a pure 3D network [8] or leverage a simple 2.5D strategy [11] to treat and fuse adjacent slices equally without considering their relative importance. Here, we view LN metastasis classification as a multi-instance learning problem, group the CT slices into multiple sets of 3-channel images, and transform a 3D LN CT image with a volume-level metastasis label in multiple 2D slice groups that can be processed by pre-trained 2D network, acting as a bag of instances in the MIL paradigm. During the prototype-guided aggregation, the fusion weight of each instance is recalibrated so that the slice with more suspicious LN metastasis features can be enhanced by assigning a higher weight to the fused feature. In implementation, we adopt MobileNetv3 [6] with pre-trained weights on ImageNet [4] as the image encoder to generate the two-streamed multiple slice-group features. To effectively aggregate two-streamed slice-group features, we propose a prototype-guided MIL Aggregator. As shown in Fig. 1, we randomly initialize two learnable prototypes: LN-global Prototype and LN-local Prototype, denoted as $P_g, P_l \in \mathbb{R}^{1\times d}$, to guide the aggregation of global stream and local stream features, respectively. Taking the global stream as an example, the learnable prototype P_g and the global slice features $F_g \in \mathbb{R}^{N\times d}$ (N represents the number of slice groups) are fed into the aggregator, which is implemented as a cross-attention layer. Formally, the learnable global prototype P_g serves as the query vector Q_g to search for slice-group features that are most relevant to LN metastasis, while the global-stream slice features F_g act as both the key K_g and the value V_g. The enhanced global prototype is then updated as follows:

$$P'_g = \text{Softmax}(\frac{Q_g K_g^T}{\sqrt{d}} V_g) + P_g \quad (1)$$

The operation for the local stream is the same as the global stream. Under the guidance of P_g and P_l, the prototype-guided MIL aggregator can effectively capture slices that are highly correlated with LN metastasis status, thereby enabling better aggregation. The enhanced prototypes P'_g, P'_l are subsequently used as a compact, metastasis-aware representation for final classification.

2.2 Multi-scale Multi-source Fusion

In LN metastasis classification, clinic-relevant meta-imaging features, such as round morphology, intensity heterogeneity, are shown to be effective. However, these meta-imaging features may not be easily captured by direct end-to-end trained deep learning models. Therefore, we explicitly compute these meta-imaging features and integrate these heterogeneous features into one deep learning framework by a new multi-scale multi-source fusion module. Specifically, we measure the long and short diameters of each 3D LN instance, calculate its radiomic first-order features, and concatenate them as the meta-imaging feature. Then, a linear layer is used to map the meta-imaging feature to the same dimensionality as the deep learning feature, denoting it as $F_m \in \mathbb{R}^{1\times d}$. Next, we leverage the multi-scale deep features to further enhance the meta-imaging features. As shown in Fig. 1(c), this process is achieved by the cross-attention mechanism. Specifically, enhanced local and global deep prototypes, P'_g, P'_l, and original slice features F_g, F_l are concatenated together as key $K_{g,l}$ and value $V_{g,l}$. The meta-imaging feature F_m is taken as query Q_m. The interaction operation is as follows:

$$F'_m = \text{Softmax}(\frac{Q_m K^T_{g,l}}{\sqrt{d}} V_{g,l}) + F_m \tag{2}$$

After deriving the enhanced meta feature F'_m, along with previous P'_g and P'_l, we simply fuse them through channel-wise concatenation. The fused representation is then projected to a logit space for LN metastasis classification via a learnable linear transformation and optimized using binary cross-entropy (BCE) loss, formulated as:

$$\mathcal{L}_{\text{cls}} = \mathcal{L}_{\text{BCE}}\left(\mathbf{W}_p \cdot [P'_g \,\|P'_l\|\, F'_m]^T, Y\right) \tag{3}$$

where $\|$ denotes feature concatenation along the channel dimension, $\mathbf{W}_p \in \mathbb{R}^{1\times 3d}$ represents the trainable projection weights. $Y \in \{0,1\}$ is the metastasis status label (0: benign, 1: malignant).

3 Experiments and Results

3.1 Experimental Settings

Dataset: We collected a dataset of 284 esophageal cancer patients who underwent esophagectomy treatment. Each patient has a preoperative contrast-enhanced CT scan and a detailed post-operative pathological report indicating whether there are metastatic LNs in the surgical dissection area. The median CT scan size is 512 × 512 × 358 voxels with a median resolution of 0.785 × 0.785 × 1.0 mm. We first detected all visible LN instances using an ensemble of recent automatic LN detection models [21,22]. Out of these detected LNs, 809 LN instances have metastasis status labels (208 positive and 601 negative), which is confirmed by the consensus of two radiologists according to the pathology report. The LN mask, along with the CT scan, is then cropped using a

$64 \times 64 \times 32$ ROI centered on each 3D LN. Experiments are conducted using five-fold cross-validation with a 70%/10%/20% training, validation, and testing split (at the patient level) for each fold.

Implementation Details: A lightweight model MobileNetv3 [6] is used as the image encoder for each of the two streams. We considered 18 slices (6 slice groups) as the encoder's input, which exceeds 80% LN's slice number. The feature dimension d of the slice features and mapped meta-features is 960. For the meta-imaging feature input, we use 20 features in total, including two morphological features (short diameter and long diameter), and 18 radiomics features (original first-order features (skewness, elongation, etc.) calculated by Radiomics library in Python 3.9). SGD optimizer with a learning rate of 1.6e−4 and cosine annealing decay is adopted, and the network is trained by 300 epochs with a mini-batch size of 32.

Comparison Methods: We compare with other methods in the following categories. 1) *Traditional deep learning category:* four state-of-the-art LN metastasis classification work are evaluated: two 3D-based methods of MobileNetv3_3d [6] and 3D DualNet [8]; two 2.5D-based methods of MobileNetv3_2.5D [6] and dual-stream 2.5D [10]. 2) *MIL category:* We compare to six popular MIL methods, including Max-pooling, Mean-pooling, ABMIL [7], GAMIL [7], TransMIL [17], and DTMIL [23]. For fair comparison, MobileNetv3_2.5D is used as the backbone in the MIL methods. 3) *Multi-source MIL category:* We also examine a leading multi-source MIL fusion method ViLa-MIL [18], which combines both deep learning and meta features.

Evaluation Metrics: To evaluate the performance comprehensively, we compute the area under the receiver operating characteristic curve (AUROC), accuracy at a specificity rate of 75% (A@S75), accuracy at a recall rate of 75% (A@R75), recall at a specificity rate of 75% (R@S75), and specificity at a recall rate of 75% (S@R75).

3.2 Experimental Results

Quantitative Comparison Results: Table 1 summarizes the quantitative comparison results. Several observations can be drawn. (1) Both 2.5D and MIL methods outperform 3D methods with an improvement of 2–5% in AUROC. This indicates that the 3D deep network fail to learn and extract the most essential LN characteristics in a 3D fashion. One reason for this may be the lack of pre-trained model weights in 3D. (2) For the methods in 3D and 2.5D categories, multi-scale approaches (DualNet [8], dual-stream [11]) achieve better performance, with an improvement of more than 2% in AUROC as compared to the single-stream counterpart. This confirms that the local and global features of LNs are both important for its metastatic classification. (3) Slice-based MIL approaches yield decreased performance ([79.89%, 80.68%] AUC) when compared to the specifically designed dual-stream 2.5D LN classification method [11] (81.09% AUC). This shows that applying previous MIL methods in

Table 1. Quantitative performance of LN metastasis classification. Four groups correspond to 3D methods, 2.5D methods, MIL methods and MIL methods combined with meta imaging features, respectively. Meta represents the meta imaging features (morphological and radiomics features).

Method	Meta	AUROC	A@R75	A@S75	S@R75	R@S75
mobilenetv3_3D [6]	×	75.61±4.34	67.99	71.47	64.94	66.16
DualNet_3D [8]	×	77.67±4.95	67.98	72.31	64.96	67.20
mobilenetv3_2.5D [6]	×	78.94±2.96	68.89	72.40	66.63	66.95
dual-stream_2.5D [11]	×	81.09±3.88	72.66	73.21	71.91	69.50
Max-pooling	×	80.10±3.16	70.05	72.55	68.59	68.60
Mean-pooling	×	80.40±3.03	71.49	72.95	70.35	67.30
ABMIL [7]	×	80.34±3.92	71.80	72.24	70.41	69.28
GAMIL [7]	×	80.51±4.31	71.66	73.12	70.29	68.67
TransMIL [17]	×	79.89±2.32	69.73	72.20	68.09	66.99
DTMIL [23]	×	80.68±4.54	71.88	72.81	70.45	69.41
Ours	×	**82.54±2.10**	**76.57**	**73.97**	**76.84**	**72.37**
		(+1.45%)	(+3.91%)	(+0.76%)	(+4.93%)	(+2.87%)
ViLa-MIL [18]	✓	82.02±2.90	73.18	73.47	72.27	71.69
Ours	✓	**84.68±2.00**	**78.53**	**74.68**	**79.48**	**76.50**
		(+2.66%)	(+5.35%)	(+1.21%)	(+7.21%)	(+4.81%)

single stream to the LN classification task would not exhibit improved accuracy. In contrast, with the proposed two-streamed prototype-guided MIL aggregation method, our method exceeds the dual-stream 2.5D method [11] (indicating that multiple slices aggregation is important) and other image-based MIL methods, and achieves the best performance of 82.54% AUC, 76.84% S@R75 and 72.37% R@S75. (4) When further incorporating meta-imaging features, our method further boosts the AUROC to 84.68% with +2.66% improvement compared to the multi-source fusion method ViLa-MIL [18].

Qualitative Results: Figure 2 illustrates the attention weight distributions of prototype-guided MIL aggregation module in the global and local streams (mean values of all attention weights in the test split), showing the importance of the positions of the slices in the classification contribution. It is observed that in both streams, the importance of the LN slices decreases from center to periphery, indicating that the malignant features are more likely to be near the central of a 3D LN instance. Figure 3 presents three malignant LN examples, each showing two different slices. The upper row presents the CT slice that has a higher attention weight, while the lower row shows the CT slice with a low attention weight. The examples demonstrate that our prototype-guided MIL aggregation method is effective in identifying the most important CT slices for metastasis status.

Table 2. Effect of the different meta-imaging features: shape and texture.

shape	texture	AUC	S@R75	R@S75
		82.54±2.10	76.84	72.37
✓		83.76±2.18	77.54	74.68
	✓	83.38±3.03	76.01	74.12
✓	✓	84.68±2.00	79.48	76.50

Table 3. Effect of different input LN slice numbers.

slice number	AUC	S@R75	R@S75
9	81.12±2.78	73.51	70.67
18	82.54±2.10	76.84	72.37
27	81.77±2.60	73.63	71.89

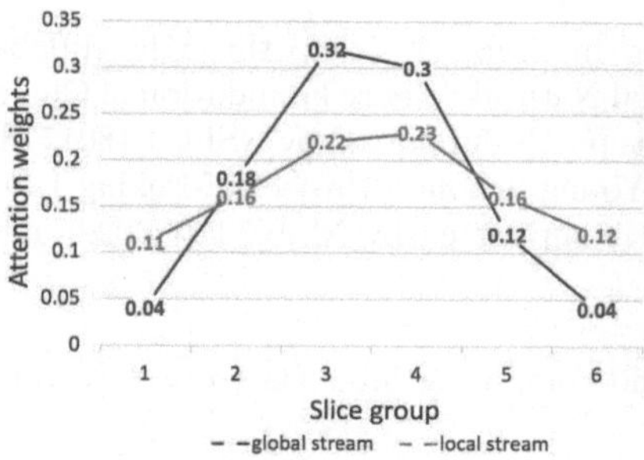

Fig. 2. Attention weight distributions of the prototype-guided MIL aggregation in two streams.

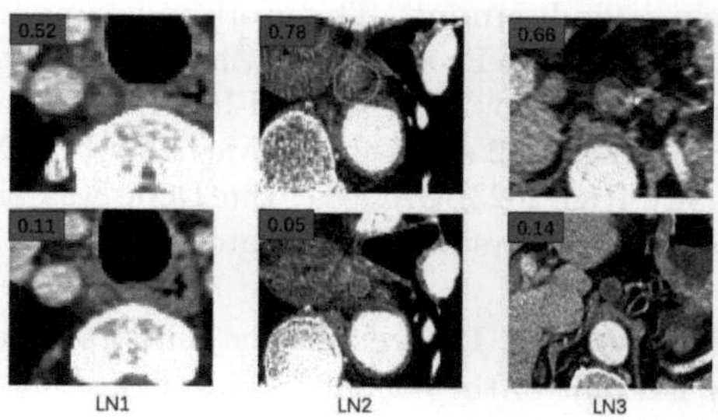

Fig. 3. Three qualitative examples of LN CT image with its attention weight assigned to the slice.

Ablation Results of Incorporating Meta-imaging Features: We separately demonstrate the effects of incorporating shape features and texture features in Table 2. We can see that individually adding each of these two features both improve the model's performance. In contrast, shape features provide a greater improvement, indicating that the global information (shape, size) plays a more important role in the task. Finally, combining the shape and texture features, our full model improves AUROC, S@R75, and R@S75 to 84.68%, 79.48%, and 76.50%, demonstrating its effectiveness.

Ablation Results of Different Numbers of Input Slices: We select 9, 18, and 27 slices of LN, respectively, to combine into slice-wise instances as input to the prototype-guided MIL aggregation module. As shown in Table 3, we observe that selecting 18 slices achieves the best results. Since LNs are relatively small, 18 slices can fully encompass most of LNs. Selecting 9 slices may miss metastatic information, and selecting 27 slices leads to feature redundancy hindering the effective fusion of key features, both leading to suboptimal performance.

4 Conclusion

In this work, we formulate the 3D LN metastasis classification as an MIL problem by extracting and combining slice-level features (instance) into a complete bag representation. Building on this, we propose a two-streamed MIL framework with a prototype-guided aggregation method that effectively captures LN characteristics at both local and global scales. Furthermore, a multi-scale multi-source fusion module is introduced to integrate the heterogeneous meta-imaging features with deep learning features, enhancing the comprehensive representation of LN. Five-fold cross-validation on a cohort of 284 esophageal cancer patients with 809 pathology-confirmed LN instances demonstrate the superiority of our methods compared to the state-of-the-art approaches with +2.66% in AUROC and +4.81% in sensitivity improvements.

Acknowledgement. This work was supported by Alibaba Group through Alibaba Research Intern Program, and Zhejiang Provincial Natural Science Foundation of China under Grant No. 2024-KYI-00I-I05. Li Zhang was partly supported by NSFC 81801778, NSFC 12090022 and Clinical Medicine Plus X-Young Scholars Project of Peking University PKU2023LCXQ041. Bin Dong was partly supported by NSFC 12090022, and the New Cornerstone Investigator Program.

Disclosure of Interests. The authors have no competing interests to declare that are relevant to the content of this article.

References

1. Ajani, J.A., et al.: Esophageal and esophagogastric junction cancers, version 2.2019, NCCN clinical practice guidelines in oncology. J. Natl. Compr. Canc. Netw. **17**(7), 855–883 (2019)
2. Bayanati, H., et al.: Quantitative CT texture and shape analysis: can it differentiate benign and malignant mediastinal lymph nodes in patients with primary lung cancer? Eur. Radiol. **25**, 480–487 (2015)
3. Boiselle, P.M., Patz, E.F., Jr., Vining, D.J., Weissleder, R., Shepard, J., McLoud, T.C.: Imaging of mediastinal lymph nodes: CT, MR, and FDG PET. Radiographics **18**(5), 1061–1069 (1998)
4. Deng, J., Dong, W., Socher, R., Li, L.J., Li, K., Fei-Fei, L.: ImageNet: a large-scale hierarchical image database. In: 2009 IEEE Conference on Computer Vision and Pattern Recognition, pp. 248–255. IEEE (2009)
5. Foley, K., Christian, A., Fielding, P., Lewis, W., Roberts, S.: Accuracy of contemporary oesophageal cancer lymph node staging with radiological-pathological correlation. Clin. Radiol. **72**(8), 693-e1 (2017)
6. Howard, A., et al.: Searching for mobilenetv3. In: ICCV, pp. 1314–1324 (2019)
7. Ilse, M., Tomczak, J., Welling, M.: Attention-based deep multiple instance learning. In: International Conference on Machine Learning, pp. 2127–2136 (2018)
8. Kann, B.H., et al.: Pretreatment identification of head and neck cancer nodal metastasis and extranodal extension using deep learning neural networks. Sci. Rep. **8**(1), 14036 (2018)
9. Lee, J.H., Ha, E.J., Kim, J.H.: Application of deep learning to the diagnosis of cervical lymph node metastasis from thyroid cancer with CT. Eur. Radiol. **29**, 5452–5457 (2019)
10. Li, B., et al.: The value of enhanced CT scanning for predicting lymph node metastasis along the right recurrent laryngeal nerve in esophageal squamous cell carcinoma. Ann. Transl. Med. **8**(24) (2020)

11. Li, H., et al.: Semi-supervised lymph node metastasis classification with pathology-guided label sharpening and two-streamed multi-scale fusion. In: Linguraru, M.G., et al. (eds.) MICCAI 2024. LNCS, vol. 15011, pp. 623–633. Springer, Cham (2024). https://doi.org/10.1007/978-3-031-72120-5_58
12. Liu, J., Wang, Z., Shao, H., Qu, D., Liu, J., Yao, L.: Improving CT detection sensitivity for nodal metastases in oesophageal cancer with combination of smaller size and lymph node axial ratio. Eur. Radiol. **28**, 188–195 (2018)
13. Roth, H.R., et al.: Improving computer-aided detection using convolutional neural networks and random view aggregation. IEEE Trans. Med. Imaging **35**(5), 1170–1181 (2015)
14. Roth, H.R., et al.: A new 2.5D representation for lymph node detection using random sets of deep convolutional neural network observations. In: Golland, P., Hata, N., Barillot, C., Hornegger, J., Howe, R. (eds.) MICCAI 2014. LNCS, vol. 8673, pp. 520–527. Springer, Cham (2014). https://doi.org/10.1007/978-3-319-10404-1_65
15. Schreuder, A., Jacobs, C., Scholten, E.T., van Ginneken, B., Schaefer-Prokop, C.M., Prokop, M.: Typical CT features of intrapulmonary lymph nodes: a review. Radiol. Card. Imaging **2**(4), e190159 (2020)
16. Schwartz, L., et al.: Evaluation of lymph nodes with recist 1.1. Eur. J. Cancer **45**(2), 261–267 (2009)
17. Shao, Z., et al.: TransMIL: transformer based correlated multiple instance learning for whole slide image classification. In: Advances in Neural Information Processing Systems, vol. 34, pp. 2136–2147 (2021)
18. Shi, J., Li, C., Gong, T., Zheng, Y., Fu, H.: ViLa-MIL: dual-scale vision-language multiple instance learning for whole slide image classification. In: CVPR, pp. 11248–11258 (2024)
19. Siegel, R.L., Miller, K.D., Fuchs, H.E., Jemal, A.: Cancer statistics, 2022. CA Canc. J. Clin. **72**(1), 7–33 (2022)
20. Wu, L., et al.: Multiple level CT radiomics features preoperatively predict lymph node metastasis in esophageal cancer: a multicentre retrospective study. Front. Oncol. **9**, 1548 (2020)
21. Yu, Q., et al.: Effective lymph nodes detection in CT scans using location debiased query selection and contrastive query representation in transformer. In: Leonardis, A., Ricci, E., Roth, S., Russakovsky, O., Sattler, T., Varol, G. (eds.) ECCV 2024. LNCS, vol. 15100, pp. 180–198. Springer, Cham (2025). https://doi.org/10.1007/978-3-031-72946-1_11
22. Yu, Q., et al.: Slice-consistent lymph nodes detection transformer in CT scans via cross-slice query contrastive learning. In: Linguraru, M.G., et al. (eds.) MICCAI 2024. LNCS, vol. 15005, pp. 616–626. Springer, Cham (2024). https://doi.org/10.1007/978-3-031-72086-4_58
23. Zhang, H., et al.: DTFD-MIL: double-tier feature distillation multiple instance learning for histopathology whole slide image classification. In: Proceedings of the IEEE/CVF Conference on Computer Vision and Pattern Recognition, pp. 18802–18812 (2022)

Lymphoma Prognosis with Lesion-Anatomy Context Fusion and Attention-Based Multi-lesion Aggregation

Song Zhang[1], Jiajin Zhang[1,2,3](✉), Liheng Qiu[4], Wei Liu[1], Dakai Jin[1], Le Lu[1], Shenmiao Yang[4], and Ke Yan[1]

[1] DAMO Academy, Alibaba Group, Hangzhou, China
[2] Shanghai Jiao Tong University, Shanghai, China
[3] Hupan Lab, 310023 Hangzhou, China
zhangjiajin.zjj@alibaba-inc.com
[4] Peking University People's Hospital, Beijing, China

Abstract. Early identification of lymphoma patients with poor prognosis is crucial to determining personalized treatment plans and improving prognosis. Currently, commonly used prognostic biomarkers include clinical variables such as International Prognostic Index. Quantitative parameters based on PET/CT and deep learning methods have also shown promising results. However, there are still several challenges in PET/CT-based prognostic studies: heterogeneity in the number and location of lesions, insufficient representation of lesion features, and the lack of anatomical context modeling of the lesions. We propose a novel framework named **LAMP**, with lesion-anatomy context fusion and attention-based multi-lesion aggregation as its two key components. The former takes into account information about the surrounding anatomical organs of the lesions to improve their representation. The latter treats each lesion region as an instance, assigning attention scores that reflect the contribution of each lesion, and aggregates them accordingly. A total of 229 lymphoma patients were collected to evaluate our model. In prediction tasks for progression-free survival and overall survival, the 5-fold cross-validation C-index is 0.791 and 0.828, respectively, outperforming existing models based on clinical variables and deep learning. **LAMP** has the potential to become a clinical auxiliary tool to differentiate patients with varying risk levels, facilitating the development of personalized treatment plans.

Keywords: PET/CT · Lymphoma Prognosis · Anatomical Context

1 Introduction

Lymphoma is a malignant neoplasm of the lymphatic system [1,2]. Unlike solid tumors, lymphoma often spread throughout the entire body. Diffuse large B-cell

J. C. Gee et al. (Eds.): MICCAI 2025, LNCS 15960, pp. 322–332, 2026.
https://doi.org/10.1007/978-3-032-04927-8_31

lymphoma (DLBCL), the most common type, accounts for 30% of non-Hodgkin's lymphoma [3,4]. Approximately 60% to 80% of patients achieve complete remission following first-line treatment with R-CHOP [5,6]. However, 20% to 40% may experience relapse or progression after initial therapy. Early identification of patients with poor prognosis is crucial, as it enables clinicians to avoid potentially ineffective therapy and timely adjust in treatment strategy [7–9].

Currently, the primary prognostic marker for DLBCL is the International Prognostic Index (IPI) and its variants [10,11]. IPI has long been the cornerstone for risk stratification in newly diagnosed DLBCL patients. However, IPI is predominantly based on clinical factors, overlooking tumor heterogeneity among individuals. Furthermore, with the development of new therapies, IPI has become increasingly inadequate to meet current clinical needs. As an essential clinical imaging tool for DLBCL prognosis, ^{18}F-FDG PET/CT provides a non-invasive method to assess tumor metabolism and capture tumor heterogeneity from a macro perspective. Prognostic biomarkers derived from PET/CT, such as metabolic tumor volume, and total lesion glycolysis, have shown promising results [14–17]. However, these metabolic parameters primarily rely on simple features, such as standard uptake value (SUV) and lesion volume, without fully capturing the complexity of tumor characteristics.

In addition to PET/CT-based quantitative parameters specifically designed for lymphoma, many other methods have been explored for prognosis and diagnosis. A common approach is radiomics feature extraction from regions of interest (ROIs) via manual or semi-automatic segmentation [12,13,18–20]. However, delineating ROIs is time-consuming, and general-purpose radiomics features may fail to provide specific and sufficient feature representation of lymphoma lesion regions. Recently, the use of deep learning techniques for PET/CT image analysis has grown [21–23]. However, these approaches of cropping fixed organ regions limit its application in diseases like lymphoma, where lesions are scattered in various anatomical regions, with heterogeneous lesion numbers, locations, and sizes across patients. Some researchers have attempted to apply the aforementioned traditional methods to lymphoma [24–26]. Meanwhile, Liu et al. [27] employed multi-task learning by integrating segmentation and prognosis prediction within a unified framework. However, they resampled whole-body PET/CT to a fixed-size input, which cannot adapt to varying lesion locations and sizes. Graph-based approaches [28,29] provide a novel perspective for analyzing DLBCL. In these methods, each lesion is treated as a node of a graph, with radiomics features extracted as the attributes of each node. Graph attention mechanisms are then employed to model the interactions between lesions. However, these methods mainly focus on local lesion features while neglecting crucial spatial context, including lesion distribution and its anatomical correlation with surrounding organs.

For DLBCL, PET/CT-based prognostic assessment still presents several challenges: **Heterogeneity in lesion number and location**. Unlike solid tumors, where the ROI of a specific lesion/organ can be cropped for analysis, DLBCL lesions are dispersed across different regions and their size varies, making it difficult to define a standardized ROI. The inherent characteristic of lymphoma necessitates a more flexible approach to analysis. **Insufficient representation of lesion regions**. Radiomics features lack specificity in representing lesion regions, while whole-image-based deep learning methods usually focus on dominant lesions and may overlook small metastatic lesions. The strategy of extracting features from each lesion region separately and then aggregating them (usually through average pooling) also requires further optimization. **Lack of modeling the anatomical context of lesions**. For example, extranodal involvement in organs like the liver and marrow in DLBCL is typically associated with poorer prognosis [30], but current methods cannot explicitly model the anatomical context of lesions.

To address the aforementioned challenges, this paper proposes a model for **L**ymphoma using **A**natomical context to lesions and **M**ulti-lesion aggregation for **P**rognosis (**LAMP**). This approach aims to enhance the accuracy of prognosis by effectively extracting and integrating features from multiple lesions while considering the anatomy of the lesions. The contributions of this work are as follows:

- We use attention-based multi-lesion aggregation to solve the issue of heterogeneous lesion distribution. The learned attention score is interpretable and can show how each lesion location contributes to prognosis prediction.
- We characterize the interaction between lesions and anatomical structures to enhance the anatomical context information of each lesion, thereby improving the representation of the lesion in its spatial and functional environment, which contributes to more accurate prognosis prediction.
- Extensive validation demonstrates that our approach outperforms existing prognosis models. Both quantitative evaluations and qualitative analyses confirmed the advantages of the proposed method.

2 Method

The overview of the proposed framework is presented in Fig. 1. For PET/CT-based prognostic prediction in lymphoma, we implement a comprehensive computational pipeline comprising three stages. Initially, a multimodal segmentation network is employed to achieve voxel-level lesion delineation, meanwhile generating high-dimensional feature maps that maintain spatial correspondence with the original imaging data. Subsequently, whole-body anatomical segmentation is obtained on CT volumes to derive organ masks, enabling lesion-anatomy context fusion through an attention-based feature interaction mechanism. Finally, multi-lesion aggregation is implemented to integrate features across distributed lesion regions, incorporating both local lesion characteristics and contextual organ information for precise prognostic stratification. We denote the N training data as $\mathbb{D} = \{X^i_{\mathrm{CT}}, X^i_{\mathrm{PET}}, C^i, S^i\}_{i=1}^N$, where X^i_{CT}, and X^i_{PET} denote the input CT image and PET image, C^i and S^i denote censorship status and survival time. Now we will delve into further details.

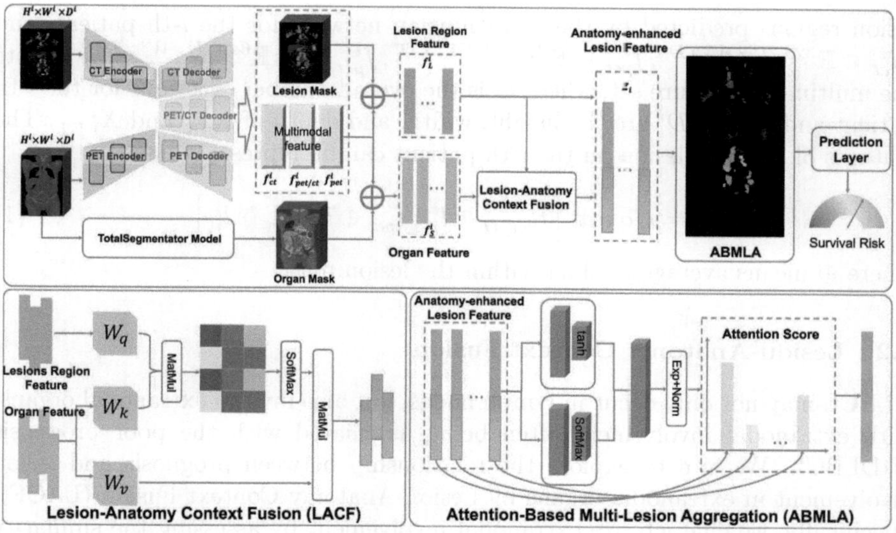

Fig. 1. Overview architecture of our proposed LAMP model.

2.1 Lesion Segmentation and Feature Extraction

Based on the nnUNet [31] framework, we designed a multimodal segmentation network that integrates PET and CT images. This network consists of two encoders with shared weights, each inputting either X_{CT} or X_{PET}. Following the encoders, three distinct decoders are introduced to process the feature representations from X_{CT}, X_{PET}, and the fusion of X_{CT} and X_{PET} features. Each decoder serves a distinct purpose: one extracts anatomical information from CT, another captures metabolic information from PET, and the third enables a deep interaction between the PET and CT information. Each decoder predicts the lesion masks separately. Finally, a pixel-wise weight map is computed to fuse the three probability maps adaptively and generate the final lesion mask. This design combines PET and CT in both feature and decision levels. On our DLBCL dataset of 229 cases under five-fold cross-validation, this lesion segmentation network achieved a Dice coefficient of 0.9199, a false positive volume (FPV) of 3.4940, and a false negative volume (FNV) of 2.0300. For lesion-level detection, it achieved a precision of 0.8931 and recall of 0.8991, significantly outperforming the plain nnUNet. It provides a solid foundation for our subsequent analysis.

To accurately differentiate lesions from normal tissues in PET/CT, a segmentation network needs to comprehensively encode each lesion's local appearance and global anatomical information. Therefore, the feature map of the segmentation network should also be valuable for downstream tasks, such as prognosis. Based on this assumption, we extracted feature maps with the same spatial dimensions as the input images from the final block of each decoder branch in the aforementioned segmentation network, and then pool lesion features from them. Formally, let $\mathbb{L}^i = \{l^i_1, l^i_2, \cdots, l^i_{i_J}\} \in \mathbb{R}^{H^i \times W^i \times D^i}$ represents the lesion regions predicted by the segmentation network for the i-th patient, and $\{f_{ct}^i \in \mathbb{R}^{32\times H^i\times W^i\times D^i}, f_{pet}^i \in \mathbb{R}^{32\times H^i\times W^i\times D^i}, f_{pet/ct}^i \in \mathbb{R}^{64\times H^i\times W^i\times D^i}\}$ denote the multimodal feature set, where i_J is the overall number of lesions for the i-th patient and H^i,W^i,D^i are the height, width, and depth of X_{CT}^i and X_{PET}^i. The features of the j-th lesion on the i-th patient can be represented as:

$$f_{i_j}^i = Concat\left[f_{pet/ct}^i \oplus l_{i_j}^i, f_{pet}^i \oplus l_{i_j}^i, f_{ct}^i \oplus l_{i_j}^i\right] \tag{1}$$

where $\oplus$ means average pooling within the lesion mask.

2.2 Lesion-Anatomy Context Fusion

DLBCL may not only occur in lymph nodes, but also involve extranodal organs, with extranodal involvement often being associated with the poor prognosis of DLBCL. We aim to explore the relationship between prognosis and lesion involvement in extranodal organs by Lesion-Anatomy Context Fusion (LACF). Specifically, we characterize extranodal involvement by assessing the similarity between the features of the lesion regions and the features of the organ regions. This allows us to capture the degree of organ involvement of each lesion. Then, we weight and aggregate these organ region features to enrich the characterization and representation of each lesion region. This approach not only enhances the lesion feature's anatomical and metabolic profiles but also incorporates crucial information regarding the organ context, providing a more comprehensive understanding of the disease's progression and prognosis.

First, we use TotalSegmentator [32] to segment major anatomical structures $\mathbb{S}^i = \{s_1^i, s_2^i, \cdots, s_K^i\}$ on the CT images (The regions belonging to the same anatomical structure on the same side will be merged, reducing the number of labels from 117 to 71.). Following the same feature extraction method as described for the lesion regions, we extract the feature representation for each anatomical organ.

$$f_{s_k}^i = Concat\left[f_{pet/ct}^i \oplus s_k^i, f_{pet}^i \oplus s_k^i, f_{ct}^i \oplus s_k^i\right] \tag{2}$$

where $f_{s_k}^i$ stands for the features extracted from the s_k-th anatomical organs of the i-th patient. Next, we employ the attention mechanism to perform the association analysis between the lesion region features and the anatomical organs features. We use the features of the lesion region $f_L^i \in \mathbb{R}^{i_J\times 128}$ as query Q, the anatomical organs' feature representation $f_S^i \in \mathbb{R}^{K\times 128}$ as key K and value V:

$$\begin{aligned} h_L^i &= \mathrm{Attention}(Q, K, V) \\ &= \mathrm{Attention}\left(f_L^i W_q, f_S^i W_k, f_S^i W_v\right) \\ &= \mathrm{softmax}\left(\frac{(f_L^i W_q)(f_S^i W_k)^T}{\sqrt{d_k}}\right) f_S^i W_v. \end{aligned} \tag{3}$$

The attention mechanism calculates the similarity between the query and key, allowing the model to focus on the most relevant anatomical features that interact with the lesion. This enables the model to establish a more precise correlation between the lesion's characteristics and the surrounding anatomical context, thus improving the overall lesion representation and enhancing the prediction of prognosis. After that, $f_{i_j}^i$ and the i_j-th row from h_L^i are concatenated to get the anatomy-enhanced lesion feature

$$z_{i_j}^i = Concat\left[f_{i_j}^i, \left(h_L^i\right)_{i_j}\right]. \tag{4}$$

2.3 Attention-Based Multi-lesion Aggregation

DLBCL presents unique challenges due to its heterogeneous anatomical distribution and variable number of lesions across patients. Conventional methodologies employed fixed-size whole-image input paradigms or naive average pooling of all lesion features. They are unable to adaptively process lesion regions with varying anatomical locations and pathological significance. To overcome these limitations, we propose a novel solution by reformulating the aggregation of the lesion feature as an attention-based multi-lesion aggregation (ABMLA) problem using gated attention mechanisms.

$$P^i = \sum_{j=1}^{i_J} a_{i_j}^i z_{i_j}^i, \quad a_{i_j}^i = \frac{exp\left\{w^T\left(tanh(V{z_{i_j}^i}^T) \oplus sigm(U{z_{i_j}^i}^T)\right)\right\}}{\sum_{j=1}^{i_J} exp\left\{w^T\left(tanh(V{z_{i_j}^i}^T) \oplus sigm(U{z_{i_j}^i}^T)\right)\right\}} \tag{5}$$

where $a_{i_j}^i$ is the attention score learned by self-attention layers with parameters U,V,and w. Through that, the relative contribution of individual lesion regions to the predictive outcome could be assessed. P^i will be input into the linear layer to obtain the final survival risk score. NLL (negative log-likelihood) survival loss is leveraged to train the model [33].

3 Experiments

Data Collection. To evaluate the performance of our method, 229 pretreatment whole-body PET/CT image pairs from patients with DLBCL are collected, with prognostic information available for 195 patients. This study focuses on two main outcomes: progression-free survival (PFS) and overall survival (OS). Compared to OS, PFS is more critical, as it is of primary concern to clinicians due to its direct relevance to disease progression in patients. Multiple preprocessing steps are applied to the PET/CT image pairs for model training to ensure proper alignment and standardization, including standardized uptake values conversion, registering the CT images to the PET images, and z-score normalization.

Implementation and Evaluation. Since the major focus of this paper is prognosis instead of lesion segmentation, we will mainly introduce the prognosis algorithm in this section. We used the Adam optimizer with an initial learning rate of 0.0001 to train our model. We utilized a batch size of 1 with a gradient accumulation step of 32 due to nonuniform lesion numbers across patients. Two metrics, the time-dependent area under the receiver operating characteristic curve (Time-AUC) and the concordance index (C-index), are used to evaluate the prognosis performance. To ensure data integrity and prevent leakage, we applied the same five-fold cross-validation split for both the segmentation network and the prognosis network training. To extract feature maps with the same spatial dimensions as the input images, we implemented patch-wise training and inference during the segmentation network training.

Table 1. Comparison results using C-index and Time-AUC metrics under five-fold cross-validation.

Methods	PFS			OS		
	C-index(↑)	1y AUC(↑)	2y AUC(↑)	C-index(↑)	1y AUC(↑)	2y AUC(↑)
IPI	$0.714_{0.007}$	$0.755_{0.005}$	$0.758_{0.004}$	$0.749_{0.014}$	$0.722_{0.025}$	$0.785_{0.016}$
Ann Arbor	$0.632_{0.001}$	$0.667_{0.000}$	$0.669_{0.007}$	$0.667_{0.003}$	$0.665_{0.016}$	$0.680_{0.005}$
Meta. Param.†	$0.705_{0.006}$	$0.734_{0.012}$	$0.736_{0.007}$	$0.766_{0.018}$	$0.791_{0.046}$	$0.795_{0.018}$
Meta. Param.*	$0.724_{0.004}$	$0.766_{0.003}$	$0.756_{0.008}$	$0.787_{0.006}$	$\mathbf{0.825}_{0.021}$	$0.830_{0.012}$
Radiomics [18]	$0.754_{0.006}$	$0.800_{0.001}$	$0.792_{0.011}$	$0.771_{0.014}$	$0.750_{0.032}$	$0.806_{0.018}$
Guo2021 [25]	$0.778_{0.004}$	$0.816_{0.002}$	$0.801_{0.008}$	$0.801_{0.002}$	$0.790_{0.007}$	$0.822_{0.003}$
Liu2022 [27]	$0.719_{0.001}$	$0.718_{0.011}$	$0.708_{0.002}$	$0.797_{0.003}$	$0.807_{0.009}$	$0.807_{0.002}$
Thiery2025 [29]	$0.758_{0.003}$	$0.839_{0.010}$	$0.782_{0.005}$	$0.809_{0.001}$	$0.820_{0.001}$	$0.843_{0.002}$
LAMP (ours)	$\mathbf{0.791}_{0.002}$	$\mathbf{0.844}_{0.002}$	$\mathbf{0.823}_{0.004}$	$\mathbf{0.828}_{0.001}$	$0.782_{0.013}$	$\mathbf{0.863}_{0.004}$

* 1y AUC and 2y AUC refer to the Time-AUC at the 1-year and 2-year time points. The metabolic parameters are computed based on manual † and automatic segmentation*. The larger numbers within each cell in the table represent the mean values, while the smaller numbers denote the standard deviations.

Table 2. Ablation studies under five-fold cross-validation.

Methods	**PFS**			**OS**		
	C-index(↑)	1y AUC(↑)	2y AUC(↑)	C-index(↑)	1y AUC(↑)	2y AUC(↑)
Baseline	$0.750_{0.003}$	$0.816_{0.004}$	$0.775_{0.003}$	$0.794_{0.008}$	$\mathbf{0.844}_{\mathbf{0.015}}$	$0.822_{0.011}$
w/ ABMLA	$0.771_{0.002}$	$0.802_{0.005}$	$0.777_{0.003}$	$0.800_{0.003}$	$0.724_{0.016}$	$0.827_{0.006}$
w/ LACF	$0.765_{0.001}$	$0.841_{0.002}$	$0.796_{0.001}$	$0.813_{0.008}$	$0.795_{0.028}$	$0.842_{0.011}$
Full model	$\mathbf{0.791}_{0.002}$	$\mathbf{0.844}_{0.002}$	$\mathbf{0.823}_{0.004}$	$\mathbf{0.828}_{0.001}$	$0.782_{0.013}$	$\mathbf{0.863}_{0.004}$

$*$ The larger numbers within each cell in the table represent the mean values, while the smaller numbers denote the standard deviations.

Method Comparison. To evaluate the performance of the proposed model, we compared it against four categories of baselines: (1) Widely-used clinical prognosis indices, including IPI and Ann Arbor staging (Ann Arbor) [10,11]; (2) Metabolic parameters (Meta. Param.) [34] extracted from segmentation masks, including those derived from both manual and automatic segmentation methods; (3) End-to-end whole image-based models, such as Guo2021 [25], Liu2022 [27]; and (4) Graph Convolutional Networks (GCNs) [29]. As shown in Table 1, the performance of the proposed model outperforms the currently used prognostic clinical indices and metabolic parameters. The C-index for the proposed model shows an improvement of over 9% and 5% for PFS and OS, respectively, compared to the best-performing model. This demonstrates the superior predictive power of our model in both endpoints. For Guo2021 [25] and Liu2022 [27], the constraint of a fixed input size may hinder model performance by limiting its ability to capture variations in lesion size and distribution. Moreover, jointly performing segmentation and prognosis prediction within a single model [27] introduces challenges in task balancing, potentially leading to suboptimal learning of both objectives. While Thiery2025 [29] incorporate inter-lesional propagation relationships, GCN lacks the integration of perilesional anatomical context, which may contribute to unfavorable performance. The last, the Kaplan-Meier curves presented in Fig. 2. A and 2. B demonstrates a statistically significant distinction ($p \leq 0.05$) between high-risk and low-risk patients. This finding further validates the model's discriminative capacity and underscores its potential clinical translational value.

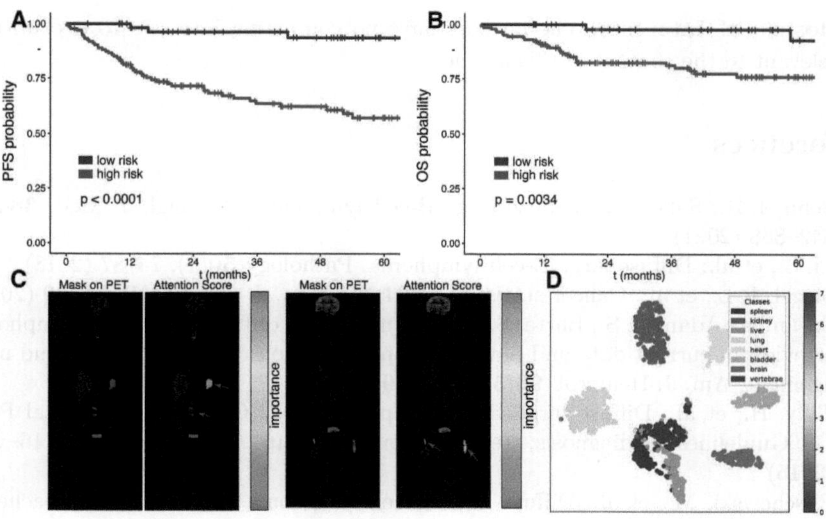

Fig. 2. (A) and (B) are the Kaplan-Meier (KM) curves for PFS and OS prognosis. (C) presents mapping the attention scores $a_{i_j}^i$ over the lesion regions. (D) shows the features of key anatomical organs by t-SNE.

Ablation Study. Table 2 highlights two key advantages of our proposed model: ABMLA and LACF, both jointly contribute to performance improvement compared with baselines (all features are aggregated by averaging). Additionally, in Fig. 2. C, we visualized the attention scores learned through ABMLA. A higher attention score is assigned to the lesion regions within the bone marrow and spleen, as involvements of these organs typically indicate a poor prognosis. The t-SNE scatter map in Fig. 2. D indicates that features extracted from the segmentation network can differentiate key organs, which is beneficial for the prognosis task via our LACF strategy.

4 Conclusion

For malignancies like lymphoma, which have multiple, variable lesion regions and challenges in feature representation and aggregation, we propose a PET/CT-based model LAMP that combines attention-based multi-lesion aggregation with contextual interaction with surrounding anatomical structures. Our approach is comprehensively evaluated on PFS and OS prognosis using multiple metrics. The results demonstrate the superiority of our model both quantitatively and qualitatively, highlighting its potential to serve as an auxiliary tool to assist clinicians in treatment planning.

Acknowledgments. This work was supported by DAMO Academy through DAMO Academy Research Intern Program.

Disclosure of Interests. The authors have no competing interests to declare that are relevant to the content of this article.

References

1. Sehn, L.H., Salles, G.: Diffuse large B-cell lymphoma. N. Engl. J. Med. **384**(9), 842–858 (2021)
2. Li, S., et al.: Diffuse large B-cell lymphoma. Pathology **50**(1), 74–87 (2018)
3. Siegel, R.L., et al.: Cancer statistics, 2024. CA Can. J. Clin. **74**(1), 12–49 (2024)
4. Susanibar-Adaniya, S., Barta, S.K.: 2021 update on diffuse large B cell lymphoma: a review of current data and potential applications on risk stratification and management. Am. J. Hematol. **96**(5), 617–629 (2021)
5. Tilly, H., et al.: Diffuse large B-cell lymphoma (DLBCL): ESMO Clinical Practice Guidelines for diagnosis, treatment and follow-up. Ann. Oncol. **26**, v116–v125 (2015)
6. Roschewski, M., et al.: Diffuse large B-cell lymphoma—treatment approaches in the molecular era. Nat. Rev. Clin. Oncol. **11**(1), 12–23 (2014)
7. Eertink, J.J., et al.: Optimal timing and criteria of interim PET in DLBCL: a comparative study of 1692 patients. Blood Adv. **5**(9), 2375–2384 (2021)
8. Cabero Martinez, A., et al.: Early identification of very high risk patients with diffuse large B cell lymphoma (DLBCL) by PET-CT. Blood **142**(Supplement 1), 4511–4511 (2023)
9. Wenzl, K., et al.: Multiomic analysis identifies a high-risk signature that predicts early clinical failure in DLBCL. Blood Cancer J. **14**(1), 100 (2024)

10. Shipp, M.A.: International Non-Hodgkin's lymphoma prognostic factors project. A predictive model for aggressive non-Hodgkin's lymphoma. New Engl. J. Med. **329**, 987–994 (1993)
11. Ruppert, A.S., et al.: International prognostic indices in diffuse large B-cell lymphoma: a comparison of IPI, R-IPI, and NCCN-IPI. Blood **135**(23), 2041–2048 (2020)
12. Wang, R., Zheng, G.: CyCMIS: Cycle-consistent cross-domain Medical Image Segmentation via diverse image augmentation. Med. Image Anal. (76), 102328 (2022)
13. Wang, R., et al.: Few-shot medical image segmentation regularized with self-reference and contrastive learning. In: International Conference on Medical Image Computing and Computer-Assisted Intervention, pp. 514–523 (2022)
14. Ceriani, L., et al.: Metabolic heterogeneity on baseline 18FDG-PET/CT scan is a predictor of outcome in primary mediastinal B-cell lymphoma. Blood **132**(2), 179–186 (2018)
15. Sasanelli, M., et al.: Pretherapy metabolic tumour volume is an independent predictor of outcome in patients with diffuse large B-cell lymphoma. Eur. J. Nucl. Med. Mol. Imaging **41**, 20172022 (2014)
16. Mikhaeel, N.G., et al.: Combination of baseline metabolic tumour volume and early response on PET/CT improves progression-free survival prediction in DLBCL. Eur. J. Nucl. Med. Mol. Imaging **43**, 12091219 (2016)
17. Esfahani, S.A., et al.: Baseline total lesion glycolysis measured with 18F-FDG PET/CT as a predictor of progression-free survival in diffuse large B-cell lymphoma: a pilot study. Am. J. Nucl. Med. Mol. Imaging **3**(3), 272 (2013)
18. Carlier, T., et al.: Prognostic value of 18F-FDG PET Radiomics features at baseline in PET-Guided consolidation strategy in diffuse large B-Cell lymphoma: a machine-learning analysis from the GAINED study. J. Nucl. Med. **65**(1), 156–162 (2024)
19. Eertink, J.J., et al.: 18F-FDG PET/CT baseline radiomics features improve the prediction of treatment outcome in diffuse large B-cell lymphoma patients. Blood **136**, 27–28 (2020)
20. Eertink, J.J., et al.: Baseline PET radiomics outperforms the IPI risk score for prediction of outcome in diffuse large B-cell lymphoma. Blood **141**(25), 3055–3064 (2023)
21. Zhong, Y., et al.: PET/CT based cross-modal deep learning signature to predict occult nodal metastasis in lung cancer. Nat. Commun. **14**(1), 7513 (2023)
22. Meng, M., et al.: DeepMTS: deep multi-task learning for survival prediction in patients with advanced nasopharyngeal carcinoma using pretreatment PET/CT. IEEE J. Biomed. Health Inform. **26**(9), 4497–4507 (2022)
23. Cheng, N.M., et al.: Deep learning for fully automated prediction of overall survival in patients with oropharyngeal cancer using FDG-PET imaging. Clin. Cancer Res. **27**(14), 3948–3959 (2021)
24. Häggström, I., et al.: Deep learning for [18F] fluorodeoxyglucose-PET-CT classification in patients with lymphoma: a dual-centre retrospective analysis. Lancet Digital Health **6**(2), e114–e125 (2024)
25. Guo, R., et al.: Weakly supervised deep learning for determining the prognostic value of 18 F-FDG PET/CT in extranodal natural killer/T cell lymphoma, nasal type. Eur. J. Nuclear Med. Mol. Imaging 1–11 (2021)
26. Yuan, C., et al.: Multimodal deep learning model on interim [18F] FDG PET/CT for predicting primary treatment failure in diffuse large B-cell lymphoma. Eur. Radiol. **33**(1), 77–88 (2023)

27. Liu, P., et al.: Joint lymphoma lesion segmentation and prognosis prediction from baseline FDG-PET images via multitask convolutional neural networks. IEEE Access **10**, 81612–81623 (2022)
28. Lv, W., et al.: Functional-structural sub-region graph convolutional network (FSGCN): application to the prognosis of head and neck cancer with PET/CT imaging. Comput. Methods Programs Biomed. **230**, 107341 (2023)
29. Thiery, O., et al.: PET-based lesion graphs meet clinical data: An interpretable cross-attention framework for DLBCL treatment response prediction. Comput. Med. Imaging Graph. **120**, 102481 (2025)
30. Yang, S., et al.: A matched comparison of Pola-R-CHP Vs R-CHOP for DLBCL with extranodal involvement. Blood **144**, 6478 (2024)
31. Isensee, F., et al.: nnU-Net: a self-configuring method for deep learning-based biomedical image segmentation. Nat. Methods **18**(2), 203–211 (2021)
32. Wasserthal, J., et al.: TotalSegmentator: robust segmentation of 104 anatomic structures in CT images. Radiol. Artif. Intell. **5**(5), e230024 (2023)
33. Chen, R.J., et al.: Multimodal co-attention transformer for survival prediction in gigapixel whole slide images. In: Proceedings of the IEEE/CVF International Conference on Computer Vision, vol. 2021, 1–13 (2021)
34. Frood, R., et al.: Baseline PET/CT imaging parameters for prediction of treatment outcome in Hodgkin and diffuse large B cell lymphoma: a systematic review. Eur. J. Nuclear Med. Mol. Imaging **48**, 31983220 (2021)
35. Huang, W., et al.: LIDIA: precise liver tumor diagnosis on multi-phase contrast-enhanced CT via iterative fusion and asymmetric contrastive learning. In: International Conference on Medical Image Computing and Computer-Assisted Intervention, pp. 394–404 (2024)
36. Li, C., et al.: Improved esophageal varices assessment from non-contrast CT scans. In: International Conference on Medical Image Computing and Computer-Assisted Intervention, pp. 349–359 (2024)

MambaMER: Adaptive EEG-Guided Multimodal Emotion Recognition with Mamba

Xiangle Ping, Wenhui Huang(✉), and Yuanjie Zheng

School of Information Science and Engineering, Shandong Normal University, Jinan, China

whhuang.sdu@gmail.com

Abstract. In recent years, multimodal emotion recognition has gradually become a research hotspot. Although existing methods have achieved significant results by integrating information from different modalities, irrelevant or conflicting emotional information across modalities often limits performance improvement. Inspired by Mamba's ability to effectively filter irrelevant information and model long-range dependencies with linear complexity, we propose a new paradigm for EEG-guided adaptive multimodal emotion recognition with Mamba. This paradigm effectively addresses the interference caused by cross-modal information conflicts, enhancing the performance of multimodal emotion recognition. Firstly, to alleviate the interference caused by conflicts between different modalities, we design a multi-scale EEG-guided conflict suppression module. Guided by multi-scale EEG features, this module uses a selective cross state space model to suppress irrelevant information and conflicts in eye movement features, thereby obtaining enhanced eye movement features. Secondly, to deeply integrate the complementary features between the EEG modality and the enhanced eye movement modality, we propose a novel cross-modal fusion mechanism, consisting of Mutual-Cross-Mamba and Merge-Mamba, which effectively captures long-range dependencies in the fused features, thereby enhancing the integration and utilization of cross-modal information. Experimental results on the SEED, SEED-IV, and SEED-V datasets demonstrate that our method significantly surpasses current state-of-the-art methods.

Keywords: EEG · Eye Movement · Multimodal Emotion Recognition · Mamba

1 Introduction

Multimodal emotion recognition (MER) has gained attention for its ability to decode complex human emotions [9]. By integrating physiological signals and behavioral data, multimodal methods improve emotion recognition accuracy. Electroencephalography (EEG) captures dynamic changes linked to emotional

J. C. Gee et al. (Eds.): MICCAI 2025, LNCS 15960, pp. 333–343, 2026.
https://doi.org/10.1007/978-3-032-04927-8_32

states, making it a key modality [12,28]. Eye movement data offers valuable behavioral and cognitive cues, especially in response to external stimuli [4,18, 27]. Combining EEG with eye movement data provides both neurological and behavioral insights, leading to a more comprehensive understanding of emotional states [5,6,21].

In recent years, with the continuous advancement of multimodal emotion recognition research, many methods have been proposed that integrate information from different modalities and explore their complementarity and individual characteristics to achieve more accurate emotion recognition [1,24,25]. For example, Fu et al. [5] designed a multimodal feature fusion neural network to capture complementary eye movement and EEG information, while Wang et al. [22] applied an attention mechanism to integrate EEG and eye movement signals for emotion recognition. However, these methods often fail to fully account for irrelevant or conflicting emotional information between modalities. Additionally, using attention mechanisms for cross-modal fusion significantly increases computational complexity.

To address the aforementioned issues, this paper proposes a Mamba-based adaptive EEG-guided multimodal emotion recognition model that effectively suppresses emotion-irrelevant interference and conflicting information in eye movement signals, thereby enhancing the performance of multimodal emotion recognition. The main contributions of this work are as follows (Fig. 1):

1. We propose an adaptive EEG-guided multimodal emotion recognition model based on Mamba, which effectively mitigates interference from cross-modal conflicts and enhances recognition accuracy. To the best of our knowledge, this is the first study to apply the Mamba model to resolve cross-modal conflicts in emotion recognition.
2. We designed a multi-scale EEG-guided conflict suppression module, which learns to suppress irrelevant information and conflicts in the eye movement features under the guidance of multi-scale EEG features, thereby enhancing the eye movement features.
3. We propose a novel cross-modal fusion mechanism consisting of Mutual-Cross-Mamba and Merge-Mamba, which aims to realize the deep interaction of multimodal features and effectively capture the remote dependencies in the fused features.

2 Method

2.1 Preliminaries

The state-space model (SSM) [8] consists of a state equation describing the system's internal dynamics and an observation equation linking the system state to the observed values. For an input $x(t) \in \mathbb{R}$ and a hidden state $y(t) \in \mathbb{R}$, the system is represented by a linear ordinary differential equation (ODE) as follows:

$$h'(t) = \mathbf{A}h(t) + \mathbf{B}x(t), \quad y(t) = \mathbf{C}h(t). \tag{1}$$

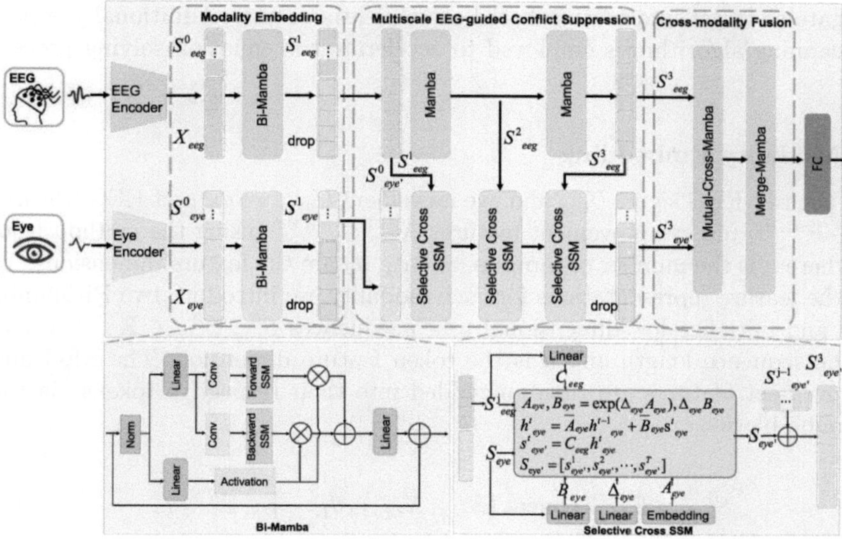

Fig. 1. Our framework consists of three modules: modality embedding, multi-scale EEG-guided conflict suppression, and cross-modal fusion. The modality embedding module utilizes Bi-Mamba to extract unified modality features from EEG and eye movement data. The multi-scale EEG-guided conflict suppression module extracts multi-scale EEG features using two Mamba blocks and uses these features to guide the Selective Cross-SSM in generating enhanced eye movement feature representations. The cross-modal fusion module integrates multimodal information through Mutual-Cross-Mamba and Merge-Mamba, and outputs the final emotion classification result via a fully connected layer.

Here, $\mathbf{A} \in \mathbb{R}^{N \times N}$ is the state matrix, $\mathbf{B} \in \mathbb{R}^{N \times 1}$ is the input matrix, and $\mathbf{C} \in \mathbb{R}^{1 \times N}$ is the output matrix. Mamba integrates continuous systems with deep learning algorithms by applying zero-order hold (ZOH) [19] to transform the continuous parameters $\mathbf{A}$ and $\mathbf{B}$ into their discrete counterparts $\bar{\mathbf{A}}$ and $\bar{\mathbf{B}}$, incorporating the time scaling parameter Δ. The conversion formula is as follows:

$$\overline{\mathbf{A}} = \exp(\Delta\mathbf{A}), \quad \overline{\mathbf{B}} = (\Delta\mathbf{A})^{-1}(\exp(\Delta\mathbf{A}) - \mathbf{I}) \cdot \Delta\mathbf{B}. \tag{2}$$

In practice, following the approach of [7], we approximate $\overline{\mathbf{B}}$ using the first-order Taylor series as follows:

$$\overline{\mathbf{B}} = (e^{\Delta\mathbf{A}} - \mathbf{I})\mathbf{A}^{-1}\mathbf{B} \approx (\Delta\mathbf{A})(\Delta\mathbf{A})^{-1}\Delta\mathbf{B} = \Delta\mathbf{B}. \tag{3}$$

This approach converts the continuous ODE into a discrete form. The expression is as follows:

$$h_t = \overline{\mathbf{A}}h_{t-1} + \overline{\mathbf{B}}x_t, \quad y_t = \mathbf{C}h_t. \tag{4}$$

Building on the aforementioned discrete state-space equations, mamba introduces data dependencies into the model parameters, enabling it to selectively propagate or forget information based on sequential inputs. Additionally, a parallel scanning algorithm is employed to accelerate the equation solving process [7].

2.2 Modality Embedding

Given the raw EEG signal U_{eeg} and eye movement U_{eye}, we extract EEG features $X_{\text{eeg}} \in \mathbb{R}^{n\times d_1}$ and eye movement features $X_{\text{eye}} \in \mathbb{R}^{n\times d_2}$ using the method from [11], where n is the number of samples, and d_1, d_2 are the feature dimensions. To unify the feature representations for each modality, we introduce two Bi-Mamba blocks and initialize low-dimensional token sequences $S^0_{\text{eeg}}, S^0_{\text{eye}} \in \mathbb{R}^{T\times d}$, where T is the sequence length and d is the token feature dimension. The EEG and eye movement features are then embedded into their respective tokens via the Bi-Mamba blocks:

$$S^1_{eeg} = \mathrm{E}^0_{eeg}\left(\text{concat}\left(S^0_{eeg}, X_{eeg}\right), \theta_{E^0_{eeg}}\right) \in \mathbb{R}^{T\times d}, \tag{5}$$

$$S^1_{eye} = \mathrm{E}^0_{\text{eye}}\left(\text{concat}\left(S^0_{eye}, X_{eye}\right), \theta_{E^0_{eye}}\right) \in \mathbb{R}^{T\times d}. \tag{6}$$

Here, $\mathrm{E}^0_{\text{eeg}}$ and $\mathrm{E}^0_{\text{eye}}$ denote the Bi-Mamba layers with parameters $\theta_{\mathrm{E}^0_{\text{eeg}}}$ and $\theta_{\mathrm{E}^0_{\text{eye}}}$, respectively. The operation concat$(\cdot)$ represents concatenation. Bi-Mamba embeds the modality features into tokens, generating the unified feature representations S^1_{eeg} and S^1_{eye}.

2.3 Multi-Scale EEG-Guided Conflict Suppression

After achieving a unified representation transformation for the features of each modality, our goal is to filter out irrelevant or conflicting interference information between modalities. Specifically, we define the S^1_{eeg} as low-scale EEG features and further extract medium-scale and high-scale EEG features (i.e., S^2_{eeg} and S^3_{eeg}) by introducing two Mamba layers:

$$S^i_{eeg} = \mathrm{E}^i_{eeg}\left(S^{i-1}_{eeg}, \theta_{E^i_{eeg}}\right) \in \mathbb{R}^{T\times d}, \tag{7}$$

where $i \in \{2, 3\}$ represents EEG features at different scales, with $\mathrm{E}^i_{\text{eeg}}$ and $\theta_{E^i_{\text{eeg}}}$ denoting the i-th Mamba layer and its parameters.

Next, we initialize the enhanced eye movement feature $S^0_{eye'} \in \mathbb{R}^{T\times d}$ and update it by calculating the relationship between EEG and eye movement features using selective cross-state-space computation. The process is as follows:

$$\overline{\mathbf{A}}_{\text{eye}}, \overline{\mathbf{B}}_{eye} = \exp(\Delta_{\text{eye}}\mathbf{A}_{\text{eye}}), \Delta_{\text{eye}}\mathbf{B}_{\text{eye}}, \tag{8}$$

$$h^t{}_{eye} = \overline{\mathbf{A}}_{eye} h^{t-1}{}_{eye} + \overline{\mathbf{B}}_{eye} s^t{}_{eye}, \tag{9}$$

$$s^t{}_{eye'} = \mathbf{C}^i_{eeg} h^t{}_{eye}, \tag{10}$$

$$S_{eye'} = \left[s^1{}_{eye'}, s^2{}_{eye'}, ..., s^T{}_{eye'}\right]. \tag{11}$$

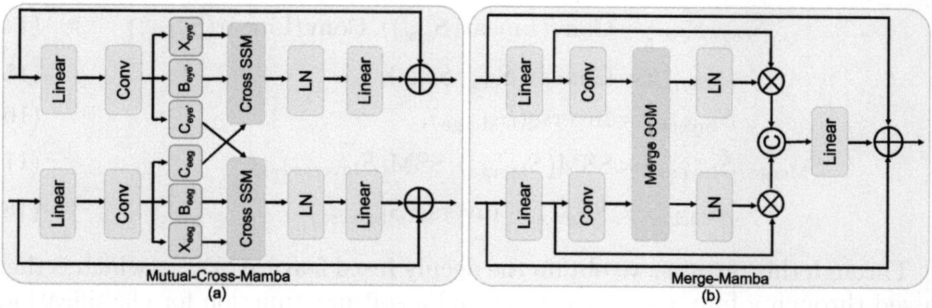

Fig. 2. Cross-modality fusion. (a) Mutual-Cross-Mamba and (b) Merge-Mamba.

Here, $\mathbf{C}_{\text{eeg}}$ is computed from the EEG features at different scales, and the enhanced eye movement modality feature $S^j_{eye'}$ can be updated as follows:

$$S^j_{eye'} = S^{j-1}_{\text{eye}'} + S_{\text{eye}'}, \tag{12}$$

where $j \in \{1,2,3\}$ and $S^j_{eye'} \in \mathbb{R}^{T\times d}$ represent the j-th selective cross-state-space model and its corresponding output.

2.4 Cross-Modal Fusion

The proposed cross-modal fusion mechanism consists of the Mutual-Cross-Mamba (MutCroMB) block and the Merge-Mamba (MerMB) block. MutCroMB enhances inter-modal interactions by applying the SSM in both directions across modalities. MerMB uses a selective scanning mechanism to fuse interacting features, producing the final fusion result. The fusion process is described below:

$$\begin{aligned} \widehat{S}_{\text{eeg}}, \widehat{S}_{\text{eye}'} &= \text{MutCroMB}\big(S^3_{\text{eeg}}, S^3_{\text{eye}'}\big), \\ S_{\text{fusion}} &= \text{MerMB}\big(\widehat{S}_{\text{eeg}}, \widehat{S}_{\text{eye}'}\big). \end{aligned} \tag{13}$$

Specifically, MutCroMB enhances modal feature interaction via a cross-multiplication mechanism. As shown in Fig. 2 (a), the two input features are processed by convolution and then passed into the cross SSM. According to Eq. (4), the matrix $\mathbf{C}$ decodes information from the hidden state h_t to generate the output y_t.

In MutCroMB, features from both modalities interact to generate cross-modal enhanced features. To capture long-range dependencies, the MerMB module integrates MutCroMB's output. As shown in Fig. 2 (b), $\widehat{S}_{\text{eeg}}$ and $\widehat{S}_{\text{eye}'}$ are first processed through linear layers and convolutional layers, and then enter the Merge SSM block. To ensure comprehensive information capture, we perform a reverse scan on the merged sequence S_{Merge}, producing S_{Inverse}, which undergoes further processing to yield $\hat{S}_{\text{Merge}}$ and $\hat{S}_{\text{Inverse}}$. Finally, the inverted output is reversed and subsequently added to the merged sequence. This process can be represented as:

$$\widetilde{S}_{\text{eeg}}, \widetilde{S}_{\text{eye}'} = \text{Conv}(\text{Linear}(\hat{S}_{\text{eeg}})), \text{Conv}(\text{Linear}(\hat{S}_{\text{eye}'})), \tag{14}$$

$$S_{\text{Merge}} = \text{Concat}(\widetilde{S}_{\text{eeg}}, \widetilde{S}_{\text{eye}'}), \tag{15}$$

$$S_{\text{Inverse}} = \text{Inverse}(S_{\text{Merge}}), \tag{16}$$

$$\hat{S}_{\text{Merge}}, \hat{S}_{\text{Inverse}} = \text{SSM}(S_{\text{Merge}}), \text{SSM}(S_{\text{Inverse}}), \tag{17}$$

$$S_{\text{fusion}} = \hat{S}_{\text{Merge}} + \text{Inverse}(\hat{S}_{\text{Inverse}}). \tag{18}$$

Through this process, we obtain the deeply fused feature S_{fusion}, which is then passed through a fully connected layer and a softmax function for classification. The model is optimized by minimizing the cross-entropy loss:

$$\mathcal{L} = -\frac{1}{N}\sum_{i}\sum_{c=1}^{M} y_{ic}\log\left(p_{ic}\right), \tag{19}$$

where M is the total number of categories, y_{ic} is a binary indicator (0 or 1) for sample i belonging to category c, and p_{ic} is the predicted probability of sample i belonging to category c.

3 Experiment

3.1 Datasets

We conducted experiments on three datasets: SEED [15], SEED-IV [26], and SEED-V [14]. The SEED dataset includes data from 12 Chinese subjects watching 15 clips covering happy, neutral, and sad emotions. The SEED-IV contains data from 15 subjects, involving four emotions (happy, sad, fear, and neutral). The SEED-V includes data from 16 subjects, covering five emotions (happy, sad, fear, neutral, and disgust).

3.2 Implementation Details

We followed the training/testing protocols from the original papers for each dataset, using the same data splits as previous studies [14,15,26]. The model was trained for 100 epochs with a batch size of 16, using the Adam optimizer and a learning rate of $1e^{-4}$. A dropout layer with a rate of 0.5 was added to prevent overfitting. All experiments were performed on an NVIDIA RTX 3090 with CUDA 11.8, using Python 3.10.13 and PyTorch 2.1.1.

3.3 Experiment Results

We evaluated the proposed method on the SEED, SEED-IV, and SEED-V datasets and conducted a comparative analysis with other multimodal methods. The experimental results in Table 1 show that, on the SEED dataset, the average recognition accuracy of our model reaches 96.82%, an improvement of 1.74% over the best-performing model; on the SEED-IV dataset, the average recognition accuracy is 94.93%, an increase of 3.33%; and on the SEED-V dataset, the

Table 1. Comparison of the average accuracy and standard deviation (%) of various multimodal methods across different datasets. (Bold indicates the best accuracy.)

Methods	Modality	SEED	SEED-IV	SEED-V
Fuzzy Intergral [14,17]	EEG and EM	87.60 ± 19.9	73.60 ± 16.7	73.20 ± 8.70
BDAE [16,26]	EEG and EM	91.01 ± 8.91	85.11 ± 11.79	79.70 ± 4.76
DCCA [14]	EEG and EM	94.60 ± 6.12	87.50 ± 9.20	85.30 ± 5.6
DCCA-FCP [23]	EEG and EM	95.08 ± 6.42	-	84.51 ± 5.11
CAN [20]	EEG and EM	94.03 ± 6.62	87.71 ± 9.74	–
ECO-FET [10]	EEG and EM	93.69 ± 8.22	87.76 ± 9.19	77.13 ± 4.16
ATAM [13]	EEG and EM	94.80 ± 7.50	91.60 ± 10.0	–
Ours	EEG and EM	**96.82 ± 5.20**	**94.93 ± 6.12**	**86.95 ± 5.89**

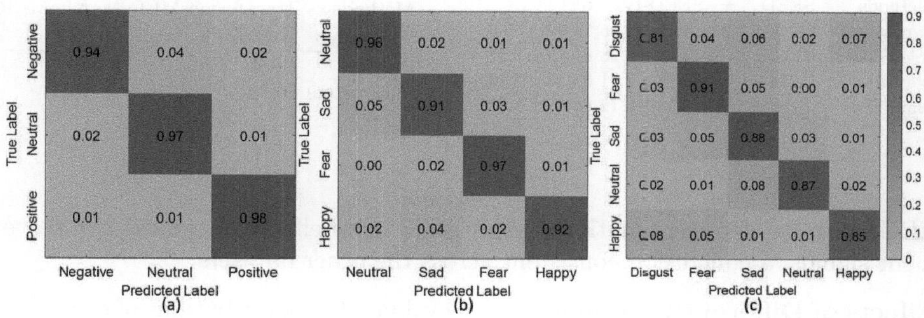

Fig. 3. Confusion matrix (a) SEED, (b) SEED-IV and (c) SEED-V.

average recognition accuracy is 86.95%, surpassing the best model by 1.65%. These results thoroughly demonstrate the model's outstanding performance in emotion recognition tasks.

Figure 3 presents the confusion matrices of our proposed method on the SEED, SEED-IV, and SEED-V datasets. It is clear that on the SEED dataset neutral and positive emotions are easy to distinguish. On the SEED-IV, neutral and fear emotions are easier to recognise than sadness and happiness. On the SEED-V, fear is the easiest emotion to recognise, while disgust is the most difficult.

3.4 Ablation Study and Analysis

Effects of Different Modalities and Dominant Modalities. We conducted ablation experiments to assess the impact of different modalities. Table 2 presents results on the SEED, SEED-IV, and SEED-V datasets, comparing EEG, eye movement, and both modalities, with each signal as the dominant modality. On SEED, the model's accuracy with only eye movement signals is 83.18%, while using eye movement as the dominant modality improves accuracy to 87.34%. However, eye movement as the dominant modality performs worse than EEG alone (94.21%) or EEG as the dominant modality (96.82%). This indicates that

Table 2. Effects of different modalities and dominant modalities (mean ± std. dev (%)).

Modality	SEED	SEED-IV	SEED-V
EM	83.18 ± 7.05	80.35 ± 7.23	71.68 ± 8.11
EEG	94.21 ± 5.66	91.29 ± 6.03	83.62 ± 6.41
EEG+EM (EM Dominant)	87.34 ± 6.89	83.68 ± 6.47	75.39 ± 7.15
EEG+EM (EEG Dominant)	96.82 ± 5.20	94.93 ± 6.12	86.95 ± 5.89

Table 3. Effects of different components (mean ± std. dev (%).)

Methods	SEED	SEED-IV	SEED-V
w/o MECS	86.76 ± 7.38	85.45 ± 7.96	78.69 ± 5.63
w/o CMF	91.52 ± 6.59	89.03 ± 6.28	81.72 ± 7.04
w/o ME	92.18 ± 6.93	90.52 ± 7.21	83.26 ± 7.16
MambaMER	96.82 ± 5.20	94.93 ± 6.12	86.95 ± 5.89

Table 4. Performance comparison of mamba and transformer.

Methods	TransformerMER	MambaMER
Accuracy	96.21%	96.82%
FLOPs	1.58B	0.86B
Parameters	86.72M	28.56M

EEG, with its stronger emotional expression, more effectively guides eye movement signals, a conclusion consistent across the other datasets.

Effects of Different Components. To evaluate the contributions of each component in the proposed model, we conducted ablation experiments on the SEED, SEED-IV, and SEED-V datasets, with the results presented in Table 3. Removing the EEG-guided conflict suppression (MECS) module led to a significant drop in performance, highlighting its crucial role in capturing cross-modal relationships. Similarly, removing the cross-modal fusion (CMF) module also resulted in a performance decline, further emphasizing its importance in integrating complementary information and enhancing inter-modal interactions. Additionally, removing the modality embedding (ME) module also reduced performance. These results thoroughly demonstrate the key role of the MECS, CMF, and ME modules in enhancing the model's expressive power and robustness.

Performance Comparison of Mamba and Transformer. By replacing Mamba with Transformer [2,3] in the model, the TransformerMER model was constructed. We compared the accuracy, FLOPs, and parameters of MambaMER and the deformation model on the SEED dataset to validate the effectiveness and superiority of the proposed MambaMER model. Table 4 summarizes the experimental results of the two methods. The results indicate that MambaMER slightly outperforms the deformation model in terms of accuracy, while demonstrating a significant advantage in FLOPs and parameters. Therefore, the experimental results provide strong evidence for the feasibility of the MambaMER model and its advantages over TransformerMER.

4 Conclusion and Discussion

In this paper, we propose an adaptive EEG-guided multimodal emotion recognition paradigm based on Mamba. To mitigate interference caused by conflicts between different modalities, we design a multi-scale EEG-guided conflict suppression module. Guided by EEG features at different scales, a selective cross-state space model extracts conflict-independent representations from eye movement features, effectively filtering out irrelevant or conflicting information. Additionally, we introduce a novel cross-modal fusion mechanism to enhance feature interaction and capture long-range dependencies. Experiments on SEED, SEED-IV, and SEED-V datasets show that our method outperforms state-of-the-art approaches. In the future, we will focus on addressing the issue of missing modalities in cross-modal fusion and explore more adaptive mechanisms to effectively compensate for the impact of information loss on emotion recognition performance.

Acknowledgements. This work is supported by funds from the National Natural Science Foundation of China (62476156, 62003196, 62076249, 62072289, and 62073201), the Young Talent of Lifting Engineering for Science and Technology in Shandong (SDAST2024QTA091), the Youth Innovation Technology Project of Higher Education in Shandong Province (2023KJ193) and the Provincial Natural Science Foundation of Shandong Province of China (ZR2020QF032).

Disclosure of Interests. The authors have no competing interests to declare that are relevant to the content of this article.

References

1. Che, T., et al.: SDOF-GAN: symmetric dense optical flow estimation with generative adversarial networks. IEEE Trans. Image Process. **30**, 6036–6049 (2021)
2. Chen, Z., Zheng, Y., Gee, J.C.: Transmatch: a transformer-based multilevel dual-stream feature matching network for unsupervised deformable image registration. IEEE Trans. Med. Imaging **43**(1), 15–27 (2023)
3. Dosovitskiy, A., et al.: An image is worth 16x16 words: transformers for image recognition at scale. arXiv preprint arXiv:2010.11929 (2020)
4. Fan, X., Xu, P., Zhao, Q., Hao, C., Zhao, Z., Wang, Z.: A domain adaption approach for EEG-based automated seizure classification with temporal-spatial-spectral attention. In: Linguraru, M.G., et al. (eds.) MICCAI 2024. LNCS, vol. 15005, pp. 14–24. Springer, Cham (2024). https://doi.org/10.1007/978-3-031-72086-4_2
5. Fu, B., Gu, C., Fu, M., Xia, Y., Liu, Y.: A novel feature fusion network for multimodal emotion recognition from EEG and eye movement signals. Front. Neurosci. **17**, 1234162 (2023)
6. Gong, X., Chen, C.P., Zhang, T.: Cross-cultural emotion recognition with EEG and eye movement signals based on multiple stacked broad learning system. IEEE Trans. Comput. Soc. Syst. (2023)
7. Gu, A., Dao, T.: Mamba: linear-time sequence modeling with selective state spaces. arXiv preprint arXiv:2312.00752 (2023)

8. Gu, A., Goel, K., Gupta, A., Ré, C.: On the parameterization and initialization of diagonal state space models. Adv. Neural. Inf. Process. Syst. **35**, 35971–35983 (2022)
9. Hazmoune, S., Bougamouza, F.: Using transformers for multimodal emotion recognition: taxonomies and state of the art review. Eng. Appl. Artif. Intell. **133**, 108339 (2024)
10. Jiang, W.B., Li, Z., Zheng, W.L., Lu, B.L.: Functional emotion transformer for EEG-assisted cross-modal emotion recognition. In: ICASSP 2024-2024 IEEE International Conference on Acoustics, Speech and Signal Processing (ICASSP), pp. 1841–1845. IEEE (2024)
11. Li, L., et al.: STM-net based spatial-temporal multi-modal fusion network for emotion recognition. In: International Conference on Image, Signal Processing, and Pattern Recognition (ISPP 2024), vol. 13180, pp. 1250–1255. SPIE (2024)
12. Liu, H., et al.: EEG-based multimodal emotion recognition: a machine learning perspective. IEEE Trans. Instrument. Measur. (2024)
13. Liu, W., Luo, Y., Lu, Y., Lu, Y.: A multitask framework for emotion recognition using eeg and eye movement signals with adversarial training and attention mechanism. In: 2023 IEEE International conference on bioinformatics and biomedicine (BIBM). pp. 2551–2558. IEEE (2023)
14. Liu, W., Qiu, J.L., Zheng, W.L., Lu, B.L.: Comparing recognition performance and robustness of multimodal deep learning models for multimodal emotion recognition. IEEE Trans. Cogn. Dev. Syst. **14**(2), 715–729 (2021)
15. Liu, W., Zheng, W.L., Li, Z., Wu, S.Y., Gan, L., Lu, B.L.: Identifying similarities and differences in emotion recognition with EEG and eye movements among Chinese, German, and French people. J. Neural Eng. **19**(2), 026012 (2022)
16. Liu, W., Zheng, W.-L., Lu, B.-L.: Emotion recognition using multimodal deep learning. In: Hirose, A., Ozawa, S., Doya, K., Ikeda, K., Lee, M., Liu, D. (eds.) ICONIP 2016. LNCS, vol. 9948, pp. 521–529. Springer, Cham (2016). https://doi.org/10.1007/978-3-319-46672-9_58
17. Lu, Y., Zheng, W.L., Li, B., Lu, B.L.: Combining eye movements and EEG to enhance emotion recognition. In: IJCAI, vol. 15, pp. 1170–1176. Buenos Aires (2015)
18. Nordfält, J., Ahlbom, C.P.: Utilising eye-tracking data in retailing field research: a practical guide. J. Retail. **100**(1), 148–160 (2024)
19. Pechlivanidou, G., Karampetakis, N.: Zero-order hold discretization of general state space systems with input delay. IMA J. Math. Control. Inf. **39**(2), 708–730 (2022)
20. Qiu, J.L., Li, X.Y., Hu, K.: Correlated attention networks for multimodal emotion recognition. In: 2018 IEEE International Conference on Bioinformatics and Biomedicine (BIBM), pp. 2656–2660. IEEE (2018)
21. Wang, J., et al.: Information bottleneck-based interpretable multitask network for breast cancer classification and segmentation. Med. Image Anal. **83**, 102687 (2023)
22. Wang, Y., Jiang, W.B., Li, R., Lu, B.L.: Emotion transformer fusion: complementary representation properties of EEG and eye movements on recognizing anger and surprise. In: 2021 IEEE International Conference on Bioinformatics and Biomedicine (BIBM), pp. 1575–1578. IEEE (2021)
23. Wu, X., Zheng, W.L., Li, Z., Lu, B.L.: Investigating EEG-based functional connectivity patterns for multimodal emotion recognition. J. Neural Eng. **19**(1), 016012 (2022)
24. Yang, D., Huang, S., Kuang, H., Du, Y., Zhang, L.: Disentangled representation learning for multimodal emotion recognition. In: Proceedings of the 30th ACM International Conference on Multimedia, pp. 1642–1651 (2022)

25. Zhao, Y., Gu, J.: Feature fusion based on mutual-cross-attention mechanism for EEG emotion recognition. In: Linguraru, M.G., et al. (eds.) MICCAI 2024. LNCS, vol. 15011, pp. 276–285. Springer, Cham (2024)
26. Zheng, W.L., Liu, W., Lu, Y., Lu, B.L., Cichocki, A.: Emotionmeter: a multimodal framework for recognizing human emotions. IEEE Trans. Cybern. **49**(3), 1110–1122 (2018)
27. Zheng, Y., et al.: SYMREG-GAN: symmetric image registration with generative adversarial networks. IEEE Trans. Pattern Anal. Mach. Intell. **44**(9), 5631–5646 (2021)
28. Zheng, Y., et al.: Image matting with deep gaussian process. IEEE Trans. Neural Networks Learn. Syst. **34**(11), 8879–8893 (2022)

Meta-learning Physics-Informed Neural Networks for Personalized Cardiac Modeling

Maryam Toloubidokhti(✉), Ryan Missel, Shichang Lian, and Linwei Wang

Rochester Institute of Technology (RIT), Rochester, NY 14623, USA
{mt6129,rxm7244,sl5074,lxwast}@rit.edu

Abstract. The advancement of personalized cardiac modeling, particularly through digital cardiac twins, enables tailored treatments based on the physiology of the individual patient. Traditional physics-based methods for optimizing the parameters of these cardiac models face challenges in clinical adoption due to their computational cost. Recent shifts towards data-driven approaches offer improved efficiency, but struggle with generalization and integration of core electrophysiological principles. The emerging use of physics-informed neural networks (PINNs) has the potential to combine the advantages of these two approaches, although still requiring retraining from scratch for each subject. This paper introduces a novel framework for meta-learning PINNs to overcome these challenges, enabling rapid personalization of a PINN to new subjects' data via simple feedforward computation. We instantiate this meta-PINN framework using the Eikonal model as the governing physics, demonstrating its efficacy in significantly reducing computational demands while improving the predictive accuracy of personalized cardiac models. Source code available at https://github.com/temporary-repos/MICCAI2025

Keywords: PINNs · meta-learning · cardiac electrophysiology

1 Introduction

The progress in cardiac modeling, particularly through the lens of digital twins, marks a significant leap in cardiac care. These models, when properly personalized for an individual, allow the planning of treatments [1,6,18] and prediction of interventional outcomes [8,22] for individual subjects. However, rapidly personalizing the functional parameters of these models remains an open challenge.

Earlier progress in estimating cardiac model parameters is deeply rooted in physics-based methodologies, primarily involving iterative optimization of a cardiac model to fit its outputs to available measurements [5,22,23]. Due to the need for repeated simulations of complex physics-based equations over a large parameter space, these approaches are associated with significant time and

J. C. Gee et al. (Eds.): MICCAI 2025, LNCS 15960, pp. 344–354, 2026.
https://doi.org/10.1007/978-3-032-04927-8_33

computational cost, resulting in limited applicability for real-time or near-real-time clinical use. Furthermore, because these optimizations are independently conducted for each personalization task, they miss the opportunity to accumulate knowledge about *how to rapidly personalize* to different individuals' data.

To address these limitations, recent advancements have pivoted towards data-driven approaches. Techniques such as reinforcement learning [16,17] and direct modeling of input-output relationships [9,10] have been explored. The emergence of deep learning has further enabled neural surrogates capable of approximating complex physiological phenomena. Examples include neural surrogates for cardiovascular hemodynamics that leverage transfer learning for rapid adaptation to new subjects [25], the application of conditional variational autoencoders for predicting ventricular activation using 3D bi-ventricular meshes and electrocardiogram simulations [15], and the meta-learning of spatiotemporal graph convolutional neural networks as a patient-specific surrogate of cardiac electrophysiological (EP) models [14]. While these approaches prioritize computational efficiency, their interpretability is often obscured by neglecting the underlying physical principles of cardiac behavior. Furthermore, their training often requires time-consuming generation of a large quantity of high-fidelity simulation data, while their ability to generalize from simulation to clinical data remains unclear.

Physics-informed neural networks (PINNs) recently emerged as a potential solution to bridge the gap between physics-based and data-driven approaches [19]. PINNs integrate known physical laws – in the form of a partial differential equation (PDE) – through a PDE residual loss to encourage the neural network to approximate PDE solutions without direct supervision from these solutions [7,19]. Recent studies have successfully utilized PINNs in personalized cardiac EP modeling [13,20,24]. In EP-PINN [13], a PINN is trained to generate action potentials in a 2D grid while estimating the PDE's key EP parameters. In [24], a spatial-temporal adaptive strategy is introduced to enhance the ability of PINNs to handle complex geometry, long time domain, and sharp gradients in PDE solutions. In FiberNet [20], a PINN is learned to solve for the Eikonal equation when inferring its spatially-varying fiber orientations from electroanatomical maps.

Despite this progress, a significant challenge remains for the widespread application of PINNs in personalized cardiac modeling: because the optimization of PINN is governed (in part) by a known PDE with its associated physics-based parameters, to personalize it requires the PINN to be re-trained from scratch along with estimating patient-specific parameters for the governing PDE. This results in a bottleneck similar to traditional physics-based approaches: the computational burden for clinical adoption, and a missed opportunity to *learn how to personalize* a PINN to shift this burden from deployment to offline development.

To address this critical gap, we introduce a novel framework for meta-learning the personalization of PINNs, enabling the rapid adaptation of a PINN to new patient-specific data without the need for retraining. It includes three key components: 1) an *adaptive* PINN whose weight parameters are generated from a hypernetwork from inferred patient-specific embeddings, 2) a conduction velocity network that personalizes the parameters of the governing PDE in a similar fashion, and 3) a meta-learning approach that learns to extract patient-specific embeddings from few-shot context observations in a rapid feedforward fashion. As a proof of concept, we developed the proposed meta-PINN framework for the isotropic Eikonal PDE [21] with spatially varying tissue properties. In *in-silico* and *in-vivo* experiments, we evaluated meta-PINN against physics-based and neural approaches to personalized EP modeling, as well as state-of-the-art PINN approaches where each PINN is individually optimized for each individual heart. We show that, at deployment time given sparse measurements from a patient, meta-PINN is able to deliver a personalized and physics-informed neural surrogate for cardiac EP via real-time feedforward computation, making an important stride towards fast adaptation of cardiac digital twins.

2 Background

Eikonal PDE: The Eikonal PDE is widely used to model the rapid propagation of electrical currents across the myocardium. We consider the isotropic Eikonal:

$$|\nabla T(\boldsymbol{x})|F(\boldsymbol{x}) = 1 \tag{1}$$

where $T(\boldsymbol{x})$ is a scalar field representing the arrival time of the electrical wavefront at spatial position $\boldsymbol{x} \in \mathbb{R}^3$, and $F(\boldsymbol{x})$ is a scalar field with units of speed (mm/ms) representing the local conduction velocity. While $F(\boldsymbol{x})$ denotes speed in an isotropic medium, we use the term *conduction velocity* following standard cardiac electrophysiology terminology. This equation effectively captures the isotropic propagation of electrical signals through the heterogeneous medium of the heart, accounting for varying conductivity in space.

PINNs: PINNs embed known physical laws into the learning algorithm of neural networks. In the context of the Eikonal PDE, a PINN $T_{NN_\phi}(\boldsymbol{x})$ parameterized by ϕ can be designed to output the arrival time of the electrical wavefront at any input spatial position $\boldsymbol{x}$. Instead of conventional supervised training, the key to PINN training is a PDE residual loss L_{PDE} that encapsulates the adherence of $T_{NN_\phi}(\boldsymbol{x})$ to the governing Eikonal physics as:

$$L_{PDE}(T_{NN_\phi}; F(\boldsymbol{x})) = \frac{1}{N_{PDE}} \sum_{j=1}^{N_{PDE}} \left(|\nabla T_{NN_\phi}(\boldsymbol{x}_j)|F(\boldsymbol{x}_j) - 1\right)^2 \tag{2}$$

where N_{PDE} is the number of *residual points* on which the PDE constraint is enforced. When observations of the arrival time $Y = T_{\text{obs}}(\boldsymbol{x})$ are available at N_{data} number of locations $\boldsymbol{x}_i$, a data-fidelity loss L_{data} can also be enforced:

$$L_{data}(T_{NN_\phi}, Y) = \frac{1}{N_{data}} \sum_{i=1}^{N_{data}} \left(T_{NN_\phi}(\boldsymbol{x}_i) - Y(\boldsymbol{x}_i)\right)^2 \tag{3}$$

If the conduction velocity $F(\boldsymbol{x})$ is unknown, it must be simultaneously estimated when $\boldsymbol{\phi}$ is optimized, resulting in the following optimization objective:

$$\{\hat{\boldsymbol{\phi}}, \hat{F}(\boldsymbol{x})\} = \arg \min_{\phi, F(\boldsymbol{x})} \{L_{data}(T_{NN_\phi}, Y) + \alpha L_{PDE}(T_{NN_\phi}; F(\boldsymbol{x}))\} \tag{4}$$

where the hyperparameter α balances the two losses. In state-of-the-art PINNs, $T_{NN_{\hat{\phi}}}(\boldsymbol{x})$ and $\hat{F}(\boldsymbol{x})$ must be optimized from scratch for each set of measurements.

3 Methodology

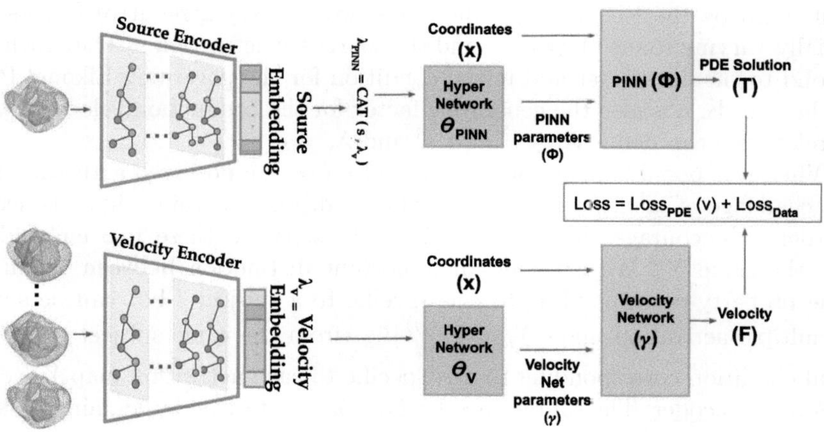

Fig. 1. Overview of meta-PINN. At inference time, sparse context AT samples on the ventricular geometry are used to extract a patient-specific embedding that adapts the PINN and velocity network to estimate personalized conduction velocity and the full volumetric AT map of a query sample, given only the first 20% of its activation.

Figure 1 outlines the proposed meta-PINN framework for personalized cardiac EP modeling. Instead of optimizing T_{NN_ϕ} and $F(\boldsymbol{x})$ from scratch for each set of observations Y using Eq. (4), we aim to *learn to personalize* T_{NN_ϕ} and $F(\boldsymbol{x})$ via a meta-learning approach to automatically adapt them to patient-specific Y in real time. This is achieved with two key components:

- Learn-to-personalize meta-inference: a feedforward meta-learner that extracts patient-specific embeddings from context observations of a subject.
- Adaptive generative models: a PINN $T_{NN_\phi}(\boldsymbol{x})$ and velocity network $F_{NN_\gamma}(\boldsymbol{x})$, with $\boldsymbol{\phi}$ and $\boldsymbol{\gamma}$ adaptable to patient-specific embeddings through hypernetworks, to predict for any other samples for the same patient.

Adaptable PINN and Velocity-Net: To model the spatially-varying conduction velocity to capture local abnormality, we model $F(\boldsymbol{x})$ with a neural network $F_{NN_\gamma}(\boldsymbol{x})$ and term it the *velocity-net*, similar to the framework presented in [20,21]. To avoid repeating the optimization of $\boldsymbol{\phi}$ and $\boldsymbol{\gamma}$ for T_{NN_ϕ} and F_{NN_γ} for every observation Y, we model $\boldsymbol{\phi}$ and $\boldsymbol{\gamma}$ with hypernetworks [2,12] as:

$$\boldsymbol{\phi} = H_{\boldsymbol{\theta}_{\text{PINN}}}(\boldsymbol{\lambda}_{\text{PINN}}), \quad \boldsymbol{\gamma} = H_{\boldsymbol{\theta}_{\text{v}}}(\boldsymbol{\lambda}_{\text{v}}) \tag{5}$$

where $\boldsymbol{\lambda}_{\text{PINN}}$ or $\boldsymbol{\lambda}_{\text{v}}$ are learnable patient-specific embeddings that will be described in the next section, while $\boldsymbol{\theta}_{\text{PINN}}$ and $\boldsymbol{\theta}_{\text{v}}$ represent the learnable weight parameters of the hypernetworks. These adaptive generative models mitigate the need for inefficient optimization-from-scratch for each new set of observations.

Learning to Identify the PINN and Velocity-Net: To identify the patient-specific embedding $\boldsymbol{\lambda}_{\text{PINN}}$ and $\boldsymbol{\lambda}_{\text{v}}$ that can adapt T_{NN_ϕ} and F_{NN_γ}, we consider their respective generative factors. For an observed activation sequence to be modeled by the PINN T_{NN_ϕ}, there are two primary generative factors: the spatially-varying tissue property $\boldsymbol{c}$ and the sources of activation $\boldsymbol{s}$. They each correspond to the parameter and initial condition for the governing Eikonal PDE: in other words, $\boldsymbol{c}$ is also the generative factor for the conduction velocity F_{NN_γ}. Therefore, we can define $\boldsymbol{\lambda}_{\text{PINN}} = (\boldsymbol{c}, \boldsymbol{s})$ and $\boldsymbol{\lambda}_{\text{v}} = \boldsymbol{c}$.

While it is possible to simply infer $\boldsymbol{c}$ and $\boldsymbol{s}$ from an observed activation map Y, a challenge of *identifiability* arises: what additional learning signal is needed in order to encourage the model to learn to separate these two embeddings from the same Y? We consider an important distinction between them: the tissue property corresponding to $\boldsymbol{c}$ is specific to a subject i but can be shared by multiple activation maps $\mathcal{Y}_i = \{Y_i^{(j)}\}_{j=1}^k$ from the same subject, while the initial condition corresponding to $\boldsymbol{s}$ is specific to each activation map $Y_i^{(j)}$.

Source Encoder: The source encoder $\mathcal{E}_{\psi_\text{s}}$, parameterized by $\boldsymbol{\psi}_\text{s}$, aims to identify the representation of the source of activation unique to each $Y_i^{(j)}$ as:

$$\boldsymbol{s}_i^{(j)} = \mathcal{E}_{\psi_\text{s}}(Y_i^{(j)}) \tag{6}$$

Velocity Encoder: The velocity encoder $\mathcal{E}_{\psi_\text{c}}$, parameterized by $\boldsymbol{\psi}_\text{c}$, aims to identify the representation for the heterogeneous tissue property by extracting the shared embedding from k activation maps $\{Y_i^{(j)}\}_{j=1}^k$ from the same subject i. To provide an explicit learning signal for a shared $\boldsymbol{c}_i$ for subject i, we obtain an embedding from each $Y_i^{(j)}$ and then aggregate them via a simple averaging:

$$c_i = \frac{1}{K}\sum_{k=1}^{K} \mathcal{E}_{\psi_\text{c}}(Y_i^{(k)}) \tag{7}$$

The Learn-to-Identify Meta-objectives: Given a dataset of observed activation maps for N subjects $\mathcal{Y} = \{\mathcal{Y}_i\}_{i=1}^N$, where M activation time maps with varying sources of activation are available for each subject $\mathcal{Y}_i = \{Y_i^{(j)}\}_{j=1}^M$, we now formulate a meta-learning objective to learn to personalize T_{NN_ϕ} and F_{NN_γ}. More specifically, for each subject i, we consider k activation maps as context samples $\mathcal{Y}_i^c = \{Y_i^{c,(j)}\}_{j=1}^k$ while the remaining as query samples $\mathcal{Y}_i^q = \{Y_i^{q,(j)}\}_{j=1}^{M-k}$.

For each query sample Y_i^q, the model extracts a subject-specific embedding $\boldsymbol{c}$ from the context set $\mathcal{Y}_i^c$ and activation-specific embedding $\boldsymbol{s}$ from only the first 20% activation of Y_i^q. These embeddings are then used to adapt the PINN and velocity-net for predicting the full Y_i^q while minimizing the PDE residual loss as:

$$\hat{\boldsymbol{\Theta}} = \arg\min_{\boldsymbol{\Theta}} \sum\nolimits_{i=1}^{N} \sum\nolimits_{Y_i^q \in \mathcal{Y}_i^q} \{L_{data}(T_{NN_\phi}, Y_i^q) + \alpha L_{PDE}(T_{NN_\phi}; F_{NN_\gamma})\} \tag{8}$$

where $\boldsymbol{\Theta} = \{\boldsymbol{\psi}_s, \boldsymbol{\psi}_c, \boldsymbol{\theta}_{\text{PINN}}, \boldsymbol{\theta}_v\}$. This optimization is carried out in an episodic training scheme, where the division of context and query samples is changed at each epoch for each subject. The outcome is a meta-model that, given k (which can be variable) observed activation maps for a subject, can obtain a subject-specific PINN along with an estimate of the spatially-varying conduction velocity for that subject—all via simple and rapid feedforward computation.

4 Experiments and Results

In all experiments, meta-PINN employed encoders with six SplineConv layers (kernel size = 5, dim = 3) followed by adaptive max pooling, effectively capturing spatial relations in the ventricular geometry. The PINN hypernetwork consisted of four fully connected layers, generating parameters for the six-layer PINN with 20 neurons per layer and Tanh activation. Similarly, the velocity hypernetwork had five hidden layers, producing parameters for the eight-layer velocity network with ReLU activation. Optimization was carried out using the AdamW optimizer, starting with a learning rate of 2×10^{-5} that we adjusted using a cosine annealing scheduler. We gradually increased the PDE loss hyper-parameter, α, from 0 to 1 throughout the training epochs. Each training episode, executed on an NVIDIA RTX 3090 with 24GB of memory, took $\sim$ 3 seconds.

4.1 Synthetic Data Experiments

Data: We created activation-time (AT) maps using the Eikonal model in Eq. (1), utilizing the fast marching method from the *fim-python* library [11]. We obtained four unique human ventricular geometries from MRI images, aligned via non-rigid registration. Two tissue classes were modeled: scar (0.1 m/s) and healthy (0.6 m/s), with scars placed at five distinct locations per geometry. For each scar-geometry combination, we generated 20 AT maps with random placements of the site of origin, resulting in 400 AT maps, 200 for training and 200 for test. Train and test samples were split by activation origin per task; all tasks appear in both splits. We set $k = 5$ for the context set. To reflect practice, we use endocardial AT measurements (200-400 nodes), to predict AT and conduction velocity throughout the ventricular myocardium (1600-2400 nodes).

Baselines and Metrics: We evaluated meta-PINN against three baselines: 1) direct optimization of $F(\boldsymbol{x})$ for the Eikonal PDE using the gradient-free BOBYQA algorithm [4] (*physics-based*); 2) data-driven meta-learning of T_{NN_ϕ} but without PDE-residual loss, similar to [14] – this can also be seen as an ablation of the presented meta-PINN (*meta-neural*); and 3) standalone PINN models where T_{NN_ϕ} is optimized for each observed activation time and F_{NN_γ} is shared across PINNs of the same subject, similar to that described in [20] (*multi-PINN*). Because *meta-neural* and *meta-PINN* have predictive abilities once personalized from context samples, they were tested for that purpose. *Physics-based* optimization and *multi-PINN* were optimized to each test sample instead.

For quantitative metrics, we considered 1) relative mean squared error (RMSE) and 2) spatial correlation coefficient (SCC) between predicted and true AT maps and velocity fields, as well as 3) computational cost to achieve personalization.

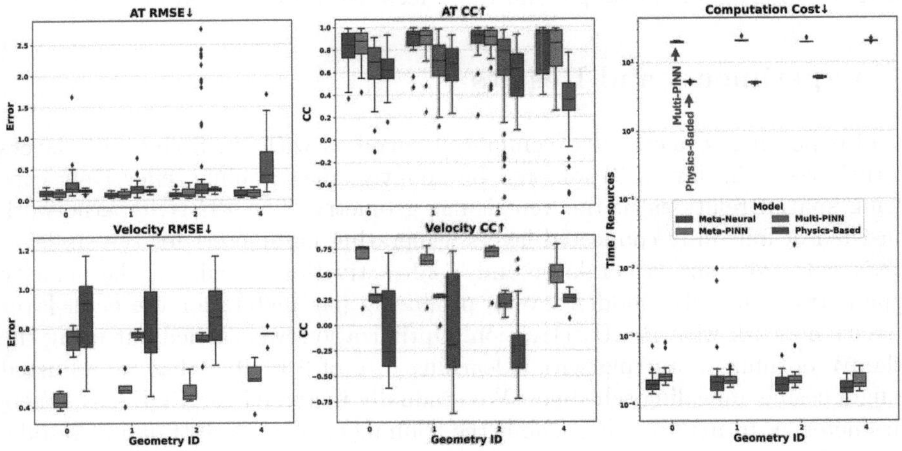

Fig. 2. Quantitative results on synthetic data. Meta-PINN gained significant improvements in AT and velocity accuracy compared to physics-based optimization and multi-PINN at a fraction of computation. Its performance and computation were comparable to meta-neural, with added benefit of interpretable personalized velocity estimation.

Results: Figure 2 summarizes the quantitative results while visual examples are shown in Fig. 3. Compared to *physics*-based optimization (red) and *multi-PINN* (green), meta-PINN (orange) obtained significantly improved accuracy in both AT map prediction and velocity estimation, *e.g.*, $22.3 \pm 4.0\%$ and $68.9 \pm 10.0\%$ improvements in SCC respectively for AT and velocity maps, compared to physics; and $19.7 \pm 4.7\%$ and $36.2 \pm 4.1\%$ improvements compared to multi-PINN. These improvements were obtained with a fraction of computation time to personalize ($0.009 \pm 0.007\%$ of physics-based, and $0.003 \pm 0.002\%$ of multi-PINN). Compared to meta-neural (blue), meta-PINN achieved comparable accuracy in predicting unseen AT maps with slightly higher computation time. It however had the benefit of delivering subject-specific conduction velocity maps as shown in Fig. 3, leading to interpretable clinical utility.

Ablation and Analysis: We conducted three ablation-style analyses. First, we evaluated the model's sensitivity to the number of context observations k. Meta-PINN showed stable performance with fewer context maps: for k=1, AT/velocity SCC $= 0.86 \pm 0.17/0.56 \pm 0.26$; for k=3, $0.86 \pm 0.16/0.59 \pm 0.20$; and for k=5,

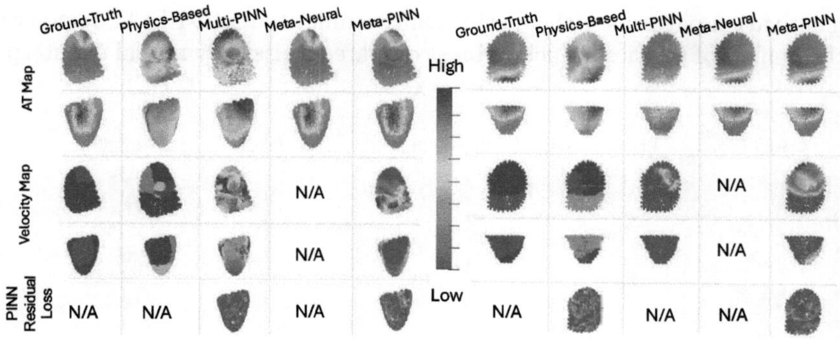

Fig. 3. Visual examples on synthetic data. Meta-PINN showed significantly improved AT and velocity compared to multi-PINN and physics-based optimization. Meta-neural showed similar AT maps without an ability to deliver personalized conduction velocity.

$0.87 \pm 0.15/0.60 \pm 0.19$ suggesting strong robustness with sparse supervision. Second, to assess whether accuracy differs between observed and unobserved regions, we conducted a preliminary stratified analysis: meta-PINN achieved AT/velocity SCC of 0.87/0.57 on the endocardium (where measurements are) and 0.87/0.60 on the remaining myocardium; given the minimal difference, full regional analysis is left to future work. Finally, we evaluated meta-PINN's adherence to the Eikonal PDE. As shown in the last row of Fig. 3, residual heatmaps indicate smoothly minimized PDE loss. Without any test-time retraining, meta-PINN achieved a per-node residual of 0.31 ± 0.18, comparable to individually trained PINNs (0.29 ± 0.04), confirming its ability to satisfy PDE constraints.

4.2 Real Data Experiments

Data: We considered *in-vivo* data from an animal model experiment [3], which included cardiac activation sequences driven by bipolar stimulation via intramural plunge needles at six different left-ventricular locations. Epicardial potentials were recorded using a 247-electrode sock, and geometric surfaces were then generated from the electrode positions acquired during the experiment.

Baselines and Metrics: We directly tested meta-PINN and meta-neural trained from synthetic data, each time considering five epicardial maps as the context to predict the left-out activation map. Note the large generalization gap due to both the sim-to-real gap and the change of endocardial to epicardial measurements. Physics-based and multi-PINN were optimized from scratch.

Results: Figure 4 provides several visuals of predicted AT maps. Note the relatively stronger performance of physics-based optimization and multi-PINN as they were optimized from scratch to all six available maps. Meta-neural, in comparison, completely failed in this generalization task by producing similar outputs across all test cases. Meta-PINN, in comparison, was able to capture the main activation patterns with a performance similar to multi-PINN, demonstrating the benefit of physics inductive bias compared to its fully neural counterpart.

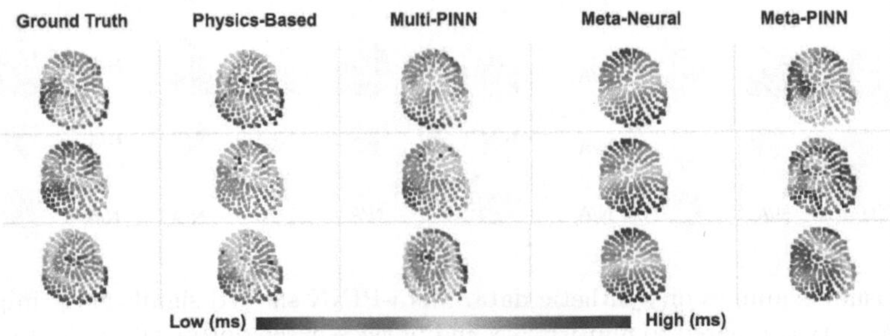

Fig. 4. Visual examples on real-data AT maps.

4.3 Conclusion and Discussion

We presented an innovative meta-PINN for learning to rapidly adapt a personalized and physics-informed surrogate to sparse surface AT measurements from a subject. Results showed improved accuracy and computational efficiency over existing approaches. As a proof-of-concept, this study can be strengthened by a systematic generalization study considering unseen geometry or scar settings in synthetic experiments, as well as evaluation on a larger number of real-data experiments. Future works will extend meta-PINN to anisotropic Eikonal PDE, higher-fidelity cardiac EP models, as well as beyond the domain of cardiac EP.

Acknowledgments. This study was supported by NIH National Heart, Lung, And Blood Institute (NHLBI) grant R01HL145590, NIH National Institute of Nursing Research (NINR) grant R01NR0183 and NSF OAC-2212548.

Disclosure of Interests. The authors have no competing interests to declare that are relevant to the content of this article.

References

1. Arevalo, H.J., et al.: Arrhythmia risk stratification of patients after myocardial infarction using personalized heart models. Nat. Commun. **7**(1), 11437 (2016)
2. de Avila Belbute-Peres, F., Chen, Y.f., Sha, F.: Hyperpinn: learning parameterized differential equations with physics-informed hypernetworks. The symbiosis of deep learning and differential equations **690** (2021)
3. Bergquist, J.A., Good, W.W., Zenger, B., Tate, J.D., Rupp, L.C., MacLeod, R.S.: The electrocardiographic forward problem: a benchmark study. Comput. Biol. Med. **134**, 104476 (2021)

4. Cartis, C., Fiala, J., Marteau, B., Roberts, L.: Improving the flexibility and robustness of model-based derivative-free optimization solvers. ACM Trans. Math. Softw. (TOMS) **45**(3), 1–41 (2019)
5. Chinchapatnam, P., et al.: Estimation of volumetric myocardial apparent conductivity from endocardial electro-anatomical mapping. In: 2009 Annual International Conference of the IEEE Engineering in Medicine and Biology Society, pp. 2907–2910. IEEE (2009)
6. Corral-Acero, J., et al.: The 'digital twin'to enable the vision of precision cardiology. Eur. Heart J. **41**(48), 4556–4564 (2020)
7. Cuomo, S., Di Cola, V.S., Giampaolo, F., Rozza, G., Raissi, M., Piccialli, F.: Scientific machine learning through physics-informed neural networks: where we are and what's next. J. Sci. Comput. **92**(3), 88 (2022)
8. Gerach, T., et al.: Electro-mechanical whole-heart digital twins: a fully coupled multi-physics approach. Mathematics **9**(11), 1247 (2021)
9. Giffard-Roisin, S., et al.: Sparse bayesian non-linear regression for multiple onsets estimation in non-invasive cardiac electrophysiology. In: Pop, M., Wright, G.A. (eds.) FIMH 2017. LNCS, vol. 10263, pp. 230–238. Springer, Cham (2017). https://doi.org/10.1007/978-3-319-59448-4_22
10. Giffard-Roisin, S., et al.: Noninvasive personalization of a cardiac electrophysiology model from body surface potential mapping. IEEE Trans. Biomed. Eng. **64**(9), 2206–2218 (2016)
11. Grandits, T.: A fast iterative method python package. J. Open Source Softw. **6**(66), 3641 (2021). https://doi.org/10.21105/joss.03641
12. Ha, D., Dai, A., Le, Q.V.: Hypernetworks. arXiv preprint arXiv:1609.09106 (2016)
13. Herrero Martin, C.: Ep-pinns: cardiac electrophysiology characterisation using physics-informed neural networks. Front. Cardiovascular Med. **8**, 768419 (2022)
14. Jiang, X., Li, Z., et al.: Few-shot generation of personalized neural surrogates for cardiac simulation via bayesian meta-learning. In: International Conference on Medical Image Computing and Computer-Assisted Intervention, pp. 46–56. Springer (2022). https://doi.org/10.1007/978-3-031-16452-1_5
15. Li, L., Camps, J., Banerjee, A., Beetz, M., Rodriguez, B., Grau, V.: Deep computational model for the inference of ventricular activation properties. In: International Workshop on Statistical Atlases and Computational Models of the Heart, pp. 369–380. Springer (2022). https://doi.org/10.1007/978-3-031-23443-9_34
16. Neumann, D., Mansi, T.: Machine learning methods for robust parameter estimation. In: Artificial Intelligence for Computational Modeling of the Heart, pp. 161–181. Elsevier (2020)
17. Neumann, D., et al.: A self-taught artificial agent for multi-physics computational model personalization. Med. Image Anal. **34**, 52–64 (2016)
18. Personalized virtual-heart technology for guiding the ablation of infarct-related ventricular tachycardia. Nat. Biomed. Eng. **2**(10), 732–740 (2018)
19. Raissi, M., Perdikaris, P., Karniadakis, G.E.: Physics-informed neural networks: a deep learning framework for solving forward and inverse problems involving nonlinear partial differential equations. J. Comput. Phys. **378**, 686–707 (2019). https://doi.org/10.1016/j.jcp.2018.10.045
20. Ruiz Herrera, C., Grandits, T., Plank, G., Perdikaris, P., Sahli Costabal, F., Pezzuto, S.: Physics-informed neural networks to learn cardiac fiber orientation from multiple electroanatomical maps. Eng. Comput. **38**(5), 3957–3973 (2022)
21. Sahli Costabal, F., Yang, Y., Perdikaris, P., Hurtado, D.E., Kuhl, E.: Physics-informed neural networks for cardiac activation mapping. Front. Phys. **8**, 42 (2020)

22. Sermesant, M., et al.: Patient-specific electromechanical models of the heart for the prediction of pacing acute effects in crt: a preliminary clinical validation. Med. Image Anal. **16**(1), 201–215 (2012)
23. Wong, K.C., et al.: Velocity-based cardiac contractility personalization from images using derivative-free optimization. J. Mech. Behav. Biomed. Mater. **43**, 35–52 (2015)
24. Ye, Y., Liu, H., Jiang, X., Toloubidokhti, M., Wang, L.: A spatial-temporally adaptive pinn framework for 3d bi-ventricular electrophysiological simulations and parameter inference. In: International Conference on Medical Image Computing and Computer-Assisted Intervention, pp. 163–172. Springer (2023). https://doi.org/10.1007/978-3-031-43990-2_16
25. Zhou, Y., He, Y., Wu, J., Cui, C., Chen, M., Sun, B.: A method of parameter estimation for cardiovascular hemodynamics based on deep learning and its application to personalize a reduced-order model. Inter. J. Numerical Methods Biomed. Eng. **38**(1), e3533 (2022)

MG-UNet: A Memory-Guided UNet for Lesion Segmentation in Chest Images

Shuaipeng Ding, Mingyong Li(✉), and Chao Wang

School of Computer and Information Science, Chongqing Normal University, Chongqing 401331, China
limingyong@cqnu.edu.cn

Abstract. Lesion segmentation in medical images is a key task for the intelligent diagnosis of lung diseases. Although existing multimodal methods have achieved significant progress in medical image segmentation by combining image and text information, these methods still rely on textual input during the inference phase, limiting their applicability in real-world scenarios. To address this limitation, this paper proposes an innovative Memory-Guided UNet model (MG-UNet). MG-UNet introduces a learnable memory bank that automatically extracts and stores textual information during the training phase. In the decoding stage, the proposed memory-guided decoder retrieves knowledge relevant to the current image from the memory bank, thereby eliminating the need for textual input during inference. Extensive experiments were conducted on the QaTa-Cov19 and MosMedData+ datasets to validate the effectiveness of MG-UNet. The experimental results demonstrate that MG-UNet not only outperforms existing unimodal and multimodal methods in terms of segmentation performance but also excels in text-free inference scenarios using only 15% of the training data, surpassing the current best unimodal methods. This characteristic significantly reduces the reliance on annotated data for medical image segmentation, offering greater flexibility and scalability for practical clinical applications. The code will be available soon.

Keywords: Chest CT · Multi-modal · Memory Bank · Medical image segmentation

1 Introduction

Chest imaging modalities, such as X-rays and CT scans, play a crucial role in the diagnosis and monitoring of various pulmonary diseases, including infectious diseases and neoplastic disorders. With the rapid development of deep learning, deep neural networks have been widely applied to radiological image analysis, supporting auxiliary diagnostic tasks such as disease classification, lesion detection, and segmentation. Among these tasks, lesion segmentation is particularly critical, as it not only enables precise localization and delineation of pathological

J. C. Gee et al. (Eds.): MICCAI 2025, LNCS 15960, pp. 355–365, 2026.
https://doi.org/10.1007/978-3-032-04927-8_34

regions within the thorax but also provides quantitative foundations for disease staging and treatment planning, making it a key component of clinical diagnosis. Existing medical image segmentation methods [1–3], primarily based on the UNet architecture [4], have achieved significant progress. However, due to their reliance on large amounts of high-quality annotated data and the high cost and inefficiency associated with expert involvement in the annotation process, their widespread application in real clinical settings remains severely limited.

To mitigate similar issues available in natural image processing, CLIP [5] capitalizes on the complementary information provided by accompanying text, thereby reducing the dependency on high-quality annotated image data and maximizing the utilization of available information. Inspired by CLIP, approaches like MedCLIP [6] and GLoRIA [7] have extended these ideas to medical image-report pairs situations, still showcasing excellent performance. Motivated by the impressive performance gains achieved through the integration of textual information, Li et al. [8] proposed a novel CNN-Transformer structure named LViT to integrate multimodal information in the early stage. GuideDecoder [9] implemented a novel approach that focuses on improving the decoder using both image and text features. MMI-UNet [10] achieved state-of-the-art (SOTA) performance on the QaTa-COV19 [11] dataset by integrating visual and linguistic features during the encoder stage.

Although the above research has significantly improved segmentation performance by leveraging text, these multimodal methods require textual input during both training and inference, limiting their practicality in real clinical scenarios. To address this issue, we introduce a learnable Memory Bank to store textual information, serving as a bridge between textual and visual information. This allows us to enhance visual representation during inference by retrieving relevant historical information from the Memory Bank, eliminating the need for text input. Our contributions are as follows:

- We propose Memory-Guided UNet (MG-UNet), an innovative multimodal learning method for lesion segmentation in chest images, achieving excellent segmentation results during the inference phase without relying on text input.
- We introduce an intermittent memory bank updating (IMBU) strategy that allows the model to progressively transition from reliance on text input to operating without text input during the training phase, thereby achieving greater adaptability.
- Dual inference mechanism: The model dynamically updates the Memory Bank through corresponding text, utilizing the updated Memory Bank to extract knowledge relevant to the current image. In the absence of text input, the model can directly retrieve knowledge from the historical Memory Bank. This mechanism not only enhances the model's adaptability but also significantly improves the efficiency of knowledge utilization, thereby increasing its operational viability in clinical diagnosis.

2 Method

The overall architecture of our proposed Memory-Guided UNet model (MG-UNet) is illustrated in Fig. 1. The model consists of three main components: the image encoder, the Memory Bank Updating (MBU) mechanism, and the Memory-Guided Decoder (MG-Decoder). The image encoder encodes the input images into feature representations, while the MBU mechanism automatically learns and stores textual knowledge into the Memory Bank during training. Subsequently, the image features are processed by the MG-Decoder, which extracts knowledge relevant to the current image from the Memory Bank and fuses it with the encoder features. The iterative process of the MG-Decoder effectively utilizes different levels of image features and knowledge from the Memory Bank, resulting in a more efficient fused representation. Finally, a segmentation head is used to generate the final segmentation predictions.

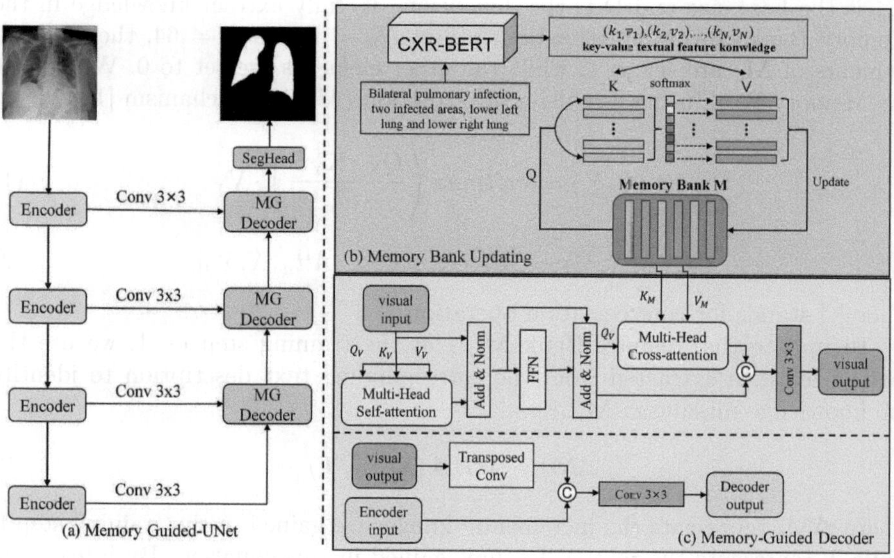

Fig. 1. The overview of the proposed (a) Memory-Guided Unet with (b) Memory Bank Updating and (c) Memory-Guided Decoder.

We will elaborate on these components in the following sections and introduce an intermittent memory bank updating strategy to enable the model to better adapt to the text-free inference process.

2.1 Image Encoder

We utilize ConvNeXt tiny [12] as the image encoder to extract multiple visual features from the input image. Given an input image of size $H \times W \times 3$, the encoder extracts feature maps with dimensions $\frac{H}{4} \times \frac{W}{4} \times 96$, $\frac{H}{8} \times \frac{W}{8} \times 192$, $\frac{H}{16} \times \frac{W}{16} \times 384$, and $\frac{H}{32} \times \frac{W}{32} \times 768$, respectively. These feature maps are then adjusted to a uniform number of channels through 3×3 convolution operations, set to 64 in this paper to balance model performance and efficiency. The resulting features are denoted as E_4, E_3, E_2, and E_1.

2.2 Memory Bank Updating

We first adopt the pre-trained CXR-BERT [13] as the text encoder to extract textual features, which are then reduced in dimensionality through a linear mapping, denoted as $\mathbf{T}$. The dimensionality of these textual features is $L \times C$, where C represents the dimensionality of the extracted features and L signifies the length (number of tokens) of the text description. We set C to 64 and freeze the text encoder during training.

Then we initialize the Memory Bank $\mathbf{M}_0 \in \mathbb{R}^{N_m \times D}$ through a matrix, where N_m is the base size and D is the dimension. To fully extract knowledge in the Memory Bank, during initialization, we set $N_m = L$ and $D = 64$, the diagonal elements of $\mathbf{M}_0$ are set to 1, while the other elements are set to 0. We update the Memory Bank using a multi-head attention (MHA) mechanism [14]:

$$Att_i(X, Y) = softmax\left(\frac{\mathrm{Q_X} \cdot \mathrm{K_Y}^T}{\sqrt{d_n}}\right) \cdot \mathrm{V_Y} \tag{1}$$

$$MHA(X, Y) = [Att_1(X, Y); \ldots; Att_n(X, Y)] \tag{2}$$

where[;] stands for concatenation operation.

To update the Memory Bank $\mathbf{M}_{t-1}$ at the training step $\mathrm{t}-1$, we use the textual features extracted from the corresponding text description to identify the knowledge missing in $\mathbf{M}_{t-1}$.

$$\Delta\mathbf{M}_t = MHA\left(\mathbf{M}_{t-1}, \mathbf{T}\right) \tag{3}$$

where $\Delta\mathbf{M}_t$ represents the incremental knowledge gained in the training step t while T represents the textual features gained in this equation. By integrating incremental knowledge, we can update the Memory Bank during the training step.

$$\mathbf{M}_t = \mathbf{M}_{t-1} + Norm\left(\Delta\mathbf{M}_t\right) \tag{4}$$

where $Norm$ refers to layer normalization to normalize the incremental knowledge.

2.3 Memory-Guided Decoder

For the decoder feature D_i, we first capture internal relationships and dependencies through multi-head self-attention and a feed-forward network (FFN).

Then, we extract image-related knowledge from the Memory Bank using multi-head cross-attention. Finally, the fused features are refined through a 3×3 convolutional layer. The process can be described as follows:

$$D_i^{sa} = Norm\left(MHA\left(D_i, D_i\right) + D_i\right) \tag{5}$$

$$F_i^{sa} = Norm\left(FFN\left(D_i^{sa}\right) + D_i^{sa}\right) \tag{6}$$

$$F_i^{ca} = Norm\left(MHA\left(F_i^{sa}, \mathbf{M}_t\right)\right) \tag{7}$$

$$F_i = Conv_{3\times 3}\left[F_i^{ca}; F_i^{sa}\right] \tag{8}$$

Then, the processed features F_i are upsampled using a transposed convolution and concatenated with the feature E_{i+1} from the encoder, and finally processed using a 3×3 convolution to obtain the decoder features D_{i+1}, as shown in Fig. 1(c).

After three iterations, the decoder feature is first upsampled to match the resolution of the original input image. Then, a 1×1 convolution and a sigmoid activation function are applied to generate the segmentation output.

2.4 Intermittent Memory Bank Updating

During the training phase, since text corresponding to the images is used to update the memory bank, directly utilizing a fixed-parameter memory bank for text-free inference may result in feature degradation, leading to inaccurate segmentation. To mitigate this potential issue, we propose a straightforward training strategy (intermittent memory bank updating). Specifically, we train MG-UNet over several rounds, each consisting of two epochs. In the first epoch, the memory bank is updated using the text associated with the images. In the second epoch, we simulate the inference phase by omitting the text update process and instead directly retrieving relevant historical information from the memory bank for the current image.

3 Experiments and Results

3.1 Datasets and Implementation

Datesets: To evaluate the performance of our proposed MG-UNet model, we conducted experiments on two publicly available datasets: QaTa-COV19 [11] and MosMedData+ [15]. The first dataset, compiled through a collaborative effort between researchers at Qatar University and Tampere University, consists of 9258 chest X-ray images depicting COVID-19 cases. The second dataset, MosMedData+, includes 2729 CT scan slices specifically depicting lung infections. Notably, both datasets feature similar textual annotations that focus on key clinical aspects such as the presence of infection in both lungs, the number of lesions, and their approximate locations. These annotations are illustrated in Fig. 1(a) for reference.

Implementation: Following [8], we split the QaTa-COV19 dataset into training, validation, and testing sets, containing 5716, 1429, and 2113 samples, respectively. The MosMedData+ dataset is divided into a training set of 2183 images, a validation set of 273 images, and a testing set of 273 samples. All images are cropped to a size of 224 × 224. For data augmentation, a random zoom technique with a probability of 10% is applied. The implementation utilizes PyTorch [16], PyTorch Lightning, and MONAI [17]. The entire training and testing process is conducted on a Nvidia GeForce RTX 3090 with 24 GB VRAM. During training, we use a combined loss function consisting of Dice loss and cross-entropy loss, with the network optimized using the AdamW optimizer and a batch size of 32. A cosine annealing learning rate schedule is employed, starting from 3e−4 and decaying to 1e−6.

3.2 Performance Comparison with Existing Methods

We compared MG-UNet and the text-free reasoning model MG-UNet* with commonly used single-modal methods and the latest multi-modal medical image segmentation methods. As shown in Table 1, MG-UNet outperformed all evaluation methods on the QaTa-COV19 and MosMedData+ datasets. Notably, MG-UNet* improved the DSC (Dice Similarity Coefficient) by 7.68% and 3.80%, respectively, compared to the best single-modal method nnunet on the two datasets. It also demonstrated better performance compared to some multi-modal methods. Although there is still a gap compared to the latest multi-modal method MMI-UNet, but MG-UNet* achieved a roughly 50% reduction in FLOPs compared to MMI-UNet, and it does not rely on text input during the inference phase, providing greater flexibility and efficiency for practical clinical applications.

The qualitative experimental results compared with ground truth are shown in Fig. 2. We conducted qualitative comparisons between the proposed MG-UNet method and other approaches on the QaTa-COV19 and MosMedData+ datasets, respectively. In the visualization results, yellow indicates true positive regions (correctly identified infected areas), red denotes false negative regions (missed infections), and green represents false positive regions (normal tissues misclassified as infections). The experimental results demonstrate that due to limited feature extraction capabilities, traditional unimodal methods exhibit significant performance limitations in complex scenarios, leading to substantial false negative segmentation results (noticeable increase in red regions). Compared with other multimodal fusion methods, MG-UNet not only effectively suppresses over-segmentation phenomena (indicated by blue annotations) through its innovative feature fusion mechanism but, more importantly, maintains excellent true positive recognition capability. This advantage enables our model to accurately segment infected regions while effectively filtering out irrelevant areas.

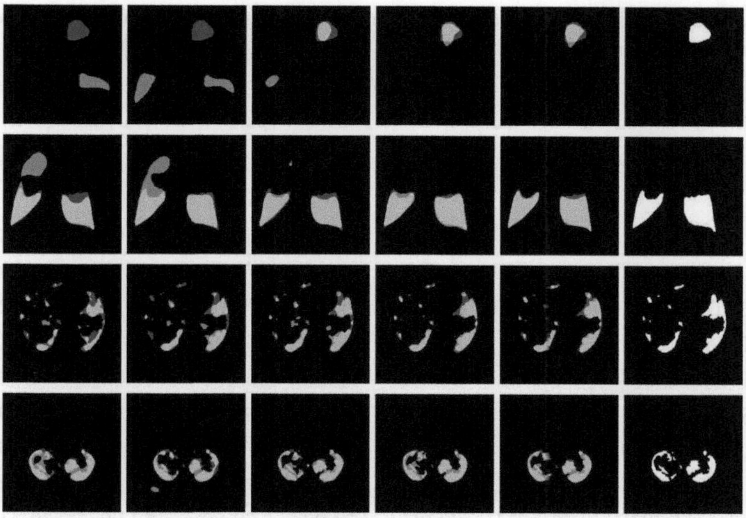

Fig. 2. Qualitative results on the QaTa-COV19 dataset and the MosMedData+ dataset. Yellow, red, and green indicate true positive, false negative, and false positive, respectively. (Color figure online)

3.3 Ablation Study

To validate the effectiveness of the proposed module, we conducted an ablation study on the QaTa-COV19 dataset. As shown in Table 2, compared to the baseline model that uses only images, MG-UNet achieved 6.72% improvement in Dice Similarity Coefficient (DSC), confirming the effectiveness of the Memory-Guided Decoder. MG-UNet* showed 3.72% decrease in DSC compared to MG-UNet, likely due to the model's reliance on a fixed-parameter memory bank, which resulted in knowledge confusion and affected segmentation accuracy. By incorporating our proposed training strategy, the model gradually transitioned from text-based training to text-free inference, reducing the knowledge confusion caused by the rigid memory bank, leading to 0.92% improvement in DSC scores.

Furthermore, we performed extensive experiments to assess the impact of varying training data sizes on the model's segmentation performance. As illustrated in Table 3, even with limited training data, MG-UNet demonstrated comparable performance to nnUNet. Specifically, when trained with only 15% of the dataset, MG-UNet and MG-UNet* achieved 7.60% and 4.88% higher DSC scores, respectively, compared to nnUNet, the best performing unimodal model trained on the full dataset.

Table 1. Quantitative comparison on segmentation results with uni-modal and previous multi-modal learning methods. optimal and suboptimal performance is highlighted, ↑(↓) denotes the higher (lower) the better. MG-UNet results are averaged over five runs.

Method	Params↑	FLOPs↓	QaTa-COV19		MosMedData+	
			DSC↑	IoU↑	DSC↑	IoU↑
UNet[4]	14.8M	50.3G	79.02	69.46	64.60	50.73
UNet++ [3]	74.5M	94.6G	79.62	70.25	71.75	58.39
AttUNet [18]	34.9M	101.9G	79.31	70.04	66.34	52.82
TransUNet [2]	105M	56.7G	78.63	69.13	71.24	58.44
UCTransNet [19]	65.6M	63.2G	79.15	69.60	65.90	52.69
Swin-UNet [1]	82.3M	67.3G	78.07	68.34	63.29	50.19
nnUNet [20]	19.1M	412.7G	80.42	70.81	72.59	60.36
CLIP [5]	87.0M	105.3G	79.81	70.66	71.97	59.64
LAVT [21]	118.6M	83.8G	79.28	69.89	73.29	60.41
LViT [8]	29.7M	54.1G	83.66	75.11	74.57	61.33
GuideDecoder [9]	44.0M	22.4G	89.78	81.45	77.75	63.60
TGCAM [22]	-	-	90.60	82.81	77.82	63.69
MMI-UNet [10]	56.2M	22.1G	90.88	83.28	78.42	64.50
MAdapter [23]	-	-	90.22	82.16	78.62	64.78
MG-UNet	30.5M	16.19G	90.90	83.30	78.65	64.80
MG-UNet*	30.5M	10.95G	88.10	77.80	76.39	61.79

Table 2. The ablation study of the QaTa-COV19 test set. "w/o text" indicates that there is no text, and the model uses only the UNet decoder. "w/text" means that text is used during both the training and inference phases, referring to MG-UNet. "Test w/o text" indicates that text is used only during the training phase, referring to MG-UNet*. TS represents the IMBU strategy we proposed during the training phase.

Model		QaTa-COV19	
		DSC↑	IoU↑
w/o text	Baseline	84.18	72.62
w/ text	MG-UNet	90.90	83.30
Test w/o text	MG-UNet*	87.18	77.27
	MG-UNet*+TS	88.10	77.80

Table 3. Impact of the training data size on segmentation performance. The best results are shown in bold.

Method	DSC↑	IoU↑
nnUNet (100% training data)	80.42	70.81
MG-UNet (15% training data)	88.02	78.60
MG-UNet (25% training data)	88.91	80.03
MG-UNet (50% training data)	89.94	81.87
MG-UNet (100% training data)	**90.90**	**83.30**
MG-UNet* (15% training data)	85.30	74.37

4 Conclusion

This paper proposes the MG-UNet model for the segmentation of infected regions in chest images. By designing a learnable memory bank, we establish a bridge between visual and textual information, ensuring that the model can perform inference and maintain segmentation performance even in the absence of textual input. Extensive evaluations on the QaTa-COV19 and MosMedData+ datasets demonstrate that MG-UNet outperforms the best-performing unimodal and multimodal methods. Notably, even with limited training data during the text-free inference phase, MG-UNet surpasses the best unimodal method, highlighting its potential to significantly reduce the reliance on extensive data annotation while offering greater flexibility and scalability for practical applications.

Acknowledgements. This work was supported by the Nature Science Foundation Project of Chongqing Science and Technology Bureau (No. 2024NSCQ-MSX3120).

Disclosure of Interests. The authors have no competing interests to declare that are relevant to the content of this article.

References

1. Cao, H., et al.: Swin-Unet: Unet-like pure transformer for medical image segmentation. In: Karlinsky, L., Michaeli, T., Nishino, K. (eds.) ECCV 2022. LNCS, vol. 13803, pp. 205–218. Springer, Cham (2022). https://doi.org/10.1007/978-3-031-25066-8_9
2. Chen, J., et al.: TransUNet: transformers make strong encoders for medical image segmentation. arXiv preprint arXiv:2102.04306 (2021)
3. Zhou, Z., Rahman Siddiquee, M.M., Tajbakhsh, N., Liang, J.: UNet++: a nested U-net architecture for medical image segmentation. In: Stoyanov, D. (ed.) DLMIA/ML-CDS -2018. LNCS, vol. 11045, pp. 3–11. Springer, Cham (2018). https://doi.org/10.1007/978-3-030-00889-5_1
4. Ronneberger, O., Fischer, P., Brox, T.: U-Net: convolutional networks for biomedical image segmentation. In: Navab, N., Hornegger, J., Wells, W.M., Frangi, A.F. (eds.) MICCAI 2015, Part III. LNCS, vol. 9351, pp. 234–241. Springer, Cham (2015). https://doi.org/10.1007/978-3-319-24574-4_28

5. Radford, A., et al.: Learning transferable visual models from natural language supervision. In: International Conference on Machine Learning, pp. 8748–8763. PMLR (2021)
6. Wang, Z., Wu, Z., Agarwal, D., Sun, J.: MedClip: contrastive learning from unpaired medical images and text. arXiv preprint arXiv:2210.10163 (2022)
7. Huang, S.C., Shen, L., Lungren, M.P., Yeung, Gloria, S.: A multimodal global-local representation learning framework for label-efficient medical image recognition. In: Proceedings of the IEEE/CVF International Conference on Computer Vision, pp. 3942–3951 (2021)
8. Li, Z., Li, Y., Li, Q., et al.: LviT: language meets vision transformer in medical image segmentation. IEEE Trans. Med. Imaging **43**(1), 96–107 (2023)
9. Zhong, Y., Xu, M., Liang, K., Chen, K., Wu, M.: Ariadne's thread: using text prompts to improve segmentation of infected areas from chest X-ray images. In: Greenspan, H., et al. (eds.) MICCAI 2023. LNCS, vol. 14223, pp. 724–733. Springer, Cham (2023). https://doi.org/10.1007/978-3-031-43901-8_69
10. Bui, P.-N., Le, D.-T., Choo, H.: Visual-textual matching attention for lesion segmentation in chest images. In: Linguraru, M.G., et al. (eds.) MICCAI 2024. LNCS, vol. 15009, pp. 702–711. Springer, Cham (2024). https://doi.org/10.1007/978-3-031-72114-4_67
11. Degerli, A., Kiranyaz, S., Chowdhury, M.E., Gabbouj, M.: OsegNet: operational segmentation network for COVID-19 detection using chest x-ray images. In: 2022 IEEE International Conference on Image Processing (ICIP), pp. 2306 2310. IEEE (2022)
12. Liu, Z., Mao, H., Wu, C.Y., Feichtenhofer, C., Darrell, T., Xie, S.: A convnet for the 2020s. In: Proceedings of the IEEE/CVF Conference on Computer Vision and Pattern Recognition, pp. 11976–11986 (2022)
13. Boecking, B., et al.: Making the most of text semantics to improve biomedical vision language processing. In: Avidan, S., Brostow, G., Cissé, M., Farinella, G.M., Hassner, T. (eds.) ECCV 2022. LNCS, vol. 13696, pp. 1–21. Springer, Cham (2022). https://doi.org/10.1007/978-3-031-20059-5_1
14. Vaswani, A., et al.: Attention is all you need. Adv. Neural Inf. Process. Syst. **30** (2017)
15. Morozov, S.P., Andreychenko, A.E., Pavlov, N.A., et al.: MosMedData: chest CT scans with COVID-19 related findings dataset. arXiv preprint arXiv:2005.06465 (2020)
16. Imambi, S., Prakash, K.B., Kanagachidambaresan, G.: PyTorch. Programming with TensorFlow: Solution for Edge Computing Applications, pp. 87–104 (2021)
17. Cardoso, M.J., et al.: MONAI: an open-source framework for deep learning in healthcare. arXiv preprint arXiv:2211.02701 (2022)
18. Oktay, O., et al.: Attention U-net: learning where to look for the pancreas. arxiv 2018. arXiv preprint arXiv:1804.03999 (1804)
19. Wang, H., Cao, P., Wang, J., Zaiane, O.R.: UCTransNet: rethinking the skip connections in U-Net from a channel-wise perspective with transformer. In: Proceedings of the AAAI Conference On Artificial Intelligence, vol. 36, pp. 2441–2449 (2022) (2021)
20. Isensee, F., Jaeger, P.F., Kohl, S.A., Petersen, J., Maier-Hein, K.H.: nnU-Net: a self-configuring method for deep learning-based biomedical image segmentation. Nat. Methods **18**(2), 203–211 (2021)

21. Yang, Z., Wang, J., Tang, Y., Chen, K., Zhao, H., Torr, P.H.: LAVT: language-aware vision transformer for referring image segmentation. In: Proceedings of the IEEE/CVF Conference on Computer Vision and Pattern Recognition, pp. 18155–18165 (2022)
22. Guo, Y., et al.: Common vision-language attention for text-guided medical image segmentation of pneumonia. In: Linguraru, M.G., et al. (eds.) MICCAI 2024. LNCS, vol. 15009, pp. 192–201. Springer, Cham (2024). https://doi.org/10.1007/978-3-031-72114-4_19
23. Zhang, X., et al.: MAdapter: a better interaction between image and language for medical image segmentation. In: Linguraru, M.G., et al. (eds.) MICCAI 2024. LNCS, vol. 15009, pp. 425–434. Springer, Cham (2024). https://doi.org/10.1007/978-3-031-72114-4_41

MIBF-Net: Multi-Modal Information Balanced Fusion Network for Clinical Diagnosis via Patient Narratives and Lesion Image

Zixuan Tang[1,5], Bai Sun[1,5], Shidan He[1,5], Yuan Hong[2], Dongdong Yu[3], Zhenzhong Liu[4], Mengtang Li[1], Bin Chen[3], and Shen Zhao[1,5](✉)

[1] School of Intelligent Systems Engineering, Sun Yat-Sen University, Shenzhen 518107, China
z-s-06@163.com
[2] Zhejiang Normal University, Hangzhou 321004, China
[3] Orthopedics Department, The First Affiliated Hospital of Zhejiang University, Hangzhou 310003, China
[4] Tianjin Key Laboratory for Advanced Mechatronic System Design and Intelligent Control, School of Mechanical Engineering, Tianjin University of Technology, Tianjin 300384, China
[5] Guangdong Key Laboratory of Big Data Analysis and Processing, Guangzhou 510006, China

Abstract. Accurate clinical diagnosis requires comprehensive analysis of medical imaging and patient narratives. However, current computer-aided diagnosis methods focus primarily on imaging modalities while neglecting the integration of patient-reported clinical narratives, due to the scarcity of high-quality patient narratives and the limitations in multimodal information fusion. To address these issues, we propose a dual-component framework consisting of: 1) a Retrieval Augmented Patient Narratives Generation Module (RANGM) that employs a retrieval-enhanced mechanism to guide pre-trained large language models in generating clinically plausible patient narratives; and 2) a Multimodal Information Balanced Fusion Network (MIBF-Net) incorporating our novel Information Balanced Fusion Attention (IBFA) module for effective cross-modal integration, along with a Modal Prediction-Divergent Loss (MP-Loss) to enhance the model's ability to diagnose samples that have ambiguous single modal prediction distribution. Owing to the plug-and-play design, our MIBF-Net can integrate with existing imaging-based state-of-the-art methods. Extensive experiments demonstrate significant performance improvements of 2.3%–4.6% on the HAM10000 dataset and 3.8%–6.4% on the ISIC2019 dataset. Our code is publicly available at https://github.com/sysu19351118/MIBF-Net.

Keywords: Multi-modal · Medical Image Classification · Deep Learning

J. C. Gee et al. (Eds.): MICCAI 2025, LNCS 15960, pp. 366–375, 2026.
https://doi.org/10.1007/978-3-032-04927-8_35

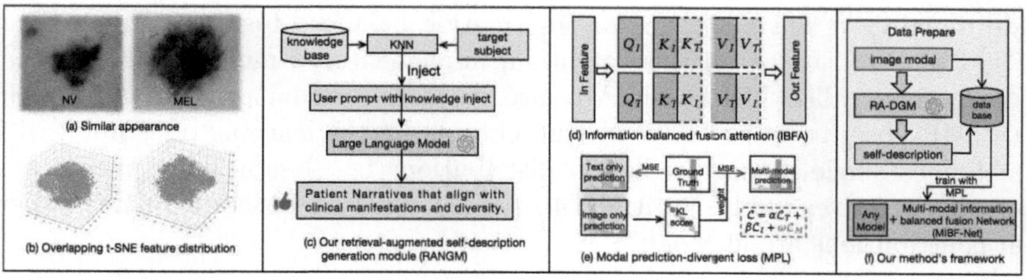

Fig. 1. (a)-(b) Some diseases, such as NV and MEL, exhibit highly similar visual appearances with overlapping feature distributions. (c) Our RANGM module enhances LLM generation capabilities by retrieving relevant medical knowledge through K-Nearest Neighbors (KNN) matching, enabling the generation of highly realistic patient narratives. (d)-(e) The meticulously designed Information Balanced Fusion Attention (IBFA) module achieves effective integration across modalities. Through the Modal Prediction-Divergent Loss (MP-Loss), we reinforce the model's focus on analyzing challenging samples with ambiguous modality information. (f) Our approach can serve as a plug-and-play solution that can be seamlessly integrated with any existing model to enhance their understanding of patient narratives and improve diagnostic accuracy.

1 Introduction

The integration of linguistic modalities (such as patient narratives) with medical imaging is essential for accurate disease diagnosis [24], as visual data alone often inadequately captures complex medical conditions. For instance, early-stage melanomas and nevi are challenging to differentiate visually [12,20,21] , as both exhibit irregular shapes and brownish-black pigmentation (Fig. 1(a)). This difficulty is further compounded by the significant overlap in their t-SNE distributions (Fig. 1(b)). Patient narratives, such as localized pain or lifestyle factors, provide critical contextual information that enhances diagnostic precision. Thus, combining linguistic and visual modalities is vital for robust disease diagnosis [3,6,18,29].

However, obtaining and using patient narratives is challenging [16] [9]. Clinically, there is a lack of high-quality patient narratives data with image-text matching. Additionally, the use of existing text generation models is limited by poor medical knowledge generation quality and low diversity [2,23,27], making it challenging to simulate real clinical diagnostic scenarios. Meanwhile, there is a lack of effective multi-modal fusion methods and training approaches that focus on samples with predictive discrepancies [7,10,13]. Existing multi-modal information fusion methods exhibit information weighting bias when attending to multiple modalities, failing to balance attention between the two types of information [8,14,26,28,30]. For samples with significant discrepancies in single-modal predictions, there is no effective method to specifically supervise these samples [1].

To address these issues, we propose a novel probability recall augmented generation method (Fig. 1(c)) capable of generating realistic patient narratives.

Additionally, we have designed an information balanced fusion attention-IBFA (Fig. 1(d)) and, based on this attention, implemented a multimodal classification method called MIBF-Net. We also devised a modal prediction-divergent loss (MP-Loss) to assist the model in focusing on the learning of samples with ambiguous single modal prediction distribution (i.e., lesion image modal and patient narratives modal are trend to predict different disease). In this paper, our contributions are fourfold:

- We are the first to combine patient narratives with lesion images for computer-aided diagnosis. Patient narratives provide information beyond the lesion images, such as pain perception. Compared to using lesion images alone for diagnosis, our method better meets clinical needs and aligns more closely with the logic of clinical diagnosis.
- To generate high quality patient narratives, we propose the Retrieval Augmented Patient Narratives Generation Module (RANGM), which can retrieve corresponding knowledge to enhance the professionalism and diversity of the patient narratives generated by the LLM.
- To utilize patient narratives and lesion images efficiently, we propose the Multi-modal Information Balanced Fusion Network (MIBF-Net). Thanks to our Information Balanced Fusion Attention (IBFA), MIBF-Net can fuse information from the two modalities in a balanced manner, without being dominated by single-modal judgments.
- To focus more on hard samples that have ambiguous single modal prediction distribution, we propose a meticulously designed modal prediction-divergent loss (MP-Loss) to supervise network training. This loss function guides the model pay more attention on these samples using KL divergency score, thereby improving classification performance for challenging samples.

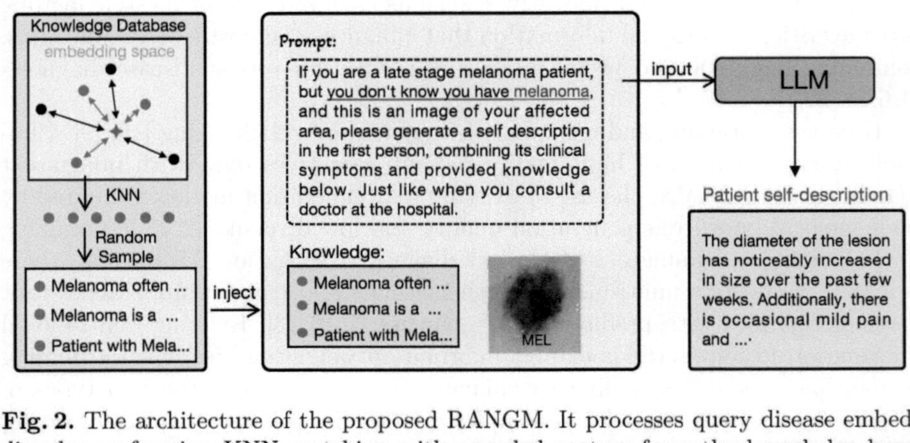

Fig. 2. The architecture of the proposed RANGM. It processes query disease embedding by performing KNN matching with encoded vectors from the knowledge base, randomly selecting M matches from the top K candidates to inject into the prompt, thereby guiding the LLM to generate high-quality patient narratives.

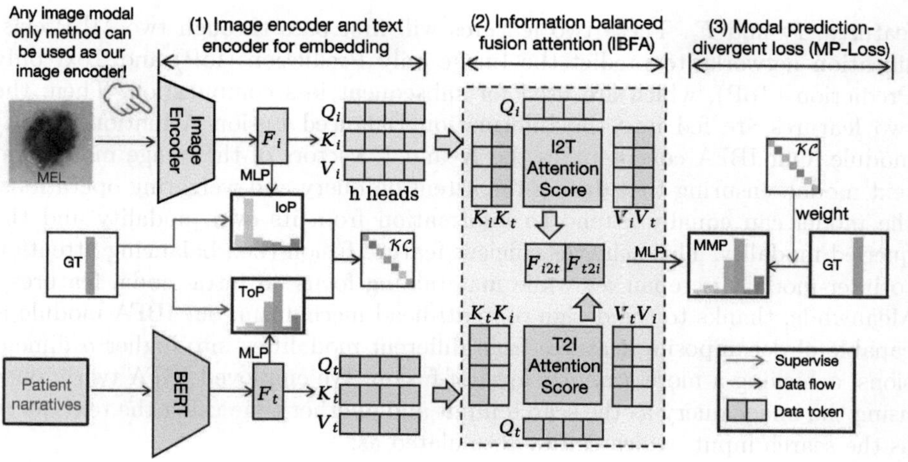

Fig. 3. The architecture of the proposed MIBF-Net: the input images and patient narratives are first encoded through encoders to obtain individual modality predictions (Image only prediction (IoP) and Text only Prediction ToP), then the encoded features are fused through IBFA for multimodal feature integration, with MLP subsequently predicting multimodal results (MMP), while MP-Loss loss optimizes network parameters by leveraging the distribution discrepancy between IoP and ToP along with MMP prediction distribution bias.

2 Method

Retrieval Augmented Patient Narratives Generation Module (RANGM) overcomes the limitations of existing large language models (LLM) in downstream professional knowledge, enabling more comprehensive and diverse generation. We have compiled a substantial knowledge base by gathering information related to the diseases covered in our dataset. This knowledge base is then hashed using a BERT encoder as the hash function. To retrieve relevant knowledge, we employ the same BERT encoder for the disease names and their basic definitions that we wish to query. Through KNN (k-nearest neighbors) hashing matching with the knowledge base, we obtain the retrieval knowledge. To enhance diversity, we randomly select the top M pieces of knowledge with the highest matching scores as the final retrieval knowledge, which is then integrated into our meticulously designed prompt. Assume that the patient narratives, query tokens, and knowledge tokens are y, q, $\mathcal{H}$, the above process can be formulated as:

$$y = \arg\max_{y} \mathcal{LLM}\left(y \mid p, \mathcal{R}_{\text{final}} = \text{RandomSelect}\left(\underset{h_i \in \mathcal{H}}{\text{argtopK}}\left(\frac{q \cdot h_i}{\|q\|\|h_i\|}\right), M\right)\right) \tag{1}$$

Multi-modal Information Balanced Fusion Network (MIBF-Net). The pipeline of our MIBF-Net is shown in Fig. 3. The text modal and image modal input will firstly passing through their respective encoders to obtain their own features, F_t and F_i. These two features will first pass through two MLP classification networks to predict the Image only Prediction (IoP) and Text only Prediction (ToP), which are used for subsequent loss computation. Then, the two features are fed into the Information Balanced Fusion Attention (IBFA) module. Our IBFA concatenates the K and V vectors of the image modal and text modal, ensuring that during the attention query and weighting operations, the model can equally attend to information from its own modality and the queried modality. This achieves efficient feature fusion (i.e., balancing attention to inter-modal discrepancies while maintaining focus on intra-modal features). Meanwhile, thanks to the design of multi-head mechanism, our IBFA module is capable of decomposing features from different modalities into higher h dimensions, achieving a more granular level of fusion. We employed IBFA twice: once using the image query as the search input and another time using the text query as the search input, which can be formulated as:

$$\mathbf{F}_{i2t} = \mathcal{IBFA}(\mathbf{Q}_i, \mathbf{K}, \mathbf{V}) = \left[\sigma \left(\frac{\mathbf{Q}_i \mathbf{W}_j^{(q)} \cdot ([\mathbf{K}_i, \mathbf{K}_t] \mathbf{W}_j^{(k)})^\top}{\sqrt{d_k}} \right) [\mathbf{V}_i, \mathbf{V}_t] \cdot \mathbf{W}_j^{(v)} \right]_{j=1}^{h} \mathbf{W}^0 \tag{2}$$

$$\mathbf{F}_{t2i} = \mathcal{IBFA}(\mathbf{Q}_t, \mathbf{K}, \mathbf{V}) = \left[\sigma \left(\frac{\mathbf{Q}_t \mathbf{W}_j^{(q)} \cdot ([\mathbf{K}_t, \mathbf{K}_i] \mathbf{W}_j^{(k)})^\top}{\sqrt{d_k}} \right) [\mathbf{V}_t, \mathbf{V}_i] \cdot \mathbf{W}_j^{(v)} \right]_{j=1}^{h} \mathbf{W}^0 \tag{3}$$

where $[\cdot]$ means concatenate, i represents image modal, t represents text modal and $\sigma(z) = \frac{e^{z_i}}{\sum_{j=1}^{n} e^{z_j}}$ is softmax function used to normalize the probability distribution of attention scores, emphasizing important information. Then the fused information $[\mathbf{F}_{i2t}, \mathbf{F}_{t2i}]$ from the two modalities is concatenated and fed into the final MLP classifier to get multi-modal predictions (MMP).

Modal Prediction-Divergent Loss. We propose a loss function that leverages the average Kullback-Leibler (KL) divergence [17] between the classification predictions of the image and text modalities to dynamically weight the supervision of the final multimodal output. We denote the network parameters for predicting these three outputs as θ_i, θ_t, and $\theta_{i,t}$, respectively. As illustrated in Fig. 3, we employ two Multi-Layer Perceptrons (MLP) to generate the *image-only prediction* (i.e., IoP $= f(\mathbf{x}; \theta_i)$) and the *text-only prediction* (i.e., ToP $= f(\mathbf{x}; \theta_t)$). Simultaneously, the IBFA module fuses the information from both modalities to produce the *multi-modal prediction* (i.e., MMP $= f(\mathbf{x_i}, \mathbf{x_t}; \theta_{i,t})$). Based on this framework, the modal prediction-divergent loss can be formulated as follows:

$$\mathcal{L}_{\text{MP-Loss}} = \mathbb{E}_{(\mathbf{x_i}, \mathbf{x_t}, y) \sim \mathcal{D}} \begin{bmatrix} \alpha \cdot ||y - f(\mathbf{x_i}; \theta_i)||_2 \\ + \beta \cdot ||y - f(\mathbf{x_t}; \theta_t)||_2 \\ + \gamma \cdot \mathcal{KL} \cdot ||y - f(\mathbf{x_i}, \mathbf{x_t}; \theta_{i,t})||_2 \end{bmatrix} \tag{4}$$

where, $\mathcal{KL}$ denotes the Kullback-Leibler divergence score calculated between the IoP and the ToP:

$$\mathcal{KL} = \frac{1}{2}\left(\sum_{x} f(\mathbf{x};\theta_i)\log\frac{f(\mathbf{x};\theta_i)}{f(\mathbf{x};\theta_t)} + \sum_{x} f(\mathbf{x};\theta_t)\log\frac{f(\mathbf{x};\theta_t)}{f(\mathbf{x};\theta_i)}\right) \tag{5}$$

This approach enhances learning for challenging samples where there is a significant information discrepancy between the image and text modalities.

3 Experiment

Datasets. We conduct experiment on ISIC2019[1] and HAM10000 [22] datasets to prove the effectiveness of proposed method. The ISIC2019 dataset contains 25,531 dermatological images, while the HAM10000 dataset comprises 10,015 dermatological images. Both datasets cover seven skin disease categories, including melanoma (MEL), nevus (NV), basal cell carcinoma (BCC), actinic keratosis (AK), benign keratosis (BKL), dermatofibroma (DF), and vascular lesions (VASC). Additionally, the ISIC2019 dataset includes an extra category of squamous cell carcinoma (SCC). These datasets are characterized by significant class imbalance and high visual similarity between different categories, making them challenging benchmarks. For the experimental setup, we adopt a random split strategy, using 80% as the training set and 20% as the test set. we use Top-1 accuracy and F1-score to evaluate the model's performance. All images are resized to 224 pixles.

Implementation Details. We use BERT [4] as text encoder. For the image encoder, our approach can incorporate existing image-only diagnostic methods as our image encoder, enabling these state-of-the-art models to seamlessly understand patient narratives and perform clinical scenario-compliant diagnostic tasks. We employ GPT-4 as our Large Language Model (LLM), and API access can be obtained through an official subscription[2]. Since both datasets include records of the lesion locations, we have also integrated the lesion locations into the outputs generated by the LLM. For the loss function, we set $\alpha = 0.6$, $\beta = 0.4$, $\gamma = 0.1$. We utilize the Adam optimizer with a learning rate of 5e−5 to train our network with 50 epochs. Experiment is executed on four Nvidia RTX 3090 GPUs.

Comparative Experiments. Since our work is the first to utilize patient narratives as an auxiliary modality for image-based disease classification, meanwhile, our method does not specify a particular image encoder, and thus, it can be considered as a plug-and-play module applicable to any existing state-of-the-art image classification method, such as Resnet50 [11], DaViT [5] , HiFuse [15], Med-Mamba [25] and ConvNeXt [19]. The effectiveness of our proposed method has been demonstrated through experiments on two datasets as shown in Tab. 1. For the HAM10000 dataset, our method achieves a top accuracy of 0.9295. Notably, the HiFuse method demonstrates a 4.6% improvement after integrating our module. On the ISIC2019 dataset, our approach attains a maximum accuracy of

[1] https://challenge.isic-archive.com/landing/2019/.

[2] https://openai.com/api/.

Table 1. Comparison of Different Methods on Two Datasets with Improvements

Experiment on HAM10000 Dataset											
Method	Acc	F1	MEL	NV	BCC	AK	BKL	DF	VASC	SCC	Improv.
Resnet50 [11]	0.8915	0.8911	0.6552	0.9763	0.8600	0.5714	0.7328	0.6875	0.8320	-	+3.15%
w/ ours	0.9230	0.9490	0.7034	0.9824	0.8979	0.7200	0.8384	0.8260	0.7812	-	
DaViT [5]	0.8740	0.8720	0.6453	0.9664	0.7449	0.6533	0.6332	0.6957	0.9062	-	+3.20%
w/ ours	0.9060	0.9053	0.6667	0.9704	0.8600	0.6571	0.8190	0.7500	0.9500	-	
HiFuse [15]	0.8540	0.8499	0.5407	0.9555	0.7449	0.6133	0.6245	0.6522	0.875	-	+4.60%
w/ ours	0.9000	0.8751	0.7703	0.9023	0.9259	0.9642	0.9432	0.9047	0.8928	-	
MedMamba [25]	0.8940	0.8814	0.3448	0.9911	0.8800	0.8857	0.7856	0.7500	0.9500	-	+3.00%
w/ ours	0.9240	0.9198	0.7500	0.9788	0.8571	0.7066	0.8384	0.7826	0.9687	-	
ConvNeXt [19]	0.9065	0.9059	0.5747	0.9719	0.9000	0.8286	0.8190	0.8750	0.9500	-	+2.30%
w/ ours	0.9295	0.8899	0.7707	0.9524	0.9021	0.857	0.9090	0.9523	0.9227	-	
Experiment on ISIC2019 Dataset											
Method	Acc	F1	MEL	NV	BCC	AK	BKL	DF	VASC	SCC	Improv.
Resnet50 [11]	0.8488	0.8488	0.7580	0.9188	0.9034	0.5783	0.7395	0.7111	0.9400	0.5752	+6.44%
w/ ours	0.9132	0.9116	0.8527	0.9757	0.9083	0.7617	0.8355	0.7010	0.9162	0.71264	
DaViT [5]	0.8934	0.8917	0.7947	0.9589	0.9271	0.7283	0.8069	0.6957	0.9355	0.7288	+3.40%
w/ ours	0.9274	0.9263	0.8613	0.9761	0.9484	0.7774	0.8860	0.7783	0.9359	0.7339	
HiFuse [15]	0.8741	0.8740	0.7859	0.9410	0.8480	0.7283	0.8340	0.7391	0.9335	0.6610	+4.17%
w/ ours	0.9158	0.914	0.8585	0.9766	0.8821	0.7888	0.8584	0.7886	0.8620	0.7475	
MedMamba [25]	0.8867	0.8847	0.7439	0.9581	0.9331	0.7717	0.8263	0.7826	0.9032	0.6441	+3.86%
w/ ours	0.9253	0.9290	0.8571	0.9706	0.9208	0.8288	0.9061	0.8195	0.8916	0.7825	
ConvNeXt [19]	0.9061	0.9045	0.8366	0.9697	0.9301	0.8043	0.7683	0.6522	0.9355	0.7627	+3.82%
w/ ours	0.9443	0.9430	0.8807	0.9877	0.9568	0.8316	0.8865	0.8865	0.9458	0.8660	

0.9445, while ResNet50 shows a significant 6.44% enhancement when equipped with our module. Furthermore, substantial improvements are observed in both the F1-score and classification accuracy across individual disease categories.

Ablation Study. The ablation experiments conducted on the ISIC dataset further demonstrate the effectiveness of our proposed method. First, we show that our RANGM module generates more diverse textual narratives. As illustrated in Fig. 4(a), compared to the baseline without RANGM, the samples generated with RANGM exhibit significant improvements across evaluation metrics including Distinct-N, Vocabulary Usage, and Self-BLEU. We have also randomly selected some samples, as shown in Fig. 5, and found that after applying RANGM, the generated patient narratives align more closely with clinical scenarios, and the relevant knowledge reflected is more accurate. Additionally, separate ablation studies on individual modules validate their contributions. Using ResNet-50 as the image encoder, the baseline achieves a Top-1 accuracy of 0.8488 without any specialized design. After integrating the RANGM module (which supplements patient narratives), the accuracy increases to 0.8696, demonstrating that incorporating non-visual information (e.g., pain symptoms, lifestyle habits) through clinical narratives effectively enhances classification performance. Further replacing conventional cross-attention with our IBFA module for modality fusion elevates accuracy to 0.9015, confirming that IBFA's dual attention mechanism

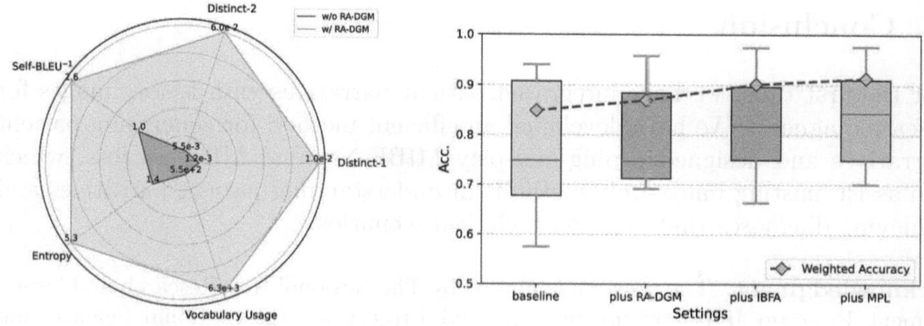

(a) Comparison of Text Generation Methods with and without RANGM

(b) Comparison of Text Generation Methods with and without our proposed module.

Fig. 4. Ablation experiment results. (a) After employing our RANGM, the metrics used to evaluate the diversity of model generation have shown significant improvement. (b) The box plot indicates the accuracy rates of various categories under different settings, with the line representing the change in accuracy across all samples. As our designs are sequentially incorporated, the classification performance continuously improves, which verifies the effectiveness of our method.

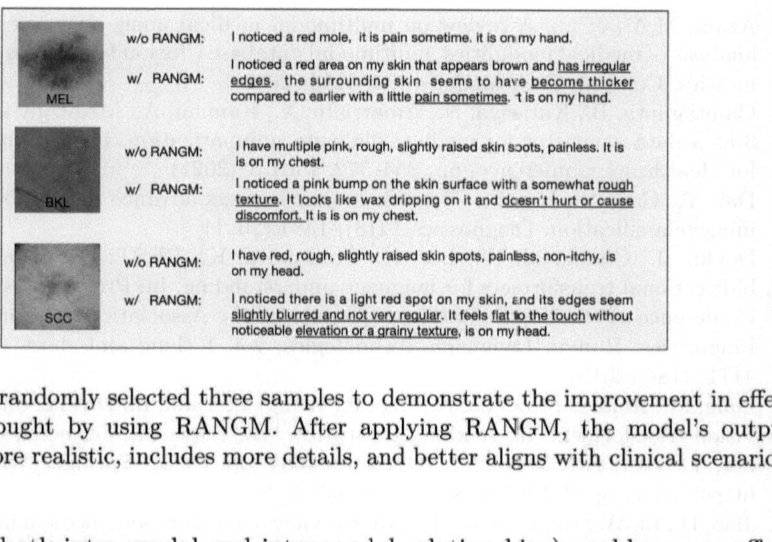

Fig. 5. We randomly selected three samples to demonstrate the improvement in effectiveness brought by using RANGM. After applying RANGM, the model's output becomes more realistic, includes more details, and better aligns with clinical scenarios.

(modeling both intra-modal and inter-modal relationships) enables more effective information aggregation. Finally, employing the MP-Loss loss to strengthen supervision on ambiguous single-modality prediction distribution samples let accuracy reaches 0.9132. Meanwhile, as shown in Fig. 4(b), the box plot analysis reveals improvements in both per-class accuracy and median accuracy, indicating MP-Loss's capability to better supervise challenging samples.

4 Conclusion

For the first time, we have integrated patient narratives with lesion images for disease diagnosis. We have developed an efficient method for generating patient narratives and designed a plug-and-play MIBF-Net and MP-Loss loss, which can assist existing image-only methods in understanding patient narratives and achieving diagnoses that align with clinical scenarios.

Acknowledgment. The research is funded by The National Key Research and Development Program Inter-governmental Special Project for International Science and Technology Innovation Cooperation (2022YFE0112500), Shenzhen Science and Technology Program (JCYJ20240813151224032, JCYJ20240813151102004), and Shenzhen Medical Research Fund (B2402030).

Disclosure of Interests. The authors have no competing interests to declare that are relevant to the content of this article.

References

1. Azam, M.A., et al.: A review on multimodal medical image fusion: compendious analysis of medical modalities, multimodal databases, fusion techniques and quality metrics. Comput. Biol. Med. **144**, 105253 (2022)
2. Chintagunta, B., Katariya, N., Amatriain, X., Kannan, A.: Medically aware GPT-3 as a data generator for medical dialogue summarization. In: Machine Learning for Healthcare Conference, pp. 354–372. PMLR (2021)
3. Dai, Y., Gao, Y., Liu, F.: TransMed: transformers advance multi-modal medical image classification. Diagnostics **11**(8), 1384 (2021)
4. Devlin, J., Chang, M.-W., Lee, K., Toutanova, K.: BERT: pre-training of deep bidirectional transformers for language understanding. In: Proceedings of the 2019 Conference of the North American Chapter of the Association for Computational Linguistics: Human Language Technologies, vol. 1 (long and short papers), pp. 4171–4186 (2019)
5. Ding, M., Xiao, B., Codella, N., Luo, P., Wang, J., Yuan, L.: DaViT: dual attention vision transformers. In: Avidan, S., Brostow, G., Cissé, M., Farinella, G.M., Hassner, T. (eds.) ECCV 2022. LNCS, vol. 13684, pp. 74–92. Springer, Cham (2022). https://doi.org/10.1007/978-3-031-20053-3_5
6. Jiao, D., Li, W., Ke, L., Xiao, B.: An overview of multi-modal medical image fusion. Neurocomputing **215**, 3–20 (2016)
7. Duan, J., Xiong, J., Li, Y., Ding, W.: Deep learning based multimodal biomedical data fusion: an overview and comparative review. Inf. Fusion **112**, 102536 (2024)
8. Gao, J., Li, P., Chen, Z., Zhang, J.: A survey on deep learning for multimodal data fusion. Neural Comput. **32**(5), 829–864 (2020)
9. Grob, R., et al.: What words convey: the potential for patient narratives to inform quality improvement. Milbank Q. **97**(1), 176–227 (2019)
10. Guo, Z., Li, X., Huang, H., Guo, N., Li, Q.: Medical image segmentation based on multi-modal convolutional neural network: study on image fusion schemes. In: 2018 IEEE 15th International Symposium on Biomedical Imaging (ISBI 2018), pp. 903–907. IEEE (2018)

11. He, K., Zhang, X., Ren, S., Sun, J.: Deep residual learning for image recognition. In: Proceedings of the IEEE Conference on Computer Vision and Pattern Recognition, pp. 770–778 (2016)
12. Hessler, M., et al.: Melanoma biomarkers and their potential application for in vivo diagnostic imaging modalities. Int. J. Mol. Sci. **21**(24), 9583 (2020)
13. Huang, B., Yang, F., Yin, M., Mo, X., Zhong, C.: A review of multimodal medical image fusion techniques. Comput. Math. Methods Med. **2020**(1), 8279342 (2020)
14. Huang, Z., Wang, X., Huang, L., Huang, C., Wei, Y., Liu, W.: CCNet: criss-cross attention for semantic segmentation. In: Proceedings of the IEEE/CVF International Conference on Computer Vision, pp. 603–612 (2019)
15. Huo, X., et al.: Hifuse: Hierarchical multi-scale feature fusion network for medical image classification. Biomed. Signal Process. Control **87**, 105534 (2024)
16. Hutchison, C., McCreaddie, M.: The process of developing audiovisual patient information: challenges and opportunities. J. Clin. Nurs. **16**(11), 2047–2055 (2007)
17. Kullback, S., Leibler, R.A.: On information and sufficiency. Ann. Math. Stat. **22**(1), 79–86 (1951)
18. Liu, C., et al.: Visual–language foundation models in medicine. Vis. Comput., 1–20 (2024)
19. Liu, Z., Mao, H., Wu, C.-Y., Feichtenhofer, C., Darrell, T., Xie, S.: A convnet for the 2020s. In: Proceedings of the IEEE/CVF Conference on Computer Vision and Pattern Recognition, pp. 11976–11986 (2022)
20. Matthews, T.E., Piletic, I.R., Selim, M.A., Simpson, M.J., Warren, W.S.: Pump-probe imaging differentiates melanoma from melanocytic nevi. Sci. Transl. Med. **3**(71), 71ra15–71ra15 (2011)
21. Rastgoo, M., Garcia, R., Morel, O., Marzani, F.: Automatic differentiation of melanoma from dysplastic nevi. Comput. Med. Imaging Graph. **43**, 44–52 (2015)
22. Tschandl, P., Rosendahl, C., Kittler, H.: The HAM10000 dataset, a large collection of multi-source dermatoscopic images of common pigmented skin lesions. Sci. Data **5**(1), 1–9 (2018)
23. Wang, B., et al.: Pre-trained language models in biomedical domain: a systematic survey. ACM Comput. Surv. **56**(3), 1–52 (2023)
24. Wang, Z., Wu, Z., Agarwal, D., Sun, J.: MedCLIP: contrastive learning from unpaired medical images and text. In: Proceedings of the Conference on Empirical Methods in Natural Language Processing. Conference on Empirical Methods in Natural Language Processing, vol. 2022, p. 3876 (2022)
25. Yue, Y., Li, Z.: MedMamba: vision mamba for medical image classification. arXiv preprint arXiv:2403.03849 (2024)
26. Zhang, C., Yang, Z., He, X., Deng, L.: Multimodal intelligence: representation learning, information fusion, and applications. IEEE J. Sel. Topics Signal Process. **14**(3), 478–493 (2020)
27. Zhang, H., Song, H., Li, S., Zhou, M., Song, D.: A survey of controllable text generation using transformer-based pre-trained language models. ACM Comput. Surv. **56**(3), 1–37 (2023)
28. Zhang, R., Guo, H., Zhang, Z., Yan, P., Zhao, S.: Gamed-snake: gradient-aware adaptive momentum evolution deep snake model for multi-organ segmentation (2025)
29. Zhang, R., et al.: MARL-MambaContour: unleashing multi-agent deep reinforcement learning for active contour optimization in medical image segmentation (2025)
30. Zhao, F., Zhang, C., Geng, B.: Deep multimodal data fusion. ACM Comput. Surv. **56**(9), 1–36 (2024)

MiCo: Multiple Instance Learning with Context-Aware Clustering for Whole Slide Image Analysis

Junjian Li[1], Jin Liu[1,2(✉)], Hulin Kuang[1], Hailin Yue[1], Mengshen He[1], and Jianxin Wang[1,2]

[1] Hunan Provincial Key Lab on Bioinformatics, School of Computer Science and Engineering, Central South University,, Changsha 410083 Hunan, China
liujin06@csu.edu.cn

[2] Xinjiang Engineering Research Center of Big Data and Intelligent Software, School of Software, Xinjiang University, Urumqi 830091, China

Abstract. Multiple instance learning (MIL) has shown significant promise in histopathology whole slide image (WSI) analysis for cancer diagnosis and prognosis. However, the inherent spatial heterogeneity of WSIs presents critical challenges, as morphologically similar tissue types are often dispersed across distant anatomical regions. Conventional MIL methods struggle to model these scattered tissue distributions and capture cross-regional spatial interactions effectively. To address these limitations, we propose a novel Multiple instance learning framework with Context-Aware Clustering (MiCo), designed to enhance cross-regional intra-tissue correlations and strengthen inter-tissue semantic associations in WSIs. MiCo begins by clustering instances to distill discriminative morphological patterns, with cluster centroids serving as semantic anchors. To enhance cross-regional intra-tissue correlations, MiCo employs a Cluster Route module, which dynamically links instances of the same tissue type across distant regions via feature similarity. These semantic anchors act as contextual hubs, propagating semantic relationships to refine instance-level representations. To eliminate semantic fragmentation and strengthen inter-tissue semantic associations, MiCo integrates a Cluster Reducer module, which consolidates redundant anchors while enhancing information exchange between distinct semantic groups. Extensive experiments on two challenging tasks across nine large-scale public cancer datasets demonstrate the effectiveness of MiCo, showcasing its superiority over state-of-the-art methods. The code is available at https://github.com/junjianli106/MiCo.

Keywords: Multiple Instance Learning · Whole Slide Image · Context-Aware Clustering

1 Introduction

Whole Slide Images (WSIs) provide comprehensive and detailed representations of cellular morphology and the tumor microenvironment, both of which are

J. C. Gee et al. (Eds.): MICCAI 2025, LNCS 15960, pp. 376–385, 2026.
https://doi.org/10.1007/978-3-032-04927-8_36

crucial for accurate diagnostic evaluation and prognostic assessment [2,3,6,21]. However, histopathological WSIs often exhibit extremely high resolutions, reaching up to 100,000 × 100,000 pixels. The high resolution of WSIs, along with the absence of pixel-level annotations, presents significant challenges in their analysis and modeling [8,9,12,14,17]. Although Multiple Instance Learning (MIL) has emerged as a paradigm for weakly supervised WSI analysis [1,20], existing methods still struggle to handle the spatial complexity of WSIs.

Despite their capability in extracting discriminative features from local image patches, current foundation models still struggle with the challenge of spatial heterogeneity in pathological structures. This core limitation manifests at two interrelated levels: a failure to capture both fine-grained cellular arrangements and broader tissue architecture. At the macroscopic level, key structures of the same histological type are often discontinuously distributed as multi-focal sites across distant regions. For instance, lesions of lymphovascular invasion or perineural invasion may span multiple anatomical planes [11,13]. This spatial separation hinders the model from learning the intrinsic correlations among instances that belong to the same pathological class but are located in different regions. At the microscopic level, each of these distributed structures interacts with a highly heterogeneous cellular microenvironment, forming complex local pathological patterns that require simultaneous modeling of local discriminative features and cross-regional semantic dependencies.

Despite significant progress, existing MIL methods still struggle to model long-range dependencies between pathologically related regions that are morphologically distant, a challenge arising from the inherent spatial heterogeneity of WSIs. For instance, traditional attention-based MIL methods (e.g., AMIL [5]) leverage attention mechanisms to dynamically identify and aggregate critical instances, which tend to focus on instance-level feature learning but fail to capture the semantic relationships between distant pathological regions. Transformer-based methods (e.g., TransMIL [15]) use transformers to learn the global contextual information of WSIs but treat all interactions as homogeneous, failing to capture dynamic heterogeneity interactions across the tumor microenvironment. Graph-based methods (e.g., PatchGCN [1]) rely on fixed neighborhood definitions, while hierarchical methods like HVTSurv [16] restrict local learning to pre-determined sub-regions, both inadequate for modeling tissue-level morphological continuities across spatial distributions.

To overcome these limitations, we propose MiCo, a novel Multiple Instance Learning framework with Context-Aware Clustering. MiCo employs a Cluster Route module to enhance intra-tissue semantic associations by aggregating and propagating information from dispersed patches of the same tissue type, refining instance-level representations. Additionally, MiCo integrates a Cluster Reducer module to eliminate semantic fragmentation and strengthen inter-tissue semantic associations, consolidating semantically redundant anchors while preserving pathological diversity. Extensive experiments on two challenging tasks across nine large-scale public cancer datasets validate MiCo's effectiveness, showcasing

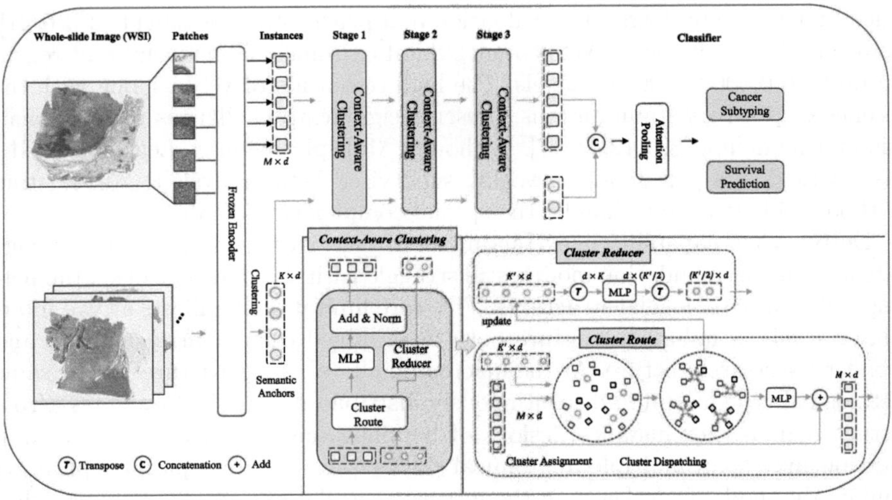

Fig. 1. Overview of MiCo. MiCo consists of multi-layered context-aware clustering modules. Each module is organized by a Cluster Route module, which aggregates and propagates semantic information, and a Cluster Reducer module, which consolidates redundant anchors while enhancing information exchange between semantically distinct anchors.

its superior performance over state-of-the-art methods in survival prediction and cancer subtyping tasks.

2 Methodology

2.1 Overview of MiCo

To address the challenges of spatial heterogeneity in histopathological WSIs, we propose MiCo, a novel MIL framework with Context-Aware Clustering. As illustrated in Fig. 1, MiCo consists of multi-layered context-aware clustering modules, each organized by two core components: the Cluster Route (CluRoute) module and the Cluster Reducer (CluReducer) module. MiCo begins by clustering instances to distill discriminative morphological patterns, with cluster centroids serving as semantic anchors. The CluRoute dynamically links instances of the same tissue type across distant regions based on feature similarity, aggregating and propagating semantic information to refine instance-level features. The CluReducer consolidates redundant anchors while facilitating information exchange between distinct semantic groups, thereby eliminating semantic fragmentation and strengthening inter-tissue semantic associations. Then, the refined instances and anchor features are fused, and an attention pooling mechanism is applied to enable the model to focus on the most task-relevant information.

2.2 WSI Preprocessing

Given a WSI $\boldsymbol{W}_j$, we first crop the tissue regions into non-overlapping patches $\{p_{j,m}\}_{m=1}^{M}$, where the total number of patches varies per WSI. Each patch is encoded into a feature vector $h_{j,m} \in \mathbb{R}^d$ using the CONCH v1.5 [4], a multi-modal whole-slide foundation model pre-trained with self-supervised and vision-language objectives. This process produces a feature matrix $H_j \in \mathbb{R}^{M \times d}$ for each WSI. To simplify notation, we drop the subscript j. To initialize semantic anchors, we apply K-means clustering on the training instances, obtaining cluster centers $S = \{s_k\}_{k=1}^{K}$. These centers serve as semantic anchors representing distinctive tissue patterns.

2.3 Context-Aware Clustering

Morphologically similar tissue types are often dispersed across distant anatomical regions, making it challenging to capture their contextual relationships and semantic consistency. MiCo addresses this challenge through its multi-layered context-aware clustering modules, which consist of two core components: the Cluster Route and the Cluster Reducer. The Cluster Route aggregates and propagates semantic information by linking instances of the same tissue type across distant regions, while the Cluster Reducer consolidates redundant anchors and facilitates information exchange between distinct semantic groups.

Cluster Route. Given a WSI containing M instances, each patch p_m is represented by a feature embedding $h_m \in \mathbb{R}^d$. To establish cross-regional intra-tissue correlations, we compute the cosine similarity between each instance feature h_i and K' learnable semantic anchors $\{s_k\}_{k=1}^{K'}$:

$$A_{l(m,k)} = \frac{h_m^\top s_k}{|h_m| \cdot |s_k|}, \tag{1}$$

where $A_l \in \mathbb{R}^{M \times K'}$ quantifies the semantic alignment between instances and semantic anchors.

The argmax operation ensures that each instance is assigned to its most relevant semantic anchor, preserving semantic consistency across pathological structures. However, argmax is non-differentiable, which disrupts gradient flow to the anchors. To address this, we use a straight-through estimator [19] to enable instance-to-anchor assignments while maintaining gradient propagation:

$$\hat{A}_l = \text{one-hot}\left(\arg\max(A_l)\right) + A_l - \text{sg}(A_l), \tag{2}$$

where sg$(\cdot)$ denotes the stop-gradient operator. During forward propagation, $\hat{A}_l \in \mathbb{R}^{M \times K'}$ is a binary matrix where each row is one-hot encoded, assigning each instance to its most similar cluster. During backward propagation, gradients bypass the non-differentiable arg max operation and flow through A_l, enabling end-to-end training.

For each semantic anchor s_k, we aggregate the features of its assigned instances to generate a context-aware representation:

$$\tilde{s}_k = \frac{1}{N_k} \sum_{m=1}^{M} \hat{A}_{l(m,k)} \cdot h_m, \quad N_k = \sum_{m=1}^{M} \hat{A}_{l(m,k)}. \tag{3}$$

where $\tilde{s}_k \in \mathbb{R}^d$ captures shared morphological semantics (e.g., tumor stroma or immune infiltration) within the anchor group, and N_k denotes the number of instances assigned to the k-th anchor. This aggregation step unifies spatially dispersed instances sharing similar semantics, directly mitigating spatial heterogeneity. We collect all $\tilde{s}_k$ into a matrix $\tilde{S} \in \mathbb{R}^{K' \times d}$, which serves as the input to the Cluster Reducer module.

To propagate inter-region context, we align each instance feature h_m with its assigned semantic anchor semantics:

$$h_m' = h_m + MLP(h_m + \hat{A}_{l(m,k)} \cdot \tilde{s}_k). \tag{4}$$

The updated feature h_m' combines instance-level features with cross-regional semantic context, creating a unified representation of spatially dispersed tissue structures. This enriched feature is then passed to the next context-aware clustering module for further refinement, enabling the model to progressively capture complex anatomical relationships.

Cluster Reducer. To reduce semantic redundancy, the Cluster Reducer merges anchors with similar characteristics by modeling inter-anchor relationships. Given initial semantic anchors $\tilde{S} \in \mathbb{R}^{K' \times d}$, we transpose $\tilde{S}$ to $\tilde{S}^\top \in \mathbb{R}^{d \times K'}$ and apply a MLP to model nonlinear interactions:

$$\tilde{S}' = MLP(\tilde{S}^\top), \tag{5}$$

where $\tilde{S}' \in \mathbb{R}^{d \times K'/2}$ represents the refined anchors after redundancy reduction. The MLP captures semantic dependencies, such as merging scattered tumor-infiltrating lymphocyte clusters, by consolidating semantically similar anchors. Transposing $\tilde{S}'$ back to $\tilde{S} \in \mathbb{R}^{K'/2 \times d}$ yields compact and meaningful anchors that preserve essential tissue characteristics.

3 Experiments and Results

3.1 Datasets

Survival Prediction. We use seven publicly available cancer datasets from TCGA[1] (**BLCA**, **BRCA**, **GBMLGG**, **HNSC**, **KIRC**, **KIRP**, and **LUAD**) in our experiments. These datasets collectively include data from 3,523 patients and 4,091 H&E diagnostic WSIs.

[1] https://portal.gdc.cancer.gov.

Cancer Subtyping. We conduct comparative experiments on two challenging public datasets: **TCGA-BRCA** and **TCGA-NSCLC**. The TCGA-BRCA dataset contains 1,034 H&E slides of two invasive cancer subtypes: invasive ductal carcinoma (IDC) and invasive lobular carcinoma (ILC). The TCGA-NSCLC dataset includes 1,030 H&E slides from two subtypes: Lung Squamous Cell Carcinoma (TCGA-LUSC) and Lung Adenocarcinoma (TCGA-LUAD).

3.2 Implementation Details

We compare MiCo with eight state-of-the-art methods: AMIL [5], TransMIL [15], DeepGraphConv [10], Patch-GCN [1], ILRA [20], HVTSurv [16], RRTMIL [18], and WiKG [7]. WSIs are preprocessed using CONCH v1.5 [4] to extract 448×448 patch features at 20× magnification. All experiments adopt 4-fold cross-validation, with datasets split into training, validation, and test sets at a ratio of 60:15:25. The semantic anchor number is set to 64. Training configurations are unified across methods: 200 epochs, batch size 1, learning rate 2e-4, and early stop 8. Survival prediction performance is evaluated using the Concordance Index (C-Index) with standard deviation, while cancer subtyping is assessed via Accuracy (ACC), F1-score (F1), and Area Under the Curve (AUC) metrics.

Table 1. Survival prediction performance comparison across seven cancer types.

Models	BLCA	BRCA	GBMLGG	HNSC	KIRC	KIRP	LUAD	MEAN
AMIL [5]	$0.551_{0.10}$	$0.576_{0.11}$	$0.736_{0.15}$	$0.590_{0.02}$	$0.681_{0.05}$	$0.778_{0.04}$	$0.584_{0.04}$	0.642
TransMIL [15]	$0.610_{0.07}$	$0.555_{0.04}$	$0.760_{0.14}$	$0.565_{0.02}$	$0.694_{0.05}$	$0.773_{0.07}$	$0.585_{0.09}$	0.649
DeepGraphConv [10]	$0.563_{0.08}$	$0.581_{0.10}$	$0.739_{0.14}$	$0.591_{0.02}$	$0.681_{0.02}$	$0.738_{0.04}$	$0.581_{0.02}$	0.639
PatchGCN [1]	$0.607_{0.05}$	$0.600_{0.12}$	$0.736_{0.16}$	$0.584_{0.01}$	$0.694_{0.05}$	$0.783_{0.05}$	$0.582_{0.04}$	0.655
ILRA [20]	$0.612_{0.04}$	$0.551_{0.07}$	$0.722_{0.16}$	$0.600_{0.01}$	$0.663_{0.04}$	$0.688_{0.05}$	$0.590_{0.04}$	0.632
HVTSurv [16]	$0.606_{0.05}$	$0.579_{0.09}$	$0.792_{0.01}$	$0.575_{0.03}$	$0.692_{0.04}$	$0.778_{0.06}$	$0.591_{0.03}$	0.658
RRTMIL [18]	$0.588_{0.05}$	$0.542_{0.08}$	$0.738_{0.16}$	$0.597_{0.02}$	$0.699_{0.04}$	$0.770_{0.06}$	$0.579_{0.04}$	0.644
WiKG [7]	$0.607_{0.07}$	$0.582_{0.09}$	$0.752_{0.16}$	$0.579_{0.02}$	$0.704_{0.06}$	$0.763_{0.09}$	$0.595_{0.06}$	0.654
MiCo	$\mathbf{0.619}_{0.03}$	$\mathbf{0.608}_{0.12}$	$\mathbf{0.813}_{0.02}$	$\mathbf{0.605}_{0.01}$	$\mathbf{0.715}_{0.05}$	$\mathbf{0.800}_{0.07}$	$\mathbf{0.604}_{0.04}$	**0.680**

3.3 Results and Discussion

Survival Prediction. The experimental results demonstrate the superior performance of MiCo across seven cancer types for survival prediction tasks. As shown in Table 1, MiCo achieves a mean C-index of 0.680, outperforming all other methods. This performance underscores the effectiveness of MiCo's context-aware clustering in addressing the challenges posed by spatial heterogeneity in histopathological WSIs. MiCo's ability to model cross-regional dependencies and enhance semantic associations across dispersed tissue regions enables it to capture complex morphological patterns and improve survival prediction accuracy.

Table 2. Cancer subtyping performance comparison across two cancer types.

Method	TCGA-BRCA			TCGA-NLCSC			MEAN		
	ACC	F1	AUC	ACC	F1	AUC	ACC	F1	AUC
AMIL [5]	$0.909_{0.01}$	$0.858_{0.02}$	$0.948_{0.01}$	$0.923_{0.02}$	$0.923_{0.02}$	$0.979_{0.01}$	0.916	0.890	0.964
TransMIL [15]	$0.906_{0.01}$	$0.857_{0.02}$	$0.945_{0.02}$	$0.919_{0.02}$	$0.919_{0.02}$	$0.977_{0.02}$	0.913	0.889	0.961
DeepGraphConv [10]	$0.913_{0.01}$	$0.859_{0.01}$	$0.947_{0.01}$	$0.913_{0.02}$	$0.913_{0.02}$	$0.969_{0.02}$	0.913	0.886	0.958
PatchGCN [1]	$0.917_{0.01}$	$0.871_{0.02}$	$0.944_{0.02}$	$0.919_{0.02}$	$0.919_{0.02}$	$0.979_{0.01}$	0.918	0.895	0.962
ILRA [20]	$0.899_{0.01}$	$0.836_{0.03}$	$0.931_{0.02}$	$0.918_{0.02}$	$0.918_{0.02}$	$0.973_{0.02}$	0.909	0.885	0.952
RRTMIL [18]	$0.910_{0.01}$	$0.855_{0.01}$	$0.945_{0.02}$	$0.915_{0.02}$	$0.915_{0.02}$	$0.976_{0.01}$	0.913	0.890	0.961
WiKG [7]	$0.915_{0.02}$	$0.859_{0.03}$	$0.948_{0.01}$	$0.920_{0.03}$	$0.920_{0.03}$	$0.980_{0.01}$	0.917	0.889	0.964
MiCo	$\mathbf{0.922}_{0.01}$	$\mathbf{0.875}_{0.02}$	$\mathbf{0.952}_{0.01}$	$\mathbf{0.931}_{0.02}$	$\mathbf{0.931}_{0.02}$	$\mathbf{0.981}_{0.01}$	**0.927**	**0.903**	**0.967**

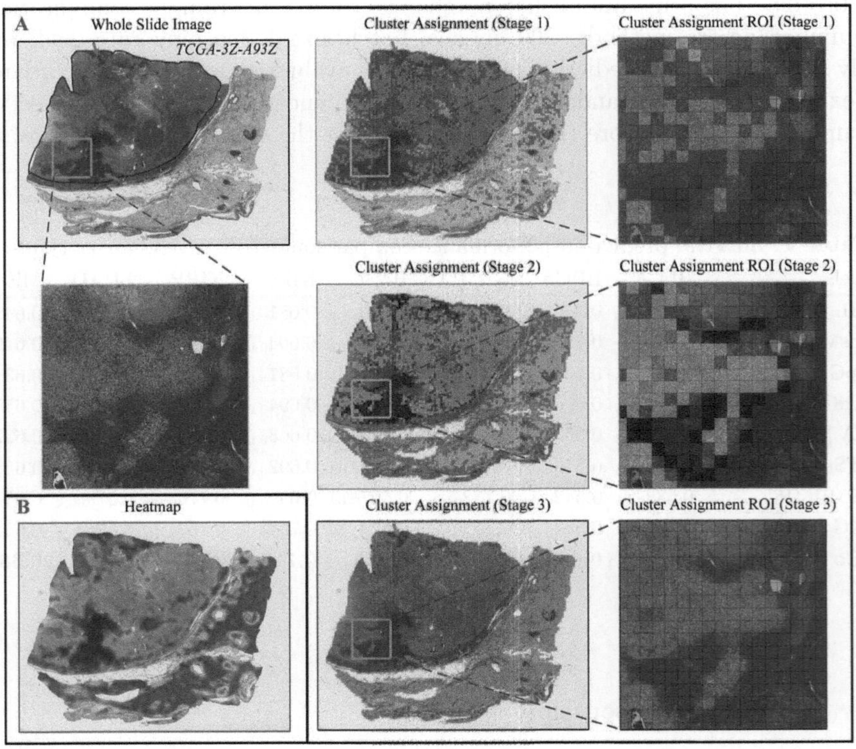

Fig. 2. A. Qualitative results of cluster assignment, showcasing WSI thumbnails, cluster assignment results, and corresponding regions of interest (ROIs) at different stages. Blue outlines denote ground truth tumor regions, while areas assigned to the same cluster are displayed in identical colors. The green arrows highlight how spatially distant regions are progressively aggregated into a single, semantically consistent cluster. **B.** Interpretability analysis of MiCo, with red regions in the heatmaps indicating areas of high attention. (Color figure online)

Cancer Subtyping. The experimental results for cancer subtyping, as presented in Table 2, demonstrate MiCo's superior performance across two main datasets, TCGA-BRCA and TCGA-NLCSC. MiCo achieves state-of-the-art results in terms of accuracy, F1 score, and AUC, outperforming all other methods. With a mean ACC of 0.927, MiCo significantly surpasses the best-performing baseline, PatchGCN, which achieves a mean ACC of 0.918. These results highlight MiCo's ability to accurately classify cancer subtypes by leveraging its context-aware clustering mechanism and dynamic semantic anchors, effectively alleviating the challenges of spatial heterogeneity and semantic fragmentation in histopathological WSIs.

Ablation Study. To further determine the efficacy of each crucial component in MiCo, we conduct a series of ablation studies on survival prediction and cancer subtyping tasks. The average results, as summarized in Table 3, underscore the critical importance of each module in enhancing the overall performance of MiCo. The ablation experiments evaluate the impact of removing key components, including semantic anchor initialization, the CluReducer, and the CluRoute. Removing semantic anchor initialization reduces the C-index to 0.668, underscoring its importance in capturing discriminative morphological patterns. Omitting the CluReducer further decreases the C-index to 0.659, highlighting its role in consolidating redundant anchors and enhancing inter-anchor information exchange. Similarly, removing the CluRoute results in a C-index of 0.656, demonstrating its effectiveness in establishing cross-regional semantic relationships and refining instance-level representations.

Interpretability Analysis. As shown in Fig. 2, we analyze MiCo's interpretability through cluster assignments and heatmaps. Figure 2.A demonstrates that cluster assignments of semantically related regions are progressively merged across stages, preventing overly coarse representations in later phases. The Cluster Reducer module halves cluster numbers at each stage of the MiCo. This progressive reduction aims to intelligently refine and consolidate semantic information, rather than causing coarseness or indiscriminately mixing distinct regional information. The heatmap results further reveal strong alignment with ground truth tumor annotations in Fig. 2.B.

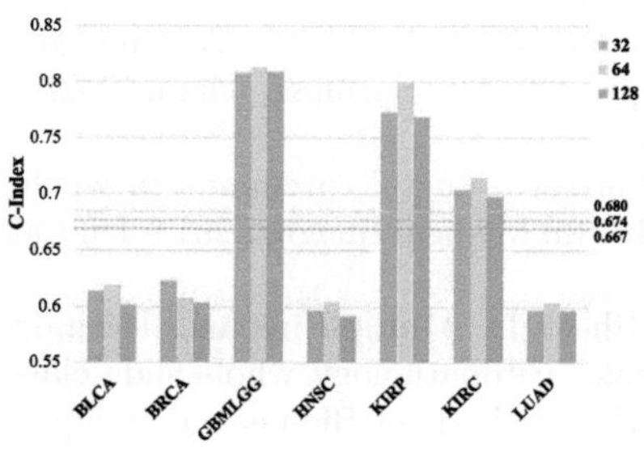

Fig. 3. Analysis of semantic anchors' impact.

Table 3. Average results of the ablation study for MiCo.

Method	Survival Prediction	Cancer Subtyping		
	C-Index	ACC	F1	AUC
w/o Semantic Anchors Init	0.668	0.922	0.896	0.964
w/o CluReducer	0.659	0.921	0.897	0.965
w/o CluRoute	0.656	0.910	0.890	0.955
MiCo	**0.680**	**0.927**	**0.903**	**0.967**

Semantic Anchors Analysis. As shown in Fig. 3, we evaluate MiCo's survival prediction performance with different numbers of semantic anchors (32, 64, and 128). The mean C-index values are 0.674 (32 anchors), 0.680 (64 anchors), and 0.667 (128 anchors). The 64-anchor configuration achieves the highest mean C-index, balancing morphological diversity and redundancy.

4 Conclusion

In this paper, we propose MiCo, a novel Multiple Instance Learning framework with Context-Aware Clustering, designed to tackle spatial heterogeneity in histopathological WSIs. MiCo incorporates a Cluster Route module to strengthen cross-regional intra-tissue correlations and a Cluster Reducer module to consolidate redundant anchors while promoting information exchange between distinct semantic groups. Extensive experiments on nine cancer datasets validate MiCo's superiority over state-of-the-art methods in both survival prediction and cancer subtyping tasks.

Acknowledgement. This work was supported in part by Xinjiang Uygur Autonomous Region Key R&D program (No. 2024B03039-3), the National Natural Science Foundation of China (No. U24A20256), the Scientific Research Fund of Hunan Provincial Education Department (No. 23A0020), the Science and Technology Major Project of Changsha (No. kh2502004), the central government guides local funds for scientific and technological development of China (No. ZYYD2025QY25), the Central South University Innovation-Driven Research Programme (No. 2023CXQD018), the Fundamental Research Funds for the Central Universities of Central South University (No. 2024ZZTS0108), and High-Performance Computing Center of Central South University.

Disclosure of Interests. The authors have no competing interests to declare that are relevant to the content of this article.

References

1. Chen, R.J., Lu, M.Y., Shaban, M., Chen, C., Chen, T.Y., Williamson, D.F.K., Mahmood, F.: Whole slide images are 2D point clouds: context-aware survival prediction using patch-based graph convolutional Networks. In: de Bruijne, M., et al. (eds.) MICCAI 2021. LNCS, vol. 12908, pp. 339–349. Springer, Cham (2021). https://doi.org/10.1007/978-3-030-87237-3_33
2. Cruz-Roa, A., et al.: Automatic detection of invasive ductal carcinoma in whole slide images with convolutional neural networks. In: Medical Imaging 2014: Digital Pathology, vol. 9041, p. 904103. SPIE (2014)
3. Das, K., Conjeti, S., Roy, A.G., Chatterjee, J., Sheet, D.: Multiple instance learning of deep convolutional neural networks for breast histopathology whole slide classification. In: 2018 IEEE 15th International Symposium on Biomedical Imaging (ISBI 2018), pp. 578–581. IEEE (2018)
4. Ding, T., et al.: Multimodal whole slide foundation model for pathology (2024). https://arxiv.org/abs/2411.19666

5. Ilse, M., Tomczak, J., Welling, M.: Attention-based deep multiple instance learning. In: International Conference on Machine Learning, pp. 2127–2136. PMLR (2018)
6. Lee, M.: Recent advancements in deep learning using whole slide imaging for cancer prognosis. Bioengineering **10**(8), 897 (2023)
7. Li, J., et al.: Dynamic graph representation with knowledge-aware attention for histopathology whole slide image analysis. In: Proceedings of the IEEE/CVF Conference on Computer Vision and Pattern Recognition, pp. 11323–11332 (2024)
8. Li, J., Kuang, H., Liu, J., Yue, H., Wang, J.: Ca2cl: cluster-aware adversarial contrastive learning for pathological image analysis. IEEE J. Biomed. Health Inform. **29**, 5095–5108 (2025)
9. Li, J., et al.: DARC: deep adaptive regularized clustering for histopathological image classification. Med. Image Anal. **80**, 102521 (2022)
10. Li, R., Yao, J., Zhu, X., Li, Y., Huang, J.: Graph CNN for survival analysis on whole slide pathological images. In: Frangi, A.F., Schnabel, J.A., Davatzikos, C., Alberola-López, C., Fichtinger, G. (eds.) MICCAI 2018. LNCS, vol. 11071, pp. 174–182. Springer, Cham (2018). https://doi.org/10.1007/978-3-030-00934-2_20
11. Lia, M., Horn, L.C., Sodeikat, P., Höckel, M., Aktas, B., Wolf, B.: The diagnostic value of core needle biopsy in cervical cancer: A retrospective analysis. PLoS ONE **17**(1), e0262257 (2022)
12. Lipkova, J., et al.: Deep learning-enabled assessment of cardiac allograft rejection from endomyocardial biopsies. Nat. Med. **28**(3), 575–582 (2022)
13. Medvedev, O., et al.: Perineural spread in head-and-neck malignancies: imaging findings-an updated literature review. Bosn. J. Basic Med. Sci. **22**(1), 22 (2021)
14. Qu, L., Liu, S., Liu, X., Wang, M., Song, Z.: Towards label-efficient automatic diagnosis and analysis: a comprehensive survey of advanced deep learning-based weakly-supervised, semi-supervised and self-supervised techniques in histopathological image analysis. Phys. Med. Biol. **67**(20), 20TR01 (2022)
15. Shao, Z., Bian, H., Chen, Y., Wang, Y., Zhang, J., Ji, X., et al.: Transmil: transformer based correlated multiple instance learning for whole slide image classification. Adv. Neural. Inf. Process. Syst. **34**, 2136–2147 (2021)
16. Shao, Z., Chen, Y., Bian, H., Zhang, J., Liu, G., Zhang, Y.: HVTSurv: hierarchical vision transformer for patient-level survival prediction from whole slide image. In: Proceedings of the AAAI Conference on Artificial Intelligence, vol. 37, pp. 2209–2217 (2023)
17. Song, C., et al.: LVPnet: a latent-variable-based prediction-driven end-to-end framework for lossless compression of medical images (2025). https://arxiv.org/abs/2506.17983
18. Tang, W., Zhou, F., Huang, S., Zhu, X., Zhang, Y., Liu, B.: Feature re-embedding: Towards foundation model-level performance in computational pathology. In: Proceedings of the IEEE/CVF Conference on Computer Vision and Pattern Recognition, pp. 11343–11352 (2024)
19. Van Den Oord, A., Vinyals, O., et al.: Neural discrete representation learning. In: Advances in Neural Information Processing Systems, vol. 30 (2017)
20. Xiang, J., Zhang, J.: Exploring low-rank property in multiple instance learning for whole slide image classification. In: The Eleventh International Conference on Learning Representations (2023)
21. Zarella, M.D., et al.: A practical guide to whole slide imaging: a white paper from the digital pathology association. Arch. Pathol. Lab. Med. **143**(2), 222–234 (2019)

MicroMIL: Graph-Based Multiple Instance Learning for Context-Aware Diagnosis with Microscopic Images

JongWoo Kim[1], Bryan Wong[1], Huazhu Fu[2], Willmer Rafell Quiñones Robles[1], Young Sin Ko[3], and Mun Yong Yi[1(✉)]

[1] Korea Advanced Institute of Science and Technology, Daejeon, South Korea
{gsds4885,bryan.wong,munyi}@kaist.ac.kr
[2] Institute of High Performance Computing, Agency for Science, Technology and Research (A*STAR), Singapore, Singapore
[3] Seegene Medical Foundation, Pathology Center, Seoul, South Korea

Abstract. Cancer diagnosis has greatly benefited from the integration of whole-slide images (WSIs) with multiple instance learning (MIL), enabling high-resolution analysis of tissue morphology. Graph-based MIL (GNN-MIL) approaches have emerged as powerful solutions for capturing contextual information in WSIs, thereby improving diagnostic accuracy. However, WSIs require significant computational and infrastructural resources, limiting accessibility in resource-constrained settings. Conventional light microscopes offer a cost-effective alternative, but applying GNN-MIL to such data is challenging due to extensive redundant images and missing spatial coordinates, which hinder contextual learning. To address these issues, we introduce **MicroMIL**, the first weakly-supervised MIL framework specifically designed for images acquired from conventional light microscopes. MicroMIL leverages a representative image extractor (RIE) that employs deep cluster embedding (DCE) and hard Gumbel-Softmax to dynamically reduce redundancy and select representative images. These images serve as graph nodes, with edges computed via cosine similarity, eliminating the need for spatial coordinates while preserving contextual information. Extensive experiments on a real-world colon cancer dataset and the BreakHis dataset demonstrate that MicroMIL achieves state-of-the-art performance, improving both diagnostic accuracy and robustness to redundancy. The code is available at https://github.com/kimjongwoo-cell/MicroMIL.

Keywords: Digital Pathology · Conventional Light Microscopes · Microscopy Images · Multiple Instance Learning

1 Introduction

Cancer remains a leading global cause of mortality, necessitating advancements in diagnostic technologies to enhance early detection and improve survival rates.

J. Kim and B. Wong—Co-first authors with equal contribution.

J. C. Gee et al. (Eds.): MICCAI 2025, LNCS 15960, pp. 386–396, 2026.
https://doi.org/10.1007/978-3-032-04927-8_37

Whole-slide imaging (WSI) has emerged as a transformative tool in digital pathology, offering high-resolution insights into tissue morphology and disease-related anomalies [6,13]. However, WSIs require substantial infrastructure due to their high acquisition costs, memory demands, and lengthy processing times, making them less practical in resource-limited settings [2,23].

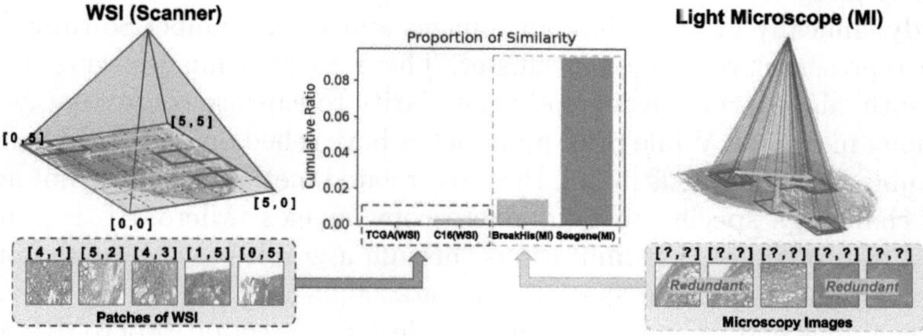

Fig. 1. Left: Valid patches from WSI (scanner) are acquired using a sliding-window approach and have absolute positions. **Right:** Light microscopy images lack known positions and contain many redundancies due to subjective capture by pathologists. **Middle:** Ratio of image pairs exceeding the redundancy threshold (0.995).

While WSI scanners are becoming more accessible, light microscopes remain far more widely used, especially in low-resource settings. Microscopy images thus offer a practical and cost-effective alternative for enabling AI-driven diagnostic solutions in diverse healthcare contexts [21]. Low-cost optical microscopes tailored for low- and middle-income countries continue to be developed and microscopy-based diagnostics remain the clinical standard [19,26].

Despite these advantages, as illustrated in Fig. 1, light microscopy images pose unique challenges, including the **absence of absolute spatial coordinates** due to manual acquisition by pathologists and **significant redundancy** caused by multiple image captures. To highlight these issues, we compare microscopic and WSI datasets (TCGA NSCLC[1] and Camelyon16 [3]) and show that light microscopy images exhibit significantly higher redundancy, with the highest observed in real-world microscopic datasets.

Recent advancements in weakly-supervised multiple instance learning (MIL) have facilitated the use of WSIs for cancer diagnosis by requiring only slide-level labels, thereby reducing the need for exhaustive annotations [7]. Within this paradigm, graph-based MIL (GNN-MIL) models [4,5] have shown promise by leveraging spatial relationships among patches to capture contextual information, representing patches as nodes and their interactions as edges. However, these methods are inherently designed for WSIs and cannot be directly applied to light microscopy images, where spatial coordinates are unknown and image

[1] https://www.cancer.gov/tcga.

redundancy is prevalent. Overcoming these limitations requires a specific approach that accommodates the unique characteristics of light microscopy images while preserving the benefits of graph-based contextual modeling.

To address the challenges of absent spatial coordinates and high redundancy, we propose **MicroMIL**, the first weakly-supervised MIL framework specifically designed for conventional light microscopy images. MicroMIL introduces a representative image extractor (RIE) that leverages deep cluster embedding (DCE) [11] to dynamically group redundant images and hard Gumbel-Softmax [14] to select a representative image per cluster. These selected images serve as graph nodes, with edges formed using cosine similarity to capture contextual information among instances. While prior approaches have relied on statistical heuristics or ensemble-based methods [8,20], the most related method [15] does not address critical challenges specific to light microscopy images. MicroMIL is explicitly designed to overcome these limitations through a graph-based formulation that operates without relying on spatial metadata.

By enabling end-to-end representative feature selection, MicroMIL jointly optimizes clustering and instance selection within a unified framework, ensuring that the most informative representations contribute to the final prediction. To achieve this, we propose an online redundancy-aware learning strategy that dynamically refines instance selection while maintaining feature diversity. The graph-based representation further enhances structural preservation by connecting similar nodes, mitigating the loss of spatial information. Extensive experiments on a real-world colon cancer dataset and the BreakHis dataset validate MicroMIL's effectiveness, demonstrating significant gains in diagnostic accuracy and redundancy robustness compared to state-of-the-art MIL methods.

2 Methodology

Each patient is treated as a *bag*, and the corresponding light-microscope images as *instances*, following the MIL paradigm. The goal is to predict patient diagnosis in a weakly-supervised setting without instance-level labels. We propose MicroMIL, a MIL framework for microscopic image analysis (see Fig. 2). The framework consists of three components: (1) a frozen pre-trained feature extractor for generating image features, (2) a RIE that reduces redundancy by clustering similar images using DCE and selecting representatives via hard Gumbel-Softmax, and (3) a graph-based aggregation module, where nodes represent the selected represented images and edges are constructed using an upper triangular cosine similarity matrix to link similar nodes. Finally, a GNN captures contextual information for analysis.

2.1 Representative Image Extractor

In the embedding-based MIL framework, a frozen pre-trained feature extractor E maps each microscopic image I_s to a d-dimensional feature vector $f_s = E(I_s)$, forming the feature set $F = \{f_1, f_2, \ldots, f_S\}$, where S varies across patients. To

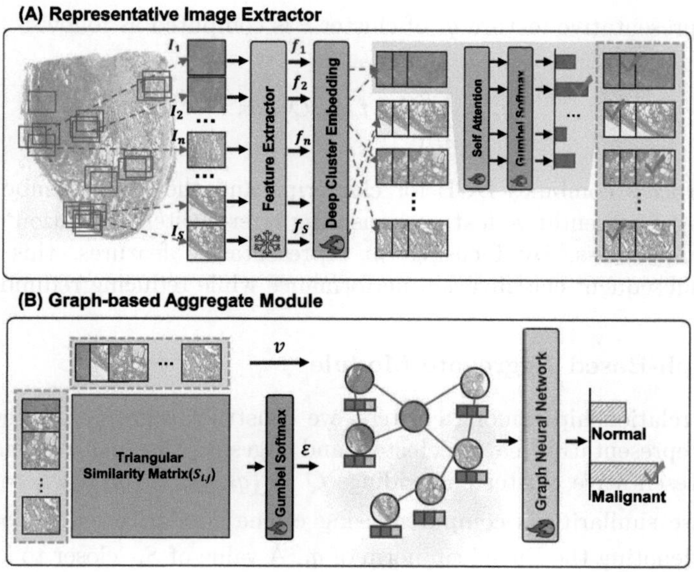

Fig. 2. Proposed end-to-end MicroMIL framework.

group redundant images, we employ deep cluster embedding (DCE) [25], which iteratively assigns data points to clusters and refines cluster centers to minimize intra-cluster distances. Let $\mu_c \in \mathbb{R}^d$ be the centroid of the c-th cluster, where $c = 1, \ldots, C$. The soft assignment probability $z_{s,c}$, indicating the likelihood of f_s belonging to the c-th cluster, is defined as:

$$z_{s,c} = \frac{\left(1 + \|f_s - \mu_c\|^2\right)^{-1}}{\sum_{j=1}^{C} \left(1 + \|f_s - \mu_j\|^2\right)^{-1}}, \quad Z \in \mathbb{R}^{S \times C} \tag{1}$$

The DCE algorithm alternates between updating the centroids μ_c and refining cluster assignments Z until convergence.

To select the most representative feature from each cluster, we use the hard Gumbel-Softmax function [14], which allows for end-to-end differentiability. Given logits X_x and Gumbel noise $g_x \sim \text{Gumbel}(0, 1)$, the hard Gumbel-Softmax function is defined as:

$$\text{HardGumbel}(X) = \text{one\_hot}\left(\arg\max_x (X_x + g_x)\right) \tag{2}$$

Applying this function to feature-cluster interactions, the hard cluster assignments $\tilde{Z}$ are determined as:

$$\tilde{z}_{s,c} = \text{HardGumbel}(s_{s,c}), \quad \tilde{Z} \in \mathbb{R}^{S \times C} \tag{3}$$

where $s_{s,c} = w^\top (f_s \odot z_{:,c})$, with $w \in \mathbb{R}^d$ being a learnable weight vector and $\odot$ denoting element-wise multiplication.

The representative feature q_c of cluster c is computed as follows:

$$q_c = \sum_{s=1}^{S} \tilde{z}_{s,c} f_s, \quad Q \in \mathbb{R}^{C \times d} \tag{4}$$

This process combines DCE for clustering and the hard Gumbel-Softmax for selecting representative features, ensuring inter-cluster separation and intra-cluster compactness. By focusing on representative features, this approach improves subsequent classification performance while reducing redundancy.

2.2 Graph-Based Aggregate Module

To model relationships among clusters, we construct a graph G, where nodes represent representative feature clusters and edges capture pairwise similarities. Given representative cluster embeddings $Q = \{q_1, q_2, \ldots, q_C\}$, where $q_c \in \mathbb{R}^d$, the pairwise similarity is computed using cosine similarity as $S_{ij} = \frac{q_i^\top q_j}{\|q_i\|\|q_j\|}$, with $\|q_i\|$ denoting the Euclidean norm of q_i. A value of S_{ij} closer to 1 indicates higher similarity between clusters.

To retain only the most important relationships, we apply the same hard Gumbel-Softmax function to the similarity matrix:

$$\tilde{m}_{i,j} = \text{HardGumbel}(S_{ij}), \quad \tilde{M} \in \mathbb{R}^{C \times C} \tag{5}$$

The resulting graph $G = (V, E)$ is defined by nodes $V = \{1, 2, \ldots, C\}$ and edges $E = \{(i, j) \mid \tilde{m}_{i,j} > 0\}$.

Once the graph is constructed, a GNN propagates and refines the cluster embeddings. The initial node features are $H^{(0)} = R$ and through L GNN layers, node embeddings are updated by aggregating information from neighboring nodes. The entire process is represented as:

$$y = \sigma\left(W_{\text{class}} \cdot \text{mean}\left(\text{GNN}(G, R)\right)\right) \tag{6}$$

where $\text{GNN}(G, R)$ represents the L-layer operations on G, W_{class} is the classification weight matrix, $\text{mean}(\cdot)$ aggregates node embeddings, and σ is the activation function. The entire framework, including DCE, RIE, and GNN, is trained end-to-end using binary cross-entropy (BCE) loss.

3 Experiments and Results

3.1 Datasets

BreakHis. BreakHis [24] is a widely used benchmark dataset for microscopy image analysis and cancer research. It comprises 7,909 images from 81 patients, with an average of 96.4 images per patient. Among these, 2,480 images (from 24 patients) are labeled as normal, while 5,429 images (from 57 patients) are labeled as malignant. The dataset spans multiple magnifications (40×, 100×,

200×, and 400×) and is collected under controlled conditions, which may limit its applicability to real-world scenarios.

Real-world (Seegene) Dataset. Our real-world dataset, collected from the Seegene Medical Foundation, consists of 135,100 images from 899 patients, averaging 150.3 images per patient. The dataset was approved by both the foundation's IRB (SMF-IRB-2020-007) and the KAIST IRB (KAIST-IRB-23-214). It includes 52,339 images (from 493 patients) labeled as normal and 82,761 images (from 406 patients) labeled as malignant, primarily at 100× and 200× magnifications, aligning with common clinical practices.

3.2 Implementation Details

For a fair comparison, we use ResNet18 pre-trained on ImageNet as the feature extractor for all models, with a hidden dimension of 128 for consistency. MicroMIL is tuned with a dropout rate of 0.5, a learning rate of 1×10^{-3}, and the Adam optimizer. For the final results, we use 36 clusters for the real-world (Seegene) dataset and 16 for BreakHis, given its smaller image count per patient. We experiment with 16 (4^2), 25 (5^2), and 36 (6^2) clusters and report the best-performing configuration. Performance differences across these settings are minimal and MicroMIL consistently outperforms all baselines. While online clustering (DCE) requires a predefined cluster count, we aim to explore automatic cluster number selection in future work. All models, implemented in PyTorch, are trained with two graph layers on an NVIDIA GeForce RTX 3080 GPU.

3.3 Baselines Comparison

Table 1 shows that MicroMIL consistently surpasses baseline MIL models across all evaluated metrics on both real-world and public datasets. This performance advantage stems from the MicroMIL's specific design to address two key challenges of light microscopy datasets: image redundancy and no absolute position. In contrast, existing MIL models are designed for WSIs obtained from scanners, thereby without the need to account for these characteristics. By effectively tackling the unique characteristics present, MicroMIL proves to be exceptionally well-suited for patient diagnosis using light microscopy images.

3.4 Impact of Representative Image Extractor

In histopathology, patient-level prediction requires information exchange across diverse tissue regions [1,9], but redundant images hinder this process. To address this, we propose the Representative Image Extractor (RIE), which selects representative images from visually similar clusters to enhance diversity and improve predictive performance. Figure 3 (*without* RIE) shows that redundancy limits diverse interactions and leads to performance degradation (Fig. 4), which RIE successfully mitigates by filtering redundant instances. The effect varies depending on dataset redundancy, with smaller improvement in BreakHis due to lower redundancy, whereas the real-world dataset shows a larger improvement, demonstrating RIE's effectiveness in handling highly redundant data.

Table 1. Performance metrics of baselines and MicroMIL on Real-world (Seegene) and BreakHis datasets. Best results are in **bold** and second-best results are underlined.

Model	Real-world (Seegene)			BreakHis		
	ACC	AUC	F1	ACC	AUC	F1
ABMIL [ICML'18] [12]	0.9444	0.9764	0.9433	0.8929	0.8947	0.8805
MS-DA-MIL [CVPR'20] [10]	0.9556	0.9829	0.9514	0.8929	0.9591	<u>0.9268</u>
DSMIL [CVPR'21] [16]	0.9444	0.9829	0.9440	0.8214	0.8947	0.8155
CLAM [Nat BioMed'21] [18]	0.9556	0.9873	0.9552	0.9286	0.9298	0.9181
TransMIL [NeurIPS'21] [22]	<u>0.9778</u>	0.9873	<u>0.9776</u>	0.8929	<u>0.9825</u>	<u>0.9268</u>
DTFD-MIL [CVPR'22] [27]	0.9611	<u>0.9901</u>	0.9607	<u>0.9286</u>	0.9766	0.9222
IBMIL [CVPR'23] [17]	0.9611	0.9894	0.9606	<u>0.9286</u>	0.9532	0.9181
ACMIL [ECCV'24] [28]	0.9611	0.9893	0.9606	0.8929	0.9474	0.8857
MicroMIL	**0.9922**	**0.9994**	**0.9925**	**0.9643**	**0.9942**	**0.9730**

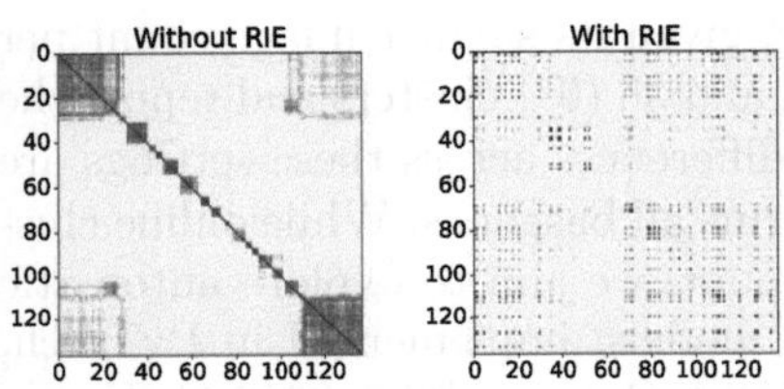

Fig. 3. Similarity-based edge formation probabilities (≥ 0.8) heatmaps for real-world (Seegene) data, comparing cases *without (Left)* and *with (Right)* the RIE.

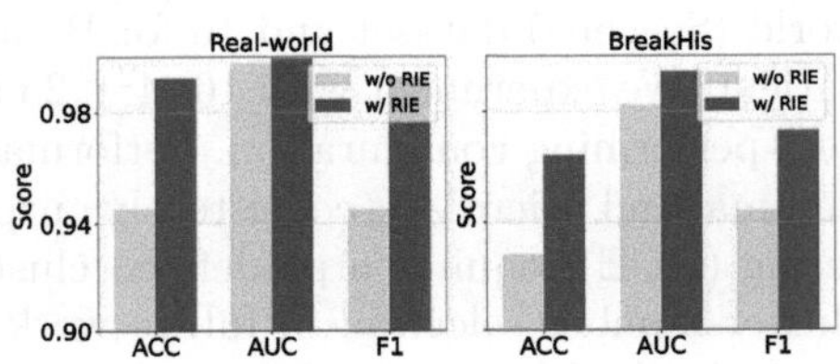

Fig. 4. Performance metrics *without (w/o)* and *with (w/)* the RIE on Real-world (Seegene) and BreakHis datasets.

3.5 Connecting Relative Neighborhood Nodes Method

Light microscopy images lack spatial information, necessitating effective edge construction. We utilize a GNN-based method that connects nodes to their most similar neighbors. As shown in Fig. 5, performance drops without connections due to a lack of contextual relationships. Random connections lead to weak interactions, while cosine similarity-based edges, linking highly similar nodes, capture meaningful relationships and outperform random or reverse-similarity (1/similarity) edges. This highlights the importance of leveraging similarity to enhance context integration and predictive performance.

3.6 Ablation Studies

Representative Image Extractor Methods. To enhance feature learning, we adopt an online approach using DCE and Gumbel-Softmax for selecting influential representative images. As shown in Fig. 6, a comparison of clustering

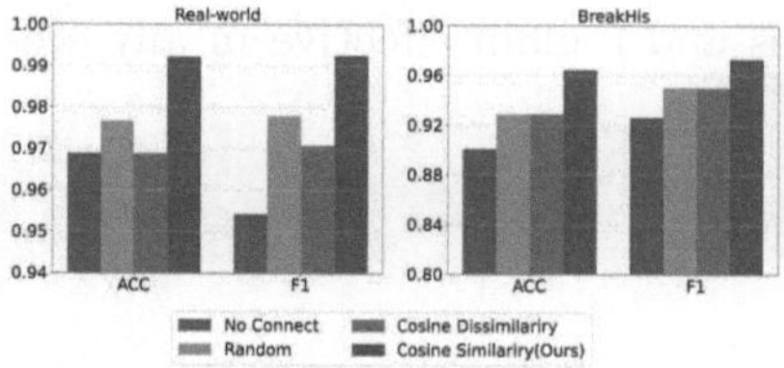

Fig. 5. Performance metrics of different edge generation methods.

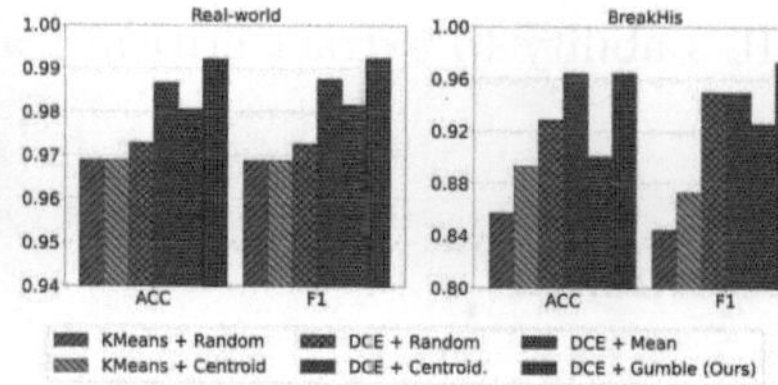

Fig. 6. Performance metrics of different representative image extractor methods.

methods (KMeans, DCE) and selection strategies (Random, Mean, Centroid, Gumbel-Softmax) reveals that offline KMeans underperforms, while online DCE + Gumbel-Softmax achieves superior results. Notably, Gumbel-Softmax outperforms Mean and Centroid by better highlighting key images, demonstrating the effectiveness of online clustering and selection in optimizing feature learning. This online approach enables dynamic adaptation, ensuring robust feature representations and reducing the risk of suboptimal cluster assignments.

Table 2. Robustness to redundancy in baseline models and MicroMIL. Arrows indicate data flow direction (train → test). We count images exceeding the 0.995 redundancy threshold (Fig. 1, middle) per patient, then select the top 10% (**T10**) and bottom 10% (**B10**) of patients. Best results are in **bold**.

Model	(1) T10 → B10			(2) B10 → T10			(3) T10 → T10		
	ACC	AUC	F1	ACC	AUC	F1	ACC	AUC	F1
ABMIL	0.8090	0.8592	0.7966	0.9213	0.9592	0.9210	0.9091	0.9229	0.9089
MS-DA-MIL	0.9101	0.9526	0.9248	0.9213	0.9642	0.9248	0.9091	0.9348	0.9209
DSMIL	0.9101	0.9474	0.9092	0.9326	0.9755	0.9319	0.9091	0.9438	0.9089
CLAM	0.9326	0.9796	0.9315	0.9213	0.9770	0.9194	0.9318	0.9521	0.9318
TransMIL	0.9213	0.9776	0.9212	0.8989	0.8526	0.8963	0.9318	0.8854	0.9309
DTFD-MIL	0.9438	0.9658	0.9428	0.9213	0.9750	0.9203	0.9318	0.9583	0.9318
IBMIL	0.9326	0.9709	0.9319	0.9326	0.9719	0.9319	0.9318	0.9458	0.9315
ACMIL	0.9434	0.9704	0.9431	0.9213	0.9689	0.9210	0.9318	0.9521	0.9315
MicroMIL	**0.9663**	**0.9842**	**0.9630**	**0.9551**	**0.9801**	**0.9542**	**0.9545**	**0.9958**	**0.9524**

Robustness on Image Redundancy Shift. To evaluate MicroMIL's robustness to image redundancy, we set a similarity threshold of 0.995 to identify redundant image pairs. Table 2 shows that baseline MIL methods degrade under extreme redundancy settings (B10→T10, T10→T10), while MicroMIL consistently maintains high performance. Even in low-redundancy simulations (T10→B10), MicroMIL outperforms all baselines. These results confirm

MicroMIL's ability to extract critical features and remain effective in any scenario.

4 Conclusion

We introduce MicroMIL, the first weakly-supervised MIL framework specifically designed for conventional light microscopy images, addressing the limitations of GNN-MIL approaches that rely on spatial coordinates and exhibit low redundancy tolerance. By integrating deep cluster embedding (DCE) and hard Gumbel-Softmax, MicroMIL effectively reduces redundancy and selects representative instances, enabling a graph-based representation without requiring absolute spatial positioning while explicitly modeling contextual cues. Experiments on the real-world and BreakHis datasets demonstrate state-of-the-art performance, improving diagnostic accuracy while maintaining robustness to redundancy. MicroMIL offers a scalable, spatially agnostic solution that advances weakly-supervised MIL for microscopy imaging in resource-constrained settings.

Acknowledgments. This research was supported by the Seegene Medical Foundation, South Korea, under the project "Research on Implementing a Next Generation Medical Diagnosis System based on AI" (Grant Number: G01210159) and by the National Research Foundation of Korea (NRF) grant funded by the Korea government (MSIT) (No.RS-2022-NR068758).

Disclosure of Interests. The authors have no competing interests to declare that are relevant to the content of this article.

References

1. Ahn, S., Huang, H.: Multiregion sequence analysis to predict intratumor heterogeneity and clonal evolution. In: Deep Sequencing Data Analysis, pp. 283–296 (2021)
2. Alkassar, S., Jebur, B.A., Abdullah, M.A., Al-Khalidy, J.H., Chambers, J.A.: Going deeper: a magnification-invariant approach for breast cancer classification using histopathological images. IET Comput. Vision **15**(2), 151–164 (2021)
3. Bejnordi, B.E., et al.: Diagnostic assessment of deep learning algorithms for detection of lymph node metastases in women with breast cancer. JAMA **318**(22), 2199–2210 (2017)
4. Bontempo, G., Bartolini, N., Lovino, M., Bolelli, F., Virtanen, A., Ficarra, E.: Enhancing PFI prediction with GDS-MIL: a graph-based dual stream mil approach. In: Foresti, G.L., Fusiello, A., Hancock, E. (eds.) ICIAP 2023. LNCS, vol. 14233, pp. 550–562. Springer, Cham (2023). https://doi.org/10.1007/978-3-031-43148-7_46
5. Chan, T.H., Cendra, F.J., Ma, L., Yin, G., Yu, L.: Histopathology whole slide image analysis with heterogeneous graph representation learning. In: Proceedings of the IEEE/CVF Conference on Computer Vision and Pattern Recognition, pp. 15661–15670 (2023)

6. Evans, A.J., Krupinski, E.A., Weinstein, R.S., Pantanowitz, L.: 2014 American telemedicine association clinical guidelines for telepathology: Another important step in support of increased adoption of telepathology for patient care. J. Pathol. Inform. **6** (2015)
7. Gadermayr, M., Tschuchnig, M.: Multiple instance learning for digital pathology: a review of the state-of-the-art, limitations & future potential. Comput. Med. Imaging Graph., 102337 (2024)
8. Gandomkar, Z., Brennan, P.C., Mello-Thoms, C.: Mudern: multi-category classification of breast hikumartopathological image using deep residual networks. Artif. Intell. Med. **88**, 14–24 (2018)
9. Halperin, R.F., et al.: Leveraging spatial variation in tumor purity for improved somatic variant calling of archival tumor only samples. Front. Oncol. **9**, 119 (2019)
10. Hashimoto, N., et al.: Multi-scale domain-adversarial multiple-instance CNN for cancer subtype classification with unannotated histopathological images. In: Proceedings of the IEEE/CVF Conference on Computer Vision and Pattern Recognition, pp. 3852–3861 (2020)
11. Huang, P., Huang, Y., Wang, W., Wang, L.: Deep embedding network for clustering. In: 2014 22nd International Conference on Pattern Recognition, pp. 1532–1537. IEEE (2014)
12. Ilse, M., Tomczak, J., Welling, M.: Attention-based deep multiple instance learning. In: International Conference on Machine Learning, pp. 2127–2136. PMLR (2018)
13. Iyengar, J.N.: Whole slide imaging: the futurescape of histopathology. Indian J. Pathol. Microbiol. **64**(1), 8 (2021)
14. Jang, E., Gu, S., Poole, B.: Categorical reparameterization with gumbel-softmax. arXiv preprint arXiv:1611.01144 (2016)
15. Kim, J.W., Jeong, M.K., Park, H.M., Ko, Y.S., Yi, M.: Leveraging spatial relationships in microscopic images for patient cancer diagnosis. In: 2024 IEEE International Symposium on Biomedical Imaging (ISBI), pp. 1–5. IEEE (2024)
16. Li, B., Li, Y., Eliceiri, K.W.: Dual-stream multiple instance learning network for whole slide image classification with self-supervised contrastive learning. In: Proceedings of the IEEE/CVF Conference on Computer Vision and Pattern Recognition, pp. 14318–14328 (2021)
17. Lin, T., Yu, Z., Hu, H., Xu, Y., Chen, C.W.: Interventional bag multi-instance learning on whole-slide pathological images. In: Proceedings of the IEEE/CVF Conference on Computer Vision and Pattern Recognition, pp. 19830–19839 (2023)
18. Lu, M.Y., Williamson, D.F., Chen, T.Y., Chen, R.J., Barbieri, M., Mahmood, F.: Data-efficient and weakly supervised computational pathology on whole-slide images. Nat. Biomed. Eng. **5**(6), 555–570 (2021)
19. McDermott, S.: Multi-modal microscopy imaging with the openflexure delta stage. Opt. Express **30**(15), 26377–26395 (2022)
20. Nguyen, H.G., Blank, A., Dawson, H.E., Lugli, A., Zlobec, I.: Classification of colorectal tissue images from high throughput tissue microarrays by ensemble deep learning methods. Sci. Rep. **11**(1), 1–11 (2021)
21. Sangameswaran, R.: MAIScope: a low-cost portable microscope with built-in vision AI to automate microscopic diagnosis of diseases in remote rural settings. arXiv preprint arXiv:2208.06114 (2022)
22. Shao, Z., Ji, X., et al.: TransMIL: transformer based correlated multiple instance learning for whole slide image classification. Adv. Neural. Inf. Process. Syst. **34**, 2136–2147 (2021)

23. Spanhol, F.A., Oliveira, L.S., Petitjean, C., Heutte, L.: Breast cancer histopathological image classification using convolutional neural networks. In: 2016 International Joint Conference on Neural Networks (IJCNN), pp. 2560–2567 (2016)
24. Spanhol, F.A., Oliveira, L.S., Petitjean, C., Heutte, L.: A dataset for breast cancer histopathological image classification. IEEE Trans. Biomed. Eng. **63**(7), 1455–1462 (2015)
25. Xie, J., Girshick, R., Farhadi, A.: Unsupervised deep embedding for clustering analysis. In: International Conference on Machine Learning, pp. 478–487. PMLR (2016)
26. Zhang, H., Zhang, W., Zuo, Z., Yang, J.: Towards ultra-low-cost smartphone microscopy. Microsc. Res. Tech. **87**(7), 1521–1533 (2024)
27. Zhang, H., et al.: DTFD-MIL: double-tier feature distillation multiple instance learning for histopathology whole slide image classification. In: Proceedings of the IEEE/CVF Conference on Computer Vision and Pattern Recognition, pp. 18802–18812 (2022)
28. Zhang, Y., Li, H., Sun, Y., Zheng, S., Zhu, C., Yang, L.: Attention-challenging multiple instance learning for whole slide image classification. In: Leonardis, A., Ricci, E., Roth, S., Russakovsky, O., Sattler, T., Varol, G. (eds.) ECCV 2024. LNCS, vol. 15111, pp. 125–143. Springer, Cham (2024). https://doi.org/10.1007/978-3-031-73668-1_8

MNM: Multi-level Neuroimaging Meta-analysis with Hyperbolic Brain-Text Representations

Seunghun Baek, Jaejin Lee, Jaeyoon Sim, Minjae Jeong, and Won Hwa Kim(✉)

Pohang University of Science and Technology, Pohang, South Korea
{habaek4,jlee4923,simjy98,minjaetidtid,wonhwa}@postech.ac.kr

Abstract. Various neuroimaging studies suffer from small sample size problem which often limit their reliability. Meta-analysis addresses this challenge by aggregating findings from different studies to identify consistent patterns of brain activity. However, traditional approaches based on keyword retrieval or linear mappings often overlook the rich hierarchical structure in the brain. In this work, we propose a novel framework that leverages hyperbolic geometry to bridge the gap between neuroscience literature and brain activation maps. By embedding text from research articles and corresponding brain images into a shared hyperbolic space via the Lorentz model, our method captures both semantic similarity and hierarchical organization inherent in neuroimaging data. In the hyperbolic space, our method performs multi-level neuroimaging meta-analysis (MNM) by 1) aligning brain and text embeddings for semantic correspondence, 2) guiding hierarchy between text and brain activations, and 3) preserving the hierarchical relationships within brain activation patterns. Experimental results demonstrate that our model outperforms baselines, offering a robust and interpretable paradigm of multi-level neuroimaging meta-analysis via hyperbolic brain-text representation.

Keywords: Neuroimaging Meta-analysis · Hyperbolic Representation

1 Introduction

Recent research in neuroscience is rapidly expanding our understanding of brain function and structure, uncovering diverse aspects of neural organization across various regions [10,19]. However, neuroimaging studies are often underestimated due to small sample sizes of their experiments [3,20]. Brain meta-analysis tackles these challenges by integrating neuroscientific descriptions (i.e., body text) with corresponding brain activation maps from multiple studies to identify consistent patterns in the brain. Traditional approaches [7,11,21] have relied on predefined keywords and regression models to discover these patterns by predicting brain activation maps from textual descriptions. More recent approaches utilize large language models (LLMs) to handle longer sequences [15], and employ contrastive learning to map textual descriptions and activation coordinates into a joint representation space [13] for text-to-brain activation prediction.

J. C. Gee et al. (Eds.): MICCAI 2025, LNCS 15960, pp. 397–406, 2026.
https://doi.org/10.1007/978-3-032-04927-8_38

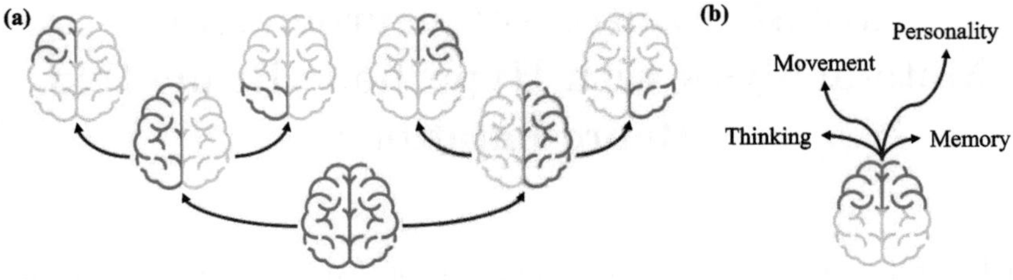

Fig. 1. Motivation of MNM. (a) The brain exhibits a spatial hierarchy, where broad regions can be further subdivided into more specific regions. (b) Each brain region can correspond to various articles covering diverse topics, such as functionalities.

Notice that the human brain exhibits a naturally nested organization, where broad regions are gradually subdivided into more specialized areas as in Fig. 1(a). Capturing these *multi-level relationships* is essential for robustly linking textual descriptions from a coarse activation (i.e., involved hemisphere) to fine details (i.e., the most relevant region of interests (ROIs)). Also, as depicted in Fig. 1(b), each brain region is defined by its unique functionality and can be linked to numerous articles that explore diverse topics related to these functions. Such insights on hierarchy are not included in previous approaches, where they map brain activation and text pairs to the identical point in the Euclidean space [13].

To better jointly represent brain activation and text pairs, we rely on hyperbolic spaces with exponentially expanding geometry, which are suited for representing tree-like structures. Therefore, we propose to embed the corresponding pairs into a shared hyperbolic space, together with an angle-based contrastive learning to capture mutual context between brain activation and text. Our framework, **M**ulti-level **N**euroimaging **M**eta-analysis with Hyperbolic Brain-Text Representations (MNM), preserves these inherent hierarchical relationships and overcomes the limitations of the Euclidean method, which lead to the following **major contributions: 1)** We introduce a hyperbolic embedding framework for neuroimaging meta-analysis that naturally captures semantic associations between brain and text and brain's hierarchical structure. **2)** We propose a brain structural hierarchy guidance such that brain structural hierarchy is preserved in the hyperbolic space. **3)** MNM is validated through extensive experiments on neuroimaging meta-analysis, such as cross-modal retrieval and brain activation prediction, demonstrating enhanced semantic alignment and interpretability in the same setting in [13]. Cross-modal retrieval and brain activation map prediction demonstrate that our approach not only improves semantic alignment between brain and text modalities but also preserves the multi-level organization of brain regions. Hierarchical structure analyses and ablation studies further confirm that incorporating both brain structural and brain-text hierarchies leads to more interpretable analyses and enhanced retrieval performance.

Related Works on Neuroimaging Meta-Analysis. Previous approaches for brain meta-analysis have transitioned from manual to automated methods. BrainMap [11] manually curated neuroimaging articles, a strategy that struggles to scale with the growing size of literature. NeuroSynth [21] and NeuroQuery [7] automate the extraction of brain activation coordinates using fixed keywords and statistical techniques, yet they are limited by bag-of-words representations that fail to capture complex semantics. Text2Brain [15] addresses this by using SciBERT [2] embeddings and convolutional neural networks to generate 3D activation maps, though it faces challenges with longer texts. Recently, NeuroConText [13] employs a contrastive framework that leverages advanced language models such as Mistral-7B [9] alongside the Dictionary of Functional Mode (DiFuMo) atlas [5] to enhance the alignment between paired modalities. While we adopt identically processed dataset from [13], MNM outperforms it by leveraging hyperbolic geometry to capture the brain's intrinsic hierarchical structure, thereby achieving superior semantic alignment and more flexible multi-level meta-analysis.

2 *Preliminary*: Hyperbolic Spaces and Lorentz Model

Unlike Euclidean space, hyperbolic space expands exponentially under constant negative curvature, making it highly effective for representing hierarchical structures. This enables capturing multi-level traits in both brain structure and text.

Lorentz Model. Lorentz model $\mathbb{L}^n$ expresses the n-dimensional hyperbolic space with curvature c. It is defined in a $(n+1)$-dimensional Euclidean space $\mathbb{R}^{n+1}$ as $\mathbb{L}^n = \{\mathbf{x} \in \mathbb{R}^{n+1} \mid \langle \mathbf{x}, \mathbf{x} \rangle_{\mathbb{L}} = -1/c, x_0 > 0\}$, where Lorentzian inner product is defined as $\langle \mathbf{x}, \mathbf{y} \rangle_{\mathbb{L}} = -x_0 y_0 + \sum_{i=1}^{d} x_i y_i$. Following [14], $\mathbf{x} \in \mathbb{L}^n$ is expressed as $[x_{\text{time}} \in \mathbb{R}, \mathbf{x}_{\text{space}} \in \mathbb{R}^n]$. Then, x_{time} is determined as:

$$x_{\text{time}} = \sqrt{1/c + \sum_{i=1}^{d} x_i^2}. \tag{1}$$

Geodesics. The shortest distance between two points $\mathbf{x}$, $\mathbf{y}$ on the hyperboloid is called Lorentzian distance $d_{\mathbb{L}}(\mathbf{x}, \mathbf{y})$ and given by:

$$d_{\mathbb{L}}(\mathbf{x}, \mathbf{y}) = \sqrt{1/c}\ \cosh^{-1}(-c\langle \mathbf{x}, \mathbf{y} \rangle_{\mathbb{L}}). \tag{2}$$

Tangent Space. Tangent space at a point $\mathbf{u} \in \mathbb{L}^n$ is an Euclidean space spanned by a set of vectors orthogonal to $\mathbf{u}$. For $\mathbf{z} \in \mathbb{R}^{n+1}$ in the tangent space, *exponential map* $\mathbf{Exp}_{\mathbf{u}}(\mathbf{z})$ transforms $\mathbf{z}$ into $\mathbf{x} \in \mathbb{L}^n$ in the hyperbolic space as

$$\mathbf{x} = \mathbf{Exp}_{\mathbf{u}}(\mathbf{z}) = \cosh\left(\sqrt{c}\sqrt{\langle \mathbf{z}, \mathbf{z} \rangle_{\mathbb{L}}}\right)\mathbf{u} + \frac{\sinh(\sqrt{c}\sqrt{\langle \mathbf{z}, \mathbf{z} \rangle_{\mathbb{L}}})}{\sqrt{c}\sqrt{\langle \mathbf{z}, \mathbf{z} \rangle_{\mathbb{L}}}}\mathbf{z}. \tag{3}$$

With the hyperboloid origin $\mathbf{O} = [1/c, \mathbf{0}]$ introduced in [6], $\mathbf{x}_{\text{space}}$ is defined as

$$\mathbf{x}_{\text{space}} = \mathbf{Exp}_{[1/c,0]}(\mathbf{z}) = \frac{\sinh(\sqrt{c}\sqrt{\langle \mathbf{z}, \mathbf{z} \rangle_{\mathbb{L}}})}{\sqrt{c}\sqrt{\langle \mathbf{z}, \mathbf{z} \rangle_{\mathbb{L}}}}\mathbf{z} \tag{4}$$

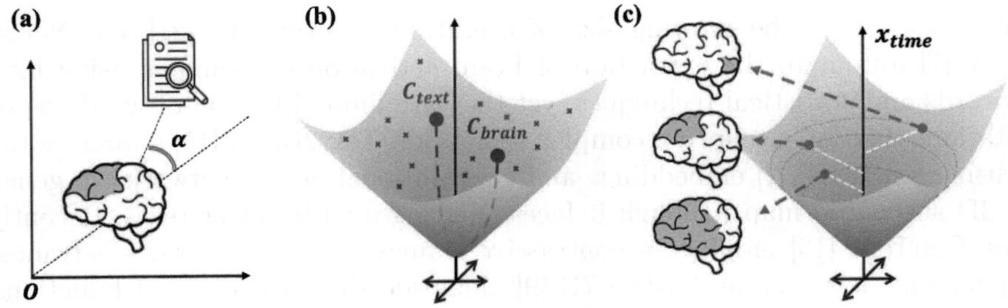

Fig. 2. Concept of Our Objective Functions. (a) The angle-based contrastive loss minimizes the exterior angle between a text embedding and its corresponding brain activation embedding. (b) The centroid loss encourages the centroid of brain embeddings $\mathbf{C}_{\text{brain}}$ to be positioned closer to the hyperboloid origin than the centroid of text embeddings $\mathbf{C}_{\text{text}}$. (c) The hierarchical loss aligns brain embeddings along the hyperbolic time axis by the number of activated regions for brain structural hierarchy.

Centroid. In the hyperbolic space, the *centroid* of N points in Lorentz space can be computed in the Klein space $\mathbb{K}^n$. Conversion between two spaces are

$$T_{\mathbb{L}\to\mathbb{K}}(\mathbf{x}) = \frac{\mathbf{x}_{\text{space}}}{x_{\text{time}}}, \quad T_{\mathbb{K}\to\mathbb{L}}(\mathbf{k}) = \frac{[1, \mathbf{k}]}{\sqrt{c(1 - \|\mathbf{k}\|^2)}}, \tag{5}$$

where $\mathbf{k} \in \mathbb{R}^n$ is a point in $\mathbb{K}^n$. Lorentzian centroid is then defined as

$$\text{centroid}_{\mathbb{L}}\left(\{\mathbf{x}_j\}_{j=1}^N\right) = T_{\mathbb{K}\to\mathbb{L}}\left(\frac{\sum_{j=1}^N \gamma_j T_{\mathbb{L}\to\mathbb{K}}(\mathbf{x}_j)}{\sum_{j=1}^N \gamma_j}\right), \tag{6}$$

where the Lorentz factors $\gamma_j \in \mathbb{R}$ are defined as $1/\sqrt{1 - c\|T_{\mathbb{L}\to\mathbb{K}}(\mathbf{x}_j)\|^2}$.

3 Hyperbolic Brain-Text Representation

Given a neuroscience article indexed by i, let (b_i, t_i) denotes a pair of brain activation data and LLM-representation of its corresponding neuroscientific text. For neuroimaging meta-analysis, which seeks to identify the most relevant brain regions from a given textual description or retrieve corresponding brain activations, our goal is to embed b_i and t_i into a shared space while preserving both the brain-text hierarchy as well as the structural hierarchy of the brain (Fig. 2).

Angle-Based Contrastive Learning. Unlike contrastive learning [6] that minimizes the spatial distances in a hyperbolic space, we adopt an angle-based similarity measure [17] since regularizing geodesic distance weakens the hierarchical dependency. We employ two parallel encoders: a brain activation encoder $\mathcal{E}_{\text{brain}}$ and a text encoder $\mathcal{E}_{\text{text}}$. After mapping the b_i and t_i into their respective Euclidean latent spaces, we project each of these onto d-dimensional hyperbolic space $\mathbb{L}^d$ via Eq. (4) as $\mathbf{z}_i^{\text{text}} = \mathcal{E}_{\text{text}}(t_i)$ and $\mathbf{z}_i^{\text{brain}} = \mathcal{E}_{\text{brain}}(b_i)$. Since each brain region can be characterized by different functionalities as depicted in Fig. 1(b), we assume that brain activations have higher hierarchy (i.e., broader context) than textual descriptions. Denoting $z^{\cdot}_{i,\text{time}}$ and $\mathbf{z}^{\cdot}_{i,\text{space}}$ as the time and space component of $\mathbf{z}^{\cdot}_i$, exterior angle of $\mathbf{z}^{\text{text}}_i$ from $\mathbf{z}^{\text{brain}}_i$ is then defined as

$$\text{ext}(\mathbf{z}_{\text{brain}}, \mathbf{z}_{\text{text}}) = \cos^{-1}\left(\frac{z^{\text{text}}_{i,\text{time}} + z^{\text{brain}}_{i,\text{time}} \cdot c\langle \mathbf{z}^{\text{brain}}, \mathbf{z}^{\text{text}}\rangle_{\mathbb{L}}}{\|\mathbf{z}^{\text{brain}}_{i,\text{space}}\| \sqrt{(c\langle \mathbf{z}^{\text{brain}}, \mathbf{z}^{\text{text}}\rangle_{\mathbb{L}})^2 - 1}}\right). \tag{7}$$

As exterior angle is defined in $[-\pi, \pi]$, smaller absolute angle indicates higher correspondence. To increase similarity between corresponding pair and discrepancies between non-corresponding pairs, we utilize InfoNCE loss [16] to define our angle-based loss by minimizing the absolute angles of positive pairs as

$$\mathcal{L}_{\text{angle}} = -\frac{1}{N}\sum_{i=1}^{N} \log \frac{\exp\left(-|\text{ext}(\mathbf{z}^{\text{img}}_i, \mathbf{z}^{\text{text}}_i)|/\tau\right)}{\sum_{j=1}^{N} \exp\left(-|\text{ext}(\mathbf{z}^{\text{img}}_i, \mathbf{z}^{\text{text}}_j)|/\tau\right)}. \tag{8}$$

Centroid Regularization. To reflect the hierarchy between the brain activations and text, $\mathbf{z}^{\text{brain}}$ needs to be positioned closer to the hyperboloid origin $\mathbf{O}$ than $\mathbf{z}^{\text{text}}$ as a smaller distance from $\mathbf{O}$ corresponds to a higher hierarchy. However, $\mathcal{L}_{\text{angle}}$ alone does not have explicit control over the spatial position of the embeddings. To explicitly enforce this ordering, we first compute the centroids of the brain activation embeddings and the article text embeddings, denoted by $\mathbf{c}_{\text{brain}}$ and $\mathbf{c}_{\text{text}}$ each, as described in Eq. (6). We then impose a constraint that guides these centroids toward predetermined target distances respectively as

$$\mathcal{L}_{\text{cent}} = \left\| d_{\mathbb{L}}(\mathbf{O}, \mathbf{c}_{\text{text}}) - \frac{1}{\sqrt{c}}\cosh^{-1}(cp)\right\| + \left\| d_{\mathbb{L}}(\mathbf{O}, \mathbf{c}_{\text{img}}) - \frac{1}{\sqrt{c}}\cosh^{-1}(cq)\right\|, \tag{9}$$

where $p, q \in \mathbb{R}^+$ are hyperparameters satisfying $p > q$ and $\|\cdot\|$ is the Euclidean norm. This condition ensures that the 'broader' brain activation embeddings are positioned closer to the hyperboloid origin than the 'detailed' text embeddings.

Brain Structural Hierarchy Guidance. Brain activation map exhibits varying levels of specificity from global regions (e.g., left hemisphere) to local regions (e.g., superior temporal gyrus) as illustrated in Fig. 1(a). To refine this structural hierarchy among brain embeddings, we introduce a brain structural hierarchy guidance based on the number of the activation regions. Specifically, we set R_i as how many ROIs in b_i have values greater than predefined threshold δ. The brain structural hierarchy guidance between two brain embeddings is defined as

$$\mathcal{L}_{\text{hier}} = \frac{1}{N^2}\sum_{i=1}^{N}\sum_{j=1}^{N} \mathbb{1}_{ij}(R_i - R_j) \cdot \max(\log(z^{\text{brain}}_{i,\text{time}}/z^{\text{brain}}_{j,\text{time}}), 0), \tag{10}$$

where $\mathbb{1}_{ij}$ is an indicator function that equals 1 if $R_i > R_j$. This penalizes when a broader brain activation (i.e., a more general pattern) is wrongly placed at a lower hierarchical level (i.e., with a larger time value) than a localized activation.

Joint Optimization. The final training loss is a weighted combination as:

$$\mathcal{L} = \mathcal{L}_{\text{angle}} + \lambda_1 \mathcal{L}_{\text{cent}} + \lambda_2 \mathcal{L}_{\text{hier}}, \tag{11}$$

where $\lambda_1, \lambda_2 \in \mathbb{R}^+$ are hyperparameters that balance the contributions of centroid regularization and brain hierarchical loss. By jointly minimizing this objective, the model learns to capture both the semantic alignment and the hierarchical structure necessary for robust neuroimaging meta-analysis.

4 Experimental Settings

Data Preparation. Our work follows the setting established by NeuroConText [13], which processed 20,674 neuroscientific articles from PubMed along with their corresponding brain activation peak coordinates. With a pretrained LLM (i.e., Mistral-7B), text representation t_i is derived from textual descriptions (e.g., abstract or body). A corresponding discrete brain activation peak coordinate is converted into continuous brain activation voxel maps via Kernel Density Estimation (KDE), and subsequently projected onto a 512-dimensional coefficients using the Dictionary of Functional Modes (DiFuMo) atlas as b_i.

Implementation Details. For $\mathcal{E}_{\text{text}}$ and $\mathcal{E}_{\text{brain}}$, we use a two-layer and three-layer residual MLP, respectively, followed by layer normalization. We train the model for 200 epochs using the AdamW optimizer [12] with a learning rate of 1×10^{-4}, a batch size of 4096, and a weight decay of 0.05. We set $\lambda_1 = 0.5$, $\lambda_2 = 30$, $\delta = 5$, $p = 2.0$, and $q = 0.5$. We use pretrained weights for Text2Brain [15] and NeuroQuery [7], while NeuroConText [13] is trained under the same settings as our model. *The code and dataset will be publicly available.*

Table 1. Comparison of Brain-Text Cross-Modal Retrieval Performance using two types of textual descriptions. The best performance is indicated in **bold**.

	Retrieval	Text→Brain			Brain→Text		
	Metric [%]	Recall@5	Recall@10	Recall@100	Recall@5	Recall@10	Recall@100
Abstract	Text2Brain [15]	0.247±0.026	0.469±0.022	4.837±0.075	-	-	-
	NeuroQuery [7]	4.266±0.262	7.560±0.389	36.055±0.761	-	-	-
	NeuroConText [13]	9.964±0.663	15.033±0.808	45.120±1.098	10.046±0.683	15.183±0.790	45.168±1.077
	MNM-Reverse	8.852±1.224	13.486±42.585	42.585±3.000	10.332±0.562	15.096±0.964	44.636±1.270
	MNM *w/o* $\mathcal{L}_{\text{hier}}$	10.699±0.650	15.807±0.653	45.734±1.286	10.458±0.590	16.218±0.764	46.406±1.057
	MNM	**10.728±0.673**	**15.860±0.668**	**45.864±1.417**	**10.593±0.627**	**16.320±0.822**	**46.590±1.217**
Body	Text2Brain [15]	0.237±0.014	0.479±0.015	4.832±0.079	-	-	-
	NeuroQuery [7]	4.421±0.374	7.696±0.458	36.055±0.562	-	-	-
	NeuroConText [13]	13.660±0.520	20.228±0.742	51.945±0.961	13.747±0.579	19.517±0.449	51.572±0.771
	MNM-Reverse	13.756±0.575	20.136±0.485	51.969±0.736	14.087±0.556	21.176±0.575	53.134±0.845
	MNM *w/o* $\mathcal{L}_{\text{hier}}$	14.288±0.747	20.649±0.850	52.670±0.735	14.569±0.648	21.302±0.396	53.342±0.797
	MNM	**14.434±0.618**	**20.741±0.574**	**52.709±0.928**	**14.806±0.449**	**21.360±0.414**	**53.623±0.698**

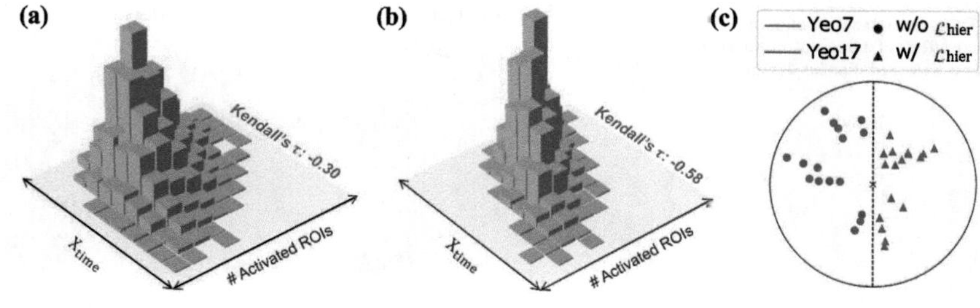

Fig. 3. Effect of Brain Hierarchical Loss. (a) Histogram between x_{time} and the number of activated ROIs without $\mathcal{L}_{\text{hier}}$, showing -0.30 of Kendall's τ [18] (b) Same histogram with $\mathcal{L}_{\text{hier}}$, decreasing Kendall's τ to -0.58. (c) 2D Poincaré disk projection of Yeo7 and Yeo17 network [22] embeddings (left: w/o $\mathcal{L}_{\text{hier}}$, right: w/ $\mathcal{L}_{\text{hier}}$).

5 Results and Analysis

Brain-Text Cross Modal Retrieval. To evaluate our semantic alignment performance, we perform text-to-brain and brain-to-text retrievals. In the text-to-brain setting, the model ranks candidate brain activations by their relevance to given textual descriptions (e.g., abstracts or full-body text). The brain-to-text setting reverses it, from a brain activation to the most relevant article. We adopt a 10-fold cross-validation to rigorously evaluate recall@K, $K \in \{5, 10, 100\}$. Recall@K evaluates whether the true pair appears among the top-K similar candidates, making it suitable for ranking tasks. For comparison, the following baselines are adopted. Since Text2Brain [15] and NeuroQuery [7] directly generate 3D brain activations from text, mean squared error (MSE) is used as similarity measure between the predicted brain activation and each candidate. Unlike [13] using cosine similarity, MNM adopts the negation of the absolute exterior angle.

For MNM, we employ two additional settings: (1) MNM-Reverse that reverses the hierarchy of MNM by flipping the brain and text relationship in $\mathcal{L}_{\text{ext}}$ and $\mathcal{L}_{\text{cent}}$, and (2) MNM *w/o* $\mathcal{L}_{\text{hier}}$. As shown in Table 1, our method consistently outperforms all baselines in both text-to-brain and brain-to-text retrieval, with the largest improvement observed in brain-to-text retrieval when using full-body text. Degraded performances in two additional settings demonstrate the rationality of brain-text hierarchy and the need for explicit hierarchy guidance.

Brain Hierarchy Analysis. To show the effect of $\mathcal{L}_{\text{hier}}$, we analyze how samples are distributed in the latent space. One of our goals is to model the inherent hierarchy of brain activation patterns, such that more general (i.e., widely activated) regions lie closer to the hyperbolic origin than more specific (i.e., smaller or functionally narrower) ones. To verify the assumption with the trained embeddings using $\mathcal{L}_{\text{hier}}$, we plot histograms of x_{time} against the number of activated ROIs, along with Kendall's τ [18] to quantify the ordinal correlation in a range $[-1, 1]$. A negative τ indicates that embeddings with broader activations tend to

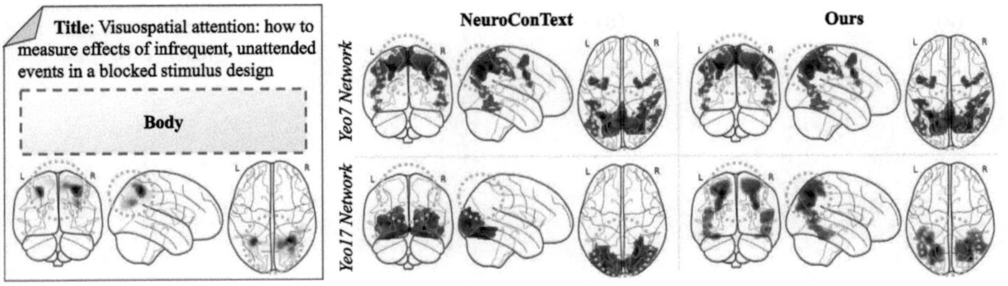

Fig. 4. Reconstructed Brain Activation Map from a Neuroscience Article [8]. While NeuroConText [13] fails to capture activation in the intraparietal sulcus using Yeo17 networks, MNM assigns high similarity scores to them as in the ground truth.

reside at a higher "level" (i.e., smaller x_{time}) in the hyperbolic space. In Fig. 3, while embeddings are dispersed (a) without $\mathcal{L}_{\text{hier}}$ resulting in -0.30 of Kendall's τ, they are more inversely aligned (b) with $\mathcal{L}_{\text{hier}}$ resulting in Kendall's $\tau = -0.58$.

We further examine two well-known cortical parcellation schemes, Yeo7 and Yeo17 networks [22]. The Yeo7 network segments the brain into seven broad functional systems, while the Yeo17 network refines this into 17 detailed subdivisions. To visualize these divisions, we map them into the learned hyperbolic space and project them onto a 2D Poincaré disk [4]. Due to page limitations, we visualize only half of the Poincaré disk, varying the presence of $\mathcal{L}_{\text{hier}}$. In Fig. 3(c), the embeddings from both networks exhibit no discernible order in the left half when $\mathcal{L}_{\text{hier}}$ is absent. However, with $\mathcal{L}_{\text{hier}}$, Yeo7 embeddings cluster near the origin in the right half, enhancing interpretability. This organization supports a coarse-to-fine analysis by selecting x_{time} from brain activation map candidates.

Brain Activation Map Prediction. To evaluate text-brain alignment beyond the task-specific decoder, we use the embeddings of Yeo7 and Yeo17 networks [22] as our brain activation bases. With body text of given articles, we calculate the similarity of the text embedding and brain activation bases of two networks. NeuroConText [13] is a unique baseline, since it aligns both modalities in the shared latent space as we do. We employ the same similarity metric as in the brain-text cross-modal retrieval experiment. The resulting similarities are transformed into probabilities using a softmax function and assigned to the corresponding voxels in the 3D brain. For better visualization, we threshold the top 10% of voxels and visualize using brain glass schematics via the Nilearn [1] software.

Figure 4 illustrates how embeddings can be recovered using the Yeo7 and Yeo17 networks for a study [8] that investigates the neural mechanisms underlying visuospatial attention reorientation. The study demonstrates that both blocked and event-related analyses consistently activate the intraparietal sulcus and superior parietal cortex. In coarse prediction using the Yeo7 networks, both NeuroConText and MNM exhibit a high correlation (i.e., top 10% similarity) with the activated regions. However, with the finer-grained Yeo17 network, NeuroConText fails to capture the target regions, whereas our method successfully identifies them. Since most detected regions overlap across different parcellation levels, MNM demonstrates greater consistency compared to NeuroConText. These results indicate that our hyperbolic representation can flexibly adapt to various parcellation schemes when inferring activation maps from articles.

6 Conclusion

In this work, we propose MNM, a novel framework that leverages hyperbolic geometry to capture the inherent hierarchical structure of brain activations while aligning them more effectively with corresponding textual descriptions. Through extensive experiments, we demonstrate that MNM not only surpasses baselines in cross-modal retrieval but also provides greater interpretability in activation map prediction. By providing robust performances across various parcellation schemes, MNM paves the way for multi-level neuroimaging meta-analysis.

Acknowledgments. This research was supported by NRF-2022R1A2C2092336 (50%), RS-2025-02216257 (20%), RS-2022-II220290 (20%), and RS-2019-II191906 (AI Graduate Program at POSTECH, 10%).

Disclosure of Interests. The authors have no competing interests to declare that are relevant to the content of this article.

References

1. Abraham, A., Pedregosa, F., Eickenberg, M., et al.: Machine learning for neuroimaging with scikit-learn. Front. Neuroinform. **8**, 71792 (2014)
2. Beltagy, I., Lo, K., Cohan, A.: SciBERT: a pretrained language model for scientific text. In: EMNLP-IJCNLP, pp. 3615–3620. ACL (2019)
3. Button, K.S., Ioannidis, J.P., Mokrysz, C., et al.: Power failure: why small sample size undermines the reliability of neuroscience. Nat. Rev. Neurosci. **14**(5), 365–376 (2013)
4. Cvetkovski, A., Crovella, M.: Multidimensional scaling in the poincaré disk. arXiv preprint arXiv:1105.5332 (2011)
5. Dadi, K., Varoquaux, G., Machlouzarides-Shalit, A., et al.: Fine-grain atlases of functional modes for FMRI analysis. Neuroimage **221**, 117126 (2020)
6. Desai, K., Nickel, M., Rajpurohit, T., et al.: Hyperbolic image-text representations. In: ICML, pp. 7694–7731. PMLR (2023)
7. Dockès, J., Poldrack, R.A., Primet, R., et al.: Neuroquery, comprehensive meta-analysis of human brain mapping. elife **9**, e53385 (2020)
8. Giessing, C., Thiel, C.M., Stephan, K.E., et al.: Visuospatial attention: how to measure effects of infrequent, unattended events in a blocked stimulus design. Neuroimage **23**(4), 1370–1381 (2004)
9. Jiang, A.Q., Sablayrolles, A., Mensch, A., et al.: Mistral 7b (2023). https://arxiv.org/abs/2310.06825
10. Kim, W., et al.: Multi-resolutional shape features via non-Euclidean wavelets: applications to statistical analysis of cortical thickness. NeuroImage **93**, 107–123 (2014)

11. Laird, A.R., Lancaster, J.J., Fox, P.T.: Brainmap: the social evolution of a human brain mapping database. Neuroinformatics **3**, 65–77 (2005)
12. Loshchilov, I., Hutter, F.: Decoupled weight decay regularization. In: ICLR (2019)
13. Meudec, R., Ghayem, F., Dockès, J., et al.: Neurocontext: contrastive text-to-brain mapping for neuroscientific literature. In: Linguraru, M.G., et al. (eds.) Medical Image Computing and Computer Assisted Intervention. MICCAI 2024. LNCS, vol. 15003, pp. 325–335. Springer, Cham (2024). https://doi.org/10.1007/978-3-031-72384-1_31
14. Minkowski, H.: Raum und zeit. In: Aichelburg, P.C. (ed.) Zeit im Wandel der Zeit. Facetten der Physik, vol. 23, pp. 123–136. Vieweg+Teubner Verlag, Wiesbaden (1988). https://doi.org/10.1007/978-3-322-89451-9_11
15. Ngo, G.H., Nguyen, M., Chen, N.F., Sabuncu, M.R.: Text2Brain: synthesis of brain activation maps from free-form text query. In: de Bruijne, M., et al. (eds.) MICCAI 2021. LNCS, vol. 12907, pp. 605–614. Springer, Cham (2021). https://doi.org/10.1007/978-3-030-87234-2_57
16. Oord, A.v.d., Li, Y., Vinyals, O.: Representation learning with contrastive predictive coding. arXiv preprint arXiv:1807.03748 (2018)
17. Ramasinghe, S., Shevchenko, V., Avraham, G., et al.: Accept the modality gap: an exploration in the hyperbolic space. In: CVPR, pp. 27263–27272 (2024)
18. Sen, P.K.: Estimates of the regression coefficient based on Kendall's tau. J. Am. Stat. Assoc. **63**(324), 1379–1389 (1968)
19. Sim, J., Lee, M., Wu, G., Kim, W.H.: Multi-modal graph neural network with transformer-guided adaptive diffusion for preclinical alzheimer classification. In: Linguraru, M.G., et al. (eds.) Medical Image Computing and Computer Assisted Intervention. MICCAI 2024. LNCS, vol. 15005, pp. 511–521. Springer, Cham (2024). https://doi.org/10.1007/978-3-031-72086-4_48
20. Szucs, D., Ioannidis, J.P.: Empirical assessment of published effect sizes and power in the recent cognitive neuroscience and psychology literature. PLoS Biol. **15**(3), e2000797 (2017)
21. Yarkoni, T., Poldrack, R.A., Nichols, T.E., et al.: Large-scale automated synthesis of human functional neuroimaging data. Nat. Methods **8**(8), 665–670 (2011)
22. Yeo, B.T., Krienen, F.M., Sepulcre, J., et al.: The organization of the human cerebral cortex estimated by intrinsic functional connectivity. J. Neurophysiol. (2011)

MrTrack: Register Mamba for Needle Tracking with Rapid Reciprocating Motion During Ultrasound-Guided Aspiration Biopsy

Yuelin Zhang[1], Qingpeng Ding[1], Long Lei[2], Yongxuan Feng[1], Raymond Shing-Yan Tang[3], and Shing Shin Cheng[1](✉)

[1] Department of Mechanical and Automation Engineering, The Chinese University of Hong Kong, Shatin, Hong Kong
sscheng@cuhk.edu.hk

[2] Department of Computer Science and Engineering, The Chinese University of Hong Kong, Shatin, Hong Kong

[3] Department of Medicine and Therapeutics and Institute of Digestive Disease, The Chinese University of Hong Kong, Shatin, Hong Kong

Abstract. Ultrasound-guided fine needle aspiration (FNA) biopsy is a common minimally invasive diagnostic procedure. However, an aspiration needle tracker addressing rapid reciprocating motion is still missing. MrTrack, an aspiration needle tracker with a mamba-based register mechanism, is proposed. MrTrack leverages a Mamba-based register extractor to sequentially distill global context from each historical search map, storing these temporal cues in a register bank. The Mamba-based register retriever then retrieves temporal prompts from the register bank to provide external cues when current vision features are temporarily unusable due to rapid reciprocating motion and imaging degradation. A self-supervised register diversify loss is proposed to encourage feature diversity and dimension independence within the learned register, mitigating feature collapse. Comprehensive experiments conducted on both robotic and manual aspiration biopsy datasets demonstrate that MrTrack not only outperforms state-of-the-art trackers in accuracy and robustness but also achieves superior inference efficiency. Project page: https://github.com/PieceZhang/MrTrack.

Keywords: Ultrasound Needle Tracking · Mamba · Temporal Context

1 Introduction

Ultrasound (US)-guided fine needle aspiration (FNA) has emerged as a pivotal minimally invasive diagnostic procedure for percutaneous interventions [22],

---

Supplementary Information The online version contains supplementary material available at https://doi.org/10.1007/978-3-032-04927-8_39.

J. C. Gee et al. (Eds.): MICCAI 2025, LNCS 15960, pp. 407–417, 2026.
https://doi.org/10.1007/978-3-032-04927-8_39

facilitating the retrieval of cellular material by rapid reciprocating movement [23]. Despite its growing clinical importance, a dedicated needle tracker addressing the rapid needle aspiration is still missing. In practice, the imaging environment during FNA presents unique challenges for a US needle tracker, where reciprocating rapid needle motion during the aspiration [23] induces abrupt nonlinear tip displacement and degraded imaging quality. Although recent models enhance needle tracking by deep learning [13], they were mainly designed for stable environments, yet the nonlinear needle motion and artifact-prone imaging of aspiration biopsy can render these methods ineffective.

Recent developments in tracking reveal that capturing long-range temporal dependencies can enhance tracking [7,14,36]. However, conventional CNN-based [7] or transformer-based [36] trackers struggle with computational constraints when integrating temporal context. In contrast, Mamba [9], a recently developed framework based on state space models (SSMs) [12], has shown its superiority and efficiency in modeling long-term temporal features with satisfactory complexity growth. Although there exist Mamba trackers with temporal modeling [14,28,33,35], their performance and efficiency suffer from inefficient temporal integration. The register then offers a promising solution [4], which is a learnable token that aggregates global context into a compact representation for efficient downstream processing. It calls for an efficient register mechanism to model a spatiotemporal profile for the aspiration needle. The global context from historical search maps should be sequentially *extracted* and stored using the register. The stored latent features from registers should also be effectively *retrieved* to provide external cues for the tracking of the current frame, thus compensating for rapid motion and abrupt appearance changes of the target being tracked.

In this work, **MrTrack**, an ultrasound needle tracker for FNA based on Mamba and register, is proposed with a **register extractor** and a **register retriever**. To the best of our knowledge, it is the first US needle tracker designed for the FNA procedure to robustly track the needle tip undergoing fast reciprocating motion. A **register diversify loss (RD loss)** is also proposed as a self-supervised regularization to prevent collapse in the learned register by encouraging register diversity and dimension independence. The experimental results on both robotic and manual aspiration biopsies show state-of-the-art performance and efficiency of MrTrack. The contributions of MrTrack are fourfold: **1)** The proposed Mamba-based register extractor sequentially condenses historical frames into compact tokens using a trainable register, storing spatiotemporal profiles to provide external cues from past frames when the needle undergoes appearance variation. **2)** The proposed Mamba-based register retriever retrieves temporal cues from historical registers and integrates them with the template map, assisting in tracking an abruptly moving needle using temporal context. **3)** The self-supervised register diversify loss (RD Loss) is proposed to prevent feature collapse by promoting diversity and dimension independence, ensuring an informative register that representationally stores the varying temporal context. **4)** MrTrack achieves highly efficient inferencing, which enables real-time tracking of rapid needle motion under high-speed US imaging.

2 Related Work

Ultrasound Needle Tracking. By free-hand [17] or robotics-driven [19], US is commonly adopted in many clinical practices [2,18], including US-guided needle insertion. To track a US needle, recent learning-based methods have bee proposed, including discriminative correlation filters [24], convolutional neural networks [21], and transformer [29,30]. However, these methods were largely developed for relatively stable needle dynamics. In contrast, the dynamic environment of aspiration biopsy, where rapid needle motion and significant imaging artifacts frequently obscure fine needle features [23], exposes the limitations of existing trackers. Although some needle trackers incorporate historical motion [29] to compensate for temporary tip loss, the reciprocating and abrupt movements in aspiration can limit the effectiveness of such motion integration, where the motion can be highly nonlinear and unpredictable.

Mamba. Recently, based on SSMs [12], Mamba [9] outperforms transformers, leveraging its selective scan mechanism to integrate information over extended temporal ranges with a linear complexity [32]. Mamba-based trackers have demonstrated their capability in tracking tasks and prove the importance of modeling long-term temporal dependencies [14,28]. In contrast, transformer-based [36] or CNN-based [7] trackers with temporal modeling tend to struggle under similar computational constraints due to higher complexity. The existing Mamba trackers with temporal-context modeling face challenges due to the lack of an effective and efficient temporal extraction and integration paradigm [14,28]. As a trainable prompt, the register has been applied in transformer [4] and Mamba [25] for its ability to suppress artifacts and, more critically, function as a compact global representation. The learnable register can thus be helpful for aspiration needle tracking by adapting rapid motion with temporal context.

Self-supervised Regularization. Advances in self-supervised learning have demonstrated that enforcing feature invariance can yield powerful representations [10,34]. Recent approaches like BYOL [8], Barlow Twins [31], and VICReg [1] promote feature diversity and prevent collapse through variance preservation and decorrelation of embedding dimensions. However, the self-supervised regularization for training learnable prompts like the register in Mamba is still missing.

3 Method

3.1 Preliminaries on Mamba

The advancement of Mamba is based on the structured state-space models (SSMs), which rely on a continuous linear time-invariant (LTI) system. It can be expressed with the following linear ordinary differential equation (ODE)

$$\begin{aligned} h'(t) &= \mathbf{A}h(t) + \mathbf{B}u(t), \\ y(t) &= \mathbf{C}h(t), \end{aligned} \tag{1}$$

where $\mathbf{A} \in \mathbb{R}^{N \times N}$, $\mathbf{B} \in \mathbb{R}^{N \times 1}$, and $\mathbf{C} \in \mathbb{R}^{1 \times N}$. Following the equations, the input $u(t) \in \mathbb{R}$ is mapped to the output $y(t) \in \mathbb{R}$ by a linear transformation with $h(t) \in \mathbb{R}^{N \times 1}$. A discretization is then required for this continuous system. With Δ denoting the timescale step size, the discretized system is given by

$$\begin{aligned} h_t &= \bar{\mathbf{A}} h_{t-1} + \bar{\mathbf{B}} u_t, \\ y_t &= \mathbf{C} h_t, \end{aligned} \tag{2}$$

where $\mathbf{A}$ and $\mathbf{B}$ is discretized to be $\bar{\mathbf{A}} = exp(\Delta \cdot \mathbf{A})$ and $\bar{\mathbf{B}} = (\Delta \cdot \mathbf{A})^{-1}(exp(\Delta \cdot \mathbf{A}) - I) \cdot \Delta \cdot \mathbf{B}$. In Mamba [9], the selective scan mechanism is incorporated to learn $\mathbf{B}, \mathbf{C}, \Delta$ from input x by an input-dependent way, which are given by $\mathbf{B} = \text{Linear}(x)$, $\mathbf{C} = \text{Linear}(x)$, $\Delta = \text{Softplus}(\tilde{\Delta} + \text{Linear}(x))$, where $\tilde{\Delta}$ is learnable.

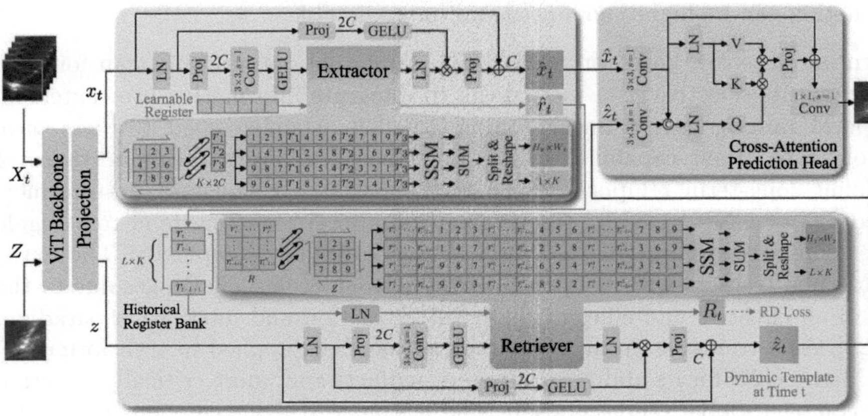

Fig. 1. Overview of MrTrack. The notes (e.g. $3 \times 3, s = 1$) on convolution layers represent the kernel size and stride.

3.2 Overview of MrTrack

As shown in Fig. 1, a ViT-Base [5] is adopted as the backbone, where the embedded features of the template map $z \in \mathbb{R}^{H_z \times W_z \times C}$ and search map $x_t \in \mathbb{R}^{H_x \times W_x \times C}$ ($C = 256$) are obtained, where t denotes the current time step. The x_t and z are then passed to the Mamba-based register extractor and retriever, respectively, to extract registers from the search map and retrieve information into the template map. The register is trained in a self-supervised manner by the register diversify loss $\mathcal{L}_{\text{RD}}$. With the obtained $\hat{x}_t$ and $\hat{z}_t$, the tracking result is predicted by the cross-attention prediction head.

3.3 Mamba-Based Register Extractor and Register Retriever

The proposed Mamba-based register extraction-retrieval framework benefits aspiration needle tracking by aggregating long-term visual features at minimal overhead. The extractor learns compact global representation from a search map x_t through a trainable register $r \in \mathbb{R}^{k \times 2C}$ $(k = 8)$, where r is shareable for all frames. Specifically, after a series of layer normalization, projection, and convolution, x_t is passed to the extractor together with r, in which a cross-map scanning is performed to evenly insert the register *behind* each image segment throughout the x_t, as shown in Fig. 1. This insertion ensures a thorough feature translation. The fused vectors are then learned by the SSM, where not only the feature from x_t is conveyed into the output r_t but also the x_t itself is modeled to be $\hat{x}_t$ with richer representation and suppressed artifacts [25]. By the extractor, the dedicated compact descriptor for each frame is efficiently obtained, then stored in the register bank $R_t = [r_t, ..., r_{t-L+1}] \in \mathbb{R}^{L \times k \times 2C}$, $L = 300$.

The retriever retrieves the information from the register bank R_t to the template z. Through cross-map scanning, the descriptors r_i in R_t are evenly inserted back into the sequence of image tokens of z, yet *before* each image segment. After SSM modeling, the temporal context in R_t is retrieved into $\hat{z}_t$, which is a dynamic template containing global temporal context from time t to $t-L$. Through the retriever, the feature of the initial template is aggregated with the latest transient cues, which can be helpful during aspiration when abrupt and rapid motion happens.

3.4 Register Diversify Loss

To ensure that the learned register does not collapse and to encourage distinctiveness in the register bank, the register diversify loss (RD loss) is proposed. RD loss comprises two components: a variance promoting term $\mathcal{L}_{var}$ and a cross-register diversify term $\mathcal{L}_{div}$. The variance promoting term $\mathcal{L}_{var}$ penalizes feature dimensions whose standard deviation falls below a threshold τ ($\tau = 1$ in this paper) to encourage diversity. The *softplus* is applied here as a smooth penalty with differentiability, given by softplus$(v) = ln(1 + e^v)$. $\mathcal{L}_{var}$ is then defined by

$$L_{var} = \frac{1}{d}\sum_{i=1}^{d} \text{softplus}(\tau - \sqrt{V_i + \epsilon}), \tag{3}$$

where d is the dimensionality, ϵ is a small constant. V_i is variance of the i_{th} dimension. The variance of the learnable register is then promoted with $\mathcal{L}_{var}$.

Except for promoting variance of the register, the cross-register diversify term $\mathcal{L}_{div}$ is proposed to encourage the diversity within the register bank, thus preventing the register extractor from encoding similar features for different frames. Assume each descriptor r_i in $R_t = [r_t, ..., r_{t-L}]$ is defined by $r_i = [r_{i,1}, ..., r_{i,d}] \in \mathbb{R}^{k \times d}$, where $d = 2C$, $\mathcal{L}_{div}$ is presented with a covariance format given by

$$\mathcal{L}_{\text{div}} = \frac{1}{dL(L-1)}\sum_{p=1}^{d}\sum_{\substack{i,j=1 \\ i \neq j}}^{L} \left[(r_{i,p} - \bar{r})\,(r_{j,p} - \bar{r})\right]^2, \text{ where } \bar{r} = \frac{1}{L}\sum_{n=1}^{L} r_{n,p}. \tag{4}$$

$\mathcal{L}_{div}$ drives the off-diagonal elements of the covariance close to zero thus promoting decorrelation. Different from the covariance regularization in [1] that decorrelates the dimension within the same sample, $\mathcal{L}_{div}$ decorrelates dimensions between different samples r_i from the register bank, thus encouraging the model to learn independently varying features for different frames. By combining Eq. 3 and Eq. 4 with scalar $\alpha = \beta = 0.01$, the RD loss $\mathcal{L}_{\text{RD}}$ is given by

$$\mathcal{L}_{\text{RD}} = \alpha\mathcal{L}_{\text{var}} + \beta\mathcal{L}_{\text{div}}. \tag{5}$$

This self-supervised regularization prevents collapse by encouraging dimension independence in the learnable register and promoting diversity in register bank.

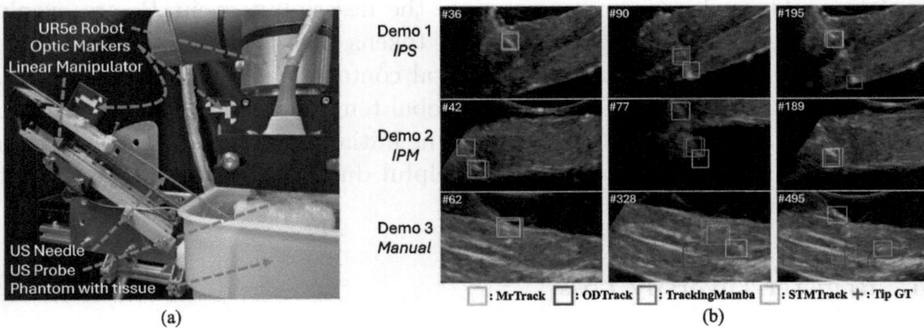

Fig. 2. Experiment setup (a) and tracking demonstration (b). More demonstrations are shown in the supplementary video.

4 Experiments and Results

4.1 Experiment Setup and Data Collection

As shown in Fig. 2(a), a robotic US needle insertion platform and a US imaging system with a Wisonic Clover 60 US machine and a Wisonic C5-1 convex transducer (sampling frequency 4.5 MHz) was adopted for the experiments. A 20 gauge fine aspiration needle [6] was used to insert into a phantom made with fresh pork supported by solidified agar. The ground truth position of the needle tip was collected by an optic tracker (ClaroNav MicronTracker 3), which has an RMSE of 0.189 mm in our setup. Datasets for both robotic and manual needle manipulation experiments simulating the aspiration biopsy procedure were collected. During the robotic insertion, the needle was installed on a linear manipulator, while the US probe was held by a UR5e robot arm. The needle was first inserted at a constant velocity of 20 mm/s to reach the target with a depth of more than 50 mm. After reaching the target, the needle would move back and forth five times at a velocity of 30 mm/s and a depth of 15 mm to perform aspiration. Both in-plane-static (IPS) and in-plane-moving (IPM) techniques [13,29] were performed at three needle insertion angles (30°, 45°, 60°). The robotic insertion dataset includes 54,428 frames in videos (1920 × 1080, 30 FPS) from 239 trials (105 IPS, 134 IPM). After sampling it by 5 to remove redundancy, the dataset was divided into training (7,619 samples), validation (1,088 samples), and testing (2,177 samples) sets in the ratio 7:1:2. During the manual insertion, both the needle and probe were held by the user. Four users were asked to perform four aspirations each, with the needle being moved back and forth two to three times per second. It resulted in a dataset that includes 16 videos (10,948 frames), which is only for testing to validate generalization.

During the evaluation, all models were implemented using PyTorch and trained on the same dataset with four NVIDIA RTX 4090 GPUs. All models adopt the same training strategy (350 epochs, batch size 32, AdamW optimizer) and the same input size of 384 × 384 and 192 × 192 for search and template map. Scaling, blur, and position shifting were adopted for image augmentation. The learning rate was set to 5e-4 and dropped by 10 after 200 epochs.

Table 1. Evaluation and ablation study results. The methods with the best and the second best performance are in red and cyan. AUC and P are reported in percentage (%), and Err and SD are reported in mm. In ablation studies, the performance difference relative to the baseline has been highlighted.

Method	Robotic In-plane-static (IPS)			Robotic In-plane-moving (IPM)			Robotic Mean			Manual			FPS
	AUC$_\uparrow$	$P_\uparrow$	Err±SD$_\downarrow$	AUC$_\uparrow$	$P_\uparrow$	Err±SD$_\downarrow$	AUC$_\uparrow$	$P_\uparrow$	Err±SD$_\downarrow$	AUC$_\uparrow$	$P_\uparrow$	Err±SD$_\downarrow$	
SiamRPN++ [15]	44.1	60.0	4.71±3.10	58.2	75.0	3.28±3.00	50.3	66.7	4.14±3.06	42.0	59.3	5.50±3.73	31.3
SiamCAR [11]	45.3	68.8	4.20±2.38	59.7	80.5	2.84±2.25	52.0	74.6	3.66±2.33	40.3	57.1	5.72±3.49	48.2
SiamBAN [3]	52.9	73.1	3.90±2.51	61.5	89.8	2.39±2.09	56.2	79.6	3.30±2.34	49.0	69.5	3.60±3.18	33.2
SwinTrack [16]	51.0	72.6	3.93±1.83	64.4	93.1	1.91±1.72	57.1	80.6	3.13±1.79	48.2	70.6	3.62±2.85	60.1
STMTrack [7]	54.7	76.9	3.80±2.00	64.1	92.5	2.10±1.98	59.3	84.6	3.14±2.00	51.5	76.8	3.05±2.42	24.4
TrackingMamba [26]	55.2	79.7	3.62±1.71	63.9	92.9	1.90±1.62	59.5	86.3	2.93±1.72	54.1	77.0	3.39±2.98	63.1
ODTrack [36]	57.6	78.6	3.76±1.73	63.5	94.5	1.81±1.69	60.8	85.0	2.98±1.71	55.4	76.2	3.20±2.11	53.2
MrTrack	63.6	92.0	2.51±1.31	67.2	93.2	1.88±1.10	65.4	92.3	2.29±1.22	59.0	83.9	2.80±1.59	73.9

Ablations	Robotic (Mean)			Manual (Mean)			FPS
	AUC$_\uparrow$	$P_\uparrow$	Err±SD$_\downarrow$	AUC$_\uparrow$	$P_\uparrow$	Err±SD$_\downarrow$	
Baseline: MrTrack	65.4	92.3	2.29±1.22	59.0	83.9	2.80±1.59	73.9
v_{r1}: $L = 100, k = 8$	63.5(-1.9)	92.2(-0.1)	2.27(-0.02)±1.73(+0.51)	59.7(+0.7)	82.1(-1.8)	2.94(+0.14)±1.72(+0.13)	76.2(+2.3)
v_{r2}: $L = \infty, k = 8$	65.9(+0.5)	94.0(+1.7)	2.19(-0.10)±1.34(+0.12)	57.8(-1.2)	80.6(-3.3)	2.94(+0.14)±1.80(+0.21)	58.0(-15.9)
v_{r3}: $L = 300, k = 2$	65.1(-0.3)	92.2(-0.1)	2.24(-0.05)±1.25(+0.03)	58.1(-0.9)	82.9(-1.0)	2.92(+0.12)±1.55(-0.04)	73.7(-0.2)
v_{r4}: $L = 300, k = 32$	62.4(-3.0)	91.2(-1.1)	2.31(+0.02)±1.39(+0.17)	53.8(-5.2)	81.9(-2.0)	3.19(+0.39)±1.99(+0.40)	68.0(-5.9)
v_{M1}: *Trans E & R*	59.0(-6.4)	85.8(-6.5)	3.10(+0.81)±1.40(+0.18)	55.3(-3.7)	78.8(-5.1)	2.99(+0.19)±1.65(+0.06)	61.3(-12.6)
v_{M2}: w/o Register	61.2(-4.2)	88.5(-3.8)	2.49(+0.20)±1.79(+0.57)	54.2(-4.8)	79.4(-4.5)	3.38(+0.58)±2.03(+0.44)	76.1(+2.2)
v_{RDL}: w/o $\mathcal{L}_{RD}$	64.1(-1.3)	90.5(-1.8)	2.40(+0.11)±1.65(+0.43)	56.4(-2.6)	83.5(-0.4)	2.88(+0.08)±1.77(+0.18)	73.9(-0.0)

4.2 Results and Discussion

The evaluation was performed between MrTrack and seven trackers, including classic Siamese trackers [3,11,15], a tracker with motion integration [16], a Mamba-based tracker [26], and trackers with temporal modeling [7,36]. The results are reported by four metrics, including average error (Err) and standard deviation (SD) in mm, and commonly used tracking evaluation metrics area under curve (AUC) [27] and precision (P) [20] in percentage (%). The inference speed (FPS) is also reported. As shown in Table 1, MrTrack achieves the best performance for most metrics while maintaining the best efficiency with an inference speed of 73.9 FPS. In Err for robotic and manual insertion, it outperforms the second-best method by 21.8% and 8.2%, respectively. For SD, the improvements are even more substantial, reaching 28.7% and 24.6%. For manual insertion, which involves more abrupt motion and imaging disturbances, MrTrack still achieves consistent SOTA performance, which shows its robustness and generalizability to unseen images and motion, as the manual insertion data was not used for training. The demonstration in Fig. 2(b) also shows that MrTrack achieves the most accurate tracking of the aspiration needle with reciprocating motion. More examples are provided in the supplementary video.

This result demonstrates the effectiveness of MrTrack with the Mamba-based register, which captures the global spatiotemporal profile of a rapidly moving needle during aspiration. Among the comparison methods, ODTrack [36] also adopts temporal learning yet with worse performance and lower FPS. It may be due to its *token propagation* strategy, which with a fixed-length token to record the temporal context and pass through the whole network, leading to error accumulation. In contrast, MrTrack extracts and stores each frame individually, ensuring that every historical token possesses the same importance, thereby preventing error accumulation. Although TrackingMamba [26] also adopts a Mamba framework, it performs poorly because it only utilizes the vanilla Mamba without temporal context modeling. STMTrack [7] also learns temporal features by simply concatenating historical frames but has lower FPS and worse performance due to the lack of an effective temporal integration strategy. SwinTrack [16] adopts a historical motion fusion strategy to model temporal context, but since the motion in aspiration is nonlinear and abrupt, this strategy may lead to worse performance for this task. Currently, MrTrack still lacks a mechanism to update the template when needle visibility deteriorates. Future work will introduce an explicit template updating strategy to address this issue.

4.3 Ablation Study

Ablation studies were carried out on 7 variations. v_{r1} to v_{r4} explore the temporal integration capacity of the register, where the results indicate that the baseline structure ($L = 300, k = 8$) with a moderate register bank length L and a reasonable register scale k is the most suitable for the aspiration needle tracking task. Although v_{r2} somewhat improves performance by integrating infinite registers, it also leads to a significant decrease in inference speed, as well as a loss of robustness caused by error accumulation. v_{M1} replaces the proposed Mamba-based register extractor and retriever with a fundamental self-attention transformer [5] similar to [4]. This results in degradation across all metrics, indicating that the Mamba-based register framework is more effective than transformer. v_{M2} further shows the effectiveness of the register-based temporal integration, where the register and register bank are removed while keeping Mamba only. v_{RDL} only applies supervised training without $\mathcal{L}_{\mathrm{RD}}$, where the result shows the effectiveness, i.e., register diversity and dimension independence are promoted, thus enhancing the whole register framework and aspiration needle tracking performance. The robustness of $\mathcal{L}_{\mathrm{RD}}$ is also showed by avoiding tuning the value of hyper-parameters α and β.

The results demonstrate the effectiveness of the proposed modules under rapid motion and degraded image quality, where both the learnable register mechanism and the register diversify loss are crucial for robust temporal integration and feature discrimination. It underscores the importance of capturing long-term temporal context, which not only improves tracking accuracy but also enhances stability across various insertion techniques.

5 Conclusion

In this paper, a US needle tracker, MrTrack, has been proposed for FNA which features fast reciprocating needle motion. It leverages a Mamba-based register mechanism and a self-supervised register diversify loss to ensure robust and accurate needle tip tracking. In the future, more experiments will be conducted on large-scale manual insertion data and real clinical video, as well as on handling needle out-of-plane caused by deflection.

Acknowledgments. Research reported in this work was supported in part by Research Grants Council (RGC) of Hong Kong (T45-401/22-N, CUHK 14217822, CUHK 14207823, and AoE/E-407/24-N) and in part by Innovation and Technology Commission of Hong Kong (ITS/234/21, ITS/235/22, ITS/224/23, ITS/225/23, and Multi-scale Medical Robotics Center, InnoHK). The content is solely the responsibility of the authors and does not necessarily represent the official views of the sponsors.

Disclosure of Interests. The authors have no competing interests to declare that are relevant to the content of this article.

References

1. Bardes, A., et al.: VicReg: variance-invariance-covariance regularization for self-supervised learning. arXiv preprint arXiv:2105.04906 (2021)
2. Cai, Y., et al.: Autonomous robotic ultrasound approach for fetoscope tracking by fusing optical and 2D ultrasound data. IEEE Robot. Autom. Lett. (2024)
3. Chen, Z., et al.: SIAMBAN: target-aware tracking with SIAMESE box adaptive network. IEEE Trans. Pattern Anal. Mach. Intell. **45**(4), 5158–5173 (2022)

4. Darcet, T., et al.: Vision transformers need registers. arXiv preprint arXiv:2309.16588 (2023)
5. Dosovitskiy, A.: An image is worth 16x16 words: transformers for image recognition at scale. arXiv preprint arXiv:2010.11929 (2020)
6. D'Onofrio, M., et al.: Ultrasound-guided percutaneous fine-needle aspiration of solid pancreatic neoplasms: 10-year experience with more than 2,000 cases and a review of the literature. Eur. Radiol. **26**, 1801–1807 (2016)
7. Fu, Z., et al.: STMTrack: template-free visual tracking with space-time memory networks. In: Proceedings of the IEEE/CVF Conference on Computer Vision and Pattern Recognition, pp. 13774–13783 (2021)
8. Grill, J.B., et al.: Bootstrap your own latent-a new approach to self-supervised learning. Adv. Neural. Inf. Process. Syst. **33**, 21271–21284 (2020)
9. Gu, A., Dao, T.: Mamba: Linear-time sequence modeling with selective state spaces. arXiv preprint arXiv:2312.00752 (2023)
10. Gui, J., et al.: A survey on self-supervised learning: Algorithms, applications, and future trends. IEEE Trans. Pattern Anal. Mach. Intell. (2024)
11. Guo, D., et al.: SIAMCAR: Siamese fully convolutional classification and regression for visual tracking. In: Proceedings of the IEEE/CVF Conference on Computer Vision and Pattern Recognition, pp. 6269–6277 (2020)
12. Kalman, R.E.: A new approach to linear filtering and prediction problems (1960)
13. Kimbowa, A., et al.: Advancements in needle visualization enhancement and localization methods in ultrasound: a literature review. Artif. Intell. Surg. **4**(3), 149–169 (2024)
14. Lai, S., et al.: MambaVT: spatio-temporal contextual modeling for robust RGB-T tracking. arXiv preprint arXiv:2408.07889 (2024)
15. Li, B., et al.: SiamRPN++: evolution of SIAMESE visual tracking with very deep networks. In: Proceedings of the IEEE/CVF Conference on Computer Vision and Pattern Recognition, pp. 4282–4291 (2019)
16. Lin, L., et al.: Swintrack: a simple and strong baseline for transformer tracking. Adv. Neural. Inf. Process. Syst. **35**, 16743–16754 (2022)
17. Luo, M., et al.: Recon: online learning for sensorless freehand 3D ultrasound reconstruction. Med. Image Anal. **87**, 102810 (2023)
18. Ma, X., et al.: Feasibility of pointcloud-based ultrasound-CT registration towards automated, robot-assisted image-guidance in spine surgery. In: 2024 International Symposium on Medical Robotics (ISMR), pp. 1–7. IEEE (2024)
19. Ma, X., et al.: Cross-modality registration using bone surface pointcloud for robotic ultrasound-guided spine surgery. J. Med. Robot. Res. 2540004 (2025)
20. Müller, M., Bibi, A., Giancola, S., Alsubaihi, S., Ghanem, B.: TrackingNet: a large-scale dataset and benchmark for object tracking in the wild. In: Ferrari, V., Hebert, M., Sminchisescu, C., Weiss, Y. (eds.) ECCV 2018. LNCS, vol. 11205, pp. 310–327. Springer, Cham (2018). https://doi.org/10.1007/978-3-030-01246-5_19
21. Mwikirize, C., Nosher, J.L., Hacihaliloglu, I.: Single shot needle tip localization in 2D ultrasound. In: Shen, D., et al. (eds.) MICCAI 2019. LNCS, vol. 11768, pp. 637–645. Springer, Cham (2019). https://doi.org/10.1007/978-3-030-32254-0_71
22. Richardson, MD, R.R., Richardson, R.R.: Imaging modalities: advantages and disadvantages. In: Atlas of Acquired Cardiovascular Disease Imaging in Children, pp. 1–4 (2017)
23. Rumack, C.M., Levine, D.: Diagnostic Ultrasound. Elsevier Health Sciences (2017)
24. Shen, C., He, J., Huang, Y., Wu, J.: Discriminative correlation filter network for robust landmark tracking in ultrasound guided intervention. In: Shen, D., et al.

(eds.) MICCAI 2019. LNCS, vol. 11768, pp. 646–654. Springer, Cham (2019). https://doi.org/10.1007/978-3-030-32254-0_72
25. Wang, F., et al.: Mamba-r: vision mamba also needs registers. arXiv preprint arXiv:2405.14858 (2024)
26. Wang, Q., et al.: Trackingmamba: visual state space model for object tracking. IEEE J. Sel. Top. Appl. Earth Observat. Remote Sens. (2024)
27. Wu, Y., et al.: Online object tracking: a benchmark. In: Proceedings of the IEEE Conference on Computer Vision and Pattern Recognition, pp. 2411–2418 (2013)
28. Xie, J., et al.: Robust tracking via mamba-based context-aware token learning. arXiv preprint arXiv:2412.13611 (2024)
29. Yan, W., et al.: Learning-based needle tip tracking in 2D ultrasound by fusing visual tracking and motion prediction. Med. Image Anal. **88**, 102847 (2023)
30. Yan, W., et al.: Task-oriented network design for visual tracking and motion filtering of needle tip under 2d ultrasound. IEEE Trans. Med. Imaging (2024)
31. Zbontar, J., et al.: Barlow twins: self-supervised learning via redundancy reduction. In: International Conference on Machine Learning, pp. 12310–12320. PMLR (2021)
32. Zhang, H., et al.: A survey on visual mamba. Appl. Sci. **14**(13), 5683 (2024)
33. Zhang, Y., et al.: Motion-guided dual-camera tracker for endoscope tracking and motion analysis in a mechanical gastric simulator. arXiv preprint arXiv:2403.05146 (2024)
34. Zhang, Y., et al.: A unified framework for microscopy defocus deblur with multi-pyramid transformer and contrastive learning. In: Proceedings of the IEEE/CVF Conference on Computer Vision and Pattern Recognition, pp. 11125–11136 (2024)
35. Zhang, Y., et al.: Mambaxctrack: mamba-based tracker with SSM cross-correlation and motion prompt for ultrasound needle tracking. IEEE Robot. Autom. Lett. (2025)
36. Zheng, Y., et al.: Odtrack: online dense temporal token learning for visual tracking. In: Proceedings of the AAAI Conference on Artificial Intelligence, vol. 38, pp. 7588–7596 (2024)

MSGFlowNet: Learning Effective Connectivity Network Based on Sparse Generative Flow Network from fMRI and EEG Data

Zhihao Su, Jihao Zhai, Junzhong Ji, and Jinduo Liu(✉)

Beijing Municipal Key Laboratory of Multimedia and Intelligent Software Technology, Beijing Institute of Artificial Intelligence, College of Computer Science, Beijing University of Technology, Beijing, China
jinduo@bjut.edu.cn

Abstract. Brain effective connectivity (EC) is key to understanding causal neural interactions and brain organization. However, learning EC from single-modal brain data, such as functional magnetic resonance imaging (fMRI) or electroencephalography (EEG), is limited by the inability to simultaneously capture sparse temporal and spatial information. This paper proposes a novel multimodal sparse generative flow network (MSGFlowNet), which integrates fMRI and EEG data through an attention-guided encoder and employs a multi-head self-attention sparse Transformer to extract features from the fused data. These features are then processed by two output heads of the generative flow network: one computes state transition probabilities and updates the mask, while the other determines the probability of generating a termination state. Experiments on synthetic and real-world datasets demonstrate that MSGFlowNet significantly outperforms state-of-the-art methods.

Keywords: Brain Effective Connectivity · Generative Flow Network · Sparse Transformer · functional Magnetic Resonance Imaging · Electroencephalography

1 Introduction

Brain effective connectivity (EC) describes the causal relationships between brain regions and is crucial for understanding brain function [7,17,30]. Currently, EC learning methods fall into two categories: traditional assumption-based statistical methods and data-driven deep learning methods [21]. Traditional methods like structural equation modeling (SEM) [18], dynamic causal modeling (DCM) [8], and Granger causality analysis (GCA) are widely used but struggle to model complex brain networks due to their strong assumptions.

Deep learning has advanced data-driven EC learning [4]. Graph neural networks (GNNs) leverage brain network topology to model causal relationships,

J. C. Gee et al. (Eds.): MICCAI 2025, LNCS 15960, pp. 418–428, 2026.
https://doi.org/10.1007/978-3-032-04927-8_40

with models such as dynamic diffusion-variational GNN [15] and directed structure learning GNN [2] making progress in brain disease classification. Recurrent neural networks (RNNs) are well-suited for time-series modeling, with a Granger causality estimator based on the recurrent neural network [29] capturing time lags and nonlinear dependencies.Recurrent neural networks (RNNs) are well-suited for time-series modeling, with a Granger causality estimator based on the recurrent neural network [27]capturing time lags and nonlinear dependencies. Several approaches have been developed to address time-varying causal relationships [5] and developing more rigorous model selection frameworks [13]. Generative adversarial networks (GANs) have shown promise in data augmentation and missing data imputation. Furthermore, the framework incorporates a diffusion model-based EC learning method [12] and integrates dynamic EC with structural EC for brain network classification [3]. However, these approaches primarily infer EC from functional magnetic resonance imaging (fMRI) [25], neglecting the potential of multimodal integration.

fMRI provides high spatial resolution but low temporal resolution, whereas electroencephalography (EEG) offers high temporal resolution [22, 31] but lower spatial resolution. Combining both modalities enhances EC learning accuracy but presents challenges [26]. Anwar et al. inferred EC using GCA but did not fully integrate fMRI and EEG information [19]. Tu et al. proposed a linear state-space model that combines temporal information from both modalities but struggles with nonlinear causal relationships [28]. Liu et al. proposed a novel method to bridge modality differences but failed to address high-dimensional sparsity after data fusion [16]. Efficient integration of fMRI and EEG, while overcoming high-dimensional sparsity, remains a key challenge.

To address these issues, we propose a novel multimodal sparse generative flow network (MSGFlowNet). The model consists of three main stages: data fusion, sparse feature extraction, and state generation. MSGFlowNet redefines the EC learning problem as a generative task and effectively fuses fMRI and EEG data using an attention-based encoder. Subsequently, a sparse Transformer with multi-head self-attention is employed to extract features from the fused data, and two output heads in the generative flow network (GFN) compute the EC state and state transition probabilities. Finally, the EC is updated and generated.

Our contributions are summarized as follows:

1. This is the first study that develops a sparse generative flow network to learn brain effective connectivity from fMRI and EEG data.

2. We utilize fMRI and EEG data as complementary modalities, effectively leveraging the high spatial resolution of fMRI and the high temporal resolution of EEG to overcome the limitations of using a single modality.

3. We propose a novel sparse Transformer architecture that effectively extracts key features from high-dimensional sparse fused fMRI-EEG data, significantly enhancing EC learning accuracy.

2 Related Work

Brain Effective Connectivity. The brain EC can represent as a directed graph $G = (V, E, W)$, containing $|V|$ nodes and $|E|$ directed edges, which reflect the regions of interest (ROIs) in the brain and the causal relationships between these regions, respectively. The weight or adjacency matrix W is an asymmetric matrix used to represent the causal effects between brain regions of brain data.

Generative Flow Network. GFN is a probabilistic model over discrete sample spaces with a compositional structure [6], which is a recently introduced method to train energy-based generative models. They have now been successfully applied to a variety of settings such as biological sequences, causal discovery, discrete latent variable modeling, and computational graph scheduling.

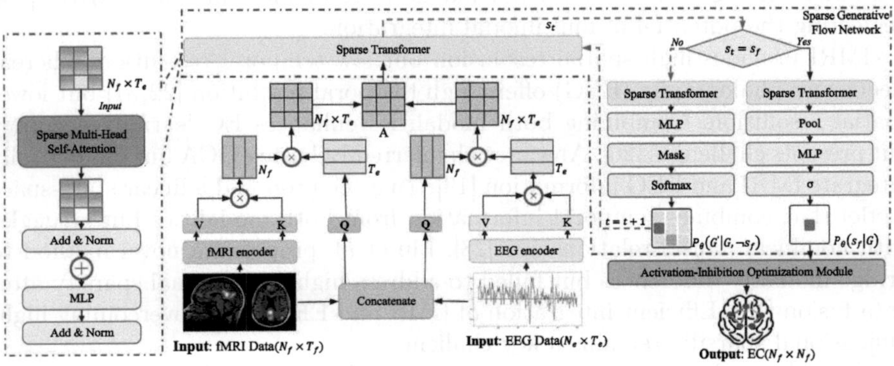

Fig. 1. Flowchart of MSGFlowNet. The framework processes multimodal data through three stages: (1) fusion of fMRI and EEG signals, (2) sparse Transformer for feature extraction, and (3) generation of EC states with GFN.

3 Method

In this section, we present MSGFlowNet, a novel three-stage method for learning brain effective connectivity networks from complementary fMRI and EEG data.

3.1 Main Idea

To accurately learn EC from fMRI and EEG data, we propose MSGFlowNet model, as shown in Fig. 1. MSGFlowNet redefines the EC learning problem as a generative task and effectively fuses fMRI and EEG data using an attention-based encoder. Subsequently, a sparse Transformer with multi-head self-attention is employed to extract features from the fused data, and two output heads in the generative flow network compute the EC state and state transition probabilities. Finally, the EC is updated and generated.

3.2 fMRI and EEG Feature Fusion

We propose attention-guided a multimodal encoder for integrating fMRI and EEG signals into a unified representation, thereby enabling efficient cross-modal information integration [20]. By combining the fMRI signal x_f and EEG signal x_e, we introduce fMRI and EEG source encoding to align their spatial and temporal resolutions. These encoding modules solve the inverse problems for both modalities, generating fMRI and EEG source-level time series (s_f and s_e). For fMRI, the BOLD inverse problem is defined by the equation:

$$s_f = x_f H + \epsilon_f, \tag{1}$$

where $x_f \in \mathbb{R}^{N_f \times T_f}$ is the measured fMRI signal of N_f brain regions during T_f instants, $s_f \in \mathbb{R}^{N_f \times T_e}$ is the fMRI sources-level time series, and T_e is the duration of the EEG signal. $H \in \mathbb{R}^{T_f \times T_e}$ is the BOLD response matrix consisting of trainable neural network parameters. ϵ_f is the Gaussian noise. For EEG, the inverse problem estimates the neural sources as:

$$s_e = I x_e + \epsilon_e, \tag{2}$$

where $x_e \in \mathbb{R}^{N_e \times T_e}$ is the measured EEG signal of N_e electrodes during T_e instants. $s_e \in \mathbb{R}^{N_f \times T_e}$ is the EEG sources-level time series. $I \in \mathbb{R}^{N_e \times N_f}$ is a leadfield matrix consisting of trainable neural network parameters. ϵ_e is the Gaussian noise.

In the attention-guided layer, the query(Q_m), key(K_m), and value(V_m) vectors are generated from the input data through fully connected layers. Specifically, Q_m is derived from the input s_m, while K_m and V_m are extracted from the fMRI source-level time series s_f and EEG source-level time series s_e. The attention map A is generated by computing the dot product between Q_m and K_m. Then, A is multiplied by V_m to obtain the fused feature time series, enabling efficient capture and fusion of intersecting information across modalities.

The fusion data is represented as a two-dimensional matrix A:

$$A^{N_f \times T_e} = \begin{bmatrix} a_{11} & \cdots & a_{1N_f} \\ \vdots & \ddots & \vdots \\ a_{T_e,1} & \cdots & a_{T_e,N_f} \end{bmatrix}, \tag{3}$$

where N_f is the number of brain regions, T_e is the number of time points, and a_{ij} represents the activity level of the i-th brain region in the j-th time point.

3.3 Sparse Transformer for Feature Extraction

The fMRI data ($N_f \times T_f$) and EEG data ($N_e \times T_e$) have high spatial and temporal resolutions, respectively, but their fusion results in high-dimensional sparsity. To address this issue, we propose a sparse feature extraction model.

A causal graph is a mathematical model based on a directed acyclic graph (DAG), represented as $G = \langle N, E \rangle$, where N is the set of brain regions and E

is the set of edges. These edges represent the connectivity relationships between two brain regions. Causal graphs focus not merely on probabilistic dependencies between variables, but on causal mechanisms. Their joint probability distribution is decomposed via the causal structure into:

$$P(X_1, X_2, X_3, \cdots X_n) = \prod_{i=1}^{n} P(X_i|Pa(X_i)), \tag{4}$$

where $P_a(X_i)$ is the set of parent nodes for X_i. The task is to learn the hidden EC $G = \langle N, E \rangle$ behind the high-dimensional sparse data.

Due to the sparsity of the fused data, direct input often leads to less accurate learned EC. Therefore, we design a sparse Transformer module, which uses the sparse multi-head self-attention mechanism to extract important features from the sparse data and integrates them into a global representation. This mechanism is more advantageous than the conventional single-head self-attention in learning long-term dependencies. The sparse Transformer module mainly relies on the sparse multi-head self-attention mechanism, which is defined as follows:

$$Atten_i(q_i, K, V) = \sum_j \frac{k(q_i, k_j)}{\sum_l k(q_i, k_l)} v_j = E_{p(k_j|q_i)}[v_j]. \tag{5}$$

The self-attention combines the value matrix V by computing the conditional probability $p(k_j|q_i)$. Sparsity is introduced by limiting the attention probability distribution, ensuring that only a portion of the keys interact with the current query matrix. The sparsity measure $M(q_i, K)$ is given by:

$$M(q_i, K) = \ln \sum_{j=1}^{L} exp(q_i k_j^T / \sqrt{d_k}) - \frac{1}{L} \sum_{j=1}^{L} (\frac{q_i k_j^T}{\sqrt{d_k}}). \tag{6}$$

Due to the high complexity of the dot product operation and potential instability caused by the log-sum-exp (LSE) function, we use the maximum average method to reduce complexity, yielding:

$$\widetilde{M} = max_j \left\{ \frac{q_i k_j^T}{\sqrt{d_k}} \right\} - \frac{1}{L} \sum_{j=1}^{L} (\frac{q_i k_j^T}{\sqrt{d_k}}). \tag{7}$$

Finally, the sparse multi-head self-attention is optimized by traversing each key and adopting a Top-u query, as shown in:

$$Atten^{sparse}(\overline{Q}, K, V) = softmax\left(\frac{\overline{Q}K^T}{\sqrt{d_k}} V\right). \tag{8}$$

This approach significantly improves the efficiency of the Transformer in handling long sequences, reducing both computational and storage overheads while maintaining the expressive power of the attention mechanism. It is especially suitable for tasks that involve large-scale data and high-dimensional features.

3.4 GFN-Based EC State Generation

After extracting sparse features, MSGFlowNet initiates the dynamic construction process of EC through GFN. The goal is to model the posterior of the explanatory graph $Q(G|D)$ given the data and jointly learn the distribution over parameters θ that parameterize the latent drift function $f(x)$. The generative process of the GDN proceeds as follows::

$$p(G, \theta, D) = p(G)p(\theta|G)p(D|G, \vartheta). \tag{9}$$

The terminal state s_f indicates the termination of the sequential construction. Some states $s \in Y$ are connected to s_f, called complete states. Each complete state has a reward $R(s) \geq 0$, and for incomplete states $s \in S \setminus Y$, $R(s) = 0$. The goal is to find a flow that satisfies the flow-matching condition for all states s':

$$\sum_{s \in Pa(s')} F_\theta(s \to s^{'}) - \sum_{s'' \in Ch(s')} F_\theta(s^{'} \to s^{''}) = R(s^{'}). \tag{10}$$

When this condition is satisfied, MSGFlowNet samples from the complete states $s \in Y$ with probability proportional to $R(s)$. To fit the parameters θ of the forward transition probabilities, we minimize the detailed-balance loss:

$$Loss(\theta) = \mathbb{E}_\pi \left[\left[\log \frac{R(G^{'})P_B(G|G^{'})P_\theta(s_f|G)}{R(G)P_\theta(G^{'}|G)P_\theta(s_f|G^{'})} \right]^2 \right]. \tag{11}$$

The forward transition probabilities are parameterized by neural network. We model these transitions with two parts: one for the termination probability $P_\theta(s_f|G)$, and another for the non-termination transition $P_\theta(G'|G, \neg s_f)$:

$$P_\theta(G'|G) = (1 - P_\theta(s_f|G))P_\theta(G'|G, \neg s_f). \tag{12}$$

The transition $G \to G'$ involves adding edges to the graph, and a mask Mask_0 is used to filter out invalid actions. The reward for graph G is given by:

$$R(G) = P(G)P(D|G). \tag{13}$$

Finally, after sampling transitions randomly and updating the parameters, we obtain the probability distribution over the graph G as the output of MSGFlowNet after several iterations.

4 Experiments

4.1 Simulated Dataset Analysis

We generated three simulated datasets to evaluate MSFlowNet and other EC learning methods. For simulated fMRI data, we followed prior work [16,32] and used the forward dynamic causal model (DCM) to generate time series data with 5 brain regions (200 time points). For simulated EEG data, we used a state-space model [28] to generate synchronized time series data with 3 channels (20,000 time points). Each dataset contains data from 10 subjects. To assess performance under sparse conditions, we set the network edges to 3, 5, and 7, and uniformly assigned edge weights of 0.4 to test the methods' ability to detect weak connectivity. We employ four metrics: F1 score (F1), precision (Pre), recall (Rec), and structural Hamming distance (SHD) [1,10] to evaluate the performance of EC learning.

Table 1. Performance comparison of different methods on simulation datasets

Methods	sim1				sim2				sim3			
	Prec↑	Rec↑	F1↑	SHD↓	Prec↑	Rec↑	F1↑	SHD↓	Prec↑	Rec↑	F1↑	SHD↓
pwLiNGAM [11]	0.50	0.81	0.59	2.00	0.30	0.29	0.28	5.62	0.36	0.56	0.42	5.22
Two-Step [24]	0.55	0.53	0.53	3.60	0.51	0.51	0.51	**1.80**	0.34	0.41	0.37	4.90
GFN [14]	**0.76**	0.74	0.70	1.00	0.78	0.54	0.63	2.90	0.77	0.51	0.61	2.55
DiffAN [23]	0.59	0.59	0.59	1.20	0.28	0.24	0.27	5.20	0.32	0.39	0.35	5.50
MetaRLEC [32]	0.33	0.25	0.28	3.00	0.32	0.28	0.31	5.25	0.26	0.29	0.27	5.83
MCAN [16]	0.33	0.54	0.39	3.00	**1.00**	0.25	0.40	5.00	**1.00**	0.35	0.51	3.00
SGFlowNet	0.76	0.74	0.73	**0.90**	0.78	0.54	0.63	2.90	0.73	**0.64**	0.63	3.50
MSGFlowNet	0.66	**1.00**	**0.80**	1.00	0.98	0.54	**0.70**	2.01	0.82	0.52	**0.64**	**2.26**

As shown in Fig. 2, MSGFlowNet outperforms other EC learning methods, including SGFlowNet (the MSGFlowNet using single-modal fMRI), across various datasets. In particular, MSGFlowNet performs well on the sim1 and sim2 datasets. As presented in Table 1, this underscores MSGFlowNet's ability to effectively capture the intricate relationships in sparse and complex data, demonstrating its robustness in learning EC under challenging conditions.

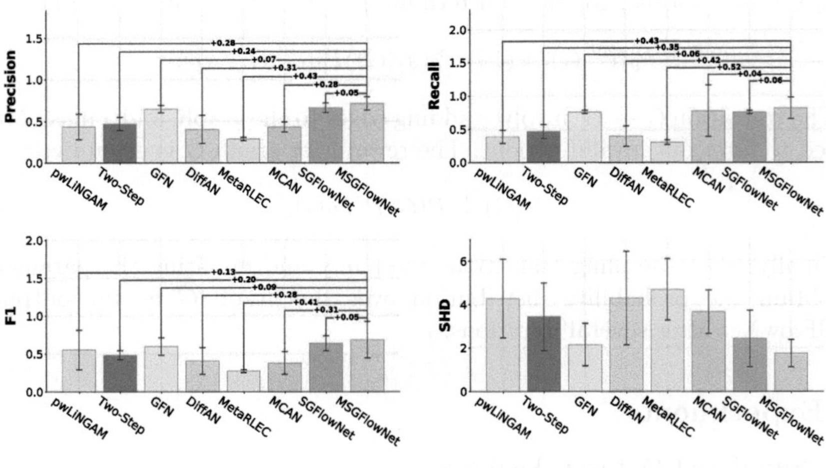

Fig. 2. Average performance comparison of various methods on 3 simulated datasets. The height of each bar represents the model's performance, and the performance differences relative to MSGFlowNet are indicated by connecting lines.

Additionally, MSGFlowNet outperforms MCAN (also employs fMRI and EEG data) across all core evaluation metrics. These results validate MSGFlowNet's effective integration of fMRI and EEG data, successfully addressing high-dimensional sparsity and outperforming other multimodal methods.

4.2 Real Dataset Analysis (Downstream Classification Task)

To further validate the effectiveness of MSDlowNet, we utilize simultaneous from the OpenNeuro dataset [9] for downstream classification tasks, and compare its performance against state-of-the-art methods.

The raw dataset is preprocessed using statistical parametric mapping (for fMRI) and EEGLAB (for EEG) toolboxes, yielding 55 resting-state and 183 sleep-state simultaneous fMRI-EEG recordings. We employ 15 functionally critical brain regions as the ROIs based on the research of Enzo Tagliazucchi et al. [27] and retain 30 EEG channels after excluding electrooculography (EOG) and electrocardiography (ECG) artifacts.

We evaluated EC learning performance by classifying rest-state vs. sleep-state brain activity using ground truth labels. All methods were tested with the same polynomial kernel SVM classifier, and assessed using four metrics: accuracy, F1-score, precision, and recall.

Figure 3 shows MSGFlowNet obtains better classification results by effectively combining fMRI's spatial resolution and EEG's temporal resolution to capture EC characteristics. Our method outperforms traditional approaches [27] using fewer regions and less data.

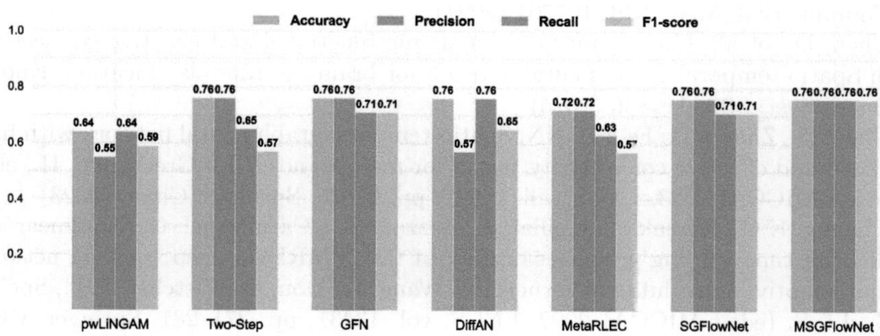

Fig. 3. Performance comparison of different methods on classification tasks

In addition to classification metrics, we also examine the neurobiological interpretability of the learned EC networks. EC analyses revealed increased directed links from the right supramarginal gyrus to the left precuneus during rest compared to sleep. These regions are associated with attention, self-awareness, and sensory integration. The enhanced connectivity during rest reflects heightened cognitive readiness and environmental monitoring, which diminish during sleep as consciousness declines.

5 Conclusion and Limitation

This study introduces MSGFlowNet, for learning EC from fMRI and EEG data. The model combines an attention-guided encoder with a multi-head attention-guided sparse Transformer to efficiently fuse multimodal information, addressing the high-dimensional sparsity issue after data fusion. The experimental results indicate that MSGFlowNet performs well on multiple brain analysis tasks. A limitation of the current work is the computational efficiency for large-scale brain network analysis. Future work will address this by integrating advanced neuroimaging modalities to enable deeper insights into brain information processing mechanisms.

Acknowledgements. This work was sponsored by Beijing Nova Program (20240484635), supported by National Natural Science Foundation of China (62106009, 62276010, 62171297).

Disclosure of Interests. The authors have no competing interests to declare that are relevant to the content of this article.

References

1. Cai, R., Qiao, J., Zhang, Z., Hao, Z.: Self: structural equational likelihood framework for causal discovery. In: Proceedings of the AAAI Conference on Artificial Intelligence, vol. 32 (2018)
2. Cao, J., Yang, L., Sarrigiannis, P.G., Blackburn, D., Zhao, Y.: Dementia classification using a graph neural network on imaging of effective brain connectivity. Comput. Biol. Med. **168**, 107701 (2024)
3. Chen, D., et al.: Guiding fusion of dynamic functional and effective connectivity in Spatio-temporal graph neural network for brain disorder classification. Knowl.-Based Syst. **309**, 112856 (2025)
4. Chen, D., Zhang, L.: Fe-STGNN: Spatio-temporal graph neural network with functional and effective connectivity fusion for mci diagnosis. In: Greenspan, H., et al. (eds.) MICCAI 2023. LNCS, vol. 14227, pp. 67–76. Springer, Cham (2023)
5. Chuang, K.C., Ramakrishnapillai, S., Bazzano, L., Carmichael, O.: Nonlinear conditional time-varying granger causality of task FMRI via deep stacking networks and adaptive convolutional kernels. In: Wang, L., Dou, Q., Fletcher, P.T., Speidel, S., Li, S. (eds.) MICCAI 2022. LNCS, vol. 13431, pp. 271–281. Springer, Cham (2022). https://doi.org/10.1007/978-3-031-16431-6_26
6. Deleu, T., et al.: Bayesian structure learning with generative flow networks. In: Uncertainty in Artificial Intelligence, pp. 518–528. PMLR (2022)
7. Friston, K.J.: Functional and effective connectivity: a review. Brain Connect. **1**(1), 13–36 (2011)
8. Friston, K., Harrison, L., Penny, W.: Dynamic causal modelling. Neuroimage **19**(4), 1273–1302 (2003)
9. Gu, Y., et al.: An orderly sequence of autonomic and neural events at transient arousal changes. Neuroimage **264**, 119720 (2022)
10. Huang, B., Zhang, K., Lin, Y., Schölkopf, B., Glymour, C.: Generalized score functions for causal discovery. In: Proceedings of the 24th ACM SIGKDD International Conference on Knowledge Discovery & Data Mining, pp. 1551–1560 (2018)

11. Hyvarinen, A.: Pairwise measures of causal direction in linear non-gaussian acyclic models. In: Proceedings of 2nd Asian Conference on Machine Learning. pp. 1–16. JMLR Workshop and Conference Proceedings (2010)
12. Ji, J., Fan, J., Liu, J.: MSAD-LEC: estimating large-scale brain effective connectivity network based on multi-subgraph attention diffusion. Knowl.-Based Syst. **309**, 112858 (2025)
13. Li, F., Wang, X., Lin, Q., Hu, Z.: Unified model selection approach based on minimum description length principle in granger causality analysis. IEEE Access **8**, 68400–68416 (2020)
14. Li, W., Li, Y., Zhu, S., Shao, Y., Hao, J., Pang, Y.: Gflowcausal: generative flow networks for causal discovery (2023). https://arxiv.org/abs/2210.08185
15. Liang, G., Tiwari, P., Nowaczyk, S., Byttner, S., Alonso-Fernandez, F.: Dynamic causal explanation based diffusion-variational graph neural network for spatiotemporal forecasting. In: IEEE Transactions on Neural Networks and Learning Systems (2024)
16. Liu, J., Han, L., Ji, J.: MCAN: multimodal causal adversarial networks for dynamic effective connectivity learning from FMRI and EEG data. IEEE Trans. Med. Imaging **43**(8), 2913–2923 (2024)
17. Liu, J., Ji, J., Xun, G., Zhang, A.: Inferring effective connectivity networks from FMRI time series with a temporal entropy-score. IEEE Trans. Neural Networks Learn. Syst. **33**(10), 5993–6006 (2021)
18. Mclntosh, A.: Structural equation modeling and its application to network analysis in functional brain imaging. Hum. Brain Mapp. **2**(1–2), 2–22 (1994)
19. Muthuraman, M.: Effective connectivity of cortical sensorimotor networks during finger movement tasks: a simultaneous fNIRS, fMRI, EEG study. Brain Topogr. **29**, 645–660 (2016)
20. Noorzadeh, S., Maurel, P., Oberlin, T., Gribonval, R., Barillot, C.: Multi-modal EEG and fMRI source estimation using sparse constraints. In: Descoteaux, M., Maier-Hein, L., Franz, A., Jannin, P., Collins, D.L., Duchesne, S. (eds.) MICCAI 2017. LNCS, vol. 10433, pp. 442–450. Springer, Cham (2017). https://doi.org/10.1007/978-3-319-66182-7_51
21. Rossini, P.M., et al.: Methods for analysis of brain connectivity: an IFCN-sponsored review. Clin. Neurophysiol. **130**(10), 1833–1858 (2019)
22. Sakkalis, V.: Review of advanced techniques for the estimation of brain connectivity measured with eeg/meg. Comput. Biol. Med. **41**(12), 1110–1117 (2011)
23. Sanchez, P., Liu, X., O'Neil, A.Q., Tsaftaris, S.A.: Diffusion models for causal discovery via topological ordering. In: The Eleventh International Conference on Learning Representations (2022)
24. Sanchez-Romero, R., Ramsey, J.D., Zhang, K., Glymour, M.R., Huang, B., Glymour, C.: Estimating feedforward and feedback effective connections from fMRI time series: Assessments of statistical methods. Network Neurosci. **3**(2), 274–306 (2019)
25. Smith, S.M.: The future of fMRI connectivity. Neuroimage **62**(2), 1257–1266 (2012)
26. Song, X., et al.: Multicenter and multichannel pooling GCN for early ad diagnosis based on dual-modality fused brain network. IEEE Trans. Med. Imaging **42**(2), 354–367 (2022)
27. Tagliazucchi, E., Laufs, H.: Decoding wakefulness levels from typical fMRI resting-state data reveals reliable drifts between wakefulness and sleep. Neuron **82**(3), 695–708 (2014)

28. Tu, T., Paisley, J., Haufe, S., Sajda, P.: A state-space model for inferring effective connectivity of latent neural dynamics from simultaneous EEG/fMRI. In: Advances in Neural Information Processing Systems, vol. 32 (2019)
29. Wang, Y., et al.: Estimating brain connectivity with varying-length time lags using a recurrent neural network. IEEE Trans. Biomed. Eng. **65**(9), 1953–1963 (2018)
30. Xiong, W., Liu, J., Ji, J., Ma, F.: Brain effective connectivity estimation via fourier spatiotemporal attention. In: Proceedings of the 31st ACM SIGKDD Conference on Knowledge Discovery and Data Mining, vol. 1, pp. 1657–1668 (2025)
31. Zhang, X., Shi, E., Yu, S., Zhang, S.: DTCA: dual-branch transformer with cross-attention for EEG and eye movement data fusion. In: Linguraru, M.G., et al. (eds.) MICCAI 2024. LNCS, vol. 15002, pp. 141–151. Springer, Cham (2024). https://doi.org/10.1007/978-3-031-72069-7_14
32. Zhang, Z., Ji, J., Liu, J.: Metarlec: meta-reinforcement learning for discovery of brain effective connectivity. In: Proceedings of the AAAI Conference on Artificial Intelligence, vol. 38, pp. 10261–10269 (2024)

Multi-agent Reasoning for Cardiovascular Imaging Phenotype Analysis

Weitong Zhang[1(✉)], Mengyun Qiao[2,3,4(✉)], Chengqi Zang[5], Steven Niederer[6], Paul M. Matthews[3,8,9], Wenjia Bai[1,3,4], and Bernhard Kainz[1,7]

[1] Department of Computing, Imperial College London, London, UK
wz1820@ic.ac.uk

[2] Department of Mechanical Engineering, University College London, London, UK
mq21@ic.ac.uk

[3] Department of Brain Sciences, Imperial College London, London, UK

[4] Data Science Institute, Imperial College London, London, UK

[5] University of Tokyo, Tokyo, Japan

[6] National Heart and Lung Institute, Imperial College London, London, UK

[7] FAU Erlangen-Nürnberg, Nürnberg, Germany

[8] UK Dementia Research Institute, Imperial College London, London, UK

[9] Rosalind Franklin Institute, Harwell Science and Innovation Campus, Didcot, UK

Abstract. Identifying the associations between imaging phenotypes and disease risk factors and outcomes is essential for understanding disease mechanisms and improving diagnosis and prognosis models. However, traditional approaches rely on human-driven hypothesis testing and selection of association factors, often overlooking complex, non-linear dependencies among imaging phenotypes and other multi-modal data. To address this, we introduce a **M**ulti-agent **E**xploratory **S**ynergy for the **H**eart (MESHAgents) framework that leverages large language models as agents to dynamically elicit, surface, and decide confounders and phenotypes in association studies, using cardiovascular imaging as a proof of concept. Specifically, we orchestrate a multi-disciplinary team of AI agents, which spontaneously generate and converge on insights through iterative, self-organizing reasoning. The framework dynamically synthesizes statistical correlations with multi-expert consensus, providing an automated pipeline for phenome-wide association studies (PheWAS). We demonstrate the system's capabilities through a population-based study of imaging phenotypes of the heart and aorta. MESHAgents autonomously uncovered correlations between imaging phenotypes and a wide range of non-imaging factors, identifying additional confounder variables beyond standard demographic factors. Validation on diagnosis tasks reveals that MESHAgents-discovered phenotypes achieve performance comparable to expert-selected phenotypes, with mean AUC differences as small as $-0.004_{\pm 0.010}$ on disease classification tasks. Notably, the recall score improves for 6 out of 9 disease types. Our framework provides clinically relevant imaging phenotypes with transparent reasoning, offering a scalable alternative to expert-driven methods.

Weitong Zhang and Mengyun Qiao:—Equal contribution.

J. C. Gee et al. (Eds.): MICCAI 2025, LNCS 15960, pp. 429–439, 2026.
https://doi.org/10.1007/978-3-032-04927-8_41

Keywords: Multi-agent system · Phenome-wide association studies · Cardiovascular imaging · Integration of imaging and non-imaging data

1 Introduction

Medical imaging is essential for deriving quantitative phenotypes of anatomical structure and function and understanding their associations with various non-imaging factors, such as genetic predispositions [1], environmental conditions [2] and physiological characteristics [3] etc. How to better integrate imaging with non-imaging data to perform multi-modal data analysis has received increasing attention in the imaging community. To ensure reliable and unbiased associations between imaging and non-imaging data, it is crucial to identify associations affecting both risk factors and imaging phenotypes [4]. A confounder is an association variable that influences both the independent variable and the dependent variable, creating a spurious association between them. Ignoring confounders can lead to misinterpretations in phenome-wide association studies (PheWAS) and genome-wide association studies (GWAS), impacting clinical analyses. Confounding factors also need to be considered when evaluating bias in machine learning algorithms [5], particularly in medical imaging research [6,7]. Conventional methods rely on established empirical criteria or expert selection, not accounting for complicated non-linear relationships between imaging phenotypes and health outcomes. Performing automated reasoning on large-scale, multi-modal datasets and integrating the multi-disciplinary expertise is not well investigated in the past.

Recent advances in large language models (LLMs) [8,9] have exhibited notable reasoning abilities across a wide range of tasks and applications [10,11], with these capabilities stemming from extensive training on vast comprehensive corpora covering diverse topics. In real-world scenarios, LLMs tend to encounter domain-specific tasks (*e.g.*, strategy [12], safety [13]) that necessitate a combination of domain expertise and complex reasoning abilities [14,15]. This backdrop makes the adoption of LLMs in the medical field a noteworthy research topic, which has recently gained growing attention [16,17].

Challenges. Two major challenges prevent LLMs from handling tasks in multi-modal medical imaging data analysis: (i) Imaging phenotype analysis demands multi-disciplinary domain expertise [19,20] spanning medicine, biomechanics, and statistics, yet LLMs suffer from catastrophic forgetting [21,22] in maintaining comprehensive knowledge across disciplines. (ii) The analysis requires phenotype-aware reasoning to identify robust correlations between imaging phenotypes and a wide range of non-imaging factors [4,18], whereas LLMs tend to produce *hallucination* [23,24] that undermines phenotype discovery.

Contributions. (i) We introduce the first multi-agent system for medical imaging PheWAS where domain-specific LLM agents achieve association consensus through structured interactions, mitigating catastrophic forgetting through a multi-disciplinary team. (ii) We demonstrate that role-based collaborative reasoning enhances imaging phenotype discovery without retrieval augmentation, maintaining interpretable decision trails while reducing hallucination. (iii)

Expert-Selected[18, 7]	MESHAgents-Discovered
Imaging phenotypes	
RVEDV (mL)	RVEDV (mL)
LVEDV (mL)	RVESV (mL)
LVEF (%)	RVSV (mL)
LVM (g)	LVCO (L/min)
RVEF (%)	LAV min (mL)
LAV max (mL)	LASV (mL)
LAEF (%)	RAV max (mL)
RAV max (mL)	RAV min (mL)
RAEF (%)	AAo max area (mm2)
AAo max area (mm2)	AAo min area (mm2)
AAo distensibility	DAo min area (mm2)
DAo max area (mm2)	WT_AHA_2 (mm)
DAo distensibility	Ell_4 (%)
Confounders	
Sex	Sex
Age	Age
Weight	Weight
Height	Height
	Alcohol intake
	Systolic BP
	Diastolic BP

(a) Parameter comparison

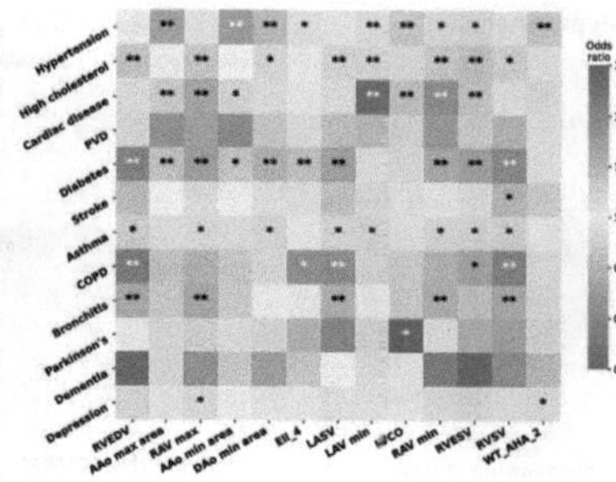

(b) Disease association heatmaps

Fig. 1. MESHAgents-discovered cardiovascular parameters and their disease associations. (a) Comparison between expert-selected in literature [7,18] and MESHAgents-discovered parameters, with overlapping parameters highlighted in green. MESHAgents autonomously identified clinically relevant phenotypes and additional confounders without prior domain knowledge. (b) Heatmaps showing significant associations between discovered imaging phenotypes and disease outcomes on 38,309 participants (**p < 0.01, *p < 0.05).

Through multi-faceted evaluation, MESHAgents shows superior phenotype discovery over existing single-agent and multi-agent approaches on population imaging data[1], while achieving diagnostic performance comparable to expert-selected parameters across three classification models (AUC difference $-0.004_{\pm 0.010}$)

Related Work. Recently, medical AI agents have shown promising progress, from single-agent systems like Med-PaLM [25] to collaborative frameworks such as MedAgents [26] and RareAgents [27] for medical question answering and diagnosis. These systems have demonstrated the potential of LLM-based agents in clinical scenarios, including triaging [28] and medical image analysis [29]. Meanwhile, multi-agent collaboration has emerged as a powerful paradigm in LLM applications [30], with approaches like role playing [31] and structured communication [32] enabling more sophisticated reasoning. Recent works have explored adversarial collaboration through debates [33] and negotiation [34] to enhance decision quality. However, these approaches lack consensus mechanisms for orchestrating distributed medical expertise and validating phenotype associations, making them insufficient for faithful medical imaging PheWAS.

2 Multi-agent Exploratory Synergy for the Heart

Problem Formulation. (i) For phenotype association analysis, given a large-scale imaging dataset containing N participants, let $\mathcal{X} = \{x_1, ..., x_N\}$ represent the set of imaging-derived phenotypes where each $x_i \in \mathbb{R}^d$ consists of d structural and functional phenotypes. For each participant, we also observe a set of

[1] We follow the evaluation setting from literature [18].

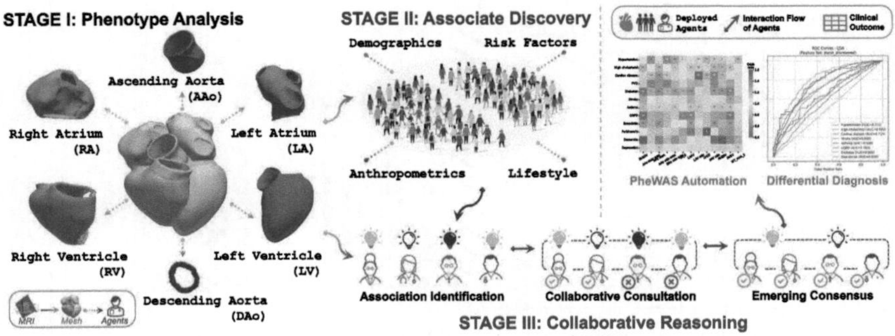

Fig. 2. `MESHAgents` **framework for cardiac imaging phenotype analysis through multi-agent reasoning.** Operating and validating in three stages: (i) specialist agents analyze cardiac structures from imaging measurements; (ii) agents discover confounders via distributed analysis; (iii) collaborative reasoning generates phenome-wide associations, validated through clinical metrics.

non-imaging factors $\mathcal{C} = \{c_1, c_2, ..., c_M\}$ spanning demographics, anthropometrics, lifestyle, and risk factors. Our objective is to discover a set of phenotypes $\mathcal{P} = \{p_1, p_2, ..., p_K\}$ where each phenotype p_k exhibits a strong and meaningful association with certain factors. MESHAgents identifies phenome-wide associations $\mathcal{A} = \{(p_k, c_m, a_{km})\}$ between image phenotype p_k and factor c_m. (ii) For diagnosis, we apply MESHAgents-discovered features to a disease classification model. Each patient profile is represented using MESHAgents-identified parameters, including both imaging phenotypes and non-imaging factors: $\mathcal{X}_i^{MESH} = \{p_k(x_i), c_m\}$ where $p_k \in \mathcal{P}$ and $c_m \in \mathcal{C}$. Standard classifiers then classify these features to a disease type: $f_{classifier}(\mathcal{X}_i^{MESH}) \rightarrow \mathcal{D} = \{d_1, d_2, ..., d_9\}$. This provides a more stringent evaluation than query-based tasks [27], though MESHAgents could support direct diagnostic querying with information retrieval (Fig 2).

MESHAgents Overview. MESHAgents orchestrates an expert team of agents α_i to analyze imaging phenotypes and discover associative factors through three progressive stages: $\mathcal{R}$ is medical imaging data containing structural and functional measurements, and $\mathcal{A}$ represent the identified phenotype-factor relationships contributing to diagnosis.

Dynamic Memory. To enable collective analysis, we design memory-augmented agents that integrates domain expertise with historical experience across local analysis and group reasoning. Each specialist agent α_i maintains a structured, long-term memory bank $\mathcal{M}_i$ that encodes imaging phenotype patterns and associative factors from past analyses. For a given phenotype set $\mathcal{X}$, the framework employs a similarity-based mechanism: $\mathcal{R}_i = \arg\max_{h \in \mathcal{H}_i} \text{Sim}\,(Emb(\mathcal{X}), Emb(h))$, where $\mathcal{H}_i$ represents agent α_i's historical cases and $\mathcal{R}_i$ is the retrieved relevant experience. $Emb(\cdot)$ maps imaging phenotypes to an embedding space. This memory mechanism enables agents to leverage past experiences and statistical evidence when evaluating phenotypes $\mathcal{P}$ and factors $\mathcal{C}$.

Consensus Mechanism. To foster independent reasoning while preventing premature convergence, MESHAgents employ a sequential discussion protocol. In each round t, expert agent α_i formulates its opinion independently based on three essential components: $O_i^t = g_i(\mathcal{V}_i, \mathcal{H}_{t-1}, \mathcal{M}_i)$, where $g_i(\cdot)$ is the opinion generation of agent α_i that synthesizes domain-specific insights, discussion history, and group context. Domain analysis $\mathcal{V}_i$ ensures specialty-specific insights, discussion history $\mathcal{H}_{t-1}$ enables progressive refinement, and memory bank $\mathcal{M}_i$ provides group context. This sequential consensus building offers advantages over simultaneous discussion: (i) it reduces mutual influence during initial assessment, allowing each expert to form an unbiased domain-specific opinion [35], (ii) it enables integration and update of previous insights through $\mathcal{H}_{t-1}$ while avoiding false consensus [13], and (iii) it provides transparent decision trails instead of black-box models. The coordinator agent guides this process until reaching agreement (typically within 10 rounds), ensuring both independent expertise contribution and coherent consensus formation. Critically, this sequential mechanism creates traceable emergence patterns and transparent validation, whether phenotype analysis, associative factor discovery, or collective reasoning process.

Tools. Each agent use a suite of tools $\mathcal{T} = \{T_1, ..., T_J\}$ that process imaging phenotypes $\mathcal{R}$. Each tool T_j implements specific analytical functions: $\mathcal{TR} = \bigcup_{j=1}^{J} T_j(\mathcal{R})$, where this represents the collective evidence generated by applying agent-selected tools to the data. These tools perform statistical significance tests, effect size calculation, and population distribution analysis. The final decision $\mathcal{A}$ emerges through synthesis of agent opinions $\mathcal{O}^{(t)}$ at discussion round t, memory retrievals $\mathcal{R}_i$, and tool-derived evidence $\mathcal{TR}$: $\mathcal{A} = f_{AP}(\{\mathcal{O}^{(t)}\}_{t=1}^{T}, \{\mathcal{R}_i\}_{i=1}^{N}, \mathcal{TR})$, where f_{AP} represents an aggregation function that weights different information sources based on their statistical confidence (< 0.05) and on-topic relevance (> 0.3). This integrated framework provides the foundation for the three-stage analysis pipeline, enabling phenotype valuation $\mathcal{V}_i$, factor effect size assessment $\mathcal{E}_i$, and collaborative reasoning across three stages:

`Stage I: Phenotype Analysis.` In the initial stage, each agent α_i analyses the imaging phenotypes within a specific domain: {left ventricle (LV), right ventricle (RV), left atrium (LA), right atrium (RA), ascending aorta (AAo) and descending aorta (DAo)}. Here, we define the specialty by different anatomical structures, similar to specialty training in medical education. Given a phenotype set $\mathcal{P} = p_1, p_2, ..., p_K$, each agent evaluates: $\mathcal{V}_i = \{\phi_i(p_k, \mathcal{R}_i) | p_k \in \mathcal{P}\}$ where $\phi_i(\cdot)$ represents agent α_i's phenotype valuation function incorporating statistical significance assessment, clinical relevance evaluation, population distribution analysis, and stability among the data. Each agent is able to use tools to extract features from statistical pools within the scope of prior knowledge for passing to the discussion. The collective phenotype valuations $\mathcal{V}_i$ form the fidelity of representation before the discussion.

`Stage II: Associative Factor Discovery.` Then, agents examine a wide range of non-imaging factors $\mathcal{C} = c_1, c_2, ..., c_M$ through a similar two-level analysis: For local analysis, each agent α_i evaluates factor effects within its domain: $\mathcal{E}_i = \{(p_n, c_m, \psi_i(p_n, c_m, \mathcal{V}_i)) | p_n \in \mathcal{P}, c_m \in \mathcal{C}\}$ where $\psi_i(\cdot, \cdot)$ quantifies the phenotype-factor association strength. For global analysis, the coordinator agent aggregates local assessments $\{\mathcal{E}_i\}_{i=1}^N$ to establish cross-domain factor effects $\mathcal{E}_G = h(\{\mathcal{E}_i\}_{i=1}^N)$, where $h(\cdot)$ is a consensus aggregation function. The agents are set to reveal highly significant associations with a wide range of non-imaging factors, such as early-life factors, lifestyle and cognitive functions etc.

`Stage III: Collaborative Reasoning.` The final stage facilitates collaborative reasoning among agents to reach final consensus and report generation. This is inspired by multidisciplinary team (MDT) panel meetings in clinical decision making. We design a dynamic long-term memory mechanism for the agents in MESHAgents, enabling them to store, retrieve, and update memories like human experts. Agents can facilitate personalized analysis and diagnosis based on historical interactions. Meanwhile, agents interact through structured dialogues following protocols of evidence-based argumentation. The consensus formation follows an iterative process until convergence across all decision-making nodes (*i.e.,* recommended phenotype, top associative factor, top diagnosis).

3 Experiments and Results

Dataset and Setting. The UK Biobank data is used, which are available from UK Biobank via a standard application procedure[2]. Cardiac imaging phenotypes are derived from cardiac MR images using a deep learning-based pipeline [18] and released by the UK Biobank[3]. For generalizability and cross-validation, `MESHAgents` mine phenotypes and factors on a training set of 26,893 participants, and tested on the expanded dataset of 38,309 participants, which includes the initial cohort plus additional participants. Meanwhile, we perform hypothesis testing with existing expert knowledge in Fig. 1, and performance comparison with automatic systems in Table 1 and Fig. 3. All experiments are performed on two Nvidia RTX 6000 GPUs (48 GB).

Benchmarks. (i) The LLMs we evaluate are the latest version of GPT-4o and GPT3.5 [36]. All models are set in a zero-shot Chain-of Thought (CoT) [38] setting. (ii) For medical multi-agent frameworks: we select MedAgents [26] and RareAgents [27] as the latest medical frameworks. For those LLMs that originally cannot complete the PheWAS task, we introduce the task definition into their agent initialization process for fair comparison.

Evaluation Metrics. (i) In auto PHeWAS: We evaluate automatically discovered phenotypes through two complementary metrics: *Dependency:* Measures both phenotypic redundancy and hallucination tendencies. For phenotype set $\mathcal{P}$, calculated as $Q(\mathcal{P}) = [1 - \frac{2}{K(K-1)} \sum_{i=1}^{K-1} \sum_{j=i+1}^{K} |corr(p_i, p_j)|] \times \frac{|\mathcal{P}_{valid}|}{|\mathcal{P}|}$, where $|\mathcal{P}_{valid}|$ is the number of valid phenotypes in set $\mathcal{P}$. Lower values indicate greater phenotypic independence and fewer hallucinated features. *Coverage:* Assesses anatomical comprehensiveness across six anatomical structures (LV, RV, LA, RA, AAo, DAo) in [18,39] . We define coverage as $C(\mathcal{P}) =$

[2] http://www.ukbiobank.ac.uk/register-apply.

[3] https://biobank.ndph.ox.ac.uk/showcase/label.cgi?id=157.

Table 1. Performance comparison of different Agent-based models on two metrics for Auto PheWAS.

Model	Auto PheWAS	
	Dependency↓	Coverage↑
Independent LLMs (zero-shot CoT)		
GPT-3.5 [36]	0.642	0.524
GPT-4omini [36]	<u>0.356</u>	0.767
Claude-3.5 [37]	0.520	0.841
Medical Multi-agent Frameworks		
MedAgents [26]	0.509	0.829
RareAgents [27]	0.363	0.629
w/o Mem. & Tool	0.492	<u>0.847</u>
w/o Stage III	0.385	**0.871**
MESHAgents	**0.350**	**0.871**

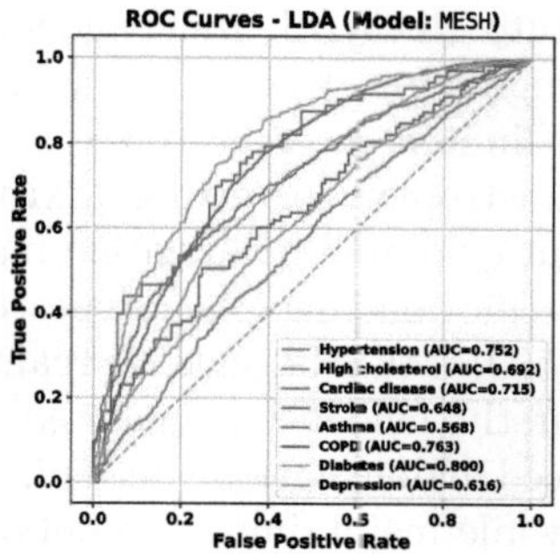

Fig. 3. ROC curves (LDA) for classification across cardiovascular conditions using discovered features.

Table 2. Performance comparison between experts-selected and MESHAgentsdiscoved on disease classification tasks. (Green and bold values indicate where MESHAgents outperformed analyses using human-selected phenotypes. Summary: For SVM, MESHAgents showed $\overline{AUC} = -0.013_{\pm 0.076}$ and $\overline{Recall} = -0.005_{\pm 0.042}$; for LDA, $\overline{AUC} = -0.004_{\pm 0.010}$ and $\overline{Recall} = +\mathbf{0.020}_{\pm \mathbf{0.033}}$; for AdaBoost, $\overline{AUC} = -0.014_{\pm 0.036}$ and $\overline{Recall} = -0.009_{\pm 0.040}$.)

	Hypertension		High cholesterol		Cardiac disease		PVD		Stroke		Asthma		COPD		Diabetes		Depression	
	AUC	Recall	AUC	Recall	AUC	Recall	AUC	Recall	AUC	Recall	AUC	Recall	AUC	Recall	AUC	Recall	AUC	Recall
SVM																		
Experts	0.739	0.708	0.645	0.672	0.671	0.548	0.316	0.462	0.398	0.805	0.549	0.691	0.561	0.521	0.730	0.670	0.573	0.667
MESHAgents	0.726	0.706	0.635	**0.716**	0.645	0.513	0.265	0.462	**0.619**	0.779	0.549	0.676	0.549	0.438	0.678	**0.678**	0.568	**0.679**
LDA																		
Experts	0.753	0.687	0.692	0.655	0.722	0.614	0.479	0.308	0.644	0.628	0.570	0.567	0.760	0.630	0.801	0.732	0.619	0.625
MESHAgents	0.752	**0.691**	0.692	**0.660**	0.715	**0.624**	**0.521**	**0.462**	**0.648**	0.619	0.568	**0.568**	**0.763**	**0.685**	0.800	0.732	0.616	0.616
AdaBoost																		
Experts	0.747	0.705	0.676	0.670	0.701	0.609	0.487	0.538	0.587	0.602	0.548	0.542	0.697	0.644	0.775	0.716	0.595	0.611
MESHAgents	0.743	0.700	0.673	0.658	0.689	0.601	**0.538**	0.462	**0.619**	0.566	**0.554**	**0.552**	0.649	0.644	**0.777**	**0.724**	0.580	0.598

$w_s \times \frac{|\Omega_{covered}|}{|\Omega|} + w_f \times \frac{|F_{covered}|}{|F_{total}|}$, where $\Omega_{covered}$ represents covered anatomical structures and $F_{covered}$ represents covered structure-function combinations. Higher coverage indicates global representation. (ii) In disease diagnosis: We assess the clinical utility of the MESHAgents-discovered phenotype-factor associations $\mathcal{A}$, using these features to train classification models for nine disease types. We implement three classification algorithms (AdaBoost [40], LDA [41], and SVM [42]) and report two evaluation metrics for classification performance: AUC, which measures overall discriminative ability, and recall, which quantifies the model's capacity to correctly identify positive disease cases. All results are obtained through five-fold cross-validation to ensure statistical reliability, and the practical value of automatically discovered phenotype-factor set.

Results. (i) Auto PheWAS: Independent LLMs generated unreliable phenotype sets with high dependency (GPT-3.5: 0.642), trade-off (GPT-4omini: 0.356 but 0.767), and limited coverage (GPT-3.5: 0.524), reflecting hallucination tendencies and catastrophic forgetting. Existing multi-agent frameworks showed improvement but exhibited polarization biases (RareAgents: 0.629), either overfocusing on specific imaging phenotypes or generating redundant phenotypes (MedAgents: 0.509). MESHAgents overcame these limitations through three key components: Firstly, evidence-augmented analysis with tools and memory significantly reduced hallucination; Second, the consensus mechanism promoted cross-checking while maintaining diversity; and the distributed expertise analysis prevented knowledge distortion. These components combined to achieve the best dependency score (0.350) and strong coverage (0.871) among all automated methods for PheWAS.

(ii) Disease Association Study: The comparison with imaging phenotypes selected by human in literature [7,18]. While MESHAgents do not fully match human patterns, it independently discovered significant relevant phenotypes without domain-specific knowledge. Figure 1(a) shows explainable overlap between MESHAgents-discovered and expert-selected parameters, with MESHAgents identifying alternative markers with diagnostic value. Importantly, MESHAgents achieves this automation while maintaining validity, as evidenced by its disease identification in Fig. 1(b). This shows MESHAgents' ability to effectively balance automated exploration with clinical relevance, offering complementary advantages to traditional expert-driven phenotype association studies.

(iii) Diagnosis: Table 2 shows a comparison between MESHAgent-discovered and expert-selected imaging phenotypes in disease classification across three classifiers. Notably, MESHAgents shows comparable overall performance with mean AUC differences of only $-0.013_{\pm 0.076}$ (SVM), $-0.004_{\pm 0.010}$ (LDA), and $-0.014_{\pm 0.036}$ (AdaBoost). For recall metrics—MESHAgent achieves $\overline{\text{Recall}}$ of $-0.005_{\pm 0.042}$ (SVM), $+0.020_{\pm 0.033}$ (LDA, outperforming human-selected features), and $-0.009_{\pm 0.040}$ (AdaBoost). The superior recall stems from MESHAgents' ability to identify complementary phenotypes that better capture disease variance in linear discriminant space in Fig. 3. MESHAgents-discovered phenotypes achieved superior results in specific conditions like Stroke (AUC +0.221 in SVM), PVD (AUC +0.042 in LDA), and Diabetes. This suggests that MESHAgents' automatic phenotype selection captures clinically relevant patterns without requiring extensive domain knowledge. The classification results across Fig. 3 further validate that MESHAgents-identified parameters maintain robust diagnostic value across different conditions, demonstrating the generalizability and scalability of our approach. These findings suggest that automatic phenotype discovery through multi-agent reasoning can effectively complement expert feature selection, potentially reducing subjective bias while maintaining clinical relevance in cardiovascular studies.

4 Conclusion

The framework could be further enhanced through the integration of vector embedding bases. For knowledge integration, an embedding library of medical imaging literature could be incorporated into our existing agent communication protocols, allowing phenotype-confounder evaluations to reference specific studies. In general, MESHAgents framework demonstrates advantages of multi-agent reasoning for medical imaging phenotype analysis compared to traditional approaches that rely on predefined criteria.

Acknowledgements. W.Z. is supported by the JADS programme and the UKRI Centre for Doctoral Training in AI for Healthcare (EP/S023283/1). M.Q. is supported by the EPSRC DeepGeM Grant (EP/W01842X/1) and the Dame Julia Higgins Postdoc Collaborative Research Fund. W.B. is supported by the EPSRC DeepGeM Grant (EP/W01842X/1), CVD-Net Programme Grant (EP/Z531297/1) and the BHF New Horizons Grant (NH/F/23/70013). S.N. is supported by the National Institutes of Health (R01-HL152256), the European Research Council (PREDICT-HF 864055), the British Heart Foundation (RG/20/4/34803), the EPSRC (EP/X012603/1 and EP/P01268X/1), the Technology Missions Fund under the EPSRC (EP/X03870X/1), and the Alan Turing Institute. P.M.M. acknowledges generous personal support from the Edmond J. Safra Foundation and Lily Safra, an NIHR Senior Investigator Award, Rosalind Franklin Institute, and the UK Dementia Research Institute, which is funded predominantly by the UKRI Medical Research Council. B.K. acknowledges the HPC resources provided by NHR@FAU under the NHR projects b143dc and b180dc. NHR funding is provided by federal and Bavarian state authorities. NHR@FAU hardware is partially funded by the DFG - 440719683. Additional support was received by the ERC - project MIA-NORMAL 101083647, DFG 513220538, 512819079, and by the state of Bavaria (HTA). This research was conducted using the UK Biobank Resource under Application Number 18545. We thank all UK Biobank participants and staff.

Disclosure of Interests. The authors have no competing interests to declare that are relevant to the content of this article.

References

1. Sing, C.F., Stengård, J.H., Kardia, S.L.: Genes, environment, and cardiovascular disease. Arteriosclerosis Thrombosis Vascular Biol. **23**(7) (2003)
2. Cosselman, K.E., Navas-Acien, A., Kaufman, J.D.: Environmental factors in cardiovascular disease. Nature Rev. Cardiol. **12**(11) (2015)
3. Rozanski, A., et al.: Impact of psychological factors on the pathogenesis of cardiovascular disease and implications for therapy. Circulation (1999)
4. Skelly, A.C., Dettori, J.R., Brodt, E.D.: Assessing bias: the importance of considering confounding. Evidence-based Spine-care J. **3**(01), 9–12 (2012)
5. Mukherjee, P., et al.: Confounding factors need to be accounted for in assessing bias by machine learning algorithms. Nature Med. **28**(6) (2022)
6. Qiao, M., et al.: CHeart: a conditional spatio-temporal generative model for cardiac anatomy. IEEE Trans. Med. Imaging (2023)

7. Qiao, M., et al.: A personalized time-resolved 3D mesh generative model for unveiling normal heart dynamics. Nature Mach. Intell., 1–12 (2025)
8. Touvron, H., Lavril, T., et al.: Llama: Open and efficient foundation language models. ArXiv preprint. 2023;abs/2302.13971
9. OpenAI. GPT-4 Technical report. ArXiv preprint. 2023;abs/2303.08774
10. Lu, P., et al.: Chameleon: Plug-and-play compositional reasoning with large language models. arXiv preprint arXiv:2304.09842 (2023)
11. Park, J.S, et al.: Generative agents: interactive simulacra of human behavior. In: UIST. Association for Computing Machinery (2023)
12. Zhang, W., Zang, C., Schmidt, M., Blythman, R.: BiD: Behavioral agents in dynamic auctions (2024)
13. Zhang, W., Zang, C., Kainz, B.: Truth or deceit? A Bayesian decoding game enhances consistency and reliability (2024)
14. Moor, M., et al.: Foundation models for generalist medical artificial intelligence. Nature. **616**(7956) (2023)
15. Singhal, K., Azizi, S., Tu, T., Mahdavi, S., Wei, J., et al.: Large language models encode clinical knowledge. Nature **07**, 620 (2023)
16. Zhang, X., et al.: AlpaCare: Instruction-tuned Large Language Models for Medical Application (2023)
17. Bao, Z, et al.: DISC-MedLLM: Bridging general large language models and real-world medical consultation (2023)
18. Bai, W., et al.: A population-based phenome-wide association study of cardiac and aortic structure and function. Nature Med. **26**(10) (2020)
19. Liévin, V., Hother, C.E., Winther, O.: Can large language models reason about medical questions? arXiv preprint arXiv:2207.08143 (2022)
20. Schmidt, H.G., Rikers, R.M.: How expertise develops in medicine: knowledge encapsulation and illness script formation. Med. Educ. **41**(12) (2007)
21. Tarale, P., Rietman, E., Siegelmann, H.T.: Distributed multi-agent lifelong learning. Trans. Mach. Learn. Res. (2025)
22. Zheng, G., et al.: Towards a collective medical imaging AI: enabling continual learning from peers. In: MIDL (2024)
23. Harris, E.: Large language models answer medical questions accurately, but can't match clinicians' knowledge. JAMA (2023)
24. Ji, Z.: Survey of hallucination in natural language generation. ACM Comput. Surv. **55**(12), 1–38 (2023)
25. Singhal, K., et al.: Large language models encode clinical knowledge. Nature. **620**(7972) (2023)
26. Tang, X., et al.: Medagents: Large language models as collaborators for zero-shot medical reasoning. arXiv preprint arXiv:2311.10537 (2023)
27. Chen X, et al.: RareAgents: Autonomous multi-disciplinary team for rare disease diagnosis and treatment. arXiv:2412.12475 (2024)
28. Lu, M., Ho, B., Ren, D., Wang, X.: TriageAgent: Towards better multi-agents collaborations for large language model-based clinical triage. In: Findings of the Association for Computational Linguistics: EMNLP, p. 5747-64 (2024)
29. Li, B., et al.: MMedAgent: Learning to use medical tools with multi-modal agent. arXiv:2407.02483 (2024)
30. Li, Y., Zhang, Y., Sun, L.: MetaAgents: Simulating interactions of human behaviors for LLM-based task-oriented coordination via collaborative generative agents. arXiv:2310.06500 (2023)
31. Wang Z, et al.: Unleashing the emergent cognitive synergy in large language models: a task-solving agent through multi-persona self-collaboration. NAACL (2024)

32. Wu, Q., et al.: AutoGen: Enabling next-gen LLM applications via multi-agent conversation framework. arXiv:2308.08155 (2023)
33. Du, Y., Li, S., Torralba, A., Tenenbaum, J.B., Mordatch, I.: Improving factuality and reasoning in language models through multiagent debate. In: ICML (2023)
34. Fu, Y., Peng, H., Khot, T., Lapata, M.: Improving language model negotiation with self-play and in-context learning from AI feedback. arXiv:2305.10142 (2023)
35. Yang, Z., et al. Oasis: Open agents social interaction simulations on one million agents. arXiv preprint arXiv:2411.11581 (2024)
36. Achiam, J., et al.: GPT-4 Technical Report. arXiv:2303.08774 (2023)
37. Anthropic. Claude 3.5 (2023). https://www.anthropic.com/claude
38. Wei, J., Wang, X., et al.: Chain-of-thought prompting elicits reasoning in large language models. NeurIPS. **35**, 24824–37 (2022)
39. Strocchi, M.: A publicly available virtual cohort of four-chamber heart meshes for cardiac electro-mechanics simulations. PLoS ONE **15**(6), e0235145 (2020)
40. Freund, Y., Schapire, R.E.: A decision-theoretic generalization of on-line learning and an application to boosting. J. Comput. Syst. Sci. **55**(1), 119–39 (1997)
41. Fisher, R.A.: The use of multiple measurements in taxonomic problems. Ann. Eugen. **7**(2), 179–88 (1936)
42. Cortes, C., Vapnik, V.: Support-vector networks. Mach. Learn. **20**, 273–97 (1995)

Multi-expert Collaboration and Knowledge Enhancement Network for Multimodal Emotion Recognition

Kun Wang, Junyong Zhao, Liying Zhang, Qi Zhu(✉), and Daoqiang Zhang(✉)

College of Artificial Intelligence, Nanjing University of Aeronautics and Astronautics, Key Laboratory of Brain-Machine Intelligence Technology, Ministry of Education, Nanjing 211106, China
zhuqinuaa@163.com, dqzhang@nuaa.edu.cn

Abstract. Emotion recognition leveraging multimodal data plays a pivotal role in human-computer interaction and clinical applications, such as depression, mania, Parkinson's Disease, etc. However, existing emotion recognition methods are susceptible to heterogeneous feature representations across modalities. Additionally, complex emotions involve multiple dimensions, which presents challenges for achieving highly trustworthy decisions. To address these challenges, in this paper, we propose a novel multi-expert collaboration and knowledge enhancement network for multimodal emotion recognition. First, we devise a cross-modal fusion module to dynamically aggregate complementary features from EEG and facial expressions through attention-guided. Second, our approach incorporates a feature prototype alignment module to enhance the consistency of multimodal feature representations. Then, we design a prior knowledge enhancement module that injects original dynamic brain networks into feature learning to enhance the feature representation. Finally, we introduce a multi-expert collaborative decision module designed to refine predictions, enhancing the robustness of classification results. Experimental results on the DEAP dataset demonstrate that our proposed method surpasses several state-of-the-art emotion recognition techniques.

Keywords: Emotion recognition · EEG · Multimodal fusion · Knowledge enhancement · Multi-expert collaboration

1 Introduction

Affective computing shows significant potential in healthcare applications, especially for assisting in diagnosing emotional disorders and psychiatric conditions [5,13]. For example, quantitative emotional assessment can help in the early diagnosis of Parkinson's Disease [12]. However, mental health disorders often exhibit dynamic emotional changes in complex environments, which can be reflected through variations in facial expressions or EEG signals. Therefore, how to develop a multimodal data emotion recognition system to improve the accuracy of disease diagnosis is the key.

J. C. Gee et al. (Eds.): MICCAI 2025, LNCS 15960, pp. 440–449, 2026.
https://doi.org/10.1007/978-3-032-04927-8_42

As affective states and distinct neurophysiological patterns are established associations, particularly in frequency bands (alpha, beta, theta) and spatial characteristics like frontal asymmetry [16,19]. Emotion recognition based on EEG signals has gained wide attention in recent years [9,21]. For example, Wu *et al.* [21] proposed a graph orthogonal purification network designed to capture both emotion-relevant and emotion-irrelevant features, which address distribution discrepancies across different emotion feature spaces. Li *et al.* [9] proposed a cross-attention-based dilated causal convolutional neural network to extract more discriminative features related to emotions. This method introduces the domain discriminator to reduce individual variability. However, these methods rely solely on single-modality data, which may not provide sufficient discriminative features for complex affective computing tasks. Multimodal emotion analysis can integrate complementary multi-dimensional information (*i.e.*, EEG and facial expressions) to enhance model adaptability in complex scenarios. Therefore, multimodal-based emotion recognition methods are proposed to improve emotion recognition performance [15,25]. For instance, Sun *et al.* [15] proposed a mutual information-based disentangled multimodal representation learning framework to balance and integrate the contributions of these diverse types of information. Meanwhile, Yin *et al.* [25] proposed a multimodal and channel attention fusion transformer to model inter-channel correlations and enhance emotion recognition accuracy by emphasizing global temporal dependencies. However, existing multimodal emotion recognition methods often yield suboptimal classification performance due to the lack of prior knowledge integration [27].

Brain networks based on EEG can reveal the topological associations of activation between different brain regions, which not only enhance feature discriminability for emotion recognition but also provide spatiotemporal prior information for understanding emotional fluctuations [22]. For example, Zheng *et al.* [27] proposed a prior-driven dynamic functional connectivity network to extract complex spatial-temporal features to aid emotion recognition. In addition, Chen *et al.* [3] proposed a comprehensive multi-source learning network to incorporate the information from multi-source data. However, existing methods still ignore both the time-frequency characteristics of raw EEG signals and the detailed features of facial expressions. In addition, the static weighting mechanisms in multimodal fusion algorithms may lead to misclassification in complex emotion recognition scenarios due to their limited adaptability.

To solve the above challenges, we propose a novel multi-expert collaboration and knowledge enhancement network for multimodal emotion recognition. Specifically, we propose a cross-modal fusion module (CMF) to dynamically aggregate complementary features from EEG and facial expressions through attention-guided interaction. Furthermore, we calculate the feature prototype alignment (FPA) loss to align the multimodal feature Then, we design a prior knowledge enhancement module (PKE) that injects original dynamic brain networks into feature learning to enhance the feature representation. Finally, we propose a multi-expert collaborative decision module (MCD) to obtain robustness classification results.

2 Method

As shown in Fig. 1, the proposed framework architecture consists of four interconnected core components that collectively address the challenges of multimodal emotion recognition. The cross-modal fusion module establishes deep feature interactions between EEG signals and facial expressions. The prior knowledge enhancement module incorporates domain-specific knowledge to enrich feature representations. The multi-expert collaborative decision module integrates complementary expertise, while the feature prototype alignment loss minimizes distribution discrepancies between multimodal feature representations, preserving their distinctive characteristics.

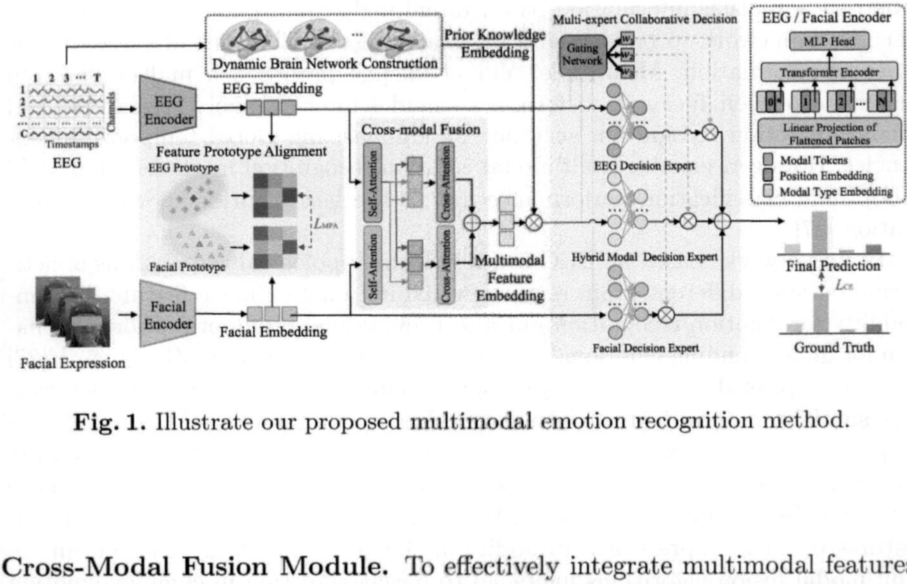

Fig. 1. Illustrate our proposed multimodal emotion recognition method.

Cross-Modal Fusion Module. To effectively integrate multimodal features, we propose the cross-modal fusion module. Specifically, the EEG signals $\mathbf{X}_{\mathrm{EEG}} \in \mathbb{R}^{T \times C_E \times D}$ and facial expression images $X_{Facial} \in \mathbb{R}^{T \times C_F \times D}$ are fed into the EEG encoder and facial encoder to obtain the EEG embedding emb_{EEG} and facial expression embedding emb_{facial}, respectively, where C_E and C_F denote the number of channels, T indicates the number of time windows, and D represents the feature dimension. It is worth noting that the EEG encoder and facial encoder use the EEGViT [24] and ViT [4] to extract the EEG feature and facial feature, respectively. Then, these modality-specific embeddings are then fused through our cross-modal fusion module, which comprises two self-attention blocks for intra-modal refinement and two cross-attention blocks for inter-modal interaction, ultimately yielding a unified multimodal feature representation emb_{MF}. Therefore, the cross-modal fusion module is defined as,

$$\begin{aligned} Q_{EEG}, K_{EEG}, V_{EEG} &= ATT_{self}(emb_{EEG}) \\ Q_{Facial}, K_{Facial}, V_{Facial} &= ATT_{self}(emb_{facial}) \\ emb_{MF} &= ATT_{cross}(Q_{Facial}, K_{EEG}, V_{EEG}) \\ &\oplus ATT_{cross}(Q_{EEG}, K_{Facial}, V_{Facial}) \end{aligned} \tag{1}$$

where ATT_{self} and ATT_{cross} represent the self-attention block and the cross-attention block, respectively. $\oplus$ denotes the concatenate operation.

Prior Knowledge Enhancement Module. In EEG recordings, each electrode not only captures electrophysiological activity within its corresponding brain region, but also contributes to emotional state recognition [18,23]. Therefore, we propose to construct a dynamic brain network to establish prior knowledge for improving feature discriminability. In our paper, EEG electrode channels are simplified as nodes in constructing an EEG network to characterize information in brain regions, and signal linkages between them form edges connecting these nodes. The connection strength is quantified as edge weights in the network. Specifically, we obtain each subject's EEG data $X_{EEG} \in \mathrm{R}^{T \times C_E \times D}$. To establish the functional connectivity network, we compute the Pearson correlation coefficient between EEG from a pair of channels within the t-th time window,

$$PCC_{ij}(t) = \frac{Cov(x_i(t), x_j(t))}{\sigma_i(t)\sigma_j(t)} \tag{2}$$

where Cov denotes the covariance between two vectors, σ represents the standard deviation operation, $x_i(t)$ and $x_j(t)$ denotes the EEG of a pair of channels i and j within the t-th time window, respectively. Thus, for each subject, the dynamic brain network $DFCN_{ori} \in \mathrm{R}^{T \times C \times C}$ consists of T transient functional connectivity network $[C(1), C(2), \cdots, C(t)]^T$, where each matrix captures time-varying functional interactions between brain regions. Finally, we use the dynamic brain network as prior knowledge to guide the fusion process of multimodal features and help the model better capture key emotion-relevant features. The fusion process is defined as,

$$F_{prior} = emb_{MF} \otimes DFCN_{ori} \tag{3}$$

where $\otimes$ represents the element-wise multiplication operation.

Multi-expert Collaborative Decision Module. To reduce decision error in complex emotion recognition tasks, we proposed a multi-expert collaborative decision module to integrate cross-domain expert knowledge to improve the decision-making accuracy of the model. Specifically, the multi-expert collaborative decision module consists of three same-expert decision blocks (*i.e.*, EEG decision expert, hybrid modal decision expert, and facial decision expert) and a gating network. Furthermore, the different embedding feature is fed into the different expert decision blocks to obtain classification results. Therefore, each expert decision block is defined as,

$$\text{expert}_m = \text{W}_{o,m} \cdot GELU(\text{W}_{h,m} \cdot emb_m + \text{b}_{h,m}) + \text{b}_{o,m} \tag{4}$$

where $m \in M = [EEG, facial, MF]$ represent different modal sets. $\text{W}_{o,m}$, $\text{b}_{o,m}$, $\text{W}_{h,m}$, $\text{b}_{h,m}$ are the learnable weights and biases of the m-th expert network. The concatenated embedding features are subsequently fed into a gating network to dynamically determine weighting coefficients for each expert's output. This gating network employs an average pooling layer followed by the SoftMax function, enabling adaptive weight allocation according to diverse feature representations. Thus, the gating weights are defined as,

$$\text{weight}_m = Softmax(Avgpool(\text{W}_m \cdot emb_m + \text{b}_m)) \tag{5}$$

The outputs of all experts are computed and stacked together to form a tensor $\text{expert}_{EEG}(emb_{EEG})$, $\text{expert}_{facial}(emb_{facial})$, $\text{expert}_{MF}(emb_{MF})$. It follows that the output of the multi-expert collaborative decision module is the weighted sum of all expert outputs,

$$outputs = \sum_{m \in M} \text{weight}_m \cdot \text{expert}_m(emb_m) \tag{6}$$

where W_m is the m-th element of the weights vector. $\text{expert}_m(emb_m)$ represents the m-th expert's output. Finally, we obtain the final emotion recognition classification result.

Feature Prototype Alignment. Different modalities (*i.e.*, EEG signals and facial expressions) exhibit inherent structural and representational disparities that challenge their effective alignment within a shared embedding space. To address this issue, we propose a feature prototype alignment mechanism to leverage the complementary features of multimodal emotions, which can preserve their distinctive characteristics. We employed K-means clustering to capture global patterns in emotion distribution, initializing prototypes as modality-specific emotion category centroids during the feature embedding process. Thus, the prototype vector of the modal $u \in [EEG, faical]$ is defined as,

$$P_u = \frac{1}{N_u} \sum_{i=1}^{N_u} emb_u \tag{7}$$

where N_u is the number of samples for modal u, and emb_u is the feature representation of the modal u. Minimizing the distance between projected features and their respective modal prototypes ensures cross-modal alignment.

$$\mathcal{L}_P = \frac{1}{N} \sum_{i=1}^{N} \left(\left\| emb_{EEG}^i - P_{EEG} \right\|_2^2 + \left\| emb_{facial}^i - P_{facial} \right\|_2^2 \right) \tag{8}$$

where $|| \cdot ||$ represents the L2_norm.

3 Experiment Results

Multimodal Emotion Databases. The DEAP dataset is a multimodal dataset designed for the analysis of human emotional states [7]. The dataset comprises a 32-channel electroencephalogram (EEG) and peripheral physiological signals collected from 32 participants, with supplementary facial expression recordings obtained from the first 22 subjects. During the experiment, each participant viewed 40 standardized one-minute video stimuli, followed by comprehensive affective evaluations using a validated assessment scale. Post-stimulus ratings encompassed five discrete dimensions: arousal, valence, dominance, liking, and familiarity.

Data Processing. A comprehensive data preprocessing pipeline was implemented for the DEAP dataset, wherein the 32-channel EEG signals were downsampled to 128 Hz. Power spectral density (PSD) features were subsequently extracted using the Welch method with a 3-second non-overlapping window segmentation. Following established methodologies in the field [26], we formulated the emotion recognition task as a binary classification problem by implementing a threshold of 5 to dichotomize the continuous rating scales. For facial expression analysis, we employed OpenFace [1] to extract facial features from video frames, ensuring temporal synchronization with corresponding physiological data.

Experiment settings. In our experiment, we employed the leave-one-subject-out (LOSO) cross-validation strategy, which is particularly suitable for subject-independent emotion recognition tasks. In each validation iteration, the samples from a single subject were held out as the testing set, while the remaining subjects' samples from the dataset constituted the training set. The final performance metrics were computed as the average values across all iterations, thereby providing a comprehensive assessment of the model's generalization capability across different individuals. The identification performance was evaluated using accuracy (ACC) and F1-score. The proposed model was implemented in PyTorch and trained on an NVIDIA GeForce RTX3090 GPU. The Adam optimizer was used to optimize our method, and the learning rate and batch size were set as 0.001 and 40, respectively. We employed the cross-entropy loss to supervise the multimodal learning process. The source code of this work is available at https://github.com/EEGBrainNet/Emotion-Recognition.

Results and Discussions. To evaluate the effectiveness of our proposed method, we compared our method with EEG-based methods (*i.e.*, DGCNN [14] and EEGNet [8]), facial-based methods (*i.e.*, ViT [4] and GAT [17]), and multimodal-based methods (*i.e.*, MKL [2], LSTM [6], DCCA [10], MoGE [11], DFCNs [27], and Milmer [20]) in DEAP dataset. The experimental results are reported in Table 1. As shown in Table 1, our proposed method achieves the best emotion recognition performance. For example, on valence and arousal, the average ACC and F1 of our method achieved 70.01%, 74.82%, 71.67%, and 76.15%, respectively. The main reason for the superiority of our method is that our method not only uses the dynamic brain network as prior knowledge to enhance

Table 1. Experimental results of our method with comparison methods (Mean±Std%).

Modality	Method	Valence		Arousal	
		ACC	F1	ACC	F1
EEG	DGCNN [14]	62.54/09.32	61.54/11.25	62.56/09.64	68.09/11.67
	EEGNet [8]	61.11/06.19	60.41/20.47	61.52/09.59	66.07/12.37
Facial	ViT [4]	62.51/10.17	65.11/10.66	62.52/08.23	69.39/13.94
	GAT [17]	61.53/06.98	63.92/18.62	66.38/09.47	74.69/13.82
EEG+Facial	MKL [2]	56.94/08.38	64.31/13.90	55.14/09.25	65.39/10.03
	LSTM [6]	64.86/09.52	62.95/14.95	64.30/11.78	76.28/09.96
	DCCA [10]	57.08/08.79	68.42/08.04	53.75/08.59	65.04/10.26
	MoGE [11]	65.01/08.93	66.67/10.74	65.08/09.35	69.57/16.24
	DFCNs [27]	67.36/05.58	69.17/09.01	68.47/08.65	74.68/13.36
	Milmer [20]	67.77/09.49	68.06/15.43	67.78/08.49	75.02/11.52
	Ours	**70.01/09.42**	**74.82/16.59**	**71.67/07.50**	**76.15/13.35**

Table 2. Ablation studies of our proposed method (Mean±Std%).

Method	Valence		Arousal	
	ACC	F1	ACC	F1
Baseline	61.66/09.50	64.89/16.29	63.19/09.68	65.58/13.25
w/o CMF	66.51/08.25	67.28/12.45	67.44/08.44	67.51/12.13
w/o PKE	68.19/09.08	68.89/15.51	69.16/09.75	70.17/10.86
w/o MCD	67.50/08.45	69.76/12.94	68.33/08.53	68.29/14.50
w/o FPA	69.02/09.34	70.50/16.25	70.27/08.89	74.74/11.66
Ours	**70.01/09.42**	**74.82/16.59**	**71.67/07.50**	**76.15/13.35**

the feature representation, but also introduces multi-expert collaborative decision to improve the classification performance of emotion recognition.

To evaluate the effectiveness of our proposed module, we conduct several ablation studies on the DEAP dataset. Besides, we establish a baseline that simply fuses two modal features through concatenate operations. The experiment results are reported in Table 2. As shown in Table 2, our proposed module achieved effective improvement with the baseline method, with an increase of 8.35%, 9.93%, 8.48%, and 10.57% for ACC and F1 on valence and arousal, respectively. Specifically, we proposed the cross-modal fusion module and the prior knowledge enhancement module to obtain better improvement with the baseline. The possible reason may be that the two modules not only can enhance emotion-relevant feature representation, but also help the model focus on key emotion-relevant brain regions, reducing redundant feature learning. Furthermore, the removal of the MCD module leads to performance degradation compared to our method, resulting in accuracy reductions of 2.51% and 3.34% for

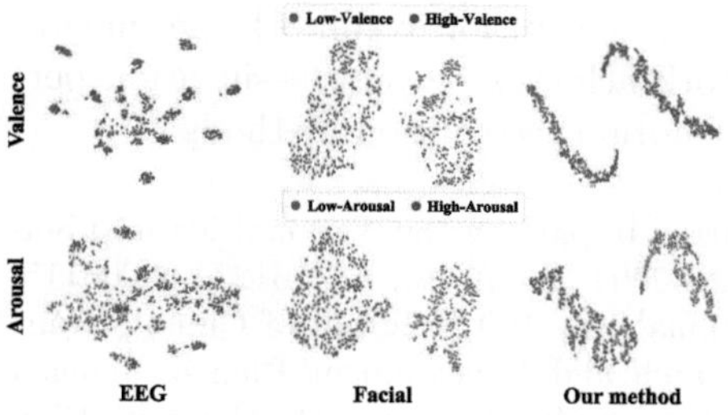

Fig. 2. The visualization of the distribution of features extracted by each modal and multimodal of the proposed model.

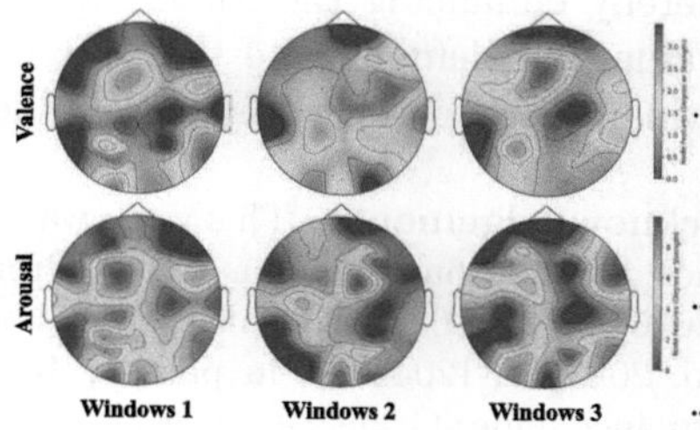

Fig. 3. Dynamic brain network topography across time windows.

valence and arousal classification, respectively. The reason lies in that MCD effectively reduces decision bias and further improves classification performance. Moreover, when the FPA module is removed from our proposed method, performance degradation occurs. Experimental results demonstrate that the FPA module can effectively reduce inter-modal heterogeneity and enhance cross-modal semantic consistency.

To analyze the performance differences among emotion recognition methods, we applied t-distributed Stochastic Neighbor Embedding (t-SNE) to visualize high-dimensional feature distributions. As shown in Fig. 2, our method demonstrates clearer cluster separation in the embedded space, indicating enhanced discriminative power for affective states. Experimental results confirm the framework's superiority in multimodal emotion recognition, particularly in valence-arousal classification accuracy. In addition, we also display the spatiotemporal characteristics of brain activity patterns through topographic mapping in Fig. 3. It is worth noting that the color intensity of brain regions reflects EEG-facial expression correlation levels, with red hues indicating higher correlations and blue hues denoting lower ones. These findings validate that our method effectively identifies emotion-relevant electrodes and establishes corresponding brain networks as prior structural information, thereby enhancing model performance in emotion recognition tasks.

4 Conclusion

In this paper, we propose a novel multi-expert collaboration and knowledge enhancement network for multimodal emotion recognition. Specifically, we develop a cross-modal fusion module to integrate EEG and facial expression features. Second, feature prototype alignment is employed to enhance cross-modal consistency. Then, we also proposed a prior knowledge enhancement module that incorporates dynamic brain network topology as prior knowledge to strengthen feature representation. To reduce the uncertainty in decision fusion across different modalities, we develop a multi-expert collaborative strategy that systematically integrates complementary insights from experts across various modalities, thereby enhancing the robustness of the decision fusion results. Experimental evaluations demonstrate that our framework achieves better classification performance compared with state-of-the-art emotion recognition methods.

Acknowledgements. This work was supported in part by the National Natural Science Foundation of China under Grants 62136004, 62276130, 62376123, 62006115, 62076129, and 62371234, in part by the National Key RD Program of China (Grant No. 2023YFF1204803), in part by Key Research and Development Plan of Jiangsu Province under Grant BE2022842, in part by the Key Research and Development Plan in Jiangsu (Social Development) under Grant BE2022677, in part by Natural Science Foundation of Jiangsu Province (Grant No. BK20231438).

Disclosure of Interests. The authors have no competing interests to declare that are relevant to the content of this article.

References

1. Baltrušaitis, T., Robinson, P., Morency, L.P.: Openface: an open source facial behavior analysis toolkit. In: 2016 IEEE winter conference on applications of computer vision (WACV), pp. 1–10. IEEE (2016)
2. Cai, Q., Cui, G.C., Wang, H.X.: EEG-based emotion recognition using multiple kernel learning. Mach. Intell. Res. **19**(5), 472–484 (2022)
3. Chen, C., et al.: Comprehensive multisource learning network for cross-subject multimodal emotion recognition. IEEE Trans. Emerg. Top. Comput. Intell. **9**(1), 2471–285X (2025)
4. Dosovitskiy, A., et al.: An image is worth 16x16 words: Transformers for image recognition at scale. arXiv preprint arXiv:2010.11929 (2020)
5. Gandhi, A., Adhvaryu, K., Poria, S., Cambria, E., Hussain, A.: Multimodal sentiment analysis: a systematic review of history, datasets, multimodal fusion methods, applications, challenges and future directions. Inf. Fusion **91**, 424–444 (2023)
6. Hochreiter, S., Schmidhuber, J.: Long short-term memory. Neural Comput. **9**(8), 1735–1780 (1997)
7. Koelstra, S., et al.: Deap: a database for emotion analysis; using physiological signals. IEEE Trans. Affect. Comput. **3**(1), 18–31 (2011)
8. Lawhern, V.J., et al.: EEGNet: a compact convolutional neural network for EEG-based brain–computer interfaces. J. Neural Eng. **15**(5), 056013 (2018)
9. Li, C., Bian, N., Zhao, Z., Wang, H., Schuller, B.W.: Multi-view domain-adaptive representation learning for EEG-based emotion recognition. Inf. Fusion **104**, 102156 (2024)
10. Liu, W., Qiu, J.L., Zheng, W.L., Lu, B.L.: Comparing recognition performance and robustness of multimodal deep learning models for multimodal emotion recognition. IEEE Trans. Cogn. Dev. Syst. **14**(2), 715–729 (2021)
11. Liu, X.H., Jiang, W.B., Zheng, W.L., Lu, B.L.: Moge: Mixture of graph experts for cross-subject emotion recognition via decomposing EEG. In: 2024 IEEE International Conference on Bioinformatics and Biomedicine (BIBM), pp. 3515–3520. IEEE (2024)
12. Pepa, L., Spalazzi, L., Ceravolo, M.G., Capecci, M.: Supervised learning for automatic emotion recognition in Parkinson's disease through smartwatch signals. Expert Syst. Appl. **249**, 123474 (2024)

13. Saganowski, S., Perz, B., Polak, A.G., Kazienko, P.: Emotion recognition for everyday life using physiological signals from wearables: a systematic literature review. IEEE Trans. Affect. Comput. **14**(3), 1876–1897 (2022)
14. Song, T., Zheng, W., Song, P., Cui, Z.: EEG emotion recognition using dynamical graph convolutional neural networks. IEEE Trans. Affect. Comput. **11**(3), 532–541 (2018)
15. Sun, H., et al.: Multimodal sentiment analysis with mutual information-based disentangled representation learning. IEEE Trans. Affect. Comput., 1–12 (2025)
16. Tan, H., et al.: Intracranial EEG signals disentangle multi-areal neural dynamics of vicarious pain perception. Nat. Commun. **15**(1), 5203 (2024)
17. Velickovic, P., et al.: Graph attention networks. Stat **1050**(20), 10–48550 (2017)
18. Wan, Z., Yu, Q., Dai, W., Li, S., Hong, J.: Data generation for enhancing EEG-based emotion recognition: extracting time-invariant and subject-invariant components with contrastive learning. IEEE Trans. Consum. Electron., 1–14 (2024)
19. Wang, Y., Zhang, B., Tang, Y.: DMMR: Cross-subject domain generalization for EEG-based emotion recognition via denoising mixed mutual reconstruction. In: Proceedings of the AAAI Conference on Artificial Intelligence., vol. 38, pp. 628–636 (2024)
20. Wang, Z., et al.: Milmer: a framework for multiple instance learning based multimodal emotion recognition. arXiv preprint arXiv:2502.00547 (2025)
21. Wu, M., Chen, C.P., Chen, B., Zhang, T.: Grop: Graph orthogonal purification network for EEG emotion recognition. IEEE Trans. Affect. Comput., 1–14 (2024)
22. Wu, Y., Meng, T., Li, Q., Xi, Y., Zhang, H.: Study on multidimensional emotion recognition fusing dynamic brain network features in EEG signals. Biomed. Signal Process. Control **100**, 107054 (2025)
23. Xu, J., Qian, W., Hu, L., Liao, G., Tian, Y.: EEG decoding for musical emotion with functional connectivity features. Biomed. Signal Process. Control **89**, 105744 (2024)
24. Yang, R., Modesitt, E.: VIT2EEG: leveraging hybrid pretrained vision transformers for EEG data. arXiv preprint arXiv:2308.00454 (2023)
25. Yin, J., et al.: Research on multimodal emotion recognition based on fusion of electroencephalogram and electrooculography. IEEE Trans. Instrum. Meas. **73**, 6502012 (2024)
26. Zhang, Z., Liu, Y., Zhong, S.H.: Ganser: a self-supervised data augmentation framework for EEG-based emotion recognition. IEEE Trans. Affect. Comput. **14**(3), 2048–2063 (2022)
27. Zheng, C., Shao, W., Zhang, D., Zhu, Q.: Prior-driven dynamic brain networks for multi-modal emotion recognition. In: International Conference on Medical Image Computing and Computer-Assisted Intervention, pp. 389–398. Springer (2023)

Multimodal Fusion Network with Distribution-Based Tumor-Marker Imputation for Multi-origin Metastatic Cervical Lymphadenopathy Classification

Rui Li[1], Chunyan Li[2], Xi Lin[2(✉)], Jinfeng Xu[3], Fang Li[4], and Yao Lu[1(✉)]

[1] School of Computer Science and Engineering, Sun Yat-sen University, Guangzhou 510275, China
luyao23@mail.sysu.edu.cn

[2] Department of Ultrasound State Key Laboratory of Oncology in South China, Guangdong Provincial Clinical Research Center for Cancer, Sun Yat-sen University Cancer Center, Guangzhou 510060, China
linxi@sysucc.org.cn

[3] Department of Biostatistics, City University of Hong Kong, Hong Kong 999077, China

[4] Department of Ultrasound, Chongqing Key Laboratory for Intelligent Oncology in Breast Cancer, Chongqing University Cancer Hospital, Chongqing 400030, China

Abstract. Accurate identification the primary tumor of metastatic cervical lymphadenopathy (CLA) is crucial for guiding clinical treatment, yet clinical diagnosis remains challenging due to the complexity of tracing multi-potential origins using ultrasound images and incomplete clinical information. Existing deep learning methods typically utilize the imaging semantic features from B-mode ultrasound (BUS) and color Doppler flow imaging (CDFI), or incorporate basic clinical information, neglecting the importance of patient-specific features such as tumor markers (TMs) in clinical diagnosis. To address these limitations, we propose a new multimodal imaging-features and distribution-based tumor-marker fusion network (MDFN) for five categories of CLA metastatic origins (thyroid, head and neck, respiratory, female reproductive, and digestive). First, a distribution-based TM imputation method is proposed to reconstruct missing TMs, which treats the available clinical information of each patient (such as sex, age, neck region, etc.) as a vector to construct data distributions between TMs and avoid the data bias issues. Building on these personalized TMs, we propose the first population-personalized fusion framework, which integrates semantic features related to lymph node morphology from BUS images, semantic features related to vascular distribution from CDFI images, and TM features consistent with individualized patient data, thereby simulating clinical reasoning patterns. The effectiveness of the proposed MDFN method was evaluated using extensive experimental results from 3,100 multi-origin metastatic CLA cases, achieving an area under the receiver operating characteristic

R. Li, C. Li—Equal contribution.

J. C. Gee et al. (Eds.): MICCAI 2025, LNCS 15960, pp. 450–459, 2026.
https://doi.org/10.1007/978-3-032-04927-8_43

(AUC) of 0.891, with corresponding accuracy, sensitivity, specificity, and F_1 of 0.863, 0.604, 0.913, and 0.661, respectively, outperforming other state-of-the-art methods.

Keywords: Metastatic cervical lymphadenopathy · Multimodal imaging-features · Distribution-based tumor-marker

1 Introduction

Metastatic cervical lymphadenopathy (CLA), a systemic manifestation of malignancy, represents one of the most prevalent secondary cancers in patients over 40 years of age [12]. It originates from diverse primary tumors, including thyroid carcinoma, head and neck neoplasms, nasopharyngeal cancer, and pulmonary malignancies, etc., whose accurate identification becomes crucial for formulating appropriate therapeutic strategies [2]. Current clinical protocols typically require extensive diagnostic workups to localize the primary tumor sites of metastatic CLA [9]. This underscores the imperative need to develop effective, non-invasive computer-aided diagnosis (CAD) methods for identifying primary tumor sites.

Current non-invasive diagnostic approaches predominantly utilize conventional ultrasound modalities, encompassing B-mode ultrasound (BUS) and color Doppler flow imaging (CDFI). Radiologists primarily pay more attention to tumor morphology and anatomical features based on BUS and vascular structure features based on CDFI [3]. Nevertheless, the inherent pathological heterogeneity of metastatic CLA poses significant challenges in determining primary tumor origins through conventional imaging alone. Existing research predominantly focuses on distinguishing benign from malignant lymphadenopathy [7] or detecting metastatic involvement [16], with only one published study attempting metastatic CLA classification using BUS [19]. This preliminary investigation achieved an AUC of 0.822 through multi-modality integration of BUS, CDFI, ultrasound elastography (UE), contrast-enhanced ultrasound (CEUS), and clinical parameters in a limited cohort (n=280). However, the clinical applicability of the study was hampered by the small sample size and the omission of mapping the relationships between patient-specific characteristics, such as tumor markers (TMs) [17], which are critical for clinical diagnosis.

To better align with clinical diagnosis, we first proposed a new multimodal fusion network, which integrates distribution-based TMs with morphological features from BUS and vascular patterns from CDFI, enabling the classification of five-category primary tumor origins (head and neck, thyroid, respiratory system, female reproduction system, and digestive system), including 90% metastatic CLA [6]. Our contributions are summarized below:

(1) Distribution-based TM imputation: A new imputation method introduces the idea of adversarial learning to reconstruct missing TM data from each patient's clinical information, explicitly maintaining the multivariate correlation structures among TMs through cycle-consistency regularization.

(2) Multimodal imaging features and distribution-based TM fusion: A new multimodal and distribution-based fusion method is proposed, which integrates anatomically related features extracted from BUS images, hemodynamic related features extracted from CDFI images, and distribution-based TMs profiles after imputation, thereby simulating the clinical diagnosis workflow.

(3) Large-scale multi-origin metastatic classification: Establishment of the first multi-modality population-based features with personalized TMs for five-category primary origins determination in 3,100 histologically confirmed cases, significantly outperforming previous attempts in both scope and accuracy.

2 Related Work

Ultrasound-based Related work: Due to the relevant literature on multi-origin CLA prediction, this mainly introduces the classification of medical lesions similar to it. Notably, recent research [18] [10] demonstrates benign-malignant classification accuracy rates of 85%-90% when integrating multi-modality ultrasound features (incorporating BUS, CDFI, and CEUS, etc.). Although convolutional neural network architecture has shown effectiveness in deep semantic feature extraction, its feature space remains population-based, and the integration of the mapping relationship between each patient's features is still an area to be explored.

Data imputation-based related work: Missing clinical parameters represent a pervasive challenge in medical artificial intelligence, where accurate data imputation critically determines diagnostic model performance. Current imputation methodologies bifurcate into statistical approaches (e.g., Multiple Imputation by Chained Equations [14], matrix decomposition [11]) and deep learning methods. The former leverage mathematical optimization to preserve feature correlations, whereas the latter construct nonlinear mapping spaces through generative architectures like adversarial networks [5]. However, the essence of the above research is still based on a population-based image feature space, overlooking the interaction between different clinical data of the same patient.

3 Method

As shown in Fig. 1, the proposed multimodal imaging-features and distribution-based tumor-marker fusion network (MDFN)[1] primarily comprises BUS branch, CDFI branch, distribution-based TMs imputation branch, and feature fusion branch.

3.1 BUS Branch

For clinicians focusing on morphological information in BUS images, this branch employs the InceptionResNetV2 architecture [13] to extract population-based,

[1] The code can be found at https://github.com/lr256/MDFN.

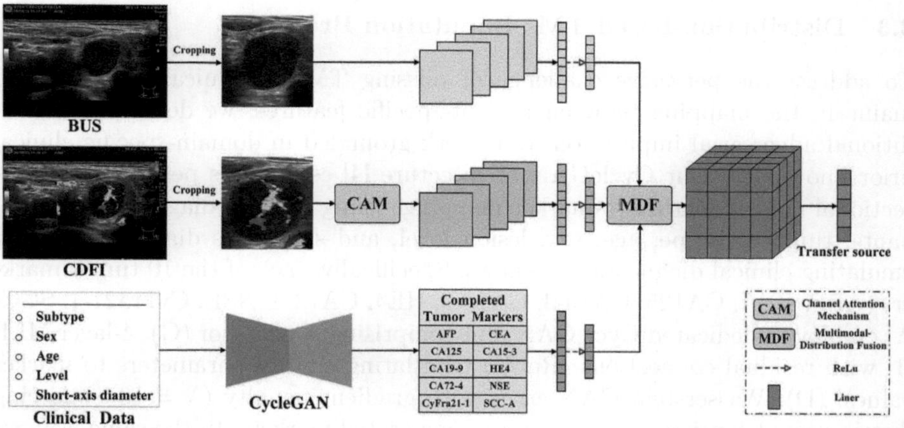

Fig. 1. Proposed framework of the multimodal imaging-features and distribution-based tumor-marker fusion network (MDFN). BUS images were analyzed to extract anatomical semantic features, while CDFI images were analyzed to extract vascular distribution features indicated by colorful dots.

morphology-based semantic feature. This multiscale convolutional backbone synergistically integrates Inception-style parallel pathways (with 1×1, 3×3, and 5×5 convolutional filters) and residual skip connections, enabling hierarchical learning of discriminative morphological patterns while mitigating gradient vanishing in deep networks. The architecture's cross-channel normalization and batch-aware feature recalibration mechanisms enhance clinically relevant feature discovery, particularly for subtle textural variations critical in BUS interpretation. Following depthwise feature aggregation through a nonlinear projection module, we apply dimensionality reduction via a learnable projection layer (1536-d to 1024-d), generating fixed-dimensional semantic embeddings.

3.2 CDFI Branch

The CDFI branch employs a two-stage attention guidance architecture to extract population-based semantic features related to vascular patterns. First, raw Doppler signals amplify hemodynamically salient color components through a 3-channel (RGB) channel attention mechanism module while suppressing acquisition artifacts [8]. This channel-adaptive feature recalibration precedes Inception-ResNetV2 processing to preserve directional flow signatures critical for microvascular assessment. The backbone network then extracts directional hemodynamic features through its multiscale receptive fields, with particular emphasis on vessel tortuosity and branching patterns via depthwise separable convolutions. Following feature aggregation, a nonlinear projection module (two fully-connected layers with ReLU) reduces dimensionality from 1536-d to 1024-d.

3.3 Distribution-based TMs Imputation Branch

To address the pervasive challenge of missing TMs in clinical oncology and maintain the mapping between patient-specific features, we developed a conditional adversarial imputation framework grounded in domain-specific clinical prior knowledge. Our CycleGAN architecture [4] establishes personalized bidirectional clinical-feature-to-marker mappings using five key diagnostic determinants: tumor subtype, age, sex, lesion level, and short-axis diameter, thereby emulating clinical diagnostic workflows. Specifically, each of the 10 tumor markers (AFP, CEA, CA125, CA15-3, CA19-9, HE4, CA72-4, NSE, CyFra21-1, SCC-A) employs a dedicated CycleGAN unit comprising: Generator (G): 4-layer MLP [1] with residual connections (ReLU) translating clinical parameters to marker values. (D): Wasserstein GAN critic with gradient penalty ($\lambda = 10$) enforcing distributional consistency between real/generated markers. In the same way, we also map it to a unified 1024-d using a nonlinear projection module.

3.4 Multimodal-Distribution Fusion Branch

To achieve an effective fusion of multimodal and distribution-based features (morphological-related semantic features extracted from BUS, hemodynamic features extracted from CDFI, and personalized TM features), we employ hierarchical tensor fusion network [15]. It constructs a joint representation space of cross-modal feature interactions by calculating the outer product of three feature vectors, capturing nonlinear and higher-order correlation between different modes. Finally, the low-dimensional features, after tensor decomposition, are mapped to the probability of primary tumor origin of CLA metastasis through a linear connection layer.

4 Results and Discussion

This study retrospectively analyzed the data of 3100 metastatic cervical lymph nodes from Sun Yat-sen University Cancer Center, of which 280 were from thyroid, 218 were from head and neck, 1249 were from respiratory system, 910 were from female reproductive system, and 443 were from digestive system. The missing ratios for these TMs were 54.5%, 3.6%, 57.6%, 39.7%, 36.9%, 84.9%, 60.1%, 36.1%, 53.3%, and 85.9%, respectively. We divided the dataset into training and test sets with an 8:2 ratio and performed 3-fold cross-validation on the training. We conducted comparative experiments to evaluate the method's effectiveness.

4.1 Analysis of Classification Performance

The experimental results validate the classification performance of the proposed MDFN framework compared with different modalities, as shown in Table 1. Comparative analysis of two single-modality methods revealed a 3.9% improvement in AUC for M_{BUS} over M_{CDFI} (0.743 vs 0.704), aligning with radiologists' predominant reliance on morphological features in clinical practice. Although the AUC of $M_{BUS+CDFI}$ increased by 0.3% (0.746 vs 0.743), the F_1 was significantly improved by 10.4% (0.445 vs 0.341), indicating enhanced precision-recall balance. Crucially, integration of imputed TM with imaging data yielded significant performance gains across all modality. The AUC of baseline M_{BUS} improved by 9.7% when combined with TM (M_{BUS+TM}). Similarly, CDFI combined with TM ($M_{CDFI+TM}$) elevated AUC from 0.704 to 0.806 (increased by 10.2%), validating the complementary diagnostic value of TMs. Compared with other methods, the proposed method MDFN effectively fused the heterogeneous characteristics of multimodal imaging and distribution-based TMs, achieving the optimal value among five evaluation indicators (AUC, Accuracy, Sensitivity, Specificity, F_1): AUC and F_1 are 5.1% and 11.1% higher, respectively, than the next best method (M_{BUS+TM}). Figure 2 (a) depicts the receiver operating characteristic curve of the corresponding AUC value, where proposed MDFN achieves an optimal AUC of 89.1%, surpassing comparator methods across all specificity thresholds. These findings collectively demonstrate MDFN's capacity to leverage cross-modal synergies through tensor fusion, greatly improving classification accuracy for metastatic cervical lymphadenopathy.

Table 1. The comparison of classification performance under different modality data.

Method	AUC (95% CI)	Accuracy (95% CI)	Sensitivity (95% CI)	Specificity (95% CI)	F_1
M_{BUS}	0.743 (0.718-0.765)	0.792 (0.777-0.809)	0.413 (0.383-0.443)	0.861 (0.851-0.872)	0.341
M_{CDFI}	0.704 (0.676-0.729)	0.751 (0.736-0.767)	0.293 (0.261-0.324)	0.823 (0.813-0.834)	0.244
$M_{BUS+CDFI}$	0.746 (0.721-0.768)	0.786 (0.771-0.802)	0.399 (0.369-0.432)	0.858 (0.848-0.868)	0.445
M_{BUS+TM}	0.840 (0.816-0.862)	0.848 (0.833-0.862)	0.542 (0.495-0.59)	0.901 (0.891-0.910)	0.550
$M_{CDFI+TM}$	0.806 (0.778-0.831)	0.823 (0.809-0.839)	0.481 (0.439-0.528)	0.880 (0.871-0.891)	0.489
MDFN	0.891 (0.872-0.908)	0.863 (0.847-0.878)	0.604 (0.561-0.647)	0.913 (0.904-0.923)	0.661

4.2 Comparison with Different Imputation Method

The CycleGAN-based tumor marker imputation module constitutes a core module of proposed MDFN method, enabling robust prediction of multi-origin

Table 2. The performance comparison under different imputation methods.

Method	AUC (95% CI)	Accuracy (95% CI)	Sensitivity (95% CI)	Specificity (95% CI)	F_1
MICE	0.858 (0.837-0.878)	0.860 (0.845-0.874)	0.558 (0.515-0.604)	0.906 (0.897-0.915)	0.586
Average padding	0.866 (0.846-0.883)	0.825 (0.809-0.842)	0.523 (0.481-0.568)	0.886 (0.876-0.896)	0.491
Zero padding	0.870 (0.849-0.891)	0.845 (0.831-0.861)	0.592 (0.547-0.640)	0.900 (0.891-0.911)	0.585
Logistic regression	0.871 (0.850-0.890)	0.851 (0.837-0.865)	0.593 (0.544-0.637)	0.904 (0.895-0.913)	0.588
Our-CycleGAN	0.891 (0.872-0.908)	0.863 (0.847-0.878)	0.604 (0.561-0.647)	0.913 (0.904-0.923)	0.661

metastases. Figure 2 (b) and Table 2 provides the ROC and a quantitative comparison of the different imputation methods used in predicting multiple transfer origins. Compared to the three traditional imputation methods, fixed-value padding (average padding and zero pading) yields better classification performance (0.886 vs 0.870 vs 0.858) than the Multiple Imputation by Chained Equations (MICE) method. This may be attributed to their conservative preservation of original TM distributions rather than introducing spurious correlations through statistical assumptions. In contrast, the proposed personalized imputation method based on each patient's clinical information effectively simulates the complex relationship between TMs and achieves the best classification performance among all evaluation indicators, outperforming other methods.

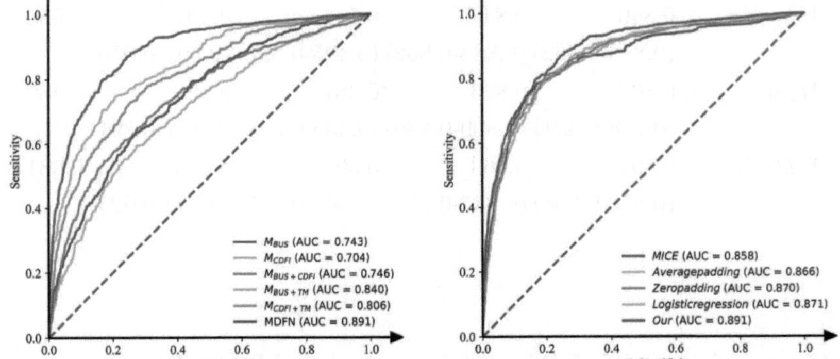

Fig. 2. ROCs used to evaluate the performance of different methods.

Table 3. Performance of the model in ablation study

Method	AUC (95% CI)	Accuracy (95% CI)	Sensitivity (95% CI)	Specificity (95% CI)	F_1
MDFN (None CAM)	0.876 (0.858-0.894)	0.849 (0.834-0.864)	0.573 (0.535-0.613)	0.904 (0.895-0.914)	0.536
MDFN (None MDF)	0.874 (0.855-0.891)	0.847 (0.832-0.862)	0.567 (0.526-0.608)	0.902 (0.892-0.912)	0.530

4.3 Ablation Study

Building on the comprehensive ablation studies in Sect. 4.1 (which compare different modalities and the presence/absence of tumor markers), this analysis focuses on the two remaining innovative components: 1) the Channel Attention Mechanism (CAM) for context-aware feature recalibration, and 2) Multimodal-Distribution Fusion (MDF) for cross-modal interaction modeling. As empirically demonstrated in Table 3, both modules showed consistent performance improvements: the CAM implementation increased the AUC from 0.876 to 0.891 (Δ+1.5%), enhancing sensitivity to vascular pattern variations through adaptive channel weighting. The addition of TF module raised the AUC from 0.874 to 0.891 and increased the F_1 by 13.1%, confirming its effectiveness in fusion imgaing and distribution-based features.

Table 4. The comparison of state-of-the art method

Reference	Method	AUC (95% CI)	Accuracy (95% CI)	Sensitivity (95% CI)	Specificity (95% CI)	F_1
Zhu et al., 2024 [19]	MSMFN	0.633 (0.605-0.658)	0.794 (0.779-0.809)	0.290 (0.274-0.304)	0.847 (0.838-0.857)	0.229
Our	MDFN	0.891 (0.872-0.908)	0.863 (0.847-0.878)	0.604 (0.561-0.647)	0.913 (0.904-0.923)	0.661

4.4 Comparison with Previous State-of-the-Art Method

Given the limited prior research on ultrasound-based multi-origin metastasis prediction in cervical lymphadenopathy (CLA), we conducted a comparative analysis with the only existing method (MSMFN) presented in Table 4. Although MSMFN originally incorporates UE and CEUS modalities alongside BUS/CDFI, we strategically disabled these components in our implementation to ensure a fair comparison based solely on core ultrasound modalities. In comparison, the proposed MDFN method aligns more closely with the diagnostic approach due to the integration of distribution-based completed TMs and dual-modality semantic features, thus improving the classification performance.

5 Conclusion

We proposed a population-personalized fusion network for identifying five-category metastatic CLA. Evaluating its effectiveness on a dataset of 3,100 malignant CLA cases from multiple metastatic origins, the method outperforms single-modality methods and those without TM imputation, achieving an AUC of 0.891, accuracy of 0.863, sensitivity of 0.604, specificity of 0.913, and F_1 score of 0.661. Additionally, the distribution-based TM imputation method from existing clinical data to preserve interaction relationships and integrates dual-modality image features, to improve multi-class classification performance.

Acknowledgments. This work was funded in part by the National Key R&D Program of China under Grant 2023YFE0204300; in part by the R&D project of Pazhou Lab (HuangPu) under Grant 2023K0606; in part by the National Natural Science Foundation of China under Grants 82441027, 62371476, and 82171955; in part by the Guangzhou Science and Technology Bureau under Grant 2023B03J1237.

Disclosure of Interests. The authors have no competing interests to declare that are relevant to the content of this article.

References

1. Abbas, D.K., Rashid, T.A., Bacanin, K.H., Alsadoon, A.: Using fitness dependent optimizer for training multi-layer perceptron. arXiv preprint arXiv:2201.00563 (2022)
2. Alečković, M., McAllister, S.S., Polyak, K.: Metastasis as a systemic disease: molecular insights and clinical implications. Biochimica et Biophysica Acta (BBA) Rev. Cancer **1872**(1),pp. 89–102 (2019)
3. Cheung, Y.C., Wan, Y.L., Lui, K.W., Lee, K.F.: Sonographically guided core-needle biopsy in the diagnosis of superficial lymphadenopathy. J. Clin. Ultrasound **28**(6), 283–289 (2000)
4. Chu, C., Zhmoginov, A., Sandler, M.: Cyclegan, a master of steganography. arXiv preprint arXiv:1712.02950 (2017)
5. Goodfellow, I., et al.: Generative adversarial networks. Commun. ACM **63**(11), 139–144 (2020)
6. Han, F., et al.: Efficacy of ultrasound-guided core needle biopsy in cervical lymphadenopathy: a retrospective study of 6,695 cases. Eur. Radiol. **28**, 1809–1817 (2018)
7. Lee, J.H., et al.: Deep learning-based computer-aided diagnosis system for localization and diagnosis of metastatic lymph nodes on ultrasound: a pilot study. Thyroid **28**(10), 1332–1338 (2018)
8. Li, H., Qiu, K., Chen, L., Mei, X., Hong, L., Tao, C.: Scattnet: semantic segmentation network with spatial and channel attention mechanism for high-resolution remote sensing images. IEEE Geosci. Remote Sens. Lett. **18**(5), 905–909 (2020)

9. Li, Y., Zhang, Q., Xiang, T., Lin, Y., Zhang, Q., Li, X.: Few-Shot lymph node metastasis classification meets high performance on whole slide images via the informative non-parametric classifier . In: proceedings of Medical Image Computing and Computer Assisted Intervention – MICCAI 2024. vol. LNCS 15012. Springer Nature Switzerland (2024)
10. Liu, Y., et al.: Ultrasound-based radiomics can classify the etiology of cervical lymphadenopathy: a multi-center retrospective study. Front. Oncol. **12**, 856605 (2022)
11. Ni, Z., Zheng, X., Zheng, X., Zou, X.: sclrtd: a novel low rank tensor decomposition method for imputing missing values in single-cell multi-omics sequencing data. IEEE/ACM Trans. Comput. Biol. Bioinf. **19**(2), 1144–1153 (2020)
12. Pynnonen, M.A., et al.: Clinical practice guideline: evaluation of the neck mass in adults. Otolaryngology–Head and Neck Surgery **157**(2_suppl), S1–S30 (2017)
13. Szegedy, C., Ioffe, S., Vanhoucke, V., Alemi, A.: Inception-v4, inception-resnet and the impact of residual connections on learning. In: Proceedings of the AAAI conference on artificial intelligence. vol. 31 (2017)
14. Van Buuren, S., Groothuis-Oudshoorn, K.: mice: multivariate imputation by chained equations in r. J. Stat. Softw. **45**, 1–67 (2011)
15. Wang, T., Chen, H., Chen, Z., Li, M., Lu, Y.: Prediction model of early recurrence of multimodal hepatocellular carcinoma with tensor fusion. Phys. Med. Biol. **69**(12), 125003 (2024)
16. You, J., et al.: Automated and reusable deep learning (autoRDL) framework for predicting response to neoadjuvant chemotherapy and axillary lymph node metastasis in breast cancer using ultrasound images: a retrospective, multicentre study. EClinicalMedicine **69** (2024)
17. Yuan, C., Wang, Y., Li, Y.: Tmdfm: a data fusion model for combined detection of tumor markers. In: 2015 IEEE International Conference on Bioinformatics and Biomedicine (BIBM), pp. 657–660. IEEE (2015)
18. Zhu, Y., et al.: Deep learning radiomics of dual-modality ultrasound images for hierarchical diagnosis of unexplained cervical lymphadenopathy. BMC Med. **20**(1), 269 (2022)
19. Zhu, Y., et al.: Deep learning radiomics of multimodal ultrasound for classifying metastatic cervical lymphadenopathy into primary cancer sites: a feasibility study. Ultraschall in der Medizin-European Journal of Ultrasound **45**(03), 305–315 (2024)

Non-invasive TB Detection Using Acoustic and Semantic Features from Cough Sounds

Yasmeena Akhter, Rishabh Ranjan, Bikash Dutta, Mayank Vatsa, and Richa Singh(✉)

Indian Institute of Technology Jodhpur, Jodhpur, India
{akhter.1,ranjan.4,d22cs051,mvatsa,richa}@iitj.ac.in

Abstract. We present a novel dual-stream deep learning architecture, *AcouSem-AFNet*, for automated tuberculosis (TB) detection using acoustic analysis of respiratory sounds. The proposed architecture utilizes two complementary pathways to extract distinct semantic and acoustic characteristics essential for identifying TB-related respiratory patterns. Specifically, the semantic stream employs a Whisper encoder to model structured patterns in respiratory events, while the acoustic stream leverages WavLM to capture detailed temporal dynamics characteristic of TB cough sounds. These distinct features are fused through a specialized backbone with squeeze-excitation mechanisms and residual connections, designed explicitly to maintain discriminative capabilities and mitigate overfitting challenges typical of limited medical datasets. Evaluated on the CODA-TB challenge dataset, our approach achieves state-of-the-art performance with an accuracy of 78.10% and an AUC of 0.79, demonstrating improvements of 3% in AUC and 2% in accuracy over leading baseline methods. Our framework enables rapid, non-invasive TB screening, particularly beneficial for resource-limited settings, demonstrating the feasibility of deep learning-based acoustic analysis as a scalable, preliminary diagnostic tool to enhance global TB screening accessibility. The code and models are publicly available at https://github.com/IAB-IITJ/AcouSem-AFNet

Keywords: Tuberculosis · CAD · Cough · Acoustic Features

1 Introduction

Tuberculosis (TB), caused by Mycobacterium tuberculosis, remains a critical global health challenge, with 9.9 million cases and 1.3 million deaths reported in 2020 [1]. TB, one of humanity's oldest infectious diseases [2], has claimed an estimated one billion lives throughout history [3]. Currently ranking as the leading cause of death among infectious diseases globally, TB continues to pose a significant public health challenge with approximately 10 million new active cases reported annually [4]. Despite its severity, approximately 40% of TB cases go undiagnosed due to limited healthcare access and diagnostic barriers [5]. This

J. C. Gee et al. (Eds.): MICCAI 2025, LNCS 15960, pp. 460–470, 2026.
https://doi.org/10.1007/978-3-032-04927-8_44

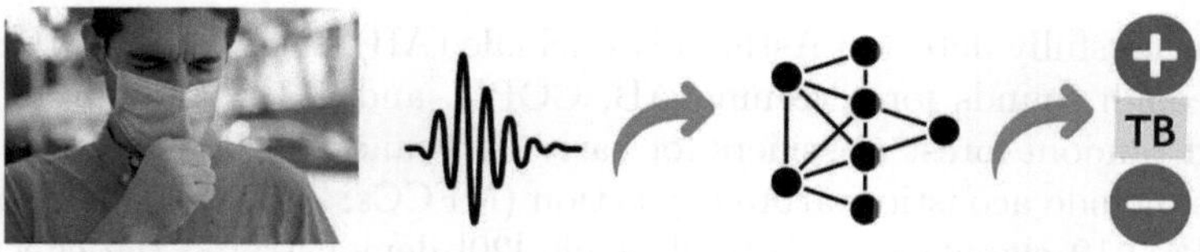

Fig. 1. Highlights the visual abstract, addressing the AI-based automatic TB detection from the cough sounds.

significant gap in detection highlights the urgent need for accessible, non-invasive screening methods.

Traditional TB screening through self-reported symptoms presents critical limitations: low specificity leading to over-testing, and stigma-induced under-reporting resulting in inadequate care [6,7]. This necessitates the development of objective, low-cost point-of-care screening methods. Acoustic-based TB detection emerges as a promising specimen-free, automated screening solution that could optimize resource allocation for molecular testing [8].

Cough, being a primary symptom of TB, presents a promising acoustic biomarker for automated disease detection [9]. The scientific basis for acoustic TB detection lies in the pathophysiology of respiratory diseases. Different pathological conditions alter airway dynamics and glottal behaviour, producing distinctive acoustic signatures during cough episodes [10]. Previous research has demonstrated the feasibility of distinguishing various respiratory conditions (asthma, bronchitis, pertussis) through cough acoustics, suggesting the viability of automated TB detection through acoustic pattern analysis [11,12], as also demonstrated in Fig. 1 as the visual abstract for this problem statement.

Recent advances in deep learning and acoustic signal processing have enabled the development of sophisticated audio analysis systems. AI models offer a transformative approach to respiratory illness diagnosis by providing non-invasive, cost-effective, and real-time data interpretation capabilities, addressing the limitations of traditional diagnostic methods and presenting particular value for resource-constrained settings. While preliminary studies [13,14]have demonstrated the potential of cough-based TB screening, these efforts were limited by small sample sizes and restricted settings. Our work addresses these limitations by introducing a novel deep-learning framework, *AcouSem-AFNet*, that leverages raw audio-based features for robust TB classification from cough sounds. This research advances the field in three key aspects: (1) we propose a new architecture optimized for acoustic TB biomarker detection, (2) we demonstrate state-of-the-art performance on the comprehensive CODA-TB dataset, and (3) we provide an extensive comparison with different models.

2 Related Work

Respiratory disease detection through cough analysis has shown promising results across multiple conditions. Using cough frequency analysis, Marsden et al. [15] successfully detected Asthma Bronchiale (AB). Infante et al. [16] demonstrated cough sounds for screening AB, COPD, and TB. Windmon et al. [17] employed random forest classifiers for early CHF and COPD detection. Recent advances include acoustic feature extraction (MFCCs, ZCR) for pneumonia [18] and COVID-19 classification [14,19], while [20] demonstrated the effectiveness of generative adversarial networks for respiratory disease classification.

In the existing literature on TB detection through cough sounds, Tracey et al. [21] and Larson et al. [22] linked cough frequency to TB recovery. Pahar et al. [14] achieved high specificity (95%) and sensitivity (93%) using logistic regression on a small dataset (16 TB patients, 35 healthy controls). In follow-up work, Pahar et al. [19] tested multiple deep learning models on 47 TB patients, with ResNet50 [23] achieving 92.59% accuracy in TB vs. COVID-19 classification and 86.31% in three-class discrimination. Frost et al. [24] implemented Bi-LSTM on 74 subjects (28 TB patients, 46 controls), achieving 75% specificity and 89% sensitivity. These studies demonstrate that cough acoustics contain significant discriminative information for the classification of respiratory diseases, suggesting a potential for developing automated screening tools.

3 Methodology

The proposed methodology, *AcouSem-AFNet*, for TB detection from respiratory audio recordings employs a dual-stream architecture integrated with a RawNet3-based backbone [25] network specifically optimized for medical audio analysis. The system processes raw audio inputs through parallel semantic and acoustic pathways to extract complementary features crucial for identifying TB-related respiratory patterns, including distinctive cough characteristics, breathing anomalies, and associated acoustic signatures. The semantic stream, powered by the Whisper-large encoder, captures structured patterns in respiratory events, while the acoustic stream, utilizing WavLM [26], preserves fine-grained temporal characteristics specific to TB-related sounds. These complementary features are integrated through a specialized backbone network incorporating squeeze-excitation mechanisms and residual connections, designed to maintain discriminative power while preventing overfitting on limited medical datasets. The complete processing pipeline consists of two main components: (1) a front-end feature extraction module utilizing dual-stream processing for comprehensive respiratory sound analysis and (2) a backend network featuring AFMS-Res2MP blocks with targeted regularization for robust TB detection.

3.1 Audio Encoders and Modality Adapters

We propose a dual-stream design that processes semantic and acoustic cues in parallel to capture acoustic features for TB detection from audio cough recordings. Several works [26,27] demonstrate that Semantic Features $\mathcal{F}_{\text{semantic}}$ and Acoustic Features $\mathcal{F}_{\text{acoustic}}$ are critical for speech processing tasks. This design choice helps focus on acoustic information, including $\mathcal{P}$ (pitch), $\mathcal{T}$ (timbre) and content information extracted from semantic tokens $\mathbf{S} = \{s_1, s_2, ..., s_n\}$.

For semantic information processing, we leverage the Whisper-large encoder [28] $\Phi_{\text{whisper}} : \mathcal{X} \rightarrow \mathcal{H}_s$, pre-trained through weak supervision, enabling robust extraction of linguistic features with zero-shot capabilities. The acoustic stream utilizes WavLM [29] $\Phi_{\text{wavlm}} : \mathcal{X} \rightarrow \mathcal{H}_a$, which specializes in capturing speaker-specific characteristics and timbral qualities. This complementary combination allows our model to analyze content-level semantic features and fine-grained acoustic properties crucial for TB detection.

The processing pipeline operates on batched input signals $\mathbf{X} \in \mathbb{R}^{B \times T}$, where B represents the batch size and T the temporal dimension. For semantic analysis, the input is transformed into log-mel spectrograms $\mathbf{S} \in \mathbb{R}^{B \times F \times T}$, with F frequency bins. For acoustic analysis, we use raw audio waveforms to pass in the WavLM encoder. These representations are processed through their respective encoders to produce feature spaces:

$$\mathbf{H}_s = \Phi_{\text{whisper}}(\mathbf{S}) \in \mathbb{R}^{B \times T_s \times D_s}$$

$$\mathbf{H}_a = \Phi_{\text{wavlm}}(\mathbf{X}) \in \mathbb{R}^{B \times T_a \times D_a}$$

where T_s, T_a represent temporal dimensions and D_s, D_a denote feature dimensions for semantic and acoustic features respectively. This parallel processing enables our model to capture both semantic content and acoustic characteristics simultaneously. We pass the outputs $\mathbf{H}_s$ and $\mathbf{H}_a$ into semantic and acoustic adapters, respectively:

$$\mathbf{A}_s = f_s(\mathbf{H}_s) = \text{Adapter}_s(\mathbf{H}_s)$$

$$\mathbf{A}_a = f_a(\mathbf{H}_a) = \text{Adapter}_a(\mathbf{H}_a)$$

The semantic and acoustic adapters each comprise two sequential 1-D convolutional layers designed for temporal downsampling and alignment, followed by a down-up bottleneck adapter. Each adapter concludes with a linear projection layer $W_s \in \mathbb{R}^{D_s \times 1024}$ and $W_a \in \mathbb{R}^{D_a \times 1024}$ that maps the features to a shared dimensional space, enabling effective integration of both information streams. These adapters align both feature streams to a common representation space $\mathbb{R}^{B \times m \times 1024}$. We concatenate the outputs of both adapters to get:

$$\mathbf{x}_m = [\mathbf{A}_s; \mathbf{A}_a] \in \mathbb{R}^{B \times m \times 2048}$$

before passing them to the backbone for final classification.

3.2 Backend Network Architecture

The backend network of our TB detection system is built upon the RawNet3 architecture [25]. This architecture was selected for its proven effectiveness in capturing nuanced acoustic characteristics while maintaining computational efficiency crucial for medical applications with limited dataset availability.

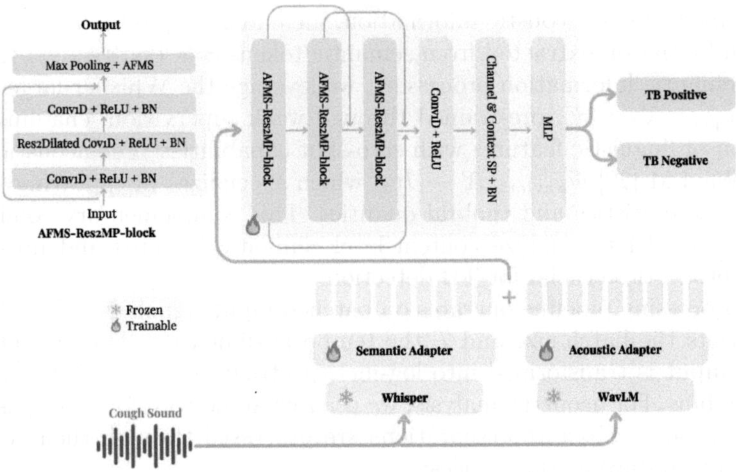

Fig. 2. Illustrates the architecture of the *AcouSem-AFNet* model. The proposed framework uses Whisper and WavLM encoders for semantic and acoustic processing, combined through adapters and AFMS blocks for final classification.

At the core of our backend network are three AFMS-Res2MP blocks [25] arranged sequentially. Each AFMS-Res2MP block, as illustrated in Fig. 2, consists of a sophisticated arrangement of convolutional layers with residual connections and feature map scaling. The input to each block first passes through a Conv1D layer, followed by ReLU activation and batch normalization (BN). This is succeeded by a Res2Dilated Conv1D layer with ReLU and BN, which employs dilated convolutions to expand the receptive field without increasing the parameter count essential for capturing the broad temporal patterns in respiratory sounds indicative of TB. Another Conv1D layer with ReLU and BN follows, with its output added to the block input via a residual connection. The final component is a max pooling layer with an α Filter-wise feature Map Scaling (AFMS) module.

The AFMS module builds upon the filter-wise Feature Map Scaling (FMS) technique, which independently scales each filter of a feature map to derive more discriminative representations of respiratory characteristics. Let $c = [c_1, c_2, \cdots, c_F]$ represent a feature map output from a residual block, where $c_f \in \mathbb{R}^T$, T is the sequence length in time, and F denotes the number of filters. The AFMS module first performs global average pooling on the time axis, followed by feed-forwarding through a fully-connected layer with sigmoid activation to derive a scale vector $s = [s_1, s_2, \cdots, s_F]$, where $s_f \in \mathbb{R}^1$.

In our implementation for TB detection, we utilize both multiplicative and additive scaling methods to enhance the discriminative power of respiratory sound features. The multiplicative method applies the scale vector as $c'_f = c_f \cdot s_f$, while the additive method implements $c'_f = c_f + s_f$. For optimal performance in capturing the subtle acoustic signatures of TB, we employ a sequential application of both methods, expressed as:

$$c'_f = (c_f + s_f) \cdot s_f \quad (1)$$

This approach provides several advantages for TB detection from respiratory sounds. The multiplicative scaling functions similarly to an attention mechanism in the filter domain, allowing the network to emphasize specific frequency components particularly relevant to TB-related acoustic patterns. Unlike conventional softmax-based attention, our sigmoid-based scaling prevents excessive information removal, preserving the complementary features captured by different filters indicating TB presence.

The additive scaling introduces data-driven perturbation to the feature maps, potentially increasing their discriminative power for distinguishing TB-related sounds from normal respiratory patterns or other conditions. This concept is particularly relevant for TB detection, where subtle variations in acoustic characteristics can significantly impact diagnostic accuracy. In the first two AFMS-Res2MP blocks, we incorporate max-pooling operations that serve as effective regularization techniques, crucial for preventing overfitting on our limited TB audio dataset. This design enhances the model's generalization capabilities across diverse patient populations and variable recording conditions while maintaining sensitivity to TB-specific respiratory signatures.

The three AFMS-Res2MP blocks process the concatenated features from the front-end dual streams, progressively refining the representation to capture the complex acoustic patterns associated with TB. A fully connected layer follows these blocks, mapping the extracted features to the final binary classification output indicating the presence or absence of TB. This carefully designed backend architecture enables robust TB detection from respiratory audio recordings while maintaining exceptional performance despite limited training data availability. The complete architecture of the *AcouSem-AFNet* model is shown in Fig. 2.

4 Experimental Setup

Dataset: To assess the performance of the *AcouSem-AFNet* model for automated diagnosis of TB from cough sounds, we used the Cough Diagnostic Algorithm for Tuberculosis (CODA TB) challenge dataset] [30].The CODA-TB challenge, organized by Sage Bionetworks, focused on developing algorithms to detect pulmonary tuberculosis (TB) through acoustic analysis of cough recordings collected over a two-week period. This initiative aimed to advance automated TB screening by leveraging machine learning approaches to identify distinctive acoustic signatures in TB-positive cough episodes. The subject age is above 18 or older, with a cough sound of 0.5 s, collected from health centers of 7 countries (India, Philippines, South Africa, Uganda, Vietnam, Tanzania, Madagascar). The task associated is a TB-positive or TB-negative cough. The CODA-TB dataset contains 9,772 cough recordings from 1,105 patients, including detailed demographic, clinical, and microbiological diagnostic metadata. Initially developed for the CODA-TB DREAM Challenge competition, the training dataset is now publicly accessible[1]. Researchers can utilize this data to develop acoustic-based TB screening models, with model evaluation performed on a separate test set. To build our proposed approach, we performed the 70:10:20 train-val-test split on the publicly available training set.

Baselines: We consider eight different existing architectures to compare the proposed *AcouSem-AFNet* model. We consider both variations of models that take raw audio as input and spectrogram-like features. We use models such as RawNet3 that take input as raw audio signals, whereas models such as MesoNet [31] and SpecRNet [32] take spectrogram features as input. We use the variant proposed in the paper [33] for the implementation of SpecRNet and MesoNet. For the spectrogram-based models, we use LFCC and the output of the Whisper ASR encoder.

Implementation Details: PyTorch is used to implement the proposed model. WavLM and Whisper are used from the Hugging Face repository. We use a variant of Whisper and WavLM. The proposed model is trained for 50 epochs, with AdamW as the optimizer and a batch size of 32. The learning rate and weight decay are set to 1e-3 and 5e-4, respectively. The training, validation, and evaluation data consist of the 6841, 977, and 1954 samples, respectively.

Evaluation Metrics: We evaluate our proposed *AcouSem-AFNet* approach using the AUROC and overall accuracy as the evaluation metrics.

5 Results

To assess the classification capabilities for TB diagnosis from cough acoustics, we compared the proposed *AcouSem-AFNet* model with the baseline models. The quantitative results are tabulated in Fig. 3 (Left), representing the AUC score and overall accuracy for baselines and the proposed model. From the results, the proposed approach outperforms the baseline models with an overall improvement from (3–18)% and (2–7)% in terms of the AUC and accuracy scores, respectively. To demonstrate performance, we provide the Receiver Operating Characteristics (ROC) Curve plot, shown in Fig. 3.

Our evaluation demonstrates that the dual-stream architecture of *AcouSem-AFNet* outperforms all baseline models in tuberculosis detection from respiratory audio recordings. As shown in Fig. 3 (Left), the proposed model achieves superior performance with 78.10% accuracy and a mean AUC of 0.79, representing significant improvements over other baseline architectures.

The performance gap between our *AcouSem-AFNet* approach and the nearest competitor (SpecRNet with 76.61% accuracy and 0.76 AUC) highlights the effectiveness of our architectural innovations. While SpecRNet demonstrates strong performance as a single-stream approach, it fails to capture the full spectrum

[1] https://www.synapse.org/Synapse:syn31472953/wiki/619711.

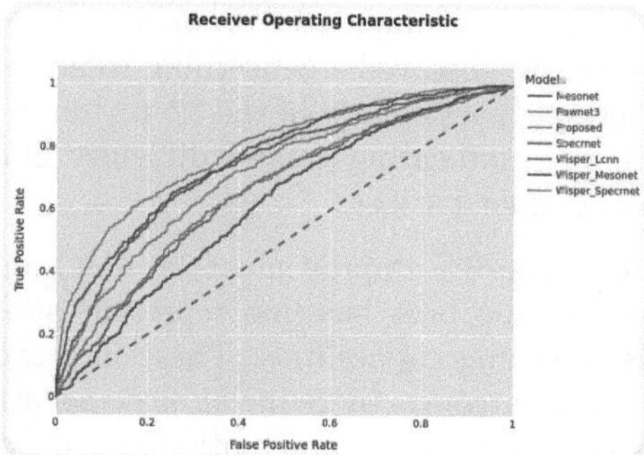

Fig. 3. TB detection model evaluation. **Left**: Table showcases the performance metrics across architectures with AcouSem-AFNet (bold) achieving the best accuracy (78.10%), and AUC (0.79), second best underlined. **Right**: Showcases the ROC curves, comparing the proposed model's (green) with existing. The results show that the proposed algorithm provides better sensitivity-specificity balance compared to the baseline algorithms.

of TB-related acoustic signatures. Similarly, RawNet3 performs reasonably well (73.69% accuracy, 0.71 AUC) but lacks the specialized dual pathway design that enables comprehensive feature extraction for semantics and acoustic characteristics. The Whisper variants (SpecRNet and MesoNet) consistently underperform with accuracies around 71.6%, and AUC scores between 0.61–0.66, indicating their limitations in capturing the subtle acoustic patterns associated with TB respiratory sounds. This underscores the insufficiency of relying solely on general audio representation models for this specialized medical diagnostic task. The superior performance of *AcouSem-AFNet* model can be attributed to several key innovations: (1) The dual-stream architecture effectively leverages complementary information pathways, with the semantic stream (Whisper-large encoder) capturing structured respiratory patterns while the acoustic stream (WavLM) preserves critical temporal characteristics specific to TB-related sounds, (2) the specialized integration backbone with squeeze-excitation mechanisms and residual connections successfully fuses these complementary features while maintaining discriminative power, even with limited medical training data, and (3) the AFMS-Res2MP blocks with targeted regularization in the backend network effectively prevent overfitting while preserving the model's ability to detect subtle TB acoustic signatures.

Our AcouSem-AFNet model demonstrates balanced gender performance with an overall accuracy of 78.10%, achieving 76.01% for male subjects and 80.14% for female subjects. This relatively small performance gap suggests that the model effectively captures TB-related acoustic signatures which are independent of gender-based physiological differences in cough characteristics. Regarding age-related performance, we created four age bins to assess the generalizability of the model: 18–34, 35–51, 52–68, and 69–85. Performance analysis reveals interesting variations, with middle-aged groups (35–51 and 52–68) showing higher accuracy

(80.4% and 80.7%, respectively) compared to younger (75.6%) and elderly populations (69.7%). However, we observed a concerning trend in precision and recall metrics for older age groups, particularly in the 52–68 bin, where precision drops to 40.6% and recall to 40%, indicating greater difficulty in correctly identifying TB-positive cases among older patients.

Clinical Significance: The 3% improvement in AUC and 2% gain in overall accuracy compared to the best baseline represent substantial progress in respiratory-based TB screening capabilities. These improvements could translate to meaningful clinical outcomes by reducing false negatives in TB screening protocols, potentially enabling earlier intervention and treatment.

Our experimental results conclusively demonstrate that the proposed *AcouSem-AFNet*, a dual-stream architecture with specialized feature integration, outperforms existing approaches for TB detection from respiratory audio recordings. The performance gains validate our hypothesis that capturing both semantic and acoustic features in cough sounds is essential for accurate TB diagnosis, opening promising avenues for non-invasive, cost-effective screening tools in resource-limited settings.

6 Conclusion

In this paper, we presented a novel deep-learning framework for automated TB detection through cough sound analysis. Our approach demonstrated superior performance on the CODA-TB challenge dataset, achieving state-of-the-art results in classification accuracy and robustness. The key innovations include our spectral representation technique and architecture optimization for acoustic biomarker detection. Experimental results validate our method's effectiveness as a rapid, non-invasive TB screening tool, particularly valuable for resource-limited settings. While our work shows promising results, future research directions include: (1) investigating multi-modal integration with clinical metadata, (2) improving model interpretability for clinical adoption, and (3) validating performance across diverse demographic populations. Our contribution establishes a strong foundation for acoustic-based TB screening, potentially addressing critical gaps in global TB detection and management.

Acknowledgment. This work is supported by Srijan: Center of Excellence on GenAI at IIT Jodhpur, India, IndiaAI Mission, and Meta.

Disclosure of Interests. The authors declare no competing interests.

References

1. WHO. Accessed 23 Feb 2025
2. Comas, I., et al.: Out-of-Africa Migration and Neolithic Coexpansion of Mycobacterium Tuberculosis with Modern Humans. Nat. Genet. **45**(10), 1176–1182 (2013)

3. T. Paulson.: "Epidemiology: a mortal foe.," Nature, vol. 502, no. 7470, 2013
4. Organization, W. H.: Global tuberculosis report 2018. World Health Organization, 2018
5. Sulis, G., Pai, M.: Missing tuberculosis patients in the private sector: business as usual will not deliver results. Public Health Action **7**(2), 80–81 (2017)
6. Nathavitharana, R.R., et al.: Guidance for studies evaluating the accuracy of tuberculosis triage tests. J. Infect. Dis. **220**, S116–S125 (2019)
7. Akhter, Y., Singh, R., Vatsa, M.: AI-based radiodiagnosis using chest X-rays: a review. Frontiers in Big Data **6**, 1120989 (2023)
8. Naidoo, P., et al.: "The south african tuberculosis care cascade: estimated losses and methodological challenges," J. Infect. Dis. vol. 216, no. suppl_7, pp. S702–S713, 2017
9. Simonsson, B., et al.: Role of autonomic nervous system and the cough reflex in the increased responsiveness of airways in patients with obstructive airway disease. J. Clin. Investig. **46**(11), 1812–1818 (1967)
10. Chung, K.F., Pavord, I.D.: Prevalence, pathogenesis, and causes of chronic cough. The Lancet **371**(9621), 1364–1374 (2008)
11. Korpáš, J., Sadloňová, J., Vrabec, M.: Analysis of the cough sound: an overview. Pulm. Pharmacol. **9**(5–6), 261–268 (1996)
12. Sharan, R.V., et al.: Predicting spirometry readings using cough sound features and regression. Physiol. Meas. **39**(9), 095001 (2018)
13. Pathri, R., et al.: "Acoustic epidemiology of pulmonary tuberculosis (TB) and COVID-19 leveraging AI/ML,"medRxiv, 2022
14. Pahar, M., et al.: Automatic Cough Classification for Tuberculosis Screening in a Real-world Environment. Physiol. Meas. **42**, 105014 (2021)
15. Marsden, P.A., et al.: Objective cough frequency, airway inflammation, and disease control in asthma. Chest **149**(6), 1460–1466 (2016)
16. Infante, C., et al.: "Use of cough sounds for diagnosis and screening of pulmonary disease," In: 2017 IEEE Global Humanitarian Technology Conference (GHTC), pp. 1–10, 2017
17. Windmon, A., et al.: Tussiswatch: a smart-phone system to identify cough episodes as early symptoms of chronic obstructive pulmonary disease and congestive heart failure. IEEE JBHI **23**(4), 1566–1573 (2018)
18. Sotoudeh, H., Tabatabaei, M., et al.: Artificial intelligence empowers radiologists to differentiate pneumonia induced by COVID-19 versus influenza viruses. Acta Inf. Medica **28**(3), 190 (2020)
19. Pahar, M., et al.: "Automatic tuberculosis and COVID-19 cough classification using deep learning," In: 2022 International Conference on Electrical, Computer and Energy Technologies, pp. 1–9, 2022
20. Ramesh, V., et al.: "CoughGAN: generating synthetic coughs that improve respiratory disease classification," In: IEEE EMBC, pp. 5682–5688, 2020
21. Tracey, B.H., et al.: "Cough detection algorithm for monitoring patient recovery from pulmonary tuberculosis," In: IEEE engineering in medicine and biology society, pp. 6017–6020, 2011
22. Larson, S., et al.: "Validation of an automated cough detection algorithm for tracking recovery of pulmonary tuberculosis patients," PLOS ONE, vol. 7, pp. 1–10, 10 2012
23. He, K., Zhang, X., Ren, S., Sun, J.: "Deep residual learning for image recognition," In: "Deep residual learning for image recognition CVPR," pp. 770–778, 2016
24. Frost, G., Theron, G., Niesler, T.: "TB or not TB? acoustic cough analysis for tuberculosis classification," ArXiv:2209.00934, 2022

25. weon Jung, J., et al.: "Pushing the limits of raw waveform speaker recognition," In: Proc. Interspeech, pp. 2228–2232, 2022
26. Hu, S., et al.: "WavLLM: towards robust and adaptive speech large language model," In: Findings of the Association for Computational Linguistics: EMNLP, pp. 4552–4572, 2024
27. Ye, Z., et al.: "Codec does matter: exploring the semantic shortcoming of codec for audio language model," ArXiv:2408.17175, 2024
28. Radford, A., et al.: "Robust speech recognition via large-scale weak supervision," CoRR, vol. abs/2212.04356, 2022
29. Chen, S., et al.: Wavlm: Large-scale Self-supervised Pre-training for full stack speech processing. IEEE J. Sel. Top. Signal Proc. **16**, 1505–1518 (2021)
30. Huddart, S., et al.: A dataset of solicited cough sound for tuberculosis triage testing. Scientific Data **11**(1), 1149 (2024)
31. Afchar, D., et al.: "MesoNet: a compact facial video forgery detection network," In: 2018 IEEE International Workshop on Information Forensics and Security (WIFS), pp. 1–7, 2018
32. Kawa, P., Plata, M., Syga, P.: "SpecRNet: towards faster and more accessible audio deepfake detection," In: 2022 IEEE International Conference on Trust, Security and Privacy in Computing and Communications (TrustCom), pp. 792–799, 2022
33. Kawa, P., et al.: Improved deepfake detection using whisper features. Proc. INTERSPEECH **2023**, 4009–4013 (2023)

Omni-Fusion of Spatial and Spectral for Hyperspectral Image Segmentation

Qing Zhang, Guoquan Pei, and Yan Wang(✉)

Shanghai Key Laboratory of Multidimensional Information Processing, East China Normal University, Shanghai 200241, China

ywang@cee.ecnu.edu.cn

Abstract. Medical Hyperspectral Imaging (MHSI) has emerged as a promising tool for enhanced disease diagnosis, particularly in computational pathology, offering rich spectral information that aids in identifying subtle biochemical properties of tissues. Despite these advantages, effectively fusing both spatial-dimensional and spectral-dimensional information from MHSIs remains challenging due to its high dimensionality and spectral redundancy inherent characteristics. To solve the above challenges, we propose a novel spatial-spectral omni-fusion network for hyperspectral image segmentation, named as Omni-Fuse. Here, we introduce abundant cross-dimensional feature fusion operations, including (1) a cross-dimensional enhancement module that refines both spatial and spectral features through bidirectional attention mechanisms; (2) a spectral-guided spatial query selection to select the most spectral-related spatial feature as the query; and (3) a two-stage cross-dimensional decoder which dynamically guide the model's attention towards the selected spatial query. Despite of numerous attention blocks, Omni-Fuse remains efficient in execution. Experiments on two microscopic hyperspectral image datasets show that our approach can significantly improve the segmentation performance compared with the state-of-the-art methods, with over 5.73% improvement in DSC. Code available at: https://github.com/DeepMed-Lab-ECNU/Omni-Fuse.

1 Introduction

Hyperspectral Imaging (HSI) is an emerging technology in pathological diagnosis, reflecting the physiological and biochemical properties of different tissue components of tissues in a non-invasice manner [1,2]. As shown in Fig. 1(a), each single-band image in a microscopic hyperspectral image (MHSI) represents the reflectance spectrum information within a specific wavelength range, with different wavelengths highlighting various pathological regions. Therefore, each pixel has a corresponding spectral curve, ensuring that the granularity of both spatial and spectral features is aligned [11,23]. Mining the correlations and complementary information between spectral and spatial data enhance the model's ability to understand and analyze complex pathological scenes.

Q. Zhang and G. Pei—Denotes equal contribution.

J. C. Gee et al. (Eds.): MICCAI 2025, LNCS 15960, pp. 471–481, 2026.
https://doi.org/10.1007/978-3-032-04927-8_45

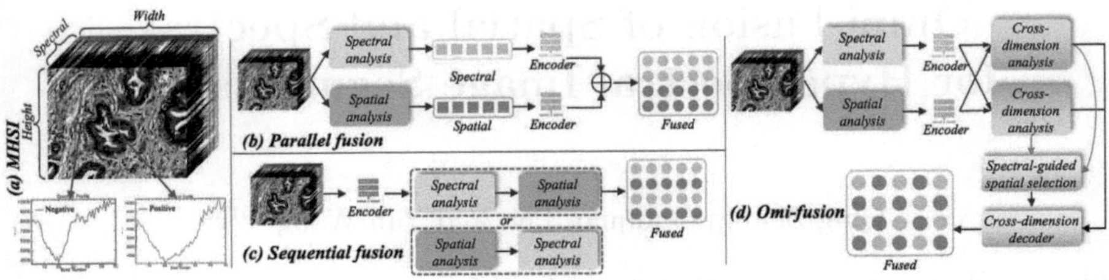

Fig. 1. (a) Microscopic hyperspectral image (MHSI) of Multi-Dimensional Choledoch Dataset (MDC) dataset. Positive and Negative regions have different spectral curves and single band images at various wavelengths can highlight different texture information. (b)–(d) Structures for spatial-spectral feature fusion.

MHSIs contain asymmetric information in spectral and spatial dimensions. The key challenge for MHSI segmentation is how to effectively align this cross-dimensional information. Recent attempts either learn spectral and spatial features in a sequential or a parallel manner [6], and try to fuse features in the encoder. Since the former explicitly processes one dimension at each stage [19–21], it is hard to enable the two dimensional features to simultaneously interact with each other. The latter always designs a dual stream to extract spectral and spatial features, respectively, and then followed by post-fusion, either in the late stage of the encoder or in the decoder [16,22]. Spectral and spatial features do not fully guide each other in this way, leading to the loss of complementary information. Though FL [6] integrates spatial and spectral features for a spatiospectral balance, they simply apply deformable cross-attention twice between the encoder and the decoder. We summarize the network architecture design in Fig. 1(b) and (c). It shows that none of the above networks provides abundant multi-dimensional feature fusion, thereby leading to sub-optimal results.

In this paper, we are the first to achieve abundant multi-dimensional feature fusion for MHSI segmentation. Similar with parallel fusion, we separately extract primary spectral and spatial features by a dual stream. Afterwards, we design our spatial-spectral omni-fusion, which makes fusion omnipresent. Our fusion lies in the **feature enhancer** ("Cross-dimensional analysis" in Fig. 1), the **query selection** ("Spectral-guided spatial selection" in Fig. 1) and the **decoder** ("Cross-dimension decoder" in Fig. 1). For feature enhancer, the primary features mutually enhance each other through two designed bidirectional cross-dimensional analysis blocks, extracting the most important and relevant spatial-spectral feature (see "Cross-Dimensional Feature Enhancement" in Fig. 2). The enhanced spatial features are further compacted according to their correlation with the enhanced spectral feature (see "Spectral-Guided Spatial Query Selection" in Fig. 2). For decoder, we design a coarse-to-fine learning strategy to

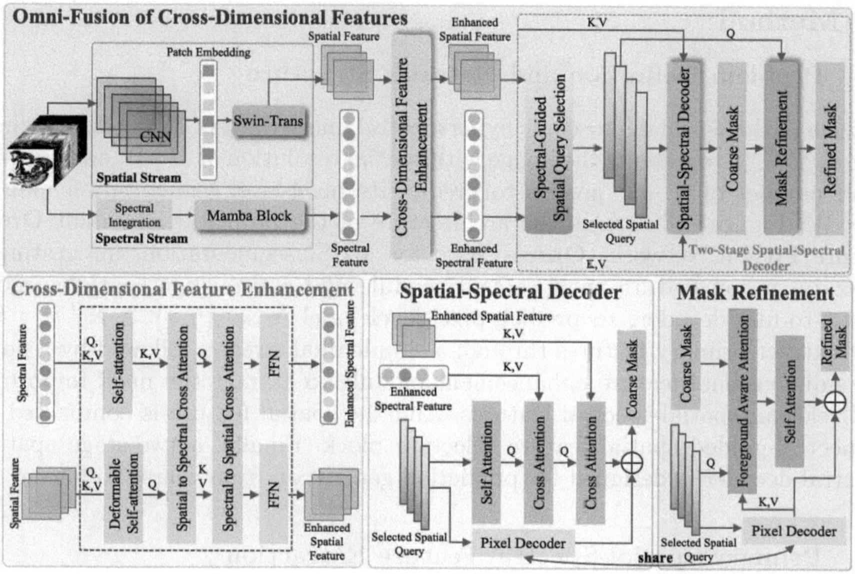

Fig. 2. Framework architecture of Omni-Fuse. The MHSI is primarily processed by a dual steam, followed by a cross-dimensional attention based feature enhancer. And the enhanced spatial features are further compacted by a spectral-guided spatial selection. Subsequently, the compacted features are decoded cooperate with enhanced spatial and spectral features to generate a coarse mask. Finally, the coarse mask is further refined by mask attention mechanism.

extract fine-grained localized features. Specifically, we propose a spectral-spatial decoder for coarse segmentation, using enhanced spectral and spatial features to dynamically guide the model's attention towards the selected spatial query (see "Spatial-Spectral Decoder" in Fig. 2). Then, we refine the coarse mask by incorporating with the selected spatial query (see "Mask Refinement" in Fig. 2). Through iterative interactions, spatial and spectral features are deeply and fully fused in this paper. Our contributions are as follows:

- We propose an extensive Omni feature fusion network (Omni-Fuse) for microscopic hyperspectral image segmentation, introducing a cross-dimensional enhancement to bidirectionally integrate spatial-spectral features.
- We design a two-stage spatial-spectral decoder to coarsely-then-finely segment tumor regions with the assistance of SAM decoder.
- Comprehensive experiments are conducted on a public MDC Dataset and a private GPCC Dataset, proving the effectiveness and efficiency of the proposed multi-modal feature fusion based segmentation method.

2 Method

2.1 Problem Definition and Network Structure

Given a pathological microscopic hyperspectral image dataset $\mathcal{D} = \{\mathbf{X}, \mathbf{Y}\}$ where $\mathbf{X} \in \mathbb{R}^{H\times W\times S}$ represents the image with spatial resolution of $H\times W$ and spectral band number of S, our goal is to predict its pixel-level segmentation map of $H \times W$. Figure 2 illustrates the architecture of the proposed abundant **Omni** feature **Fus**ion Network (**Omni-Fuse**) for MHSI segmentation, integrating a cross-dimension feature enhancement module and a two-stage spatial-spectral coarse-to-fine decoder, to predict pixel-level label $\hat{\mathbf{Y}} = \{0, 1\} \in \mathbb{R}^{H\times W}$. We first extract primary features through a simple dual stream, followed by a novel cross-dimensional feature enhancement module to extract the most important and relevant spatial-spectral features. And the spatial feature is compacted by a spectral-guided spatial feature selection block. Finally, a two-stage spatial-spectral decoder is designed for prediction generation from coarse to fine.

2.2 Primary Spatial-Spectral Feature Extraction

Due to the fact that microscopic hyperspectral image is a three-dimensional data cube, its data size is significantly larger compared to traditional RGB or grayscale image. In hyperspectral pathology analysis tasks, how to efficiently extract and fuse these multi-dimensional features is a key challenge. To address this issue, we adopt a dual-branch architecture to separately extract spatial information and spectral information. For spatial feature extraction, we apply a Convolutional Neural Network (CNN) to capture local spatial feature $F_{\text{CNN}} \in \mathbb{R}^{H'\times W'\times C}$, where C is the channel number, followed by Swin-Transformer encoder to capture long-range dependencies in the spatial domain, refining the spatial feature to $F_{\text{swin}} \in \mathbb{R}^{H'\times W'\times C}$. For Spectral feature extraction, inspired by MDN [12], we employ a bi-directional mamba based block specifically designed for capturing long-distance dependencies across the spectral domain, exporting $X_{\text{spec}} \in \mathbb{R}^{B\times H\times W\times S\times L_{\text{spec}}}$, where B is the batch size and L_{spec} is the dimension of the spectral features. For the following feature fusion of spatial and spectral information, the spectral feature is flattened to continuous spectral tokens $T_{prispec} = \{[T_0, T_1, ..., T_s] \mid T_j \in \mathbb{R}^{L_{spec}}\}$.

2.3 Cross-Dimensional Feature Enhancement

Primary spatial and spectral features are extracted separately using simple CNN or Mamba-based algorithms, the three-dimensional information of the hyperspectral image remains inadequately analysed. Therefore, we introduce a cross-dimensional enhancement module based on bidirectional attention mechanisms to integrate the aforementioned features. Each pixel in the spatial dimension has its own spectral curve, consistent with the curve of the same category area and differing from the curve of areas belonging to different categories. Since then, the granularity of spatial and spectral features is considered to be aligned.

To fully utilize the complementary information between spectral features and the structural information of spatial features, we propose a cross-dimensional feature enhancement module. For spatial feature, considering that tissues and lesions in histopathological images often exhibit irregular and deformed shapes, deformable self-attention module is selected in this module to flexibly capture local details while efficiently modeling long-range morphological correlations, exporting updated spatial feature $F'_{spa} = \texttt{DeformableSelfAttn}(F_{swin})$. We get updated spectral feature F'_{spec} through regular self-attention mechanism to capture dependencies within the spectral domain. Then, two-layer bidirectional cross-attention module is designed here to deeply achieve cross-dimensional communication:

$$F''_{spa} = \texttt{CrossAttn}(F'_{spa}, F'_{spec}) = \texttt{softmax}(\frac{F'_{spa}{F'_{spec}}^{\top}}{\sqrt{d_{spec}}})F'_{spec}, \tag{1}$$

$$F''_{spec} = \texttt{CrossAttn}(F'_{spec}, F'_{spa}) = \texttt{softmax}(\frac{F'_{spec}{F'_{spa}}^{\top}}{\sqrt{d_{spa}}})F'_{spa}, \tag{2}$$

where $\{F''_{spa}, F''_{spec}\}$ are the enhanced features and $\texttt{CrossAttn}(\cdot)$ represents the vanilla cross attention operation. $\{d_{spa}, d_{spec}\}$ are the dimension number of $\{F'_{spa}, F'_{spec}\}$ respectively. The extensively fused features greatly enhance the compatibility of feature expression in hyperspectral data. This enhancer not only makes the learned visual features spectral-aware and vice versa, but also provides spectral prompts for downstream segmentation tasks.

Usually, positive regions occupy only a small portion in pathological images, indicating that redundant information exists in the spatial features. Therefore, we introduce a spatial feature selection module, exploring features that are strongly correlated with spectral information, for the following two-stage decoder. For feature selection, we convert the enhanced features $\{F''_{spa}, F''_{spec}\}$ into tokens through patch embedding, *i.e.*, spatial token $T_{spa} \in \mathbb{R}^{N_{spa} \times d}$ and spectral token $T_{spec} \in \mathbb{R}^{N_{spec} \times d}$. Here, N_{spa} is the number of spatial tokens, N_{spec} is the number of spectral tokens, and d is the number of feature dimension. We select the most relevant N_q spatial tokens T'_{spa} by calculating the correlation between spatial and spectral tokens:

$$T'_{spa} = \texttt{Top}_{N_q}(\max^{(-1)}(T_{spa}T_{spec}^{\top})), \tag{3}$$

where $\texttt{Top}_{N_q}(\cdot)$ is the operation to select the most relevant N_q spatial tokens. $\max^{(-1)}$ means the max operation along the -1 dimension.

2.4 Two-Stage Spatial-Spectral Decoder

To precisely segment tumor regions from the microscopic hyperspectral image, we design a two-stage cross-dimension decoder to further combine enhanced spatial and spectral features, *i.e.*, the spatial-spectral decoder and the mask refinement. In the first stage, we combine attention-based cross-dimension decoder and

SAM's pixel decoder [10] to create a coarse mask. Specifically, the spatial query T'_{spa} is fed into a self-attention layer, followed by a spatial cross-attention layer and a spectral cross-attention layer to fuse the enhanced features $\{F''_{spa}, F''_{spec}\}$. We then apply a linear layer for the combination of deeply fused features and pixel decoder's output feature, to generate a coarse mask $P_c \in \mathbb{R}^{H/4 \times W/4}$.

To refine the segmentation, in the second stage, we follow Mask2Former [4] to enhance the object details with a foreground aware attention operation as calculated in Eq. 4. Through multiplying the mask with the attention map, the model can focus more on foreground regions and prove the tumor boundary:

$$F_o = \texttt{SelfAttn}(\texttt{FgroundAttn}(T'_{spa}, F_{pd}, P_c)), \tag{4}$$

where $\texttt{SelfAttn}(\cdot)$ is the self-attention operation, $\texttt{FgroundAttn}(\cdot)$ is the foreground aware attention operation with 3 layers of mask attention, and F_{pd} is the feature at the middle layer of pixel decoder in SAM. Similar with the first stage, we predict a refined segmentation mask $P_f \in \mathbb{R}^{H \times W}$ by employing a linear layer to the final features with F_o and the last layer of pixel decoder feature.

The training loss consists of two components, *i.e.*, pixel-wise classification loss and binary mask loss: $\mathcal{L} = \lambda_{ce}\mathcal{L}_{ce} + \lambda_{dice}\mathcal{L}_{dice}$. The total loss of the two-stage spatial-spectral decoder is computed as a weighted sum of the losses from both stages, *i.e.*, $\mathcal{L}_{\text{total}} = \lambda_w \mathcal{L}_{\text{stage1}} + (1 - \lambda_w)\mathcal{L}_{\text{stage2}}$. Here, $\{\lambda_{ce}, \lambda_{dice}, \lambda_w\}$ are weight values: $\lambda_{ce} = 0.25, \lambda_{dice} = 0.75, \lambda_w = 0.8$.

3 Experiments and Results

3.1 Experimental Setup

Datasets. To evaluate the model performance of our proposed method, we conduct all experiments on a private Gastric Poorly Cohesive Carcinoma (GPCC) dataset and a public Multi-Dimensional Choledoch (MDC) dataset. The **Multi-Dimensional Choledoch (MDC) Dataset** [24] is a publicly available dataset used for hyperspectral image segmentation tasks. 538 scenes are analyzed in this paper, each with 60 spectral bands from 450nm to 750nm and spatial resolution is resized to 256 × 256 pixels. The dataset focuses on the task of binary Microscopic Hyperspectral Image (MHSI) segmentation. For experimentation purposes, the MDC dataset is divided into training, validation, and test subsets using a patient-centric hard split method, maintaining a ratio of 3:1:1 for these subsets. The **Gastric Poorly Cohesive Carcinoma (GPCC) Dataset** is a private dataset containing 600 scenes with 40 spectral bands. The spatial resolution of each scene is resized to 256 × 256 pixels. This dataset focuses on a different set of medical imaging challenges, specifically targeting gastric cancer tissues. Similar to the MDC dataset, the GPCC dataset is also partitioned into training, validation, and test subsets using a patient-centric hard split method, maintaining a ratio of 3:1:1 for these subsets.

The proposed Omni-Fuse extracts primary feature utilizing Swin-Transformer encoder [14] for spatial and MDN [12] for spectral, and select SAM

Table 1. Evaluation results of the proposed method and SOTA methods on both MDC and GPCC datasets. The optimal metric values are highlighted in bold.

Method	MDC				GPCC			
	DSC↑	IoU↑	HD↓	p-value ↓	DSC↑	IoU↑	HD↓	p-value ↓
3DUNet [5]	72.48	58.92	73.99	3.06	70.76	53.79	92.26	3.05
BYOL [7]	72.83	60.18	68.54	3.19	71.73	55.64	86.70	3.15
DMVL [13]	71.97	59.49	72.75	2.96	70.36	54.37	90.07	3.00
HyperNet [18]	72.31	59.54	75.36	3.13	69.52	54.83	92.99	3.07
SimSiam [3]	73.42	60.78	68.61	3.25	71.34	55.48	85.89	3.25
SpectralFormer [8]	82.68	66.75	56.27	3.76	79.47	64.57	74.63	3.61
nnUNet [9]	74.06	61.33	71.69	3.58	71.87	55.32	86.37	3.54
Swin-UNETR [17]	72.04	58.69	72.80	3.30	70.93	54.93	89.66	3.11
FSS [22]	74.62	61.61	68.67	3.64	71.45	54.99	87.48	3.68
Spec-Tr [21]	73.06	60.36	67.96	3.22	71.60	55.01	86.84	3.27
DF-S^3R [19]	75.38	63.83	69.09	3.95	72.29	56.52	86.13	3.93
QSQL-FL [6]	78.39	65.78	63.01	4.24	75.76	60.44	80.92	4.26
VM-UNet [15]	81.53	67.23	59.75	4.63	77.53	59.54	77.12	4.66
Ours	**84.12**	**69.39**	**54.84**	-	**79.78**	**66.84**	**72.96**	-

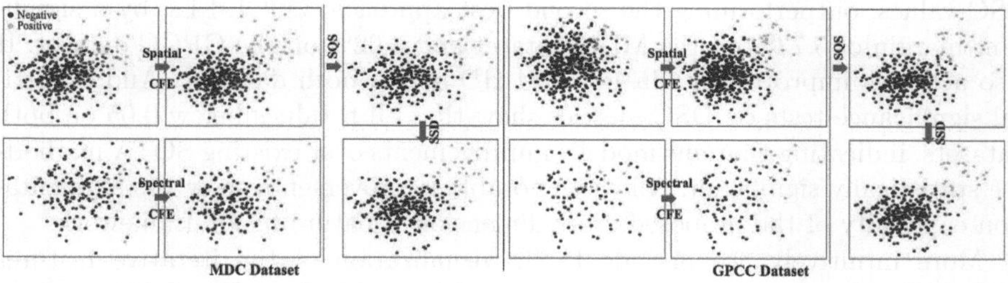

Fig. 3. tSNE visualization of the spatial-spectral features processed by several feature fusion modules. "CFE" is cross-dimensional feature enhancer, "SQS" is spectral-guided spatial query selector, and "SSD" is spatial-spectral decoder.

decoder [10] as the decoder. Data augmentation is employed in Omni-Fuse, combining elastic deformation, rotation and scaling. The batch size is set to 8 and the maximal training epoch to 300. AdamW optimizer and weight decay scheduler are utilized with initial learning rate of 0.005. The training loss is a combination of Cross-Entropy loss and Dice loss. We select 3 metrics, *i.e.*, Dice Similarity Coefficient (DSC) [%], Intersection over Union (IoU) [%], and Hausdorff Distance (HD) [px], for the assessment of model performance. All experiments are implemented using PyTorch 2.2.0 on an NVIDIA Geforce RTX 3090 GPU.

3.2 Performance Evaluation

We compare segmentation results of the proposed Omni-Fuse with state-of-the-art (SOTA) methods on both MDC and GPCC datasets. The experimental results in Table 1 show that our proposed method outperforms all SOTA methods

Table 2. Evaluation results of the proposed method with various combinations of its modules. "CNN" is CNN-based spatial feature extraction, "Mamba" is mamba-based spectral feature extraction, "CFE" is cross-dimensional feature enhancer, "SQS" is spectral-guided spatial query selctor, "SSD" os spatial-spectral decoder and "MR" is mask refiner. The optimal metric values are highlighted in bold.

CNN	Mamba	CFE	SQS	SSD	MR	MDC			GPCC		
						DSC↑	IoU↑	HD↓	DSC↑	IoU↑	HD↓
×	×	×	×	×	×	70.54	63.54	59.67	67.43	62.47	76.97
✓	×	×	×	×	×	72.68	65.75	58.34	70.89	63.28	75.76
✓	✓	×	×	×	×	76.61	66.95	57.23	74.65	64.44	74.32
✓	✓	✓	×	×	×	79.39	68.64	55.36	77.23	65.83	73.26
✓	✓	✓	✓	×	×	82.23	68.95	54.98	77.95	65.43	74.32
✓	✓	✓	✓	✓	×	83.47	68.46	54.49	78.47	66.43	72.85
✓	✓	✓	✓	✓	✓	**84.12**	**69.39**	**54.84**	**79.78**	**66.84**	**72.96**

in all evaluation metrics. The proposed Omni-Fuse method achieves the highest DSC values, outperforming the second best approach, QSQL-FLs, by a significant margin of 5.73% on the MDC dataset and 4.02% on the GPCC dataset. It also achieves improvements in IoU and HD across both datasets. And statistical significance tests on DSC metrics show that all p-values below 0.05 on both datasets, indicating that our model's improvements over existing SOTA methods are statistically significant. All these notable improvements show the segmentation capability of the proposed cross-dimensional feature fusion framework.

More intuitively, we provide tSNE visualization of the iterative features updated by the abundant multi-dimensional fusion blocks in Fig. 3. It is obvious that cross-dimensional feature enhancer can significantly increase the inter-class distance and decrease intra-class distance in both spatial and spectral dimensions. And this block effectively reduces the spectral redundancy from 0.5755 to 0.46328 on MDC and 0.6450 to 0.5154 on GPCC, indicating its ability to suppress redundant information across adjacent spectral bands. Spectral-guided spatial query selector and spatial-spectral decoder can help incrementally differentiate cancer area from the background. Although several attentions are employed for feature fusion, Omni-Fuse maintains low computational complexity and operates at high speed, as the spatial-spectral features are refined progressively.

3.3 Ablation Study

We conduct an ablation study to evaluate each component in Omni-Fuse as shown in Table 2. The baseline is conducted by removing the CNN, Mamba block, cross-dimension feature enhancer, and two-stage spatial-spectral decoder in the proposed Omni-Fuse. With CNN and Mamba blocks extracting primary spatial and spectral features, DSC has a significant increase of 6.07%. Additionally, applying feature enhancement module yields a 2.78% improvement in DSC on MDC and 2.58% on GPCC, proving its inter-class discriminative ability.

On top of this, the following spatial query selector, spatial-spectral decoder and mask refinement all make contributions for more accurate tumor area segmentation. From Table 2, if the proposed two-stage spatial-spectral decoder (SSD + MR) is replaced with SAM decoder, with enhanced spatial and spectral features added as input, the model performance drops from 84.12% to 79.39% on MDC dataset. Finally, we explore Omni-Fuse using pseudo-color image (420nm, 495nm, and 625nm mapped to RGB) to represent pure spatial information. DSC value decreases 9.74% on MDC and 8.21% on GPCC, further highlighting the superiority of hyperspectral image and the proposed omni fusion strategy.

4 Conclusion

In this paper, we introduce an omni-fusion strategy of spatial and spectral dimensions for microscopic hyperspectral image segmentation, including a novel cross-dimensional enhancer, a spectral-guided spatial query selector and a two-stage cross-dimensional decoder. All the deigned modules progressively reinforce the communication between spatial and spectral dimensions, increasing the inter-class distance and decreasing the intra-class distance. Experiments show our proposed Omni-Fuse achieves much better segmentation performance compared with the SOTA methods with 5.73% improvement in DSC on MDC dataset and 4.02% on GPCC dataset.

Acknowledgments. This work was supported by the National Natural Science Foundation of China (Grant No. 62471182), Shanghai Rising-Star Program (Grant No. 24QA2702100), and the Science and Technology Commission of Shanghai Municipality (Grant No. 22DZ2229004).

Disclosure of Interests. The authors have no competing interests in the paper as required by the publisher.

References

1. Banu, K.S., Lerma, M., Ahmed, S.U., Gardea-Torresdey, J.L.: Hyperspectral microscopy-applications of hyperspectral imaging techniques in different fields of science: a review of recent advances. Appl. Spectrosc. Rev. **59**(7), 935–958 (2024)
2. Bengs, M., et al.: Spectral-spatial recurrent-convolutional networks for *In-Vivo* hyperspectral tumor type classification. In: Martel, A.L., et al. (eds.) MICCAI 2020. LNCS, vol. 12263, pp. 690–699. Springer, Cham (2020). https://doi.org/10.1007/978-3-030-59716-0_66
3. Chen, X., He, K.: Exploring simple siamese representation learning. In: Proceedings of the IEEE/CVF Conference on Computer Vision and Pattern Recognition, pp. 15750–15758 (2021)
4. Cheng, B., Schwing, A.G., Kirillov, A.: Per-pixel classification is not all you need for semantic segmentation (2021)

5. Çiçek, Ö., Abdulkadir, A., Lienkamp, S.S., Brox, T., Ronneberger, O.: 3D U-Net: learning dense volumetric segmentation from sparse annotation. In: Ourselin, S., Joskowicz, L., Sabuncu, M.R., Unal, G., Wells, W. (eds.) MICCAI 2016. LNCS, vol. 9901, pp. 424–432. Springer, Cham (2016). https://doi.org/10.1007/978-3-319-46723-8_49
6. Dong, H., et al.: Multi-stage multi-granularity focus-tuned learning paradigm for medical HSI segmentation. In: Proceedings of Medical Image Computing and Computer-Assisted Intervention, pp. 456–466. Springer (2024)
7. Grill, J.B., et al.: Bootstrap your own latent-a new approach to self-supervised learning. Adv. Neural. Inf. Process. Syst. **33**, 21271–21284 (2020)
8. Hong, D., et al.: Spectralformer: rethinking hyperspectral image classification with transformers. IEEE Trans. Geosci. Remote Sens. **60**, 1–15 (2021)
9. Isensee, F., Jaeger, P.F., Kohl, S.A., Petersen, J., Maier-Hein, K.H.: nnU-Net: a self-configuring method for deep learning-based biomedical image segmentation. Nat. Methods **18**(2), 203–211 (2021)
10. Kirillov, A., et al.: Segment anything. In: Proceedings of the IEEE/CVF International Conference on Computer Vision, pp. 4015–4026 (2023)
11. Li, W., Gao, Y., Zhang, M., Tao, R., Du, Q.: Asymmetric feature fusion network for hyperspectral and SAR image classification. IEEE Trans. Neural Netw. Learn. Syst. **34**(10), 8057–8070 (2022)
12. Lin, S., Yun, B., Shen, W., Li, Q., Yang, A., Wang, Y.: MDN: mamba-driven dual-stream network for medical hyperspectral image segmentation. In: ICASSP 2025-2025 IEEE International Conference on Acoustics, Speech and Signal Processing (ICASSP), pp. 1–5. IEEE (2025)
13. Liu, B., Yu, A., Yu, X., Wang, R., Gao, K., Guo, W.: Deep multiview learning for hyperspectral image classification, vol. 59, no. 9, pp. 7758–7772 (2020)
14. Liu, Z., et al.: Swin transformer: hierarchical vision transformer using shifted windows. In: Proceedings of the IEEE/CVF International Conference on Computer Vision, pp. 10012–10022 (2021)
15. Ruan, J., Li, J., Xiang, S.: Vm-unet: vision mamba unet for medical image segmentation. arXiv preprint arXiv:2402.02491 (2024)
16. Song, L., Feng, Z., Yang, S., Zhang, X., Jiao, L.: Interactive spectral-spatial transformer for hyperspectral image classification. IEEE Trans. Circuits Syst. Video Technol. (2024)
17. Tang, Y., et al.: Self-supervised pre-training of swin transformers for 3D medical image analysis. In: Proceedings of the IEEE/CVF Conference on Computer Vision and Pattern Recognition, pp. 20730–20740 (2022)
18. Wang, Q., et al.: Identification of melanoma from hyperspectral pathology image using 3D convolutional networks. IEEE Trans. Med. Imaging **40**(1), 218–227 (2021)
19. Xie, X., Jin, T., Yun, B., Li, B., Wang, Q.: Exploring hyperspectral histopathology image segmentation from a deformable perspective. In: Proceedings of the 31st ACM International Conference on Multimedia, pp. 242–251 (2023)
20. Yao, Z., Liu, S., Yuan, X., Fang, L.: Specat: spatial-spectral cumulative-attention transformer for high-resolution hyperspectral image reconstruction. In: Proceedings of the IEEE/CVF Conference on Computer Vision and Pattern Recognition, pp. 25368–25377 (2024)
21. Yun, B., et al.: Spectral transformer for microscopic hyperspectral pathology image segmentation. IEEE Trans. Circuits Syst. Video Technol. (2023)
22. Yun, B., Li, Q., Mitrofanova, Q., Zhou, C., Wang, Y.: Factor space and spectrum for medical hyperspectral image segmentation. In: Proceedings of Medical Image Computing and Computer-Assisted Intervention, pp. 152–162. Springer (2023)

23. Zhang, J., et al.: Hyperspectral image classification based on dense pyramidal convolution and multi-feature fusion. Remote Sens. **15**(12), 2990 (2023)
24. Zhang, Q., Li, Q., Yu, G., Sun, L., Zhou, M., Chu, J.: A multidimensional choledoch database and benchmarks for cholangiocarcinoma diagnosis. IEEE Access **7**, 149414–149421 (2019)

PDF-Net: Prototype-Aware Dynamic Fusion Network for Nasopharyngeal Carcinoma T-Staging Classification with Epstein-Barr Virus DNA

Wantong Lu[1,2,3], Xu Han[1], Yibo Wei[1,2,3], Zanting Ye[1,2,3], and Lijun Lu[1,2,3,4](✉)

[1] School of Biomedical Engineering, Southern Medical University, Guangzhou, China
ljlubme@gmail.com

[2] Guangdong Provincial Key Laboratory of Medical Image Processing, Southern Medical University, Guangzhou, China

[3] Guangdong Province Engineering Laboratory for Medical Imaging and Diagnostic Technology, Southern Medical University, Guangzhou, China

[4] Pazhou Lab, Guangzhou, China

Abstract. Accurate T-staging classification of nasopharyngeal carcinoma (NPC) is crucial for guiding individualized treatment strategies and predicting patient prognosis. However, this task remains challenging due to the limitations of unimodal approaches, which often fail to capture the full complexity of NPC progression, and the severe class imbalance in clinical datasets, where early-stage cases (T1/T2 stage) are significantly underrepresented. In this paper, we propose a Prototype-Aware Dynamic Fusion Network (PDF-Net), a novel multimodal framework that integrates MR images with Epstein-Barr virus (EBV) DNA tabular data to improve NPC T-staging classification. Our framework introduces two key components: (1) the Dynamic Multi-Modal Alignment (DMMA) module, which aligns MR imaging features with EBV DNA data to capture complementary information across modalities, and (2) the Optimal Prototype-Aware Transport (OPAT) module, which incorporates a Prototypical Constraint to enhance the representation of T2-staging features and mitigate class imbalance. To the best of our knowledge, PDF-Net is the first framework to leverage EBV DNA data as an auxiliary tool for T-staging classification, significantly improving accuracy and robustness. Experimental results in a real clinical dataset demonstrate that our approach outperforms state-of-the-art methods, achieving an accuracy of 0.8006 ± 0.0488 and an AUC of 0.8191 ± 0.0551 for T1C images, highlighting its potential to advance NPC diagnosis and personalized treatment strategies.

Keywords: Nasopharyngeal Carcinoma · T-staging · Epstein-Barr Virus · Multimodal Fusion · Few-shot Learning · Optimal Transport

1 Introduction

Nasopharyngeal carcinoma (NPC) is a malignancy originating from the epithelial lining of the nasopharynx, with magnetic resonance imaging (MRI) serving as the pri-

W. Lu, X. Han—Co-first authors.

J. C. Gee et al. (Eds.): MICCAI 2025, LNCS 15960, pp. 482–491, 2026.
https://doi.org/10.1007/978-3-032-04927-8_46

mary modality for tumor assessment and staging due to its superior soft tissue contrast and noninvasive characteristics [1,2]. In the current, the clinical staging of NPC predominantly relies on the UICC/AJCC tumor-node-metastasis (TNM) staging system, in which the T-staging (T1–T4), reflecting the size, location, and extent of the primary tumor invasion, plays a pivotal role in treatment planning [3, 4]. Since the treatment approach for NPC varies significantly across different stages, accurate T-staging is critical for guiding individualized precision treatment and predicting prognosis.

In clinical practice, T-staging of NPC primarily depends on manual slice-by-slice inspection of MR images, which is time-consuming, labor-intensive, and highly dependent on the expertise of radiologists [1]. These challenges make it difficult to obtain large-scale annotated MR image datasets, resulting in a scarcity of labeled data. Consequently, this manual approach is neither efficient nor reliable enough to be considered the gold standard for staging. These limitations underscore the urgent need to develop effective computer-aided diagnosis (CAD) systems to improve the accuracy and efficiency of NPC staging, ultimately supporting individualized precision treatment.

The clinical manifestations of NPC are highly diverse, with early-stage symptoms often being subtle and nonspecific. As a result, the majority of patients are diagnosed at locally advanced stages (T3/T4 stage), while early-stage diagnoses (T1/T2 stage) remain relatively rare. This disparity creates a significant class imbalance in clinical datasets, posing a major challenge for accurate staging. Models trained on such imbalanced data tend to exhibit a bias toward predicting advanced stages, resulting in poor recognition of early-stage cases. Addressing this class imbalance is critical for improving the accuracy of early-stage diagnosis and enabling timely, personalized treatment strategies. In recent years, few-shot learning (FSL) has demonstrated considerable potential in medical image analysis, particularly in scenarios with limited sample sizes, by enabling efficient feature extraction and classification with minimal data [5,6]. Consequently, FSL holds significant promise for enhancing early diagnosis in NPC, aligning with our goal of early detection, timely treatment, and the implementation of individualized precision medicine.

Although deep learning has made significant progress in NPC staging, most existing studies focus solely on unimodal imaging data, overlooking the rich information contained in clinical tabular data [7,8]. Vision-only approaches often misclassify T1/T2 cases due to their subtle features or fail to differentiate between tumor invasion, inflammation, and edema, leading to staging inaccuracies. Alongside imaging, plasma circulating cell-free EBV DNA has emerged as a sensitive and specific biomarker for EBV-associated NPC, which consists of short DNA fragments released by NPC cells, can be detected through highly sensitive polymerase chain reaction (PCR) [9,10]. However, the measurement of EBV DNA is challenging due to substantial interlaboratory variability. Despite this, dynamic changes in EBV DNA levels during treatment have been linked to therapeutic efficacy and prognosis, with complete biological response (cBR) after multiple chemotherapy cycles serving as a strong predictor of long-term outcomes [11]. Given the close association between cfEBV DNA levels and NPC prognosis, integrating this biomarker with imaging data can significantly enhance the accuracy of T-stage prediction, providing a more comprehensive assessment of disease progression and treatment response.

Motivated by the above discussion, we propose a **Prototype-Aware Dynamic Fusion Network (PDF-Net)**, a novel multimodal framework that integrates MR images with EBV DNA tabular data for the classification of NPC T-staging. Our main contributions are as follows:

- We propose the **Dynamic Multi-Modal Alignment (DMMA)** module, which enables structured fusion of MR images and EBV tabular data, allowing the model to learn clinically relevant progression patterns beyond vision-only staging.
- We introduce the **Optimal Prototype-Aware Transport (OPAT)** module, which incorporates Prototypical Constraint into the Optimal Transport framework to enhance T2 feature representation and mitigate class imbalance.
- To the best of our knowledge, PDF-Net is the first framework to incorporate EBV DNA data as an auxiliary tool for T-staging classification, effectively leveraging complementary information across modalities to significantly improve classification accuracy.

2 Method

Figure 1 presents the overall framework of our proposed method for integrating MR images and EBV DNA tabular data. The framework consists of three key components: the PDF-Net, the DMMA module, and the OPAT module. These components work collaboratively to enhance T-staging classification by leveraging multi-modal data fusion and representation learning.

2.1 Dynamic Multi-modal Alignment (DMMA) Module

Muti-modal Feature Extraction. We extract features separately from two modalities: two sequences of MR images and EBV Dynamics tabular data. For images, we utilize a Vision Transformer (ViT) to extract image features $F_I \in \mathbb{R}^{d_I}$ from each MRI sequence [12]. The extracted image features capture spatial and structural characteristics of the tumor region, which are essential for accurate T-staging classification. For tabular data, we employ the FT-Transformer to process the data and extract tabular features $F_T \in \mathbb{R}^{d_T}$ that represent the temporal dynamics and numerical relationships in the EBV values across the different chemotherapy stages [13].

Since these extracted features from different modalities and have different dimensions, we project them into a unified 256-dimensional alignment space $\mathcal{A}$ using a linear transformation layer before further processing. After projection, both sets of features are transformed into $F_I^{'} \in \mathbb{R}^{1\times256}$ and $F_T^{'} \in \mathbb{R}^{1\times256}$. The alignment is achieved through the following linear transformations:

$$F_I^{'} = \mathcal{A}(F_I) = linear(F_I), \mathcal{A} \in \mathbb{R}^{d_I\times256}, \tag{1}$$

$$F_T^{'} = \mathcal{A}(F_T) = linear(F_T), \mathcal{A} \in \mathbb{R}^{d_T\times256}. \tag{2}$$

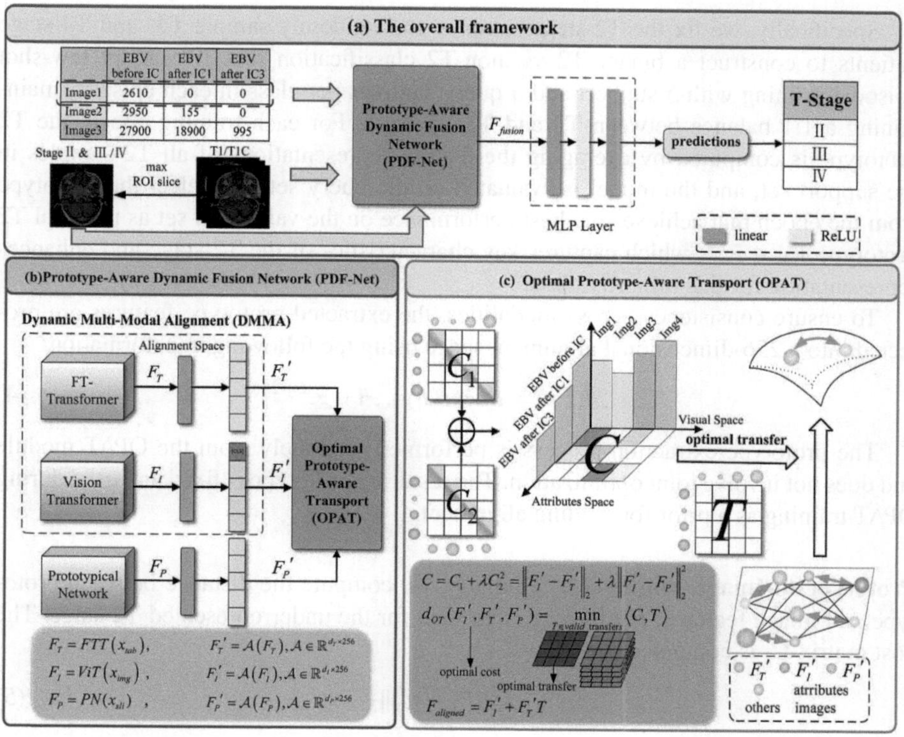

Fig. 1. Overview of the proposed framework for T-staging classification. (a) The overall framework integrating MR images and EBV DNA tabular data; (b) The PDF-Net; (c) The OPAT module. In (c), color represents modality: blue for image, green for EBV tabular data, and orange for the T2 prototype. Red arrows indicate alignment flows via optimal transport. Ball size reflects different patients and is purely for visualization—not quantitative. (Color figure online)

Tabular-to-Image Alignment. To align the image and tabular features, we employ Optimal Transport (OT) to compute a cost matrix based on Euclidean distance, which measures the pairwise similarity between the image features and the tabular features [14,15]. Smaller distances indicate greater similarity in the shared feature space. The cost matrix C_1 is computed as follows:

$$C_1 = \|F_I^{'} - F_T^{'}\|_2. \tag{3}$$

2.2 Optimal Prototype-Aware Transport (OPAT) Module

Prototype Extraction. To address the class imbalance, particularly the underrepresentation of T2-stage patients, we introduce a FSL approach [16,17]. We use a Prototypical Network to extract prototype features for the T2 stage [18]. The input to the net is the axial slice with the largest Region of Interest (ROI) area from the 3D MR volume for each patient. This slice is selected as it contains the most significant pathological information relevant to the T-staging task.

Specifically, we fix the T2-stage samples and randomly sample T3- and T4-stage patients to construct a binary T2 vs. non-T2 classification task. We adopt few-shot episodic training with 5 support and 5 query samples per class in each episode, maintaining a 1:1 balance between T2 and T3/T4 cases. For each training epoch, the T2 prototype is computed by averaging the feature representations of all T2 samples in the support set, and the model is evaluated on the query set. We select the prototype from the epoch that achieves the best performance on the validation set as the final T2 prototype $F_P \in \mathbb{R}^{d_P}$, which captures key characteristics of the T2 stage and enhances representation for this minority class.

To ensure consistency across modalities, the extracted prototype features are projected into a 256-dimensional alignment space using the following transformation:

$$F'_P = \mathcal{A}(F_P) = linear(F_P), \mathcal{A} \in \mathbb{R}^{d_P \times 256}. \quad (4)$$

The prototype extraction process is performed separately from the OPAT module and does not involve joint optimization. The resulting prototype is fixed and used during OPAT training as a prior for guiding alignment.

Prototype-to-Image Alignment. Similarly, we compute the distance between prototype and image features, reinforcing alignment for the underrepresented T2 stage. The cost matrix C_2 is computed as follows:

$$C_2 = ||F'_I - F'_P||_2. \quad (5)$$

Prototypical Constraint. The final cost matrix integrates T2 prototype constraints to mitigate bias towards majority classes (T3 & T4 stage), improving T2-stage recognition:

$$C = C_1 + \lambda C_2^2 = ||F'_I - F'_T||_2 + \lambda ||F'_I - F'_P||_2^2, \quad (6)$$

where λ is a weighting parameter that controls the influence of the T2 prototype constraint. The addition of this constraint helps to mitigate the bias towards the majority classes (T3 and T4), improving the model's ability to predict the T2 stage more accurately.

Optimal Prototype-Aware Transport. After computing the cost matrix,the Sinkhorn Algorithm is applied to compute the optimal transport plan that minimizes the cost of aligning tabular features to image features [19]. The optimization problem is formulated as:

$$d_{OT}(F'_I, F'_T, F'_P) = \min_{T \in \text{valid plans}} \langle C, T \rangle, \quad (7)$$

where C is the total cost matrix, T is the transport matrix. The objective of the problem is to minimize the cost associated with transporting the tabular features to the image features. The result is the optimal transport matrix T, which indicates the best alignment between the features from both modalities, taking into account both the spatial and temporal aspects of the data. This optimal transport framework ensures that the multimodal features are aligned in a way that minimizes the overall cost, leading to improved feature fusion and more accurate T-stage classification.

Muti-modal Feature Fusion. After obtaining the transport matrix, the aligned features are computed by combining the image features $F_I^{'}$ and the tabular features $F_T^{'}$ according to the transport matrix $T^{'}$:

$$F_{aligned} = F_I^{'} + F_T^{'}T. \tag{8}$$

Finally, the aligned features are passed through a multi-layer perceptron (MLP) classifier for the final T-stage classification. The addition of prototype constraints ensures that T2 patients are better represented in the feature space, thereby improving classification performance for this underrepresented group.

3 Experiment

3.1 Experimental Settings

Dataset. To validate the effectiveness of our method, we conducted a study using data from a single medical center, comprising a total of 1,634 NPC patients. For T-staging classification, the dataset includes 113 T2-stage cases, 893 T3-stage cases, and 628 T4-stage cases. We collected T1-weighted imaging (T1w) and contrast-enhanced T1-weighted imaging (T1Cw) sequence, along with the corresponding primary tumor mask images. To prepare the input data, we multiplied the T1w and T1Cw images with their respective mask images, effectively highlighting the ROIs corresponding to the primary tumor. These masked images were then used as inputs to our model, ensuring that the focus remained on the clinically relevant areas for accurate T-staging classification. Additionally, we utilized clinical information in the form of EBV DNA levels, measured at three time points: before chemotherapy, after one cycle of chemotherapy, and after three cycles of chemotherapy. Among the 1,634 patients, only 802 had complete EBV DNA measurements across all three time points.

Implementation Details. The MR images were resampled to a resolution of (2, 2, 2) and resized to a uniform dimension of (112, 112, 112). All images were normalized to ensure consistency. The dataset was split into training, validation, and testing sets in an 7:1.5:1.5 ratio, ensuring that the division was consistent for both MR images and EBV DNA data.

The training process was divided into three independent stages. First, we trained modality-specific encoders: a ViT for MR images and an FT-Transformer for EBV DNA data. Features were extracted before the final classification layers and fixed for downstream use. Second, Prototypical Network was also trained independently to extract a fixed T2 prototype, which was used in the OPAT module to guide the alignment process. Finally, the OPAT module was trained using the extracted image and tabular features along with the fixed prototype. During this stage, only the OPAT parameters were updated.

Models were trained using the Adam optimizer with a learning rate schedule: 1e−4 initially, decayed to 5e−5 at epoch 50 and 2e−5 at epoch 100, for a total of 150 epochs. The best checkpoint was selected based on validation ACC. All code was implemented in PyTorch and executed on a NVIDIA RTX 4090 GPU.

Table 1. Comparisons between our proposed method and other SOTA approaches.

Method	Images	EBV	T1		TIC	
			ACC	AUC	ACC	AUC
ResNet18 [20]	✓		0.6497 ± 0.0279	0.7453 ± 0.0165	0.6850 ± 0.0346	0.7558 ± 0.0382
ResNet34 [20]	✓		0.6791 ± 0.0485	0.7454 ± 0.0246	0.7162 ± 0.0.371	0.7524 ± 0.0138
DenseNet121 [21]	✓		0.6871 ± 0.0340	0.7768 ± 0.0285	0.7122 ± 0.0336	**0.7536 ± 0.0357**
ViT [12]	✓		**0.7170 ± 0.0368**	**0.7870 ± 0.2222**	**0.7243 ± 0.0320**	0.7424 ± 0.0362
BERT [22]		✓	0.5949 ± 0.0346	0.5780 ± 0.3250		
FT-Transformer [13]		✓	**0.6203 ± 0.0295**	**0.5762 ± 0.0271**		
CA [23]	✓	✓	0.6703 ± 0.0314	0.6900 ± 0.0321	0.6209 ± 0.0345	0.6626 ± 0.0335
Concat	✓	✓	0.7403 ± 0.0580	0.7443 ± 0.0389	0.7431 ± 0.0442	0.7485 ± 0.0770
OT [14]	✓	✓	0.7659 ± 0.0431	0.7802 ± 0.0495	0.7736 ± 0.0524	0.7522 ± 0.0551
GMU [24]	✓	✓	0.7535 ± 0.0313	0.778 ± 0.0290	0.7592 ± 0.0368	0.7645 ± 0.0200
3MT [25]	✓	✓	0.7614 ± 0.0385	0.7805 ± 0.0453	0.7778 ± 0.0669	0.7876 ± 0.0684
AGGN [26]	✓	✓	0.7432 ± 0.0320	0.7495 ± 0.0218	0.7615 ± 0.0403	0.7868 ± 0.0186
Ours	✓	✓	**0.7762 ± 0.0317**	**0.7924 ± 0.0183**	**0.8006 ± 0.0488**	**0.8191 ± 0.0551**

Evaluation Metrics. We conducted comparative experiments and ablation studies to demonstrate the advantages of our method. The performance was evaluated using several metrics, including accuracy (ACC), area under the curve (AUC). These metrics were chosen to comprehensively assess the model's ability to classify the T stage of NPC accurately and reliably.

3.2 Result and Analysis

We performed comparative experiments and ablation studies to demonstrate the advantages of our proposed framework.

Comparison to Other Methods. For single image modality validation, we compared our method with classical classification networks, including ResNet18, ResNet34 [20], DenseNet121 [21], and ViT [12]. For clinical tabular data, we compared FT-Transformer [13] with BERT [22] to assess the performance of structured tabular data versus textual data representation. To validate the effectiveness of multimodal fusion, we compared our method with several state-of-the-art approaches, including Cross Attention (CA) [23], Concatenation, OT [14,15], GMU [24], 3MT [25], and AGGN [26]. All multimodal baselines use the same ViT and FT-Transformer encoders as our method to ensure a fair comparison.

As shown in Table 1, our model achieved the best classification performance. For T1 images, the ACC and AUC were 0.7762 ± 0.0371 and 0.7924 ± 0.0138, respectively. For T1C images, the performance further improved, with an ACC of 0.8006 ± 0.0488 and an AUC of 0.8191 ± 0.0551. For single image modality validation, ViT achieved the best performance when using masked images as input, demonstrating its superior

Table 2. Ablation results of our method.

Method	T1		T1C	
	ACC	AUC	ACC	AUC
w/o OPAT, w/o DMMA	0.7170 ± 0.0368	0.7870 ± 0.0222	0.7243 ± 0.0302	0.7424 ± 0.0326
w/o DMMA	0.7432 ± 0.0488	0.7458 ± 0.0549	0.7436 ± 0.0629	0.7501 ± 0.0637
w/o OPAT	0.7403 ± 0.0580	0.7443 ± 0.0879	0.7431 ± 0.0422	0.7485 ± 0.0770
w/o PC	0.7569 ± 0.0431	0.7802 ± 0.0915	0.7736 ± 0.0524	0.7522 ± 0.0951
Ours	**0.7762 ± 0.0371**	**0.7924 ± 0.0138**	**0.8006 ± 0.0488**	**0.8191 ± 0.0551**

ability to capture subtle features in the ROIs. For clinical data, results demonstrated that methods based on structured tabular data outperformed text-based approaches, underscoring the advantages of structured numerical data over textual representations for NPC T-stage classification. However, single-modality approaches, whether based on masked images or EBV DNA tabular data, consistently underperformed compared to multimodal methods, underscoring the importance of integrating complementary information from multiple modalities. Additionally, other multimodal fusion methods either ignore the significant dimensional differences between modalities or are designed for prognostic tasks, which may not be directly applicable to the unique diagnostic scenarios of NPC. These limitations often result in suboptimal alignment between image and non-image feature spaces. In contrast, our PDF-Net effectively bridges the gap between MR images and EBV DNA tabular data, enabling the network to learn more discriminative features and significantly improving classification performance.

Ablation Study. We evaluated the importance of key components in our framework through ablation experiments (Table 2). The results demonstrate that removing any module leads to a decline in both ACC and AUC, highlighting the critical role of each component in the overall performance of our model. Specifically, the removal of the OPAT module caused the most significant performance drop, particularly for early-stage (T2) cases, as it weakens the model's ability to handle class imbalance and enhance the representation of underrepresented T2-stage features. Similarly, removing the DMMA module led to a substantial decrease in performance, as it disrupts the alignment of MR image features with EBV DNA tabular data, limiting the model's ability to capture complementary information across modalities.

These results underscore the importance of both the OPAT and DMMA modules in improving the overall accuracy and robustness of NPC T-staging classification. The OPAT module is particularly critical for addressing class imbalance and enhancing early-stage detection, while the DMMA module enables effective multimodal fusion by aligning MR image features with EBV DNA tabular data. Additionally, the Prototypical Constraint (PC) contributes to refining feature representation, further supporting the model's performance.

4 Conclusion

We propose a PDF-Net for the multimodal T-staging classification of NPC for the first time. This method effectively integrates MR images and EBV DNA tabular data, achieving state-of-the-art classification performance. By leveraging the DMMA module, we align MR image features with EBV DNA data to capture complementary information across modalities. Additionally, the OPAT module enhances the representation of under-represented T2-stage cases, mitigating class imbalance and improving early-stage NPC classification. Our framework pioneers a promising direction for multimodal diagnosis of NPC, offering significant potential for improving diagnostic accuracy and personalized treatment strategies.

Acknowledgments. This research was funded by the National Natural Science Foundation of China (62371221, 12326616 and 62201245), National High end Foreign Experts Recruitment Plan (G2023030025L), the Science and Technology Program of Guangdong Province 2022A0505050039 and the Open Project Program of Yunnan Key Laboratory of Intelligent Systems and Computing ISC23Y02.

Disclosure of Interests. The authors have no competing interests to declare that are relevant to the content of this article.

References

1. Chen, Y.P., Chan, A.T., Le, Q.T., Blanchard, P., Sun, Y., Ma, J.: Nasopharyngeal carcinoma. Lancet **394**(10192), 64–80 (2019)
2. King, A.D.: MR imaging of nasopharyngeal carcinoma. Magn. Reson. Imaging Clin. N. Am. **30**(1), 19–33 (2022)
3. Pan, J.J., et al.: Ninth version of the AJCC and UICC nasopharyngeal cancer TNM staging classification. JAMA Oncol. (2024)
4. Lydiatt, W.M., et al.: Head and neck cancers—major changes in the American Joint Committee on cancer eighth edition cancer staging manual. CA Can. J. Clin. **67**(2), 122–137 (2017)
5. Zheng, F., Cao, J., Yu, W., Chen, Z., Xiao, N., Lu, Y.: Exploring low-resource medical image classification with weakly supervised prompt learning. Pattern Recogn. **149**, 110250 (2024)
6. Cheng, Z., Wang, S., Xin, T., Zhou, T., Zhang, H., Shao, L.: Few-shot medical image segmentation via generating multiple representative descriptors. IEEE Trans. Med. Imaging **43**(6), 2202–2214 (2024)
7. Yang, Q., Guo, Y., Ou, X., Wang, J., Hu, C.: Automatic T staging using weakly supervised deep learning for nasopharyngeal carcinoma on MR images. J. Magn. Reson. Imaging **52**(4), 1074–1082 (2020)
8. Feng, Q., et al.: Radiomics analysis and correlation with metabolic parameters in nasopharyngeal carcinoma based on PET/MR imaging. Front. Oncol. **10**, 1619 (2020)
9. Lo, Y.D.: Molecular prognostication of nasopharyngeal carcinoma by quantitative analysis of circulating Epstein-Barr virus DNA. Can. Res. **60**(24), 6878–6881 (2000)
10. Lv, J., et al.: Improving on-treatment risk stratification of cancer patients with refined response classification and integration of circulating tumor DNA kinetics. BMC Med. **20**(1), 268 (2022)

11. Lv, J., et al.: Liquid biopsy tracking during sequential chemo-radiotherapy identifies distinct prognostic phenotypes in nasopharyngeal carcinoma. Nat. Commun. **10**(1), 3941 (2019)
12. Dosovitskiy, A., et al.: An image is worth 16x16 words: transformers for image recognition at scale. arXiv preprint arXiv:2010.11929 (2020)
13. Gorishniy, Y., Rubachev, I., Khrulkov, V., Babenko, A.: Revisiting deep learning models for tabular data. Adv. Neural. Inf. Process. Syst. **34**, 18932–18943 (2021)
14. Caffarelli, L. A., McCann, R. J.: Free boundaries in optimal transport and Monge-Ampere obstacle problems. Ann. Math. 673–730 (2010)
15. Montesuma, E. F., Mboula, F. M. N., Souloumiac, A.: Recent advances in optimal transport for machine learning. IEEE Trans. Pattern Anal. Mach. Intell. (2024)
16. Lu, J., Gong, P., Ye, J., Zhang, J., Zhang, C.: A survey on machine learning from few samples. Pattern Recogn. **139**, 109480 (2023)
17. Li, J., Li, M.: Few-shot learning: methods and applications. In: ITM Web of Conferences, vol. 70, p. 02012. EDP Sciences (2025)
18. Snell, J., Swersky, K., Zemel, R.: Prototypical networks for few-shot learning. Adv. Neural Inf. Process. Syst. **30** (2017)
19. Knight, P.A.: The Sinkhorn-Knopp algorithm: convergence and applications. SIAM J. Matrix Anal. Appl. **30**(1), 261–275 (2008)
20. He, K., Zhang, X., Ren, S., Sun, J.: Deep residual learning for image recognition. In: Proceedings of the IEEE Conference on Computer Vision and Pattern Recognition, pp. 770–778 (2016)
21. Huang, G., Liu, Z., Van Der Maaten, L., Weinberger, K. Q.: Densely connected convolutional networks. In: Proceedings of the IEEE Conference on Computer Vision and Pattern Recognition, pp. 4700–4708 (2017)
22. Devlin, J., Chang, M. W., Lee, K., Toutanova, K.: Bert: pre-training of deep bidirectional transformers for language understanding. In: Proceedings of the 2019 Conference of the North American Chapter of the Association for Computational Linguistics: Human Language Technologies, Volume 1 (Long and Short Papers), pp. 4171–4186 (2019)
23. Gheini, M., Ren, X., May, J.: Cross-attention is all you need: adapting pretrained transformers for machine translation. arXiv preprint arXiv:2104.08771 (2021)
24. Qin, R., et al.: Fine-grained lung cancer classification from PET and CT images based on multidimensional attention mechanism. Complexity **2020**(1), 6153657 (2020)
25. Liu, L., Liu, S., Zhang, L., To, X.V., Nasrallah, F., Chandra, S.S.: Cascaded multi-modal mixing transformers for Alzheimer's disease classification with incomplete data. Neuroimage **277**, 120267 (2023)
26. Wu, P., Wang, Z., Zheng, B., Li, H., Alsaadi, F.E., Zeng, N.: AGGN: Attention-based glioma grading network with multi-scale feature extraction and multi-modal information fusion. Comput. Biol. Med. **152**, 106457 (2023)

Physics-Informed Neural ODEs for Temporal Dynamics Modeling in Cardiac T_1 Mapping

Nuno Capitão[1,2], Yi Zhang[1], Yidong Zhao[1], and Qian Tao[1(✉)]

[1] Department of Imaging Physics, Delft University of Technology, Delft, The Netherlands
up201709363@up.pt, {y.zhang-43,y.zhao-8,q.tao}@tudelft.nl
[2] University of Porto, Porto, Portugal

Abstract. Spin-lattice relaxation time (T_1) is an important biomarker in cardiac parametric mapping for characterizing myocardial tissue and diagnosing cardiomyopathies. Conventional Modified Look-Locker Inversion Recovery (MOLLI) acquires 11 breath-hold baseline images with interleaved rest periods to ensure mapping accuracy. However, prolonged scanning can be challenging for patients with poor breathholds, often leading to motion artifacts that degrade image quality. In addition, T_1 mapping requires a voxel-wise nonlinear fitting to a signal recovery model involving an iterative estimation process. Recent studies have proposed deep-learning approaches for rapid T_1 mapping using shortened sequences to reduce acquisition time for patient comfort. Nevertheless, existing methods overlook important physics constraints, limiting interpretability and generalization. In this work, we present an accelerated, end-to-end T_1 mapping framework leveraging Physics-Informed Neural Ordinary Differential Equations (ODEs) to model temporal dynamics and address these challenges. Our method achieves high-accuracy T_1 estimation from a sparse subset of baseline images and ensures efficient null index estimation at the test time. Specifically, we develop a continuous-time LSTM-ODE model to enable selective Look-Locker (LL) data acquisition with arbitrary time lags. Experimental results show superior performance in T_1 estimation for both native and post-contrast sequences and demonstrate the strong benefit of our physics-based formulation over direct data-driven T_1 priors.

Keywords: Quantitative Cardiac MRI · Physics-Informed Neural Networks · Modified Look-Locker Inversion Recovery

1 Introduction

Quantitative Magnetic Resonance Imaging (qMRI) is a widely adopted noninvasive technique for cardiac parametric mapping, where spin-lattice relaxation

N. Capitão and Y. Zhang—Contributed equally to this work.

J. C. Gee et al. (Eds.): MICCAI 2025, LNCS 15960, pp. 492–501, 2026.
https://doi.org/10.1007/978-3-032-04927-8_47

time (T_1) is routinely used as a biomarker to clinically assess myocardial tissue and diagnose cardiomyopathies [5]. Cardiac T_1 maps are generated by fitting a parametric model to a series of baseline images that encode signal recovery governed by spin-lattice energy exchange after an inversion pulse.

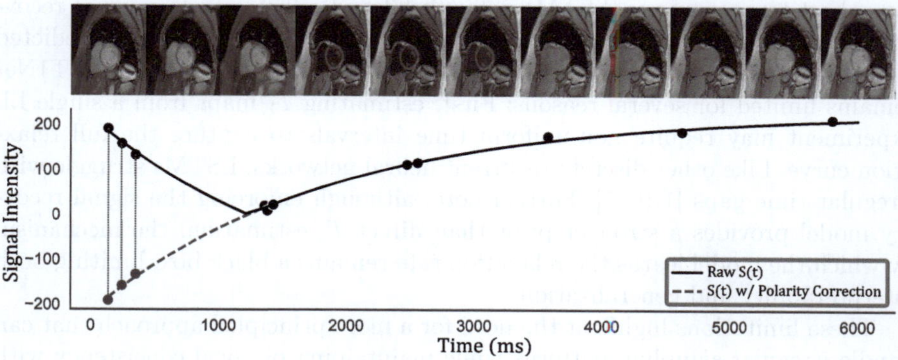

Fig. 1. MOLLI 3(3)3(3)5 reconstruction scheme illustrating signal recovery with and without polarity recovery. The corresponding temporally ordered MOLLI T_1-weighted images are shown above.

The Modified Look-Locker Inversion Recovery (MOLLI) [13] sequence has become the standard protocol in cardiac qMRI due to its high precision and reduced motion-induced artifacts. During a single breath hold, MOLLI acquires T_1-weighted images at multiple inversion times using distinct Look-Locker (LL) experiments interleaved with rest periods. In a MOLLI 3(3)3(3)5 sequence, three LL experiments collect images in sets of 3, 3, and 5, each separated by three heartbeats. These acquisitions form a image series of 11 baseline images that span the full relaxation curve, as illustrated in Fig. 1. While multiple LL experiments at different inversion times improve T_1 estimation accuracy, they also extend the breath-hold to 17 heartbeats, which can be strenuous for patients with respiratory issues. Alternative schemes such as Shortened MOLLI (ShMOLLI) [14] and Saturation Recovery Single-Shot Acquisition (SASHA) [3] reduce the total number of acquisitions to accelerate imaging, but inherently trade off precision and accuracy [16,18].

Traditional T_1 map estimation requires voxel-wise nonlinear curve fitting to a signal recovery model derived from MRI physics. This process is further complicated by the need to identify signal polarity transitions (null points) in MOLLI sequences, requiring additional iterative calculations for each pixel. The entire procedure is not only sensitive to image noise but also computationally intensive - processing a complete 3D volume over time has $\mathcal{O}(n^4)$ complexity, resulting in significant processing delays in clinical settings without specialized acceleration techniques [20].

Recent studies have shown that deep learning methods can accurately and efficiently perform cardiac T_1 mapping using as few as three to five baseline images along the inversion recovery curve, achieving results comparable to conventional MOLLI. [4] introduced MyoMapNet, a fast myocardial T_1 mapping model that uses a fully connected neural network (FCNN) to directly map pairwise LL readouts to T_1 values. However, it neglects signal polarities and is constrained to a data-driven T_1 prior. In contrast, T1Net [11] employs a bidirectional long short-term memory (LSTM) network [6] and adheres to the signal recovery model to enforce consistency between acquired MRI signals and predicted T_1 values through its loss function. Nonetheless, we hypothesize that T1Net remains limited for several reasons: First, estimating T_1 maps from a single LL experiment may require non-uniform time intervals to capture the full relaxation curve. Like other discrete recurrent neural networks, LSTMs struggle with irregular time gaps [1,9,17]. Furthermore, although enforcing the signal recovery model provides a stronger prior than direct T_1 estimation, the mechanism by which the model learns the relaxation rate remains a black box, limiting both interpretability and generalization.

These limitations highlight the need for a more principled approach that can handle irregular sampling patterns while maintaining physical consistency with the underlying MR signal dynamics. To address these challenges, we introduce a robust and accelerated T_1 mapping technique using physics-informed neural networks (PINNs) [15]. We employ a continuous-time LSTM-ODE [12] that can naturally handle LL sequences with non-uniform acquisition intervals, overcoming the time-gap limitations of traditional recurrent networks. Our Neural Ordinary Differential Equation (NODE) [2] approach explicitly models the temporal dynamics of the relaxation process, providing a continuous representation of signal evolution rather than discrete approximations. This dynamic modeling not only adheres to the signal recovery model but also explicitly formulates the recovery rate along the relaxation trajectory, ensuring that T_1 estimation follows the actual physics of spin-lattice interactions. Furthermore, the framework seamlessly incorporates signal polarity correction during inference, eliminating the need for separate processing steps.

2 Methods

2.1 MR Physics-Informed Signal Modeling and Loss Functions

To develop our physics-informed approach, we first formalize the MR signal model that governs T_1 relaxation. In qMRI, N baseline images are collected via selective sampling, with each image capturing the signal at a different point along the relaxation curve. The signal intensity $S(x, y, z)$ at a given spatial coordinate (x, y, z) is described by a parametric function that reflects the underlying physics. In a MOLLI sequence, the signal recovery follows a 3-parameter model expressed as:

$$S_{(x,y,z)}(t_i) = c_{(x,y,z)} \cdot (1 - k_{(x,y,z)} \cdot \exp(-\frac{t_i}{T^*_{1,(x,y,z)}})), \tag{1}$$

where $i \in \{1, 2, \ldots, N\}$ and t_i is the inversion time of the i-th image. The parameter set $\{c, k, T_1^*\}$ is estimated at each voxel $v \in \mathbb{R}^{x \times y \times z}$ to determine T_1 biomarker using the following equation:

$$T_{1(x,y,z)} = T^*_{1,(x,y,z)} \cdot (k_{(x,y,z)} - 1). \tag{2}$$

While Eqs. (1) and (2) provide the foundation for conventional T_1 mapping, they represent a static view of the relaxation process. Our approach incorporates the underlying physical dynamics governing spin-lattice relaxation by explicitly modeling the temporal evolution of the signal. By differentiating Eq. 1, we derive a first-order differential expression that captures the instantaneous rate of change at each inversion time t_i:

$$\frac{dS_{(x,y,z)}(t_i)}{dt} = \frac{c_{(x,y,z)} \cdot k_{(x,y,z)}}{T^*_{1,(x,y,z)}} \cdot \exp\left(-\frac{t_i}{T^*_{1,(x,y,z)}}\right). \tag{3}$$

This differential formulation enables our model to learn not just the signal values but also their temporal dynamics, providing a more complete physical representation of the relaxation process.

To incorporate both data-driven priors and physics-based constraints into our deep-learning framework, we define a physics-informed loss function with two components: a T_1 consistency loss $\mathcal{L}_{T_1}$ and a physics-based loss $\mathcal{L}_{\text{physics}}$.

$$\mathcal{L}_{T_1} = \frac{1}{HWD} \sum_{(x,y,z)} \|T_1 - \hat{T_1}\|_2^2, \tag{4}$$

where T_1 is derived from Eq. (2), enforcing consistency with our end-goal parametric mapping.

$$\mathcal{L}_{\text{physics}} = \frac{1}{NHWD} \sum_{(x,y,z)} \sum_{i=1}^{N} \|S(t_i) - \hat{S}(t_i)\|_2^2 + \lambda \|\gamma \frac{dS(t_i)}{dt} - \frac{d\hat{S}(t_i)}{dt}\|_2^2. \tag{5}$$

The first term ensures adherence to the signal recovery model (Eq. (1)), while the second term penalizes deviations from the expected signal dynamics (Eq. (3)). The scaling factor λ adjusts for the disparity in orders of magnitude observed at $\lim_{t \to 0} \frac{dS(t)}{dt}$ relative to the other loss terms. The factor γ is derived from the chain rule and accounts for the normalization ratio applied to signals and inversion times. By combining these complementary loss components, we formulate our final loss function as:

$$\mathcal{L}_{\text{total}} = \mathcal{L}_{T_1} + \mathcal{L}_{\text{physics}}. \tag{6}$$

2.2 Continuous-Time Neural ODEs for Temporal Dynamics Modeling

To effectively capture the temporal dynamics of spin-lattice relaxation with non-uniform sampling intervals, we implement a continuous-time LSTM-ODE framework that naturally handles the irregular time gaps in MOLLI sequences. Let

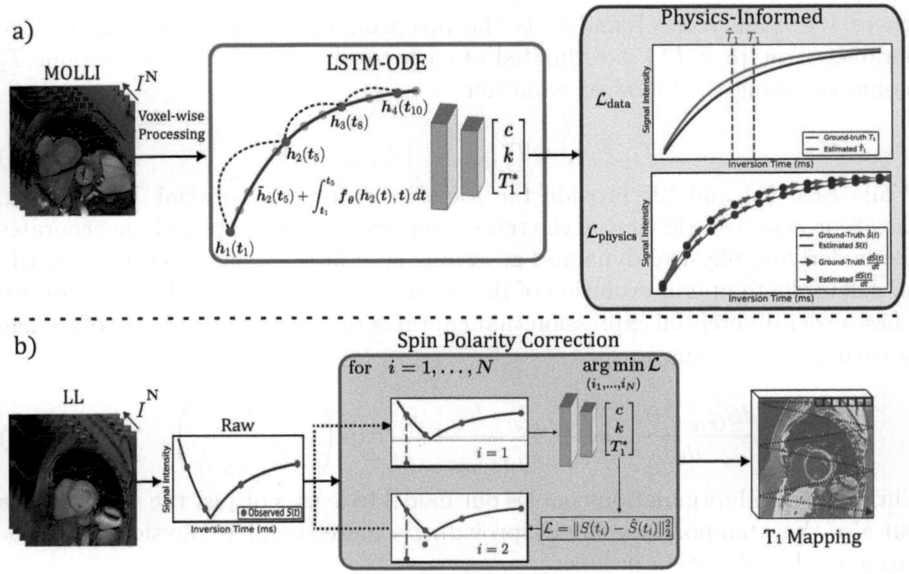

Fig. 2. Overview of the proposed physics-informed LSTM-ODE framework (a) **Training:** The model processes a subset of the 11 MOLLI baseline images in a voxel-wise manner through a continuous-time LSTM-ODE. The physics-informed loss is computed using $\{c, k, T_1^*\}$, estimated by the model. (b) **Inference:** For a series of I^N LL images, each voxel undergoes spin polarity correction. Every i-th image is sequentially inverted, and the resulting sequence is evaluated using the signal reconstruction loss. The index that minimizes the loss, arg min $\mathcal{L}$, determines the null index and the optimal parameters $\{c, k, T_1^*\}$.

$\{(S_i, t_i)\}_{i=1}^N$ represent a MOLLI voxel sequence, where S_i is the measured signal at inversion time t_i. At the i-th image, the LSTM's discrete gating update is determined by

$$(\tilde{h}_i,\, c_i) = \text{LSTMCELL}\big((S_i,\, t_i),\, (h_{i-1},\, c_{i-1})\big), \tag{7}$$

where h_{i-1} and c_{i-1} are the previous hidden and cell states, respectively, $\tilde{h}_i$ denotes the candidate hidden state for the current image, and c_i represents the updated cell state. The candidate hidden state $\tilde{h}_i$ is then evolved continuously over the interval $[t_{i-1}, t_i]$ by a NODE whose dynamics are governed by

$$\frac{d\,h(t)}{dt} = f_\theta\big(h(t),\, t\big), \tag{8}$$

where f_θ is a small, learnable neural network. We initialize the ODE at $h(t_{i-1}) = \tilde{h}_i$ and integrate forward using a black-box 5th-order DormandâĂŞPrince solver [10]. The cell states $\{c_i\}_{i=1}^N$ remain discrete to preserve stable long-term gradient propagation [12]. The final hidden state, $h_N(t_N)$, provides a continuous-time encoding of the relaxation trajectory, capturing the temporal variations in signal recovery across the N images. A multi-layer decoder then maps this encoding to the signal recovery model described by Eqs. (1–3). Formally, the mapping notation can be defined as $f_\theta : \{(S_i, t_i)\}_{i=1}^{N} \in \mathbb{R}^{N \times d_{\text{emb}} \times 2} \rightarrow \mathbb{R}^3$, where d_{emb} denotes the feature dimension of the embedded pairwise acquisitions, and the output represents the set of parameters $\{C, k, T_1^*\}$.

This continuous-time modeling approach enables our framework to accurately represent the underlying physics of relaxation processes while accommodating the practical constraints of clinical MRI acquisition protocols with varying inversion times.

2.3 Post-inversion Spin Polarity Correction

At test time, raw signal acquisitions retain their absolute magnitudes, which conflicts with our model's assumptions about post-inversion signal recovery. However, since our method is formulated around the signal recovery model, it can also act as a proxy for retrieving the spin state polarity encoded in each image. To estimate the null index at each voxel, we evaluate all possible polarity permutations across the sequence by performing N forward passes, leveraging vectorized, batch-wise inference over the full field-of-view (FoV). Figure 2 illustrates our end-to-end T_1 parametric mapping framework, which includes the post-inversion spin polarity correction. During the polarity recovery step, each i-th signal is inverted, and the entire sequence undergoes a model evaluation to compute the signal reconstruction loss term in Eq. 5. For each voxel $v \in \mathbb{R}^{x \times y \times z}$, the $\arg\min \mathcal{L}$ across trials identifies the null index, providing the best estimate of $\{c, k, T_1^*\}$ for T_1 mapping.

3 Experiments

Dataset: We used a cardiac MRI dataset of 50 subjects acquired with a 3.0T Ingenia MR scanner (Philips Healthcare). Each subject underwent both pre- and post-contrast MOLLI sequences, each consisting of 11 baseline images using the same 3-3-5 scheme provided by the manufacturer. One to three axial slices were acquired at the base, mid-ventricular, and apical levels, and the left-ventricle myocardium was manually annotated as the region of interest (ROI). To prevent data leakage, we performed a subject-wise random split: 30 subjects for training, 5 for validation, and 15 for testing. To evaluate out-of-domain generalization in signal recovery patterns at higher relaxation rates induced by a contrast agent, the training set included only native sequences, while the validation and testing sets contained both pre- and post-contrast sequences.

Implementation Details: We pre-train our LSTM-ODE model on all 11 MOLLI images and fine-tune on subsets of 3-5 images to simulate accelerated acquisition. During training, sparse readouts are strategically selected to ensure coverage across the relaxation curve: early phase ($\{t_1\text{-}t_3\}$), intermediate phase

($\{t_4\text{-}t_8\}$), and convergence phase ($\{t_9\text{-}t_{11}\}$). All models are trained using ground-truth $\{c, k, T_1^*\}$ parameters fitted to the 11-image data with the Levenberg-Marquardt (LM) algorithm. For the physics-based loss, we generate 1 000 interpolated points with 75% linearly spaced up to 2000 ms to emphasize relaxation dynamics. Our ODE solver uses tolerances of $\epsilon = 0.001$ and runs on an NVIDIA GTX 1070 GPU. We employ standard backpropagation with a derivative loss hyperparameter of $\lambda = 0.01$. The code is publicly available at https://github.com/nunomiguelc18/pinn-node-cardiac-t1mapping.

Table 1. Mean T_1 bias and SD values from Monte Carlo simulations, expressed as mean(std). Methods: TRF, MMNet (MyoMapNet), T1N (T1Net), P-MM (PINN MyoMapNet), P-LSTM (PINN LSTM-ODE). The best-performing methods are in **bold** and second-best are <u>underlined</u>. For bias values, * indicates no statistical significance ($p > 0.05$).

Method	Mean Bias (ms)			Fitting SD (ms)		
	LL_3	LL_4	LL_5	LL_3	LL_4	LL_5
Native						
TRF	-9.63(38.21)	-12.90(31.25)	15.64(18.80)	**58.12**(37.87)	**49.66**(27.16)	**46.83**(24.12)
MMNet	51.50(11.44)	21.49(10.13)	11.29(9.26)	91.57(36.67)	67.49(26.94)	<u>50.47</u>(24.36)
T1N	<u>-11.28</u>(8.83)	<u>-8.77</u>(11.75)	<u>3.11</u>*(9.74)	94.00(34.81)	69.17(29.87)	53.33(26.82)
P-MM	**0.15***(6.62)	5.39*(11.70)	7.71(8.90)	<u>82.74</u>(37.10)	<u>62.40</u>(31.39)	51.19(27.60)
P-LSTM	7.65(9.93)	**2.62***(7.74)	**0.75***(8.68)	91.18(38.74)	70.04(30.09)	50.65(27.43)
Post-Gd						
TRF	16.68(24.57)	7.64(14.57)	11.83(12.65)	**46.49**(37.58)	**39.95**(17.48)	<u>43.27</u>(16.95)
MMNet	<u>1.18</u>(22.29)	14.23(13.86)	11.40(11.58)	<u>73.67</u>(17.32)	<u>58.34</u>(15.55)	42.65(20.45)
T1N	**8.75**(10.62)	<u>8.75</u>(10.42)	20.44(7.82)	79.99(21.78)	53.97(17.18)	41.00(16.05)
P-MM	-6.52(18.47)	5.94(11.82)	<u>11.83</u>(9.65)	86.93(22.67)	59.24(18.94)	**40.29**(16.93)
P-LSTM	-12.37(18.12)	**-1.12***(11.48)	**3.38***(11.11)	110.75(52.40)	70.55(18.75)	42.92(20.47)

Comparative Studies: We benchmark our baseline model against three methods: MyoMapNet, T1Net (both designed for accelerated T_1 mapping), and SciPy's Trust Region Reflective (TRF) algorithm for bounded parameter constraints. To evaluate whether our physics-based priors improve performance relative to direct data-driven T_1 priors, we also integrate MyoMapNet's architecture into our framework for direct comparison.

Evaluation Metrics: We perform 100 Monte Carlo simulations per subject using random sampling to assess generalization across LL subsets. We evaluate two metrics: (1) mean T_1 bias relative to ground truth, and (2) fitting standard deviation (SD), which measures curve fitting quality in terms of T_1 at each voxel from the fitting residuals [7,8,19]. Both metrics are calculated within myocardial regions annotated by radiologists. Pixel-wise T_1 estimates are averaged over simulations to generate subject-level distributions, and systematic biases are evaluated using paired t-tests.

4 Results and Discussion

Table 1 presents our Monte Carlo simulation results, summarizing the mean T_1 bias and SD distributions for the Native and Post-Gd testing sets. Complementary qualitative comparisons for cardiac T_1 mapping are provided in Fig. 3. The PINN MyoMapNet model improved upon conventional MyoMapNet, demonstrating the benefits of incorporating physics during training. Our PINN LSTM-ODE method achieved the overall lowest mean bias for both Native and Post-Gd settings, indicating stronger generalization. However, its higher SD maps suggest that sparse representations may lead to inconsistent NODE-computed trajectories. The TRF algorithm performs subpar in terms of mean bias, highlighting the potential of deep learning for higher accuracy in T_1 mapping given a limited number of baseline images. Nevertheless, TRF reports the lowest SD map statistics, suggesting that deep learning methods still compromise precision compared to traditional nonlinear curve-fitting approaches.

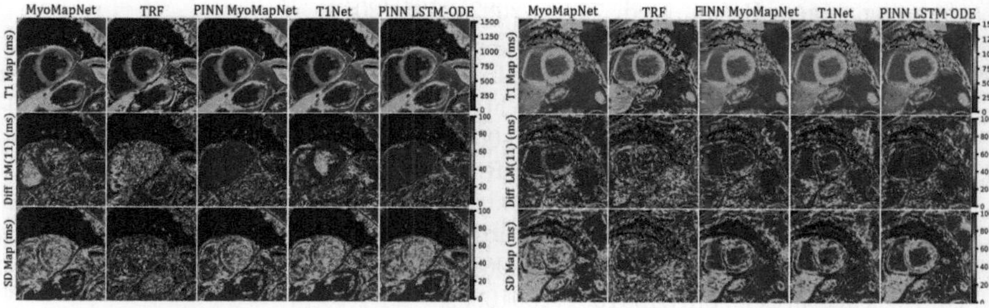

Fig. 3. Estimated mean values of T_1 and SD maps. Left: native pre-contrast sequence. Right: post-contrast sequence.

5 Conclusions

We proposed a novel deep learning framework to accelerate MOLLI-based cardiac T_1 mapping by integrating PINNs. Our experimental results show that the proposed PINN LSTM-ODE model outperforms alternative methods in both accuracy and precision, particularly with limited LL samples. Furthermore, we demonstrate that physics-informed formulations improve the performance of models that traditionally rely on direct data-driven T_1 priors, such as MyoMapNet. This highlights the benefits of embedding physics-based constraints into model training. The proposed framework enables accurate T_1 estimation and efficient FoV mapping from just 3 to 5 heartbeats, matching the performance of the conventional 17-heartbeat MOLLI sequence while significantly reducing acquisition time.

Acknowledgments. We gratefully acknowledge the Dutch Research Council (NWO) and Amazon Science for financial and computing support.

Disclosure of Interests. The authors have no competing interests to declare that are relevant to the content of this article.

References

1. Baytas, I.M., Xiao, C., Zhang, X., Wang, F., Jain, A.K., Zhou, J.: Patient subtyping via time-aware lstm networks. In: Proceedings of the 23rd ACM SIGKDD International Conference on Knowledge Discovery and Data Mining, KDD 2017, pp. 65–74. Association for Computing Machinery, New York (2017). https://doi.org/10.1145/3097983.3097997, https://doi.org/10.1145/3097983.3097997
2. Chen, R.T.Q., Rubanova, Y., Bettencourt, J., Duvenaud, D.: Neural ordinary differential equations (2019), https://arxiv.org/abs/1806.07366
3. Chow, K., Flewitt, J.A., Green, J.D., Pagano, J.J., Friedrich, M.G., Thompson, R.B.: Saturation recovery single-shot acquisition (SASHA) for myocardial t(1) mapping. Magn. Reson. Med. **71**(6), 2082–2095 (2014)
4. Guo, R., et al.: Accelerated cardiac t1 mapping in four heartbeats with inline myomapnet: a deep learning-based t1 estimation approach. J. Cardiovascular Magnet. Resonance **24**(1), 6 (2022). https://doi.org/10.1186/s12968-021-00834-0, https://doi.org/10.1186/s12968-021-00834-0
5. Haaf, P., Garg, P., Messroghli, D.R., Broadbent, D.A., Greenwood, J.P., Plein, S.: Cardiac t1 mapping and extracellular volume (ecv) in clinical practice: a comprehensive review. J. Cardiovascular Magnetic Resonance **18**(1), 89 (2016). https://doi.org/10.1186/s12968-016-0308-4, https://www.sciencedirect.com/science/article/pii/S109766472301030X
6. Hochreiter, S., Schmidhuber, J.: Long short-term memory. Neural Comput.**9**(8), 1735–1780 (1997). https://doi.org/10.1162/neco.1997.9.8.1735, https://doi.org/10.1162/neco.1997.9.8.1735
7. Kellman, P., Arai, A.E., Xue, H.: T1 and extracellular volume mapping in the heart: estimation of error maps and the influence of noise on precision. J. Cardiovasc. Magn. Reson. **15**(1), 1–12 (2013)
8. Kellman, P., Hansen, M.S.: T1-mapping in the heart: accuracy and precision. J. Cardiovasc. Magn. Reson. **16**, 1–20 (2014)
9. Kidger, P., Morrill, J., Foster, J., Lyons, T.: Neural controlled differential equations for irregular time series (2020). https://arxiv.org/abs/2005.08926
10. Konstantinov, M., Petkov, P., Christov, N.: Synthesis of linear systems with desired equivalent form. J. Comput. Appl. Math. **6**(1), 27–35 (1980). https://doi.org/10.1016/0771-050X(80)90014-5, https://www.sciencedirect.com/science/article/pii/0771050X80900145
11. Le, J.V., Mendes, J.K., McKibben, N., Wilson, B.D., Ibrahim, M., DiBella, E.V.R., Adluru, G.: Accelerated cardiac T1 mapping with recurrent networks and cyclic, model-based loss. Med. Phys. **49**(11), 6986–7000 (2022)
12. Lechner, M., Hasani, R.: Learning long-term dependencies in irregularly-sampled time series (2020). https://arxiv.org/abs/2006.04418
13. Messroghli, D.R., Radjenovic, A., Kozerke, S., Higgins, D.M., Sivananthan, M.U., Ridgway, J.P.: Modified look-locker inversion recovery (molli) for high-resolution t1 mapping of the heart. Magnet. Resonance Med. **52**(1), 141–146

(2004). https://doi.org/10.1002/mrm.20110, https://onlinelibrary.wiley.com/doi/abs/10.1002/mrm.20110
14. Piechnik, S.K., Ferreira, V.M., Dall'Armellina, E., Cochlin, L.E., Greiser, A., Neubauer, S., Robson, M.D.: Shortened modified look-locker inversion recovery (shmolli) for clinical myocardial t1-mapping at 1.5 and 3 t within a 9 heartbeat breathhold. Journal of Cardiovascular Magnetic Resonance **12**(1), 69 (Nov 2010). https://doi.org/10.1186/1532-429X-12-69, https://doi.org/10.1186/1532-429X-12-69
15. Raissi, M., Perdikaris, P., Karniadakis, G.: Physics-informed neural networks: a deep learning framework for solving forward and inverse problems involving nonlinear partial differential equations. J. Comput. Phys. **378**, 686–707 (2019). https://doi.org/10.1016/j.jcp.2018.10.045, https://www.sciencedirect.com/science/article/pii/S0021999118307125
16. Roujol, S., Weingärtner, S., Foppa, M., Chow, K., Kawaji, K., Ngo, L.H., Kellman, P., Manning, W.J., Thompson, R.B., Nezafat, R.: Accuracy, precision, and reproducibility of four T1 mapping sequences: a head-to-head comparison of MOLLI, ShMOLLI, SASHA, and SAPPHIRE. Radiology **272**(3), 683–689 (2014)
17. Schirmer, M., Eltayeb, M., Lessmann, S., Rudolph, M.: Modeling irregular time series with continuous recurrent units (2022). https://arxiv.org/abs/2111.11344
18. Weingärtner, S., et al.: Myocardial t1-mapping at 3t using saturation-recovery: reference values, precision and comparison with molli. J. Cardiovascular Magnet. Resonance **18**(1), 84 (2016). https://doi.org/10.1186/s12968-016-0302-x, https://www.sciencedirect.com/science/article/pii/S1097664723010256
19. Zhang, Y., Zhao, Y., Huang, L., Xia, L., Tao, Q.: Deep-learning-based groupwise registration for motion correction of cardiac t 1 mapping. In: International Conference on Medical Image Computing and Computer-Assisted Intervention, pp. 586–596. Springer (2024)
20. Zhang, Y., Zhao, Y., Shao, Y., Capitão, N.M.F., van den Bogert, F., Tao, Q.: Torcht_1: GPU-accelerated cardiac t_1 mapping with deep learning framework. In: Medical Imaging with Deep Learning (2024). https://openreview.net/forum?id=8ONihf0Fh9

PolarDETR: Enhancing Interpretability in Multi-modal Methods for Jawbone Lesion Detection in CBCT

Yuxuan Yang[1], Chen Zhong[1], Xinyue Zhang[2], Ruohan Ma[2], Gang Li[2], Yong Guo[1], and Jupeng Li[1](✉)

[1] School of Electronic and Information Engineering, Beijing Jiaotong University, Beijing 100044, China
lijupeng@bjtu.edu.cn

[2] Department of Oral and Maxillofacial Radiology, Peking University School and Hospital of Stomatology, Beijing 100081, China

Abstract. Rapidly advancing multi-modal learning shows great promise in medical image analysis, but challenges remain in the detection of jawbone lesions. Existing general-purpose models fail to capture the relationships between anatomical contexts and spatial locations in CBCT images, and the complexity of these models hinders interpretability. We propose PolarDETR, a novel framework combining anatomical priors and multi-modal alignment through: 1) Polar Text-Position Encoding (PTPE), which links text to spatial coordinates via polar mapping, 2) Anatomical Constraint Learning, ensuring lesion detection within anatomically plausible regions, and 3) Position Matching Optimization for spatial consistency. Evaluated on 180 clinical cases (6929 CBCT slices), our method achieves a state-of-the-art mAP of 93.66%, outperforming both single-modal (e.g., DETR at 89.35%) and multi-modal models (e.g., CORA at 91.52%). Additionally, PolarDETR excels in interpretability, with an ACS of 84.12% and PMS of 80.45%, demonstrating its potential to enhance both detection performance and clinical usability in real-world applications. Our code is available at https://github.com/Cxxxsky/PolarDETR.

Keywords: Medical Image Analysis · Multi-modal Learning · PolarDETR · Anatomical Constrain

1 Introduction

Jawbone lesion detection is crucial for early diagnosis and treatment planning in dental and maxillofacial medicine [17]. Lesions such as periapical lesions or cysts often appear in various forms on CBCT images, making their identification difficult. Accurate identification and localization are essential to prevent complications and improve outcomes, which makes advanced computational methods vital to clinicians.

J. C. Gee et al. (Eds.): MICCAI 2025, LNCS 15960, pp. 502–511, 2026.
https://doi.org/10.1007/978-3-032-04927-8_48

Recent advancements in multi-modal learning have enhanced medical image analysis, particularly for lesion detection. By integrating visual data from CBCT images with textual information from clinical reports, multi-modal methods outperform traditional single-modal approaches, improving feature representation and enabling clinically relevant interpretation. Key innovations include ConVIRT [24], which used contrastive learning for cross-modal alignment; GLoRIA [8] and MGCA [12], which employed spatial attention for feature correspondence; and MedCLIP [20], which incorporated anatomical constraints to improve consistency. Additionally, BioViL [1] redefined text encoding with hybrid models, and MRM [5] introduced a low-level feature attenuation mechanism, benefiting downstream tasks. These advancements emphasize the growing significance of multi-modal learning in medical imaging.

Despite these advances, challenges remain in applying traditional multi-modal fusion methods to specialized tasks like jawbone lesion detection. 1) Ineffectiveness of General-Purpose Models: Common fusion methods, such as feature concatenation or Cartesian coordinate mapping [4], fail to consider the jawbone's anatomical geometry. These models, often pre-trained on natural images, struggle to capture specific features like trabecular structures, bone density variations, and lesion orientation [3], leading to incomplete lesion localization. 2) Reduced Interpretability with Increasing Complexity: The complexity of modern multi-modal models will decrease interpretability. Radiologists report difficulties in understanding how deep models reach conclusions, with 38% of Artificial Intelligence (AI)-generated findings distrusted due to lack of transparency [10]. This "black-box" issue obscures critical diagnostic information, such as whether a lesion is caused by bone density changes or adjacent tooth morphology.

To address these challenges, we propose PolarDETR, a novel method for jawbone lesion detection that enhances both accuracy and interpretability. By embedding clinical text-derived location information into the detection model's query space, PolarDETR aligns anatomical knowledge with CBCT images in a polar coordinate framework, improving lesion localization while maintaining interpretability. We also introduce an interpretability-as-a-service feature that pairs anatomical heatmaps with textual data, enabling clinicians to understand the rationale behind AI diagnoses and increasing trust in the model's output.

2 Methodology

2.1 Overall Framework

An overview of the proposed framework is shown in Fig. 1. Our model is based on the DEtection TRansformer (DETR) architecture for end-to-end object detection [2]. It includes three key components for jawbone lesion detection: **P**olar **T**ext-**P**osition **E**ncoding (**PTPE**), Anatomical Constraint Learning, and Position Matching Optimization. PTPE maps clinical text to spatial coordinates in the CBCT image, aligning anatomical locations with geometric features. Anatomical priors guide the detection process to ensure lesions are within plausible regions, while Position Matching Optimization refines spatial alignment, improving accuracy and interpretability.

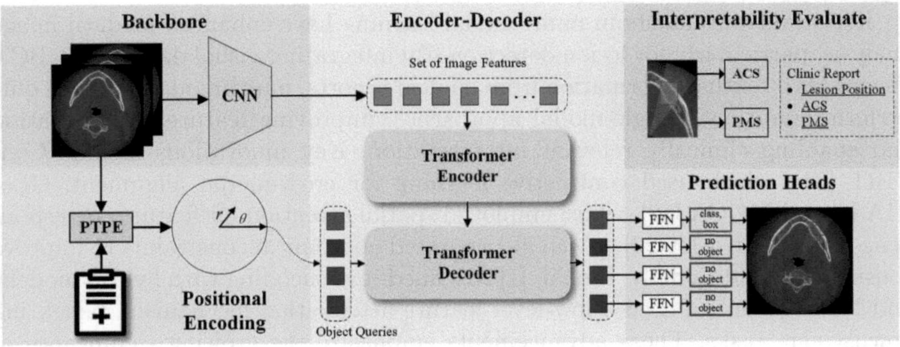

Fig. 1. The Overall Framework of Our PolarDETR.

2.2 Polar Text-Position Encoding

To effectively incorporate anatomical location information from clinical text into image-based lesion detection in CBCT images, we introduce PTPE. This approach aligns the text and image representations in a shared feature space by leveraging the spatial geometry of anatomical regions described in the clinical text.

Anatomical Consistency with the Jaw Structure. The polar coordinate system demonstrates superior congruence with jaw anatomy compared to Cartesian frameworks. Oriented on radial anatomical patterns (e.g., orientation of the trabeculae, neurovascular bundles of the mental foramen), polar parameters (r, θ) inherently encode biomechanical characteristics lost in rectangular grids [13]. Clinically, this spatial representation directly maps to diagnostic descriptions (e.g., "5 mm distal to mental foramen, third quadrant"), enhancing AI-output interpretability through native clinical lexicon alignment [14].

Design of PTPE. To construct PTPE, we first extract key positional information from the clinical text using a Named Entity Recognition (NER) model [11]. The PTPE pipeline includes an offline NER fine-tuning stage and an online inference stage integrated with PolarDETR. This model identifies anatomical terms (e.g., tooth numbers, quadrants) and their relationships. Critical elements like tooth number, quadrant, and distance from reference points (e.g., mental foramen), are parsed and converted into polar coordinates. The resulting PTPE is a 3D vector represented by θ (angle from the position of the main symptom) and r (radius from the description of the distance):

$$\text{PTPE} = [\sin(\theta), \cos(\theta), \log(r + 1)] \in \mathbb{R}^3 \tag{1}$$

To ensure geometrical consistency, a standard coordinate system with the chin foramen as the origin is defined, creating a polar coordinate system. To align image-based queries with anatomical information, PTPE vectors are integrated into DETR's object queries. This approach narrows the model's search space, focusing on anatomically relevant regions, accelerating convergence, and improving accuracy. The enhanced queries are computed as:

$$Q^{'} = Q + W_p \tag{2}$$

2.3 Anatomical Constraint and Position Matching Loss(AC-PML)

In this section, we introduce the AC-PML to improve the spatial accuracy of object localization and anatomical consistency in jawbone CBCT images. By combining anatomical understanding and precise alignment, we improve the localization of lesions and objects within the oral cavity.

Anatomical Constraint Learning. Anatomical constraint learning focuses on aligning image features with predefined anatomical regions. Most methods are handcrafted or use artificial prior information, lacking interpretability and ignoring the anatomical structure of complex regions, such as jawbone [19]. We define K anatomical regions, such as the alveolar bone and mandibular canal, using binary masks $\{M_k\}_{k=1}^{k}$ [22]. These masks serve as anatomical references for the model. For each image, we capture the relationship between image features and anatomical regions by normalizing the similarity between the query vector q and the average pooling of the anatomical mask M_k denoted as a_k. The association for each anatomical region is computed as follows:

$$\alpha_k = \frac{\exp(q^T a_k)}{\sum_{j=1}^{K}\exp(q^T a_k)}, a_k = \text{AvgPool}(M_k) \tag{3}$$

Anatomical Loss Function. To enforce anatomical consistency, we introduce an anatomical loss function based on Kullback-Leibler (KL) divergence [6] to minimize the distance between the predicted anatomical distribution α and the prior reflecting expected lesion locations α_{prior}. This approach uses anatomical priors to guide the model toward clinically plausible regions, ensuring detection focuses on anatomically relevant areas—crucial in jawbone CBCT images, where mislocalization will cause diagnostic errors. The anatomical loss is defined as:

$$L_{anatomy} = \text{KL}(\alpha || \alpha_{prior}) \tag{4}$$

Position Matching Learning. Previous research emphasizes the need to align text and visuals in tasks like lesion localization, as misalignment reduces performance [7]. Position-matching learning ensures spatial alignment between textual descriptions and corresponding image regions. Given the extracted polar coordinates (r, θ) from the text, we map these into a sector region R_{text} on the image.

The radial distance r is computed based on the pixel size and the text distance, while θ is the polar angle corresponding to the mapped tooth position. To align the predicted bounding box B_{pred} with the projected region R_{text}, we introduce a position-matching loss based on the Intersection over Union (IoU) [23] metric:

$$L_{position} = 1 - \text{IoU}(B_{pred}, R_{text}) \tag{5}$$

The final loss function combines anatomical constraint learning and position matching learning, allowing more accurate and anatomically consistent predictions. By jointly optimizing these losses, the model is trained to focus on anatomical regions and spatial alignment, improving localization and anatomical understanding in jawbone CBCT image analysis.

2.4 Interpretability Metrics

In medical image analysis, integrating multi-modal clinical data requires careful evaluation of interpretability. To assess our lesion detection framework, we introduce two evaluation metrics: the **A**natomical **C**onsistency **S**core (**ACS**) and the **P**osition **M**atching **S**core (**PMS**). These metrics quantify the anatomical plausibility and spatial alignment between the clinical text and image regions. As shown in Fig. 2, after generating detection results, the model evaluates interpretability by comparing the detected lesions with both predefined anatomical regions and clinical semantic mappings.

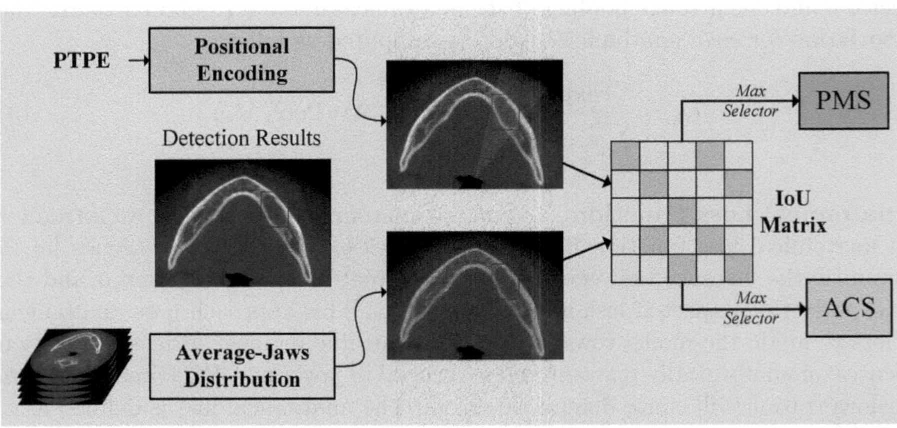

Fig. 2. Evaluation of Interpretability: Anatomical Consistency and Position Alignment.

Anatomical Consistency Score. The ACS quantifies the alignment between the predicted bounding boxes and anatomical regions. Specifically, ACS measures the IoU between each predicted bounding box and its corresponding anatomical region, taking the maximum IoU across all possible anatomical regions. The ACS is computed as follows:

$$\mathrm{ACS} = \frac{1}{N_{TP}} \sum_{N_{TP}}^{i=1} \max_{k} \mathrm{IoU}(B_i, M_k) \tag{6}$$

where N_{TP} is the number of true positive predictions, B_i is the i-th predicted bounding box and M_k represents the k-th anatomical region corresponding to the i-th predicted box. $\mathrm{IoU}(B_i, M_k)$ denotes the IoU between the predicted bounding box and the anatomical region. A higher ACS value indicates better anatomical plausibility of the model's predictions.

Position Match Score. The PMS measures the spatial consistency between the text descriptions and image predictions. This metric evaluates how well the predicted bounding boxes align with the anatomical regions described in the clinical text. The PMS is computed as:

$$\mathrm{PMS} = \frac{1}{N_{TP}} \sum_{i=1}^{N_{TP}} \mathrm{IoU}(B_i, R_{text,i}) > 0.5 \tag{7}$$

where B_i is the i-th predicted bounding box and $R_{text,i}$ is the reference region defined by the text description for the i-th sample. $\mathrm{IoU}(B_i, R_{text,i})$ denotes the IoU between the predicted bounding box and the region specified by the text. A higher PMS value indicates better spatial alignment between the text and the detected regions.

By using both ACS and PMS, we provide a comprehensive evaluation of the anatomical consistency of the model and the spatial alignment between the clinical text and image data. These metrics contribute to the interpretability of our model, ensuring that it not only achieves high detection accuracy but also provides trustworthy and interpretable predictions that can aid clinicians in their decision-making process.

3 Experiments

3.1 Data Preparation and Preprocessing

In our experiments, we used a dataset of jawbone image-text pairs from Peking University School and Hospital of Stomatology, including CBCT slices (JPG format) and corresponding clinical texts. Each pair was linked to a unique patient. We excluded pairs with misleading information or poor quality (e.g., noise, artifacts). After quality control, the final dataset had 120 image-text pairs, each with a 512×512 pixel resolution, split into 70% for training 15% for validation, and 15% for testing. CBCT slices were standardized to 512×512 pixels, and bounding-box labels were created using LabelMe. For lesion detection, the ground truth (GT) includes the predicted bounding box (BBox), sector-shaped area (PMS), and average jaw distribution (ACS) based on the FDI tooth position method. For NER, GT includes tooth number, quadrant, and distance to anatomical reference points. To augment the clinical text data, we used a language model-based strategy to increase the number of image-text pairs, refined through expert collaboration.

3.2 Implementation Details

All experiments were conducted using the PyTorch 2.4 framework and executed on four Nvidia RTX 4090 GPUs. Our model was trained from scratch for a total of 200 epochs. A batch size of 4 was used on each GPU, with an initial learning rate set to 2e-4. We applied cosine decay, with the final learning rate reaching 2e-6 by the end of training.

3.3 Comparisons

We compared our model's performance with state-of-the-art image-only models, such as YOLOv8 [18], DETR [2], and Retina U-Net [9], as well as Image-Text models like CLIP [15] and CORA [21]. For a fair comparison, we used the default parameters from the open-source codes of competing methods, ensuring consistent data volume and training cycles. We evaluated accuracy and interpretability using mAP [16] (↑), ACS (↑), and PMS (↑).

The results in Table 1 show the detection and interpretability performance of single-modal vs. multi-modal models. Among single-modal models, DETR achieved the highest mAP of 89.35%, followed by YOLOv8 (87.26%) and Retina U-Net (85.11%). Multi-modal models outperformed single-modal ones, with CORA achieving a mAP of 91.52% and PolarDETR leading with the highest mAP of 93.66%. In terms of interpretability, PolarDETR outperformed others with an ACS of 84.12% and PMS of 80.45%, surpassing both single-modal and multi-modal models. Detection results for PolarDETR, shown in Fig. 3, highlight superior lesion localization and anatomical consistency, demonstrating the practical effectiveness and robustness of our approach in clinical applications.

Table 1. Performance comparison of different models in detection and interpretability.

Model	Model Type	mAP(%)	ACS(%)	PMS(%)
YOLOv8	Single-modal (Image-only)	87.26	72.30	N/A
DETR		89.35	75.48	N/A
Retina U-Net		85.11	70.87	N/A
CLIP	Multi-modal (Image-text)	84.37	78.56	73.21
CORA		91.52	81.43	74.58
PolarDETR		**93.66**	**84.12**	**80.45**

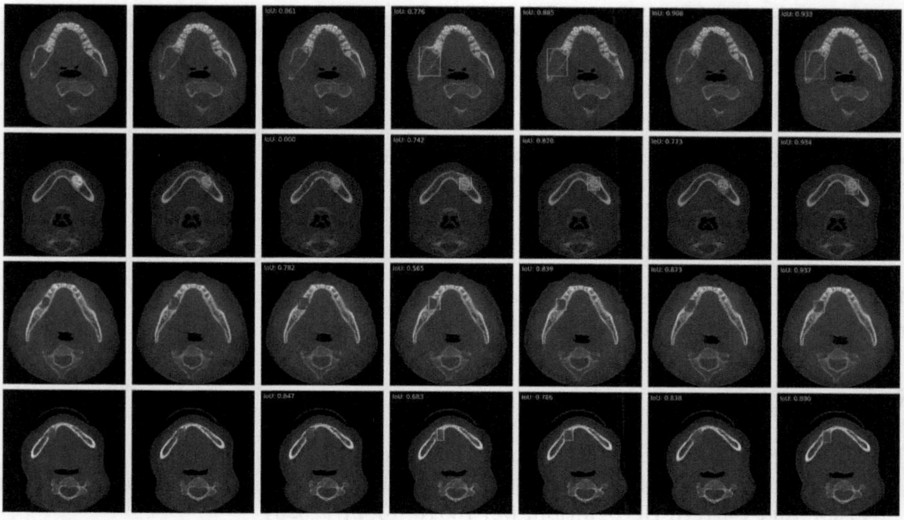

Fig. 3. Qualitative comparison of jawbone lesion detection results across different models. From left to right: original CBCT slices, GT, Retina U-Net, DETR, YOLOv8, CORA, and the proposed PolarDETR.

3.4 Ablation Study

We conducted ablation experiments to evaluate the effectiveness of the PTPE module and the AC-PML by training our model with and without these components. The first experiment compared the model with and without the PTPE module. The second experiment assessed the impact of AC-PML by comparing the full model with a version using only the standard DETR loss. Table 2 shows the results of these studies. Both the PTPE and AC-PML significantly improve detection accuracy, anatomical consistency, and position alignment.

Table 2. Ablation study results.

PTPE	AC-PML	mAP(%)	ACS(%)	PMS(%)
✗	✓	88.27	78.32	73.20
✓	✗	89.65	80.45	75.38
✓	✓	**93.66**	**84.12**	**80.45**

4 Conclusion

We proposed PolarDETR, a multi-modal approach that combines clinical text and CBCT images for jawbone lesion detection. With PTPE and AC-PML, our model enhances detection accuracy, anatomical consistency, and interpretability. Results show that PolarDETR outperforms existing models, offering greater interpretability for AI-assisted diagnosis. Future works will expand the dataset for better generalizability, explore integration with other imaging modalities, and focus on improving computational efficiency for clinical application.

Acknowledgments. This work was supported by the National Natural Science Foundation of China under Grant 62172029, the Beijing Natural Science Foundation under Grant L232029, and the Capital's Funds for Health Improvement and Research (CFH 2024-4-4107) from the Beijing Municipal Health Commission.

Disclosure of Interests. The authors have no competing interests to declare that are relevant to the content of this article.

References

1. Bannur, S., et al.: Learning to exploit temporal structure for biomedical vision-language processing. In: Proceedings of the IEEE/CVF Conference on Computer Vision and Pattern Recognition, pp. 15016–15027 (2023)
2. Carion, N., Massa, F., Synnaeve, G., Usunier, N., Kirillov, A., Zagoruyko, S.: End-to-end object detection with transformers. In: European Conference on Computer Vision, pp. 213–229 (2020)
3. Chen, R.Q., et al.: Leveraging pretrained transformers for efficient segmentation and lesion detection in cone-beam computed tomography scans. J. Endodontics **50**(10), 1505–1514 (2024)
4. Cho, H.: CNN-based autoencoder and post-training quantization for on-device anomaly detection of cartesian coordinate robots. In: 2024 IEEE 14th Annual Computing and Communication Workshop and Conference, pp. 0662–0666 (2024)
5. Colangelo, C.M., Chung, L., Bruce, C., Cheung, K.H.: Review of software tools for design and analysis of large scale MRM proteomic datasets. Methods **61**(3), 287–298 (2013)
6. Cui, J., Tian, Z., Zhong, Z., Qi, X., Yu, B., Zhang, H.: Decoupled kullback-leibler divergence loss. Adv. Neural. Inf. Process. Syst. **37**, 74461–74486 (2024)
7. Hossain, E., Sharif, O., Hoque, M.M., Preum, S.M.: Align before attend: aligning visual and textual features for multimodal hateful content detection (2024). arXiv preprint arXiv:2402.09738
8. Huang, S.C., Shen, L., Lungren, M.P., Yeung, S.: Gloria: A multimodal global-local representation learning framework for label-efficient medical image recognition. In: Proceedings of the IEEE/CVF International Conference on Computer Vision, pp. 3942–3951 (2021)
9. Jaeger, P.F., et al.: Retina u-net: Embarrassingly simple exploitation of segmentation supervision for medical object detection. In: Machine Learning for Health Workshop, pp. 171–183 (2020)
10. Lastrucci, A., Wandael, Y., Ricci, R., Maccioni, G., Giansanti, D.: The integration of deep learning in radiotherapy: exploring challenges, opportunities, and future directions through an umbrella review. Diagnostics **14**(9), 939 (2024)
11. Li, J., Sun, A., Han, J., Li, C.: A survey on deep learning for named entity recognition. IEEE Trans. Knowl. Data Eng. **34**(1), 50–70 (2020)
12. Liu, M., Liu, Y., Cui, H., Li, C., Ma, J.: Mgct: Mutual-guided cross-modality transformer for survival outcome prediction using integrative histopathology-genomic features. In: 2023 IEEE International Conference on Bioinformatics and Biomedicine, pp. 1306–1312 (2023)

13. Mei, S., Jiang, R., Ma, M., Song, C.: Rotation-invariant feature learning via convolutional neural network with cyclic polar coordinates convolutional layer. IEEE Trans. Geosci. Remote Sens. **61**, 1–13 (2023)
14. Pelé, A., Berry, P.A., Evanno, C., Jordana, F.: Evaluation of mental foramen with cone beam computed tomography: a systematic review of literature. Radiol. Res. Pract. **2021**(1), 8897275 (2021)
15. Radford, A., et al.: Learning transferable visual models from natural language supervision. In: International Conference on Machine Learning, pp. 8748–8763 (2021)
16. Shah, T.: Measuring object detection models–map–what is mean average precision. Tarang Shah-Blog **26**, 104332 (2018)
17. Shi, Y.J., et al.: Deep learning in the diagnosis for cystic lesions of the jaws: a review of recent progress. Dentomaxillofacial Radiol. **53**(5), 271–280 (2024)
18. Varghese, R., Sambath, M.: Yolov8: A novel object detection algorithm with enhanced performance and robustness. In: 2024 International Conference on Advances in Data Engineering and Intelligent Computing Systems, pp. 1–6 (2024)
19. Wang, X., Li, X., Du, R., Zhong, Y., Lu, Y., Song, T.: Anatomical prior-based automatic segmentation for cardiac substructures from computed tomography images. Bioengineering **10**(11), 1267 (2023)
20. Wang, Z., Wu, Z., Agarwal, D., Sun, J.: Medclip: Contrastive learning from unpaired medical images and text. In: Proceedings of the Conference on Empirical Methods in Natural Language Processing. Conference on Empirical Methods in Natural Language Processing. vol. 2022, p. 3876 (2022)
21. Wu, X., Zhu, F., Zhao, R., Li, H.: Cora: Adapting clip for open-vocabulary detection with region prompting and anchor pre-matching. In: Proceedings of the IEEE/CVF Conference on Computer Vision and Pattern Recognition, pp. 7031–7040 (2023)
22. Xie, Y., Yang, B., Guan, Q., Zhang, J., Wu, Q., Xia, Y.: Attention mechanisms in medical image segmentation: A survey (2023). arXiv preprint arXiv:2305.17937
23. Yu, J., Jiang, Y., Wang, Z., Cao, Z., Huang, T.: Unitbox: An advanced object detection network. In: Proceedings of the 24th ACM International Conference on Multimedia, pp. 516–520 (2016)
24. Zhang, Y., Jiang, H., Miura, Y., Manning, C.D., Langlotz, C.P.: Contrastive learning of medical visual representations from paired images and text. In: Machine Learning for Healthcare Conference, pp. 2–25 (2022)

Predicting Femoral Head Collapse Risk in Osteonecrosis Using Label Tokenization: A Multi-modality Survival Analysis Approach

Ganping Li[1(✉)], Yoshito Otake[1], Yuito Kameda[1], Keisuke Uemura[2], Kazuma Takashima[2], Hirokazu Mae[2], Sotaro Kono[2], Hidetoshi Hamada[2], Seiji Okada[2], Nobuhiko Sugano[2], and Yoshinobu Sato[1]

[1] Division of Information Science, Graduate School of Science and Technology, Nara Institute of Science and Technology, Ikoma, Nara, Japan
{li.ganping.lc2,otake,yoshi}@naist.ac.jp

[2] Department of Orthopaedics, Osaka University Graduate School of Medicine, Osaka, Japan

Abstract. Collapse of the femoral head is a critical event in osteonecrosis (ONFH) that often leads to debilitating hip pain and necessitates total hip arthroplasty. Early and accurate prediction of collapse risk is crucial for personalized treatment planning. While many studies focus on the automated diagnosis of ONFH, prognosis remains less explored. In this study, we propose a robust tri-stream deep learning framework that extracts features from T1-weighted MRI, region-of-interest (ROI) labels, and ONFH grades to estimate patient-specific collapse risk. We introduce an independent Spatial Label Encoder (SLE) module that tokenizes discrete ROI labels into dense, context-rich embeddings, thereby facilitating multi-modality model training. Experiments on 92 hips (70 patients) show that our approach performs competitively with state-of-the-art (SOTA) methods across most metrics, achieving a concordance index (CI) of 0.847 ± 0.087 and an integrated AUC of 0.884 in 5-fold cross-validation. Notably, the SLE module enhances long-term discrimination by up to 2.4% on AUC at 60 months compared to our base network. These findings highlight the potential benefits of late-fusion strategies with label tokenization for predicting femoral head collapse in ONFH, contributing to improved early intervention and prognosis.

Keywords: Survival Prediction · Osteonecrosis of the Femoral Head · Representation Learning · Deep Survival Model

1 Introduction

Osteonecrosis of the femoral head (ONFH) is a major health concern among young and middle-aged individuals, often leading to severe hip pain and dysfunction as it progresses toward femoral head collapse [2,5]. Since collapse is a

Y. Sato—Currently at Department of Mathematical and AI Medical Science, Nara Medical University.

J. C. Gee et al. (Eds.): MICCAI 2025, LNCS 15960, pp. 512–522, 2026.
https://doi.org/10.1007/978-3-032-04927-8_49

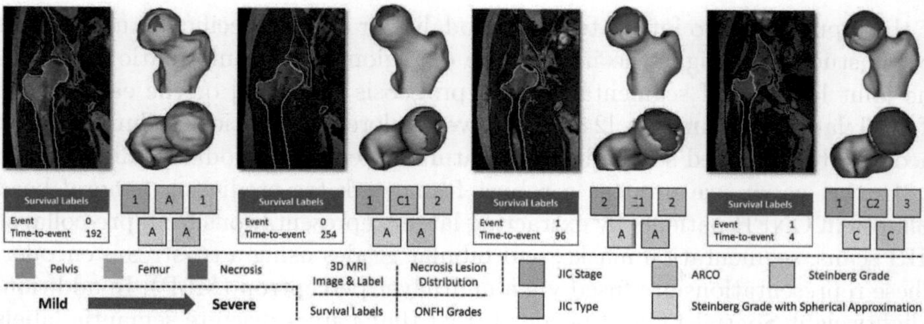

Fig. 1. Representative samples from our ONFH dataset, each annotated with multiple grade labels from different grading systems due to the lack of a universal standard. The severity of ONFH increases from left to right across the four samples.

pivotal turning point in ONFH, timely risk assessment is crucial for determining whether patients may benefit from conservative treatment or require immediate hip preservation surgery, such as total hip arthroplasty (THA) [8,12]. Figure 1 illustrates representative samples of ONFH severity progression, central to our study on femoral head collapse prediction. While many recent studies focus on diagnosing ONFH, the risk of subsequent femoral head collapse is often overlooked [11,20]. Although manual risk assessment methods have been proposed for prognostic studies of ONFH, they are time-consuming and prone to observer variability. As a result, automated deep learning approaches have gained attention for their efficiency and consistency [6].

A key challenge in predicting ONFH femoral head collapse risk is accurately modeling the relationship between the risk score and its covariates (e.g., patient data). Conventional survival prediction often relies on predefined Radiomics-based features derived from segmented regions of interest (ROI) [26], which is then processed by either statistical models (e.g., Cox Proportional Hazard (CoxPH) [3]) or machine learning models (e.g., survival support vector machine (SVM) [25]). Recent studies have proposed deep learning-based survival models with fusion mechanisms to better capture complex associations between covariates and survival outcomes. These methods employ various convolutional neural networks (CNNs)- or transformer-based encoders to leverage latent information from multimodal data such as radiological scans, electronic health records (EHR), and whole slide images (WSI) [21,22,27].

However, the presence of limited follow-up data (i.e., right-censored data) poses a challenge to the performance of survival prediction models. To address this, Qu et al. [19] introduced an auxiliary network to estimate hazards for censored intervals by leveraging statistics from uncensored and extended follow-up samples. In contrast, SurvRNC [22] captures inter- and intra-class relationships among censored and uncensored samples through representation alignments in survival-time order. Despite these advancements, both approaches rely on early-fusion strategies (e.g., concatenating multimodal images or spatial segmentations at the input stage) to integrate multi-modality or region-specific features. While recent studies highlight the importance of region-specific information, such as the joint learning of segmentation and prognosis tasks [15] or the early fusion of ROI labels and images [24], few have explored late-fusion techniques that incorporate tokenized semantic label features to enhance modality alignment.

In this paper, we propose a robust framework for predicting femoral head collapse in ONFH patients by extracting latent representations from pre-collapse MRI scans, segmentation masks, and tabular grades using a tri-stream encoder. These representations are fused via a multilayer perceptron (MLP). In addition, a lightweight Spatial Label Encoder (SLE) that parses discrete semantic labels into learnable embeddings is proposed. This design yields compact, context-rich representations of label distributions that can be seamlessly integrated with image-based features for survival analysis.

The contributions of this paper are three-fold: 1) to the best of our knowledge, this is the first general survival analysis framework for multi-modality femoral head collapse risk prediction, which could facilitate personalized surgical planning (e.g., THA) for ONFH patients; 2) the impact of incorporating the proposed SLE module is assessed on both our model and a state-of-the-art (SOTA) deep learning model, showing consistent performance gains; 3) the effectiveness of the proposed framework is demonstrated and compared with several SOTA methods using a 92-case MRI dataset of pre-collapse ONFH hips.

2 Method

2.1 Dataset and Preprocessing

We retrospectively collected T1-weighted MRI scans from the first hospital visit of 70 patients with ONFH, covering 92 hips (Japanese Investigation Committee (JIC) classification: 52 hips in stage 1, 40 in stage 2), as illustrated in Fig. 1. All MRI scans were bisected and flipped to the right for left hip targets, resulting in volumes of shape $(128, 256, n)$ with voxel spacing $(1.25, 1.25, 1)$ mm, where $n \in [54, 159]$. Among these, 63 hips were manually annotated by an experienced orthopedic surgeon, while the remaining 29 were automatically segmented using a Dynamic U-Net [1,9] trained on the labeled cases. The mean Dice coefficients for the pelvis, femur, and necrosis were 0.931, 0.945, and 0.890, respectively. Since no universal ONFH grading standard exists, each hip was assigned multiple classifications: JIC stage, JIC type, Association Research Circulation Osseous classification (ARCO), Spherical Approximation Steinberg grade, and Steinberg classification [23]. To address the modest dataset size and ensure robust findings, all experiments employed a strict patient-level 5-fold cross-validation. We are also working to secure permissions to release the data publicly in the future.

Femoral head collapse during the study period was considered the event of interest, with time-to-event measured from the first visit to either the occurrence of collapse (uncensored data) or the last available follow-up (right-censored). Approximately 47% of patients had uncensored data, with time-to-event ranging

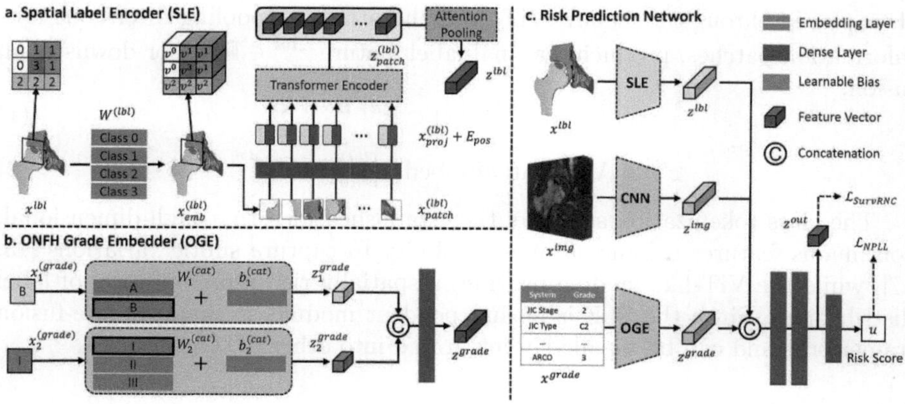

Fig. 2. Framework overview: a. **Spatial Label Encoder** maps discrete class labels to n-dim learnable vectors and extracts spatial and inter-class features in a ViT-like manner. **b. ONFH Grade Embedder** tokenizes categorical grading inputs and projects the tokens into grade features. **c. Risk Prediction Network** predicts the risk of femoral head collapse using the patient's MRI scans, ROI labels, and ONFH grades.

from 1 to 174 months. The median time-to-event was 12 months (IQR: 3.533 months), with 90% of events occurring within 82 months.

2.2 Femoral Head Collapse Risk Prediction

Problem Statement. This task aims to predict the risk of femoral head collapse in ONFH hips using multimodal data: MRI scans x_{img} of size (H, W, D), ROI labels x_{lbl}, and ONFH grades x_{grade}. We denote our dataset by $\mathcal{D} = \{\mathcal{H}_1, \ldots, \mathcal{H}_N\}$, where each hip $\mathcal{H}_i$ comprises features $X_i = (x_i^{img}, x_i^{lbl}, x_i^{grade})$, an event indicator e_i (0 for censored, 1 for uncensored), and a time-to-event T_i. The objective is to train a deep network f to predict a risk score $u = f(X)$ (with higher values indicating a worse prognosis), which can be further interpreted by survival functions $S(t \mid X)$ (e.g., via the Kaplan-Meier estimator [10]).

Spatial Label Encoder. Given a 3D image and its ROI label, a standard approach is to concatenate the scalar or one-hot label with the image to get $\mathbf{X} = [x^{img}; x^{lbl}]$ in an early fusion mechanism. While intuitive and straightforward, this approach has two drawbacks: 1) the discrete and sparse nature of categorical labels introduces redundancy, especially in multi-class settings [28,29]; and 2) early fusion can lead to insufficient intra-modality information due to entangled (concatenated) features [15]. To address these issues, we propose SLE that tokenizes the discrete multi-class label map and encodes it with a shallow vision transformer (ViT) [4], as illustrated in Fig. 2a. Given a scalar-valued label map $x^{lbl} \in \mathbb{R}^{1\times H\times W\times D}$, an embedding matrix $W^{(lbl)} \in \mathbb{R}^{C\times d}$ maps each class index c to a d-dimensional vector $v^{cls} = W^{(lbl)}[c]$, yielding a rich, learnable representation compared to sparse one-hot labels. The embedded labels $x_{emb}^{(lbl)}$ are then passed through a shallow ViT, and the attention pooling filters out low-information patches, producing a final label feature $z^{lbl} \in \mathbb{R}^{d^{lbl}}$ for downstream fusion:

$$x_{emb}^{(lbl)} = W^{(lbl)}(x^{lbl}) \in \mathbb{R}^{d \times H \times W \times D} \tag{1}$$

$$z^{lbl} = \text{ViT}(\text{PatchEmbed}(x_{emb}^{(lbl)})) \in \mathbb{R}^{d^{lbl}} \tag{2}$$

The class tokenization supports the established idea that high-dimensional, continuous features enhance a model's ability to capture subtle variations [13, 17], while the ViT-like module provides a spatially rich representation of label distributions. Since the SLE is an independent module, it supports late-fusion frameworks and can be seamlessly integrated into other SOTA methods.

ONFH Grade Embedder. This module follows the same scalar-value tokenization concept in SLE and leverages the Feature Tokenizer Transformer (FT-Transformer) [7] for categorical value embedding, as illustrated in Fig. 2.b. Given a categorical index x_j^{grade}, where $j \in \{1, 2, \ldots, G\}$ (and G is the number of grade systems), we define:

$$z_j^{(grade)} = b_j^{(cat)} + W_j^{(cat)}\left(x_j^{grade}\right) \quad \in \mathbb{R}^d, \tag{3}$$

$$z^{grade} = f_{\text{proj}}\left([z_1^{(grade)}; z_2^{(grade)}; \ldots; z_G^{(grade)}]\right) \in \mathbb{R}^{d^{grade}}, \tag{4}$$

where $b_j^{(cat)}$ represents the bias feature in the j-th categorical system, f_{proj} denotes a linear projection, and z^{grade} is the embedded grading feature of dimension d^{grade}.

Proposed Framework. Fig. 2.c demonstrates the overall risk prediction network. A tri-stream encoder, consisting of CNN-based vision encoder (shallow ResNet-18) f_{CNN}, SLE f_{SLE}, and ONFH grade embedder f_{OGE}, takes input $X = [x^{img}, x^{lbl}, x^{grade}]$ and outputs $z^{img} \in \mathbb{R}^{d^{img}}$, z^{lbl}, and z^{grade}, respectively. The three feature vectors are then fused by simple concatenation and passed to a 2-layer MLP to get feature representations z^{out}, and the final output risk score u is predicted from z^{out} through a dense output layer. To optimize model parameters, we combine the negative partial log-likelihood loss $\mathcal{L}_{\text{NPLL}}(u, e, T)$ [3] with an ordered representation alignment loss $\mathcal{L}_{\text{SurvRNC}}(\cdot)$ [22], where anchor, positive, and negative embeddings $(z_a^{out}, z_p^{out}, \{z_k^{out}\}_{k \in \mathcal{N}_{a,p}})$ enforce ordinal constraints. The total loss is:$\mathcal{L}_{\text{Total}} = \mathcal{L}_{\text{NPLL}} + \lambda * \mathcal{L}_{\text{SurvRNC}}$, with λ empirically set to 0.5.

2.3 Evaluation Metrics

Four metrics were used to thoroughly evaluate our method: the concordance index (CI), the time-dependent Area Under the Curve (AUC(t)), the Integrated AUC (iAUC), and KM analysis. CI, as the primary metric, measures the model's

Table 1. Performance of different SOTA survival prediction methods on a 5-fold cross-validation evaluation. The best prediction values are shown in bold. Note that *CoxPH was trained only by ONFH grades.

Metrics → Methods↓	CI ↑	AUC (t) ↑					iAUC ↑
		6(m)	12(m)	36(m)	60(m)	120(m)	12~120(m)
CoxPH* [3]	0.820±0.059	0.875±0.061	0.869±0.082	0.872±0.070	0.853±0.108	0.852±0.108	0.860±0.096
SurvivalSVM [25]	0.781±0.095	0.829±0.160	0.856±0.168	0.882±0.106	0.858±0.073	0.806±0.103	0.851±0.081
DeepMTS [16]	0.824±0.093	0.825±0.107	0.841±0.111	0.857±0.075	0.867±0.069	0.840±0.111	0.861±0.078
XSurv [15]	0.822±0.068	0.878±0.080	**0.912±0.064**	**0.920±0.065**	0.856±0.079	0.783±0.139	0.865±0.081
SurvRNC [22]	0.832±0.094	0.877±0.086	0.878±0.122	0.863±0.078	0.868±0.104	0.858±0.132	0.869±0.103
Ours (w/o SLE)	0.838±0.076	0.879±0.101	0.885±0.136	0.874±0.077	0.853±0.087	0.866±0.117	0.866±0.091
SurvRNC (w SLE)	0.841±0.088	0.879±0.096	0.894±0.120	0.875±0.071	0.868±0.102	0.860±0.120	0.873±0.098
Ours (w SLE)	**0.847±0.087**	**0.883±0.105**	0.885±0.122	0.880±0.068	**0.877±0.085**	**0.883±0.106**	**0.884±0.085**

overall discriminative power (ranging from 0 to 1). In contrast, AUC(t) evaluates the model's time-specific discrimination ability to distinguish individuals who experience an event before time t versus those surviving beyond t. iAUC is the aggregated version of AUC(t) that reflects overall discrimination ability over a time range $[t_{min}, t_{max}]$. Finally, KM analysis determines the stratification efficacy of the model by dividing patients into high-risk and low-risk groups.

3 Experiments and Results

3.1 Baselines and Configurations

To evaluate the validity of our proposed method, we conducted comparative experiments with several SOTA multi-modality networks that use 3D medical images as their primary input. Specifically, we assessed DeepMTS [16], XSurv [15], and SurvRNC [22] using a 5-fold patient-wise cross-validation on our femoral head collapse dataset. During training, 20% of the hip joints were set aside as the validation set for each fold. For a fair comparison, we replaced the original models' sub-modalities and clinical indicators with comparable sources of collapse-related information (as used by [6] for ROI labels and [2] for ONFH grades). Additionally, we compared our method with two conventional survival models, CoxPH and SurvivalSVM, using Radiomics features to represent both image and label modalities. Radiomics feature extraction and selection followed the preprocessing pipeline described in [6,15]. For all methods (both classical and deep learning), we report only the highest evaluation scores across different modality combinations.

All deep learning models were fine-tuned based on their default hyperparameters and survival losses ($\mathcal{L}_{\text{NPLL}}$ or $\mathcal{L}_{\text{Total}}$), trained and tested on a computing device with an Intel Xeon W-2295 @3.00GHz CPU, 125GB RAM, and an NVIDIA A6000 GPU (48 GB) using PyTorch [18] version 2.0.1 with CUDA 11.8. Each model underwent the same data augmentations and was trained for 310 epochs with a batch size of 16, a learning rate of 4×10^{-5}, and the AdamW optimizer [14] (weight decay of 10^{-4}).

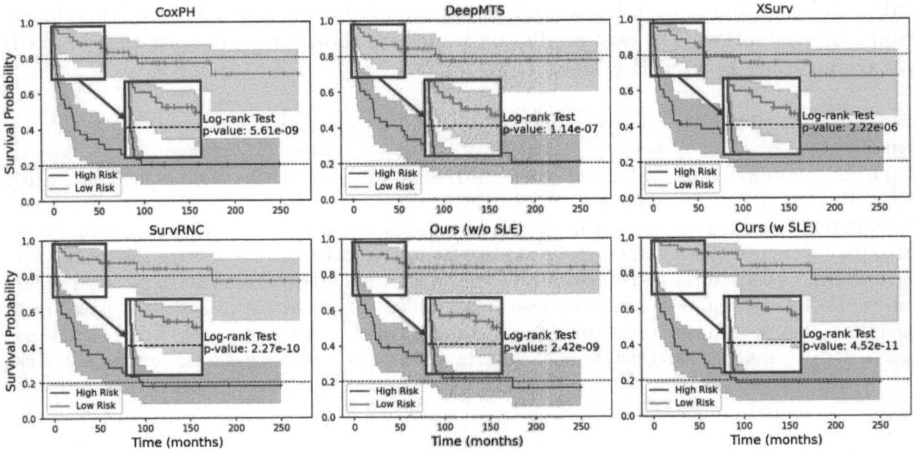

Fig. 3. KM analysis for CoxPH and deep learning methods on our ONFH dataset. Our full method with SLE achieves the best p-value in the log-rank test while our base network demonstrates the best stratification ability at the end of the censor duration.

To further validate our approach, we tested our method with the SLE module replaced by early fusion to confirm the effectiveness of the base network. Additionally, we integrated our SLE module into SurvRNC to demonstrate its adaptability and efficacy. For our proposed method, the dimensions d^{img}, d^{lbl}, and d^{grade} were empirically set to 128, 128, and 64, respectively.

3.2 Results and Discussions

Table 1 compares the performance of all methods across three metrics. Notably, CoxPH achieves its best performance when trained solely on categorical ONFH grades, whereas other methods require all modalities. Our proposed base network (w/o SLE) achieves the highest CI (0.838) among previous methods and demonstrates strong short- to mid-term AUC ($t \leq 36$), highlighting the effectiveness of tokenization-based embedding for tabular data. When integrated with SLE, our method addresses the long-term discrimination issue ($t = 60, 120$) and shows improvements over most metrics (except mid-term $t = 12, 36$). In particular, AUC(60) increases by 0.024, AUC(120) by 0.017, and iAUC by 0.018. Adding SLE to the second-best (w/o SLE) method SurvRNC also improves all metrics, suggesting enhanced feature representation while maintaining model stability.

Regarding baseline methods, CoxPH serves as a strong competitor when trained only on categorical ONFH grades, aligning with previous studies [15, 22] given the close relationship between disease grades and ONFH prognosis [6]. In contrast, SurvivalSVM performs the worst on CI, likely due to convergence failure by the limited amount of trainable data. Among deep learning approaches, DeepMTS slightly outperforms CoxPH on CI but performs worse in AUC, possibly due to its entangled volume inputs and its original multitask design. XSurv achieves the highest mid-term AUC(36) at 0.92, which may benefit from its dual symmetric encoder of the same perception field for volumes from two modalities with aligned spatial information [27]. However, the deeper network is more sensitive to a small, right-skewed dataset (as is common with uncensored data in medical scenarios), resulting in poorer long-term AUC. SurvRNC, despite employing early fusion, provides balanced results across different time windows thanks to its effective representation alignment loss $\mathcal{L}_{\text{SurvRNC}}$. Our proposed method integrates key multimodal mechanisms from previous SOTA approaches while addressing their limitations by: 1) disentangling spatial modalities via multiple lightweight encoders for late fusion, 2) facilitating implicit cross-modality alignment by maintaining tokenizers in the SLE and ONFH grade encoders, and 3) aligning representation in survival-time order through $\mathcal{L}_{\text{SurvRNC}}$.

To further assess stratification efficacy, Fig. 3 presents KM curves for CoxPH and various deep learning methods, with shaded regions indicating confidence intervals. The log-rank test p-values confirm statistically significant differences between risk groups. While our base network shows a larger discrepancy at the final time ($t > 174$), incorporating SLE results in a smaller p-value between risk groups. Our full method with SLE also yields a more stable survival curve for $t < 60$ in the low-risk group and a steeper early decline in the high-risk group, leading to a more distinct separation between risk groups earlier, which is clinically valuable for predicting femoral head collapse where timely intervention is critical. These findings align with the numerical estimates in Table 1, demonstrating a moderate overall advantage of our approach over other SOTA methods on the ONFH dataset, as reflected by lower p-values and earlier curve separation.

3.3 Limitations and Future Works

The proposed framework has two main limitations. First, our study is based on a small, single-center retrospective dataset, which may impact the generalizability of the evaluated methods. Some ground-truth labels may also contain noise due to reliance on automated segmentation. Second, all modalities in our experiments are pre-aligned temporally, and volume-based modalities (MRI and its embedded ROI labels) are also spatially aligned in advance. The impact of these alignments requires further investigation, particularly in the context of follow-up data integration of different modalities and positioning.

Future work will focus on enhancing robustness by incorporating external datasets and mitigating label noise through uncertainty-aware modeling. We also plan to integrate additional imaging modalities with diverse spatial and temporal characteristics (e.g., X-ray and follow-up Computed Tomography) to further assess the model's generalizability. Moreover, a comprehensive analysis of each modality's contribution is needed. Further refinements to the framework will aim to maximize the benefits of SLE. Lastly, ongoing follow-up data collection will support more comprehensive comparisons in future studies.

4 Conclusions

This study presents a tri-stream framework[1] that integrates MRI scans, ROI labels, and ONFH grades to predict femoral head collapse risk, while also evaluating several SOTA multi-modality methods in this context. By combining label tokenization and time-ordered representation alignments, our approach reduces the redundancy of scalar or one-hot multi-class labels while facilitating implicit cross-modality alignment. The proposed SLE has demonstrated promising and consistent improvements, suggesting its potential as an enhanced label-encoding strategy. While further validation is required, our framework provides a structured and robust baseline for ONFH prognosis, with the potential to aid early clinical interventions and advance research in personalized medicine.

Acknowledgments. This study was funded by MEXT/JSPS KAKENHI (19H01176, 20H04550, 21K16655, 23K15714) and AMED under Grant Number JP25hma322015.

Disclosure of Interests. The authors have no competing interests to declare that are relevant to the content of this article.

References

1. Cardoso, M.J., et al.: Monai: An open-source framework for deep learning in healthcare (2022). arXiv preprint arXiv:2211.02701
2. Chen, L., et al.: Predicting the collapse of the femoral head due to osteonecrosis: from basic methods to application prospects. J. Orthop. Transl. **11**, 62–72 (2017)
3. Cox, D.R.: Regression models and life-tables. J. Roy. Stat. Soc.: Ser. B (Methodol.) **34**(2), 187–202 (1972)
4. Dosovitskiy, A., et al.: An image is worth 16x16 words: transformers for image recognition at scale (2020). arXiv preprint arXiv:2010.11929
5. Fan, Y., Liu, X., Zhong, Y., Zhang, J., Liu, Y., Fang, H., He, W., Zhou, C., Chen, Z.: Evaluation of the predictive values of collapse and necrotic lesion boundary for osteonecrosis of the femoral head prognosis. Front. Endocrinol. **14**, 1137786 (2023)
6. Gao, S., et al.: Prediction of femoral head collapse in osteonecrosis using deep learning segmentation and radiomics texture analysis of MRI. BMC Med. Inform. Decis. Mak. **24**(1), 320 (2024)
7. Gorishniy, Y., Rubachev, I., Khrulkov, V., Babenko, A.: Revisiting deep learning models for tabular data. Adv. Neural. Inf. Process. Syst. **34**, 18932–18943 (2021)
8. Hindoyan, K.N., Lieberman, J.R., Matcuk, G.R., Jr., White, E.A.: A precise and reliable method of determining lesion size in osteonecrosis of the femoral head using volumes. J. Arthroplasty **35**(1), 285–290 (2020)
9. Isensee, F., et al.: NNU-net: Self-adapting framework for u-net-based medical image segmentation (2018). arXiv preprint arXiv:1809.10486
10. Kaplan, E.L., Meier, P.: Nonparametric estimation from incomplete observations. J. Am. Stat. Assoc. **53**(282), 457–481 (1958)

[1] The source code is available at: https://github.com/RIO98/FemoralCollapse Prediction.

11. Klontzas, M.E.: Deep learning enables the differentiation between early and late stages of hip avascular necrosis. Eur. Radiol. **34**(2), 1179–1186 (2024)
12. Kuroda, Y., et al.: Classification of osteonecrosis of the femoral head: who should have surgery? Bone Joint Res. **8**(10), 451–458 (2019)
13. Li, G., et al.: Prediction of disease-related femur shape changes using geometric encoding and clinical context on a hip disease CT database. In: International Conference on Medical Image Computing and Computer-Assisted Intervention, pp. 368–378. Springer (2024)
14. Loshchilov, I., Hutter, F.: Decoupled weight decay regularization (2017). arXiv preprint arXiv:1711.05101
15. Meng, M., Bi, L., Fulham, M., Feng, D., Kim, J.: Merging-diverging hybrid transformer networks for survival prediction in head and neck cancer. In: International Conference on Medical Image Computing and Computer-Assisted Intervention, pp. 400–410. Springer (2023)
16. Meng, M., Gu, B., Bi, L., Song, S., Feng, D.D., Kim, J.: Deepmts: Deep multi-task learning for survival prediction in patients with advanced nasopharyngeal carcinoma using pretreatment pet/CT. IEEE J. Biomed. Health Inform. **26**(9), 4497–4507 (2022)
17. Mildenhall, B., Srinivasan, P.P., Tancik, M., Barron, J.T., Ramamoorthi, R., Ng, R.: Nerf: Representing scenes as neural radiance fields for view synthesis. Commun. ACM **65**(1), 99–106 (2021)
18. Paszke, A., et al.: Pytorch: An imperative style, high-performance deep learning library. Adv. Neural Inf. Process. Syst. **32** (2019)
19. Qu, L., Huang, D., Zhang, S., Wang, X.: Multi-modal data binding for survival analysis modeling with incomplete data and annotations. In: International Conference on Medical Image Computing and Computer-Assisted Intervention, pp. 501–510. Springer (2024)
20. Rakhshankhah, N., Abbaszadeh, M., Kazemi, A., Rezaei, S.S., Roozpeykar, S., Arabfard, M.: Deep learning approach to femoral AVN detection in digital radiography: differentiating patients and pre-collapse stages. BMC Musculoskelet. Disord. **25**(1), 547 (2024)
21. Ramanathan, V., Pati, P., McNeil, M., Martel, A.L.: Ensemble of prior-guided expert graph models for survival prediction in digital pathology. In: International Conference on Medical Image Computing and Computer-Assisted Intervention, pp. 262–272. Springer (2024)
22. Saeed, N., Ridzuan, M., Maani, F.A., Alasmawi, H., Nandakumar, K., Yaqub, M.: Survrnc: Learning ordered representations for survival prediction using rank-n-contrast. In: International Conference on Medical Image Computing and Computer-Assisted Intervention, pp. 659–669. Springer (2024)
23. Sultan, A.A., et al.: Classification systems of hip osteonecrosis: an updated review. Int. Orthop. **43**, 1089–1095 (2019)
24. Tang, W., Zhang, H., Yu, P., Kang, H., Zhang, R.: Mmmna-net for overall survival time prediction of brain tumor patients. In: 2022 44th Annual International Conference of the IEEE Engineering in Medicine Biology Society (EMBC), pp. 3805–3808. IEEE (2022)
25. Van Belle, V., Pelckmans, K., Suykens, J.A., Van Huffel, S.: Survival SVM: a practical scalable algorithm. In: ESANN, pp. 89–94 (2008)
26. Van Griethuysen, J.J., et al.: Computational radiomics system to decode the radiographic phenotype. Can. Res. **77**(21), e104–e107 (2017)

27. Warner, E., et al.: Multimodal machine learning in image-based and clinical biomedicine: survey and prospects. Int. J. Comput. Vision **132**(9), 3753–3769 (2024)
28. Zhang, R., Kong, T., Wang, X., You, M.: Mask encoding: a general instance mask representation for object segmentation. Pattern Recogn. **124**, 108505 (2022)
29. Zhang, R., Tian, Z., Shen, C., You, M., Yan, Y.: Mask encoding for single shot instance segmentation. In: Proceedings of the IEEE/CVF Conference on Computer Vision and Pattern Recognition, pp. 10226–10235 (2020)

PRETI: Patient-Aware Retinal Foundation Model via Metadata-Guided Representation Learning

Yeonkyung Lee[1], Woojung Han[1], Youngjun Jun[1], Hyeonmin Kim[2], Jungkyung Cho[2], and Seong Jae Hwang[1(✉)]

[1] Yonsei University, Seoul, South Korea
{yeonkyung.lee,dnwjddl,youngjun,seongjae}@yonsei.ac.kr
[2] Mediwhale, Seoul, South Korea
{luke.kim,kyla.cho}@mediwhale.com

Abstract. Retinal foundation models have significantly advanced retinal image analysis by leveraging self-supervised learning to reduce dependence on labeled data while achieving strong generalization. Many recent approaches enhance retinal image understanding using report supervision, but obtaining clinical reports is often costly and challenging. In contrast, metadata (e.g., age, gender) is widely available and serves as a valuable resource for analyzing disease progression. To effectively incorporate patient-specific information, we propose PRETI, a retinal foundation model that integrates metadata-aware learning with robust self-supervised representation learning. We introduce Learnable Metadata Embedding (LME), which dynamically refines metadata representations. Additionally, we construct patient-level data pairs, associating images from the same individual to improve robustness against non-clinical variations. To further optimize retinal image representation, we propose Retina-Aware Adaptive Masking (RAAM), a strategy that selectively applies masking within the retinal region and dynamically adjusts the masking ratio during training. PRETI captures both global structures and fine-grained pathological details, resulting in superior diagnostic performance. Extensive experiments demonstrate that PRETI achieves state-of-the-art results across diverse diseases and biomarker predictions using in-house and public data, indicating the importance of metadata-guided foundation models in retinal disease analysis. Our code and pretrained model are available at https://github.com/MICV-yonsei/PRETI

Keywords: Foundation Model · Masked Image Modeling · Metadata

1 Introduction

Foundation models have recently gained attention for their remarkable adaptability across various tasks [5,24,30]. By training on extensive unlabeled data, they capture core patterns and relationships that can be fine-tuned with minimal effort for various downstream applications [11,38]. While their applications in areas like computer vision and natural language processing have shown promising results [5,21,30], their use in medical contexts remains relatively unexplored.

J. C. Gee et al. (Eds.): MICCAI 2025, LNCS 15960, pp. 523–533, 2026.
https://doi.org/10.1007/978-3-032-04927-8_50

A key motivation for developing medical foundation models lies in the scarcity of annotated medical data, as labeling clinical images requires expert knowledge and is both time-consuming and expensive. Recently, medical foundation models have shown the potential of using large-scale unlabeled data [26,40]. Especially in ophthalmology, retinal foundation models are being studied due to the discovery that clinical features associated with ocular diseases can be observed in color fundus photography (CFP) [7,35]. RETFound [40] demonstrated how image-based pre-training can serve as a strong foundation for retinal fundus analysis. Expanding on this, foundation models such as FLAIR [31], RETCLIP [13], and UrFound [36] incorporate report data, enhancing the multimodal understanding of retinal diseases. While these approaches have significantly advanced retinal image analysis through report supervision, expert descriptions of patients such as clinical diagnostic reports are difficult to obtain, thereby reducing the amount of available data. On the other hand, among various types of metadata, age and gender serve as minimal patient information that can be easily obtained. Moreover, they exert a critical influence on understanding how ocular pathologies develop, progress, and respond to treatment [22,33]. Thus, we incorporate metadata into foundation models to enable them to capture patient-specific variations.

Despite the significance of metadata, research on foundation models leveraging metadata has been relatively scarce [4,14,17]. These approaches treat metadata as static embeddings [4] or incorporate it through a contrastive loss criterion [14,17]. However, relying on metadata in such a static manner fails to fully leverage metadata for capturing complex image-metadata relationships. Instead, we introduce the *Learnable Metadata Embedding (LME)*. LME dynamically integrates metadata into the learning process, allowing it to adapt and evolve alongside visual features. In contrast to static embeddings or fixed constraints, LME enables a more flexible representation of retinal data. Specifically, we focus on the two key metadata variables (age and gender) which are fundamental to understanding retinal disease patterns [3,28].

Enhancing representation learning through metadata is a notable step forward. However, retinal foundation models still face challenges from imaging variations due to device differences, lighting, and inter-eye anatomical distinctions. To address these issues, we construct *patient-level data pairs*, grouping images from the same individual, even when captured with different cameras. Additionally, considering the clinical symmetry between the left and right eyes in ocular diseases [27,32,34], we treat them as equivalent when forming patient-level pairs. This strategy accounts for non-disease-related variations, enabling the model to learn robust representations. To effectively utilize both metadata and patient-level pairs, we build upon a self-supervised modeling framework based on Siamese Masked Autoencoder (SiamMAE) [15], an extension of MAE [16]. This architecture suits our patient-level pairs by disrupting one image and restoring it using information from the other, learning visual correspondence between both eyes.

While MAE is a self-supervised learning method that learns image features by reconstructing images from randomly masked inputs, its standard random masking strategy does not align well with retinal images, where important struc-

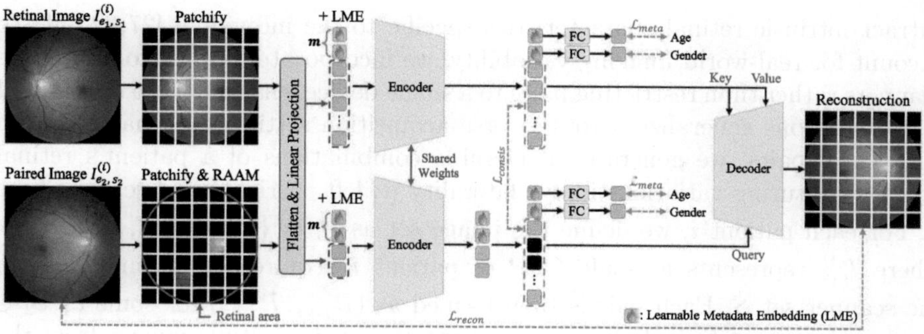

Fig. 1. The architecture of PRETI, a foundation model for retinal image representation learning, illustrates the integration of color fundus photography (CFP) and metadata through Learnable Metadata Embedding (LME). The model processes paired retinal images using Patient-level Pairs and employs shared encoders with dynamic metadata adaptation (age and gender). It also incorporates the Retina-Aware Adaptive Masking (RAAM) strategy for selective masking and reconstructing images via a decoder to enhance representation learning for robust clinical generalization.

tures are confined to the eye region. To address this, we implement *Retina-Aware Adaptive Masking (RAAM)*, a technique designed to align with the specific properties of retinal image data. RAAM selectively applies masking within the retinal region and dynamically adjusts the masking ratio during training. This refinement enhances the ability of the model to capture both global structures and fine details, making it better suited for retinal image analysis.

Based on patient-level pairs incorporating LME and RAAM, we confidently propose PRETI, a foundation model that focuses on CFP and metadata for comprehensive assessment. Through extensive experiments, we validate its state-of-the-art performance across diverse disease and biomarker prediction tasks.

2 Methods

Our foundation model, PRETI, integrates visual and metadata information for retinal image representation learning (Fig. 1). We first introduce patient-level data pairs, followed by the retina-aware adaptive masking strategy (RAAM), which is designed specifically for retinal images. Then, we present the learnable metadata embedding (LME) for capturing the complex relationship between images and metadata, along with the objectives introduced for training.

2.1 Patient-Level Data Pairs

We construct patient-level data by grouping CFPs for each individual, designing each pair to consist of two images from the same patient, potentially spanning both left and right eyes and different scanners. When pairs include left and right eye images, we leverage the natural symmetry and biological correlation to extract intrinsic retinal characteristics specific to the individual [27,32,34]. To account for real-world imaging variability, we incorporate images from multiple scanners rather than restricting pairs to a single device, ensuring that the learned representations generalize across diverse acquisition settings. To maximize the number of pairs, we generate all possible combinations of a patient's retinal images, capturing variations in eye laterality (l: left, r: right) and scanner type $\mathcal{S}$. For each patient i, we define the image set as $\mathcal{P}_i = \{I_{e,s}^{(i)} | e \in \{l, r\}, s \in \mathcal{S}\}$, where $I_{e,s}^{(i)}$ represents a e-side CFP of patient i acquired by scanner s from the scanner set $\mathcal{S}$. Each pair is then formed as $(I_{e_1,s_1}^{(i)}, I_{e_2,s_2}^{(i)})$ for some $e_1, e_2 \in \{l, r\}$ and $s_1, s_2 \in \mathcal{S}$, including cases where $s_1 = s_2$ if multiple images from the same scanner are available. This approach integrates both patient-specific retinal information and inter-scanner variability for effective model training.

2.2 Retina-Aware Adaptive Masking Strategy

To account for the distinct properties of retinal images, where the key features primarily reside within the retinal region, we propose a Retina-Aware Adaptive Masking (RAAM). This adaptive masking strategy selectively applies masking to the retinal region and dynamically adjusts the masking ratio during training.

(1) Region-Specific Retinal Masking. Traditional MAE training [16] applies random masking across the entire image. However, in retinal images, the non-retinal regions (e.g., background) do not contain meaningful information for learning. To ensure that the model focuses on critical retinal structures within the circular retinal image area, we introduce a Region-Specific Masking approach, where the unmasked patches are selectively chosen from within the retinal area (a red circle in Fig. 1). A patch is considered part of the retinal region if more than 50% of its pixels fall within the retinal mask. To implement this, we generate a retinal region mask using the automated retinal image analysis tool, Automorph [39].

(2) Cosine Decay-Based Masking Ratio Scheduling. In retinal image analysis, capturing both global structural patterns and fine-grained details is essential. To achieve this, we introduce a dynamic masking schedule that progressively reduces the masking ratio during training. Starting with a high masking ratio r_0, the model is encouraged to prioritize learning broad structural features, minimizing early reliance on localized information [8]. As training progresses, the masking ratio decreases towards the final masking ratio r_T, allowing the model to focus on fine-grained retinal structures while maintaining strong global representations. To ensure stable learning and a smooth transition, we adopt a Cosine Decay-Based Scheduling [25], which keeps the masking ratio stable early on and reduces it more rapidly as training progresses. The masking ratio r_t follows a cosine decay function: $r_t = 0.5\,(1 - \cos(\pi t/T))\,(r_T - r_0) + r_0$, where T is the total number of training epochs.

2.3 Learnable Metadata Embedding

To dynamically integrate metadata into the learning process, we introduce the Learnable Metadata Embedding (LME), which enables metadata to adapt and evolve alongside visual features. We utilize the patient's age and gender as the sole metadata, keeping the approach lightweight and focused, and introduce learnable meta embeddings, m_{age} and m_{gender}, for each attribute. Formally, we define a set of learnable embeddings $m \in \{m_{\text{age}}, m_{\text{gender}}\}$ and prepend them to the image patch embeddings, ensuring seamless integration with visual features. After the forward pass, the model produces an output embedding for each prompt, which is optimized to align with its corresponding ground-truth metadata labels (y_{age} or y_{gender}). Concretely, we optimize these embeddings through a meta loss function $\mathcal{L}_{\text{meta}}$, as outlined in Eq. (1) of the following section. By minimizing $\mathcal{L}_{\text{meta}}$, each prompt specializes in representing its corresponding clinical attribute: m_{age} adapts to capture age-related variations, and m_{gender} learns to reflect gender-specific patterns in CFPs.

2.4 Joint Optimization Objectives

To train our foundation model, we combine multiple objectives to capture both visual features and metadata. These include reconstruction loss for robust feature extraction, consistency loss to maintain stable representations across patient variations, and meta loss to guide the learnable embeddings for age and gender.

Reconstruction Loss. We include a reconstruction loss, following the objective used in Masked Autoencoders [16], to guide the encoder toward learning discriminative visual features. Given the necessity of capturing fine-grained anatomical structures in retinal images, we further incorporate perceptual loss [18] to make the model restore subtle details more effectively.

Consistency Loss. To further ensure stable learning and robust integration across patient-level variations, we use a feature consistency loss based on cosine similarity: $\mathcal{L}_{\text{consis}} = \frac{1}{N}\sum_{n=1}^{N}\left(1 - \hat{z}^n_{e_1,s_1} \cdot \hat{z}^n_{e_2,s_2}\right)$, where $\hat{z}^n_{e_1,s_1}$ are the normalized feature embeddings of an unmasked patch and $\hat{z}^n_{e_2,s_2}$ corresponds to the embedding of its paired counterpart in the opposite spatial location. Here, N denotes the total number of unmasked patch pairs across all images in a batch.

Meta Loss. To incorporate clinically relevant metadata into the learning process, we introduce a meta loss that ensures the model effectively captures both continuous and categorical patient attributes:

$$\mathcal{L}_{\text{meta}} = \text{RMSE}(\hat{y}_{\text{age}}, y_{\text{age}}) + \text{CE}(\hat{y}_{\text{gender}}, y_{\text{gender}}), \tag{1}$$

where RMSE is the root-mean-square error, applied to enforce accurate age prediction and CE is the cross-entropy loss used for gender classification.

All loss terms are combined into a weighted sum, where $\lambda_{\text{recon}}, \lambda_{\text{consis}}, \lambda_{\text{meta}}$ are fixed scaling factors: $\mathcal{L} = \lambda_{\text{recon}}\mathcal{L}_{\text{recon}} + \lambda_{\text{consis}}\mathcal{L}_{\text{consis}} + \lambda_{\text{meta}}\mathcal{L}_{\text{meta}}$.

Table 1. Comparison of various methods on disease prediction and biomarker prediction on an in-house dataset. The best AUROC (ROC) and AUPRC (PRC) scores are in **bold**, while the second-best scores are underlined. Our proposed method is highlighted in a pastel color.

Method	Disease Prediction						Biomarker Prediction			
	DR		Glaucoma		AMD		CAC		eGFR	
	ROC	PRC	ROC	PRC	ROC	PRC	ROC	PRC	ROC	PRC
train/test on in-house										
RETFound (ViT-S)	0.871	0.586	0.531	0.508	0.488	0.499	0.783	0.623	0.614	0.598
PRETI (ViT-S)	**0.977**	**0.866**	**0.717**	**0.582**	**0.751**	**0.653**	**0.898**	**0.732**	**0.687**	**0.642**
test on in-house										
MAE (ViT-B) [16]	0.965	0.818	0.697	0.590	<u>0.729</u>	0.616	0.805	0.620	0.504	0.497
RETFound (ViT-L) [40]	0.969	0.838	<u>0.747</u>	0.587	0.695	0.608	0.890	0.696	0.582	0.566
UrFound (ViT-B) [36]	<u>0.971</u>	**0.910**	0.691	<u>0.591</u>	0.700	<u>0.625</u>	0.860	0.689	0.572	0.556
RET-CLIP (ViT-B) [13]	0.857	0.579	0.642	0.547	0.516	0.504	0.795	0.630	0.522	0.543
PRETI (ViT-B)	**0.982**	<u>0.885</u>	**0.777**	**0.622**	**0.772**	**0.659**	**0.899**	**0.734**	**0.699**	**0.643**

3 Experiments

We train our foundation model PRETI on a large-scale dataset and then evaluate it through downstream classification tasks, with minimal additional training.

Datasets. PRETI was trained on an in-house dataset comprising 1,017,549 CFPs from 292,006 patients across six medical institutions. This dataset includes non-public data and UK Biobank, which was internally processed for consistency; we thus refer to the entire dataset as in-house. For downstream tasks, we evaluate disease prediction for Diabetic Retinopathy (DR), Glaucoma, and Age-Related Macular Degeneration (AMD), as well as biomarker prediction such as Coronary Artery Calcium (CAC) score and estimated Glomerular Filtration Rate (eGFR) using an in-house dataset. These biomarkers were selected due to their known associations with retinal features and strong correlation with demographic metadata such as age and sex. These downstream datasets contain 1,110 to 6,861 CFPs per disease or biomarker, separate from pre-training data, with a 70:15:15 train:val:test split. Additionally, we conduct disease prediction tasks using public datasets (IDRID [29], APTOS2019 [20], MESSIDOR-2 [9], GF [1], PAPILA [23], Retina [19], and JSIEC [6]).

Training Details. The images are resized to 224 × 224 and augmented using random crop and flip, with different augmentation ratios applied to each eye. In our pre-training setup, we use a Vision Transformer (ViT) architecture [12], with ViT-S or ViT-B for the encoder and ViT-S for the decoder. The encoder is initialized with ImageNet-1K pre-trained weights [10]. We train with a batch size of 512 per GPU on 8 H100 GPUs for 300 epochs (15 warm-up). The learning rate is 5e-5. We set the loss function weights as $\lambda_{\text{recon}} = 1.4$, $\lambda_{\text{meta}} = 0.2$, and $\lambda_{\text{consis}} = 0.4$ by default. In reconstruction loss, the original MAE loss weight is 1.0, and the perceptual loss weight is 0.4. We also use RAAM with making ratios

Table 2. Performance comparison across different datasets and foundation models on public datasets.

Method	DR				Glaucoma				Multicategory			
	APTOS2019		MESSIDOR-2		GF		PAPILA		Retina		JSIEC	
	ROC	PRC	ROC	PRC	ROC	PRC	ROC	PRC	ROC	PRC	ROC	PRC
Report Supervision												
FLAIR [31]	0.932	0.686	0.819	0.483	0.905	0.792	0.752	0.610	0.863	0.679	0.917	0.704
UrFound [36]	0.949	0.716	0.882	0.608	0.958	0.880	0.783	0.625	0.901	0.793	0.995	0.923
RET-CLIP [13]	0.951	0.748	-	-	0.958	0.889	0.853	0.754	0.942	0.871	0.999	0.972
Non-report Supervision												
MAE [16]	0.941	0.676	0.842	0.488	0.931	0.832	0.629	0.475	0.742	0.537	0.985	0.818
RETFound [40]	0.943	0.726	0.864	0.568	0.943	0.863	0.855	0.748	0.847	0.697	0.990	0.884
PRETI (ViT-S)	**0.948**	0.725	0.862	0.600	**0.955**	**0.888**	0.838	0.714	0.831	0.656	0.990	0.853
PRETI (ViT-B)	**0.948**	**0.743**	**0.866**	**0.612**	**0.955**	0.886	**0.879**	**0.764**	**0.893**	**0.771**	**0.994**	**0.912**

of $r_0 = 0.985$ and $r_T = 0.85$. Downstream tasks with our pre-trained encoder run for 50 epochs (batch size 16), saving the best validation model for testing (Table. 2).

3.1 Evaluation Results

We assess PRETI against state-of-the-art retinal foundation models, using in-house data (disease/biomarker prediction) and benchmarks (disease prediction).

Disease/Biomark Prediction. We evaluate PRETI and baselines on internal data for disease and biomarker prediction (Table 1). For fair comparison based on training data, we also train RETFound on the same data as PRETI and evaluate its performance.

As a result, PRETI outperforms baselines in disease prediction tasks and maintains strong performance regardless of encoder size (ViT-S, B). Notably, it surpasses existing foundation models in biomarker prediction tasks, demonstrating its versatility. By effectively integrating metadata, PRETI leverages patient-specific factors, ensuring robustness across various tasks.

Benchmarks. PRETI is evaluated on disease classification benchmarks consisting of seven public datasets. Unlike report-supervision models such as FLAIR [31], UrFound [36], and RET-CLIP [13], which are trained using clinical descriptions, our model uses only images and metadata. Despite this, PRETI performs comparably to or outperforms baselines, demonstrating generalizability across diverse datasets. These evaluations shows the potential of our foundation models using only metadata, even when clinical text data is unavailable.

3.2 Ablation Study

Effect of LME and RAAM. We conduct our experiments by adding RAAM and LME to the SiamMAE on public datasets (Table 3). In most datasets,

Table 3. Comparative analysis of LME and RAAM in the our PRETI model.

Method	DR				Glaucoma				Multicategory			
	APTOS2019		MESSIDOR-2		GF		PAPILA		Retina		JSIEC	
	ROC	PRC	ROC	PRC	ROC	PRC	ROC	PRC	ROC	PRC	ROC	PRC
Baseline	0.935	0.692	0.846	0.495	0.940	0.844	0.813	0.690	0.832	0.675	0.976	0.788
+ RAAM	0.943	0.713	0.864	0.585	0.950	0.875	0.845	0.736	0.861	0.717	0.989	0.857
+ LME	0.943	0.724	0.851	0.534	0.953	0.883	0.864	0.740	0.871	0.718	0.989	0.875
PRETI	**0.948**	**0.743**	**0.866**	**0.612**	**0.955**	**0.886**	**0.879**	**0.764**	**0.893**	**0.771**	**0.994**	**0.912**

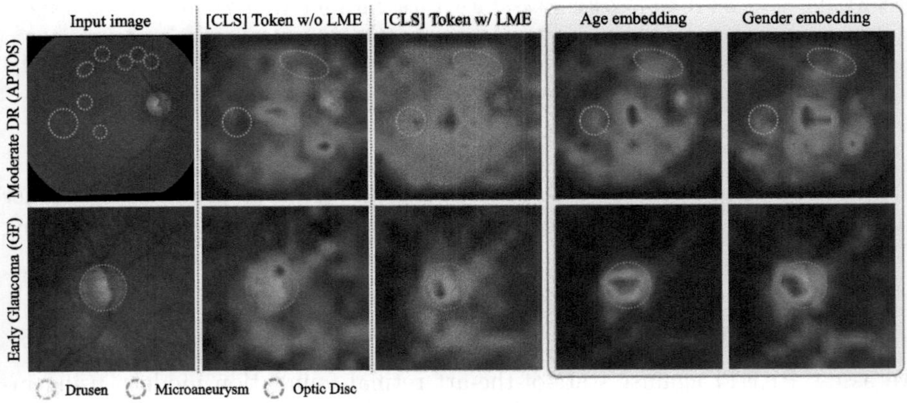

Fig. 2. Visualization of attention maps for the [CLS] token and metadata embeddings, comparing focus with and without LME. These maps highlight regions key to diagnosing eye diseases like diabetic retinopathy and glaucoma.

LME and RAAM each strictly improved AUROC and AUPRC, showing significant effects on MESSIDOR-2, PAPILA, Retina, and JSIEC datasets. Notably, PRETI, which incorporates both RAAM and LME, achieves the highest performance due to retinal area awareness and the utilization of metadata.

Impact of LME Through Attention Maps. To examine how PRETI utilizes metadata visually, we analyze attention maps. Specifically, we extract and visualize the attention maps of age and gender tokens from PRETI, along with the CLS token attention maps from both the baseline model without LME and PRETI with LME, using the APTOS2019 and GF datasets (Fig. 2). In the first row, featuring a retinal image of a patient with moderate diabetic retinopathy (DR), the attention maps of the age and gender tokens in the LME-enabled model precisely highlight structures and features critical for DR diagnosis. Notably, these metadata attention maps capture not only the macula and vessels but also drusen (yellow circles) and microaneurysms (pink circles), which the model without LME fails to detect but is important for DR [37]. This demonstrates how metadata helps the model effectively recognize key structures essential for DR prediction. Similarly, in the second row, for early glaucoma, age and gender embeddings of the LME-enabled model sharply focus on the optic disc (green circles) and the optic cup within it, both of which are critical for glaucoma diagnosis [2]. Furthermore, in both cases, the CLS token in the LME-enabled model, influenced by metadata embeddings, captures more accurate disease-relevant information for DR and glaucoma, demonstrating that metadata enhances the model's ability to target critical diagnostic features.

4 Conclusion

Our PRETI, a metadata-driven retinal foundation model that leverages LME and RAAM, achieved state-of-the-art performance in disease and biomarker prediction tasks. Ultimately, PRETI serves as a patient-specific, metadata-aware foundation model, improving the generalizability by learning ocular disease-related features. Using metadata demonstrated effectiveness for retinal analysis, suggesting its potential for broader applications. While our study focused on age and gender, future work could explore additional metadata types and extend our methodology to other medical imaging domains.

Acknowledgements. This work was supported in part by the IITP RS-2024-00457882 (AI Research Hub Project), IITP 2020-II201361, NRF RS-2024-00345806, NRF RS-2023-00219019, and NRF RS-2023-002620.

Disclosure of Interests. The authors have no competing interests to declare that are relevant to the content of this article.

References

1. Ahn, J.M., et al.: A deep learning model for the detection of both advanced and early glaucoma using fundus photography. PLoS ONE **13**(11), e0207982 (2018)
2. Almazroa, A., et al.: Optic disc and optic cup segmentation methodologies for glaucoma image detection: a survey. J. Ophthalmol. **2015**, 180972 (2015). https://doi.org/10.1155/2015/180972
3. Bai, L., Chen, S., Gao, M., Abdelrahman, L., Ghamdi, M.A., Abdel-Mottaleb, M.: The influence of age and gender information on the diagnosis of diabetic retinopathy: Based on neural networks. Proceedings of the 43rd Annual International Conference of the IEEE Engineering in Medicine and Biology Society (EMBC) **2021**, 3514–3517 (2021).https://doi.org/10.1109/EMBC46164.2021.9630601
4. Cai, G., et al.: A multimodal transformer to fuse images and metadata for skin disease classification. Vis. Comput. **39**(7), 2781–2793 (2022)
5. Caron, M., et al.: Emerging properties in self-supervised vision transformers. In: International Conference on Computer Vision, pp. 9650–9660 (2021)
6. Cen, L.P., et al.: Automatic detection of 39 fundus diseases and conditions in retinal photographs using deep neural networks. Nat. Commun. **12**(1), 4828 (2021)
7. Cheung, C.Y., et al.: A deep-learning system for the assessment of cardiovascular disease risk via the measurement of retinal-vessel calibre. Nat. Biomed. Eng. **5**(6), 498–508 (2021)

8. Ciocarlan, A., et al.: Self-supervised learning for real-world object detection: a survey (2024). arXiv preprint arXiv:2410.07442
9. Decencière, E., et al.: Feedback on a publicly distributed image database: the messidor database. Image Analysis and Stereology, pp. 231–234 (2014)
10. Deng, J., et al.: Imagenet: A large-scale hierarchical image database. In: Conference on Computer Vision and Pattern Recognition, pp. 248–255. IEEE (2009)
11. Dettmers, T., et al.: Qlora: efficient finetuning of quantized LLMLs. Adv. Neural. Inf. Process. Syst. **36**, 10088–10115 (2023)
12. Dosovitskiy, A., et al.: An image is worth 16x16 words: Transformers for image recognition at scale (2020). arXiv preprint arXiv:2010.11929
13. Du, J., et al.: Ret-clip: A retinal image foundation model pre-trained with clinical diagnostic reports. In: International Conference on Medical Image Computing and Computer-Assisted Intervention, pp. 709–719. Springer (2024)
14. Dufumier, B., et al.: Contrastive learning with continuous proxy meta-data for 3d mri classification. In: International Conference on Medical Image Computing and Computer-Assisted Intervention, pp. 58–68. Springer (2021)
15. Gupta, A., et al.: Siamese masked autoencoders. Adv. Neural. Inf. Process. Syst. **36**, 40676–40693 (2023)
16. He, K., et al.: Masked autoencoders are scalable vision learners. In: Conference on Computer Vision and Pattern Recognition, pp. 16000–16009 (2022)
17. Holland, R., et al.: Metadata-enhanced contrastive learning from retinal optical coherence tomography images. Med. Image Anal. **97**, 103296 (2024)
18. Johnson, J., Alahi, A., Fei-Fei, L.: Perceptual losses for real-time style transfer and super-resolution. In: European Conference on Computer Vision, pp. 694–711. Springer (2016)
19. jr2ngb: Cataract (2019). https://www.kaggle.com/datasets/jr2ngb/cataractdataset. kaggle
20. Karthik, Maggie, Dane, S.: Aptos 2019 blindness detection (2019). https://kaggle.com/competitions/aptos2019-blindness-detection. kaggle
21. Kirillov, A., et al.: Segment anything. In: International Conference on Computer Vision, pp. 4015–4026 (2023)
22. Korpole, N.R., et al.: Gender difference in ocular diseases, risk factors and management with specific reference to role of sex steroid hormones. J. Mid-life Health **13**(1), 20–25 (2022). https://doi.org/10.4103/jmh.jmh_28_22
23. Kovalyk, O., et al.: Papila: Dataset with fundus images and clinical data of both eyes of the same patient for glaucoma assessment. Sci. Data **9**(1), 291 (2022)
24. Liu, H., et al.: Visual instruction tuning. Adv. Neural. Inf. Process. Syst. **36**, 34892–34916 (2023)
25. Loshchilov, I., Hutter, F.: Sgdr: Stochastic gradient descent with warm restarts (2016). arXiv preprint arXiv:1608.03983
26. Moor, M., et al.: Foundation models for generalist medical artificial intelligence. Nature **616**(7956), 259–265 (2023)
27. Nam, J.W., et al.: Clinical evaluation of unilateral open-angle glaucoma: a two-year follow-up study. Chonnam Med. J. **57**(2), 144 (2021)
28. Paulus, J.K., et al.: Field synopsis of sex in clinical prediction models for cardiovascular disease. Circulation: Cardiovascular Quality and Outcomes **9**(2 Suppl 1), S8–S15 (2016). https://doi.org/10.1161/CIRCOUTCOMES.115.002473
29. Porwal, P., et al.: Indian diabetic retinopathy image dataset (idrid): a database for diabetic retinopathy screening research. Data **3**(3), 25 (2018)

30. Radford, A., et al.: Learning transferable visual models from natural language supervision. In: International Conference on Machine Learning, pp. 8748–8763. PMLR (2021)
31. Silva-Rodriguez, J., et al.: A foundation language-image model of the retina (flair): encoding expert knowledge in text supervision. Med. Image Anal. **99**, 103357 (2025)
32. Speilburg, A.M., et al.: Symmetry of peripheral retinal nonperfusion in diabetic retinopathy by ischemic index. J. optom. **11**(4), 262–267 (2018)
33. Tillmann, A., et al.: Gender differences in retinal diseases: a review. Clin. Exp. Ophthalmol. **52**(3), 317–333 (2024). https://doi.org/10.1111/ceo.14364
34. Trivizki, O., et al.: Symmetry of macular fundus features in age-related macular degeneration. Ophthalmol. Retina **7**(8), 672–682 (2023)
35. Wagner, S.K., et al.: Insights into systemic disease through retinal imaging-based oculomics. Transl. Vision Sci. Technol. **9**(2), 6–6 (2020)
36. Yu, K., et al.: Urfound: Towards universal retinal foundation models via knowledge-guided masked modeling. In: International Conference on Medical Image Computing and Computer-Assisted Intervention, pp. 753–762. Springer (2024)
37. Z, Y., et al.: The association between diabetic retinopathy and the prevalence of age-related macular degeneration - the kailuan eye study. Frontiers in Public Health **10**, 922289 (2022). https://doi.org/10.3389/fpubh.2022.922289
38. Zhang, R., et al.: Llama-adapter: Efficient fine-tuning of language models with zero-init attention (2023). arXiv preprint arXiv:2303.16199
39. Zhou, Y., et al.: Automorph: Automated retinal vascular morphology quantification via a deep learning pipeline. Transl. Vision Sci. Technol, **11**(7), 12–12 (2022)
40. Zhou, Y., et al.: A foundation model for generalizable disease detection from retinal images. Nature **622**(7981), 156–163 (2023)

Pre-to-Post Operative MRI Generation with Retrieval-Based Visual In-Context Learning

Bogyeong Kang[1], Sang-Jun Park[1], Minjoo Lim[1], Myeongkyun Kang[2], Keun-Soo Heo[1], Ji-Hye Oh[1], Hyun Jung Lee[1], and Tae-Eui Kam[1(✉)]

[1] Department of Artificial Intelligence, Korea University, Seoul, South Korea
{kangbk,kamte}@korea.ac.kr

[2] Department of Robotics and Mechatronics Engineering, Daegu Gyeongbuk Institute of Science and Technology (DGIST), Daegu, South Korea

Abstract. Glioblastoma is an aggressive brain tumor requiring precise treatment planning. Magnetic resonance imaging (MRI) is essential for pre-operative assessment, surgical resection planning, and post-operative monitoring. Therefore, generating post-operative MRI from pre-operative MRI can assist neurosurgeons in many ways, such as predicting surgical outcomes and guiding treatment planning. However, generating post-operative MRI from pre-operative MRI is challenging, as the resection extent depends on tumor location and infiltration to minimize potential complications, necessitating consideration of surgical outcomes based on tumor location and shape. Furthermore, post-operative MRI differs significantly from pre-operative MRI due to structural and visual changes, such as tissue shift, edema, hemorrhage, and the resection region. To address these challenges, we propose a novel post-operative MRI generation method that generates post-operative MRI from pre-operative MRI using tumor-aware visual in-context learning. Specifically, we provide explicit visual instruction for generating post-operative MRI from pre-operative MRI, improving the capture of structural changes. To consider tumor-specific post-operative outcomes, we propose tumor-guided retrieval, which retrieves the tumor case most similar to the query pre-operative MRI, and a tumor-aware prompt adapter that integrates tumor resection and anatomical structure information. Our proposed method achieves superior performance on publicly available dataset and is the first to generate post-operative MRI from pre-operative MRI, introducing a new approach to improving patient prognosis.

Keywords: Post-operative MRI · MRI Generation · Diffusion Model

1 Introduction

Glioblastoma is a rapidly infiltrative brain tumor with a poor prognosis, necessitating accurate diagnosis for treatment planning and improving patient outcomes [2,16,19]. Magnetic resonance imaging (MRI) is essential for assessing

J. C. Gee et al. (Eds.): MICCAI 2025, LNCS 15960, pp. 534–544, 2026.
https://doi.org/10.1007/978-3-032-04927-8_51

glioblastoma, providing critical information for pre-operative diagnosis, post-operative monitoring, and treatment planning [6,8,18,25]. Neurosurgeons determine the extent of surgical resection by analyzing past cases of pre- and post-operative MRI [27]. Generating post-operative MRI from pre-operative MRI allows neurosurgeons to estimate potential outcomes before surgery, significantly improving surgical and treatment planning.

Recent advances in generative models [10,12,23] have led to extensive research on image translation methods [13,22,32], which have primarily focused on preserving the structure of the source image while reflecting the style of the target image [13,22,32]. However, generating post-operative MRI from pre-operative MRI is far more challenging than previous methods [13,22,32] that focus only on style translation due to two challenges. First, post-operative MRI generation should consider post-operative outcomes specific to tumor location and shape [9,11], as determining the extent of resection is crucial for minimizing patient complications when a tumor is located in a critical brain region or has diffusely infiltrated surrounding tissues [4,11,26]. Second, post-operative MRI generation should capture the visual and structural changes resulting from resection, as post-operative MRI exhibits differences compared to pre-operative MRI, including the resection region, brain tissue shift, edema, and hemorrhage [7,20].

To address these challenges, we propose a novel post-operative MRI generation method that generates post-operative MRI from pre-operative MRI using tumor-aware visual in-context learning, enhancing the capture of structural changes and accurately reflecting the tumor resection region based on tumor location and shape. We provide explicit visual instruction for generating post-operative MRI from pre-operative MRI through **visual in-context learning**, improving the capture of structural changes between paired pre- and post-operative MRI and enhancing generation quality. To consider tumor-specific post-operative outcomes, we introduce **tumor-guided retrieval**, which retrieves an instruction pre- and post-operative MRI pair from another subject with a tumor location and shape closely matching those in the query pre-operative MRI. This approach provides instruction on information from similar tumor cases during post-operative MRI generation, allowing the capture of structural changes and tumor-specific resection outcomes to enhance the accuracy of post-operative MRI generation. Furthermore, we introduce a **tumor-aware prompt adapter** designed to capture not only the tumor location and anatomical structure in the pre-operative MRI but also the tumor resection region in the post-operative MRI, allowing the generated post-operative MRI to better reflect tumor-specific surgical outcomes. We demonstrate its effectiveness by achieving outstanding performance in generating post-operative MRI from pre-operative MRI using a publicly available LUMIERE [28] dataset.

The main contributions of our method can be summarized as follows:

- To the best of our knowledge, this is the first approach to generate post-operative MRI from pre-operative MRI using visual in-context learning, capturing visual and structural changes resulting from surgical resection.

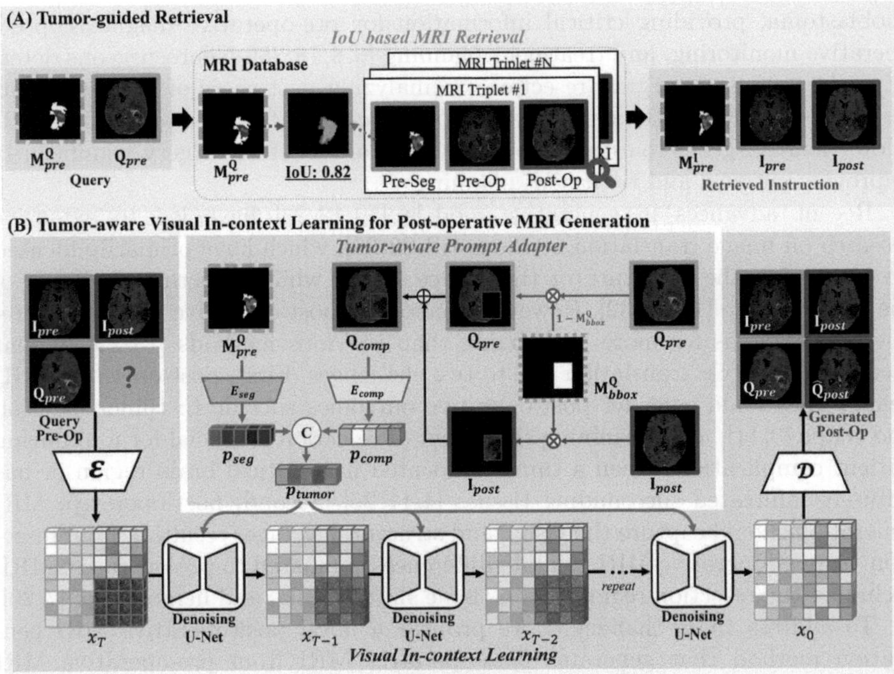

Fig. 1. Overview of the proposed method for post-operative MRI generation, which accurately reflects the tumor resection region and post-operative structural changes through two key components: (A) tumor-guided retrieval, (B) tumor-aware visual in-context learning for post-operative MRI generation.

- To consider tumor-specific post-operative outcomes, we propose tumor-guided retrieval, which retrieves pre- and post-operative MRI pairs with a tumor location and shape similar to the query, and a tumor-aware prompt adapter that integrates tumor resection and anatomical structure information.

2 Method

In this section, we describe our proposed method for post-operative MRI generation (Fig. 1), which comprises two key components: (A) tumor-guided retrieval and (B) tumor-aware visual in-context learning for post-operative MRI generation. First, we retrieve MRI scans that closely match the tumor location and shape of the query pre-operative MRI to consider tumor-specific post-operative changes through tumor-guided retrieval. Then, we generate post-operative MRI with a tumor-aware prompt adapter and visual in-context learning to better capture tumor-specific structural changes resulting from resection.

2.1 Tumor-Guided Retrieval

To consider tumor-specific post-operative outcomes, we design a tumor-guided retrieval strategy that retrieves a pre- and post-operative MRI pair with tumor location and shape closely matching those in the query pre-operative MRI. We construct an MRI database consisting of triplets: {Pre-operative MRI, Post-operative MRI, Pre-operative Segmentation Mask}, derived from the training dataset. Given the query pre-operative MRI Q_{pre} and its corresponding segmentation mask M^{Q}_{pre}, we compute the Intersection over Union (IoU) between the segmentation mask of the query pre-operative MRI and those of all triplets stored in the MRI database. We retrieve the MRI triplet with the highest IoU to ensure that the segmentation map includes a tumor region that best matches the location and shape of the tumor in the query pre-operative MRI. To provide more robust guidance, we select the top-k most similar triplets to be utilized as instructions for post-operative MRI generation, denoted as Instruction = $\{I_{pre}, I_{post}, M^{I}_{pre}\}$, where I_{pre} and I_{post} are the retrieved pre- and post-operative MRI, and M^{I}_{pre} is the corresponding segmentation mask. To further refine the precision of tumor-guided retrieval, we incorporate axial plane alignment in addition to segmentation mask similarity. Specifically, we restrict retrieval to MRI triplets whose axial plane range aligns with that of the query pre-operative scan, ensuring that retrieved images reflect tumors located within a similar axial plane.

2.2 Tumor-Aware Visual In-Context Learning for Post-operative MRI Generation

Tumor-Aware Prompt Adapter. We design a tumor-aware prompt adapter that guides the generation of tumor-specific post-operative MRI by integrating information from the tumor resection region in the retrieved post-operative MRI and the anatomical structure in the pre-operative MRI. Given a retrieved instruction's post-operative MRI I_{post}, we integrate its tumor resection region with the anatomical structure of the query pre-operative MRI Q_{pre}. This is achieved through a compositing operation, which is formulated as follows: $Q_{comp} = M^{Q}_{bbox} \cdot I_{post} + (1 - M^{Q}_{bbox}) \cdot Q_{pre}$, where Q_{pre} is the query pre-operative MRI, I_{post} is the post-operative MRI from the retrieved instruction, and M^{Q}_{bbox} refers to the bounding box mask, which is derived from the segmentation mask of the query pre-operative MRI. Next, the composite image Q_{comp} is fed into E_{comp}, which consists of a pretrained image encoder of MedSAM [17] followed by multiple convolution and pooling layers. The Q_{comp} is processed by the pre-trained image encoder to extract features, which are then passed through the subsequent layers and reshaped into a 1024-dimensional latent representation p_{comp}. Additionally, the pre-operative segmentation mask M^{Q}_{pre} is fed into E_{seg}, where features are extracted using a convolutional layer, reshaped, and projected through a fully connected layer into a 1024-dimensional latent representation p_{seg}. Finally, the p_{seg} is then concatenated with the p_{comp} to form the final prompt embedding p_{tumor}. The p_{tumor} is injected into the Stable Diffusion [24] process through cross-attention to guide tumor resection characteristics in the post-operative MRI and tumor location with anatomical structure in the pre-operative MRI.

Visual In-Context Learning. To capture the structural changes between pre- and post-operative MRI, we generate the post-operative MRI $\hat{Q}_{post}$ by adopting visual in-context learning [1,29,31]. In our visual in-context learning, we explicitly consider tumor-specific post-operative changes using the query pre-operative MRI Q_{pre} and the retrieved instruction pair $\{I_{pre}, I_{post}\}$ obtained by tumor-guided retrieval. To achieve this, we construct a grid-like image $Grid(\{I_{pre}, I_{post}, Q_{pre}, B\})$. Here, B represents a blank image, which is reconstructed into $\hat{Q}_{post}$ through the diffusion process. We employ the Stable Diffusion [24] that operates in the latent space rather than the pixel space. First, the grid-like image $Grid(\{I_{pre}, I_{post}, Q_{pre}, B\})$ is encoded into a latent representation x_0 using a variational autoencoder [15]. During the forward process, noise is gradually added only to B region over time steps t, generating noisy representations x_t based on a noise scheduler [12]. In the reverse process, a denoising network iteratively removes noise from B through a Markov process [12], ultimately reconstructing x_0, which corresponds to the generated post-operative MRI $\hat{Q}_{post}$. This process is formulated as $p_\theta(x_{t-1}|x_t, c)$ where θ represents the learnable parameters of the diffusion model, and c is the conditioning input [24]. In our task, c is the prompt p_{tumor} from the tumor-aware prompt adapter. The training objective of the diffusion model is to predict the added noise ϵ, and the loss function is formulated as follows [24]: $L_{LDM} = \mathbb{E}_{x_0,\epsilon,t}\left[\|\epsilon - \epsilon_\theta(x_t, t, c)\|^2\right]$, where ϵ_θ is the noise prediction network [24]. We adopt Classifier-Free Guidance [23], which enables the model to generate post-operative MRI by conditioning on retrieved instructions while maintaining flexibility in sampling.

3 Experiments and Results

3.1 Experiments Settings

Dataset and Implementation Details. We utilize the publicly available LUMIERE [28] dataset, containing 71 paired pre- and post-operative MRI scans, normalized to MNI space. Contrast-enhanced T1-weighted MRI is converted into 2D axial slices, from which tumor-containing slices with segmentation masks are selected. The dataset is split into 2,530 training and 205 testing slices. We train a Stable Diffusion [24] using AdamW [14] with a learning rate of 1e-5 and a cosine scheduler. We utilize the MedSAM [17] as an image encoder to extract features from the composite image and tumor-guided retrieval uses top-$k = 15$ for training and 1 for inference. During training, we randomly select one triplet from the retrieved top-15 triplets in each batch and use it as the instruction.

Competing Methods. We compare our performance against established diffusion based image editing methods [3,12,21,30,31] using SSIM, MS-SSIM, PSNR, and LPIPS as metrics. We train DDPM [12] by concatenating pre- and

Table 1. Comparison of post-operative MRI generation results with state-of-the-art image editing methods.

Method	SSIM (↑)	MS-SSIM (↑)	PSNR (↑)	LPIPS (↓)
DDPM [12]	0.5744	0.7587	31.23	0.1878
PBE [30]	0.6695	0.7295	31.09	0.1994
InstructPix2Pix [3]	0.6649	0.7594	31.53	0.1925
VISII [21]	0.6682	0.7611	31.50	0.1921
ImageBrush [31]	0.6718	0.7569	31.56	0.1928
Ours	**0.7001**	**0.7814**	**31.67**	**0.1845**

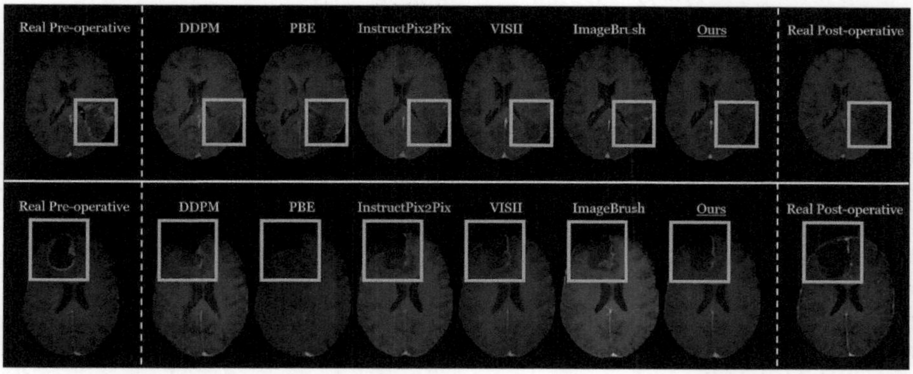

Fig. 2. Qualitative comparison of post-operative MRI generation with state-of-the-art image editing methods.

post-operative MRI. For PBE [30], we randomly select resection regions from post-operative MRI in the training dataset as exemplar images. For Instruct-Pix2Pix [3] and VISII [21], we follow the same concatenation approach as DDPM and provide the text instruction "Post-operative glioblastoma MRI". For Image-Brush [31], we provide random instruction during visual in-context learning.

3.2 Main Results

To evaluate our approach for post-operative MRI generation, we conduct a comparative study on image editing methods [3,12,21,30,31]. Table 1 shows the performance results of post-operative MRI generation. Our method outperforms all image editing methods [3,12,21,30,31] across all evaluation metrics, achieving SSIM of 0.7001, MS-SSIM of 0.7814, PSNR of 31.67, and LPIPS of 0.1845. These results demonstrate that our approach is specifically tailored for generating post-operative MRI, effectively capturing tumor-specific post-operative changes. Furthermore, Fig. 2 presents the qualitative results of post-operative MRI generation from pre-operative MRI. Our method generates images that are the most similar to real post-operative MRI among all image editing methods [3,12,21,30,31]. In

Table 2. An ablation study on different retrieval strategies and the effectiveness of the two prompts in our tumor-aware prompt adapter. We evaluate three retrieval strategies: Random (random selection), Image-retrieval (retrieval based on cosine similarity in the image embedding space), and Ours (tumor-guided retrieval). Additionally, we assess the impact of two prompts: p_{seg} (segmentation mask of the input pre-operative MRI) and p_{comp} (composite image).

Method	SSIM (↑)	MS-SSIM (↑)	PSNR (↑)	LPIPS (↓)
Random	0.6823	0.7654	31.59	0.1875
Image-retrieval	0.6846	0.7684	31.63	0.1869
Ours (*w/o.* p_{seg})	0.6835	0.7703	31.60	0.1887
Ours (*w/o.* p_{comp})	0.6784	0.7758	31.61	0.1863
Ours	**0.7001**	**0.7814**	**31.67**	**0.1845**

particular, as shown in Fig. 2, DDPM [12], PBE [30], InstructPix2Pix [3], and VISII [21] fail to reflect the post-operative tumor resection region, often generating blurred and indistinct structures instead. These findings highlight the importance of visual in-context learning for capturing structural changes caused by resection in post-operative MRI generation. Compared to ImageBrush [31], which also employs visual in-context learning, our method demonstrates visibly superior performance. This improvement can be attributed to our tumor-guided retrieval strategy, which tailors instructions based on tumor location and shape. Additionally, our tumor-aware prompt adapter explicitly incorporates the pre-operative tumor location and post-operative resection region, enhancing the post-operative MRI generation. These results demonstrate that our approach effectively captures tumor-specific post-operative changes, enabling accurate generation of post-operative MRI.

3.3 Ablation Study

Effectiveness of Tumor-Guided Retrieval. We conduct an ablation study by comparing post-operative MRI generation using random selection and image-retrieval instead of our tumor-guided retrieval. In our experiments, random selection refers to selecting an MRI instruction randomly from the training dataset. The image-retrieval strategy embeds images into latent space using a ViT encoder [5] and retrieves the top-k most similar images based on cosine similarity. As shown in Fig. 3, the random selection strategy selects the instruction that differ entirely in anatomical structure, as well as in tumor location and shape. The image-retrieval strategy selects the instruction with a similar overall anatomical structure but differing tumor location and shape. As shown in Table 2, Ours outperforms both random selection and image-retrieval, achieving the best performance. Notably, the image-retrieval strategy performs comparably to random selection, indicating its limited effectiveness for post-operative MRI generation. This limitation arises because the image-retrieval strategy primarily relies on overall anatomical similarity and does not consider the tumor

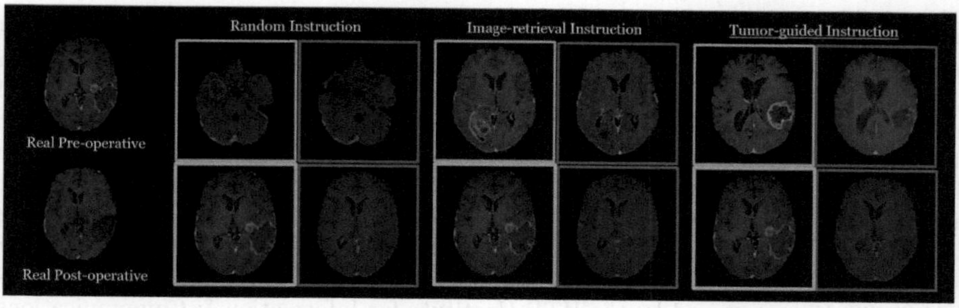

Fig. 3. Qualitative results from the ablation study comparing retrieval strategies for visual in-context learning in post-operative MRI generation.

location and shape. However, since the tumor region plays a critical role in post-operative MRI generation, considering it in retrieval is essential for achieving promising performance. These findings emphasize the importance of tumor-guided retrieval, demonstrating that explicitly considering tumor-specific characteristics is crucial for accurate post-operative MRI generation.

Effectiveness of Tumor-Aware Prompt Adapter. We further conduct an ablation study to evaluate the effectiveness of our tumor-aware prompt adapter. Specifically, we evaluate the impact of different conditioning strategies in the Stable Diffusion process by modifying the injected information. First, we use only the composite image as the prompt, without the segmentation mask (Ours ($w/o.$ p_{seg})). As shown in Table 2, this approach results in lower performance compared to Ours, which utilizes both the composite image and the segmentation mask. This highlights the importance of leveraging the input pre-operative segmentation mask as a prompt, since it provides spatial guidance for the tumor's location and shape. Notably, using only the segmentation mask as the prompt, without the composite image (Ours ($w/o.$ p_{comp})), results in even more significant performance degradation. This suggests that the composite image serves as a critical prompt that reflects tumor-specific surgical outcomes. Therefore, these results demonstrate that our tumor-aware prompt adapter captures tumor-specific post-operative outcomes, enabling more accurate post-operative MRI generation.

4 Conclusion

We propose a novel method for post-operative MRI generation from pre-operative MRI. Our proposed method integrates tumor-guided retrieval, a tumor-aware prompt adapter, and visual in-context learning to better capture post-operative structural changes and accurately reflect tumor resection based on tumor location and shape. We demonstrate the superiority of our proposed method in generating post-operative MRI on the LUMIERE dataset, highlighting its potential for improving surgical planning and patient-specific treatment strategies.

Acknowledgments. This work was supported by Institute of Information & communications Technology Planning & Evaluation (IITP) grant funded by the Korea government (MSIT) (No. RS-2019-II190079, Artificial Intelligence Graduate School Program(Korea University)), the National Research Foundation of Korea (NRF) grant funded by the Korea government (MSIT) (No. RS-2023-00212498), the Institute of Information & communications Technology Planning & Evaluation (IITP) grant funded by the Korea government (MSIT) (RS-2024-00457882, Artificial Intelligence Research Hub Project), and the MSIT (Ministry of Science and ICT), Korea, under the ITRC (Information Technology Research Center) support program (IITP-2025-RS-2024-00436857) supervised by the IITP (Institute for Information & Communications Technology Planning & Evaluation).

Disclosure of Interests. The authors have no competing interests to declare that are relevant to the content of this article.

References

1. Bar, A., Gandelsman, Y., Darrell, T., Globerson, A., Efros, A.: Visual prompting via image inpainting. In: Advances in Neural Information Processing Systems, vol. 35, pp. 25005–25017 (2022)
2. Booth, T.C., et al.: Glioblastoma post-operative imaging in neuro-oncology: current UK practice (gin cup study). Eur. Radiol. **31**, 2933–2943 (2021)
3. Brooks, T., Holynski, A., Efros, A.A.: Instructpix2pix: learning to follow image editing instructions. In: Proceedings of the IEEE/CVF Conference on Computer Vision and Pattern Recognition, pp. 18392–18402 (2023)
4. Brown, T.J., et al.: Association of the extent of resection with survival in glioblastoma: a systematic review and meta-analysis. JAMA Oncol. **2**(11), 1460–1469 (2016)
5. Dosovitskiy, A.: An image is worth 16x16 words: Transformers for image recognition at scale. arXiv preprint arXiv:2010.11929 (2020)
6. Farace, P., et al.: Early MRI changes in glioblastoma in the period between surgery and adjuvant therapy. J. Neurooncol. **111**, 177–185 (2013)
7. Farace, P., et al.: Clinical target volume delineation in glioblastomas: pre-operative versus post-operative/pre-radiotherapy MRI. Br. J. Radiol. **84**(999), 271–278 (2011)
8. Garcia-Ruiz, A., et al.: Precise enhancement quantification in post-operative MRI as an indicator of residual tumor impact is associated with survival in patients with glioblastoma. Sci. Rep. **11**(1), 695 (2021)
9. Gerritsen, J.K.W., et al.: Resection versus biopsy in patients with glioblastoma (resbiop study): study protocol for an international multicentre prospective cohort study (encram 2202). BMJ Open **14**(9) (2024)
10. Goodfellow, I., et al.: Generative adversarial nets. In: Advances in Neural Information Processing Systems, vol. 27 (2014)
11. de Haan, Y., Moussa, A., Gerritsen, J., Schnitzler, S., Dirven, C., Vincent, A.: P27. 03. b biopsy versus resection in glioblastoma treatment: a propensity score-matched analysis of a retrospective cohort. Neuro-Oncology **26**, 136–137 (2024)

12. Ho, J., Jain, A., Abbeel, P.: Denoising diffusion probabilistic models. Advances in Neural Information Processing Systems, vol. 33, pp. 6840–6851 (2020)
13. Kang, B., et al.: Target-aware cross-modality unsupervised domain adaptation for vestibular schwannoma and cochlea segmentation. Sci. Rep. **14**(1), 27883 (2024)
14. Kingma, D.P.: Adam: A method for stochastic optimization. arXiv preprint arXiv:1412.6980 (2014)
15. Kingma, D.P., Welling, M., et al.: Auto-encoding variational bayes (2013)
16. Kubben, P.L., ter Meulen, K.J., Schijns, O.E., ter Laak-Poort, M.P., van Overbeeke, J.J., van Santbrink, H.: Intraoperative MRI-guided resection of glioblastoma multiforme: a systematic review. Lancet Oncol. **12**(11), 1062–1070 (2011)
17. Ma, J., He, Y., Li, F., Han, L., You, C., Wang, B.: Segment anything in medical images. Nat. Commun. **15**(1), 654 (2024)
18. Majós, C., et al.: Early post-operative magnetic resonance imaging in glioblastoma: correlation among radiological findings and overall survival in 60 patients. Eur. Radiol. **26**, 1048–1055 (2016)
19. McGirt, M.J., Villavicencio, A.T., Bulsara, K.R., Friedman, A.H.: MRI-guided stereotactic biopsy in the diagnosis of glioma: comparison of biopsy and surgical resection specimen. Surg. Neurol. **59**(4), 279–283 (2003)
20. Miao, X., Chen, H., Tang, M., Huang, D., Gao, T., Chen, Y.: Post-operative MRI synthesis from pre-operative MRI and post-operative CT using conditional GAN for the assessment of degree of resection. Displays **83**, 102742 (2024)
21. Nguyen, T., Li, Y., Ojha, U., Lee, Y.J.: Visual instruction inversion: image editing via image prompting. In: Advances in Neural Information Processing Systems, vol. 36, pp. 9598–9613 (2023)
22. Park, T., Efros, A.A., Zhang, R., Zhu, J.Y.: Contrastive learning for unpaired image-to-image translation. In: European Conference on Computer Vision, pp. 319–345. Springer (2020)
23. Ramesh, A., Dhariwal, P., Nichol, A., Chu, C., Chen, M.: Hierarchical text-conditional image generation with clip latents. arXiv preprint arXiv:2204.06125 **1**(2), 3 (2022)
24. Rombach, R., Blattmann, A., Lorenz, D., Esser, P., Ommer, B.: High-resolution image synthesis with latent diffusion models. In: Proceedings of the IEEE/CVF Conference on Computer Vision and Pattern Recognition, pp. 10684–10695 (2022)
25. Rykkje, A.M., Larsen, V.A., Skjøth-Rasmussen, J., Nielsen, M.B., Carlsen, J.F., Hansen, A.E.: Timing of early postoperative MRI following primary glioblastoma surgerya retrospective study of contrast enhancements in 311 patients. Diagnostics **13**(4), 795 (2023)
26. Sanai, N., Berger, M.S.: Glioma extent of resection and its impact on patient outcome. Neurosurgery **62**(4), 753–766 (2008)
27. Shukla, G., et al.: Advanced magnetic resonance imaging in glioblastoma: a review. Chin. Clin. Oncol. **6**(4), 40–40 (2017)
28. Suter, Y.: The LUMIERE dataset: longitudinal glioblastoma MRI with expert RANO evaluation. Sci. Data **9**(1), 768 (2022)
29. Wang, X., Wang, W., Cao, Y., Shen, C., Huang, T.: Images speak in images: a generalist painter for in-context visual learning. In: Proceedings of the IEEE/CVF Conference on Computer Vision and Pattern Recognition, pp. 6830–6839 (2023)
30. Yang, B., et al.: Paint by example: exemplar-based image editing with diffusion models. In: Proceedings of the IEEE/CVF Conference on Computer Vision and Pattern Recognition, pp. 18381–18391 (2023)

31. Yang, Y., et al.: Imagebrush: Learning visual in-context instructions for exemplar-based image manipulation. In: Advances in Neural Information Processing Systems, vol. 36, pp. 48723–48743 (2023)
32. Zhu, J.Y., Park, T., Isola, P., Efros, A.A.: Unpaired image-to-image translation using cycle-consistent adversarial networks. In: Proceedings of the IEEE International Conference on Computer Vision, pp. 2223–2232 (2017)

RadioFormer: Integrating Radiologist Inductive Bias for Tumor Classification on Multi-Sequence MR Images

Xiaoyu Bai[1,2] and Yong Xia[1,2,3](✉)

[1] Research & Development Institute of Northwestern Polytechnical University in Shenzhen, Shenzhen 518057, China

[2] National Engineering Laboratory for Integrated Aero-Space-Ground-Ocean Big Data Application Technology, School of Computer Science and Engineering, Northwestern Polytechnical University, Xi'an 710072, China

[3] Ningbo Institute of Northwestern Polytechnical University, Ningbo 315048, China

yxia@nwpu.edu.cn

Abstract. Multi-sequence magnetic resonance imaging (MRI) plays a critical role in tumor diagnosis but relies heavily on manual interpretation, which is both labor-intensive and dependent on expert knowledge. While deep learning-based diagnostic methods show significant potential, they typically require large datasets for effective training. However, the high cost of data collection and annotation often limits the available dataset size. This highlights the need for models that can effectively train on small datasets, mitigate overfitting, and achieve reliable performance. To address these challenges, we propose RadioFormer, a novel model that incorporates radiologist inductive bias to facilitate efficient learning on small MRI datasets. Unlike traditional 2D or 3D architectures, RadioFormer emulates the radiologist's diagnostic process by explicitly parsing MRI data into three hierarchical levels: (1) single-sequence slice feature extraction, (2) multi-sequence slice information aggregation, and (3) inter-slice information (volume) aggregation. Each level builds upon the previous one, ensuring smooth information flow and a hierarchical understanding of lesion characteristics. By integrating expert knowledge into its design, RadioFormer effectively leverages inductive bias to enhance model generalization on small datasets. We evaluated RadioFormer on three public datasets for brain, breast, and liver tumor classification, where it achieved state-of-the-art performance across all tasks. The code and pre-processed data for RadioFormer are available at https://github.com/aa1234241/RadioFormer/tree/master.

Keywords: Tumor classification · Multi-sequence MRI · Inductive bias

1 Introduction

Cancer remains one of the leading causes of mortality worldwide, accounting for millions of deaths each year [6]. Magnetic resonance imaging (MRI) plays an

J. C. Gee et al. (Eds.): MICCAI 2025, LNCS 15960, pp. 545–555, 2026.
https://doi.org/10.1007/978-3-032-04927-8_52

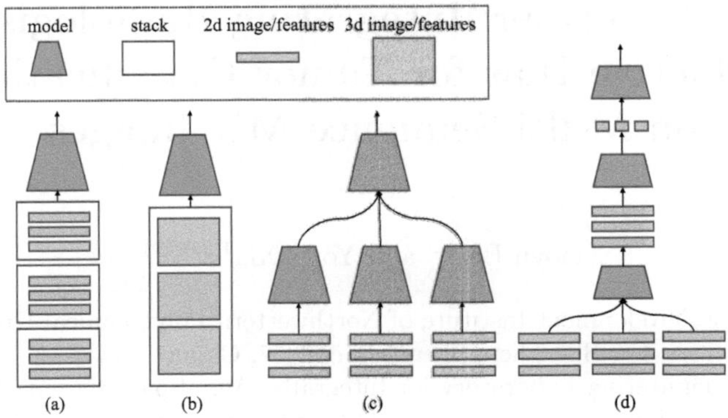

Fig. 1. Illustration of common tumor classification models: (a) The 2D model, which stacks slices from each MRI sequence as input; (b) The 3D model, which stacks all sequences together as input; (c) The mid-fusion model, which first extracts features from each sequence and then fuses them in the middle stage of the model; (d) Our RadioFormer, which divides feature extraction and fusion into three stages, inspired by the diagnostic practice of radiologists.

important role in tumor detection, characterization, and treatment planning. In particular, multi-sequence MRI provides a more detailed view of tissue properties by capturing images at different time points after contrast agent injection or across different imaging protocols. However, accurately interpreting multi-sequence MRI remains challenging, requiring extensive expertise and experience from radiologists. Additionally, manual diagnosis is time-consuming and subject to inter-observer variability, highlighting the need for automated diagnostic models to assist clinicians in tumor classification.

The advent of deep learning has catalyzed significant progress in image-based tumor diagnosis. While current methods primarily focus on specific organs, such as the liver [25,27–29,32], lung [1,19], kidney [8,26], breast [30,31], and brain [3,22], they generally follow similar design strategies. These methods typically process input 3D images in one of three ways: (a) as 2D slices, using 2D models for classification [11,18,20]; (b) by leveraging 3D models for direct classification [17,23]; or (c) by first extracting features from each sequence and fusing them at an intermediate stage of the model [16,25,27], as shown in Fig. 1(a)(b)(c). The 2D approach often utilizes pre-trained models from ImageNet to enhance performance. For instance, Swati et al. used the pre-trained VGG-16 model for brain tumor classification [20]. While this approach is straightforward and effective, it lacks a holistic interpretation of the 3D tumor volume, which can limit its ability to capture complex, inter-slice tumor characteristics. In contrast, the 3D approach classifies the entire 3D volume directly but suffers from the challenge of limited dataset size. Tumor datasets typically consist of only hundreds to a few thousand cases, making them prone to overfitting and resulting in suboptimal performance. Some researchers attempt to address this limitation by using pre-trained 3D models from video data [14,16], but the significant domain and semantic gap between medical imaging and video data often lead to unsatisfactory results. Other approaches utilize customized models to better extract and fuse features from different sequences. For example, Wang et al. proposed a method called TransLiver [25], which first uses independent Pyramid Vision Transformers (PVTs) [24] for each sequence image to extract sequence-specific features and then inputs them into a fusion module to obtain the final feature representation. While this approach improves feature fusion, it still faces the challenge of limited data size.

In this study, we introduce RadioFormer, a radiologist-inspired hierarchical Transformer model designed to effectively learn from modestly sized tumor datasets. RadioFormer departs from traditional 2D or 3D model designs by incorporating a hybrid 2D-3D data flow that mirrors the inductive bias of radiologists. Specifically, the architecture of RadioFormer is informed by the diagnostic practice of radiologists, who typically begin by inspecting each sequence image slice-by-slice, identifying key slices containing valuable diagnostic clues. These key slices are then collectively reviewed across sequences to reach a comprehensive diagnosis. RadioFormer mirrors this process through tripartite levels: the single-sequence slice information extraction level, the multi-sequence slice information aggregation level, and the inter-slice (volume) information aggregation level. At each level, RadioFormer employs pure vision Transformer blocks to encode information. The output of each level serves as the input tokens for the subsequent level, enabling a seamless flow of information and a hierarchical understanding of the lesion characteristics.

Our RadioFormer is examined on three public datasets: the LLD-MMRI liver tumor dataset [16] for classifying seven types of liver tumors, the Advanced-MRI-Breast-Lesions dataset [9] for benign and malignant breast tumor classification, and the ReMIND dataset [13] for classifying three types of brain tumors. The experimental results demonstrate that our RadioFormer achieves the best performance across all three datasets, highlighting its superior capability. The code and pre-processed data for our method are made publicly available.

2 Method

Figure 2 illustrates the image pre-processing step and overall architecture of the proposed RadioFormer model. In the data pre-processing step, given the original, unregistered multi-sequence MRI images, we employ the open-source UAE-M method [4] to register all sequence images to one sequence. Subsequently, the aligned multi-sequence tumor volumes are cropped from the registered images. These cropped tumor volumes then serve as the input to our RadioFormer model.

RadioFormer is a tripartite-level Transformer-based model designed to fully harness the diagnostic cues from volumetric data. The first level focuses on feature extraction from individual sequence slices. The second level aggregates information from aligned single-sequence slices, processing and consolidating it into a series of multi-sequence slice tokens. Finally, the third level fuses the multi-sequence slice tokens to produce the final classification result. In the following subsections, we will delve into each level in greater detail.

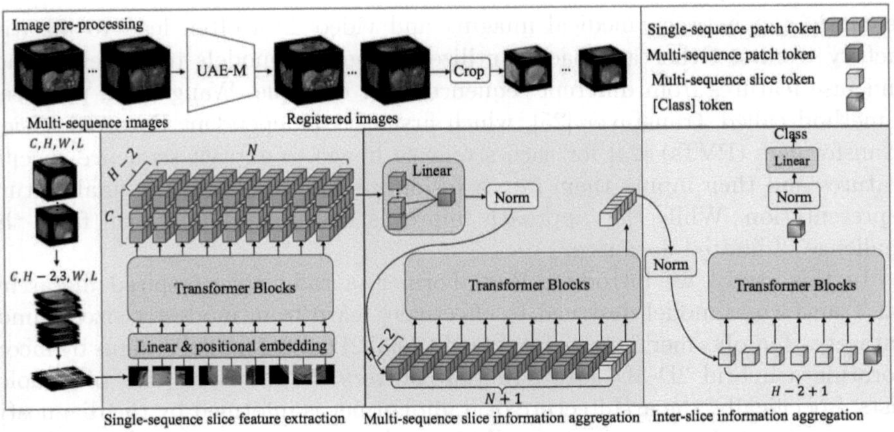

Fig. 2. The overall architecture of the proposed RadioFormer model.

2.1 Single-Sequence Slice Feature Extraction

Radiologists typically interpret MR images by first examining them slice by slice, looking for important diagnostic clues for further investigation. Our RadioFormer model emulates this approach by first extracting features from each sequence slice independently. Given a multi-sequence tumor volume with dimensions (C, H, W, L), where C denotes the number of sequences, and H, W, L represents the height, width, and length of the tumor volume, respectively. We first reorganize the volume into an array of 2D transverse images. Each transverse image contains three consecutive slices and has a shape of $(3, W, L)$.This reorganization results in an array of images with dimensions $(C, H-2, 3, W, L)$.

Subsequently, we divide each transverse image into patches of size (P, P) and flatten them, resulting in a sequence of vectors $[\mathbf{x}_{c,h}^1; \mathbf{x}_{c,h}^2; \cdots; \mathbf{x}_{c,h}^N]$, where the length of this sequence is denoted as $N = \frac{W}{P} \times \frac{L}{P}$. These vectors are then linearly embedded, and positional embeddings are added, as shown in the following equation:

$$\mathbf{z}_{c,h}^{(0)} = \left[\mathbf{x}_{c,h}^1 \mathbf{E_s}; \mathbf{x}_{c,h}^2 \mathbf{E_s}; \cdots; \mathbf{x}_{c,h}^N \mathbf{E_s}\right] + \mathbf{E}_{\text{pos}}, \tag{1}$$

where $c \in [1, C]$, $h \in [1, H-2]$, D denotes the feature length, $\mathbf{E}_{\text{s}} \in \mathbb{R}^{(P^2 \cdot 3) \times D}$ is the linear embedding layer and $\mathbf{E}_{\text{pos}} \in \mathbb{R}^{N \times D}$ is the positional embedding. Each $\mathbf{z}_{c,h}^{(0)}$ is then input into a series of ViT [10] Transformer encoder blocks. These blocks consist of alternating layers of multiheaded self-attention (MSA) and multi-layer perception (MLP) modules, with Layer Norm (LN) applied before each module. The mathematical representation is as follows:

$$\begin{aligned} \mathbf{z}'_{c,h} &= \text{MSA}\left(\text{LN}\left(\mathbf{z}_{c,h}^{(\ell-1)}\right)\right) + \mathbf{z}_{c,h}^{(\ell-1)}, \\ \mathbf{z}_{c,h}^{(\ell)} &= \text{MLP}\left(\text{LN}\left(\mathbf{z}'_{c,h}\right)\right) + \mathbf{z}'_{c,h}, \end{aligned} \tag{2}$$

where $\ell \in [1, L_s]$, L_s is the number of the Transformer blocks in this level. The output of the final Transformer block $\mathbf{z}_{c,h}^{(L_s)} = [\mathbf{y}_{c,h}^1; \mathbf{y}_{c,h}^2; \cdots; \mathbf{y}_{c,h}^N]$ are regarded as single-phase patch tokens. These tokens will serve as the input for the subsequent multi-phase slice information aggregation level.

2.2 Multi-Sequence Slice Information Aggregation

The objective of this level is to integrate the features extracted from single-phase slices into a higher-level representation that captures multi-sequence information. We begin with the spatially corresponding single-sequence patch tokens $[\mathbf{y}_{1,h}^i; \mathbf{y}_{2,h}^i; \cdots; \mathbf{y}_{C,h}^i]$. A linear layer $\mathbf{E}_m \in \mathbb{R}^{CD \times D}$ is utilized to project the concatenated tokens into a lower-dimensional multi-sequence patch token, as defined by:

$$\mathbf{m}_h^i = \mathrm{LN}(\mathrm{Concat}[\mathbf{y}_{1,h}^i; \mathbf{y}_{2,h}^i; \cdots; \mathbf{y}_{C,h}^i]\mathbf{E}_m). \tag{3}$$

Each multi-sequence slice is then represented by a series of these multi-sequence patch tokens $[\mathbf{m}_h^1; \mathbf{m}_h^2; \cdots; \mathbf{m}_h^N]$. Building upon the previous level, we process these multi-sequence patch tokens using a stack of Transformer encoder blocks. To aggregate patch-wise information into a slice-wise representation, we introduce a learnable multi-sequence slice token $\mathbf{s}$ to the patch tokens $[\mathbf{m}_h^1; \cdots; \mathbf{m}_h^N; \mathbf{s}]$. The fusion of multi-phase slice information is encapsulated in the following equations:

$$\begin{aligned}
\mathbf{v}_h^{(0)} &= \left[\mathbf{m}_h^1; \mathbf{m}_h^2; \cdots; \mathbf{m}_h^N; \mathbf{s}\right], \\
\mathbf{v}_h' &= \mathrm{MSA}\left(\mathrm{LN}\left(\mathbf{v}_h^{(\jmath-1)}\right)\right) + \mathbf{v}_h^{(\jmath-1)}, \\
\mathbf{v}_h^{(\jmath)} &= \mathrm{MLP}\left(\mathrm{LN}\left(\mathbf{v}_h'\right)\right) + \mathbf{v}_h',
\end{aligned} \tag{4}$$

where $\jmath \in [1, L_m]$, and L_m denotes the number of Transformer blocks in this level. The output from the final Transformer block retains only the multi-sequence slice token $\mathbf{s}_h$, corresponding to the initial input token $\mathbf{s}$, as it encapsulates the integrated information from all multi-sequence patches within this slice.

2.3 Inter-Slice Information Aggregation

Having obtained the multi-sequence slice tokens $\mathbf{s}_h$ from the previous level, we now aggregate information across the entire series of multi-sequence slices $[\mathbf{s}_1, \cdots, \mathbf{s}_{H-2}]$ to make the final classification. We begin by applying layer normalization to this sequence and, following the Vision Transformer (ViT) paradigm, we append a [class] token to the series of multi-sequence slice tokens. As with the previous levels, this processed sequence is then input into a series of Transformer encoder blocks.

$$\begin{aligned}
\mathbf{o}^{(0)} &= \left[\mathrm{LN}([\mathbf{s}_1; \mathbf{s}_2; \cdots; \mathbf{s}_{H-2}]); \mathbf{c}\right], \\
\mathbf{o}' &= \mathrm{MSA}\left(\mathrm{LN}\left(\mathbf{o}^{(\imath-1)}\right)\right) + \mathbf{o}^{(\imath-1)}, \\
\mathbf{o}^{(\imath)} &= \mathrm{MLP}\left(\mathrm{LN}\left(\mathbf{o}'\right)\right) + \mathbf{o}',
\end{aligned} \tag{5}$$

where $\imath \in [1, L_{\text{f}}]$, and L_{f} represents the number of Transformer blocks in this final level. The output corresponding to the [class] token $\mathbf{c}$ is then subjected to an additional layer normalization followed by a linear projection layer to produce the classification logits.

3 Experiments

Data: Our RadioFormer model was trained and evaluated on three datasets: the LLD-MMRI liver tumor dataset [16], the Advanced-MRI-Breast-Lesions (ABL) dataset [9], and the ReMIND brain tumor dataset [13]. The LLD-MMRI dataset consists of seven types of liver tumors (hepatocellular carcinoma, intrahepatic cholangiocarcinoma, hepatic metastasis, hepatic cysts, hepatic hemangiomas, focal nodular hyperplasia, hepatic abscesses), with a total of 498 cases, each having 8 MRI sequences. The ABL dataset includes 94 cases, each with 6 MRI sequences for benign and malignant breast tumor classification. The ReMIND dataset contains 71 cases, each with 2 MRI sequences for classifying three types of brain tumors (Oligodendroglioma, Astrocytoma, Glioblastoma). For the LLD-MMRI dataset, we follow the LLD-MMRI2023 challenge protocol[1], performing 5-fold cross-validation using the official train-validation splits (316 cases for training and validation, and 78 cases for testing, Stage One rule) and also report results on the test set (316 cases for training, 78 for validation, and 104 for testing, Stage Two rule). For the ABL and ReMIND datasets, we perform 5-fold cross-validation using random train-validation-test splits with a ratio of 3:1:1.

Implementation Details: The RadioFormer model was developed using PyTorch and all experiments were conducted on an RTX 4090 GPU. The architecture comprises three levels, each with a specified number of Transformer blocks: $L_{\text{s}} = 12$ for the single-sequence slice feature extraction level, $L_{\text{m}} = 2$ for the multi-sequence slice information aggregation level, and $L_{\text{f}} = 4$ for the inter-slice information aggregation level. To initialize the first level, we loaded ImageNet-pretrained ViT model weights. For training, we employed the AdamW optimizer with an initial learning rate of 1e-4, which was adjusted using a cosine learning rate schedule. The minimum learning rate was set to 1e-5, and the training spanned 300 epochs. The first 5 epochs served as a warm-up phase, during which the learning rate was progressively increased. The standard cross-entropy was used as the loss function. The batch size was fixed at 4, and each lesion volume was resized to $16 \times 128 \times 128$. Our data augmentation strategy included random rotations and flips across various anatomical axes. During the training phase, lesion volumes were randomly cropped to $14 \times 112 \times 112$. For evaluation, a central crop of the same dimensions was extracted. In line with the LLD-MMRI2023 challenge, the model's performance was evaluated using the F1-score and Cohen's Kappa metrics. On LLD-MMRI dataset, the training time of each model is about 2 h, and the inference time is 1 s per case.

[1] https://github.com/LMMMEng/LLD-MMRI2023.

Table 1. Comparative performance for tumor classification on LLD-MMRI Stage One test set. The results are formulated as F1-score / Cohen's Kappa.

	LLD-MMRI2023 Stage One train-val splits				
Methods	Fold1	Fold2	Fold3	Fold4	Fold5
Swin3D [7]	0.662/0.610	0.643/0.606	0.679/0.634	0.644/0.612	0.675/0.644
Transliver [25]	0.714/0.695	0.705/0.667	0.687/0.681	0.650/0.624	0.689/0.679
Swin-S [15]	0.755/0.698	**0.735**/0.694	0.721/0.662	0.734/0.729	0.747/0.689
Resnet18 [12] (Rank 2)	0.740/0.699	0.702/0.699	0.708/0.692	0.729/0.713	0.721/0.710
Uniformer [14] (Rank 1)	0.746/0.728	0.716/**0.724**	0.721/0.693	0.732/0.717	0.693/0.674
ViViT [2]	0.663/0.619	0.692/0.671	0.733/0.691	0.691/0.667	0.697/0.684
VideoMAE [21]	0.719/0.701	0.676/0.645	0.722/0.706	0.712/0.672	0.693/0.674
TimeSformer [5]	0.641/0.609	0.689/0.681	0.714/0.666	0.643/0.607	0.641/0.622
RadioFormer	**0.796**/**0.777**	0.710/0.713	**0.760**/**0.747**	**0.771**/**0.761**	**0.796**/**0.779**

Results: In Table 1, we present the detailed performance results for each fold under the LLD-MMRI Stage One rule. To ensure a comprehensive comparison, we evaluate recent classification methods, including Transliver [25], a Transformer-based model specifically designed for liver tumor classification; Swin-S [15]; Swin3D [7]; and video-based models such as ViViT [2], VideoMAE [21], and TimeSformer [5]. Additionally, we compare our method with the top two approaches from the LLD-MMRI2023 challenge[2], which were based on a modified Uniformer [14] (Rank 1) and a ResNet18 model [12] (Rank 2). Our RadioFormer consistently outperforms all compared methods in terms of both F1-score and Cohen's Kappa across four out of five folds, achieving competitive results on the remaining fold. This strong and consistent performance across different train-validation splits underscores the effectiveness of our approach in liver tumor classification. We further report the average five-fold cross-validation results for the ABL and ReMIND datasets, along with the results on the LLD-MMRI Stage Two test set in Table 2. In all cases, RadioFormer achieves the best overall performance, surpassing the recent SDR-Former [16], the solution provided by the official LLD-MMRI dataset team. Additionally, we report the computational complexity of the comparative methods. As shown, RadioFormer has fewer training parameters than most competing methods while maintaining moderate GFLOPS, demonstrating a favorable balance between efficiency and accuracy.

Ablation Study: To evaluate the contributions of the different levels in our RadioFormer model, we first conducted an ablation study focusing on the multi-sequence slice information aggregation (second) level and the inter-slice information aggregation (third) level. We designed two variant models for this purpose. Model1: In this variant, we replaced the second level of our RadioFormer with a global average pooling layer. This modification eliminates the explicit feature

[2] The codes of the top five teams are open-sourced.

Table 2. Comparative performance for tumor classification on the LLD-MMRI Stage Two test set and the average 5-fold cross-validation results on the ABL and ReMIND datasets. The results are reported as F1-score / Cohen's Kappa.

	Datasets			Complexity
Methods	ABL	ReMIND	LLD-MMRI Test	GFLOPS/Training Params (M)
Swin3D [7]	0.611/0.214	0.407/0.182	0.688/0.651	21.7/31.6
Transliver [25]	0.591/0.178	0.422/0.179	0.716/0.654	572.8/154.6
Resnet18 [12]	0.646/0.311	0.489/0.249	0.711/0.665	384.0/15.1
Uniformer [14]	0.542/0.103	0.491/0.258	0.712/0.666	112.6/49.5
ViViT [2]	0.621/0.303	0.364/0.102	0.733/0.691	67.1/88.5
VideoMAE [21]	0.630/0.290	0.417/0.173	0.722/0.706	64.4/88.0
TimeSformer [5]	0.589/0.181	0.433/0.178	0.714/0.666	84.2/ 122.9
H2Former [16]	-	-	0.774/0.726	-
SDR-Former [16]	-	-	0.791/0.747	-
RadioFormer	**0.695/0.362**	**0.568/0.357**	**0.806/0.745**	209.8/32.1

processing that occurs at the second level. Model2: we replaced the third level of our RadioFormer with a global average pooling layer. We then compared their average F1-score and Cohen's Kappa of the last 100 epochs on the fold1 validation subset. The results are shown in Table 3. The comparison reveals that both the second and third levels of RadioFormer contribute to the model's overall performance.

Our RadioFormer explicitly processes the input multi-sequence data in three levels, following a C-N-H order. First, it fuses the slices along the sequence dimension (C), reducing the number of channels (C) to 1. Next, it consolidates the N tokens within each slice into a single token. Finally, it aggregates the H slice tokens to produce the final classification result. This approach mirrors the clinical practice of radiologists. Additionally, we tested other parsing orders on the LLD-MMRI Stage One Fold1, with results presented in Table 4. As shown, our C-N-H parsing order achieves the best performance. This further demonstrates the importance of incorporating radiologists' inductive bias into model design.

Table 3. Ablations on the effects of second and third levels.

model	F1-score/Cohen's Kappa
Model1	0.788/0.762
Model2	0.815/0.807
RadioFormer	0.838/0.812

Table 4. Ablations on the parsing order.

order	F1-score/Cohen's Kappa
N-C-H	0.656/0.619
N-H-C	0.683/0.637
H-N-C	0.631/0.598
H-C-N	0.729/0.726
C-H-N	0.754/0.741
C-N-H	0.796/0.777

4 Conclusion

In this work, we introduce RadioFormer, a radiologist-inspired model designed to effectively learn from modestly sized multi-sequence MRI tumor datasets. Unlike traditional 2D or 3D model architectures, RadioFormer incorporates a hybrid 2D-3D data processing flow that mirrors the way radiologists interpret MRI images. Our extensive testing on three different datasets demonstrates that RadioFormer consistently delivers stable and high-performance results across all datasets.

Acknowledgments. This work was supported in part by the Shenzhen Science and Technology Program under Grants JCYJ20220530161616036, in part by the National Natural Science Foundation of China under Grants 62171377 and 92470101, in part by the "Pioneer" and "Leading Goose" R&D Program of Zhejiang, China, under Grant 2025C01201(SD2), and in part by the Ningbo Clinical Research Center for Medical Imaging under Grant 2021L003 (Open Project 2022LYKFZD06).

Disclosure of Interests. The authors have no competing interests to declare that are relevant to the content of this article.

References

1. Ardila, D., et al.: End-to-end lung cancer screening with three-dimensional deep learning on low-dose chest computed tomography. Nat. Med. **25**(6), 954–961 (2019)
2. Arnab, A., Dehghani, M., Heigold, G., Sun, C., Lučić, M., Schmid, C.: Vivit: a video vision transformer. In: Proceedings of the IEEE/CVF International Conference on Computer Vision, pp. 6836–6846 (2021)
3. Ayadi, W., Elhamzi, W., Charfi, I., Atri, M.: Deep CNN for brain tumor classification. Neural Process. Lett. **53**, 671–700 (2021)
4. Bai, X., et al.: UAE: Universal anatomical embedding on multi-modality medical images. arXiv preprint arXiv:2311.15111 (2024)
5. Bertasius, G., Wang, H., Torresani, L.: Is space-time attention all you need for video understanding? In: ICML, vol. 2, p. 4 (2021)
6. Bray, F., Ferlay, J., Soerjomataram, I., Siegel, R.L., Torre, L.A., Jemal, A.: Global cancer statistics 2018: globocan estimates of incidence and mortality worldwide for 36 cancers in 185 countries. CA Cancer J. Clin. **68**(6), 394–424 (2018)

7. Cardoso, M.J., et al.: Monai: An open-source framework for deep learning in healthcare. arXiv preprint arXiv:2211.02701 (2022)
8. Dai, C., et al.: Deep learning assessment of small renal masses at contrast-enhanced multiphase CT. Radiology **311**(2), e232178 (2024)
9. Daniels, D., et al.: Standard and delayed contrast-enhanced MRI of malignant and benign breast lesions with histological and clinical supporting data (advanced-mri-breast-lesions)(version 2). https://www.cancerimagingarchive.net/collection/advanced-mri-breast-lesions/
10. Dosovitskiy, A., et al.: An image is worth 16x16 words: Transformers for image recognition at scale. arXiv preprint arXiv:2010.11929 (2020)
11. Frid-Adar, M., Klang, E., Amitai, M., Goldberger, J., Greenspan, H.: Synthetic data augmentation using GAN for improved liver lesion classification. In: 2018 IEEE 15th International Symposium on Biomedical Imaging (ISBI 2018), pp. 289–293. IEEE (2018)
12. He, K., Zhang, X., Ren, S., Sun, J.: Deep residual learning for image recognition. In: CVPR, pp. 770–778 (2016)
13. Juvekar, P., et al.: The brain resection multimodal imaging database (ReMIND). https://www.cancerimagingarchive.net/collection/remind/
14. Li, K., et al.: Uniformer: Unified transformer for efficient spatiotemporal representation learning. arXiv preprint arXiv:2201.04676 (2022)
15. Liu, Z., et al.: Swin transformer: hierarchical vision transformer using shifted windows. In: Proceedings of the IEEE/CVF International Conference on Computer Vision, pp. 10012–10022 (2021)
16. Lou, M., et al.: SDR-former: a siamese dual-resolution transformer for liver lesion classification using 3d multi-phase imaging. Neural Networks 107228 (2025)
17. Qu, R., Xiao, Z.: An attentive multi-modal CNN for brain tumor radiogenomic classification. Information **13**(3), 124 (2022)
18. Romero, F.P., et al.: End-to-end discriminative deep network for liver lesion classification. In: 2019 IEEE 16th International Symposium on Biomedical Imaging (ISBI 2019), pp. 1243–1246. IEEE (2019)
19. Shen, S., Han, S.X., Aberle, D.R., Bui, A.A., Hsu, W.: Explainable hierarchical semantic convolutional neural network for lung cancer diagnosis. In: CVPR workshops, pp. 63–66 (2019)
20. Swati, Z.N.K., et al.: Brain tumor classification for MR images using transfer learning and fine-tuning. Comput. Med. Imaging Graph. **75**, 34–46 (2019)
21. Tong, Z., Song, Y., Wang, J., Wang, L.: VideoMAE: masked autoencoders are data-efficient learners for self-supervised video pre-training. In: Advances in neural information processing systems, vol. 35, pp. 10078–10093 (2022)
22. Tummala, S., Kadry, S., Bukhari, S.A.C., Rauf, H.T.: Classification of brain tumor from magnetic resonance imaging using vision transformers Ensembling. Curr. Oncol. **29**(10), 7498–7511 (2022)
23. Wang, R., et al.: Cross-attention guided loss-based deep dual-branch fusion network for liver tumor classification. Inf. Fusion **114**, 102713 (2025)
24. Wang, W., et al.: Pyramid vision transformer: a versatile backbone for dense prediction without convolutions. In: Proceedings of the IEEE/CVF International Conference on Computer Vision, pp. 568–578 (2021)
25. Wang, X., Ying, H., Xu, X., Cai, X., Zhang, M.: Transliver: a hybrid transformer model for multi-phase liver lesion classification. In: International Conference on Medical Image Computing and Computer-Assisted Intervention, pp. 329–338. Springer (2023)

26. Xi, I.L., et al.: Deep learning to distinguish benign from malignant renal lesions based on routine MR imaging. Clin. Cancer Res. **26**(8), 1944–1952 (2020)
27. Xu, X., et al.: A knowledge-guided framework for fine-grained classification of liver lesions based on multi-phase CT images. IEEE J. Biomed. Health Inform. **27**(1), 386–396 (2022)
28. Yasaka, K., Akai, H., Abe, O., Kiryu, S.: Deep learning with convolutional neural network for differentiation of liver masses at dynamic contrast-enhanced CT: a preliminary study. Radiology **286**(3), 887–896 (2018)
29. Ying, H., et al.: A multicenter clinical AI system study for detection and diagnosis of focal liver lesions. Nat. Commun. **15**(1), 1131 (2024)
30. Zhang, Y., et al.: Prediction of breast cancer molecular subtypes on DCE-MRI using convolutional neural network with transfer learning between two centers. Eur. Radiol. **31**, 2559–2567 (2021)
31. Zhang, Y., et al.: Deep learning-based automatic diagnosis of breast cancer on MRI using mask R-CNN for detection followed by resnet50 for classification. Acad. Radiol. **30**, S161–S171 (2023)
32. Zhou, J., et al.: Automatic detection and classification of focal liver lesions based on deep convolutional neural networks: a preliminary study. Front. Oncol. **10**, 581210 (2021)

RadiomicsRetrieval: A Customizable Framework for Medical Image Retrieval Using Radiomics Features

Inye Na[1], Nejung Rue[2], Jiwon Chung[1], and Hyunjin Park[1,2](✉)

[1] Department of Electrical and Computer Engineering, Sungkyunkwan University, Suwon 16419, South Korea

[2] Department of Artificial Intelligence, Sungkyunkwan University, Suwon 16419, South Korea

hyunjinp@skku.edu

Abstract. Medical image retrieval is a valuable field for supporting clinical decision-making, yet current methods primarily support 2D images and require fully annotated queries, limiting clinical flexibility. To address this, we propose ***RadiomicsRetrieval***, a 3D content-based retrieval framework bridging handcrafted radiomics descriptors with deep learning-based embeddings at the *tumor level.* Unlike existing 2D approaches, *RadiomicsRetrieval* fully exploits volumetric data to leverage richer spatial context in medical images. We employ a promptable segmentation model (e.g., SAM) to derive tumor-specific image embeddings, which are aligned with radiomics features extracted from the same tumor via contrastive learning. These representations are further enriched by anatomical positional embedding (APE). As a result, *RadiomicsRetrieval* enables **flexible querying** based on shape, location, or partial feature sets. Extensive experiments on both lung CT and brain MRI public datasets demonstrate that radiomics features significantly enhance retrieval specificity, while APE provides global anatomical context essential for location-based searches. Notably, our framework requires only minimal user prompts (e.g., a single point), minimizing segmentation overhead and supporting diverse clinical scenarios. The capability to query using either image embeddings or selected radiomics attributes highlights its adaptability, potentially benefiting diagnosis, treatment planning, and research on large-scale medical imaging repositories. Our code is available at https://github.com/nainye/RadiomicsRetrieval.

Keywords: Medical Image Retrieval · Radiomics Features · Anatomical Position · Flexible Query

1 Introduction

The growing volume of medical imaging has made content-based image retrieval (CBIR) a promising tool for radiologists to locate clinically relevant cases [20].

I. Na, N. Rue—Equal contribution.

J. C. Gee et al. (Eds.): MICCAI 2025, LNCS 15960, pp. 556–566, 2026.
https://doi.org/10.1007/978-3-032-04927-8_53

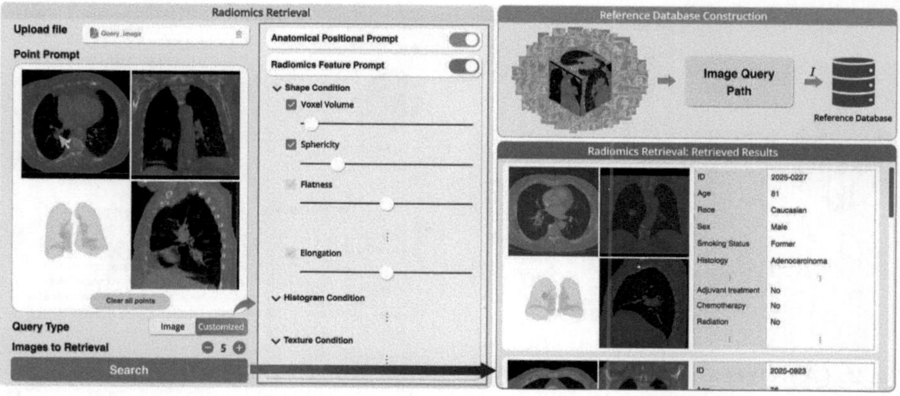

Fig. 1. Illustration of the RadiomicsRetrieval. Users can retrieve similar tumors by providing either an image with minimal point prompts or customized conditions (e.g., anatomical location or partial radiomics features). This diverse setup supports both image-based queries and more general feature-driven searches.

While CBIR methods relying on deep learning-based encoders have surpassed those based on handcrafted features, most remain limited to 2D models that lose vital 3D anatomical context essential for accurate lesion characterization [1,6,15,16,23]. Current CBIR methods emphasize global image-level similarity rather than lesion-specific relevance [22]. Although some methods accept lesion masks as queries, this is often impractical for 3D data due to the labor-intensive nature of volumetric delineation [15]. Moreover, most methods still limit queries to images, masks, or textual labels, which reduces flexibility in real-world clinical settings [8].

We propose *RadiomicsRetrieval*, a CBIR framework that unifies traditional radiomics with deep learning via tumor-level contrastive learning. Radiomics features, established handcrafted descriptors, capture detailed tumor shape and intensity information, but demand extensive manual segmentation [21]. In contrast, deep learning methods can bypass segmentation, yet often fail to retain robust tumor specificity [9,16]. To bridge this gap, we leverage promptable segmentation models such as SAM [14] to derive tumor-specific image embeddings from minimal user input. Radiomics features are separately extracted from the same tumor and contrastive learning between these *image embeddings* and *radiomics embeddings* integrates tumor-specific attributes—shape, intensity, and spatial location—into a unified representation.

A key advantage of *RadiomicsRetrieval* is **flexible querying**: users may query with an image for detailed local and global context or with radiomics features for feature-specific searches (e.g., "find highly spherical tumors in the frontal lobe"). By encoding feature names and values as text-value pairs in a language-model style, we handle varying feature sets and enable partial feature usage (Fig. 1). This design broadens clinical utility and improves search effi-

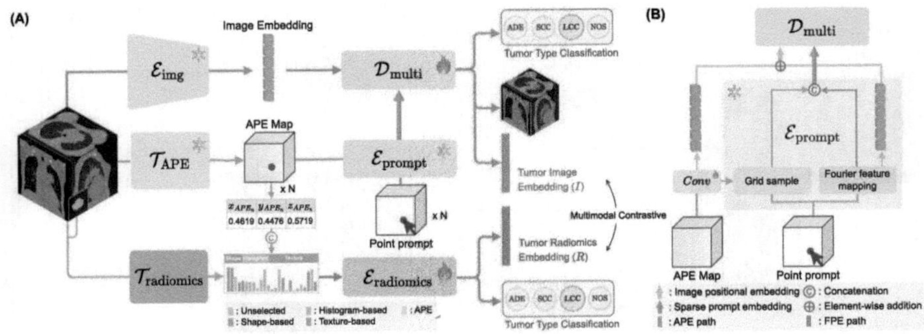

Fig. 2. (A) Overall architecture of RadiomicsRetrieval, showing parallel paths for tumor image (green), radiomics (blue), and anatomical (purple) embeddings. **(B)** Detailed view of how the anatomical positional embedding (APE) is integrated with the Fourier-based positional encoding (FPE). (Color figure online)

ciency, making *RadiomicsRetrieval* a highly adaptable CBIR solution. Our main contributions are:

1. **Hybrid Radiomics-Deep Learning Approach:** Combines traditional radiomics with deep learning via tumor-level contrastive learning for enhanced clinical relevance.
2. **Flexible Query Modes:** Supports both images for detailed context and radiomics embeddings for feature-specific or partial-feature queries.
3. **Minimizing Labeling Efforts:** Uses promptable segmentation models to derive tumor-specific image embeddings from minimal prompts, reducing the burden of manual segmentation.

2 Method

RadiomicsRetrieval operates on each tumor through two parallel paths (Fig. 2 (A)): an **image-query path** and a **radiomics-query path**. The image path generates a *tumor image embedding* (I) via a promptable segmentation model, while the radiomics path produces a *tumor radiomics embedding* (R) from hand-crafted features. These embeddings are aligned via *tumor-level contrastive learning*, yielding a shared latent space where clinically similar lesions cluster and dissimilar lesions are pushed apart.

2.1 Image-Query Path

We extend SAM-Med3D [26], which typically outputs only a segmentation mask for the prompted region of interest (ROI), by augmenting its lightweight mask decoder with additional *classification* and *contrastive* tokens—denoted $\mathcal{D}_{\text{multi}}$. Given a volumetric input ($128 \times 128 \times 128$) and a minimal prompt (e.g., a single point), the model produces a *tumor image embedding* (I), along with auxiliary mask and classification outputs. Only during training, full segmentation masks guide the network in learning tumor-specific (ROI-level, not image-level) embeddings. However, when constructing the reference database (i.e., for retrieval), the model requires only a few point prompts per tumor, thereby reducing the need for manual ROI specification. These embeddings then populate a reference database, against which future queries are compared (Fig. 1, top-right).

2.2 Radiomics-Query Path

A *tumor radiomics embedding* (R) is encoded with TransTab [27] by tokenizing each radiomics feature as a feature name-value pair. Users can specify any subset of features without retraining, supporting flexible search conditions. We extract 72 standard radiomics features per tumor, including 14 shape-based, 18 histogram, and 40 texture features via PyRadiomics [25]. During training, we randomly sample radiomics subsets ranging from 72 to 1 feature, enhancing robustness for partial feature queries (Fig. 3(**B**)).

2.3 Anatomical Positional Embedding (APE)

We incorporate global anatomical context using a 3-channel voxel-based APE [13], integrating it into both image and radiomics paths (Fig. 2(**A**)). In the **image path**, the APE map is downsampled through convolutional layers to align with SAM's Fourier-based positional encoding (FPE) [24] dimensions and then integrated element-wise (Fig. 2(**B**)). Additionally, sampled APE values at prompt coordinates are concatenated with sparse prompt embeddings to reinforce local anatomical context. In the **radiomics path**, voxel-level APE explicitly enriches tumor-intrinsic radiomics features, improving location-aware retrieval. Specifically, APE values at point-prompt locations are concatenated with radiomics subsets as input to $\mathcal{E}_{\text{radiomics}}$. During training, we randomly include APE and also train APE-only cases, enabling flexible querying (radiomics-only, APE-only, or combined) as illustrated in Fig. 3(**B**). For lung CT (NSCLC), we use an APE model pretrained on abdominal and thoracic CT [13]; for brain tumor MRI, we pre-trained the APE model with BraTS MRI data.

Training Details.

Loss Functions. We freeze large pretrained modules (SAM and APE) and fine-tune only the lightweight decoder ($\mathcal{D}_{\text{multi}}$) and the radiomics encoder. We jointly optimize: (1) $\mathcal{L}_{\text{seg}}$ (Dice + Cross-Entropy) for segmenting the prompted tumor, (2) $\mathcal{L}_{\text{contrast}}$ (InfoNCE-style [7]) for clustering embeddings of the same tumor and separating others, and (3) $\mathcal{L}_{\text{cls}}$ (Binary Cross-Entropy or Cross-Entropy) for classifying tumor type in both the image and radiomics pathways.

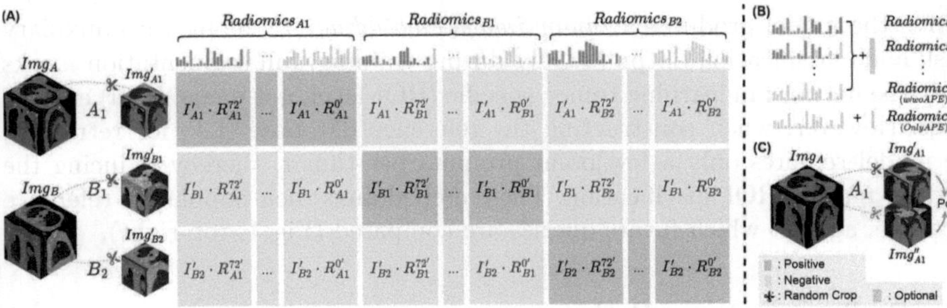

Fig. 3. Multi-positive contrastive learning. (A) Same-tumor embeddings are *positives* (yellow); different tumors are *negatives* (gray). **(B)** Radiomics feature subsets. **(C)** Paired random cropping (twice per tumor). (Color figure online)

Multi-positive Contrastive Setup. As shown in Fig. 3, we use a batch size of 80, containing 40 distinct tumors, each repeated with randomly chosen point prompts (1–10) and randomized crops. This produced two image embeddings (e.g., I' and I'') per tumor and multiple radiomics embeddings (e.g., $R^{72'} \ldots R^{0'}$, $R^{72''} \ldots R^{0''}$) incorporating matching APE values. Embeddings from the same tumor formed *positives*, while embeddings from other tumors served as *negatives*, enabling multi-positive contrastive learning.

Retrieval Inference. At inference time, a user may provide (1) an image with minimal point prompts, yielding a *tumor image embedding* via the image path, or (2) a set (or subset) of radiomics features, APE, or both, yielding a *tumor radiomics embedding*. The resulting query embedding is then compared against the reference database using cosine similarity in the shared latent space, allowing both precise image-based searches and flexible feature-driven queries (e.g., all lesions with a diameter exceeding 30 mm in the left frontal lobe).

3 Dataset

Brain Tumor MRI. We used the Adult Glioma (GLI) [3,4,19] and Meningioma (MEN) [17] subsets from the BraTS 2023 Challenge. Only the tumor core, comprising enhancing tumor and necrotic areas, was considered. Patients were split (8:1:1, stratified by tumor type) into 1440 (GLI 724, MEN 716) tumors for training, 176 (GLI 89, MEN 87) for validation, and 179 (GLI 90, MEN 89) for testing. For retrieval evaluation, ANTsPy's non-linear registration aligned anatomical masks from the ICBM T1 atlas [10–12], yielding precise left/right brain parcellations.

Lung Tumor CT. We combined non-small cell lung cancer (NSCLC) data from multiple sources (NSCLC-Radiomics [2], NSCLC-Radiomics-Interobserver1 [29], RIDER-LungCT-Seg [30], NSCLC Radiogenomics [5], and LUNG-PET-CT-Dx

[18]), focusing only on NSCLC cases. The dataset was split (7:1:2) at the patient level, stratified by subtype (adenocarcinoma [ADC], squamous cell carcinoma [SCC], large cell carcinoma [LCC], not otherwise specified [NOS]). This yielded 672 (SCC 168, LCC 85, ADC 302, NOS 48, Unlabeled 69) tumors for training, 100 (SCC 24, LCC 14, ADC 45, NOS 7, Unlabeled 10) for validation, and 192 (SCC 65, LCC 35, ADC 74, NOS 18) for testing. Unlabeled tumors were ignored in the classification loss. We used TotalSegmentator [28] to obtain anatomical masks, including five lung lobes (upper/lower on each side plus the right middle lobe).

Table 1. Location-Based Evaluation. *P@k* indicates the mean precision@k, i.e., the fraction of top-*k* retrieved samples that lie in the same region as the query. **Bold** denotes the best and underline is the second best in a given category.

Query type	Training modality	Contrastive	BraTS		NSCLC	
			P@5	P@10	P@5	P@10
Image & 1 tumor point	`<Image,FPE>`	unimodal	0.8358	0.8028	**0.9531**	**0.9167**
	`<Image,APE>`	unimodal	0.7899	0.7598	0.9271	0.8766
	`<Image,Radiomics,FPE>`	multimodal	0.5458	0.6514	0.6927	0.6807
	`<Image,Radiomics,APE>`	multimodal	**0.9117**	**0.9145**	0.9115	0.9000
	`<Image,Radiomics,FPE+APE>`	multimodal	0.9084	0.8933	0.9062	0.8984
1 point APE	`<Image,Radiomics,APE>`	multimodal	**0.8905**	**0.9000**	0.9000	**0.8891**
	`<Image,Radiomics,FPE+APE>`	multimodal	0.8603	0.8615	**0.9010**	0.8880

4 Results

We evaluated retrieval performance using the test set as queries and the training set as the reference database. Specifically, we constructed the reference database using image embeddings derived from a single tumor-center prompt per training image. To examine how radiomics features and APE influenced retrieval, we compared five training settings. We started with a baseline that used only an image path (*unimodal-contrastive*) with FPE (`<Image,FPE>`), and then either kept or replaced FPE with APE or added a radiomics path (*multimodal-contrastive*) as follows:

- `<Image,APE>`: Replaced FPE with APE.
- `<Image,Radiomics,FPE>`: Kept FPE and added a radiomics path.
- `<Image,Radiomics,APE>`: Replaced FPE with APE and added a radiomics path.
- `<Image,Radiomics,FPE+APE>`: Combined both FPE and APE alongside the radiomics path.

4.1 Evaluation of Global Context

Table 1 reports location-based retrieval performance by measuring the fraction of retrieved tumors in the same region as the query. The *Image & 1 tumor point* query type used an image with a single tumor center prompt and the resulting image embedding for retrieval. Among *unimodal-contrastive* models, `<Image,FPE>` surpassed `<Image,APE>`, indicating that FPE captured finer spatial details than the coarser, "average-patient"-style APE. Although APE alone lacked high-frequency precision, it was crucial in the *radiomics-query path*, as radiomics features solely describe tumor-internal properties. The model learned global location features by integrating APE into the radiomics path during multimodal contrastive training with image embeddings.

Turning to *multimodal-contrastive* approaches, those that used APE achieved the highest performance, showing that global anatomical cues improved location-specific matching. In contrast, **`<Image,Radiomics,FPE>`** focused on tumor-internal features and was less effective for anatomical retrieval. Finally, even a `1 point APE` query localized tumors accurately, underscoring the flexibility of our method for purely anatomical searches.

Table 2. Radiomics-Based Evaluation. *Top-k* is the average correlation between the query's radiomics features and those of the top-k retrieved samples. *Upper Bound* selects the top-k by directly comparing radiomics features in the reference set, while *Random Baseline* randomly selects samples from the reference set. **Bold** denotes the best and underline is the second best in a given category.

Query type	Training modality	Contrastive	BraTS		NSCLC	
			Top-5	Top-10	Top-5	Top-10
Image & 1 tumor point	<Image,FPE>	unimodal	0.8582	0.8476	0.8284	0.8161
	<Image,APE>	unimodal	0.8495	0.8418	0.7649	0.7543
	<Image,Radiomics,FPE>	multimodal	**0.9197**	**0.9177**	**0.8855**	**0.8818**
	<Image,Radiomics,APE>	multimodal	0.8959	0.8866	0.8316	0.8159
	<Image,Radiomics,FPE+APE>	multimodal	0.8987	0.8904	0.8403	0.8185
Radiomics features	<Image,Radiomics,FPE>	multimodal	**0.9663**	**0.9621**	**0.9448**	**0.9324**
	<Image,Radiomics,APE>	multimodal	0.9484	0.9426	0.8839	0.8847
	<Image,Radiomics,FPE+APE>	multimodal	0.9499	0.9457	0.9051	0.8956
Upper Bound			0.9794	0.9756	0.9711	0.9648
Random Baseline			0.7745	0.7806	0.6815	0.6790

4.2 Evaluation of Local Context

Table 2 shows correlations of tumor-internal radiomics features between queries and retrievals. *Multimodal-contrastive* methods substantially outperformed *unimodal-contrastive* approaches, demonstrating the benefit of combining image and radiomics paths. Notably, when radiomics features (without APE) were used as the query, the method focused more on tumor-internal attributes, yielding even higher correlation scores than image-based queries. In particular, <Image, Radiomics,FPE> achieved the highest scores, suggesting this setting prioritized local tumor details over global anatomical cues. Meanwhile, <Image,Radiomics, FPE+APE> remained competitive for local context while offering stronger location-based retrieval (see Table 1), suggesting a robust trade-off between capturing tumor-specific details and anatomical positioning.

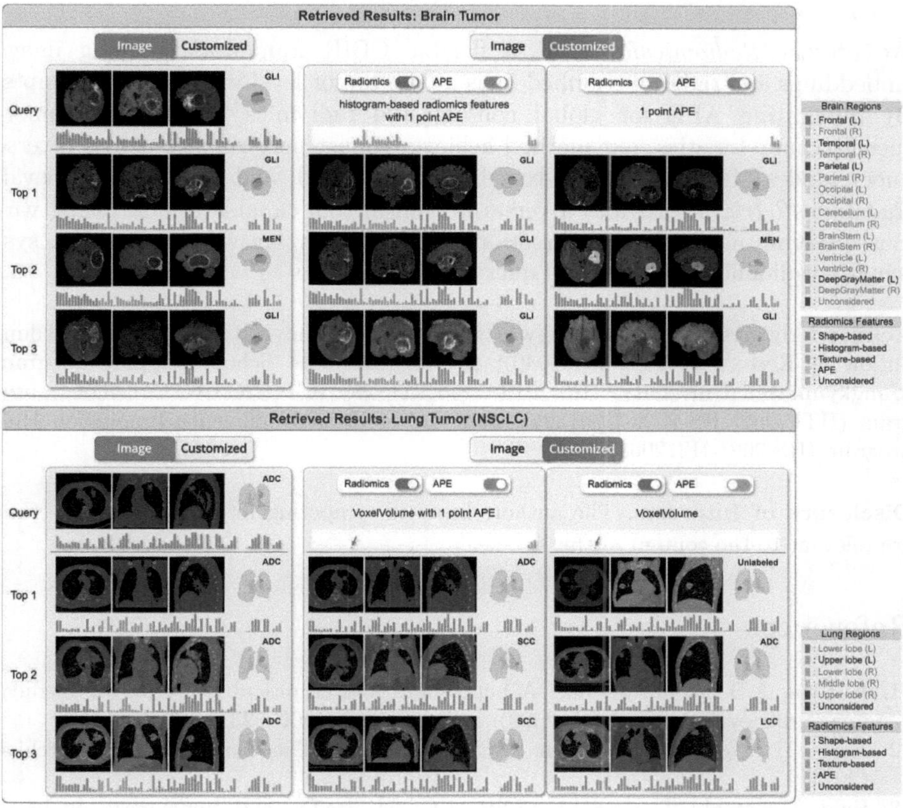

Fig. 4. Retrieved Samples Under Different Query Types. Using the same query tumor, varying the retrieval condition (image-based, location-only, or size-only, etc.) produces distinct yet valid results, highlighting *RadiomicsRetrieval*'s flexibility. The histogram under each tumor image depicts its radiomics feature values. All retrievals shown here were generated with the <Image,Radiomics,FPE+APE> configuration.

4.3 Comparisons of Retrieved Results Under Different Conditions

Figure 4 illustrates how the retrieved samples vary when using different query types on the *same* tumor. An **image query** with a single point prompt considers the tumor's overall shape, size, and intensity; a **1 point APE** query focuses solely on anatomical position, potentially returning diverse tumor appearances in that region; and a **VoxelVolume feature** query emphasizes size alone, disregarding location and shape. These examples demonstrate how *RadiomicsRetrieval* offers flexible, customizable searches that highlight different tumor attributes depending on the user's clinical or research priorities.

5 Conclusion

We present *RadiomicsRetrieval*, a flexible CBIR framework aligning image embeddings and radiomics embeddings at the tumor level via minimal prompts. By integrating APE for global context and radiomics features for tumor-specific characteristics, our method achieves robust performance across diverse query modes—including image-based, location-based, and size-based retrieval. *RadiomicsRetrieval* provides a versatile clinical tool capable of addressing various querying needs. Future directions include uncertainty quantification, systematic radiomics analyses, and query expansions.

Acknowledgments. This study was supported by the National Research Foundation of Korea (RS-2024-00408040); the AI Graduate School Support Program (Sungkyunkwan University) (RS-2019-II190421); the ICT Creative Consilience program (IITP-2025-RS-2020-II201821); and the Artificial Intelligence Innovation Hub program (RS-2021-II212068).

Disclosure of Interests. The authors have no competing interests to declare that are relevant to the content of this article.

References

1. Abacha, A.B., et al.: 3D-mir: A benchmark and empirical study on 3D medical image retrieval in radiology (2023). arXiv preprint arXiv:2311.13752
2. Aerts, H.J., et al.: Data from NSCLC-radiomics. The Cancer Imaging Archive (2014)
3. Baid, U., et al.: The RSNA-ASNR-MICCAI BraTS 2021 benchmark on brain tumor segmentation and radiogenomic classification (2021). arXiv preprint arXiv:2107.02314
4. Bakas, S., et al.: Advancing the cancer genome atlas glioma MRI collections with expert segmentation labels and radiomic features. Sci. Data **4**(1), 1–13 (2017)
5. Bakr, S., et al.: Data for NSCLC radiogenomics. The Cancer Imaging Archive (2017)
6. Böttcher, B., et al.: Evaluation of a content-based image retrieval system for radiologists in high-resolution CT of interstitial lung diseases. European Radio. Exp. **9**(1), 4 (2025)
7. Chen, T., Kornblith, S., Norouzi, M., Hinton, G.: A simple framework for contrastive learning of visual representations. In: International Conference on Machine Learning, pp. 1597–1607. PmLR (2020)

8. Chen, Y., Liu, C., Liu, X., Arcucci, R., Xiong, Z.: Bimcv-r: A landmark dataset for 3D CT text-image retrieval. In: International Conference on Medical Image Computing and Computer-Assisted Intervention, pp. 124–134. Springer (2024)
9. Choe, J., et al.: Content-based image retrieval by using deep learning for interstitial lung disease diagnosis with chest CT. Radiology **302**(1), 187–197 (2022)
10. Collins, D.L., Zijdenbos, A.P., Baaré, W.F., Evans, A.C.: Animal+ insect: improved cortical structure segmentation. In: Information Processing in Medical Imaging: 16th International Conference, IPMI99 Visegrád, Hungary, June 28–July 2, 1999 Proceedings 16, pp. 210–223. Springer (1999)
11. Fonov, V., et al.: Unbiased average age-appropriate atlases for pediatric studies. Neuroimage **54**(1), 313–327 (2011)
12. Fonov, V.S., Evans, A.C., McKinstry, R.C., Almli, C.R., Collins, D.: Unbiased nonlinear average age-appropriate brain templates from birth to adulthood. Neuroimage **47**, S102 (2009)
13. Goncharov, M., et al.: Anatomical positional embeddings. In: International Conference on Medical Image Computing and Computer-Assisted Intervention, pp. 68–77. Springer (2024)
14. Kirillov, A., et al.: Segment anything. In: Proceedings of the IEEE/CVF International Conference on Computer Vision, pp. 4015–4026 (2023)
15. Kobayashi, K., et al.: Sketch-based semantic retrieval of medical images. Med. Image Anal. **92**, 103060 (2024)
16. Kumar, A., Kim, J., Cai, W., Fulham, M., Feng, D.: Content-based medical image retrieval: a survey of applications to multidimensional and multimodality data. J. Digit. Imaging **26**, 1025–1039 (2013)
17. LaBella, D., et al.: The ASNR-MICCAI brain tumor segmentation (BraTs) challenge 2023: Intracranial meningioma (2023). arXiv preprint arXiv:2305.07642
18. Li, P., Wang, S., Li, T., Lu, J., HuangFu, Y., Wang, D.: A large-scale CT and pet/CT dataset for lung cancer diagnosis (lung-pet-CT-dx). The cancer imaging archive (2020)
19. Menze, B.H., et al.: The multimodal brain tumor image segmentation benchmark (BraTS). IEEE Trans. Med. Imaging **34**(10), 1993–2024 (2014)
20. Müller, H., Michoux, N., Bandon, D., Geissbuhler, A.: A review of content-based image retrieval systems in medical applications–clinical benefits and future directions. Int. J. Med. Inf. **73**(1), 1–23 (2004)
21. Na, I., Noh, J.J., Kim, C.K., Lee, J.W., Park, H.: Combined radiomics-clinical model to predict platinum-sensitivity in advanced high-grade serous ovarian carcinoma using multimodal MRI. Front. Oncol. **14**, 1341228 (2024)
22. Öztürk, Ş, Çelik, E., Çukur, T.: Content-based medical image retrieval with opponent class adaptive margin loss. Inf. Sci. **637**, 118938 (2023)
23. Qayyum, A., Anwar, S.M., Awais, M., Majid, M.: Medical image retrieval using deep convolutional neural network. Neurocomputing **266**, 8–20 (2017)
24. Tancik, M., et al.: Fourier features let networks learn high frequency functions in low dimensional domains. Adv. Neural. Inf. Process. Syst. **33**, 7537–7547 (2020)
25. Van Griethuysen, J.J., et al.: Computational radiomics system to decode the radiographic phenotype. Can. Res. **77**(21), e104–e107 (2017)
26. Wang, H., et al.: Sam-med3D: towards general-purpose segmentation models for volumetric medical images (2023). arXiv preprint
27. Wang, Z., Sun, J.: Transtab: Learning transferable tabular transformers across tables. Adv. Neural. Inf. Process. Syst. **35**, 2902–2915 (2022)

28. Wasserthal, J., et al.: Totalsegmentator: robust segmentation of 104 anatomic structures in CT images. Radio. Artif. Intell. **5**(5) (2023)
29. Wee, L., Aerts, H., Kalendralis, P., Dekker, A.: Data from NSCLC-radiomics-interobserver1. The Cancer Imaging Archive (2019)
30. Wee, L., Aerts, H., Kalendralis, P., Dekker, A.: Rider lung CT segmentation labels from: Decoding tumour phenotype by noninvasive imaging using a quantitative radiomics approach. The Cancer Imaging Archive (2020)

SAMASK-CLTR: A Spatial-Aware Mask Guided Learning Model for Benign and Malignant Tumor Classification in ABUS

Peirong Xu[1], Luoqian Zhu[2], Jingkun Chen[3], Xin Qian[1], Yue Sun[1], Lingyun Bao[2], and Tao Tan[1](✉)

[1] Faculty of Applied Sciences, Macao Polytechnic University, Macao, China
taotan@mpu.edu.mo
[2] Hangzhou First People's Hospital, Hangzhou, China
[3] Department of Engineering Science, University of Oxford, Oxford, UK

Abstract. Automated Breast Ultrasound (ABUS) provides three dimensional volumetric imaging that improves breast lesion detection without radiation exposure and reduces operator dependency. However, the resulting high data volume poses significant challenges for radiologists in localizing lesions accurately and distinguishing benign from malignant cases–challenges that can directly impact early diagnosis and treatment outcomes. To tackle these critical issues, we propose SAMASK-CLTR (Spatial-Aware Mask Prompting with Convolutional Transformer Architecture), a hybrid framework that combines the feature extraction power of CNNs with the global modeling capability of Transformers. In our approach, ResNet-50 extracts hierarchical, multi-scale features that are refined by a Transformer encoder-decoder to capture global context. Crucially, during decoding, a mask prompt enhanced with 3D positional encoding guides the network to focus on key tumor regions, directly addressing the challenges of precise localization and classification. Experiments on 7,073 ABUS images–including 6,973 clinical cases from Internal Datasets and 100 cases from the public ABUS Challenge Cup–demonstrate that SAMASK-CLTR achieves AUCs of 88.45% and 70.46% on internal and external datasets, respectively. These results highlight the potential of our framework to significantly enhance breast cancer diagnosis by improving the accuracy and reliability of lesion classification. Code available at: https://github.com/SAMASK-CLTR/Code.

Keywords: Auto Breast Ultrasound System · Computer Aided Diagnosis · Mask Prompt · Spatial Aware

1 Introduction

1.1 Background

Breast health is a critical concern for every woman. Among newly diagnosed female cancer cases, breast cancer accounts for up to 31% [21], making it the second leading cause of cancer death in women. Therefore, early detection of breast

J. C. Gee et al. (Eds.): MICCAI 2025, LNCS 15960, pp. 567–577, 2026.
https://doi.org/10.1007/978-3-032-04927-8_54

abnormalities, especially malignant changes, is crucial for increasing treatment success and improving quality of life, underscoring the importance of regular screenings [16].

Currently, three main methods are used for breast screening: mammography, breast ultrasound, and Automated Breast Ultrasound (ABUS). ABUS not only retains the advantages of traditional ultrasound but also captures three-dimensional ultrasonic information from multiple views, generating more comprehensive images of breast tissue [17]. This multi-view capability has increased its clinical adoption among both patients and doctors. During an ABUS examination, the device's probe automatically scans the entire breast, producing detailed three-dimensional images, which makes the process more convenient. Figure 1 shows the three views of ABUS. However, its multidimensional and multi-view nature also demands advanced interpretation skills from radiologists—especially junior physicians—as the large data volume can lead to higher rates of missed diagnoses. Although the application of Computer-Aided Detection (CAD) technology has improved diagnostic accuracy by assisting doctors in interpreting these complex images [27], the inherent challenges of 3D data remain.

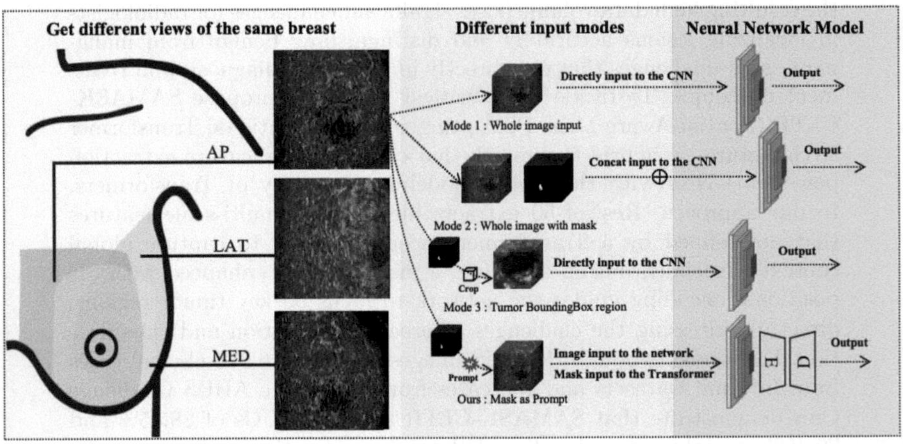

Fig. 1. This image shows different view of the same breast (AP, LAT, MED) and the framework of the four different input modes we compared.

In recent years, there has been remarkable progress in related tasks. For detection tasks, single-stage networks proposed in [14,18,19,24,26,29] are capable of fast and accurate lesion detection.For segmentation tasks, [2,3] suggest using a dynamic contrastive learning framework and a Teacher-Student Framework to improve segmentation performance. Meanwhile, models that incorporate prompt mechanisms, such as the Segment Anything Model (SAM) [9] and One-Prompt Segmentation [22], have demonstrated significant potential in the field of medical imaging. By leveraging prompt information [4,5,8,10–13,28], these models significantly boost segmentation performance or exhibit strong adaptability for multi-task processing.

Although prompt-based models achieve significant performance improvements, their primary focus remains on the segmentation or localization of suspicious lesions, rather than directly determining the pathological nature of the lesions themselves. In benign and malignant classification tasks, many strategies aim to improve classification performance by operating on the region of interest (ROI). For instance, [1,20,25,32] reduce interference from surrounding redundant information, thereby enhancing classification accuracy. However, compared to ROI manipulation, directly utilizing prompts offers a more efficient and straightforward approach. Studies such as [6,7,15,23,30] adopt Vision Transformer (ViT) architectures for ultrasound image classification, but the global attention mechanism of ViT requires large amounts of data, which can easily lead to overfitting.

To overcome these challenges and directly address the pathological classification task, we propose **SAMASK-CLTR**, a spatial-aware mask-prompted classification framework for 3D breast ultrasound imaging. This framework employs ResNet-50 as the backbone for hierarchical feature extraction, combined with a Transformer to encode multi-scale semantic representations. A key innovation is the introduction of positionally encoded mask prompts during the decoding phase, which enhances the model's spatial awareness and tightly associates the feature space with tumor regions, significantly improving classification accuracy.The contributions of our research are threefold:

- We propose a hybrid CNN-Transformer model that integrates spatial-aware mask prompts for direct benign-malignant classification, achieving substantial improvements over conventional CNNs.
- We systematically evaluated multiple input modes and thoroughly analyzed their impact on classification performance.
- We conducted extensive experiments on large-scale clinical and public datasets, validating the cross-dataset generalization capability of SAMASK-CLTR.

2 Method

SAMASK-CLTR. The SAMASK-CLTR model is a hybrid architecture that integrates convolutional neural networks with Transformer for three-dimensional data processing. Figure 2 shows the overview of our SAMASK-CLTR. The framework employs a 3D ResNet-50 network for multi-scale spatial feature extraction, where hierarchical feature maps are concatenated at the output stage and enhanced with three-dimensional positional encoding to capture spatial relationships. These encoded features are then fed into the Transformer encoder. In the decoder design, we replace the standard multi-head self-attention (MSA) mechanism with a Multi-Scale Deformable Attention module. This advanced mechanism adaptively samples features from multi-scale maps through learnable dynamic offsets, enabling the model to focus on task-relevant local regions while reducing computational cost and memory consumption, we integrate the query (Q) from the decoder with Spatial-aware Mask Prompts, combining tumor mask information with dense spatial-scale encoding, and the Key (K) propagates feature information between the decoder layers. This fusion is then fed into the network to guide attention toward lesion areas more effectively. Finally, feature fusion from multiple decoder layers produces the prediction, with the entire network optimized end-to-end using cross-entropy loss for binary classification.

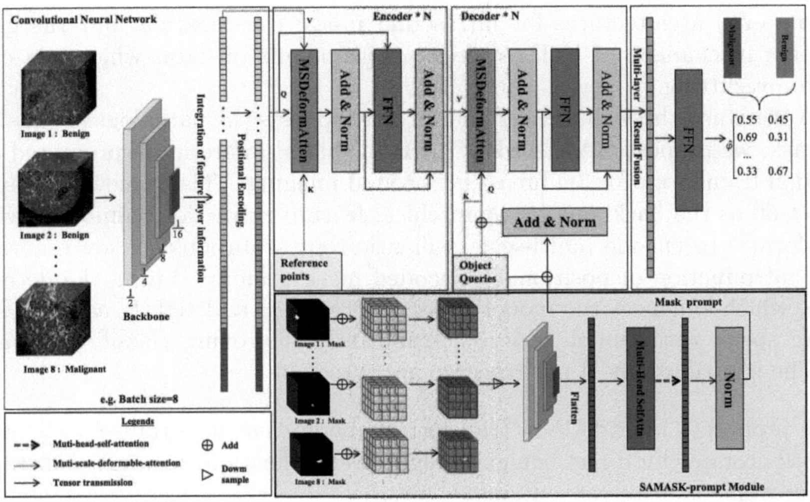

Fig. 2. SAMASK-CLTR: In this architecture, the backbone first extracts multi-scale feature information, and after applying positional encoding and flattening, feeds these features into the encoder module. Meanwhile, the Mask incorporates spatial positional encoding and downsampling before being fused with the decoder.

Deformable Attention and MSDeformable Attention. Deformable Attention addresses the computational bottleneck of standard self-attention through dynamic sparse sampling [31]. Given a query element with feature z_q and reference point coordinates $\mathbf{p}_q \in \mathbb{R}^3$, the attention operation is formulated as:

$$\text{DeformAttn}(z_q, \mathbf{p}_q, x) = \sum_{m=1}^{M} W_m \left[\sum_{k=1}^{K} A_{mqk} \cdot W'_m x \left(\mathbf{p}_q + \Delta \mathbf{p}_{mqk} \right) \right], \quad (1)$$

where M denotes the number of attention heads (typically $M = 8$), K is the number of sampled points per head (empirically set to 4), $\Delta \mathbf{p}_{mqk} \in \mathbb{R}^3$ represents learnable offsets predicted from z_q, and A_{mqk} are attention weights normalized via softmax over K points. The trilinear interpolation operation $x(\cdot)$ enables differentiable sampling from feature volume x, facilitating gradient flow.

MSDeformable Attention extends this mechanism to multi-scale feature volumes $\{x_l\}_{l=1}^{L}$ with different spatial resolutions:

$$\text{MSDeformAttn}(z_q, \{\mathbf{p}_q^l\}) = \sum_{l=1}^{L}\sum_{k=1}^{K} A_{qkl} \cdot W_l x_l\left(\phi_l(\mathbf{p}_q) + \Delta\mathbf{p}_{qkl}\right). \tag{2}$$

Here $\phi_l(\cdot)$ performs coordinate scaling to match the l-th feature level's resolution (e.g., for $L = 4$ levels with downsampling rates of $\{1/2, 1/4, 1/8, 1/16\}$ along each axis relative to the input volume).

3D Position Embedding. To extend positional information to volumetric data , we extend positional encoding from 2D to 3D to capture spatial relationships in three-dimensional space. We present the formula for 3D positional embedding here:

$$PE^{(z)}_{(pos,2i)} = \sin\left(\frac{z}{T^{\frac{2i}{d}}}\right), \quad PE^{(z)}_{(pos,2i+1)} = \cos\left(\frac{z}{T^{\frac{2i}{d}}}\right), \tag{3}$$

$$PE^{(y)}_{(pos,2i)} = \sin\left(\frac{y}{T^{\frac{2i}{d}}}\right), \quad PE^{(y)}_{(pos,2i+1)} = \cos\left(\frac{y}{T^{\frac{2i}{d}}}\right), \tag{4}$$

$$PE^{(x)}_{(pos,2i)} = \sin\left(\frac{x}{T^{\frac{2i}{d}}}\right), \quad PE^{(x)}_{(pos,2i+1)} = \cos\left(\frac{x}{T^{\frac{2i}{d}}}\right). \tag{5}$$

Spatial-Aware Mask Prompts. We propose a spatial-Aware mask prompt scheme. First, the tumor mask $M \in \mathbb{R}^{H\times W\times D}$, extracted from the input image, is enhanced via 3D sinusoidal-cosine positional encoding:

$$M_{\text{enc}}(x, y, z) = M(x, y, z) \oplus [PE_x(x) \parallel PE_y(y) \parallel PE_z(z)],$$

where PE_x, PE_y, PE_z are positional encoding functions along three axes, $\oplus$ denotes element-wise addition, and $\parallel$ represents channel concatenation. This ensures each voxel in the mask jointly encodes semantic information (tumor presence) and geometric 3D spatial coordinates. The encoded mask M_{enc} is then downsampled via multi-stage 3D convolutions to generate dimension-compatible features F, which are linearly combined with the decoder's initial query objects Q:

$$Q' = Q + F.$$

and fed into the decoder for iterative refinement. By explicitly modeling spatial-semantic correlations, this design significantly improves the model's localization accuracy and feature discriminability for tumor regions.

3 Experiment

3.1 Datasets

This study utilized an internal dataset along with the publicly available ABUS Challenge Cup external dataset. To ensure patient privacy, all data were anonymized, with patient IDs replaced by unique identifiers. Tumor region masks were annotated by junior doctors invited by the research team, while bounding box images of tumor regions were generated by cropping based on the annotation files. The dataset was divided into an internal training set, an internal validation set, and an external validation set from the ABUS Challenge Cup dataset. This hierarchical partitioning strategy not only ensures rigorous model validation but also evaluates the model's generalization ability using independent external data, thereby supporting the universality and clinical applicability of the research findings. Table 1 shows the distribution of our datasets.

Table 1. Distribution of benign and malignant cases in internal and external datasets by patients, ABUS/ABVS images, mask annotations, and lesions (external data used for validation only)

Label	Benign	Malignant	Total
Patient	2259(+42 Externals)	941(+58 Externals)	3200(+100 Externals)
ABUS/ABVS	5501(+42 Externals)	1572(+58 Externals)	7073(+100 Externals)
MASK	3888(+42 Externals)	773(+58 Externals)	4661(+100 Externals)
Lesion	4947(+42 Externals)	779(+58 Externals)	5726(+100 Externals)

3.2 Experimental Details

Our model using ResNet-50 as the backbone network for multi-scale feature extraction. Spatial features at 1/4, 1/8, and 1/16 resolutions were fused through cross-scale concatenation with 1×1 convolutions, then flattened and processed by a Transformer module containing 4 encoder-decoder layers (hidden dim 256, 6 attention heads). A set of 512 learnable object queries established feature correlations with the subsequent mask generation module, which incorporated a 3D spatial-aware positional encoding mechanism. The input volumes were preprocessed with random flipping (probability=0.3) and random contrast augmentation. The model was trained for 100 epochs with an initial learning rate of 1×10^{-4} , employing dynamic downsampling for query alignment and zero-initialized pseudo-masks to address annotation incompleteness.

In addition, we designed three input modes: Mode 1 directly inputs the whole image into the network; Mode 2 concatenates the image with its corresponding mask before inputting to the network; Mode 3 crops the lesion region from the image using the bounding box derived from the mask and inputs the cropped image into the network. Figure 1 shows illustrative diagrams of different input modes. All three modes, as well as our proposed model, adopted the same data augmentation techniques to ensure experimental consistency and comparability. Additionally, we addressed the issue of data imbalance by using random over-sampling to increase the number of malignant samples.

3.3 Comparison Experiment

We systematically evaluated three input modes (Whole Image input, Image-Lesion Mask combined input, and Lesion Bounding Box Region input) and our proposed model on both internal and external datasets. Table 2 summarizes the performance metrics of different models under various input modes. The experimental results show that in the Whole Image input mode, ResNet101 achieved AUC values of 79.28% (internal) and 56.22% (external), demonstrating the best performance; in the Image-Lesion Mask combined input mode, DenseNet121 achieved AUC values of 76.45% (internal) and 62.71% (external), achieving the highest performance; and in the Lesion Bounding Box Region input mode, DenseNet121 achieved an AUC of 72.61% on the internal test set, while

Table 2. Performance comparison of different modes on internal and external datasets

Method	Modes	Internal Dataset				External Dataset			
		AUC(%)	ACC(%)	SEN(%)	SPE(%)	AUC(%)	ACC(%)	SEN(%)	SPE(%)
Densenet121	W	76.75	83.19	54.45	<u>98.44</u>	49.74	49.44	35.14	69.41
	W+M	76.45	81.70	<u>58.48</u>	92.82	<u>62.71</u>	57.62	35.82	<u>87.97</u>
	B	72.61	75.39	51.76	90.13	49.23	49.55	33.03	67.39
Resnet18	W	76.61	83.54	53.59	97.89	53.64	55.48	43.09	72.73
	W+M	76.18	79.10	58.41	89.02	55.31	56.96	66.31	43.91
	B	72.33	75.72	49.86	86.12	50.45	55.39	44.89	62.32
Resnet50	W	76.60	81.40	55.90	93.62	49.09	56.84	71.88	36.12
	W+M	76.43	80.92	55.64	93.03	55.69	55.34	58.18	51.23
	B	69.31	74.54	53.72	84.53	45.33	52.76	59.32	42.76
Resnet101	W	<u>79.28</u>	<u>84.02</u>	54.45	98.20	56.22	56.36	<u>74.37</u>	28.31
	W+M	75.90	80.68	54.98	93.01	58.20	51.80	38.81	69.84
	B	67.93	77.76	51.13	87.92	52.75	52.33	34.29	66.14
SwimUnetr	W	78.48	82.99	53.63	97.07	48.87	49.44	30.43	75.87
	W+M	73.23	83.30	53.81	**98.92**	49.13	41.52	32.97	83.56
	B	67.22	72.43	49.88	90.34	55.42	<u>60.73</u>	22.42	71.31
3D DETR [26]	W	73.96	78.51	57.00	88.82	54.84	48.00	16.81	**91.14**
	B	68.29	72.88	47.77	82.76	52.17	45.56	24.18	88.72
SCPM-Net [14]	W	74.33	82.01	56.30	94.33	48.89	40.68	9.48	83.82
	B	66.73	80.13	53.29	87.38	45.75	41.54	13.43	82.14
Ours	W+SAMP	**88.45**	**84.13**	**79.63**	86.29	**70.46**	**70.90**	**76.68**	51.69

Mode abbreviations: W = Whole image, W+M = Whole image with mask, B = Lesion Bounding box region, W+SAMP = Whole image and spatial-aware mask prompt.

SwinUnetr achieved 55.42% on the external test set. Consequently, ResNet101, DenseNet121, and SwinUnetr were established as the state-of-the-art (SOTA) baseline models for their respective input modes. Meanwhile, we compared our proposed model with 3D DETR and SCPM-Net, and achieved AUC values of 88.45% on the internal validation set and 70.46% on the external independent validation set. A comprehensive analysis indicates that our method significantly outperforms existing models in terms of cross-dataset generalization and overall performance, with statistical significance ($p < 0.01$), further validating its robustness.

3.4 Ablation Experiment

This study systematically evaluates three architectural configurations of the SAMASK-CLTR model through module ablation experiments: (1) baseline CLTR (without mask prompting module), (2) Mask Prompt CLTR (with mask prompting mechanism), and (3) Spatial-aware MASK prompt CLTR (joint optimization of mask prompting and positional encoding). Table 3 summarizes the results of the component ablation study on internal and external datasets. The experimental results demonstrate that the mask prompting module achieves an 2.21%-point improvement in classification AUC compared to the baseline. The operates in two ways: During training, itguides feature learning by incorporating prior constraints on target regions, while during inference, it significantly enhances lesion localization accuracy. When jointly optimized with positional encoding, the model exhibits notable spatial awareness (11.43%-point additional AUC gain). This dual-optimization strategy not only strengthens spatial feature representation but also effectively improves discriminative capability for capturing critical characteristics distinguishing benign and malignant lesions.

Table 3. Component Ablation Study on Internal and External Datasets

Components			Internal Datasets				External Datasets			
CLTR	Mask	Spatial	AUC%	ACC%	SEN%	SPE%	AUC%	ACC%	SEN%	SPE%
✓			77.02	77.20	60.95	85.00	62.55	56.60	59.89	50.08
✓	✓		79.23	80.87	62.32	82.76	63.94	62.41	62.74	50.33
✓	✓	✓	**88.45**	**84.13**	**79.63**	**86.29**	**70.46**	**70.90**	**76.68**	**51.69**

The '✓ ' mark indicates the modules that have been utilized.

4 Conclusion and Discussion

This study systematically analyzed the impact of different input modes on the classification performance of the model and proposed an improved spatially-aware mask prompt module to further enhance the model's performance. In the whole image input mode, the model can capture complete contextual information, but it is also easily defocused by background , which limits the improvement of classification performance. In the whole image with mask input mode, although the mask provides a rough localization of the tumor and theoretically helps the model focus on the target area, it fails to sufficiently suppress information from non-mask regions during training, which instead introduces additional interference and leads to a decline in performance. For the lesion region input mode, directly cropping the tumor region effectively reduces background interference, but at the same time, it loses critical contextual information, which limits the model's classification capability. This study innovatively introduces a spatially-aware mask prompting module. The proposed module combines 3D positional encoding with the masked image, making the spatial contrast between the tumor region and normal tissue more pronounced. During training, this module guides the model to focus more accurately on the lesion region, suppressing noise information introduced by surrounding tissues, thereby achieving significant improvement in classification performance. Furthermore, this module demonstrates strong generalization ability in cross-dataset testing, further validating its effectiveness. Future work will focus on incorporating multimodal information prompts to further enhance the modelâĂŹs applicability in clinical settings.

Acknowledgments. This work is supported by Science and Technology Development Fund of Macao (0041/2023/RIB2), Macao Polytechnic University Grant (RP/FCA-05/2022).

Disclosure of Interests. The authors have no competing interests to declare that are relevant to the content of this article.

References

1. Alotaibi, M., et al.: Breast cancer classification based on convolutional neural network and image fusion approaches using ultrasound images. Heliyon **9**(11) (2023)
2. Chen, J., Chen, C., Huang, W., Zhang, J., Debattista, K., Han, J.: Dynamic contrastive learning guided by class confidence and confusion degree for medical image segmentation. Patt. Recogn. **145**, 109881 (2024)
3. Chen, J., et al.: From gaze to insight: bridging human visual attention and vision language model explanation for weakly-supervised medical image segmentation. arXiv preprint arXiv:2504.11368 (2025)
4. Chen, J., Huang, W., Zhang, J., Debattista, K., Han, J.: Addressing inconsistent labeling with cross image matching for scribble-based medical image segmentation. IEEE Trans. Image Process. (2025)
5. Chen, J., Zhang, J., Debattista, K., Han, J.: Semi-supervised unpaired medical image segmentation through task-affinity consistency. IEEE Trans. Med. Imaging **42**(3), 594–605 (2022)
6. Feng, H., Yang, B., Wang, J., Liu, M., Yin, L., Zheng, W., Yin, Z., Liu, C.: Identifying malignant breast ultrasound images using vit-patch. Appl. Sci. **13**(6), 3489 (2023)

7. Gheflati, B., Rivaz, H.: Vision transformers for classification of breast ultrasound images. In: 2022 44th Annual International Conference of the EMBC, pp. 480–483. IEEE (2022)
8. Hu, M., Li, Y., Yang, X.: Breastsam: adapting the segmentation anything model for breast tumor segmentation in ultrasound imaging. In: Medical Imaging 2024: Ultrasonic Imaging and Tomography. vol. 12932, pp. 182–196. SPIE (2024)
9. Kirillov, A., et al.: Segment anything. In: Proceedings of the CVPR, pp. 4015–4026 (2023)
10. Li, H., Liu, H., Hu, D., Wang, J., Oguz, I.: Promise: prompt-driven 3d medical image segmentation using pretrained image foundation models. In: ISBI, pp. 1–5. IEEE (2024)
11. Li, W., Liu, T., Feng, F., Yu, S., Wang, H., Sun, Y.: Btsspro: prompt-guided multimodal co-learning for breast cancer tumor segmentation and survival prediction. IEEE J. Biomed. Heal. Inf. (2024)
12. Li, X., Xu, M., Zhang, X., Niu, S., Zhu, J., Zhou, X.: Ceus-sam: cross-modal prompt-based sam network for breast ceus image segmentation. In: 2024 BIBM, pp. 994–999. IEEE (2024)
13. Lin, Z., et al.: Uniusnet: a promptable framework for universal ultrasound disease prediction and tissue segmentation. In: 2024 BIBM, pp. 3501–3504. IEEE (2024)
14. Luo, X., et al.: Scpm-net: an anchor-free 3d lung nodule detection network using sphere representation and center points matching. Med. Image Anal. **75**, 102287 (2022)
15. Sun, J., et al.: Classification for thyroid nodule using VIT with contrastive learning in ultrasound images. Comput. Biol. Med. **152**, 106444 (2023)
16. Sun, Y.S., et al.: Risk factors and preventions of breast cancer. Int. J. Biol. Sci. **13**(11), 1387 (2017)
17. Tan, T., Platel, B., Mus, R., Karssemeijer, N.: Detection of breast cancer in automated 3d breast ultrasound. In: Medical Imaging 2012: Computer-Aided Diagnosis. vol. 8315, pp. 56–63. SPIE (2012)
18. Tang, J., Chen, X., Fan, L., Zhu, Z., Huang, C.: Ln-detr: an efficient transformer architecture for lung nodule detection with multi-scale feature fusion. Available at SSRN 5084519
19. Tao, X., et al.: Enhancing lesion detection in automated breast ultrasound using unsupervised multi-view contrastive learning with 3d detr. Medical Image Analysis, p. 103466 (2025)
20. Wang, C., et al.: Abus-net: graph convolutional network with multi-scale features for breast cancer diagnosis using automated breast ultrasound. Expert Syst. Appl. **273**, 126978 (2025)
21. Wen, X., Guo, X., Wang, S., Lu, Z., Zhang, Y.: Breast cancer diagnosis: a systematic review. Biocybern. Biomed. Eng. **44**(1), 119–148 (2024)
22. Wu, J., Xu, M.: One-prompt to segment all medical images. In: Proceedings of the CVPR, pp. 11302–11312 (2024)
23. Xu, M., et al.: Vision transformers (vit) pretraining on 3d abus image and dual-capsvit: Enhancing vit decoding via dual-channel dynamic routing. In: 2023 BIBM, pp. 1596–1603. IEEE (2023)
24. Xu, Y., Shen, Y., Fernandez-Granda, C., Heacock, L., Geras, K.J.: Understanding differences in applying detr to natural and medical images. arXiv e-prints pp. arXiv–2405 (2024)
25. Yang, Z., Fan, T., Smedby, Ö., Moreno, R.: 3d breast ultrasound image classification using 2.5 d deep learning. In: 17th IWBI 2024. vol. 13174, pp. 443–449. SPIE (2024)

26. Zagoruyko, S.: End-to-end object detection with transformers. ECCV (2020)
27. van Zelst, J.C., et al.: Pijnappel: dedicated computer-aided detection software for automated 3d breast ultrasound; an efficient tool for the radiologist in supplemental screening of women with dense breasts. Eur. Radiol. **28**, 2996–3006 (2018)
28. Zhao, Z., et al.: One model to rule them all: towards universal segmentation for medical images with text prompts. CoRR (2023)
29. Zhou, Y.T., Yang, T.Y., Han, X.H., Piao, J.C.: Thyroid-detr: thyroid nodule detection model with transformer in ultrasound images. Biomed. Sign. Process. Control **98**, 106762 (2024)
30. Zhu, Q., Fei, L.: Cross-vit based benign and malignant classification of pulmonary nodules. PLoS ONE **20**(2), e0318670 (2025)
31. Zhu, X., Su, W., Lu, L., Li, B., Wang, X., Dai, J.: Deformable detr: deformable transformers for end-to-end object detection. arXiv e-prints pp. arXiv–2010 (2020)
32. Zhuang, Z., et al.: Tumor classification in automated breast ultrasound (abus) based on a modified extracting feature network. Comput. Med. Imaging Graph. **90**, 101925 (2021)

Self-supervised Multiview Xray Matching

Mohamad Dabboussi[1,2(✉)], Malo Huard[2], Yann Gousseau[1], and Pietro Gori[1]

[1] Telecom Paris, Palaiseau, France
dmohamad@hotmail.com
[2] Milvue, Paris, France

Abstract. Accurate interpretation of multi-view radiographs is crucial for diagnosing fractures, muscular injuries, and other anomalies. While significant advances have been made in AI-based analysis of single images, current methods often struggle to establish robust correspondences between different X-ray views, an essential capability for precise clinical evaluations. In this work, we present a novel self-supervised pipeline that eliminates the need for manual annotation by automatically generating a many-to-many correspondence matrix between synthetic X-ray views. This is achieved using digitally reconstructed radiographs (DRR), which are automatically derived from unannotated CT volumes. Our approach incorporates a transformer-based training phase to accurately predict correspondences across two or more X-ray views. Furthermore, we demonstrate that learning correspondences among synthetic X-ray views can be leveraged as a pretraining strategy to enhance automatic multi-view fracture detection on real data. Extensive evaluations on both synthetic and real X-ray datasets show that incorporating correspondences improves performance in multi-view fracture classification.

Keywords: Multi-view X-ray · DRR · Many-to-Many Correspondence · Fracture detection

1 Introduction

Accurate diagnosis in radiology often relies on the complementary information provided by multiple X-ray views that are acquired from various angles to ensure a comprehensive evaluation. In practice, radiologists examine all these views to confirm the presence and extent of lesions, thus increasing diagnostic confidence.

Motivation Multi-view imaging is essential in radiology [1,3,20], as each X-ray projection provides unique information that aids in detecting subtle abnormalities and improving overall evaluation. However, the process is time-consuming, requires expertise, and is prone to errors. These challenges highlight the need for automated methods capable of effectively handling multi-view data to support clinical decision-making.

J. C. Gee et al. (Eds.): MICCAI 2025, LNCS 15960, pp. 578–588, 2026.
https://doi.org/10.1007/978-3-032-04927-8_55

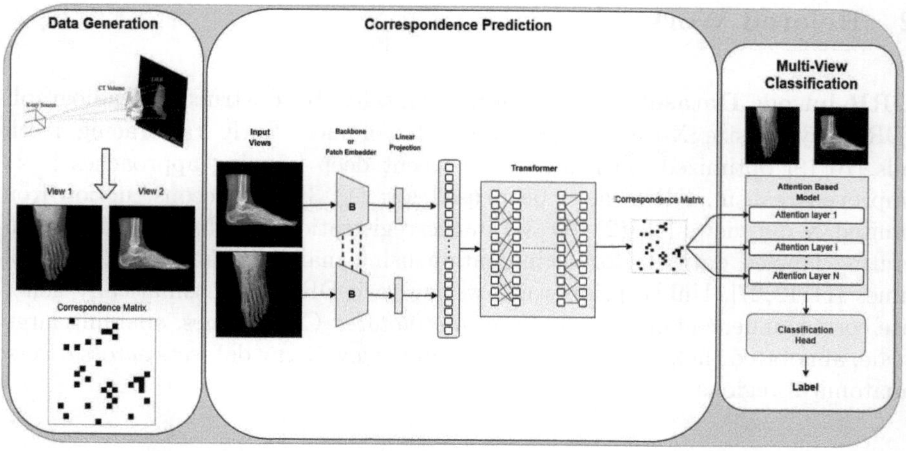

Fig. 1. End-to-end self-supervised pipeline for predicting X-ray correspondences and multi-view fracture classification. (Left) From a CT volume we generate multiple 2D DRRs (digitally reconstructed radiographs) with a correspondence matrix linking matching points. (Middle) A backbone network extracts features from each view, processed by a Transformer to predict a Correspondence Matrix via attention mechanisms. (Right) The correspondence guides attention for fracture detection.

Problem Statement While deep learning achieved impressive results in single-view medical imaging, extending it to multi-view X-ray interpretation remains challenging. Progress is hindered by two key issues: (1) the scarcity of annotated multi-view datasets, especially beyond chest X-rays, and (2) the complexity of establishing correspondences between views. Traditional image matching techniques focus on one-to-one mappings in natural images, whereas X-rays involve complex, many-to-many relationships due to the cumulative nature of pixel intensities along the X-ray path (see Fig. 2). This makes it challenging to determine whether abnormalities across views correspond to the same pathology. There is a need for an automated method to estimate accurate multi-view correspondences without extensive manual annotation, ultimately aiding radiologists and improving lesion/fracture detection.

To this end, our work (see Fig. 1 for a visual pipeline) introduces the following **contributions**:

- **DRR-based Correspondence Generation:** We generate paired simulated X-rays views and patch-level correspondence matrices from unannotated CT volumes, thereby eliminating the need for large annotated X-ray datasets.
- **Self-supervised Pre-training:** We leverage the correspondence prediction task as a self-supervised method to enhance multi-view feature learning.
- **Transformer Integration:** We incorporate correspondence information within transformer-based architectures to improve fracture classification performance.

2 Related Works

DRR-based Dataset Generation. Digitally Reconstructed Radiographs (DRR) synthesize X-ray images from CT volumes. Early ray-tracing methods [10,19] optimized efficiency, while recent deep learning approaches [5,23] improved realism. DRRs have been used for: (1) 3D CT reconstruction from limited X-ray views [4], (2) 2D/3D image registration [18,25,26], and (3) generating labeled datasets for segmentation using manually labeled 3D CT volumes [11,12,27]. Unlike prior work, we leverage DRR to automatically generate correspondence annotations from *unannotated* CT volumes, enabling large-scale, annotated (i.e., correspondences), multi-view X-ray datasets across diverse anatomical regions.

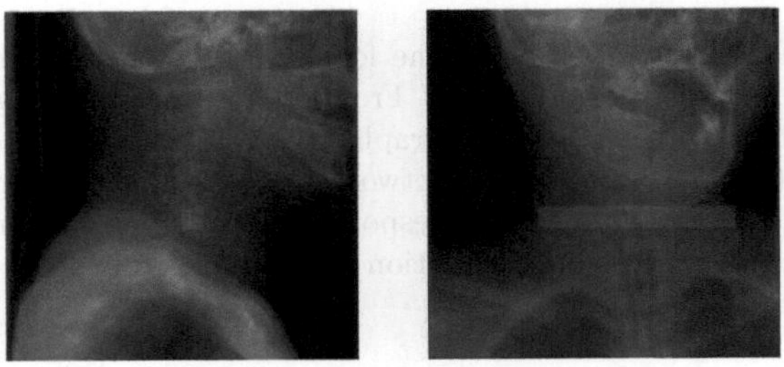

Fig. 2. Many-to-many matching between two X-ray views of the same subject. The red patch in the left view corresponds to the red bar in the right view, illustrating multiple matching pixels across views (Color figure online).

Image Matching and Correspondence Estimation. Traditional feature-based image matching [9,14] has been largely replaced by deep learning approaches. Sparse keypoint methods [13,17] and dense matching techniques [2,7,22] perform well in natural image tasks but assume one-to-one pairwise matching, which suffices for tasks involving two views with homographies, such as camera pose estimation. Multi-view X-ray imaging, however, requires many-to-many mappings due to anatomical transparency and overlapping structures. Existing methods struggle in this context, highlighting the need for specialized solutions in X-ray correspondence estimation.

To address these challenges, we propose a self-supervised pipeline that generates from unannotated CT volumes a large, diverse dataset of synthetic X-rays with automatically derived correspondence matrices, eliminating the need for manual annotations. Unlike traditional pairwise matching methods, our approach learns many-to-many correspondences across multiple views, enabling a more comprehensive understanding of their complex spatial relationships and advancing automated multi-view X-ray analysis.

3 Method

We introduce a novel approach for generating correspondence ground truth using Digitally Reconstructed Radiographs (DRR) from unannotated CT scans. Synthetic X-ray views are first created along with their corresponding correspondence matrices. These are then used as supervision for training a deep-learning model, which takes two synthetic X-ray views as input and predicts their correspondence matrix.

To evaluate the effectiveness of our approach, we conduct experiments on both synthetic and real multi-view X-ray datasets, demonstrating the effectiveness of correspondence learning not only as a pretraining strategy but also as a mechanism for attention guidance in multi-view fracture classification.

3.1 Correspondence Ground Truth Generation

Let $V \in \mathbb{R}^{H \times W \times L}$ denote a CT volume consisting of N voxels. For each non-air voxel v at position (x, y, z), obtained by thresholding the CT volume, we project it onto two distinct view directions using the Joseph method [10] to generate DRRs. This results in two projection matrices, $P_1^v, P_2^v \in \mathbb{R}^{H \times W}$, which encode the accumulated intensity values along rays traced from the X-ray source to the detector plane (see Fig. 4 for a visual explanation). We then define their flattened representations as: $\mathbf{p}_1^v = \mathit{flat}(P_1^v) \in \mathbb{R}^{HW}$ and $\mathbf{p}_2^v = \mathit{flat}(P_2^v) \in \mathbb{R}^{HW}$, where $\mathit{flat}(\cdot)$ denotes the flattening of a matrix into a vector. The voxel-specific correspondence matrix is then given by the outer product: $\mathbf{C}_{1,2}^v = \mathbf{p}_1^v \mathbf{p}_2^{v\top} \in \mathbb{R}^{HW \times HW}$. The final correspondence matrix is obtained by taking the element-wise maximum over all voxels: $\mathbf{C}_{1,2} = \max_v \left(\mathbf{C}_{1,2}^v\right)$.

To address computational challenges posed by high-resolution CT volumes (e.g., $256 \times 256 \times 256$), we perform patch-level (and not pixel-level) correspondence estimation (see Fig. 3). Specifically, we downsample the volume by a factor of $k = 16$, thereby computing the correspondence matrix at a coarser, patch-wise resolution.

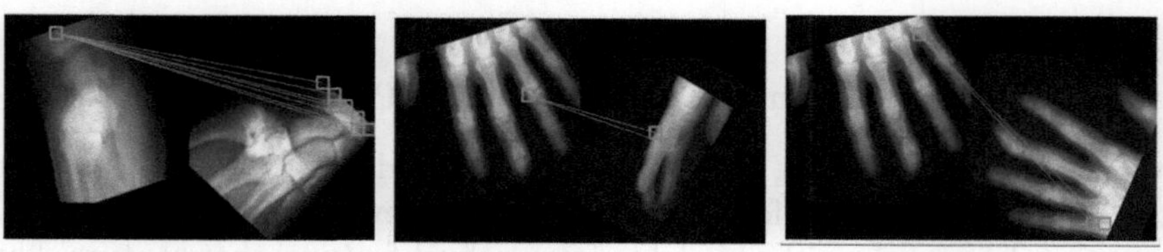

Fig. 3. Visualization of generated hand correspondences annotations.

3.2 Correspondence Prediction

We formulate correspondence prediction as a similarity assessment between features extracted from patches across different views. A pretrained backbone network, denoted as f, extracts feature maps from each image, which are then projected into a lower-dimensional embedding space. The spatial grid of these feature maps is flattened into a sequence of patch embeddings, where each embedding represents a distinct image region. Afterwards, patch embeddings from multiple views are concatenated along the sequence dimension and processed by a transformer module. This module employs self-attention to capture both intra- and inter-view interactions, effectively encoding correspondence information. Correspondences are then determined by computing a normalized dot product between feature pairs, forming a correspondence matrix. This approach naturally extends to multi-view scenarios with more than two images.

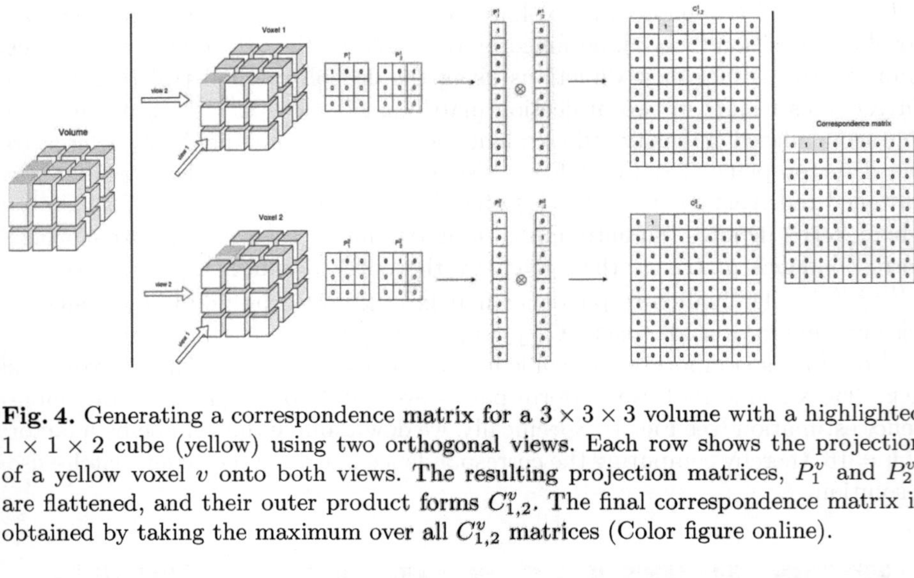

Fig. 4. Generating a correspondence matrix for a $3 \times 3 \times 3$ volume with a highlighted $1 \times 1 \times 2$ cube (yellow) using two orthogonal views. Each row shows the projection of a yellow voxel v onto both views. The resulting projection matrices, P_1^v and P_2^v, are flattened, and their outer product forms $C_{1,2}^v$. The final correspondence matrix is obtained by taking the maximum over all $C_{1,2}^v$ matrices (Color figure online).

3.3 Pre-Training Through Correspondence Prediction

As we will show in Sect. 4, we can use correspondence prediction as a self-supervised pretraining step for an auxiliary multi-view X-ray classification task. This approach encourages the model to learn robust features that capture shared anatomical information while leveraging the redundancy in multi-view data.

3.4 Attention Guidance in Multi-View X-Ray Classification

To leverage cross-view correspondence information in our multi-view X-ray classification framework, we can also integrate the correspondence matrix directly into the transformer attention mechanism.

Specifically, we concatenate the N patch tokens from both views A and B into a 2N × D sequence, producing a 2N × 2N attention map split into four N × N blocks: AA, BB (self-attention), AB, and BA (cross-attention). We add the learned correspondence matrix to AB and its transpose to BA as a soft bias before the softmax in each attention layer, guiding cross-view focus.

Let Q and K denote the query and key matrices, respectively, and let d be the dimensionality of each attention head. The standard scaled dot-product attention is computed as $A = \frac{QK^\top}{\sqrt{d}}$ (Eq. 1). Given a correspondence matrix C, the attention scores are adjusted as $A' = A + \alpha C$ (Eq. 2), where α is a learnable scaling parameter. Finally, the modified scores A' are normalized via the softmax function to yield the attention probabilities.

4 Experiments

4.1 Datasets and Implementation Details

Train Simulated Dataset. We generated correspondence matrices using our proposed method. In particular, 207,600 data samples were generated from 175 CT volumes by varying projection angles, the distance between the volume and the source, as well as by extracting appropriate crops from the CT volumes.

Test Simulated X-ray Correspondence Dataset. From 42 CT volumes of extremity, not seen during training, 16,920 pairs of views along with their corresponding correspondence matrices were generated using the same variation parameters as during training.

Real X-ray Correspondence Dataset. This dataset comprises 347 samples, each with two views. We applied a 60%/10%/30% split for training, validation, and testing. Each sample includes only three annotated positive correspondences, along with 100 negative correspondences.

MURA Public Dataset. The MURA dataset [16] comprises studies from various anatomical regions. For our purposes, we selected studies of the elbow, forearm, hand, and wrist, which correspond to the regions on which the correspondence model was trained. The final dataset includes 5,511 studies. Each study has multiple views, we used 2 views in our experiments.

Private Dataset. This dataset mainly contains multiview X-rays of the hand, forearm, foot, and knee. It has a total of 5,653 studies. Each study has multiple views, we used 2 views in our experiments.

Implementation Details. Our framework processes two input views, each of size 256 × 256 pixels. We use a pre-trained ResNet-50 [8] backbone for feature extraction, followed by a transformer with Rotary Positional Encoding (RoPE [21]) to capture spatial relationships between patches. Training utilizes the Adam optimizer with a cosine annealing learning rate scheduler. The initial learning rate is set to 1×10^{-4} for pre-training on simulated correspondence data and reduced to 1×10^{-5} when fine-tuning on partially annotated real data. The model is trained with a batch size of 16. For loss, we use mean squared error (MSE) for the correspondence task and binary cross-entropy for classification. Our classification model uses a ViT-S (6 layers, 4 heads, hidden size 384). All model modules were pre-trained on synthetic DRR correspondence data and subsequently fine-tuned for the binary fracture classification task. Each transformer layer uses a distinct learnable α for attention guidance. Importantly, our attention-bias module is modular and can be integrated into other attention-based architectures, making it broadly applicable to classification, segmentation, and detection tasks in multi-view settings. Data augmentation techniques, including random adjustments to brightness, contrast, and color inversion, are applied to enhance model robustness.

4.2 Results and Discussion

Correspondence Prediction-Simulated Data. Table 1 summarizes the performance of various models for our patch-level correspondence prediction task on a simulated test dataset. We use ResNet-50 as the backbone for feature extraction and compute the correspondence matrix using normalized dot product. Our experiments compare different attention modules: graph-based message passing methods (e.g., SuperGlue [17] and LoFTR [22]) and a standard transformer attention module. The evaluation metrics include mean squared error (MSE), precision, recall, and average precision (AP). Notably, our approach, the Standard Transformer [24] without message passing, achieved the best performance across all metrics, with the lowest MSE and the highest AP. This suggests that the transformer's flexible attention mechanism enables more effective inter-patch communication compared to the more constrained graph-based approaches. Since our goal is to predict correspondences at the patch level, acknowledging that abnormalities typically span groups of pixels rather than isolated pixels, it is natural to focus on patches. In contrast, a CNN-only baseline without any attention mechanism performed considerably worse, highlighting the importance of attention mechanisms in learning robust patch-level correspondences.

Correspondence Prediction-Real Data. Table 2 evaluates our method for correspondence matching on a real X-ray dataset with partial annotations, comparing it against existing approaches in both zero-shot and fine-tuned settings.

Table 1. Model Performance on Correspondence Simulated Test Dataset.

Attention Model	Message Pass	MSE	Precision	Recall	AP
–	–	2.25×10^{-3}	49.1	68.0	54.7
Superglue Module	✓	1.33×10^{-3}	70.3	75.6	80.1
LoFTR Module	✓	1.24×10^{-3}	74.2	78.1	83.5
Standard Transformer	✓	1.19×10^{-3}	75.5	79.4	84.1
Standard Transformer	–	$\mathbf{1.09 \times 10^{-3}}$	**77.0**	**81.5**	**85.0**

Table 2. Model Performance on the Correspondence Test Dataset of Real X-ray.

Method	Backbone	Model	Precision	Recall	AP
Zero Shot	Res50	LoFTR module	32.5	2.7	9.4
	Res50	LoFTR* module	28.3	26.3	15.9
	DinoV2-G	–	3.0	15.2	1.7
Fine-tuning	Res50	LoFTR* module	36.8	18.0	16.1
	DinoV2-G	Multi Layer Perceptron	55.23	51.6	42.2
Pre-train + Fine-tuning	Res50	Standard Transformer	**72.3**	**87.1**	**83.8**

The goal is to establish reliable point correspondences with minimal annotation. In the zero-shot setting, the pre-trained LoFTR model [22] with a ResNet50 backbone struggles due to domain shift and its one-to-one matching strategy, resulting in low recall (2.7) and modest AP (9.4). We introduce a variant LoFTR*, which incorporates a normalized dot-product and thresholding correspondence head for multi-to-multi matching, slightly improving recall (26.3) and AP (15.9), though the domain gap remains significant. Fine-tuning LoFTR* improves performance (AP: 16.1), but a stronger baseline is obtained by freezing a 1.1B-parameter DinoV2-G model [15] and fine-tuning an MLP head, achieving much better results (AP: 42.2). Our approach, a transformer-based model pre-trained and fine-tuned with only 24M parameters, outperforms all other methods, achieving the highest AP (83.8).

Table 3. Classification Performance on the MURA and the Private Dataset.

	Correspondence Pretraining	Attention Guidance	Fusion	Accuracy	Precision	Recall	Kappa
MURA	–	–	Single	73.3 ± 0.3	75.4 ± 0.2	65.4 ± 0.2	0.45
	–	–	Late	78.7 ± 0.2	83.3 ± 0.1	67.4 ± 0.2	0.56
	–	–	Early	75.8 ± 0.2	78.5 ± 0.1	65.4 ± 0.2	0.49
	✓	–	Early	80.1 ± 0.1	83.4 ± 0.1	73.6 ± 0.2	0.58
	✓	✓	Early	$\mathbf{80.6 \pm 0.1}$	$\mathbf{84.8 \pm 0.1}$	$\mathbf{74.4 \pm 0.2}$	**0.59**
Private	–	–	Single	68.8 ± 0.2	53.6 ± 0.2	41.7 ± 0.2	0.30
	–	–	Late	74.2 ± 0.2	59.5 ± 0.2	50.1 ± 0.2	0.36
	–	–	Early	71.1 ± 0.1	54.4 ± 0.1	44.2 ± 0.1	0.32
	✓	–	Early	75.0 ± 0.1	59.7 ± 0.1	49.2 ± 0.2	0.37
	✓	✓	Early	$\mathbf{76.2 \pm 0.1}$	$\mathbf{59.9 \pm 0.1}$	$\mathbf{52.6 \pm 0.2}$	**0.39**

Multi-View X-ray Classification. Table 3 evaluates the impact of correspondence pretraining and attention guidance on multi-view X-ray classification. This experiment demonstrates the benefits of pretraining a model on a correspondence task before applying it to classification, as well as the advantages of using the correspondence mask to guide attention. We used a ViT model [6] trained on both public and private datasets. We compare early fusion, where patch embeddings from multiple views are concatenated before input to the transformer, and late fusion, where scores are aggregated after independent processing. Multi-view fusion significantly outperforms single-view methods, and with correspondence pretraining, early fusion surpasses late fusion.

Pretraining on correspondence further boosts performance, despite all models being initialized with ImageNet weights. Notably, early fusion with pretraining improves accuracy from 75.8% to 80.1% on the public dataset. Additionally, incorporating the correspondence matrix as an attention bias further enhances results, yielding the highest accuracy (80.6% public, 76.2% private).

5 Conclusion

We introduced a new framework for multi-view X-ray analysis that leverages self-supervised correspondence learning to improve both correspondence matching on real X-ray images and multi-view fracture detection. Our experiments show that pretraining on correspondence significantly enhances classification accuracy, especially when integrating correspondence information into transformer-based architectures. As demonstrated in our fracture classification task, the no-annotation correspondence method we proposed opens up numerous use cases in multi-view X-ray tasks. This work provides a novel approach that can be applied to a variety of X-ray tasks, advancing the field of multi-view medical image analysis.

Disclosure of Interests. The authors have no competing interests to declare that are relevant to the content of this article.

References

1. Brandser, E.A., et al.: Contribution of individual projections alone and in combination for radiographic detection of ankle fractures. AJR Am. J. Roentgenol. **174**(6), 1691–1697 (2000)
2. Chen, H., et al.: Aspanformer: detector-free image matching with adaptive span transformer. In: European Conference on Computer Vision, pp. 20–36. Springer (2022)
3. Cheung, G.C., Lever, C.J., Morris, A.D.: X-ray diagnosis of acute scaphoid fractures. J. Hand Surg. Br. **31**(1), 104–109 (2006)
4. Corona-Figueroa, A., Frawley, J., Bond-Taylor, S., Bethapudi, S., Shum, H.P., Willcocks, C.G.: Mednerf: Medical neural radiance fields for reconstructing 3D-aware CT-projections from a single x-ray. In: 2022 44th Annual International Conference of the IEEE Engineering in Medicine & Biology Society (EMBC), pp. 3843–3848. IEEE (2022)
5. Dhont, J., Verellen, D., Mollaert, I., Vanreusel, V., Vandemeulebroucke, J.: Realdrr-rendering of realistic digitally reconstructed radiographs using locally trained image-to-image translation. Radiother. Oncol. **153**, 213–219 (2020)

6. Dosovitskiy, A., et al.: An image is worth 16x16 words: transformers for image recognition at scale. arXiv preprint arXiv:2010.11929 (2020)
7. Edstedt, J., Sun, Q., Bökman, G., Wadenbäck, M., Felsberg, M.: Roma: Robust dense feature matching. In: Proceedings of the IEEE/CVF Conference on Computer Vision and Pattern Recognition, pp. 19790–19800 (2024)
8. He, K., Zhang, X., Ren, S., Sun, J.: Deep residual learning for image recognition. In: Proceedings of the IEEE Conference on Computer Vision and Pattern Recognition, pp. 770–778 (2016)
9. Herbert, B.: Surf: speeded up robust features. Comput. Vis. Image Underst. **110**(3), 346–359 (2008)
10. Joseph, P.M.: An improved algorithm for reprojecting rays through pixel images. IEEE Trans. Med. Imaging **1**(3), 192–196 (1982)
11. Kasten, Y., Doktofsky, D., Kovler, I.: End-to-end convolutional neural network for 3d reconstruction of knee bones from bi-planar x-ray images. In: Machine Learning for Medical Image Reconstruction: Third International Workshop, MLMIR 2020, Held in Conjunction with MICCAI 2020, Lima, Peru, October 8, 2020, Proceedings 3, pp. 123–133. Springer (2020)
12. Killeen, B.D., et al.: Fluorosam: A language-aligned foundation model for x-ray image segmentation. arXiv preprint arXiv:2403.08059 (2024)
13. Lindenberger, P., Sarlin, P.E., Pollefeys, M.: Lightglue: Local feature matching at light speed. In: Proceedings of the IEEE/CVF International Conference on Computer Vision (ICCV), pp. 17627–17638 (2023)
14. Lowe, D.G.: Distinctive image features from scale-invariant keypoints. Int. J. Comput. Vision **60**, 91–110 (2004)
15. Oquab, M., et al.: Dinov2: Learning robust visual features without supervision. arXiv preprint arXiv:2304.07193 (2023)
16. Rajpurkar, P., et al.: Mura: Large dataset for abnormality detection in musculoskeletal radiographs. arXiv preprint arXiv:1712.06957 (2017)
17. Sarlin, P.E., DeTone, D., Malisiewicz, T., Rabinovich, A.: Superglue: learning feature matching with graph neural networks. In: Proceedings of the IEEE/CVF Conference on Computer Vision and Pattern Recognition (CVPR) (2020)
18. Shrestha, P., Xie, C., Shishido, H., Yoshii, Y., Kitahara, I.: X-ray to CT rigid registration using scene coordinate regression. In: International Conference on Medical Image Computing and Computer-Assisted Intervention, pp. 781–790. Springer (2023)
19. Siddon, R.L.: Prism representation: a 3D ray-tracing algorithm for radiotherapy applications. Phys. Medi. Biol. **30**(8), 817 (1985)
20. Southall, C., Friebel, T.R., Gardiner, S., Nikkhah, D.: Radiographic views for hand fractures - call for three-view national UK guidelines - a quality improvement study. Int. J. Surg. **109**(7), 1919–1922 (2023)
21. Su, J., Ahmed, M., Lu, Y., Pan, S., Bo, W., Liu, Y.: Roformer: enhanced transformer with rotary position embedding. Neurocomputing **568**, 127063 (2024)
22. Sun, J., Shen, Z., Wang, Y., Bao, H., Zhou, X.: Loftr: detector-free local feature matching with transformers. In: Proceedings of the IEEE/CVF Conference on Computer Vision and Pattern Recognition (CVPR), pp. 8922–8931 (2021)
23. Unberath, M., et al.: Deepdrr–a catalyst for machine learning in fluoroscopy-guided procedures. In: Medical Image Computing and Computer Assisted Intervention–MICCAI 2018: 21st International Conference, Granada, Spain, September 16-20, 2018, Proceedings, Part IV 11, pp. 98–106. Springer (2018)
24. Vaswani, A., et al.: Attention is all you need. In: Advances in Neural Information Processing Systems, vol. 30 (2017)

25. Wu, H., et al.: Multi-view vertebra localization and identification from CT images. In: International Conference on Medical Image Computing and Computer-Assisted Intervention, pp. 136–145. Springer (2023)
26. Zhang, B., et al.: A patient-specific self-supervised model for automatic x-ray/CT registration. In: Medical Image Computing and Computer Assisted Intervention – MICCAI 2023, pp. 515–524. Springer Nature Switzerland, Cham (2023)
27. Zhang, Y., Miao, S., Mansi, T., Liao, R.: Task driven generative modeling for unsupervised domain adaptation: Application to x-ray image segmentation. In: International Conference on Medical Image Computing and Computer-assisted Intervention, pp. 599–607. Springer (2018)

Small Lesions-aware Bidirectional Multimodal Multiscale Fusion Network for Lung Disease Classification

Jianxun Yu[1], Ruiquan Ge[2], Zhipeng Wang[2], Cheng Yang[2], Chenyu Lin[2], Xianjun Fu[3], Jikui Liu[4(✉)], Ahmed Elazab[5(✉)], and Changmiao Wang[6]

[1] Xidian University, Xi'an, China
[2] Hangzhou Dianzi University, Hangzhou, China
[3] Zhejiang College of Security Technology, School of Artificial Intelligence, Wenzhou, China
[4] Shenzhen Polytechnic University, Shenzhen, China
liujikui007@gmail.com
[5] Shenzhen University, Shenzhen, China
ahmedelazab@szu.edu.cn
[6] Shenzhen Research Institute of Big Data, Shenzhen, China

Abstract. The diagnosis of medical diseases faces challenges such as the misdiagnosis of small lesions. Deep learning, particularly multimodal approaches, has shown great potential in the field of medical disease diagnosis. However, the differences in dimensionality between medical imaging and electronic health record data present challenges for effective alignment and fusion. To address these issues, we propose the Multimodal Multiscale Cross-Attention Fusion Network (MMCAF-Net). This model employs a feature pyramid structure combined with an efficient 3D multi-scale convolutional attention module to extract lesion-specific features from 3D medical images. To further enhance multimodal data integration, MMCAF-Net incorporates a multi-scale cross-attention module, which resolves dimensional inconsistencies, enabling more effective feature fusion. We evaluated MMCAF-Net on the Lung-PET-CT-Dx dataset, and the results showed a significant improvement in diagnostic accuracy, surpassing current state-of-the-art methods. The code is available at https://github.com/yjx1234/MMCAF-Net

Keywords: Multimodal Learning · Cross-Modality · Multiscale · Interdimensional Uncertainty

1 Introduction

Computer-aided diagnosis is crucial in modern clinical applications, improving disease detection efficiency and accuracy. Deep learning methods are widely

J. Yu and R. Ge—Co-first authors.

J. C. Gee et al. (Eds.): MICCAI 2025, LNCS 15960, pp. 589–598, 2026.
https://doi.org/10.1007/978-3-032-04927-8_56

used for medical diagnosis, utilizing information from various modalities such as medical imaging and electronic health records to deliver comprehensive patient assessments [3,7]. However, detecting certain lesions, like squamous cell carcinoma and pulmonary embolism, remains challenging due to their small size in images, which can be easily missed by conventional models [2,11]. Additionally, discrepancies between modalities complicate effective multimodal fusion for clinical decision-making. These challenges highlight the need for more robust methodologies to enhance multimodal data integration and improve diagnostic accuracy.

Multi-scale methods are effective for small lesion detection by capturing local and global information, but traditional approaches often incur high computational costs. To improve efficiency, Rahman *et al.*. introduced the Multi-scale Convolutional Attention Module (MSCAM) [16], which, while beneficial, is limited to 2D image processing and lacks applicability to 3D medical imaging, highlighting a gap in multi-scale feature extraction for volumetric data. Aligning features across modalities is a persistent challenge in multimodal learning. Yao *et al.*. minimized JensenâĂŞShannon divergence for feature alignment [18], while Sanjeev *et al.*. used supervised contrastive loss to integrate tabular and image data [17]. However, these strategies often compromise the integrity of complex modalities to align with simpler ones, resulting in suboptimal performance and underscoring the need for more adaptive fusion techniques. Late fusion, known for its simplicity, is widely used but fails to capture complementary relationships between multimodal features [6,17]. To address this, Hayat *et al.*. proposed using LSTMs for aggregating multimodal representations as sequences of single-modal tokens [4], and Yao *et al.*. introduced a disease-aware attention fusion module to weigh modalities based on disease profiles [18]. Other strategies, such as Pölsterl *et al.*.'s dynamic affine feature transformation [14] and Reza *et al.*.'s MMTM module for knowledge exchange between CNNs [8], have also advanced intermediate fusion. However, they often overlook ambiguity in certain feature dimensions, which can degrade representation quality and limit downstream task effectiveness. Thus, more adaptive fusion mechanisms are necessary to enhance meaningful features while mitigating ambiguity.

To overcome these challenges, we propose a novel framework that enables simultaneous multimodal fusion while effectively capturing small lesions in medical images. The key contributions of this work are summarized as follows: 1) We introduce an E3D-MSCA module to enhance lesion detection, particularly for small lesions. 2) We propose MSCA module designed to effectively integrate multimodal data, addressing the challenges posed by significant differences between modalities. 3) We develop a Bidirectional Scale Fusion (BSF) module, which specifically addresses the challenges of integrating features across two distinct linear scales.

2 Proposed Method

As illustrated in Fig. 1, MMCAF-Net consists of three main components. The first component is an image encoder that integrates a feature pyramid with E3D-

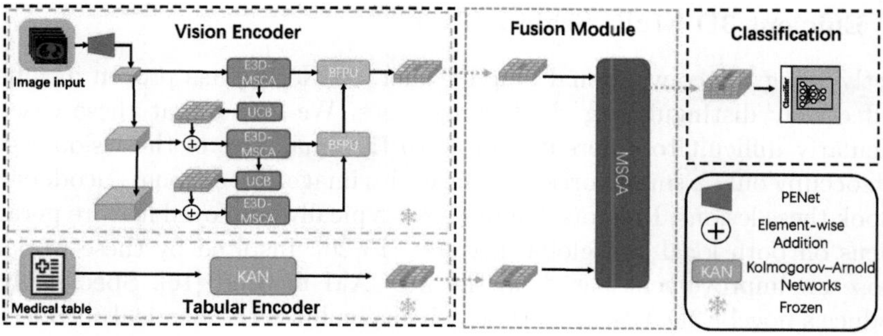

Fig. 1. Overview of the Proposed MMCAF-Net. The framework consists of an encoding module with Vision and Tabular encoders, a fusion module utilizing the MSCA module, and a classification module for final detection.

MSCA module, designed to capture both local and global features of imaging data. This allows for the effective differentiation of challenging cases. The second component employs KolmogorovâĂŞArnold Networks (KAN) [12] to encode tabular features. The advantages of KAN lie in its faster scaling laws, meaning that as the model size increases, its efficiency improves significantly. Additionally, KAN demonstrates strong expressive capability even with fewer parameters. Initially, KAN was applied to tasks such as function approximation and solving partial differential equations (PDEs). Subsequent research has expanded its use to time series forecasting tasks. In addressing our problem, we flexibly apply KAN to encode our tabular data. Lastly, the third component utilizes the MSCA module to align and fuse the features from both modalities.

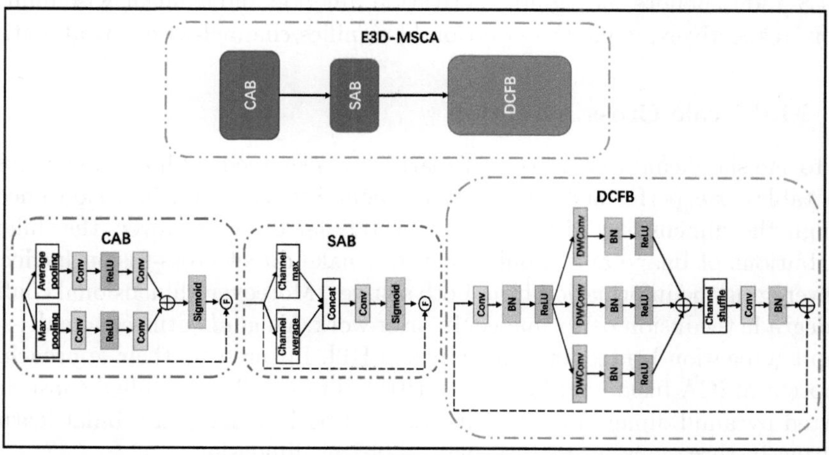

Fig. 2. Detailed structure of E3D-MSCA. It comprises 3D CAB, 3D SAB, and 3D DCFB.

2.1 Efficient 3D Multi-Scale PENet

Directly using the conventional lung encoder, PENet [5] has proven insufficient for effectively distinguishing challenging cases. We found that these cases are particularly difficult to differentiate due to the small size of the lesion regions, which occupy only a small portion of the entire image. Traditional encoders often overlook these lesions. Previous studies have typically employed feature pyramids to focus on both local and global features [13,20]. Inspired by these works, we propose an improvement based on the MSCAM method [16]. Specifically, we introduce a novel E3D-MSCA method, which combines the classical encoder with enhanced attention mechanisms. As shown in Fig. 2, E3D-MSCA consists of three modules: 3D Channel Attention Block(CAB), 3D Spatial Attention Block(SAB), and 3D Depth-Wise Convolution Fusion Block(DCFB). The formulation of E3D-MSCA is described as follows.

$$E3D - MSCA(x) = DCFB(SAB(CAB(x))), \tag{1}$$

where x represents the output features from the feature pyramid. By utilizing depth-wise convolution within the *DCFB*, we significantly enhance the performance of PENet in all metrics without incurring substantial computational costs. Finally, we incorporate the Bidirectional Feedback Propagation Unit (BFPU) proposed by Chen *et al.*. [1] to effectively fuse features from multiple scales. The formulation of BFPU is provided using:

$$F_{mid} = S\left(\text{Conv}_{3\times3}(F_a) \otimes \text{Conv}_{3\times3}(F_b)\right), \tag{2}$$

$$F_{out} = \left[F_a + F_{mid} \otimes F_a,\, F_b + F_{mid} \otimes F_b\right], \tag{3}$$

where F_a and F_b represent features extracted from two different scales. The functions S and $\otimes$ denote the sigmoid activation function and element-wise multiplication, respectively, while the notation $[\cdot]$ signifies channel-wise concatenation.

2.2 Multiscale Cross Attention

Due to the significant dimensional disparity between medical images and clinical data tables, we performed feature dimensionality reduction in vision encoder to align the dimensions with those of the tabular data. However, the differing distributions of image and tabular features make direct cross-fusion ineffective for leveraging the information from both sources. Moreover, dimensional conflicts can degrade the fusion performance. In prior work, Yu *et al.*. utilized a multi-scale group aggregation bridge for feature fusion [19]. Inspired by their approach, we propose a MSCA fusion module. As illustrated in Fig. 3, we applied consecutive inverted pyramid dimensionality reductions to both image and tabular features, resulting in three pairs of features with different dimensions.

Next, the feature pairs from each scale are fed into the Cross Attention module for feature fusion. In this module, we use the image and table features as queries (Q), while another feature is utilized as keys (K) and values (V). Following this, Q, K, and V undergo a multi-head operation. We compute the attention scores between the two modalities using Q and K, normalize these attention scores, and then multiply them with V to derive the final output. The details of this process are outlined as follows:

$$Q' = QW_q \quad (W_q \in \mathbb{R}^{D_q \times D_q}), \tag{4}$$
$$K' = KW_k \quad (W_k \in \mathbb{R}^{D_k \times D_q}), \tag{5}$$
$$V' = VW_v \quad (W_v \in \mathbb{R}^{D_v \times D_q}), \tag{6}$$

where Q has a shape of [B, N, D_q] while the keys K and values V have a shape of [B, M, D_kv].

$$Q_h = \text{reshape}(Q', B, H, N, C), \tag{7}$$
$$K_h = \text{reshape}(K', B, H, M, C), \tag{8}$$
$$V_h = \text{reshape}(V', B, H, M, C), \tag{9}$$

where C represents the focus dimension of each attention head, H denotes the number of attention heads.

$$\text{Attention}(Q_h, K_h) = \frac{Q_h K_h^T}{\sqrt{C}}, \mathrm{O_h} = \text{Attention} * \mathrm{V_h}, \tag{10}$$

Attention represents the attention weights, and O_h denotes the final output features. Subsequently, we introduce the BSF module to integrate features across multiple scales. The structure of this module is illustrated in Fig. 3. Initially, we apply a linear transformation to align the two features to a common scale. Next, we compute the dimensional importance of both features by performing an element-wise multiplication, followed by calculating the Softmax scores. This process effectively reduces the significance of the uncertain dimensions. Finally, we merge the features from the two scales based on their dimensional importance.

2.3 Loss Function

We employ a binary cross-entropy loss to quantify the distance between the predicted outcomes and the ground truths. Mathematically, the classification loss $\mathcal{L}_{cls}$ is expressed as follows:

$$\mathcal{L}_{cls}(y, \hat{y}) = -\frac{1}{N} \sum_{i=1}^{N} \left[y_i \log(\hat{y}_i) + (1 - y_i) \log(1 - \hat{y}_i) \right], \tag{11}$$

where N represents the number of samples, y_i signifies the real labels for sample i, $\hat{y}_i$ is the probability values of the model predicting sample i range between (0,1).

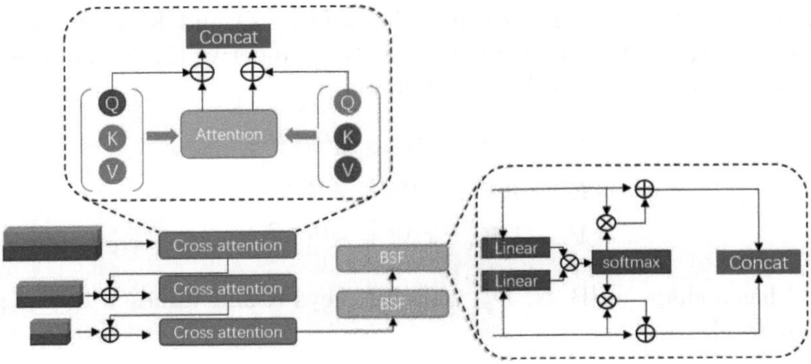

Fig. 3. Detailed structure of MSCA module.

3 Experiments

3.1 Dataset and Implementation

Dataset. We used a publicly available dataset provided by The Cancer Imaging Archive, called Lung-PET-CT-Dx [10], which includes CT or PET) scans and tabular clinical information for 355 cases. The dataset provides tumor classification labels for each patient. Most of the PET/CT data is stored in DICOM format, and all data has undergone de-identification. The dataset includes 251 adenocarcinoma samples and 61 squamous cell carcinoma samples. Since the adenocarcinoma class and squamous cell carcinoma class samples are very imbalanced in the original dataset, we decided to oversample the squamous cell carcinoma class. We use a random oversampling strategy to supplement the training set from 34 samples to 198 samples, and the number of verification sets and test sets remains the same, which are 12 and 15 respectively. The tabular data we use includes attributes such as gender, age, weight, TNM stage, and smoking history.

Implementation Details. We implemented MMCAF-Net on a device equipped with Tesla V100S-PCIE-32GB using torch-2.4.1. During the training phase, we utilized binary cross-entropy loss in conjunction with the stochastic gradient descent optimizer. The model was trained for 50 epochs, with an initial learning rate set at 0.0001 and a weight decay of 0.01. We employed a batch size of 4 and processed 12 slices per input sample. Data augmentation techniques, including rotation, sharpening, and normalization, were applied to the input images, which were subsequently reshaped to a dimension of 192 $\times$ 192 pixels.

3.2 Experiment Result

The Lung-PET-CT Result. In our study, we compared MMCAF-Net against the six state-of-the-art multi-modal approaches using the Lung-PET-CT dataset.

As shown in Table 1, MMCAF-Net outperforms all comparative approaches across several key metrics, including accuracy (ACC), F1 score (F1), specificity, sensitivity, positive predictive value (PPV), and negative predictive value (NPV). While MMCAF-Net falls short of mmtm by 1.6% in area under the receiver operating characteristic curve(AUROC), it surpasses mmtm by approximately 10% in both ACC and F1 scores, and exhibits a remarkable 15% improvement in PPV. This indicates that MMCAF-Net achieves a lower rate of false positives, demonstrating its high accuracy and reliability. Overall, it is noteworthy that all models exhibit relatively poor performance on the F1, suggesting the presence of noise or outliers in the dataset. Additionally, we visualized the classification results for challenging cases. As shown in Fig. 4, each case has 12 slices, and we selected the most representative slice of the lesion for display. The results indicate that our model outperforms the others in distinguishing subtle and difficult cases.

Ablation Study. We conducted ablation experiments on the Lung-PET-CT dataset to evaluate the effectiveness of image encoding and multi-modal fusion techniques. As demonstrated in Table 2, we compared our proposed Efficient 3D Multi-scale PENet with the Segment Anything Model (SAM) [9] combined with multi-scale techniques and the original PENet. The results indicate that our method outperforms the other two approaches across all evaluated metrics, achieving improvements of 10% and 15% in AUROC, and 10% and 14% in F1 scores, respectively. Furthermore, our method shows significant enhancements in both AUROC and NPV compared to the standalone PENet, suggesting that the improved architecture enhances the reliability of negative class sample identification. Additionally, we performed ablation experiments on different fusion methods, as presented in Table 3, comparing our proposed MSCA with Cross-Attention, Late_Fusion, and clip aligned fusion [15] techniques. The results illustrate that MSCA outperforms the other three fusion methods across most metrics, with an increase in ACC of 5%, 10%, and 12%, respectively. This indicates that our fusion method demonstrates superior overall performance compared to the other approaches.

Table 1. Quantitative comparison with other methods on the public dataset. The best and the second-best results are bolded and underlined, respectively.

Method	AUROC ↑	ACC ↑	F1 ↑	Specificity ↑	Sensitivity ↑	PPV ↑	NPV ↑
PECon [17]	0.786	0.744	0.645	0.786	<u>0.667</u>	0.625	<u>0.815</u>
MedFuse [4]	0.786	0.721	0.571	<u>0.821</u>	0.533	0.615	0.767
Drfuse [18]	0.613	0.676	0.353	0.800	0.333	0.375	0.769
MMTM [8]	**0.802**	0.698	0.581	0.750	0.600	0.562	0.778
PEfusion [6]	0.740	0.721	0.600	0.786	0.600	0.600	0.786
daft [14]	0.727	<u>0.729</u>	<u>0.667</u>	0.786	0.650	<u>0.684</u>	0.759
MMCAF-Net (Ours)	<u>0.786</u>	**0.791**	**0.690**	**0.857**	**0.667**	**0.714**	**0.828**

Table 2. Image encoder ablation experiments on the Lung-PET-CT dataset. The best and second-best results are bolded and underlined, respectively.

Method	AUROC ↑	ACC ↑	F1 ↑	Specificity ↑	Sensitivity ↑	PPV ↑	NPV ↑
SAM+E3D-MSCA	0.610	0.651	0.444	0.786	0.400	0.500	0.710
PENet [5]	0.560	0.721	0.400	<u>0.964</u>	0.267	<u>0.800</u>	0.711
PENet+E3D-MSCA	<u>0.644</u>	<u>0.721</u>	<u>0.500</u>	0.893	<u>0.400</u>	0.667	<u>0.735</u>
PENet+E3D-MSCA+drop	**0.712**	**0.767**	**0.545**	**0.964**	**0.400**	**0.857**	**0.857**

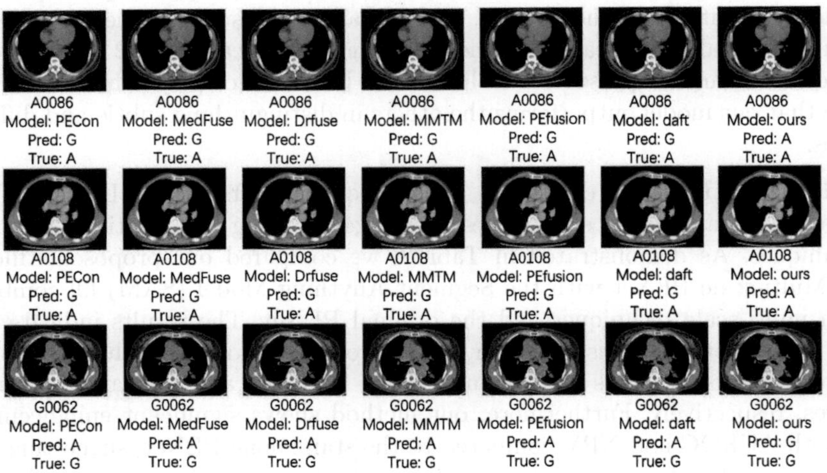

Fig. 4. Visual Representation of Classification Performance for Challenging Cases. The G denotes lung squamous cell carcinoma and A denotes lung adenocarcinoma.

Table 3. Fusion ablation experiments on the Lung-PET-CT dataset. The best and second-best results are bolded and underlined, respectively.

Method	AUROC ↑	ACC ↑	F1 ↑	Specificity ↑	Sensitivity ↑	PPV ↑	NPV ↑
Cross-Attention	<u>0.752</u>	<u>0.744</u>	<u>0.686</u>	0.714	**0.800**	0.600	**0.870**
clip_Fusion [15]	0.717	0.698	0.629	0.679	<u>0.733</u>	0.550	0.826
Late_Fusion	0.729	0.674	0.462	<u>0.821</u>	0.400	0.545	0.719
MSCA_Fusion	**0.786**	**0.791**	**0.690**	**0.857**	0.667	**0.714**	<u>0.828</u>

4 Conclusion

In this study, we introduce MMCAF-Net, a multimodal multiscale fusion model that effectively focuses on small lesions. MMCAF-Net incorporates the E3D-MSCA module to dynamically adjust feature weights across different scales, enhancing attention to small lesions. The MSCA module addresses the dimensional conflicts that arise when merging two modalities with significantly different characteristics. The BSF module tackles the impact of dimensional uncertainty between different scales on the fusion results. Comparative studies demonstrate that MMCAF-Net outperforms competing methods on most metrics, indicating its practical effectiveness. Future plans include adapting the model for the diagnosis of lung and other medical diseases in clinical settings.

Acknowledgments. This work was supported by the Open Project Program of the State Key Laboratory of CAD & CG (No. A2410), Zhejiang University; Zhejiang Provincial Natural Science Foundation of China (No. LY21F020017, 2023C03090); National Natural Science Foundation of China (No. 61702146, 62076084, U20A20386, U22A2033); Shenzhen major science and technology project (KCXFZ20240903093923031), Special projects fund in key areas of the Guangdong Provincial Department of Education (2023ZDZX2085); Guangdong Basic and Applied Basic Research Foundation (No. 2022A1515110570), Futian Healthcare Research Project (No.FTWS002); Medical Scientific Research Foundation of the Guangdong Province of China (A2023158,C2023106).

Disclosure of Interests. The authors declare no competing interests.

References

1. Chen, X., Pan, J., Dong, J.: Bidirectional multi-scale implicit neural representations for image deraining. In: Proceedings of the IEEE/CVF Conference on Computer Vision and Pattern Recognition, pp. 25627–25636 (2024)
2. Guo, Z., Gan, H.: CPP-NET: embracing multi-scale feature fusion into deep unfolding CP-PPA network for compressive sensing. In: Proceedings of the IEEE/CVF Conference on Computer Vision and Pattern Recognition, pp. 25086–25095 (2024)
3. Hager, P., Menten, M.J., Rueckert, D.: Best of both worlds: multimodal contrastive learning with tabular and imaging data. In: Proceedings of the IEEE/CVF Conference on Computer Vision and Pattern Recognition, pp. 23924–23935 (2023)
4. Hayat, N., Geras, K.J., Shamout, F.E.: Medfuse: multi-modal fusion with clinical time-series data and chest x-ray images. In: Machine Learning for Healthcare Conference, pp. 479–503. PMLR (2022)
5. Huang, S.C., et al.: Peneta scalable deep-learning model for automated diagnosis of pulmonary embolism using volumetric CT imaging. NPJ Digit. Med. **3**(1), 61 (2020)
6. Huang, S.C., Pareek, A., Zamanian, R., Banerjee, I., Lungren, M.P.: Multimodal fusion with deep neural networks for leveraging CT imaging and electronic health record: a case-study in pulmonary embolism detection. Sci. Rep. **10**(1), 22147 (2020)
7. Jiang, J.P., Ye, H.J., Wang, L., Yang, Y., Jiang, Y., Zhan, D.C.: Tabular insights, visual impacts: transferring expertise from tables to images. In: Forty-first International Conference on Machine Learning (2024)
8. Joze, H.R.V., Shaban, A., Iuzzolino, M.L., Koishida, K.: MMTM: multimodal transfer module for cnn fusion. In: Proceedings of the IEEE/CVF Conference on Computer Vision and Pattern Recognition, pp. 13289–13299 (2020)
9. Kirillov, A., et al.: Segment anything. In: Proceedings of the IEEE/CVF International Conference on Computer Vision, pp. 4015–4026 (2023)

10. Li, P., Wang, S., Li, T., Lu, J., HuangFu, Y., Wang, D.: A large-scale CT and PET/CT dataset for lung cancer diagnosis (lung-PET-CT-dx). https://doi.org/10.7937/TCIA.2020.NNC2-0461 (2020)
11. Liu, S., et al.: Rotated multi-scale interaction network for referring remote sensing image segmentation. In: Proceedings of the IEEE/CVF Conference on Computer Vision and Pattern Recognition, pp. 26658–26668 (2024)
12. Liu, Z., et al.: KAN: Kolmogorov-arnold networks. arXiv preprint arXiv:2404.19756 (2024)
13. Ma, T., Dai, X., Zhang, S., Wen, Y.: PIVIT: Large deformation image registration with pyramid-iterative vision transformer. In: International Conference on Medical Image Computing and Computer-Assisted Intervention, pp. 602–612. Springer (2023)
14. Pölsterl, S., Wolf, T.N., Wachinger, C.: Combining 3D image and tabular data via the dynamic affine feature map transform. In: International Conference on Medical Image Computing and Computer-Assisted Intervention, pp. 688–698. Springer (2021)
15. Radford, A., et al.: Learning transferable visual models from natural language supervision. In: International Conference on Machine Learning, pp. 8748–8763. PMLR (2021)
16. Rahman, M.M., Munir, M., Marculescu, R.: EMCAD: efficient multi-scale convolutional attention decoding for medical image segmentation. In: Proceedings of the IEEE/CVF Conference on Computer Vision and Pattern Recognition, pp. 11769–11779 (2024)
17. Sanjeev, S., Al Khatib, S.K., Shaaban, M.A., Almakky, I., Papineni, V.R., Yaqub, M.: PECON: contrastive pretraining to enhance feature alignment between CT and EHR data for improved pulmonary embolism diagnosis. In: International Workshop on Machine Learning in Medical Imaging, pp. 434–443. Springer (2023)
18. Yao, W., Yin, K., Cheung, W.K., Liu, J., Qin, J.: Drfuse: learning disentangled representation for clinical multi-modal fusion with missing modality and modal inconsistency. In: Proceedings of the AAAI Conference on Artificial Intelligence, vol. 38, pp. 16416–16424 (2024)
19. Yu, X., et al.: ICH-SCNET: Intracerebral hemorrhage segmentation and prognosis classification network using clip-guided SAM mechanism. In: 2024 IEEE International Conference on Bioinformatics and Biomedicine (BIBM), pp. 2795–2800. IEEE (2024)
20. Zhu, V., et al.: Low-rank continual pyramid vision transformer: incrementally segment whole-body organs in CT with light-weighted adaptation. In: International Conference on Medical Image Computing and Computer-Assisted Intervention, pp. 371–381. Springer (2024)

Surface Vision Mamba: Leveraging Bidirectional State Space Model for Efficient Spherical Manifold Representation

Rongzhao He[1], Weihao Zheng[1](✉), Leilei Zhao[2], Ying Wang[1], Dalin Zhu[3], Dan Wu[4](✉), and Bin Hu[1](✉)

[1] Lanzhou University, Lanzhou, China
{zhengweihao,bh}@lzu.edu.cn
[2] Harbin Institute of Technology, Harbin, China
[3] Gansu Maternity and Child-care Hospital, Gansu, China
[4] Zhejiang University, Hangzhou, China
danwu.bme@zju.edu.cn

Abstract. Attention-based methods have demonstrated exceptional performance in modelling long-range dependencies on spherical cortical surfaces, surpassing traditional Geometric Deep Learning (GDL) models. However, their extensive inference time and high memory demands pose challenges for application to large datasets with limited computing resources. Inspired by the state space model in computer vision, we introduce the attention-free Vision Mamba (Vim) to spherical surfaces, presenting a domain-agnostic architecture for analyzing data on spherical manifolds. Our method achieves surface patching by representing spherical data as a sequence of triangular patches derived from a subdivided icosphere. The proposed Surface Vision Mamba (SiM) is evaluated on multiple neurodevelopmental phenotype regression tasks using cortical surface metrics from neonatal brains. Experimental results demonstrate that SiM outperforms both attention- and GDL-based methods, delivering 4.8 times faster inference and achieving 91.7% lower memory consumption compared to the Surface Vision Transformer (SiT) under the Ico-4 grid partitioning. Sensitivity analysis further underscores the potential of SiM to identify subtle cognitive developmental patterns. The code is available at https://github.com/Rongzhao-He/surface-vision-mamba.

Keywords: Cortical spherical manifold · State space model · Infant magnetic resonance imaging

1 Introduction

Many methods have been developed for traditional Euclidean space data, such as Convolution Neural Networks (CNNs) and attention-based [2,20] approaches.

J. C. Gee et al. (Eds.): MICCAI 2025, LNCS 15960, pp. 599–608, 2026.
https://doi.org/10.1007/978-3-032-04927-8_57

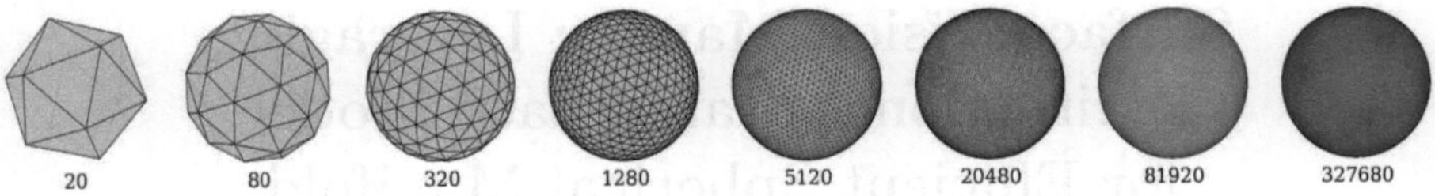

Fig. 1. Representative icosahedron discretized spherical surfaces with sequential subdivisions. The number of faces of each spherical surface is denoted under the surface.

CNNs calculate the weighted sum at each location, while attention-based methods treat the data as a sequence of patches. However, few models exist for non-Euclidean space data consist of graph, manifold and hyperbolic space data which have more complex geometries and distance metrics.

Existing models for processing non-Euclidean data can be broadly classified into attention- and Geometric Deep Learning (GDL)-based [1,3,6,10,12, 13,16,19] methods. Attention-based methods are effective in capturing long-range dependencies but are constrained in resource limited situations due to the quadratic complexity of the attention mechanism concerning sequence length, leading to higher memory consumption and slower inference time. Conversely, GDL-based methods are effective in handling complex geometric topology structure and distance metrics. However, they fail to extract global patterns, especially when applied to large-scale and highly intricate data, resulting in diminished performance. Thus, a key challenge for processing non-Euclidean data lies in improving efficiency while maintaining relatively excellent performance.

The emergence of State Space Models (SSMs) [8,11], has revitalized traditional sequence modeling, enabling efficient representation learning. A recent variant Mamba [7], has significantly surpassed traditional SSMs by integrating a selective scan mechanism that adapts parameters based on input and using a hardware-aware algorithm to parallelize scanning, thereby reducing memory I/O for more efficient inference. Motivated by ViT [5] and ViG [9], [21] adapted Mamba to computer vision, introducing a bidirectional SSM structure to address direction-sensitive challenges, termed Vision Mamba (Vim).

Non-Euclidean data, particularly spherical cortical surface, is characterized by high resolution, rich features, and intricate geometric shapes, as the cortical surface is inherently a high-dimensional manifold. While these data provide valuable insights into neurodevelopment, their effective representation poses a formidable challenge, often requiring a balance between performance and computational efficiency. Inspired by the efficiency of Vim, we extend its application to cerebral cortex analysis—an important yet underexplored area—by proposing Surface Vision Mamba (SiM). To adapt SiM to the unique characteristics of cortical surface data, we adjusted the input sequence length using various surface patching methods, as illustrated in Fig. 1.

The main contributions of this study can be summarized as follows:

1. We introduce SiM, an adaptation of Vim, as a generic backbone network for analyzing data mapped onto genus-zero surfaces.

Table 1. Demographic and clinical information of the subjects.

	dHCP (N = 526)	Replication dataset (N = 10)
Birth age [weeks $^{+days}$], median (IQR)	38^{+5} (38^{+1} - 40^{+5})	37^{+2} (35^{+4} - 39^{+0})
Scan age [weeks $^{+days}$], median (IQR)	41^{+0} (39^{+4} - 42^{+2})	39^{+4} (39^{+2} - 40^{+5})
Birth weight, mean (SD)	3.09 (0.64)	2.06 (1.00)
Head circumference at scan, mean (SD)	34.83 (3.72)	-
Radiology score (1/2/3/4/5)	270/171/51/10/24	-
Gender (M/F)	288/138	6/11

2. Leveraging the suitability of Mamba for tasks with *long-sequence* and *autoregressive* characteristics [17], we explore the impact of varying input sequence length on surface data in non-Euclidean space. We further implement autoregressive pretraining to validate the effectiveness of this approach.
3. Extensive experiments on three neurodevelopmental phenotype regression tasks, including the prediction of postmenstrual age (PMA) and long-term language and motor outcomes, demonstrate that our proposed SiM achieves promising performance compared to attention- and GDL-based models and is 4.8× faster than SiT and saves 91.7% GPU memory when performing batch inference under the Ico-4 grid partitioning.

2 Materials and Methods

Image Acquisition and Dataset. The imaging data used in this work are from the publicly available Developing Human Connectome Project (dHCP) and a private dataset collected at Gansu Provincial Maternity and Child-care Hospital (GPMCH). We used T1-weighted (T1w) and T2-weighted (T2w) images to calculate morphometric metrics of cerebral cortex.

The dHCP is approved by the United Kingdom Health Research Ethics Authority (reference number: 14/LO/1169). Additionally, we collected T1w and T2w images of 10 infants from GPMCH (2020-GSFY-05). These images were acquired in the resolution of $0.8{\times}0.8{\times}1.6\,\text{mm}^3$ with 0.8 mm overlap, and were reconstructed to 0.5 mm isotropic resolution.

Concerning the data from dHCP, a total of 526 infants covering preterm- and term-born neonates are enrolled. The neurodevelopmental assessments for these infants, conducted at 18 months of age using the Bayley-III Scales of Infant Development, can also be obtained. We used the following exclusion criteria: For PMA prediction, (*i*) we excluded the later scans of participants who were scanned twice; (*ii*) term-born neonates with focal abnormalities (radiology score $>$ 2) were excluded. The remaining infants were then split into two subsets: *Subset 1*: 408 participants who were born and scanned between 34 and 45 PMA; *Subset 2*: 16 preterm infants who were born before 34 PMA and scanned at term-equivalent

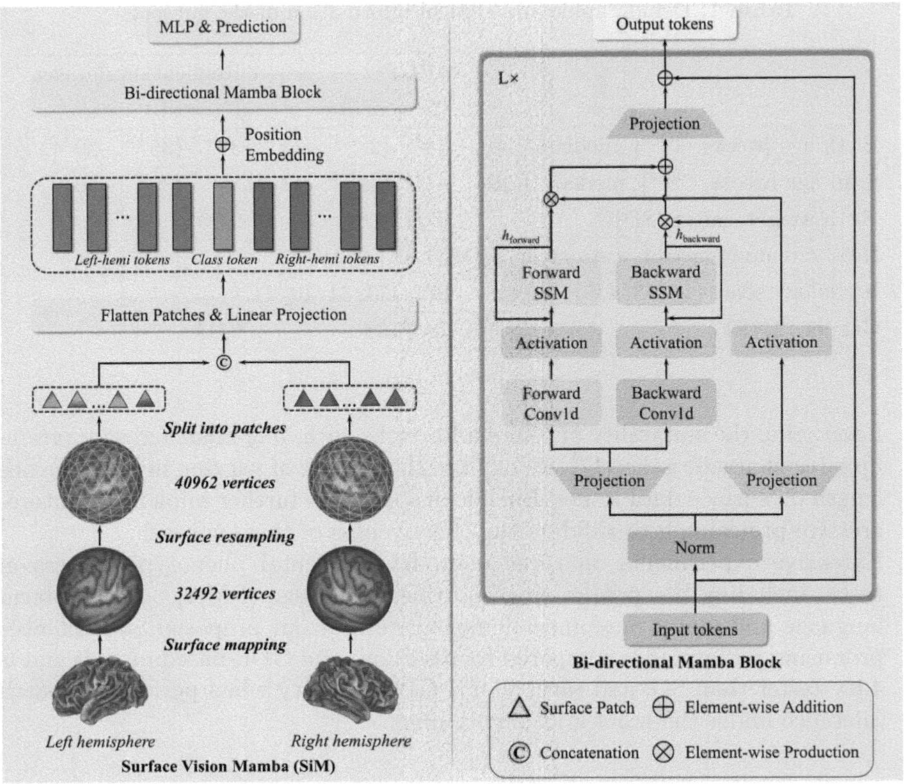

Fig. 2. Overview of the proposed Surface Vision Mamba (SiM) architecture.

age (> 37 PMA). For language and motor scores prediction, we retained the scans closest to 40 weeks for neonates scanned twice identified as *Subset 3*: 410 infants born between 23 and 43 gestational weeks (GA), with the scaled language and motor scores are 19.42 (5.26) and 20.55 (3.17), respectively, presented as mean (SD). We further utilized data from the GPMCH as a *Replication dataset*, consisting of 10 neonates to evaluate the generalization ability of the models. The demographic details are provided in Table 1[1].

Four cortical surface features (curvature, sulcal depth, cortical thickness and T1w/T2w myelination) were adopted. Each feature channel was normalized using Z-score. *Subset 1* and *3* were split into training, validation, and testing datasets in an 8:1:1 ratio. All data were registered to the dHCP 40-week spherical template, representing the cortical surface as an approximated sphere composed of triangles, with 32,492 vertices per hemisphere. We resampled the template sphere to a regular sixth-order icosphere (Ico-6) using barycentric interpolation.

[1] IQR and SD denote interquartile range and standard deviation, respectively. 20 head circumference data were missed in dHCP. The unit is centimeters.

Table 2. Summary of the parameters for icospheres of different orders.

Icosphere Order	First	Second	Third	Fourth	Fifth
The number of patches (N)	80	320	1280	5120	20480
The number of vertices (V)	561	153	45	15	6
Input Dimension (VC)	2244	612	180	60	24
Sequence Length ($2N$)	160	640	2560	10240	40960

Table 3. Hyperparameters for all training strategies in PMA prediction. Specifically, scratch means that training from scratch, fine-tuning refers to using ImageNet pre-training weights in Vim. For self-supervised pretraining, AR means autoregressive. T, S, B represent tiny-size, small-size, base-size, respectively.

	Scratch	Fine-tuning			AR Pretraining			AR Fine-tuning		
	TSB	T	S	B	T	S	B	T	S	B
Epochs	1000	600			4000	3000	3000	600		
Batch size	32	32			32			32		
Optimizer	AdamW	AdamW			AdamW			AdamW		
Adam ϵ	1e-8	1e-8			1e-8			1e-8		
Adam (β_1, β_2)	(0.9, 0.999)	(0.9, 0.999)			(0.9, 0.999)			(0.9, 0.999)		
LR	5e-5	5e-5	5e-5	3e-5	1.5e-4			1.5e-4	1e-4	8e-5
LR decay	Linear	Linear			Cosine			Cosine		
Step size	500	200			-			-		
Gamma	0.5	0.5			-			-		
Warmup epochs	None	None			10			10		
Weight decay	1e-8	1e-8			0.5			1e-6		

Surface Vision Mamba. We proposed the SiM model, as shown in Fig. 2. Specifically, the input domain is divided into $2N$ patches, represented as $\widetilde{X} = \{\widetilde{L}, \widetilde{R} | \widetilde{L} \in \mathbb{R}^{N \times V \times C}, \widetilde{R} \in \mathbb{R}^{N \times V \times C}\}$, that V is the number of vertices in a patch, and C denotes the number of feature channels. This is then flattened to $X = \{L, R | L \in \mathbb{R}^{N \times (VC)}, R \in \mathbb{R}^{N \times (VC)}\}$. Next, we projected X into D-dimensional vectors using a trainable fully connected layer. Following the design of ViT and BERT [4], a learnable class token X_{cls} is concatenated between the left and right hemispheres. To retain positional information, standard 1D position embeddings E_{pos} are added to the patch features.

$$S_0 = [X_L^1 W; \cdots X_L^N; X_{cls}; X_R^1 W; \cdots; X_R^N W] + E_{pos} \tag{1}$$

where $W \in \mathbb{R}^{(VC) \times D}$, $E_{pos} \in \mathbb{R}^{(2N+1) \times D}$, $S_0 \in \mathbb{R}^{(2N+1) \times D}$ is the initial input of SiM, X_L^1 and X_R^1 represent the first patches of the left and right hemispheres, respectively. Specifically, layer l processes the input S_{l-1} as follows:

$$S_l = SiM(S_{l-1}) + S_{l-1}, T = LayerNorm(S_l^N), \widehat{p} = MLP(T) \tag{2}$$

Table 4. Performance comparison on dHCP.

Methods	Supervised		Fine-tuning		Autoregressive		P	M
	MAE	MSE	MAE	MSE	MAE	MSE	(M)	(G)
MoNet	0.64±0.54	0.70±1.37	-	-	-	-	-	-
S2CNN	0.69±0.45	0.69±0.73	-	-	-	-	-	-
ChebNet	0.71±0.59	0.85±1.14	-	-	-	-	-	-
GConvNet	0.86±0.73	1.27±1.91	-	-	-	-	-	-
PointNet++	0.67±0.07	0.76±0.10	-	-	-	-	-	-
Spherical UNet	0.72±0.58	0.85±1.31	-	-	-	-	-	-
HRINet/1	0.75±0.67	1.05±0.05	-	-	-	-	10	-
SiT-Tiny/1	0.79±0.62	1.04±0.42	0.81±0.59	1.37±0.46	-	-	6	0.9
SiT-Small/1	0.81±0.57	1.06±0.36	0.87±0.63	0.93±0.28	-	-	22	3.6
SiT-Base/1	0.82±0.56	0.98±0.33	0.86±0.74	1.13±0.47	-	-	87	14.0
SiM-Tiny/1	0.85±0.64	0.91±0.31	0.76±0.66	1.21±0.26	1.03±0.70	1.87±0.42	7	1.9
SiM-Small/1	0.87±0.74	1.60±0.38	0.76±0.69	1.55±0.64	1.26±0.83	2.43±0.20	24	4.2
SiM-Base/1	0.86±0.69	1.20±0.04	0.84±0.63	1.24±0.19	1.06±0.67	1.34±0.30	92	15.4
HRINet/2	0.62±0.44	**0.39±0.25**	-	-	-	-	10	-
SiT-Tiny/2	0.69±0.52	0.47±0.36	0.66±0.58	0.78±0.01	-	-	6	3.5
SiT-Small/2	0.67±0.50	0.62±0.11	0.65±0.46	0.65±0.03	-	-	22	13.9
SiT-Base/2	0.64±0.57	0.57±0.55	0.72±0.47	0.58±0.20	-	-	86	55.0
SiM-Tiny/2	1.09±0.80	2.00±0.22	0.74±0.59	0.98±0.10	1.04±0.77	2.19±0.65	6	4.7
SiM-Small/2	0.98±0.80	1.89±0.38	0.60±0.49	0.52±0.10	1.12±0.92	3.50±1.79	24	16.5
SiM-Base/2	0.88±0.66	1.27±0.09	0.74±0.68	0.89±0.15	1.20±0.83	3.17±1.33	91	61.5
HRINet/3	*OOM*	*OOM*	-	-	-	-	10	-
SiT-Tiny/3	0.60±0.48	0.47±0.16	0.62±0.50	0.53±0.13	-	-	6	4.7
SiT-Small/3	0.60±0.51	0.54±0.41	0.60±0.43	0.42±0.16	-	-	24	16.5
SiT-Base/3	*OOM*	*OOM*	*OOM*	*OOM*	-	-	87	220.9
SiM-Tiny/3	1.09±0.76	2.79±1.30	0.60±0.46	0.85±0.35	0.91±0.81	3.62±2.73	7	18.9
SiM-Small/3	1.09±0.84	1.78±0.14	**0.56±0.50**	0.59±0.04	0.87±0.65	1.89±0.91	24	66.1
SiM-Base/3	1.03±0.81	2.30±0.73	0.62±0.45	0.65±0.07	1.09±0.71	2.27±0.76	93	245.5

Surface Patching Methods. We extend the sequence length by progressively subdividing the icosahedron into finer discrete levels, including first- to third-order icosphere, as summarized in Table 2. The icosphere subdivision is detailed in [18]. Different surface patching methods are visually represented in Fig. 1.

Training Methods. In this study, we explore three training strategies: (*i*) training models from scratch; (*ii*) fine-tuning pretrained weights from ImageNet (as released in Vision Mamba); and (*iii*) autoregressive pretraining due to the suitability of the Mamba for autoregressive modeling [14].

3 Results and Discussion

Model Variants. The proposed SiM configurations are built upon three variants of Vim: Vim-Tiny, Vim-Small, and Vim-Base. We adopt concise notations for model size. For instance, SiM-B/3 refers to the "Base" variant with an input size of 2560×180, using an Ico-3 grid on the sphere.

Implementation Details. All the experiments are implemented with Python 3.10.13 and PyTorch library and conducted on 4 NVIDIA A100 GPUs. The Vim-Tiny†, Vim-Small† and Vim-Base weights are adapted that Vim-Tiny† and Vim-Small† are fine-tuned under long sequence. Training details are in Table 3.

Infant Brain Age Prediction. As shown in Table 4[2], the comparison results of SiM against benchmark GDL methods [6,16] and attention-based models [2,20] on PMA prediction of *Subset 1* on the three training strategies. Notably, when fine-tuning with ImageNet pretraining weights, three variants of SiM outperform all the GDL methods using an Ico-3 grid. With comparable model parameters and MACs, SiM-S/3 achieves a performance of 0.56±0.50, surpassing SiT-S/3 (0.60±0.43). However, when training from scratch, the performance of all SiM variants decreased obviously, likely reflecting a tendency to overfit on small datasets due to the absence of strong prior weights. Although self-supervised pretraining has been shown to effectively strengthen model performance in previous studies, this benefit is less evident in our results. Actually, only the SiM-T/3 and SiM-S/3 exhibit improvements compared with training from scratch. This may be attributed to overfitting, which hampers generalization, or to the limited sample size, restricting the ability of model to capture sufficient features. In the *Subset 2*, we find that predicted brain age in preterm infants was significantly lower than chronological age, suggesting that preterm birth may delay brain development at term-equivalent age. Besides, we conducted further experiments:

1. Ablation studies on the decoder design in autoregressive pretraining, the best performance is achieved when the decoder depth is 1 and width is 256.
2. Language and motor scores prediction, which were shown in Fig. 3a and Fig. 3b, respectively. Generalization validation was displayed in Fig. 3c. The proposed SiM achieved the best performance compared to other methods.

Long Sequence and Efficiency Analysis. Figure 4 illustrates the performance and efficiency of the tiny-sized SiM model across different surface patching methods. In terms of MAE, SiM slightly outperforms SiT across all other patching methods except when using an Ico-2 grid, and both models exhibit decreasing MAE as the icosphere order increases (Fig. 4a). Regarding FPS (Frames Per

[2] Case marked as ('–') represents unspecified values. ('*OOM*') means out of memory. Bold intricates the best performance in MAE and MSE, respectively. P denotes Parameters and M refers to MACs.

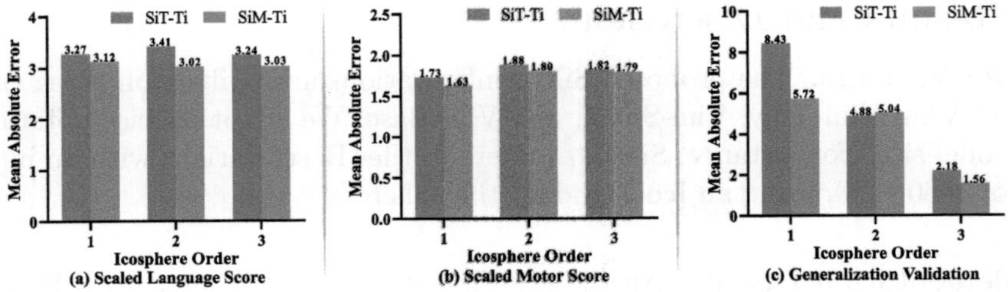

Fig. 3. Prediction performance comparison between SiT and SiM.

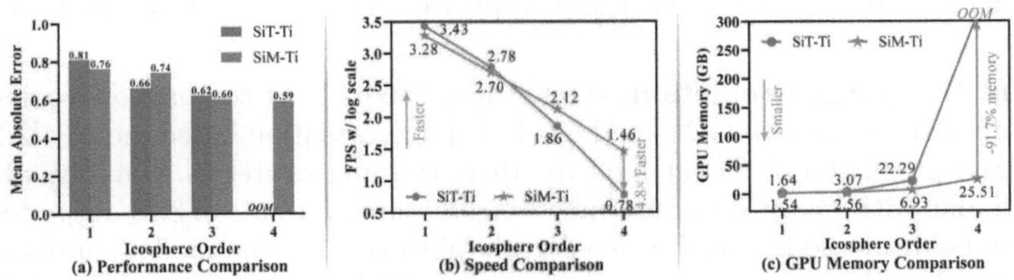

Fig. 4. PMA prediction performance and efficiency comparison between SiT and SiM.

Second), SiM is slightly slower than SiT when the icophere order is below 3 but surpasses it as the order increases (Fig. 4b). For GPU memory usage, SiM exhibits better efficiency with SiT when icosphere orders raises. Notably, when using an Ico-4 as grid, SiM is 4.8 times faster and consumes 91.7% less GPU memory compared to SiT (Fig. 4c). All experiments on efficiency analysis are conducted on a 40G A100 GPU. These results highlight the suitability of SiM for finer-grained tasks and its potential for practical clinical applications.

Cortical Regions with Significant Contributions to Age Prediction. We perform sensitive analysis [15] on the test dataset of *Subset 1* to evaluate the contribution of each vertex on cortical surface to brain age prediction, as illustrated in Fig. 5. For each vertex, we assessed four morphometric features. "All" nullifies all features, while others nullify one feature at a time per vertex.

The mean performance change for each vertex/feature was computed, normalized using Z-score. The intensity of the hot color signifies the influences of each vertex/feature. In the right hemisphere, key regions include the temporal lobe, precentral gyrus, and prefrontal and paracentral cortices. The left hemisphere shows a similar focus on the prefrontal cortex, sensory cortex, language areas, and parietal cortex when all features or only curvature are masked. Sulcal depth complements these findings by emphasizing the temporal lobe, central sulcus regions, and superior frontal and parietal areas. Cortical thickness and myelination highlight frontal regions, particularly the anterior insula, rostral middle frontal gyrus, and orbitofrontal cortex.

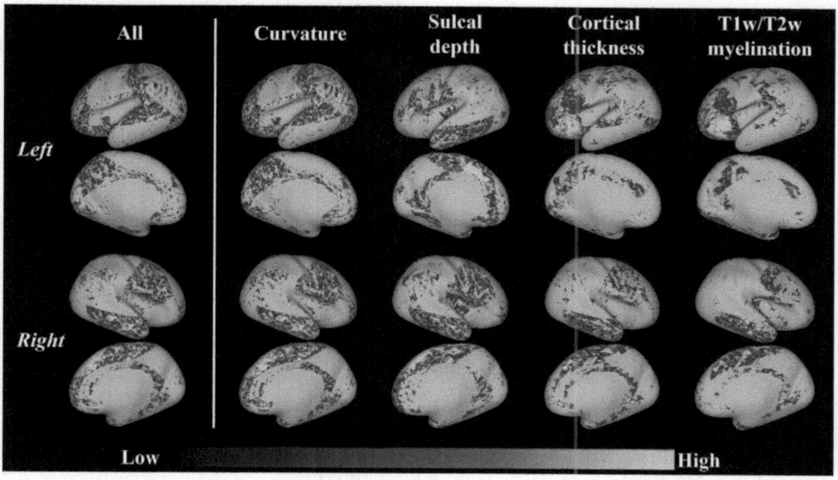

Fig. 5. Spatial distribution of informative vertices for PMA prediction.

4 Conclusion

In this study, we introduced Surface Vision Mamba (SiM), a novel vision backbone with sub-quadratic time complexity, tailored for genus-zero surfaces. We validated SiM as a more robust and efficient alternative to SiT in the challenging task of neurodevelopmental phenotype prediction from cortical surface data. Leveraging the strengths of Mamba in handling *long-sequence* and *autoregressive modeling*, we extended sequence lengths using various surface patching methods and conducted autoregressive pretraining. While SiM demonstrated sensitivity to sequence length, the benefits of autoregressive pretraining were limited, likely due to constraints of small samples. The use of longer sequences facilitated finer-grained partitioning, enhancing the ability to identify potential pathological features critical in clinical applications. Furthermore, SiM offers faster inference speeds and lower GPU memory consumption, making it both efficient and practical. Sensitivity analysis also emphasized the interpretability of SiM, highlighting its potential utility in medical research and applications.

Acknowledgments. This work was supported by the STI2030-Major Projects (2021ZD0202002), the National Natural Science Foundation of China (62202212).

Disclosure of Interests. There are no conflicts of interest to declare.

References

1. Cohen, T.S., Geiger, M., Köhler, J., Welling, M.: Spherical cnns. arXiv preprint arXiv:1801.10130 (2018)
2. Dahan, S., et al.: Surface vision transformers: attention-based modelling applied to cortical analysis. In: International Conference on Medical Imaging with Deep Learning, pp. 282–303. PMLR (2022)

3. Defferrard, M., Bresson, X., Vandergheynst, P.: Convolutional neural networks on graphs with fast localized spectral filtering. Advances in Neural Information Processing Systems 29 (2016)
4. Devlin, J.: Bert: Pre-training of deep bidirectional transformers for language understanding. arXiv preprint arXiv:1810.04805 (2018)
5. Dosovitskiy, A.: An image is worth 16x16 words: Transformers for image recognition at scale. arXiv preprint arXiv:2010.11929 (2020)
6. Fawaz, A., et al.: Benchmarking geometric deep learning for cortical segmentation and neurodevelopmental phenotype prediction. bioRxiv pp. 2021–12 (2021)
7. Gu, A., Dao, T.: Mamba: Linear-time sequence modeling with selective state spaces. arXiv preprint arXiv:2312.00752 (2023)
8. Gu, A., Goel, K., Ré, C.: Efficiently modeling long sequences with structured state spaces. arXiv preprint arXiv:2111.00396 (2021)
9. Han, K., Wang, Y., Guo, J., Tang, Y., Wu, E.: Vision gnn: an image is worth graph of nodes. Adv. Neural. Inf. Process. Syst. **35**, 8291–8303 (2022)
10. Kipf, T.N., Welling, M.: Semi-supervised classification with graph convolutional networks. arXiv preprint arXiv:1609.02907 (2016)
11. Liu, Y., et al.: Vmamba: Visual state space model (2024). https://arxiv.org/abs/2401.10166
12. Monti, F., Boscaini, D., Masci, J., Rodola, E., Svoboda, J., Bronstein, M.M.: Geometric deep learning on graphs and manifolds using mixture model cnns. In: Proceedings of the IEEE Conference on Computer Vision and Pattern Recognition, pp. 5115–5124 (2017)
13. Qi, C.R., Yi, L., Su, H., Guibas, L.J.: Pointnet++: Deep hierarchical feature learning on point sets in a metric space. Advances in neural information processing systems 30 (2017)
14. Ren, S., et al.: Autoregressive pretraining with mamba in vision. arXiv preprint arXiv:2406.07537 (2024)
15. Saltelli, A.: Sensitivity analysis for importance assessment. Risk Anal. **22**(3), 579–590 (2002)
16. Vosylius, V., et al.: Geometric deep learning for post-menstrual age prediction based on the neonatal white matter cortical surface. In: Uncertainty for Safe Utilization of Machine Learning in Medical Imaging, and Graphs in Biomedical Image Analysis: Second International Workshop, UNSURE 2020, and Third International Workshop, GRAIL 2020, Held in Conjunction with MICCAI 2020, Lima, Peru, October 8, 2020, Proceedings 2, pp. 174–186. Springer (2020)
17. Yu, W., Wang, X.: Mambaout: Do we really need mamba for vision? arXiv preprint arXiv:2405.07992 (2024)
18. Zhao, F., et al.: Spherical deformable u-net: application to cortical surface parcellation and development prediction. IEEE Trans. Med. Imaging **40**, 1217–1228 (2021). https://api.semanticscholar.org/CorpusID:231301848
19. Zhao, F., et al.: Spherical u-net on cortical surfaces: methods and applications. In: Information Processing in Medical Imaging: 26th International Conference, IPMI 2019, Hong Kong, China, June 2–7, 2019, Proceedings 26, pp. 855–866. Springer (2019)
20. Zhao, L., et al.: An attention-based hemispheric relation inference network for perinatal brain age prediction. IEEE J. Biomed. Health Inform. (2024)
21. Zhu, L., Liao, B., Zhang, Q., Wang, X., Liu, W., Wang, X.: Vision mamba: efficient visual representation learning with bidirectional state space model. In: Forty-first International Conference on Machine Learning

TemSAM: Temporal-Aware Segment Anything Model for Cerebrovascular Segmentation in Digital Subtraction Angiography Sequences

Liang Zhang[1], Xixi Jiang[2], Xiaohuan Ding[1], Zihang Huang[1], Tianyu Zhao[1], and Xin Yang[1(✉)]

[1] School of Electronic Information and Communications, Huazhong University of Science and Technology, Wuhan, China
xinyang2014@hust.edu.cn

[2] Department of Electronic and Computer Engineering, The Hong Kong University of Science and Technology, Hong Kong, China

Abstract. Digital Subtraction Angiography (DSA) is the gold standard in vascular disease imaging but it poses challenges due to its dynamic frame changes. Early frames often lack detail in small vessels, while late frames may obscure vessels visible in earlier phases, necessitating time-consuming expert interpretation. Existing methods primarily focus on single-frame analysis or basic temporal integration, treating all frames uniformly and failing to exploit complementary inter-frame information. Furthermore, existing pre-trained models like the Segment Anything Model (SAM), while effective for general medical video segmentation, fall short in handling the unique dynamics of DSA sequences driven by contrast agents. To overcome these limitations, we introduce TemSAM, a novel temporal-aware segment anything model for cerebrovascular segmentation in DSA sequences. TemSAM integrates two main components: (1) a multi-level Minimum Intensity Projection (MIP) global prompt that enhances temporal representation through a MIP-guided Global Attention (MGA) module, utilizing global information provided by MIP, and (2) a complementary information fusion module, which includes a frame selection module and a Masked Cross-Temporal Attention Module, enabling additional foreground information extraction from complementary frame. Our experimental results demonstrate that TemSAM significantly outperforms existing methods. Our code is available at https://github.com/zhang-liang-hust/TemSAM.

Keywords: Digital Subtraction Angiography · Segment Anything Model · Vessel Segmentation

1 Introduction

Digital Subtraction Angiography (DSA) [2] is highly effective in displaying vascular abnormalities, making it the gold standard for the diagnosis and treatment

J. C. Gee et al. (Eds.): MICCAI 2025, LNCS 15960, pp. 609–618, 2026.
https://doi.org/10.1007/978-3-032-04927-8_58

planning of vascular diseases such as blood flow, stenosis, and thrombosis. DSA sequences (Fig. 1-(a)) capture only transient snapshots of the dynamic vascular structure, with early frames often lacking detail in thin vessels (yellow arrow) and late frames providing poor visualization of the early phase vessels (red arrow). In contrast, its corresponding minimum intensity projection (MIP) image (Fig. 1-(b)) exhibits good visibility of the entire vessel structure. Meanwhile, unlike video object segmentation tasks in natural scenes, the apparent emergence or disappearance of vascular structures in DSA sequences reflects temporal variations in contrast agent distribution rather than physical movement, necessitating expert interpretation [4]. Consequently, the analysis of DSA sequences heavily relies on the expertise of radiologists, a process that is both time-consuming and labor-intensive. This has driven the development of fully automated methods for segmenting vascular structures in DSA sequences.

Recently, Convolutional Neural Networks (CNN)-based methods [11,13,14,16,17,19,21] have been developed for automatic vessel segmentation. Some focus on single-frame DSA analysis. For instance, Zhang et al. [21] firstly proposed a U-shaped network for cerebrovascular segmentation in single-frame DSA images. Xu et al. [17] proposed an Edge Regularization Network for cerebral vessel segmentation, using erosion-dilation for pseudo-labels generation and a Hybrid Fusion Module for refined predictions. Other studies utilize temporal information in DSA sequences. Su et al. [13] introduced CAVE, which uses a ConvGRU module to encode temporal features from 2D+time DSA series for A/V segmentation. Xie et al. [16] developed DSNet, a spatio-temporal network that incorporates MIP images as a spatial branch to enhance accuracy. However, CNN-based approaches still face challenges such as limited domain-specific training samples and restricted representation capabilities due to inadequate model capacity, making it difficult to achieve optimal vessel segmentation in DSA sequences.

The recently proposed Segment Anything Model (SAM) [6], trained on the large-scale data set SA-1B, has powerful feature extraction capabilities and can accurately focus on semantics of interest based on user prompts (e.g., points and bounding boxes). Hence, several studies have applied SAM to improve segmentation accuracy in medical scenarios [1,3,5,8,9,15,22]. For 2D medical images, adapter-based methods have been proposed to adapt SAM on medical datasets. For instance, Med-SA [15] designs a lightweight bottleneck composed of a down-projection, ReLU activation, and up-projection for fine-tuning. Several approaches have been developed to enhance the Segment Anything Model (SAM) for medical video segmentation by incorporating temporal strategies. MediViSTA-SAM [5] integrates cross-frame attention into SAM to capture temporal information. MedSAM2 [22], a generalized auto-tracking model based on SAM2 [10], treats video segmentation as an object tracking problem. Despite these advancements, existing methods either focus solely on single-frame analysis without integrating temporal information or indiscriminately treat all frames equally, overlooking the fact that certain frames, being more informative, can offer enhanced guidance for the target vessels.

To address the above challenges, we propose a novel method for vessel segmentation in DSA sequences based on SAM, named TemSAM, which adopts a local-global fusion strategy. TemSAM fully utilizes the global information provided by the MIP image and the supplementary information from complementary frames to enhance contextual representation. Specifically, *we design a multi-level MIP global prompt that enables parallel encoding of video clips and MIP image, using a MIP-guided Global Attention module to integrate global information from the MIP image.* The MIP global feature also serves as dense prompts to guide the decoding process. Additionally, *we develop a novel complementary information fusion module consisting of a frame selection module and a Masked Cross-Temporal Attention (MCTA) module to achive complementary information fusion.* The frame selection module identifies complementary frames with significant foreground differences compared to current video clip. The MCTA module then aggregates features from complementary frames and temporal information using preliminary segmentation maps, thereby facilitating high-precision vessel segmentation. In comparisons with various existing methods, we achieve state-of-the-art performance for cerebrovascular segmentation in DSA sequences, particularly excelling in the segmentation of thin vessels.

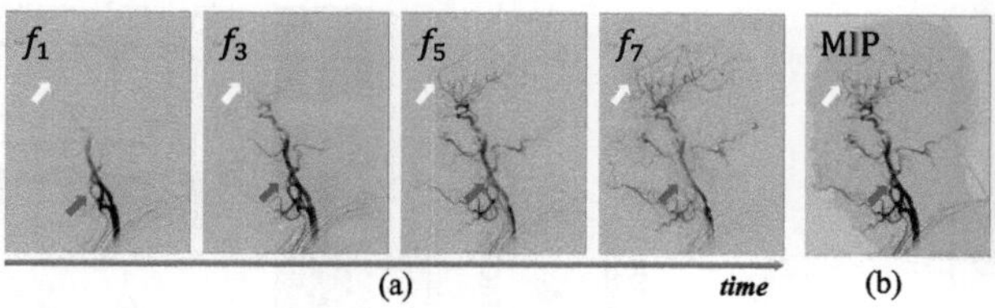

Fig. 1. Overview of the DSA sequence. (a) is a DSA sequence. (b) is the MIP image of the DSA sequence. (c) is the ground-truth of the DSA sequence.

2 Proposed Method

2.1 The Overall Architecture of TemSAM

Given a DSA sequence $x = \{x_1, x_2, \ldots, x_n\}$ and an MIP image, our objective is to predict a final segmentation map for the input sequence. As shown in Fig. 2, the DSA sequence comprises two components: current video clips $clip_t$ (in yellow bounding boxes) and remaining frames f_c (in green bounding boxes). f_c contains supplementary contextual information for $clip_t$. MIP image f_m provides comprehensive vascular structures information across all phases. We adopt a local-global fusion strategy: each $clip_t$ generate a prediction $Pred_t$ and the final prediction $Pred$ is the average of all clips' prediction, i.e., $Pred = Avg(\sum_{i=1}^{n} Pred_i)$, where n is the number of frames.

The overall architecture of TemSAM is built upon SAM. TemSAM employs a dual-branch image encoder to separately extract temporal information from local clips and global structural information from MIP images. To adapt SAM for cerebrovascular segmentation, we fine-tune it with an adapter module. In the local temporal branch and the global prompt branch, we utilize S-Adapter [15] and T-Adapter [5] separately. To effectively leverage MIP-derived global knowledge, we design a MIP-guided global attention (MGA) module between two branches, providing global context guidance for local feature extraction. Additionally, the deepest global feature F_g is fed into the mask decoder as dense prompts. Furthermore, we select the most complementary temporal information from the entire DSA sequence for the current clip and use it to refine the features of the current clip. We perform frame selection using cosine similarity, sampling the least similar one as complementary frame F_c from the remaining frames' features F_r with respect to F_t. To more effectively and precisely guide the integration of complementary information in foreground vascular regions, we introduce a two-stage hierarchical mask decoder. In the first stage, TemSAM uses SAM's original decoder to generate initial predictions M. In the second stage, we employ a Masked Cross-Temporal Attention (MCTA) module to guide feature interaction between F_t and F_c within the foreground region using mask M.

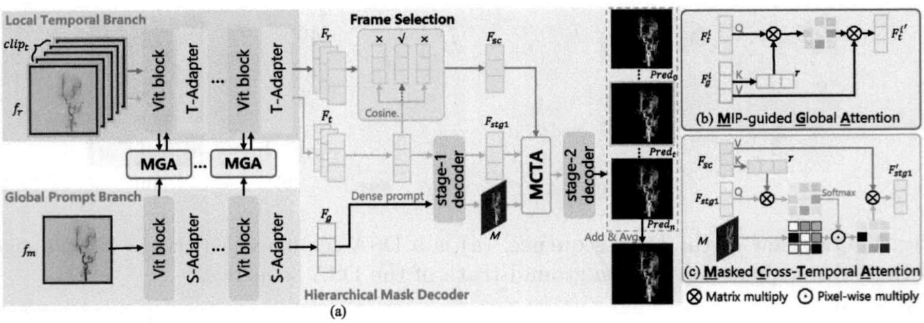

Fig. 2. (a) Overview of the proposed TemSAM. TemSAM integrates an adapter-based dual-branch image encoder coupled with a two-stage hierarchical decoder. (b) MIP-guided Global Attention module. (c) Masked Cross-Temporal Attention module.

2.2 Multi-level MIP Global Prompt

Dual-Branch Encoder. The MIP image provides vascular prior information for the DSA sequence, and thus, we leverage MIP images as prompts to enhance feature extraction of temporal information. In our method, both the MIP image and DSA sequences are fed into the encoder for parallel encoding. Specifically, the global prompt branch employs an S-Adapter [15] in 2D dimension, which sequentially performs dimension reduction, ReLU activation, and dimension restoration. The local temporal branch utilizes a T-Adapter based on multi-head self-attention (MHSA) mechanisms [5] along the temporal dimension to capture temporal information. The procedure of the two adapters can be formulated as:

$$\mathcal{A}_s(x) = RELU(xW_d)W_u \tag{1}$$

$$\mathcal{A}_t(x) = MHSA(Reshape(x \in \mathbb{R}^{(B\times T)\times H\times W\times C} \rightarrow x \in \mathbb{R}^{(B\times H\times W)\times T\times C})) \tag{2}$$

where $W_d \in \mathbb{R}^{d\times \frac{d}{4}}$ and $W_u \in \mathbb{R}^{\frac{d}{4}\times d}$ are the projection matrices, d is the dimension of the feature. Through these operations, S-Adapter and T-Adapter enable the encoder to better adapt to the feature distribution of medical image domains.

MIP-Guided Global Attention (MGA). The MGA module utilizes the global information of MIP to help aggregate the temporal information within the local window. This module establishes hierarchical interaction between global MIP features F_g^i and local temporal features F_t^i in multiple layers. MGA enables the MIP image to progressively guide and refine the extraction of local temporal information throughout the network. As shown in Fig. 2(b), we formulate F_t^i as queries while designating F_g^i as corresponding keys and values. The MGA can be formulated as:

$$F_t^{i'} = \text{softmax}(F_t^i W_q \cdot (F_g^i W_k)^T / \sqrt{d_m}) \cdot (F_g^i W_v) \tag{3}$$

where d_m is the dimension of MGA and i represents the current layer number in the encoder. $W_q \in \mathbb{R}^{d\times d_m}$, $W_k \in \mathbb{R}^{d\times d_m}$ and $W_v \in \mathbb{R}^{d\times d_m}$ are the learnable weight matrices used to project F_t^i and F_g^i to different subspaces. The output of MGA $F_t^{i'}$ is enhanced temporal feature which will be added to F_t^i and then used in subsequent encoding layer.

2.3 Complementary Information Fusion Module

Frame Selection. Local video clip often lacks some information regarding vascular visibility, while distant frames can provide necessary complementary information. Therefore, we aim to identify the most complementary frames for $clip_t$. Specifically, given current clip's mean feature $F_t^{'}$ (generated by averaging F_t) and remaining frames' features F_r , we select the feature F_c that is most dissimilar to F_t from m candidate frames. This module outputs the index of selected frame with the lowest similarity, which can be interpreted as

$$ind = argmin_{i\in[1,m]}\{Sim(F_r, F_t^{'})\} \tag{4}$$

where idx is the index of the selected complementary frame, $i \in [1, m]$ indicates the ith candidate frames, and $argmin$ is used to calculate the index of the frame with the lowest similarity, and Sim is the similarity function as follows:

$$\text{Sim}(X, Y) = (X \cdot Y) / \sqrt{dim} \tag{5}$$

According to Eq. 5, cosine distance is used to calculate the similarity between the F_r and $F_t^{'}$, and dim represents the dimension of the input feature.

Masked Cross-Temporal Attention (MCTA). The MCTA module establishes a connection between the stage-1 decoder's feature F_{stg1} and the complementary frame's feature F_c. Unlike cross-attention which attends to the global context, MCTA is guided by the predictions from the stage-1 and operates within the predicted mask, thereby further supplementing missing local features in the foreground areas. The MCTA module can be formulated as:

$$F'_{stg1} = M \odot \text{softmax}(F_{stg1}W_q \cdot (F_{sc}W_k)^T / \sqrt{d_m}) \cdot F_{stg1}W_v \tag{6}$$

which employs the probabilistic map M of stage-1 resized to the same spatial resolution as the attention map. W_q, W_k and W_v are the learnable weight matrices and d_m is the dimension of MCTA. With pixel-wise multiplication between M and the attention map, background will be ignored by multiplying a near-zero probability. The output F'_{stg1} is added to F_{stg1} and fed into the stage-2 for subsequent decoding process.

Additionally, the training loss will be applied to each stage in our mask decoder. This ensures thorough supervision and facilitates high-resolution prediction through hierarchical feature fusion.

3 Experiments

Datasets. The proposed TemSAM is evaluated on two cerebrovascular segmentation datasets DIAS [7] and DSCA [20]. DIAS annotates 60 sequences with 321 frames. DSCA includes 1792 frames from 224 sequences. All DSA sequences are resampled to a length of 8. We train on DIAS and split the training, validation, and test sets with a ratio of 3:1:2 following [7]. The DSCA dataset is used to test the generalization ability.

Evaluation Metrics. We evaluate the segmentation performance using DSC, clDice [12], IoU, Acc, and AUC. Given that accurately segmenting thin vessels presents a significant challenge, we conduct evaluation for vessels thinner than 7 pixels, following [18]. A 5-pixel search range is assigned to each thin vessel, and pixels within this range are counted for pixel-to-pixel matching. The performance is evaluated using three metrics: DSC_{thin}, Acc_{thin} and IOU_{thin}.

Implementation Details. We adopt Med-SA [15] as the baseline. The model is initialized with pre-trained ViT-B weights from SAM and implemented using Pytorch on three NVIDIA RTX A6000 GPUs. All frames are resized to 800×800 pixels with a clip length of 3. We apply data augmentation, including horizontal and vertical flipping, brightness/contrast adjustment, and random rotations. AdamW optimizer is utilized ($\beta_1 = 0.9$, $\beta_2 = 0.999$) with a StepLR scheduler that decreases the learning rate by 0.9 every 10 epochs. For a fair comparison, all models are re-implemented and trained for 200 epochs under the same settings.

Comparison with SOTA Methods. To demonstrate the superiority of our proposed method, we compare our method with some state-of-the-art (SOTA) methods, including *task-specific* methods [7,11,13], *SAM-based 2D* methods [1,8,15], and *SAM-based temporal* methods [3,5,9,22] (Tables 1 and 2).

Table 1. Comparative results on DIAS dataset.

Type	Method	DSC	Acc	IOU	AUC	clDice	DSC_{thin}	Acc_{thin}	IOU_{thin}
task-specific	VSS-Net [7]	0.7577	0.9631	0.6131	0.9762	0.6794	0.7186	0.9681	0.5641
	3D-UNet [11]	0.7618	0.9637	0.6184	0.9818	0.6879	0.7156	0.9678	0.5676
	ST-UNet [13]	0.7702	0.9630	0.6340	0.9835	0.7122	0.7147	0.9602	0.5630
SAM-based 2D	MedSAM [8]	0.6656	0.9512	0.5004	0.9339	0.5181	0.5592	0.9595	0.3907
	SAM-Med2D [1]	0.6948	0.9556	0.5344	0.9597	0.5684	0.6220	0.9599	0.4537
	Med-SA [15]	0.7107	0.9548	0.5526	0.9638	0.6277	0.6534	0.9597	0.4869
SAM-based temporal	VP-SAM [3]	0.5756	0.9406	0.4072	0.9363	0.4284	0.5188	0.9435	0.3536
	MedSAM2 [22]	0.7070	0.9549	0.5485	0.9697	0.6126	0.6321	0.9592	0.4629
	SAM+ST-Adapter [9]	0.7548	0.9605	0.6074	0.9826	0.7047	0.7011	0.9635	0.5413
	MediViSTA-SAM [5]	0.7597	0.9631	0.6142	0.9827	0.6898	0.7035	0.9659	0.5445
	ours	**0.7816**	**0.9655**	**0.6428**	**0.9836**	**0.7330**	**0.7360**	**0.9686**	**0.5834**

Table 2. Generalization comparison on DSCA Dataset.

Type	Method	DSC	Acc	IOU	AUC	clDice	DSC_{thin}	Acc_{thin}	IOU_{thin}
task specific	VSS-Net [7]	0.6464	0.9742	0.4862	0.9101	0.6596	0.5233	0.9767	0.3824
	3D-UNet [11]	0.7077	0.9739	0.5514	0.9790	0.5990	0.6394	0.9795	0.5706
	ST-UNet [13]	0.7286	0.9762	0.5766	0.9878	0.6265	0.6623	0.9789	0.5756
SAM-based 2D	MedSAM [8]	0.6055	0.9682	0.4375	0.9436	0.4062	0.4391	0.9740	0.3191
	SAM-Med2D [1]	0.6337	0.9679	0.4678	0.9577	0.4920	0.5580	0.9737	0.4665
	Med-SA [15]	0.6660	0.9702	0.5032	0.9650	0.5460	0.5948	0.9747	0.5212
SAM-based temporal	VP-SAM [3]	0.5151	0.9622	0.3500	0.9363	0.3249	0.4242	0.9654	0.3211
	MedSAM2 [22]	0.7286	0.9721	0.5753	0.9837	0.6341	0.6661	0.9766	0.6009
	SAM+ST-Adapter [9]	0.7241	0.9748	0.5703	0.9858	0.6342	0.6505	0.9769	0.6088
	MediViSTA-SAM [5]	0.7392	0.9771	0.5896	0.9882	0.6385	0.6640	0.9788	0.5865
	ours	**0.7476**	**0.9779**	**0.6044**	**0.9933**	**0.6596**	**0.6817**	**0.9800**	**0.6281**

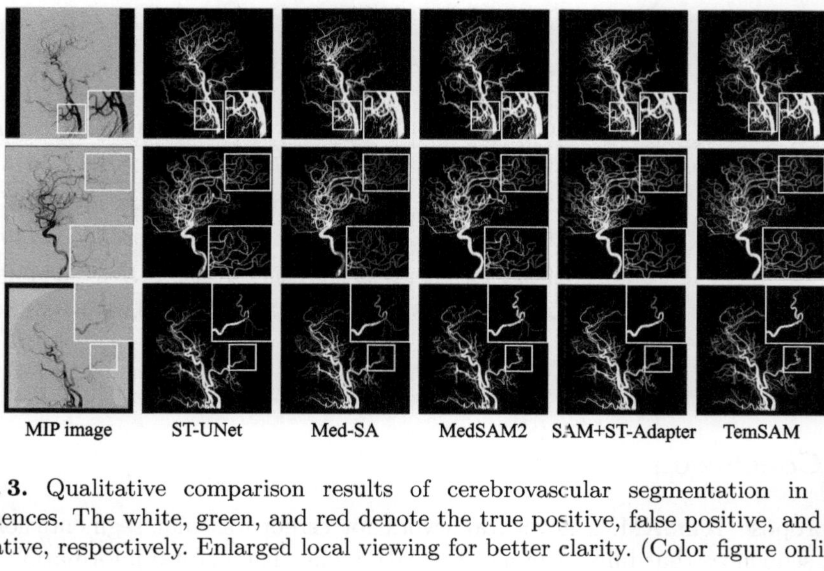

Fig. 3. Qualitative comparison results of cerebrovascular segmentation in DSA sequences. The white, green, and red denote the true positive, false positive, and false negative, respectively. Enlarged local viewing for better clarity. (Color figure online)

As quantitatively demonstrated in Tab. 1 and 2, TemSAM significantly outperforms existing SOTA task-specific methods across all evaluation metrics. These improvements can be primarily attributed to the multi-level MIP prompt, the advanced complementary information fusion module, and the effective integration of SAM's inherent design advantages. Furthermore, the 2D variant of SAM exhibits significantly inferior performance, underscoring the critical role of temporal information in DSA sequences. Notably, our approach surpasses existing SAM-based temporal methods, demonstrating superior capability in both temporal information utilization and contextual feature extraction.

Qualitative results, as shown in Fig. 3, visually demonstrate TemSAM's performance advantages over other methods. TemSAM effectively preserves vascular connectivity and enhances the detection of thin vessels in low-contrast regions. Meanwhile, TemSAM achieves an optimal balance between false positives and false negatives, demonstrating remarkable robustness in challenging clinical scenarios (1st row).

Effectiveness of Each Component in TemSAM. We conducted ablation studies by incrementally integrating each component (i.e. Dual-Branch Encoder (DBE), MGA Module, and MCTA Module) into the backbone. Additionally, we compare our Local-Global Fusion (LGF) strategy backbone with the Global Fusion (GF) strategy, which inputs the entir DSA sequence and directly outputs one final segmentation map. The results demonstrate that the LGF strategy preforms superior to GF strategy. This superiority can be attributed to the local video clip's ability to integrate local temporal information. In contrast, a global views may miss certain vessels only visible for a short time. Meanwhile, incorporating any single component of TemSAM significantly enhances the segmentation performance. Notably, combining all components achieve the optimal segmentation accuracy, especially for thin vessels (DSC_{thin}+1.92%, Acc_{thin}+0.16%, and IOU_{thin}+2.32%) (Table 3).

Table 3. Ablation study on different component combinations of TemSAM.

DBE	MGA	MCTA	DSC	Acc	IOU	AUC	clDice	DSC_{thin}	Acc_{thin}	IOU_{thin}
GIO Backbone			0.7593	0.9571	0.6003	0.9819	0.7079	0.7019	0.9612	0.5418
✗	✗	✗	0.7702	0.9644	0.6276	**0.9839**	0.7068	0.7168	0.9670	0.5602
✓	✗	✗	0.7748	0.9633	0.6335	0.9834	0.7266	0.7285	0.9665	0.5740
✓	✓	✗	0.7755	0.9634	0.6342	0.9834	0.7277	0.7300	0.9665	0.5759
✓	✓	✓	**0.7816**	**0.9655**	**0.6428**	0.9836	**0.7330**	**0.7360**	**0.9686**	**0.5834**

4 Conclusion

We propose TemSAM, a novel framework for cerebrovascular segmentation in DSA sequences that fully leverages the MIP's global structural information and complementary frames' temporal information. By introducing a multi-level MIP global prompt and a complementary information fusion module, TemSAM adaptively refines segmentation through structural priors encoded from the MIP image while aggregating contextually complementary information to enhance feature extraction. Experimental results demonstrate that TemSAM significantly outperforms existing methods.

Acknowledgments. This work was supported in part by the Natural Science Foundation of China under Grant 62472184, and in part by the Fundamental Research Funds for the Central Universities.

Disclosure of Interests. The authors have no competing interests to declare that are relevant to the content of this article.

References

1. Cheng, J., et al.: Sam-med2d. arXiv preprint arXiv:2308.16184 (2023)
2. Chng, S.M., Alvarez, H., Rodesch, G., Lasjaunias, P.: Arteriovenous malformations of the brain and spinal cord, pp. 595–608, January 2002. https://doi.org/10.1016/b978-0-323-03354-1.50048-1
3. Fang, Z., Liu, Y., Wu, H., Qin, J.: Vp-sam: taming segment anything model for video polyp segmentation via disentanglement and spatio-temporal side network. In: European Conference on Computer Vision, pp. 367–383. Springer (2024)
4. Heiss, W.D., Forsting, M., Diener, H.C.: Imaging in cerebrovascular disease. Curr. Opin. Neurol. **14**(1), 67–75 (2001)
5. Kim, S., et al.: Medivista-sam: Zero-shot medical video analysis with spatio-temporal sam adaptation. arXiv preprint arXiv:2309.13539 (2023)
6. Kirillov, A., et al.: Segment anything. In: Proceedings of the IEEE/CVF International Conference on Computer Vision, pp. 4015–4026 (2023)
7. Liu, W., et al.: Dias: a dataset and benchmark for intracranial artery segmentation in dsa sequences. Medical Image Analysis, p. 103247 (2024)
8. Ma, J., He, Y., Li, F., Han, L., You, C., Wang, B.: Segment anything in medical images. Nat. Commun. **15**(1), 654 (2024)
9. Pan, J., Lin, Z., Zhu, X., Shao, J., Li, H.: St-adapter: parameter-efficient image-to-video transfer learning. Adv. Neural. Inf. Process. Syst. **35**, 26462–26477 (2022)
10. Ravi, N., et al.: Sam 2: Segment anything in images and videos. arXiv preprint arXiv:2408.00714 (2024)
11. Ronneberger, O., Fischer, P., Brox, T.: U-net: Convolutional networks for biomedical image segmentation. In: Medical image computing and computer-assisted intervention–MICCAI 2015: 18th international conference, Munich, Germany, October 5-9, 2015, proceedings, part III 18, pp. 234–241. Springer (2015)
12. Shit, S., et al.: cldice-a novel topology-preserving loss function for tubular structure segmentation. In: Proceedings of the IEEE/CVF Conference on Computer Vision and Pattern Recognition, pp. 16560–16569 (2021)
13. Su, R., et al.: Cave: cerebral artery-vein segmentation in digital subtraction angiography. Comput. Med. Imaging Graph. **115**, 102392 (2024)
14. Van Asperen, V., et al.: Automatic artery/vein classification in 2d-dsa images of stroke patients. In: Medical Imaging 2022: Image-Guided Procedures, Robotic Interventions, and Modeling, vol. 12034, pp. 366–377. SPIE (2022)

15. Wu, J., et al.: Medical sam adapter: adapting segment anything model for medical image segmentation. arXiv preprint arXiv:2304.12620 (2023)
16. Xie, Q., et al.: Dsnet: a spatio-temporal consistency network for cerebrovascular segmentation in digital subtraction angiography sequences. In: International Conference on Medical Image Computing and Computer-Assisted Intervention, pp. 199–208. Springer (2024)
17. Xu, W., et al.: Ernet: edge regularization network for cerebral vessel segmentation in digital subtraction angiography images. IEEE J. Biomed. Health Inf. (2023)
18. Yan, Z., Yang, X., Cheng, K.T.: Joint segment-level and pixel-wise losses for deep learning based retinal vessel segmentation. IEEE Trans. Biomed. Eng. **65**(9), 1912–1923 (2018)
19. Zhang, J., Xie, Y., Wang, Y., Xia, Y.: Inter-slice context residual learning for 3d medical image segmentation. IEEE Trans. Med. Imaging **40**(2), 661–672 (2020)
20. Zhang, J., et al.: Dsca: a digital subtraction angiography sequence dataset and spatio-temporal model for cerebral artery segmentation. IEEE Trans. Med. Imaging (2025)
21. Zhang, M., et al.: A neural network approach to segment brain blood vessels in digital subtraction angiography. Comput. Methods Programs Biomed. **185**, 105159 (2020)
22. Zhu, J., Qi, Y., Wu, J.: Medical sam 2: segment medical images as video via segment anything model 2. arXiv preprint arXiv:2408.00874 (2024)

tHPM-LDM: Integrating Individual Historical Record with Population Memory in Latent Diffusion-Based Glaucoma Forecasting

Yuheng Fan[1], Jianyang Xie[1], Yimin Luo[2], Yanda Meng[3], Savita Madhusudhan[4], Gregory Y.H. Lip[5], Li Cheng[6], Yalin Zheng[1], and He Zhao[1(✉)]

[1] Eye and Vision Sciences Department, University of Liverpool, Liverpool, UK
He.Zhao@liverpool.ac.uk

[2] Department of Radiology, Weill Cornell Medicine, New York City, USA

[3] Computer Science Department, University of Exeter, Exeter, UK

[4] St Paul's Eye Unit, Liverpool University Hospitals NHS Foundation Trust, Liverpool, UK

[5] Liverpool Centre for Cardiovascular Science, University of Liverpool, Liverpool, UK

[6] Department of Electrical and Computer Engineering, University of Alberta, Edmonton, Canada

Abstract. Longitudinal medical records offer crucial insights into disease progression, including structural changes and dynamic evolution, essential for clinicians in treatment planning. However, existing disease forecasting methods are hindered by irregular data collection intervals, negligence in inter-patient relationships, and a lack of case-reference capabilities. We introduce tHPM-LDM, a glaucoma forecasting framework leveraging continuous-time attention within a historical condition module to capture disease progression from irregularly acquired records. Notably, our approach integrates population memory, enabling personalized forecasting through relevant population patterns. Empirical evaluations on the SIGF glaucoma longitudinal dataset demonstrate the significant improvements of our approach in image prediction and category consistency compared to state-of-the-art methods. Furthermore, our approach provides interpretable individual-population patterns and showcases robust performance despite missing visits.

Keywords: Longitudinal Record · Irregular-time Forecasting · Latent Diffusion Model · Population Memory

1 Introduction

Glaucoma, an irreversible and leading cause of blindness, presents significant diagnostic challenges due to its chronic, often asymptomatic nature, requiring

J. C. Gee et al. (Eds.): MICCAI 2025, LNCS 15960, pp. 619–629, 2025.
https://doi.org/10.1007/978-3-032-04927-8_59

extensive long-term monitoring [13]. Given its relatively low global prevalence (estimated 3.54% among individuals aged 40–80 [20]), developing models for accurate glaucoma forecasting from longitudinal data is essential for improving early detection and reducing healthcare costs. However, the irregular acquisition patterns inherent in longitudinal medical data present a substantial obstacle for deep learning models in capturing the dynamic nature of disease progression. While traditional methods address this by aggregating data at specific time scales [19] or estimating missing data [23], these methods may underutilize longitudinal data and weaken temporal dependencies.

Some methods focusing on architecture improvement [1], such as previous glaucoma forecasting models: DeepGF [14] employs an LSTM-based method to process fundus images for glaucoma forecasting for next visit with an indefinite time span; GLIM-Net [9] introduces a transformer that integrates image and label sequences with time intervals, enabling label predication at a given time; MSTF [21] further extends into a multi-scale framework for improved spatial dependence modeling. However, these methods primarily focus on category forecasting and lack visual prediction capabilities.

More recently, several works have been proposed for longitudinal image forecasting and completion [22,26]. C2FLDM [25], a coarse-to-fine latent diffusion model (LDM), employs a two-stage framework for glaucoma image and category forecasting. BrLP [16] incorporates prior knowledge and progression patterns through Disease Course Mapping [12] in an LDM-based Alzheimer's disease forecasting model, which are constrained by its reliance on accurate volumetric brain segmentation. These methods model the temporal relationships in a simplistic manner and ignore related disease progression pattens across the population.

In this paper, we propose a glaucoma forecasting framework called continuous-time **H**istorical and **P**opulation **M**emory-conditioned **L**atent **D**iffusion **M**odel (tHPM-LDM). We employ continuous-time modeling to address the challenge of irregularly spaced historical records over time, while incorporating population evolution to enhance personalized image and category forecasting. The main contributions are: 1) to our best knowledge, this is the first disease forecasting approach that integrates disease population information with image generation; 2) a multiscale continuous-time transformer is introduced to capture dynamic evolution within the irregularly acquired records, improves visual forecasting quality and robustness toward missing visits; 3) a retrievable memory module is proposed for individual interaction with the disease population, mimicking how clinicians reference similar cases during diagnosis.

2 Methods

Given a longitudinal dataset $D = \{R_{1:T}^{(m)} = [(x_i, l_i, t_i)]_1^T\}_{m=1}^M$ containing T visits from M participants, where $x_i \in \mathbb{R}^{W \times H \times 3}$ represents the image from the i-th visit, $l_i \in \{0, 1\}$ denotes the disease label (0 for healthy, 1 for glaucoma), and $t_i \in \mathbb{R}$ indicates the time interval between the i-th visit and the first visit. Our goal is to train a model to leverage the historical information from time steps

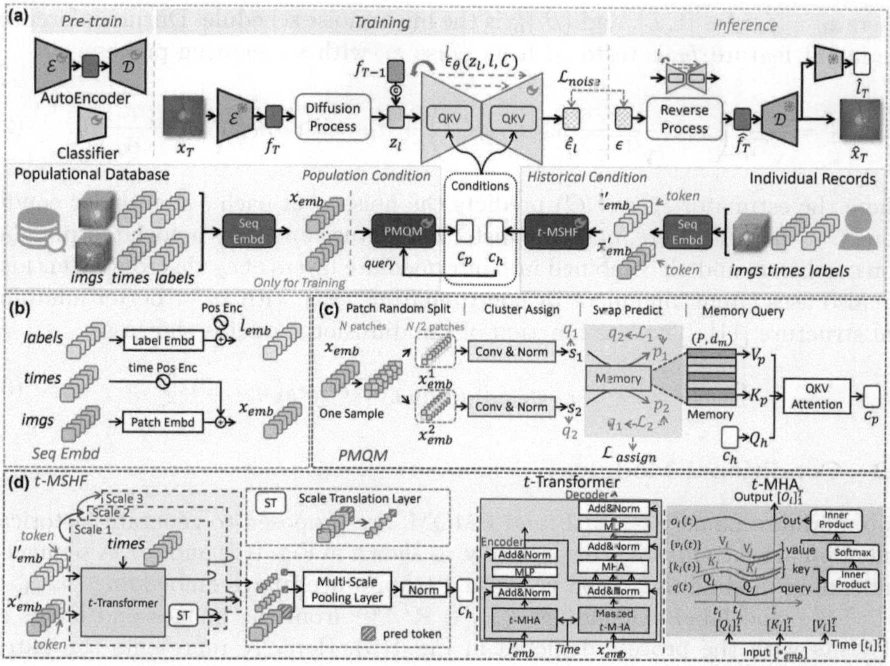

Fig. 1. Overview of our framework. (a) depicts the training and inference processes. (b) presents the embedding module, which generates embedding from longitudinal records. (c) illustrates the pipeline of PMQM in generating the population condition c_p. (d) outlines the model structure of t-MSHF in extracting the historical condition c_h.

$1 : T-1$ to forecast the participant's future image $\hat{x}_T$ and its category $\hat{l}_T$. To achieve this, we propose a LDM that generates future images, conditioned on both individual history c_h and population reference c_p. The generated images are subsequently fed into a pretrained classifier for category prediction. Figure 1 provides an overview of our approach. In the follow subsections, we introduce each of the modules in detail.

2.1 Disease Forecasting with Conditional Latent Diffusion

Our forecast framework is presented in Fig. 1(a), which contains a pre-trained auto-encoder ($\mathcal{E}$-$\mathcal{D}$) for image x_T and latent feature f_T translation, and a diffusion model for future latent prediction. In the training stage, the latent feature f_T is decomposed into noisy status $\{z_l\}_0^L$ with L-steps diffusion as:

$$z_l = \sqrt{\bar{\alpha}_l} z_0 + \sqrt{1 - \bar{\alpha}_l}\epsilon, \text{ with } \epsilon \sim \mathcal{N}(\mathbf{0}, \mathbf{I}), \; \bar{\alpha}_l = \prod_{i=1}^{l} \alpha_l, \; \alpha_l = 1 - \beta_l, \tag{1}$$

where $z_0 = f_T$, $l \in [1, L]$, and $\{\beta_i\}_0^L$ is the linear noise schedule. During inference, the latent feature f_T is restored from noise z_L with a denoising process:

$$z_{l-1} = \frac{1}{\sqrt{\alpha_l}}\left(z_l - \frac{1-\alpha_l}{\sqrt{1-\bar{\alpha}_l}}\epsilon_\theta(z_l, l, C)\right) + \sigma_l\epsilon \text{ with } \sigma_l = \sqrt{\frac{1-\bar{\alpha}_{l-1}}{1-\bar{\alpha}_l}\beta_l}, \tag{2}$$

where the estimator $\epsilon_\theta(z_l, l, C)$ predicts the noise ϵ_l at each l-step under conditions $C = \{c_h, c_p\}$. Here, c_h is the historical feature and c_p is associated population condition, and all combined into intermediate layers of ϵ_θ via cross-attention. We also uses latent alignment by concatenating f_{T-1} with z_l for better anatomical structure [11]. The loss function of the diffusion model is shown as:

$$\mathcal{L}_{\text{noise}} := \mathbb{E}_{\mathcal{E}(x_T),\epsilon\sim\mathcal{N}(0,1),l\sim\mathcal{U}(0,L)}\left[\|\epsilon - \epsilon_\theta(z_l, l, C)\|_2^2\right] \tag{3}$$

2.2 Conditional Modules

Two modules, namely t-MSHF and PMQM, are proposed to generate historical and population conditions, respectively, as shown in Fig. 1(d) and (c). A sequence embedding module `Seq Embd` is utilized to produce image embeddings $x_{\text{emb}} \in \mathbb{R}^{T\times N\times d_m}$ and label embeddings $l_{\text{emb}} \in \mathbb{R}^{T\times d_m}$ from the images and labels of all visits with the process depicted in Fig. 1(b). Here, N represents the patch number and d_m denotes the embedding dimension. These embeddings serve as inputs for t-MSHF and are used to learn the memory in PMQM. Notably, to prevent future information leakage, the last visit embeddings of x_{emb} and l_{emb} are replaced with learnable tokens for t-MSHF.

Continuous-time Multi-scale Historical Feature Module (t-MSHF). Our t-MSHF consists of t-Transformers with continuous-time attention module (t-MHA) at different scales, which captures multiscale dynamic information from individual historical records. Inspired by [4], t-MHA uses continuous functions in modeling the attention of dynamic evolution based on discrete observations. More precisely, the input embeddings at all visit times are first copied to observed sequences $[Q_i]_1^T$, $[K_i]_1^T$ and $[V_i]_1^T$. The query $q(t)$ is an interpolation function (e.g. bilinear interpolation) with observed point $q(t_i) = Q_i$. And the continuous key and value are sets of functions $\{k_i(t)\}$ and $\{v_i(t)\}$ based on each observation i. Neural networks f_{θ_k} and f_{θ_v} estimate the dynamics of key and value, where the changes of $\{k_i(t)\}$ and $\{v_i(t)\}$ described as linear ODEs [3] as follow:

$$\frac{\mathrm{d}k_i}{\mathrm{d}t} = f_{\theta_k}(k_i(t), t),\ k_i(t_i) = K_i \qquad \frac{\mathrm{d}v_i}{\mathrm{d}t} = f_{\theta_v}(v_i(t), t),\ v_i(t_i) = V_i. \tag{4}$$

Observed Q_i, K_i at time t_i are initial conditions of ODEs, and we solve these ODEs by Runge-Kutta-4 algorithm. For any visit time t, the i-th continuous correlation of query and key functions is defined as the scaled inner product of $q(t)$ and $k_i(t)$, i.e. $\alpha_i(t) := (\int_{t_i}^t q(\tau)\cdot k_i(\tau)^\mathsf{T} d\tau)/(t - t_i)$ and $\alpha_i(t_i) = q(t_i)\cdot k_i(t_i)^\mathsf{T}$ for each t_i to ensure continuity. The expected value function based on i-th observation is $\hat{v}_i(t) = (\int_{t_i}^t v_i(\tau)d\tau)/(t-t_i)$. We approximate these integrations by Gauss-Legendre Quadrature. Therefore, the output function is computed as:

$$O(t) = \sum_{i=1}^{T} \text{softmax}(\alpha_i(t)/\sqrt{d_k}) \cdot \hat{v}_i(t) \tag{5}$$

where d_k is the dimension of inputs. Discrete output sequence $[O_i]_{i=1}^{T}$ is computed based on Eq. 5 according to time $[t_i]_1^T$, which will be used for next layers.

Our t-Transformer, equipped the t-MHA, will handle label embeddings in the Transformer encoder and process image embeddings with its decoder. Additionally, the t-Transformer will be applied iteratively in a multi-stage manner with a scale translation layer to enable multi-scale learning. The final historical condition $c_h \in \mathbb{R}^{1\times d_m}$ is obtained by the average of different scales.

Population Memory Query Module (PMQM). In practice, clinicians will evaluate the patient's current status by considering both individual medical histories and the knowledge of disease progression from other cases. Motivated by this practice, we propose PMQM, which initializes a population memory $\mathcal{M}_p \in \mathbb{R}^{P\times d_m}$ to capture evolution patterns of P clusters of representative patients. Following the SwAV [2], we apply online contrastive learning with swapping strategy to update the memory from the disease population. As shown in Fig. 1(c), one image embedding x_{emb} is first divided into two sub-embeddings randomly, resulting in $x^1_{\text{emb}}, x^2_{\text{emb}} \in \mathbb{R}^{T\times N/2\times d_m}$ and further projected to $s_1, s_2 \in \mathbb{R}^{1\times d_m}$. The loss function to update the memory is shown as:

$$\mathcal{L}_{\text{assign}}(s_1, s_2) := -\frac{1}{2}\sum(q_1 \log p_2 + q_2 \log p_1), \tag{6}$$

where $p_* = \text{softmax}(s_*^{\mathsf{T}} \cdot \mathcal{M}_p)$ represents the predicted assign probability, and q_* is the optimal assign probability computed by the Sinkhorn-Knopp algorithm [5]. By minimizing the loss over the training samples, the memory $\mathcal{M}_p$ learns to represent the training participants into P clusters with centers stored in the memory. Moreover, we can build an offline cache using predicted assignment p that stores the probability of the patient from each cluster, providing a retrievable strategy to determine the top-k relevant clusters and their corresponding participants in certain patient forecasting. Finally, the condition c_p is obtained by the cross attention of historical condition c_h (Query) and the learned memory $\mathcal{M}_p$ (Key, Value) by the following equation:

$$c_p = \text{softmax}\left((Q_h \cdot K_p^{\mathsf{T}})/\sqrt{d_m}\right) V_p \tag{7}$$

2.3 Learning Conditional LDM and Memory All-in-once

After pre-training the autoencoder ($\mathcal{E}$-$\mathcal{D}$) and classifier $\mathcal{P}$, our methods can learn the LDM with condition modules all-in-once. We first perform $W = 1500$ iterations of warm-up training on the noise estimator ϵ_θ and t-MSHF with loss in Eq. 3 to prevent the random query for PMQM. After that, joint training is conducted using the total loss:

$$\mathcal{L}_{total} := \mathcal{L}_{\text{noise}} + \lambda_m \mathcal{L}_{assign} \tag{8}$$

where λ_m is the balanced hyper-parameter, empirically setting to 0.1.

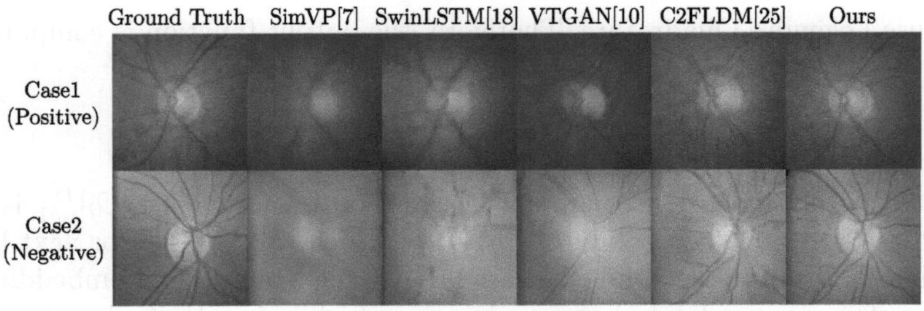

Fig. 2. Two image prediction cases using different methods.

3 Experiments

3.1 Setup Details

Dataset. We conducted experiments on the glaucoma longitudinal dataset named SIGF [14]. Following the conventional process [9], we resize fundus images to 256 × 256, compute the time interval in years, and collect $T = 6$ consecutive visits as one clip, resulting 1146 training, 16 validation and 351 testing clips.

Settings. The autoencoder ($\mathcal{E}$-$\mathcal{D}$) is a U-Net structure [17] with latent dimension 32×32×4 and pre-trained on two external glaucoma dataset named LAG [15] and ACRIMA [6]. The classifier $\mathcal{P}$ is a ResNet50 with two hidden layers and a binary head per-trained on SIGF training set, LAG and ACRIMA. The embedding dimension d_m is 128. More details can refer to our code[1].

Metrics and Baseline Methods. We take 6th image in test clip as ground truth and evaluate the visual quality and category consistency of all methods. For future image generation, we use PSNE, SSIM and MSE to measure image quality, along with semantic similarity metrics such as FID [8] and LPIPS [24]. For category forecasting, we report the performance of pre-trained classifier using accuracy, sensitivity, specificity, and AUC. Cohen's Kappa (Kappa) is utilized to compare the forecasting consistency between ground truth and generative results. We compare our approach with four baseline methods: two video models for future image generation from image sequence (CNN-based SimVP [7] and Transformer-LSTM based SwinLSTM [18]); one GAN-based image-to-image translation method VTGAN [10]; and one LDM-based sequence-to-image method C2FLDM [25]. All methods are trained using their recommended settings with identical batch sizes for a maximum of 700 epochs until convergence. Quantitative results are reported as the mean and std over five runs.

3.2 Experimental Results

Results on image prediction. Qualitative and quantitative results are shown in Fig. 2 and Table 1, respectively. Two cases generated by different methods are

[1] The implementation is available at https://github.com/yhf42/tHPM-LDM.

Table 1. Quantitative results of different methods for image prediction.

Model	PSNR ↑	SSIM ↑	MSE (10^2) ↓	FID ↓	LPIPS ↓
SimVP [7]	19.02±0.02	0.70±0.00	11.64±0.09	298.75±0.75	0.50±0.00
SwinLSTM [18]	19.01±0.04	0.67±0.00	11.35±0.06	300.23±0.86	0.51±0.00
VTGAN [10]	18.42±0.01	0.65±0.00	13.67±0.03	189.33±1.15	0.47±0.00
C2FLDM [25]	18.32±0.06	0.65±0.00	13.91±0.16	210.67±0.71	0.50±0.00
Ours	**19.30**±0.03	**0.71**±0.00	**10.40**±0.05	**141.19**±1.32	**0.44**±0.00

Table 2. Quantitative results of different methods for category forecasting.

Model	Accuracy(%)	Specificity(%)	Sensitivity(%)	AUC(%)	Kappa ↑
Ground Truth	77.09±1.48	77.31±1.79	72.50±5.00	83.03±0.35	-
SimVP [7]	29.12±1.25	26.03±1.31	93.75±0.00	59.48±1.57	0.14±0.02
SwinLSTM [18]	12.65±0.39	9.25±0.46	83.75±3.06	42.60±1.28	-0.01±0.01
VTGAN [10]	70.14±1.32	70.39±1.15	61.00±6.09	74.07±1.99	0.35±0.02
C2FLDM [25]	49.86±0.51	47.70±0.40	**95.00**±4.68	75.22±0.82	0.29±0.02
Ours	**74.02**±1.16	**74.27**±1.02	68.75±5.59	**82.06**±1.23	**0.45**±0.02

displayed in Fig. 2, where Cases 1 has positive findings in real diagnosis while Cases 2 is negative. Compared to the ground truth images shown in the left-most column, our predicted images demonstrate the best semantic similarity and clearer OC/OD and vessel structures among other methods. Although SimVP and SwimLSTM utilize historical images for image prediction, their results fail to show a clear fundus structure. VTGAN and C2FLDM generates images with better semantic information, but unstable contrast and structural variation poses challenges in identifying glaucoma. Table 1 demonstrates the superior performance of our model in both image quality and semantic consistency. Notably, while SimVP and SwinLSTM show relatively better PSNR, SSIM, and MSE since the low clarity may reduce prediction errors, their blurry structure leads to semantic dissimilarity, such as higher FID and LPIPS, making them inadequate for further analysis.

Results on category forecasting. Table 2 first summarizes the performance of pre-trained classifier $\mathcal{P}$ on test set ground truth images, which regard as the upper bound of the category forecast. Compared to other generative methods, our approach achieves a more balanced performance in glaucoma forecasting and has the highest consistency with the prediction on ground truth. The predicted images from SimVP and SwinLSTM confused the classifier's judgment due to the lack of fine-grain structure. VTGAN's results lead to a higher false-negative rate since the higher brightness increases the area of the optic cup. And the classifier

Table 3. Results of Ablation Study. LA: Latent Alignment Strategy, MS: Multi-Scale Translation, tM: t-MHA, QM: Population Memory Query Module.

LA	MS	tM	QM	PSNR ↑	SSIM ↑	MSE (10^2) ↓	FID ↓	LPIPS ↓	Kappa ↑
✓	✗	✗	✗	18.16±0.10	0.71±0.00	14.78±0.31	176.30±2.19	0.48±0.00	0.35±0.02
✓	✓	✗	✗	18.44±0.06	0.71±0.00	14.26±0.22	155.09±1.22	0.46±0.00	0.44±0.02
✓	✓	✓	✗	18.81±0.10	**0.72**±0.00	12.17±0.29	149.32±1.10	0.45±0.00	**0.45**±0.01
✓	✓	✓	✓	**19.30**±0.03	0.71±0.00	**10.40**±0.05	**141.19**±1.32	**0.44**±0.00	0.45±0.02

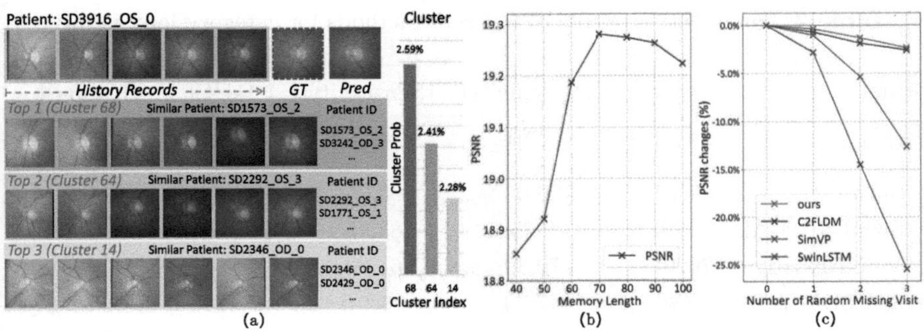

Fig. 3. (a) Memory retrieval results for forecasting of one glaucoma patient. (b) The ablation study of the memory length P in PMQM. (c) The relative changes (%) of PSNR under different number of missing visit.

tends to predict C2FLDM's results as positive due to their blurred optic disc boundaries.

3.3 Ablation Study

Our ablation study results are shown in Table 3. Latent alignment preserves basic anatomical structures, serving as a default setting. Multi-scale translation enhances performance by capturing spatial features at different scales, and t-MHA enables the LDM to better capture the dynamic progression within records. PMQM results in the best image quality by leveraging population patterns for individual prediction, though mirroring real diagnosis that considering population trends may influence the judgment. Figure 3(b) illustrates the influence of memory length P in PMQM, where $P = 70$ achieves the best result.

Population Memory Retrieve. Another benefit of our proposed population memory is that its ability to retrieve the high relevant cases. This enables our method with stronger interpretability in patient-level disease progression forecasting. Figure 3 (a) displays one example of the memory retrieval, which demonstrates that our method can successfully query groups of participants with similar structural characteristics and evolutionary trajectories.

Random Missing Visit Analysis. To evaluate the performance of sequence-based models under varying levels of missing visits, we randomly mask out different numbers of visits (2nd to 4th) to simulate records missing scenario. Figure 3(c) shows the relative changes in PSNR of SimVP, SwinLSTM, C2FLDM and our method under different levels of missing visits. The results demonstrate that conditional LDM-based methods (C2FLDM and ours) achieve better prediction performance with missing data. Furthermore, our model exhibits better robustness toward missing visits thanks to the t-MHA's capacity in modeling the irregular series and the sematic reference in population memory.

4 Conclusion

In this paper, we propose a conditional latent diffusion-based glaucoma forecasting framework. Specifically, we design t-MSHF, which incorporates continuous-time attention into multi-scale transformers to better capture the spatiotemporal dynamics of irregularly acquired records. We further enable population-assisted forecasting by querying individual-related features in PMQM. These conditions are jointly learned with the LDM to improve glaucoma image and category forecasting. Furthermore, our method enables the retrieval of participants with similar progression patterns to a given individual. This individual-population interaction does not rely on expertise in a specific disease and has potential for analyzing other diseases with longitudinal follow-up.

Acknowledgments. This work is funded by the UK Engineering and Physical Sciences Research Council (grant number EP/X01441X/1) and Medical Research Council (grant number MR/Z506175/1).

Disclosure of Interests. The authors have no competing interests to declare that are relevant to the content of this article.

References

1. Bridge, J., Harding, S., Zheng, Y.: End-to-End deep learning vector autoregressive prognostic models to predict disease progression with uneven time intervals. In: Papież, B.W., Yaqub, M., Jiao, J., Namburete, A.I.L., Noble, J.A. (eds.) MIUA 2021. LNCS, vol. 12722, pp. 517–531. Springer, Cham (2021). https://doi.org/10.1007/978-3-030-80432-9_38
2. Caron, M.: Unsupervised learning of visual features by contrasting cluster assignments. Adv. Neural. Inf. Process. Syst. **33**, 9912–9924 (2020)
3. Chen, R.T., Rubanova, Y., Bettencourt, J., Duvenaud, D.K.: Neural ordinary differential equations. Adv. Neural Inf. Process. Syst. **31** (2018)
4. Chen, Y.: Contiformer: continuous-time transformer for irregular time series modeling. Adv. Neural. Inf. Process. Syst. **36**, 47143–47175 (2023)
5. Cuturi, M.: Sinkhorn distances: lightspeed computation of optimal transport. Adv. Neural Inf. Process. Syst. **26** (2013)
6. Diaz-Pinto, A.: CNNs for automatic glaucoma assessment using fundus images: an extensive validation. Biomed. Eng. Online **18**, 1–19 (2019)

7. Gao, Z., Tan, C., Wu, L., Li, S.Z.: Simvp: simpler yet better video prediction. In: Proceedings of the IEEE/CVF conference on computer vision and pattern recognition, pp. 3170–3180 (2022)
8. Heusel, M., Ramsauer, H., Unterthiner, T., Nessler, B., Hochreiter, S.: GANs trained by a two time-scale update rule converge to a local Nash equilibrium. Adv. Neural Inf. Process. Syst. **30** (2017)
9. Hu, X.: Glim-net: chronic glaucoma forecast transformer for irregularly sampled sequential fundus images. IEEE Trans. Med. Imaging **42**(6), 1875–1884 (2023)
10. Kamran, S.A., Hossain, K.F., Tavakkoli, A., Zuckerbrod, S.L., Baker, S.A.: VTGAN: semi-supervised retinal image synthesis and disease prediction using vision transformers. In: Proceedings of the IEEE/CVF International Conference on Computer Vision (ICCV) Workshops, pp. 3235–3245 (2021)
11. Kim, M., Liu, F., Jain, A., Liu, X.: DCFace: synthetic face generation with dual condition diffusion model. In: Proceedings of the IEEE/CVF conference on computer vision and pattern recognition, pp. 12715–12725 (2023)
12. Koval, I., et al.: Ad course map charts Alzheimer's disease progression. Sci. Rep. **11**(1), 8020 (2021)
13. Leite, M.T., Sakata, L.M., Medeiros, F.A.: Managing glaucoma in developing countries (2011)
14. Li, L., Wang, X., Xu, M., Liu, H., Chen, X.: DeepGF: Glaucoma forecast using the sequential fundus images. In: Martel, A.L., et al. (eds.) Medical Image Computing and Computer Assisted Intervention – MICCAI 2020, vol. 12265, pp. 626–635. Springer International Publishing, Cham (2020). https://doi.org/10.1007/978-3-030-59722-1_60
15. Li, L., Xu, M., Wang, X., Jiang, L., Liu, H.: Attention based glaucoma detection: a large-scale database and CNN model. 2019 IEEE. In: CVF Conference on Computer Vision and Pattern Recognition (CVPR), pp. 10563–10572 (2019)
16. Puglisi, L., Alexander, D.C., Ravì, D.: Enhancing spatiotemporal disease progression models via latent diffusion and prior knowledge. In: International Conference on Medical Image Computing and Computer-Assisted Intervention, pp. 173–183. Springer (2024)
17. Rombach, R., Blattmann, A., Lorenz, D., Esser, P., Ommer, B.: High-resolution image synthesis with latent diffusion models. In: Proceedings of the IEEE/CVF conference on computer vision and pattern recognition, pp. 10684–10695 (2022)
18. Tang, S., Li, C., Zhang, P., Tang, R.: SwinLSTM: Improving spatiotemporal prediction accuracy using swin transformer and LSTM. In: Proceedings of the IEEE/CVF International Conference on Computer Vision, pp. 13470–13479 (2023)
19. Tomašev, N.: A clinically applicable approach to continuous prediction of future acute kidney injury. Nature **572**(7767), 116–119 (2019)
20. Weinreb, R.N., Aung, T., Medeiros, F.A.: The pathophysiology and treatment of glaucoma: a review. JAMA **311**(18), 1901–1911 (2014)
21. Yang, X., et al.: Multi-scale spatio-temporal transformer-based imbalanced longitudinal learning for glaucoma forecasting from irregular time series images. IEEE J. Biomed. Health Inf. (2024)
22. Yoon, J.S., Zhang, C., Suk, H.I., Guo, J., Li, X.: Sadm: Sequence-aware diffusion model for longitudinal medical image generation. In: International Conference on Information Processing in Medical Imaging, pp. 388–400. Springer (2023)
23. Yoon, J., Zame, W.R., Van Der Schaar, M.: Estimating missing data in temporal data streams using multi-directional recurrent neural networks. IEEE Trans. Biomed. Eng. **66**(5), 1477–1490 (2018)

24. Zhang, R., Isola, P., Efros, A.A., Shechtman, E., Wang, O.: The unreasonable effectiveness of deep features as a perceptual metric. In: Proceedings of the IEEE conference on computer vision and pattern recognition, pp. 586–595 (2018)
25. Zhang, Y., et al.: Coarse-to-fine latent diffusion model for glaucoma forecast on sequential fundus images. In: Linguraru, M.G., et al. (eds.) Medical Image Computing and Computer Assisted Intervention – MICCAI 2024, vol. 15005, pp. 166–176. Springer Nature Switzerland, Cham (2024). https://doi.org/10.1007/978-3-031-72086-4_16
26. Zhu, Z., et al.: Loci-diffcom: Longitudinal consistency-informed diffusion model for 3d infant brain image completion. In: International Conference on Medical Image Computing and Computer-Assisted Intervention, pp. 249–258. Springer (2024)

Time-Contrastive Pretraining for In-Context Image and Video Segmentation

Assefa Wahd(✉), Jacob Jaremko, and Abhilash Hareendranathan

Department of Radiology and Diagnostic Imaging, University of Alberta, Edmonton, Canada
{wahd,jjaremko,hareendr}@ualberta.ca

Abstract. In-context learning (ICL) has shown promise for generalizing to new visual tasks using a few examples, but current methods are limited. They typically rely on a rigid gridding strategy that restricts the number and resolution of context images. We propose **Temporal**, a novel approach that overcomes these limitations by reformulating visual ICL as a video object segmentation (VOS) problem. This VOS-based approach naturally handles a variable number of full-resolution context images. To automatically select the most relevant context for a given query, we introduce a prompt retriever pretrained on videos using a time-contrastive objective. This objective learns from the temporal coherence of video, using adjacent frames as positive examples (i.e., useful context images) and distant frames as negatives. For image segmentation, our retriever builds a pseudo-video by prepending the retrieved context images to the query image, which is then processed by the VOS model. For video segmentation, the retriever identifies keyframes, our ICL pipeline generates their masks, and these masks are propagated through the video. On the MICCAI FLARE 2022 challenge, Temporal significantly outperforms baselines, achieving a Dice score of 90.95% for image segmentation (+10.64%) and 92.45% for video segmentation (+14.88%).

Keywords: Time-Contrastive Learning · In-Context Learning · Video Object Segmentation

1 Introduction

Medical imaging generates vast quantities of unlabeled data, yet acquiring dense annotations remains costly. This challenge has motivated the development of approaches requiring fewer labels, such as *self-supervised learning* [1,2,5,9,10,13], *semi-supervised learning* [17], and *transfer learning* [20]. Recently, foundation models have introduced new paradigms such as *visual prompting* and *in-context learning* (ICL) [3,7,8,21,22], enabling adaptation to new tasks with minimal or no fine-tuning.

J. C. Gee et al. (Eds.): MICCAI 2025, LNCS 15960, pp. 630–639, 2026.
https://doi.org/10.1007/978-3-032-04927-8_60

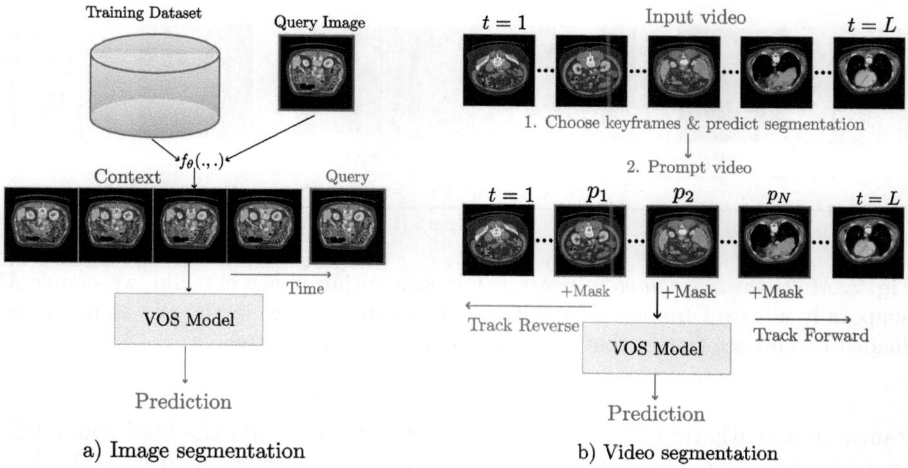

Fig. 1. Model architecture. (a) *Visual in-context learning via VOS:* The retriever f_θ selects the optimal context set P for a query image x_q. The query image x_q is appended to the context to create a video sequence, and the VOS model predicts the segmentation mask for x_q using this sequence. (b) *Video segmentation with ICL:* 1) Select keyframes (Sect. 2.3), 2) Predict masks for keyframes using step (a), 3) Propagate masks across the video.

In-context learning (ICL)—where models perform new tasks or generalize to unseen domains by conditioning on input-output examples without parameter updates—has been remarkably successful in natural language processing [4]. It offers a compelling vision for medical imaging: a single model could generalize across diverse segmentation tasks by "reasoning" over relevant annotated examples. However, adapting ICL to computer vision presents unique challenges. We highlight two critical issues: (1) selecting the most relevant context images, and (2) effectively modeling interactions between the context set and the query image. An ideal visual ICL model must support a variable number of context images and preserve their full resolution. While textual input-output pairs can be trivially concatenated, it is less clear how to best handle a variable number of images for visual ICL.

Early attempts at visual ICL relied on a *gridding* strategy, arranging input-output image pairs in a grid alongside a query image [3,18,22,23]. This strategy, however, limits the number of context images and necessitates downsampling, reducing image detail—a critical limitation for medical applications. More recent work has focused on context selection. Zhang et al. [23] found that performance is significantly impacted by the choice of context examples and proposed both unsupervised (CLIP-based [14]) and supervised retrieval techniques. Subsequent innovations included padding images with learnable vectors [22] and combining channel and spatial features for computing similarity [16].

A promising alternative emerged when Foster et al. [8] framed ICL as a video object segmentation (VOS) task. In this approach, context images are treated as

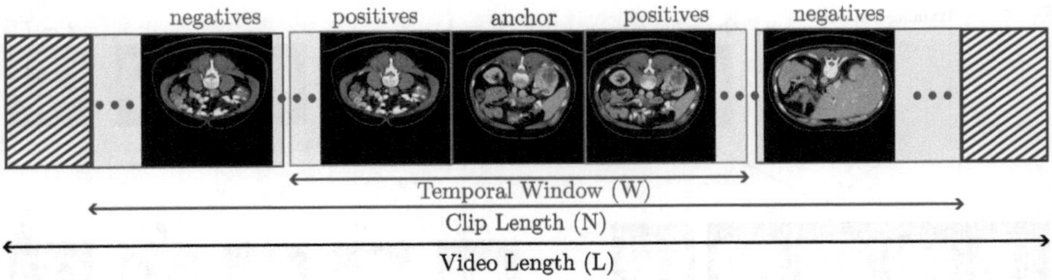

Fig. 2. *Multi-positive sampling strategy.* For each anchor image in a clip, we sample M temporally adjacent frames as positives and all other frames in the clip as negatives. Shaded regions are frames that were not sampled in a minibatch.

frames in a synthetic video, and a VOS model [6] segments the final frame (the query image). This allows for handling full-resolution images and variable-sized context sets. However, this method relied on pre-trained CLIP embeddings for context retrieval, which may not be optimal for unseen domains like medical imaging.

Our work, named **Temporal**, extends the VOS-based ICL framework by introducing a self-supervised, time-contrastive approach for context retrieval and a unified pipeline for both image and video segmentation. We pretrain a retriever on videos using a multipositive time-contrastive objective, which leverages temporal coherence as a natural supervisory signal. This avoids the need for manual labels by treating adjacent frames as positive examples. Furthermore, we provide a comprehensive analysis of context selection and an efficient keyframe-based method for video segmentation. To our knowledge, this is the first work to combine time-contrastive networks with a multipositive cross-entropy loss for self-supervised learning from videos. Our approach achieves up to 92.45% Dice on video benchmarks—a 14.88% improvement over semi-supervised VOS baselines. Figure 1 illustrates our approach. The code is available at github.com/aswahd/Temporal.

2 Method

Our method consists of three main components: (1) a formulation of visual ICL as a video object segmentation (VOS) task, (2) a prompt retriever pretrained with a multipositive time-contrastive objective, and (3) an inference pipeline for both image and video segmentation. We use SAM2 [15] as our base VOS model.

2.1 Visual In-Context Learning as VOS

Given a dataset of input-output pairs $\mathcal{D} = \{(x_n, y_n)\}_{n=1}^{N}$ and a query x_q, ICL aims to predict $\hat{y}_q$ by conditioning on a context set $P = \{(x_{c_1}, y_{c_1}), \ldots, (x_{c_K}, y_{c_K})\} \subseteq \mathcal{D}$. To overcome the limitations of gridding strategies, we frame ICL as a VOS problem. The context set P and the query x_q are treated as frames in a synthetic video, where each pair (x_{c_k}, y_{c_k}) is a labeled frame and x_q is the final, unlabeled frame. A VOS model $\mathcal{V}_\phi$ then segments x_q:

$$\mathcal{V}_\phi(P, x_q) = \mathcal{V}_\phi(x_{c_1}, y_{c_1}, \ldots, x_{c_K}, y_{c_K}, x_q). \tag{1}$$

The selection of the optimal context set P is handled by our prompt retriever, described next.

2.2 Multipositive Time-Contrastive Pretraining

Our objective is to sample multiple positive examples along with a substantial number of negatives for each anchor frame in a video, and to train the network using a contrastive loss. We formalize this as follows:

Sampling Strategy: For each video V of length L, we sample a contiguous clip $C = \{t_{\text{start}}, t_{\text{start}} + 1, \ldots, t_{\text{start}} + N - 1\}$, where N denotes the clip length and $t_{\text{start}} \sim \mathcal{U}(1, L - N)$. All frames in the clip are used in a minibatch.

Generating Label Matrix: For each anchor frame $i \in C$, positive pairs are defined as frames within a temporal window $\mathcal{W}_i = [\max(1, i - \lfloor W/2 \rfloor), \min(N, i + \lfloor W/2 \rfloor)]$, where W specifies the window size. We sample M frames from the window. This generates a binary label matrix $\mathbf{A} \in \{0, 1\}^{N \times N}$:

$$\mathbf{A}_{i,j} = \begin{cases} 1 & \text{if } |i - j| \leq \lfloor W/2 \rfloor \\ 0 & \text{otherwise} \end{cases} \tag{2}$$

To prevent trivial self-similarity matches, diagonal elements $\mathbf{A}_{i,i}$ are masked to 0.

Multi-View Augmentation: We generate two augmented views $\tilde{C}_1$ and $\tilde{C}_2$ through random geometric (e.g., translation) and photometric (e.g., Gaussian blurring) transformations. The label matrix expands to encode multi-view relationships:

$$\hat{\mathbf{A}} = \begin{bmatrix} \mathbf{A} & \mathbf{A} + \mathbf{I}_N \\ \mathbf{A} + \mathbf{I}_N & \mathbf{A} \end{bmatrix} \tag{3}$$

where $\mathbf{I}_N$ denotes the $N \times N$ identity matrix. This formulation ensures that:

- Temporal neighbors within each view remain positives
- Temporal neighbors across views are positives
- Two views of the same frame are positives (i.e., $+\mathbf{I}_N$).

For minibatch training, we sample B video clips, and extend the above label matrix to consider frames from different videos as negatives. The final label matrix is a block diagonal matrix with B blocks. See Fig. 1 for the overall architecture and Fig. 2 for an illustration of a sampled minibatch.

Multipositive Time-Contrastive Loss: Each anchor, in a minibatch of shape $(B \times N, C, H, W)$, has $2M - 1$ positives and $2B(N - M)$ negatives. The encoder f_θ projects the input batch into a set of embeddings $\mathbf{z} = \{z_1, z_2, \ldots, z_{2BN}\}$. Each

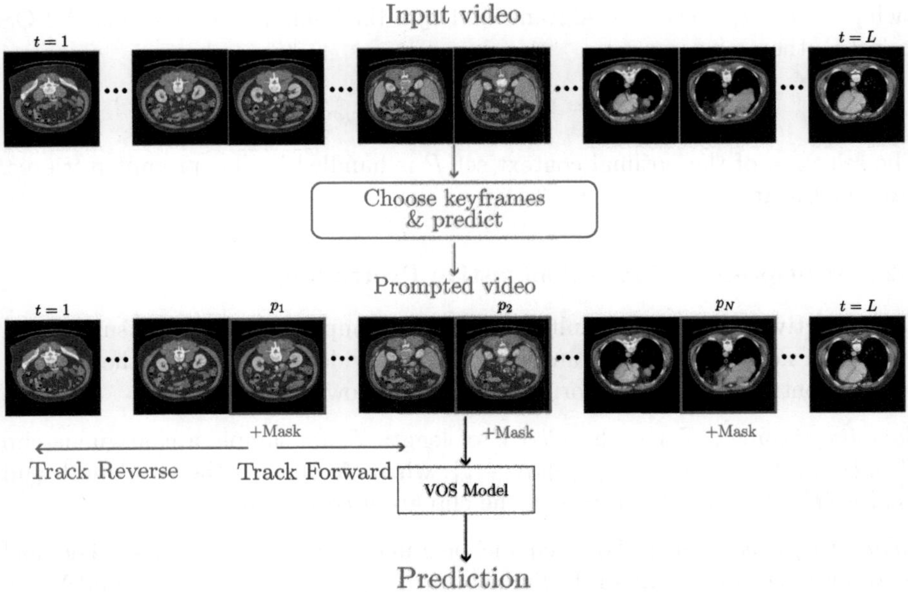

Fig. 3. Inference pipeline for video segmentation.

embedding is L2 normalized to obtain $\hat{z}_i = z_i/\|z_i\|$. The multipositive contrastive loss is defined as:

$$\mathcal{L}_{\text{contrastive}} = -\frac{1}{2BN}\sum_{i=1}^{2BN}\frac{1}{|P(i)|}\sum_{j\in P(i)}\log\frac{\exp\left(\frac{\hat{z}_i^\top \hat{z}_j}{\tau}\right)}{\exp\left(\frac{\hat{z}_i^\top \hat{z}_j}{\tau}\right)+\sum_{k\in N(i)}\exp\left(\frac{\hat{z}_k^\top \hat{z}_j}{\tau}\right)}, \tag{4}$$

where $P(i)$ denotes the set of positive samples for the i-th anchor, $N(i)$ represents the corresponding set of negatives, and τ is the temperature parameter.

2.3 Inference

Image segmentation: For a given test image x_q, we first compute its embedding $z_q = f_\theta(x_q)$ and retrieve the top-K most similar images from the training database $\mathcal{D}$ based on cosine similarity. These retrieved contexts are then used to construct a synthetic video sequence $V = \{x_{c_1}, \ldots, x_{c_K}, x_q\}$, with the query image x_q appended as the final frame. Finally, the VOS model $\mathcal{V}_\phi$ processes this sequence to produce the segmentation mask $\hat{y}_q$ for the query image, as illustrated in Fig. 1(a).

Video Segmentation: Given an input video $V = \{x_1, x_2, \ldots, x_L\}$ with L frames, our goal is to generate segmentations $\{\hat{y}_1, \hat{y}_2, \ldots, \hat{y}_L\}$. We can apply the image-based inference procedure frame-by-frame, and concatenate the results but this is computationally expensive for long sequences and it doesn't consider temporal coherence when making predictions. Instead, we propose a keyframe-based approach that leverages the temporal structure of videos. The inference proceeds as follows:

1. Choose frames that are most useful if used as prompts (keyframes). First, each frame x_t is mapped to an embedding $z_t = f_\theta(x_t) \in \mathbb{R}^d$. We then measure how "similar" each frame is to the training database by computing $s_t = \max_{d \in \mathcal{D}} \frac{z_t \cdot d}{\|z_t\| \, \|d\|}$. We form a candidate set $\mathcal{S} \subseteq V$ of size K by selecting the top-K frames with the highest similarity scores $\{s_t\}$. Intuitively, these are the frames that best match the training data distribution and thus are more likely to yield accurate segmentations using the visual ICL (Fig. 1).

2. Choose Q representative frames from $\mathcal{S}$. Although all frames in $\mathcal{S}$ are strongly matched to the training set, some may be temporally redundant. Hence, we select a subset of size Q that is more diverse. Let $\mathcal{S} = \{x_{(1)}, x_{(2)}, \dots, x_{(K)}\}$ be sorted in descending order by s_t. We impose a minimum temporal distance τ between chosen frames to encourage coverage of different time points. Starting with $\mathcal{Q} = \{x_{(1)}\}$, we add $x_{(i)}$ to $\mathcal{Q}$ if $\min_{x_j \in \mathcal{Q}} |t_i - t_j| > \tau$, where t_i is the frame index of $x_{(i)}$. This process continues until we have selected Q frames or exhausted $\mathcal{S}$.

3. Predict segmentation and filter out non-confident predictions. For each chosen frame $x_q \in \mathcal{Q}$, we apply the image inference procedure (read above or Fig. 1(b)) to obtain a predicted segmentation mask $\hat{y}_q$ and an associated confidence score c_q: $(\hat{y}_q, c_q) = \text{Temporal}(x_q, \mathcal{D})$.. Since some predictions may be unreliable, we only keep those that exceed a threshold γ: $P = \{(x_q, \hat{y}_q) \mid c_q \geq \gamma\}$. Hence, P contains high-confidence and representative prompts that will guide the subsequent video-wide segmentation.

4. Propagate predictions for the rest of the video. We feed the retained prompts P into a promptable VOS model $\mathcal{V}_\phi$. Since the selected keyframes may be in the middle of the video, we propagate the prompts in both forward and reverse directions to generate segmentations for the entire video: $\{\hat{y}_1, \hat{y}_2, \dots, \hat{y}_L\} = \mathcal{V}_\phi(V, P)$. The process is illustrated in Fig. 3.

Fine-Tuning: To bridge the domain gap between natural and medical images, we fine-tuned all parameters of the underlying VOS model using "synthetic" videos constructured from 2D images in the training set. For each image $\mathbf{x}_i$, we: 1) select top-K context images via f_θ, 2) construct video $V_i = \{\mathbf{x}_1, \dots, \mathbf{x}_K, \mathbf{x}_Q\}$ sorted by similarity, and 3) train $\mathcal{V}_\phi$ on random clips $v_i \subset V_i$ to minimize $\min_\phi \mathcal{L}_{seg}(\mathcal{V}_\phi(v_i), y)$. Fine-tuning significantly improves both image and video segmentation as shown in Tables 1 and 2.

3 Experiments

Representative Methods. We compare against grid-based [23] and VOS-based ICL [8]. We adopt SAM-2 as the video foundation model since it represents the current SOTA in generic video segmentation.

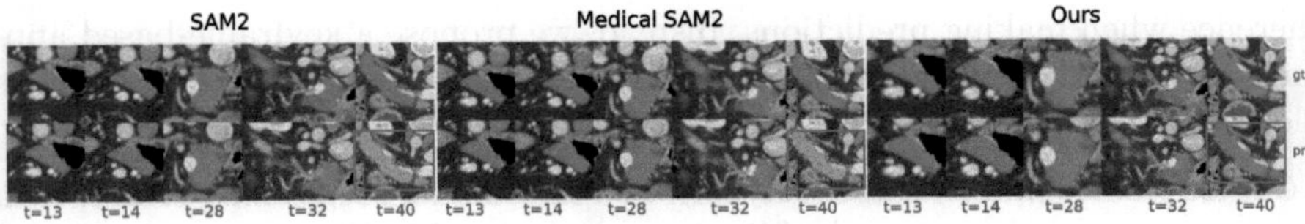

Fig. 4. *Qualitative segmentation results* on Pancreas. Our approach (right) produces more accurate organ segmentations and tracking compared to baselines. The baselines fail to track the organ after $t = 32$.

Datasets. We evaluate our approach on the MICCAI FLARE 2022 dataset, a benchmark dataset for CT scan segmentation. The dataset includes both annotated scans and a substantial collection of unlabeled data, which we utilize for self-supervised pretraining. Specifically, we use 40 scans to construct the context set and 10 scans for testing. These 10 scans comprise 922 slices containing foreground annotations.

Implementation Details. Our time-contrastive pretraining employs a ResNet-50 [11] backbone. We sample 16-frame video clips with a stride of 2 and apply random augmentations (e.g., rotation, horizontal flipping, and Gaussian blurring). For the cross entropy loss, the temperature is set to $\tau = 0.1$, and two adjacent frames are selected as positives per anchor. Training runs for 500 epochs with a minibatch of 4 clips (resulting in a batch size of $16 \times 2 \times 4 = 128$ images) using the AdamW optimizer [12] (learning rate $lr = 10^{-3}$ and weight decay 10^{-4}). The hyperparameters were chosen via grid search over $lr \in \{10^{-4}, 10^{-3}, 10^{-2}\}$ and $\tau \in \{0.1, 0.5, 1.0\}$. The VOS model utilizes a SAM2-B+ backbone [15]. For image segmentation, we empirically found that 10 context images work best (see Table 1). For video segmentation, we sample 20 keyframes. Fine-tuning is performed for 100 epochs with a learning rate of 10^{-5}, a batch size of 1, and the AdamW optimizer.

3.1 Results

Image Segmentation: The grid-based method with unsupervised context retrieval performs poorly on medical images, achieving only around 5% Dice score. So, we abandon this approach in favor of VOS-based methods. Table 1 shows that Temporal outperforms unsupervised prompt retrieval methods including DINOv2 [13], CLIP [14], and MedCLIP [19]. Our method achieves 83.55% Dice versus 80.31% for the best baseline. After fine-tuning, Temporal reaches **90.95%** Dice, a **10.64%** improvement. Performance gains are particularly significant for challenging organs like Pancreas (79.53% versus baseline's 49.10%).

Video Segmentation: We evaluate on 10 abdominal CT videos (average length: 95 frames) comparing two paradigms:

- **Semi-supervised baselines**: SAM-2 [15] and Medical SAM-2 [24] require first-frame manual prompts. So, we provide ground-truth masks for the first annotated slice of each organ.

Table 1. Image segmentation results (Dice, %) of Temporal and baselines on the FLARE 2022 dataset. "ctx" denotes the number of context images used. All baseline models are evaluated with ctx = 10. (RK: Right Kidney, Spleen, Panc: Pancreas, Aorta, IVC: Inferior Vena Cava, Gall: Gallbladder, Eso: Esophagus, Stom: Stomach, LK: Left Kidney). Fine-tuning (ft) improves performance.

Model	Liver	RK	Spleen	Panc.	Aorta	IVC	Gall.	Eso.	Stom.	LK	Avg
Grid-based methods											
Zhang et al. [23] (CLIP)	N/A										
VOS-based methods											
Foster et al. [8] (Random, 5 evals)	0.8999	0.8904	0.8561	0.3006	0.9278	0.6998	0.5378	0.4127	0.6349	0.9003	0.7060
Foster et al. [8] (CLIP)	0.9234	0.8968	0.9136	0.4886	0.9490	0.8252	0.6142	0.6434	0.7291	0.8583	0.7842
Foster et al. [8] (MedCLIP)	0.8950	0.8486	0.8605	0.4464	0.9535	0.8207	0.6647	0.6758	0.7501	0.8097	0.7724
Foster et al. [8] (DINOv2)	0.9161	0.9003	0.8815	0.4910	0.9485	0.8432	0.7519	0.6730	0.7509	0.8747	0.8031
Ours											
Temporal (ctx=5)	0.9088	0.9003	0.9222	0.6105	0.9526	0.8185	0.7579	0.7096	0.7194	0.8881	0.8188
Temporal (ctx=10)	0.9243	0.9105	0.9221	0.6211	0.9560	0.8364	0.8130	0.7228	0.7499	0.8991	0.8355
Temporal (ctx=5,ft)	0.9504	0.9115	0.9578	0.7872	0.9669	0.9204	0.8361	0.8619	0.8585	0.8980	0.8949
Temporal (ctx=10, ft, topk)	0.9506	0.9032	0.9371	0.7860	0.9669	0.9214	0.8598	0.8772	0.8671	0.8958	0.8965
Temporal (ctx=10,ft)	0.9631	0.9202	0.9650	0.7953	0.9671	0.9218	0.9073	0.8697	0.8764	0.9090	0.9095

Table 2. Video segmentation results (Dice, %). SAM2 and Medical SAM2 are evaluated in a semi-supervised setting as they require prompts to propagate. Temporal can automatically select context without supervision and achieves superior performance.

Method	Liver	RK	Spleen	Panc.	Aorta	IVC	Gall.	Eso.	Stom.	LK	Avg
Baselines (Semi-supervised)											
SAM-2 (mask prompt)	0.9121	0.9638	0.8845	0.3557	0.9497	0.8021	0.8737	0.6497	0.4395	0.9257	0.7757
Medical SAM-2 (mask prompt)	0.9070	0.8991	0.9650	0.3402	0.8932	0.7828	0.7238	0.6970	0.6786	0.8358	0.7723
Ours											
Temporal (ctx=10)	0.9263	0.8867	0.9280	0.5970	0.9526	0.8633	0.8105	0.6281	0.7149	0.8853	0.8193
Temporal (ctx=10, fine-tuned)	**0.9739**	**0.9625**	**0.9532**	**0.8132**	**0.9669**	**0.9219**	**0.8928**	**0.8686**	**0.9118**	**0.9581**	**0.9223**
Temporal (fine-tuned, mask prompt)	**0.9765**	**0.9625**	**0.9837**	**0.8203**	**0.9690**	**0.9351**	**0.8927**	**0.8875**	**0.8586**	**0.9586**	**0.9245**

– **Automated VOS**: Temporal automatically identifies the target organs without manual prompts, and tracks them throughout the entire sequence.

Table 2 highlights our method's superiority: Temporal achieves **81.93%** Dice fully automatically compared to SAM2's 77.57% with manual prompts. After fine-tuning (Sect. 2), Temporal achieves **92.23%** Dice, outperforming Medical SAM-2 (77.83%) by **14.4%**, and **92.45%** in semi-supervised segmentation. The improvement is particularly dramatic on challenging organs like Pancreas, where the best baseline achieves only 35.57% Dice compared to Temporal's **82.03%** after fine-tuning. See Fig. 4 for qualitative results.

Impact of Context Size. To evaluate the effect of the number of annotated context images, we varied the number of context examples in our experiments. As shown in Table 1, increasing the number of context images results in improved performance: for example, 74.70% Dice with 5 context images versus 81.83% Dice with 10.

Diversity-Aware Context Selection. We compare standard top-K context retrieval with a diversity-aware selection strategy. Our two-stage approach: (1) retrieves top-K most similar training images to the query using cosine similarity on SSL features, then (2) greedily selects top-Q diverse candidates. The core idea is to minimize redundancy in the context set, ensuring each selected example contributes unique information. For each candidate $\mathbf{z}_j$ in the K-subset, we iteratively select the sample maximizing:

$$\mathrm{score}(z_j) = \mathrm{sim}(z_j, z_{\mathbf{query}}) - \lambda \cdot \max_{\mathbf{z}_k \in \mathcal{S}} \mathrm{sim}(z_j, z_k),$$

where $\mathcal{S}$ contains already selected samples and λ balances relevance versus diversity. In our experiments, $\lambda = 0.7$ was determined through validation. The results in Table 1 show that diverse sampling improves Dice by $\uparrow$ 2.1% (context size 5) and $\uparrow$ 1.30% (size 10) over top-K selection, highlighting the importance of context diversity.

4 Conclusion

We introduced Temporal, a self-supervised learning objective for pretraining a prompt retriever, and we formulated visual in-context learning as a video object segmentation. When evaluated on the MICCAI FLARE 2022 dataset, our approach demonstrates substantial improvements, achieving a 9.23% and 14.88% increase in Dice scores in image and video segmentation, respectively.

Acknowledgments. Dr. Jacob Jaremko is a Canada CIFAR AI Chair, and his academic time is made available by Medical Imaging Consultants (MIC), Edmonton, Canada. This work was supported by grants from TD Ready Grant, IC-IMPACTS, One Child Every Child, the Arthritis Society, and Alberta Innovates.

Disclosure of Interests. The authors have no competing interests to declare that are relevant to the content of this article.

References

1. Assran, M., et al.: Self-Supervised Learning from Images with a Joint-Embedding Predictive Architecture (2023). https://doi.org/10.48550/arXiv.2301.08243
2. Balestriero, R., et al.: A Cookbook of Self-Supervised Learning (2023). https://doi.org/10.48550/arXiv.2304.12210
3. Bar, A., Gandelsman, Y., Darrell, T., Globerson, A., Efros, A.A.: Visual Prompting via Image Inpainting
4. Brown, T.B., et al.: Language Models are Few-Shot Learners (2020). https://doi.org/10.48550/arXiv.2005.14165

5. Chen, T., Kornblith, S., Norouzi, M., Hinton, G.: A Simple Framework for Contrastive Learning of Visual Representations (2020). https://doi.org/10.48550/arXiv.2002.05709
6. Cheng, H.K., Schwing, A.G.: XMem: Long-Term Video Object Segmentation with an Atkinson-Shiffrin Memory Model (2022). https://doi.org/10.48550/arXiv.2207.07115
7. Denner, S., et al.: Visual Prompt Engineering for Medical Vision Language Models in Radiology (2024)
8. Foster, T., Croitoru, I., Dorfman, R., Edlund, C., Varsavsky, T., Almazán, J.: Flexible visual prompts for in-context learning in computer vision (2023). https://doi.org/10.48550/arXiv.2312.06592
9. He, K., Chen, X., Xie, S., Li, Y., Dollár, P., Girshick, R.: Masked Autoencoders Are Scalable Vision Learners (2021). https://arxiv.org/abs/2111.06377v3
10. He, K., Fan, H., Wu, Y., Xie, S., Girshick, R.: Momentum Contrast for Unsupervised Visual Representation Learning (2020)
11. He, K., Zhang, X., Ren, S., Sun, J.: Deep Residual Learning for Image Recognition (2015). https://doi.org/10.48550/arXiv.1512.03385
12. Loshchilov, I., Hutter, F.: Decoupled Weight Decay Regularization (2019). https://doi.org/10.48550/arXiv.1711.05101
13. Oquab, M., et al.: DINOv2: Learning Robust Visual Features without Supervision (2024)
14. Radford, A., et al.: Learning Transferable Visual Models From Natural Language Supervision (2021). https://doi.org/10.48550/arXiv.2103.00020
15. Ravi, N., et al.: SAM 2: Segment Anything in Images and Videos
16. Sun, Y., Chen, Q., Wang, J., Wang, J., Li, Z.: Exploring Effective Factors for Improving Visual In-Context Learning (2023). https://doi.org/10.48550/arXiv.2304.04748
17. van Engelen, J.E., Hoos, H.H.: A survey on semi-supervised learning. Mach. Learn. **109**(2), 373–440 (2019). https://doi.org/10.1007/s10994-019-05855-6
18. Wang, X., Wang, W., Cao, Y., Shen, C., Huang, T.: Images Speak in Images: A Generalist Painter for In-Context Visual Learning (2023). https://doi.org/10.48550/arXiv.2212.02499
19. Wang, Z., Wu, Z., Agarwal, D., Sun, J.: MedCLIP: Contrastive Learning from Unpaired Medical Images and Text (2022). https://doi.org/10.48550/arXiv.2210.10163
20. Weiss, K., Khoshgoftaar, T.M., Wang, D.D.: A survey of transfer learning. J. Big Data **3**(1), 1–40 (2016). https://doi.org/10.1186/s40537-016-0043-6
21. Yu, B.X., et al.: Visual Tuning. ACM Comput. Surv. **56**(12), 1–38 (2024). https://doi.org/10.1145/3657632
22. Zhang, J., Wang, B., Li, L., Nakashima, Y., Nagahara, H.: Instruct Me More! Random Prompting for Visual In-Context Learning (2023). https://doi.org/10.48550/arXiv.2311.03648
23. Zhang, Y., Zhou, K., Liu, Z.: What Makes Good Examples for Visual In-Context Learning? (2023). https://doi.org/10.48550/arXiv.2301.13670
24. Zhu, J., Hamdi, A., Qi, Y., Jin, Y., Wu, J.: Medical SAM 2: Segment medical images as video via Segment Anything Model 2 (2024). https://doi.org/10.48550/arXiv.2408.00874

TMSE: Tri-Modal Survival Estimation with Context-Aware Tissue Prototype and Attention-Entropy Interaction

Ruofan Zhang[1,2], Mengjie Fang[2], Shengyuan Liu[3], Zipei Wang[1,2], Jie Tian[2,4](✉), and Di Dong[1,2](✉)

[1] School of Artificial Intelligence, University of Chinese Academy of Sciences, Beijing, China

[2] CAS Key Laboratory of Molecular Imaging, Institute of Automation, Chinese Academy of Sciences, Beijing, China

tian@ieee.org, di.dong@ia.ac.cn

[3] The Department of Electronic Engineering, The Chinese University of Hong Kong, Hong Kong, China

[4] Beijing Advanced Innovation Center for Big Data-Based Precision Medicine, Beihang University, Beijing, China

Abstract. Survival prediction plays a crucial role in clinical decision-making, enabling personalized treatments by integrating multi-modal medical data, such as histopathology images, pathology reports, and genomic profiles. However, the heterogeneity across these modalities and the high dimensionality of Whole Slide Images (WSI) make it challenging to capture survival-relevant features and model their interactions. Existing methods, typically focused on single-modal WSI, fail to leverage multimodal information, such as expert-driven pathology reports, and struggle with the computational complexity of WSI. To address these issues, we propose a novel Tri-Modal Survival Estimation framework (TMSE), which includes three components: (1) Pathology report processing pipeline, curated with expert knowledge, with both the pipeline and the processed structured report being publicly available; (2) Context-aware Tissue Prototype (CTP) module, which uses Mamba and Gaussian mixture models to extract compact, survival-relevant features from WSI, reducing redundancy while preserving histological details; (3) Attention-Entropy Interaction (AEI) module, a attention mechanism enhanced with entropy-based optimization to align and fuse three modalities: WSI, pathology reports, and genomic data. Extensive evaluation on three TCGA datasets (BLCA, BRCA, LUAD) shows that our approach achieves superior performance in survival prediction. Data and code are available: https://github.com/RuofanZhang8/TMSE.

Keywords: Multi-modal learning · Survival Prediction · Heterogeneous Biomedical Data

J. C. Gee et al. (Eds.): MICCAI 2025, LNCS 15960, pp. 640–650, 2026.
https://doi.org/10.1007/978-3-032-04927-8_61

1 Introduction

Survival prediction is vital in clinical decision-making, offering prognostic insights for personalized treatment and advancing precision medicine by integrating machine learning with multi-modal medical data [23]. As precision medicine evolves, cancer tissue pathology slides are digitized into Whole Slide Images (WSI) with expert reports, while patient genomic profiles are increasingly collected, enhancing survival analysis [1]. Integrating this multi-modal data is crucial to improving the accuracy and robustness of prognostic predictions despite complex clinical relationships [15,27].

Analyzing multi-modal information presents significant challenges due to the substantial heterogeneity [4,5,16] across modalities such as histopathology images, pathology reports, and genomic data. A common difficulties is capturing the complex relationships between these modalities. Although each modality contains rich information, identifying survival-relevant features from the vast amounts of data they contain is difficult for effective integrated survival prediction [24]. Previous studies on multi-modal fusion largely focused on pairing histopathology images with genomic data, using techniques like cross-attention [3] or optimal transport [29]. However, these methods often face challenges when models simulate the interactions of thousands of WSI patch tokens and other modality information to make patient-level predictions. It is difficult for the model to extract critical interaction information from the large number of WSI tokens, and the overwhelming amount of WSI data can easily obscure information from other modalities, posing an extremely difficult challenge. Furthermore, pathology reports, which offer expert-driven descriptive insights [14], represent a valuable complement to image data and are readily accessible. However, earlier research often neglects this expert-driven information or lacks efficient methods to integrate pathology reports with clinical knowledge, limiting their potential utility in multi-modal analysis.

To address these challenges, we propose the Tri-Modal Survival Estimation framework (TMSE). First, we design a pathology report processing pipeline that integrates expert knowledge to efficiently and uniformly extract information from pathology reports for survival prediction assistance. Second, we propose a Context-aware Tissue Prototype (CTP) module that integrates Mamba architecture and Gaussian Mixture Model (GMM) for efficient tissue representation learning. The GMM component compresses similar tissue patches into a compact set of prototype representations, effectively reducing the redundant information caused by high-resolution WSI sampling. Meanwhile, the Mamba architecture models contextual relationships among thousands of WSI patch tokens, capturing survival-relevant tumor microenvironment (TME) features that emerge from large-scale tissue interactions. This information cannot be adequately represented by GMM's tissue-level prototypes alone. This dual approach efficiently condenses extensive patch-level data into concise yet comprehensive feature representations, effectively addressing the "needle in a haystack" challenge in pathological image analysis while facilitating robust multi-modal feature alignment and fusion. Third, we propose a Attention-Entropy Interaction (AEI) module to facilitate interactions and fusion between tri-modal information. We decouple the information interaction in AEI into an information interaction mechanism and an information retention mechanism, helping the model avoid being overwhelmed by the large volume of WSI data. This enhancing multi-modal information fusion efficiency and improving the model's capacity to extract meaningful insights from diverse modalities. Our contributions can be summarized as follows:

1. We utilize an expert-knowledge-based prompt chain pipeline to organize approximately ten thousand TCGA pathology reports using large language model (LLM). The refined reports will enhance survival analysis and be made publicly available to facilitate further research.
2. We propose the CTP module, which addresses WSI redundancy through prototype-based representation learning, enabling efficient extraction and fusion of critical histological features.
3. We develop the AEI module, which mitigates multimodal alignment challenges through entropy-based fusion, achieving effective integration of WSI, pathology reports, and genomic data.
4. We validate the effectiveness of our model on three TCGA datasets: BLCA, BRCA, and LUAD, demonstrating the superior performance of our method.

2 Methods

2.1 Overview and MultiModal Data Preprocessing

We present TMSE in Fig. 1(A), a Tri-Modal Survival Estimation framework for survival prediction. It consists of two main modules: the first efficiently represents histological features using the CTP (Sect. 2.1), and the second aligns and fuses information across modalities with the AEI (Sect. 2.2). Our model innovatively leverages information from three modalities. Below, we explain how patient-level tri-modal feature representations are constructed from WSI, expert-driven pathology reports, and genomic data.

WSI. Given a WSI of one patient, we crop it into N_p patches at 20× magnification with a size of 256×256 by CLAM [18], denoted as $\{x_i\}_{i=1}^{N_w}$. Each patch image x_i is then encoded into a feature representation w_i using a pre-trained pathology foundation model PLIP [9]. The resulting representation is $W = \{w_i\}_{i=1}^{N_w}$.

Genomics. The genomic data consists of transcriptomic profiles from bulk RNA sequencing, tokenized into biological pathway vectors [10]. The resulting gene expression vectors are organized as $G = \{g_i\}_{i=1}^{N_g}$, where $g_i \in \mathbb{R}^{N_{c,g}}$ represents the expression vector for the g-th pathway, consisting of $N_{c,g}$ genes.

Pathology Report. Pathology reports, often encumbered by extraneous details and a lack of structure, pose challenges to unified feature encoding. To address this, we organize these reports using LLM (DeepSeek-V3 [17]), employing a expert-knowledge-based, chain-of-thought pipeline to extract critical information

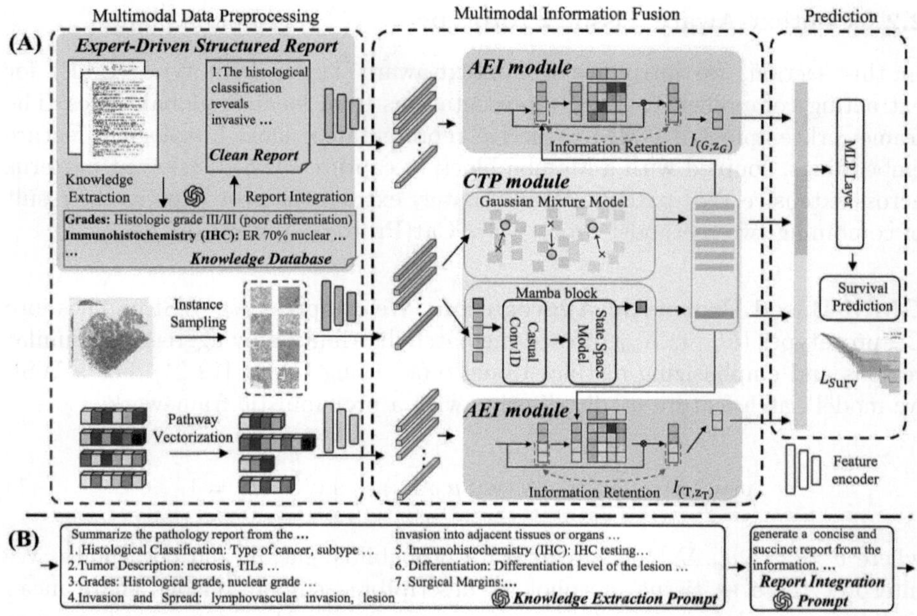

Fig. 1. (A) TMSE Overview: A tri-modal alignment pipeline featuring report preprocessing, WSI prototype extraction with the CTP module, and coupled AEI modules for information interaction. (B) Report Preprocessing Prompts: Extract report details using expert knowledge and consolidate them into a unified report representation.

and refine it into concise diagnostic reports. Our preprocessing pipeline consists of two key steps in Fig. 1(B). First, guided by clinical expertise, we devise a expert-base question prompt, which leverages seven predefined, domain-specific questions. This approach directs the LLM to focus on and extract multifaceted information from the reports, constructing a patient-level knowledge database. Second, we aggregate this patient-level knowledge database into a standardized, unified report, mitigating the risk of overlooking essential details that might occur when models directly process raw reports. We have applied this workflow to process all TCGA reports, openly sharing our pipeline and results to facilitate further research advancements. The refined report is then embedded using BiomedBERT [8], resulting in a feature $T = \{t_i\}, t_i \in \mathbb{R}^{d_T}$.

Dimension Matching. To merge features from different modalities, we align their token dimensions to a uniform length d. A linear projection layer adjusts the WSI token representations $W \in \mathbb{R}^{N_w \times d}$ and pathology reports $T \in \mathbb{R}^{1 \times d}$, while self-normalizing neural networks (SNN) [13] map variable-length genomics data via pathway-specific functions $f_{c,g}$ to $G \in \mathbb{R}^{N_g \times d}$.

2.2 Context-Aware Tissue Prototype

In this section, we introduce a Context-aware Tissue Prototype (CTP) for extracting comprehensive WSI representations from local to global scales. The framework employs a GMM to derive representative local histological feature embeddings, coupled with a Mamba block to capture survival-relevant patterns across extensive WSI patches. We ultimately extract the prototype as the result of combining two methods: $\text{Proto}_{\text{CTP}} = \text{Cat}(\text{Proto}_{\text{GMM}}, \text{Proto}_{\text{Mam}})$.

GMM-Based Prototype Aggregation. We compress N_w WSI patches into C_w prototypes ($C_w \ll N_w$), preserving patch distribution by aggregating similar regions and emphasizing distinct tissue types using GMM [12,21]. For a WSI, we model patch feature w_i distribution with a probabilistic framework:

$$p(\mathbf{w}_i;\theta) = \sum_{c=1}^{C_w} \pi_c \cdot \mathcal{N}(\mathbf{w}_i;\mu_c,\Sigma_c), \quad \text{s.t.} \sum_{c=1}^{C_w} \pi_c = 1, \tag{1}$$

where $\theta = \{\pi_c, \mu_c, \Sigma_c\}$ defines mixture probability, mean, and covariance. We interpret these as tissue morphology descriptors—similar regions share mean and variance, while π_c reflects tissue type proportions. Prototypes are formed as $\text{Proto}_{\text{GMM}} = [\pi_c, \mu_c, \Sigma_c]$. We optimize θ by initializing μ_c with k-means cluster centers, setting $\pi_c = 1/C_p$ and $\Sigma_c = I$, then maximizing $\sum_{i=1}^{N_w} \log p(\mathbf{w}_i;\theta)$ via expectation maximization.

Contextual Mamba Prototype. To model the TME information within tissue, it is essential to capture the contextual information across a large number of tissue blocks. For the feature sequence of WSI $W \in \mathbb{R}^{N_w \times d}$, we prepend a class token $t_{\text{cls}} \in \mathbb{R}^d$ to represent the entire slide patch sequence, forming $W' = [t_{\text{cls}}, W]$. After normalizing, we input the concatenated sequence into the Mamba [7] module as follows:

$$Y = \text{SSM}\left(\text{SiLU}\left(\text{Conv1D}\left(\text{Linear}\left(W'\right)\right)\right)\right) \tag{2}$$

Finally, we extract the class token from the output sequence Y as Mamba's prototype, $\text{Proto}_{\text{Mam}} = Y^{cls}$, which represents the global sequence information derived from the long feature sequence.

2.3 Attention-Entropy Interaction

In this section, we introduce the Attention-Entropy Interaction (AEI) module, which integrates attention-based interaction for bimodal alignment and fusion with entropy-based information retention. This takes into account both the fusion and preservation of multimodal information. Since WSI contain extensive information (from macro to micro details), we separately align report data (descriptive information) and genomic profiles (micro-level information) with WSI.

Cross-Modal Attention Alignment. In our approach, we replace the whole WSI patch token used in traditional attention mechanisms with CTP prototypes to compute interactive attention, improving efficiency and reducing computational complexity. For queries, we introduce learnable matrices $\mathbf{W}_Q$ and define $\mathbf{Q} = \left(\mathbf{Q}_W^{\mathrm{T}}, \mathbf{Q}_{\mathrm{G}}^{\mathrm{T}}\right)^{\mathrm{T}} = (\mathrm{Proto}_{\mathrm{CTP}}, G)^{\mathrm{T}}\, \mathbf{W}_Q$, and similarly for $\mathbf{K}$ and $\mathbf{V}$. The interactive attention is expressed as [28]:

$$\begin{pmatrix} \mathbf{Z}_W \\ \mathbf{Z}_G \end{pmatrix} = \sigma\left(\frac{1}{\sqrt{d}}\begin{pmatrix} \mathbf{Q}_W\mathbf{K}_W^{\mathrm{T}} & \mathbf{Q}_W\mathbf{K}_G^{\mathrm{T}} \\ \mathbf{Q}_G\mathbf{K}_W^{\mathrm{T}} & \mathbf{Q}_G\mathbf{K}_G^{\mathrm{T}} \end{pmatrix}\right)\begin{pmatrix} \mathbf{V}_W \\ \mathbf{V}_G \end{pmatrix} \tag{3}$$

where $\sigma(\cdot)$ denotes the softmax operation. We treat $\mathbf{Z}_G$ as the fused pathway information features after the interaction. Using the same approach, we can obtain the fused report information $\mathbf{Z}_T$.

Entropy-Based Information Retention. In this section, we propose using mutual information entropy to preserve information during the cross-modal attention alignment process. Specifically, we exclude WSI information when computing the entropy, aiming to preserve the information from the report data and genomic profiles. We apply matrix-based Rényi's α-order entropy functional [6,20] to estimate information-theoretic measures. For gene pathway data, we use pre-AEI G and post-AEI Z_G, selecting N sample pairs $\{G^m, Z_G^m\}_{m=1}^N$ from a mini-batch. Self-information and mutual Information are computed using the normalized eigenspectrum of a Gram matrix $A_G = K_G/\operatorname{tr}(K_G)$, where $K_G(m,n) = k\left(G^m, G^n\right)$ and k is a Gaussian kernel.

$$H_\alpha\left(A_G\right) = \frac{1}{1-\alpha}\log_2\left(\operatorname{tr}\left(A_G^\alpha\right)\right) \quad H_\alpha\left(A_G, A_{Z_G}\right) = H_\alpha\left(\frac{A_G \circ A_{Z_G}}{\operatorname{tr}\left(A_G \circ A_{Z_G}\right)}\right) \tag{4}$$

We set $\alpha = 1.01$ to approximate Shannon entropy, and $\circ$ denotes the element-wise product. Finally, the matrix-based Rényi's α-order mutual information entropy $I_\alpha\left(G; Z_G\right)$ used for information retention, is defined as:

$$I_\alpha\left(G; Z_G\right) = H_\alpha\left(A_G\right) + H_\alpha\left(A_{Z_G}\right) - H_\alpha\left(A_G, A_{Z_G}\right) \tag{5}$$

For textual information, $I\left(T; Z_T\right)$ can be calculated in the same manner.

2.4 Training Strategy for Survival Prediction

For a patient, the final multimodal feature is derived by applying layer normalization and mean pooling to the CTP prototypes, post-AEI text feature, and post-AEI genomic feature. For survival prediction, we employ the Negative Log-Likelihood (NLL) loss function [30], as $\mathcal{L}_{\mathrm{surv}}$. The total training loss for end-to-end model training is then calculated with weight factors α:

$$\mathcal{L} = \mathcal{L}_{\mathrm{surv}} + \alpha_1\mathcal{L}_{I(G;Z_G)} + \alpha_2\mathcal{L}_{I(T;Z_T)}. \tag{6}$$

Table 1. C-Index (mean ± std) performance over three cancer datasets. The best results are shown in **bold**, and the second best ones are underlined. G: use gene, W: use WSI, and T: use pathology report.

Model	Mod.	BLCA	BRCA	LUAD	Avg.↑
SNN [13]	G	0.649 ± 0.061	0.694 ± 0.094	0.623 ± 0.035	0.655
CLAM-MB [18]	W	0.655 ± 0.026	0.604 ± 0.052	0.594 ± 0.046	0.618
TransMIL [19]	W	0.647 ± 0.041	0.606 ± 0.054	0.627 ± 0.420	0.627
MHIM [25]	W	0.620 ± 0.031	0.572 ± 0.085	0.593 ± 0.054	0.595
R^2T-MIL [26]	W	0.651 ± 0.036	0.605 ± 0.063	0.611 ± 0.026	0.622
PANTHER [21]	W	0.612 ± 0.039	0.616 ± 0.071	0.581 ± 0.072	0.603
MCAT [3]	G+W	0.668 ± 0.065	0.723 ± 0.080	0.627 ± 0.055	0.673
MOTCAT [29]	G+W	0.673 ± 0.064	0.715 ± 0.052	0.617 ± 0.041	0.668
CMTA [31]	G+W	0.665 ± 0.044	0.710 ± 0.081	$\underline{0.661 \pm 0.035}$	$\underline{0.679}$
MMP [22]	G+W	$\underline{0.674 \pm 0.052}$	0.716 ± 0.066	0.607 ± 0.026	0.666
Pathomic [2]	G+W+T	0.657 ± 0.043	$\underline{0.727 \pm 0.028}$	0.614 ± 0.063	0.666
TMSE(Ours)	G+W+T	$\mathbf{0.687 \pm 0.057}$	$\mathbf{0.738 \pm 0.067}$	$\mathbf{0.682 \pm 0.057}$	**0.702**

3 Experiments

Datasets and Evaluation Metrics. We apply our method to three cancer survival datasets from the TCGA database, including Breast invasive carcinoma (BRCA) (n = 875), Bladder Urothelial Carcinoma (BLCA) (n = 328), and Lung Adenocarcinoma (LUAD) (n = 401). These datasets provide WSI, genomic data, patient pathology reports [11], and overall survival (OS) times. We assess the model performance using the concordance index (C-Index).

Implementation Details. In each experiment, we perform 5-fold cross validation. All models are trained for 40 epochs with a learning rate of 10^{-4}, using a cosine decay scheduler, AdamW optimizer with weight decay of 10^{-5}. We set the batch size to 1 and the number of bins to 4 in the NLL loss setting. The Gaussian prototype number is set to $C_w = 16$.

3.1 Comparison Results

We compare our model with the current state-of-the-art models in Table 1. All multimodal fusion models utilize both WSI and gene data, and we maintain this setup. Due to the scarcity of tri-modal survival prediction methods, we adapt an existing tri-modal fusion approach Pathomic [2] for comparison.

Our TMSE model outperforms all baseline methods, achieving best performance across three datasets, demonstrating its effectiveness and generalizability to various cancer datasets. By integrating three modalities, TMSE significantly outperforms unimodal models. Furthermore, compared to the contrasted tri-modal model, our model greatly enhances the efficiency of interactions. Kaplan-

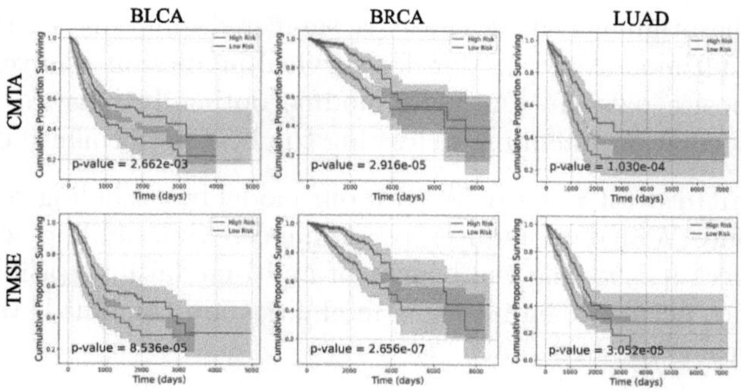

Fig. 2. Kaplan-Meier analysis for TMSE and CMTA on the three datasets.

Meier analysis in Fig. 2 further confirms TMSE's superior patient risk stratification, enhancing its p-value over the second-best model.

Table 2. Ablation results of our method.

Module	Mod.	BLCA	BRCA	LUAD	Avg.↑
w/o Proto.	W+G+T	0.663 ± 0.068	0.690 ± 0.076	0.671 ± 0.043	0.675
w/o Mamba	W+G+T	0.684 ± 0.060	0.728 ± 0.039	0.678 ± 0.060	0.697
w/o IR	W+G+T	0.668 ± 0.053	0.713 ± 0.044	0.671 ± 0.043	0.684
w/o AEI	W+G+T	0.681 ± 0.062	0.677 ± 0.053	0.682 ± 0.051	0.680
TMSE	W	0.605 ± 0.030	0.580 ± 0.066	0.600 ± 0.074	0.595
TMSE	W+T	0.637 ± 0.052	0.608 ± 0.041	0.668 ± 0.051	0.638
TMSE	W+G	0.675 ± 0.052	0.716 ± 0.044	0.648 ± 0.055	0.680
TMSE(Ours)	W+G+T	**0.687 ± 0.057**	**0.738 ± 0.067**	**0.682 ± 0.057**	**0.702**

3.2 Ablation Studies

Effect of Prototypes. We replace prototypes in the AEI module with all patch tokens, while retaining prototypes for the final survival prediction to preserve the integrity of the architecture. A C-index drop across three datasets shows that compressed prototypes boost information interaction efficiency. We drop the Mamba module, keeping only GMM-based prototypes. The C-index decreases, hinting at limitations in GMM-based feature extraction. Mamba-derived tokens, capturing broader context, prove vital for the task.

Effect of Different Modules in AEI. We explore the tradeoff in AEI between information retention and fusion by removing information retention (IR) while preserving the attention mechanism, using only the fused information, or completely removing the AEI module and relying on the original features. For BLCA and LUAD, dropping information retention significantly lowers performance, while using the original feature causes a smaller decline, indicating the value of retaining multimodal information for survival prediction. Conversely, BRCA shows an opposite pattern. Our AEI module optimally balances information fusion and information retention, delivering the best performance overall.

Effect of Multimodal Fusion. We test our model by excluding certain modalities. Using only WSI data reduces performance, while adding text and gene profiles enhances it, showing the benefit of multi-modal integration. The highest performance occurs when all three modalities are combined, validating our three-modal fusion method (Table 2).

4 Conclusion

We propose TMSE, a tri-modal survival analysis method integrating WSI, genomic profiles, and pathology reports. It uses expert-driven report extraction, a GMM-Mamba hybrid for WSI features, and entropy-combined attention to fusion modality information. Extensive tests on three cancer datasets show that our method outperforms other methods in OS survival prediction.

Acknowledgments. This work was supported by the National Key R&D Program of China (2023YFC2415200), National Natural Science Foundation of China (82441018, U24A20759, 82372053, 82361168664, 82302296), Beijing Natural Science Foundation (JQ24048), Scientific and Technological Innovation Project of China Academy of Chinese Medical Sciences (CI2023C008YG), Science and Technology Development Fund of Macao Special Administrative Region (0006/2023/AFJ).

Disclosure of Interests. The authors have no competing interests to declare that are relevant to the content of this article.

References

1. Acosta, J.N., Falcone, G.J., Rajpurkar, P., Topol, E.J.: Multimodal biomedical AI. Nat. Med. **28**(9), 1773–1784 (2022)
2. Chen, R.J., et al.: Pathomic fusion: an integrated framework for fusing histopathology and genomic features for cancer diagnosis and prognosis. IEEE Trans. Med. Imaging **41**(4), 757–770 (2020)
3. Chen, R.J., et al.: Multimodal co-attention transformer for survival prediction in gigapixel whole slide images. In: Proceedings of the IEEE/CVF International Conference on Computer Vision, pp. 4015–4025 (2021)
4. Dong, D., et al.: Deep learning radiomic nomogram can predict the number of lymph node metastasis in locally advanced gastric cancer: an international multicenter study. Ann. Oncol. **31**(7), 912–920 (2020)
5. Dong, D., et al.: Development and validation of an individualized nomogram to identify occult peritoneal metastasis in patients with advanced gastric cancer. Ann. Oncol. **30**(3), 431–438 (2019)
6. Giraldo, L.G.S., Rao, M., Principe, J.C.: Measures of entropy from data using infinitely divisible kernels. IEEE Trans. Inf. Theory **61**(1), 535–548 (2014)

7. Gu, A., Dao, T.: Mamba: linear-time sequence modeling with selective state spaces. arXiv preprint arXiv:2312.00752 (2023)
8. Gu, Y., et al.: Domain-specific language model pretraining for biomedical natural language processing. ACM Trans. Comput. Healthc. (HEALTH) **3**(1), 1–23 (2021)
9. Huang, Z., Bianchi, F., Yuksekgonul, M., Montine, T.J., Zou, J.: A visual-language foundation model for pathology image analysis using medical twitter. Nat. Med. **29**(9), 2307–2316 (2023)
10. Jaume, G., Vaidya, A., Chen, R.J., Williamson, D.F., Liang, P.P., Mahmood, F.: Modeling dense multimodal interactions between biological pathways and histology for survival prediction. In: Proceedings of the IEEE/CVF Conference on Computer Vision and Pattern Recognition, pp. 11579–11590 (2024)
11. Kefeli, J., Tatonetti, N.: TCGA-reports: a machine-readable pathology report resource for benchmarking text-based AI models. Patterns **5**(3) (2024)
12. Kim, M.: Differentiable expectation-maximization for set representation learning. In: International Conference on Learning Representations (2022)
13. Klambauer, G., Unterthiner, T., Mayr, A., Hochreiter, S.: Self-normalizing neural networks. In: Advances in Neural Information Processing Systems, vol. 30 (2017)
14. Li, H., et al.: Generalizable whole slide image classification with fine-grained visual-semantic interaction. In: Proceedings of the IEEE/CVF Conference on Computer Vision and Pattern Recognition, pp. 11398–11407 (2024)
15. Lin, D.F., et al.: Radiomic signatures associated with tumor immune heterogeneity predict survival in locally recurrent nasopharyngeal carcinoma. JNCI J. Natl. Cancer Inst. **116**(8), 1294–1302 (2024)
16. Lipkova, J., et al.: Artificial intelligence for multimodal data integration in oncology. Cancer Cell **40**(10), 1095–1110 (2022)
17. Liu, A., et al.: Deepseek-v3 technical report. arXiv preprint arXiv:2412.19437 (2024)
18. Lu, M.Y., Williamson, D.F., Chen, T.Y., Chen, R.J., Barbieri, M., Mahmood, F.: Data-efficient and weakly supervised computational pathology on whole-slide images. Nat. Biomed. Eng. **5**(6), 555–570 (2021)
19. Shao, Z., et al.: Transmil: transformer based correlated multiple instance learning for whole slide image classification. Adv. Neural. Inf. Process. Syst. **34**, 2136–2147 (2021)
20. Shi, J., et al.: Mg-trans: multi-scale graph transformer with information bottleneck for whole slide image classification. IEEE Trans. Med. Imaging **42**(12), 3871–3883 (2023)
21. Song, A.H., Chen, R.J., Ding, T., Williamson, D.F., Jaume, G., Mahmood, F.: Morphological prototyping for unsupervised slide representation learning in computational pathology. In: Proceedings of the IEEE/CVF Conference on Computer Vision and Pattern Recognition, pp. 11566–11578 (2024)
22. Song, A.H., Chen, R.J., Jaume, G., Vaidya, A.J., Baras, A., Mahmood, F.: Multimodal prototyping for cancer survival prediction. In: Forty-First International Conference on Machine Learning (2024)
23. Song, A.H., et al.: Artificial intelligence for digital and computational pathology. Nat. Rev. Bioeng. **1**(12), 930–949 (2023)
24. Steyaert, S., et al.: Multimodal data fusion for cancer biomarker discovery with deep learning. Nat. Mach. Intell. **5**(4), 351–362 (2023)
25. Tang, W., Huang, S., Zhang, X., Zhou, F., Zhang, Y., Liu, B.: Multiple instance learning framework with masked hard instance mining for whole slide image classification. In: Proceedings of the IEEE/CVF International Conference on Computer Vision, pp. 4078–4087 (2023)

26. Tang, W., Zhou, F., Huang, S., Zhu, X., Zhang, Y., Liu, B.: Feature re-embedding: towards foundation model-level performance in computational pathology. In: Proceedings of the IEEE/CVF Conference on Computer Vision and Pattern Recognition, pp. 11343–11352 (2024)
27. Wang, Z., et al.: Radiomics and deep learning in nasopharyngeal carcinoma: a review. IEEE Rev. Biomed. Eng. **17**, 118–135 (2023)
28. Xu, P., Zhu, X., Clifton, D.A.: Multimodal learning with transformers: a survey. IEEE Trans. Pattern Anal. Mach. Intell. **45**(10), 12113–12132 (2023)
29. Xu, Y., Chen, H.: Multimodal optimal transport-based co-attention transformer with global structure consistency for survival prediction. In: Proceedings of the IEEE/CVF International Conference on Computer Vision, pp. 21241–21251 (2023)
30. Zadeh, S.G., Schmid, M.: Bias in cross-entropy-based training of deep survival networks. IEEE Trans. Pattern Anal. Mach. Intell. **43**(9), 3126–3137 (2020)
31. Zhou, F., Chen, H.: Cross-modal translation and alignment for survival analysis. In: Proceedings of the IEEE/CVF International Conference on Computer Vision, pp. 21485–21494 (2023)

Top-Down Attention-Based Multiple Instance Learning for Whole Slide Image Analysis

Daniel Reisenbüchler[1(✉)], Ruining Deng[2], Christian Matek[3], Friedrich Feuerhake[4], and Dorit Merhof[5]

[1] Institute of Image Analysis and Computer Vision, University of Regensburg, Regensburg, Germany
daniel.reisenbuechler@informatik.uni-regensburg.de
[2] Weill Cornell Medicine, New York, USA
[3] Institute of Pathology, Universitätsklinikum Erlangen, Friedrich-Alexander-Universität Erlangen-Nürnberg, Erlangen, Germany
[4] Institute of Pathology, Hannover Medical School, Hannover, Germany
[5] Fraunhofer Institute for Digital Medicine MEVIS, Bremen, Germany

Abstract. Multiple instance learning (MIL) has become the de facto standard approach for whole-slide image analysis in computational pathology (CPath). While instance-wise attention tends to miss correlations between instances, self-attention can capture these interactions, but remains agnostic to the particular task. To address this issue, we introduce **T**op-**D**own **A**ttention-based **M**ultiple **I**nstance **L**earning (TDA-MIL), an architecture that first learns a general representation from the data via self-attention in an initial inference step, then identifies task-relevant instances through a feature selection module, and finally refines these representations by injecting the selected instances back into the attention mechanism for a second inference step. By focusing on task-specific signals, TDA-MIL effectively discerns subtle, yet significant, regions within each slide, leading to more precise classification. Extensive experiments on detecting lymph node metastasis in breast cancer, biomarker screening for microsatellite instability in different organs, and challenging molecular status prediction for HER2 in breast cancer show that TDA-MIL consistently surpasses other MIL baselines, underscoring the effectiveness of our proposed task-relevant refocusing and its broad applicability across CPath tasks. Our implementation is released at https://github.com/agentdr1/TDA_MIL.

Keywords: Computational Pathology · Multiple Instance Learning · Metastasis Detection · Molecular Status Prediction · Biomarker Screening

1 Introduction

Deep learning has led to significant advances in CPath, fuelled by the increasing use of whole-slide images (WSIs). However, WSIs can exceed 100,000 × 100,000

J. C. Gee et al. (Eds.): MICCAI 2025, LNCS 15960, pp. 651–660, 2026.
https://doi.org/10.1007/978-3-032-04927-8_62

pixels at 20× magnification (1.5 µm/pixel), posing significant computational challenges. A common strategy is to partition WSIs into a sequence of manageable patches (e.g., 512 × 512 pixels), which are then processed into feature representations in an offline fashion by a vision encoding model. Increasingly, self-supervised pretraining has shown promise in training domain-specific foundation models (FMs) [3], outperforming generic counterparts (e.g., ImageNet pretrained) [8]. However, transforming patch-level features into WSI-level predictions relies on MIL to identify and weight the most salient patches. Early MIL methods employ instance-wise attention [7] to highlight influential patches, but overlook contextual relationships. Recent efforts thus incorporate local neighborhood attention for stronger inductive bias [12], or apply global self-attention mechanisms to model interactions between patches [12,13,19]. Notably, a large-scale study involving a population of over 10,000 patients demonstrated that self-attention not only offers better performance but also faster convergence compared to instance-wise schemes, highlighting the importance of context in WSI analysis [19]. However, although self-attention excels at general context modeling, vanilla formulations in vision transformers tend to focus on broadly discriminative features rather than task-specific cues [14,15]. Moreover, pathologists in clinical practice first capture global information from WSIs and then focus on specific regions of interest, e.g., morphological features or objects, depending on the task. Motivated by the above technical and clinical observations, we propose *TDA-MIL*, a two-step framework that first contextualizes all patches via self-attention, then refines their representations through a dedicated feature selection module. The selected task-relevant tokens are reintroduced into the attention mechanism, further guiding the model to focus on task significant regions. This top-down infusion of task-specific context enables the network to effectively filter and contextualize up to thousands of instances, boosting the accuracy of the final WSI-level predictions. Our main contributions are summarized as follows:

(1) **Novel two-step MIL architecture.** We propose TDA-MIL, a coarse-to-fine MIL framework that (I) contextualizes all patches with self-attention and weights them through a feature-selection module, and (II) applies top-down re-attention to focus efficiently on task-specific entities.
(2) **Performance & Interpretability.** TDA-MIL consistently surpasses MIL baselines on diverse CPath benchmarks. Heat-maps show that the feature-selection module pinpoints biomarker-specific ducts overlooked by vanilla self-attention, providing clearer diagnostically relevant insights.

2 Method

We provide an overview of our TDA-MIL pipeline in Fig. 1. The framework consists of an offline *feature compression stage* and an online *aggregation stage* that leverages task-specific top-down attention strategy and the feature selection module. Below, we detail the core components of our algorithm.

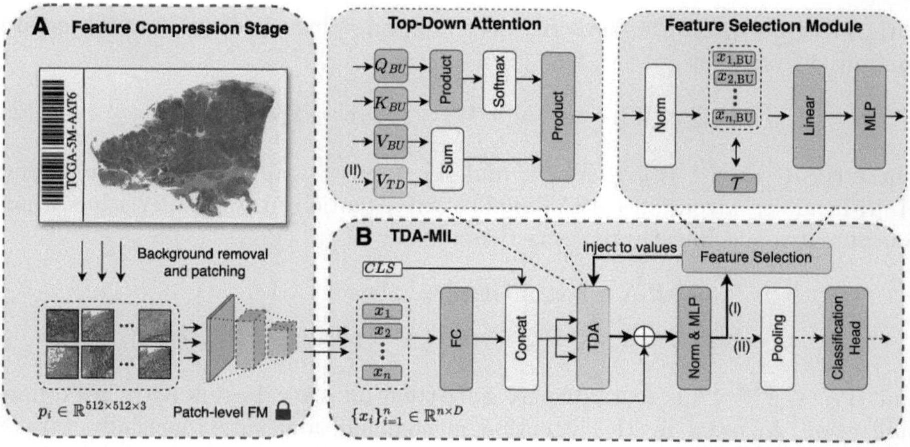

Fig. 1. Overview of the TDA-MIL pipeline. **(A)** The *offline* feature compression stage removes background areas, tessellates the WSI, and extracts patch-level features via a pathology foundation model. **(B)** The *online* aggregation stage then proceeds in two steps: (I) dimensionality reduction, self-attention-based contextualization, and feature selection, followed by (II) injecting the selected task-relevant features back into the self-attention mechanism for the final WSI-level prediction.

(A) Feature Compression Stage. Following standard preprocessing, the WSI is segmented to remove background regions using Otsu's thresholding and tessellated at 20× magnification into n smaller patches, where each $p_i \in \mathbb{R}^{512\times512\times3}$. Each patch is then passed through a vision encoding FM, extracting feature embeddings in an offline fashion. Overall, this step reduces the high-dimensional WSIs to a tractable set of patch-level feature vectors $\{x_i\}_{i=1}^n \in \mathbb{R}^{n\times D}$, where D is the resulting latent dimension depending on the used FM.

(B) TDA-MIL. In the online stage, TDA-MIL processes patch features via two sequential inference steps. Given the sequence of n features $\{x_i\}_{i=1}^n$, we first project each feature from dimension D to a lower dimension d using a fully connected (FC) layer. Subsequently, a classification token CLS $\in \mathbb{R}^{1\times D}$ is concatenated to the sequence. In the following, the CLS is treated as other tokens and we continue to denote the sequence as length n for simplicity. We denote the resulting sequence as bottom-up sequence $\{x_{i,\text{BU}}\}_{i=1}^n$.

Inference Step I:
The sequence $\{x_i\}_{i=1}^n$ is next feed through l blocks of self-attention (SA):

$$\text{SA}(Q, K, V) = \text{softmax}\left(\frac{QK^T}{\sqrt{d_k}}\right)V, \tag{1}$$

with queries $Q \in \mathbb{R}^{n \times d_k}$, keys $K \in \mathbb{R}^{n \times d_k}$, and values $V \in \mathbb{R}^{n \times d_v}$. These are computed from x by

$$Q = W_Q \cdot x, \quad K = W_K \cdot x, \quad V = W_V \cdot x, \tag{2}$$

where $W_Q \in \mathbb{R}^{d \times d_k}$, $W_K \in \mathbb{R}^{d \times d_k}$, and $W_V \in \mathbb{R}^{d \times d_v}$ are learnable parameters. Multi-head self-attention (MSA) applies self-attention in h parallel heads, then concatenates and linearly projects them:

$$\text{MSA} = \text{concat}(\text{head}_1, ..., \text{head}_h) \cdot W_O,$$
$$\text{where head}_i = \text{SA}(Q^{(j)}, K^{(j)}, V^{(j)}) \text{ for } j \in \{1, ..., h\}$$

and $W_O \in \mathbb{R}^{hd_v \times d}$ is learnable. A self-attention layer begins with layer normalization, followed by the attention mechanism, and subsequentially with a multi-layer perceptron (MLP). Next, the sequence enters the feature selection module, which refines the output sequence $\{x_i\}_{i=1}^n$ by identifying task-relevant patches (dimension n) and channel rescaling (dimension d); see Fig. 2.

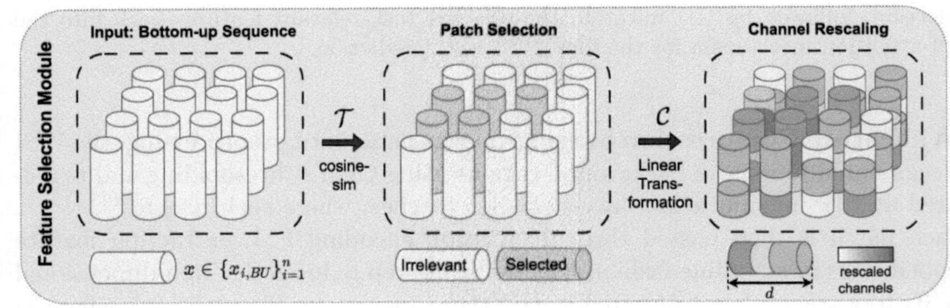

Fig. 2. Illustration of the internal behavior of the feature selection module. Each element (patch of WSI) of the input sequence is compared with a learnable task relevance token $\mathcal{T}$, and the linear transformation $\mathcal{C}$ performs tile-wise channel rescaling.

The tile selection process is defined as

$$\hat{x}_{i,\text{BU}} = \text{clamp}\left(\text{sim}\left(x_{i,\text{BU}}, \mathcal{T}\right)\right), \tag{3}$$

where sim$(\cdot, \cdot)$ denotes the cosine similarity, clamp$(\cdot)$ restricts values to $[0, 1]$ and $\mathcal{T} \in \mathbb{R}^d$ is a learnable parameter token encoding task relevance. The top-down sequence is then computed as

$$x_{i,\text{TD}} = \mathcal{C} \cdot \hat{x}_{i,\text{BU}} \cdot x_{i,\text{BU}} \text{ for } i \in \{1, ..., n\} \tag{4}$$

where $\mathcal{C} \in \mathbb{R}^{d \times d}$ is learnable. As outlined in Fig. 2, the parameter $\mathcal{T}$ acts as a task embedding to filter irrelevant tiles in a weighted fashion, while $\mathcal{C}$ performs tile-wise channel rescaling. Finally, another MLP decodes the resulting feature sequence $\{x_{i,\text{TD}}\}_{i=1}^n$ before proceeding to the second inference step.

Inference Step II: The selected tiles $x_{i,\mathrm{TD}}$ re-enter the self-attention (Eq. 1) from *Inference Step I* for a second pass. To this end, the self-attention module receives the top-down input by adding them to the values, i.e. the values in Eq. (2) are infused with information from the feature selection module (4). Specifically, we add $x_{i,\mathrm{TD}}$ to the values V while leaving the queries Q and keys K unchanged,

$$V = W_V \cdot \left(x_{\mathrm{BU}} + x_{\mathrm{TD}}\right),$$

where x_{BU} is the bottom-up sequence as in the beginning of *Inference Step I*. After self-attention, the classification token CLS is fed to a final fully connected layer, transforming $\mathbb{R}^d$ to $\mathbb{R}^c$ for the prediction, where c is the number of classes.

3 Experiments

We evaluate the effectiveness of TDA-MIL across several key tasks CPath. Below, we introduce the datasets, baselines, evaluation and implementation details.

3.1 Tasks, Datasets and Annotations

Overall, our study comprises 3,719 publicly available WSIs stained from HE stained tissue of the organs breast, colon, rectum, stomach and uterine.

Metastases Detection. We use the CAMELYON17 dataset [1] for lymph node metastasis detection in breast cancer. Following [6,18], we binarize the classes to focus on metastasis presence versus absence (N = 500; 182 positive, 318 negative). The challenge dataset is available at camelyon17.grand-challenge.org/Data/.

MSI Screening. We use TCGA-CRC [2] (N = 457; 65 positive, 392 negative) and CPTAC-COAD [4] (N = 105; 24 positive, 81 negative) to predict microsatellite instability status (MSI) in colorectal cancer. TCGA-STAD involves stomach cancer (N = 361; 60 positive, 301 negative) and TCGA-UCEC uterine endometrial carcinoma (N = 545; 117 positive, 428 negative). All TCGA datasets are publicly available at portal.gdc.cancer.gov/ and CPTAC at wiki.cancerimagingarchive.net.

Molecular Status Prediction HER2. We employ TCGA-BRCA (N = 693; 158 positive, 535 negative) for molecular status prediction of human epidermal growth factor receptor 2 (HER2) in breast cancer. We further add the Breast Cancer Needle Biopsy dataset BCNB [20] (N = 1,058; 277 positive, 781 negative) cohort, which is publicly available: paperswithcode.com/dataset/bcdalnmp.

Annotations. CAMELYON17 and BCNB provide their respective ground-truth labels. For TCGA and CPTAC, we matched clinical labels from cbioportal.org.

3.2 Comparable Methods

We benchmark TDA-MIL against comparable MIL methods, including AB-MIL [7] as a classic instance-wise attention baseline, CLAM [10] with clustering-constrained attention, DSMIL [9] employing a dual-stream attention strategy, LA-MIL [12] leveraging local attention, TransMIL [13] relying on transformer based aggregation, GPT [21] using graph transformers, RRT-MIL [17] focusing on feature re-embedding, MHIM [16] applying masked hard instance mining and S4MIL [5] adopting a structured state-space approach. Baselines are retrieved from their official repository and configured according to the published default settings. We used the same feature compression stage (Fig. 1A) for all methods.

3.3 Evaluation

We perform patient-stratified 5-fold cross-validations (CVs) for each task and report results using the area under the receiver operating curve (AUROC) and balanced accuracy (Bal. Acc), which accommodate class imbalances.

3.4 Implementation

Patch extraction is carried out with the CLAM library [10]. We employed the UNI FM [3] for feature extraction, which was not pre-trained on any dataset involved in this study and thus avoid any kind of potential data contamination from pretraining. UNI is based on a ViT-L architecture with output dimension 1024. We used a class-weighted rescaled cross-entropy loss to remedy uneven class distributions during the training process. Training is conducted with a batch size of 1 using the ADAM optimizer at a learning rate of 10^{-5} and weight decay of 10^{-2}, for a maximum of 100 epochs. The learning rate is reduced by a factor of 10 when performance plateaus for 5 subsequent epochs, and early stopping terminates training if performance fails to improve for 10 consecutive epochs. All experiments were performed on a single NVIDIA H100 GPU card.

4 Results

4.1 Performance Analysis

Table 1 reports classification performance across CPath tasks against baselines for CAMELYON17, Molecular status prediction of HER2 in TCGA-BRCA and BCNB, MSI Screening in TCGA-CRC and CPTAC-COAD. Further MSI Screening results in TCGA-STAD and TCGA-UCEC are visualized in Fig. 3a. In metastases detection, TDA-MIL achieves with 97.20 the best AUROC with +1.41 compared to the second best method CLAM. In Bal. Acc, TDA-MIL surpasses AB-MIL with a minor difference. For molecular status prediction HER2, CLAM and LA-MIL provide the second-best results in terms of AUROC and Bal. Acc,

Table 1. Performance across different CPath Tasks and Datasets. Evaluation using AUROC and balanced accuracy (Bal. Acc) metrics. Mean and standard deviation are reported over patient-stratified CV runs. Best in **bold**, second best is underlined.

Task	Metastases detection		HER2 Status		MSI Screening	
Datasets	CAMELYON17		TCGA-BRCA BCNB		TCGA-CRC CPTAC-COAD	
Model/Metric	AUROC	Bal. Acc	AUROC	Bal. Acc	AUROC	Bal. Acc
AB-MIL [7]	94.68 ± 3.9	93.35 ± 4.8	73.30 ± 1.3	64.82 ± 2.8	91.21 ± 1.5	78.54 ± 5.7
CLAM [10]	95.79 ± 2.8	93.10 ± 4.8	73.29 ± 1.2	66.29 ± 2.4	90.87 ± 2.3	77.87 ± 4.8
DSMIL [9]	82.14 ± 13	78.78 ± 12	60.75 ± 6.7	59.71 ± 5.7	68.37 ± 18	65.97 ± 13
GPT [21]	92.60 ± 4.1	89.96 ± 5.7	67.21 ± 2.4	63.29 ± 1.9	82.02 ± 5.6	74.94 ± 7.3
LA-MIL [12]	85.81 ± 4.1	79.30 ± 5.9	72.51 ± 2.7	67.27 ± 2.7	90.02 ± 3.0	83.29 ± 3.5
MHIM [16]	95.69 ± 2.5	93.12 ± 3.1	72.36 ± 1.3	66.14 ± 1.7	89.61 ± 2.1	78.71 ± 5.2
RRT-MIL [17]	92.86 ± 2.9	88.75 ± 4.0	69.91 ± 1.4	63.37 ± 2.0	89.98 ± 1.8	80.56 ± 5.1
S4MIL [5]	90.04 ± 1.0	86.43 ± 1.1	71.45 ± 2.1	64.88 ± 2.0	89.86 ± 1.6	81.73 ± 5.6
TransMIL [13]	92.11 ± 1.4	87.08 ± 1.5	71.46 ± 1.8	62.95 ± 1.8	89.61 ± 1.6	77.24 ± 2.1
TDA-MIL	**97.20** ± 1.2	**93.38** ± 1.8	**73.66** ± 1.1	**68.62** ± 2.9	**91.78** ± 2.3	**86.45** ± 1.7

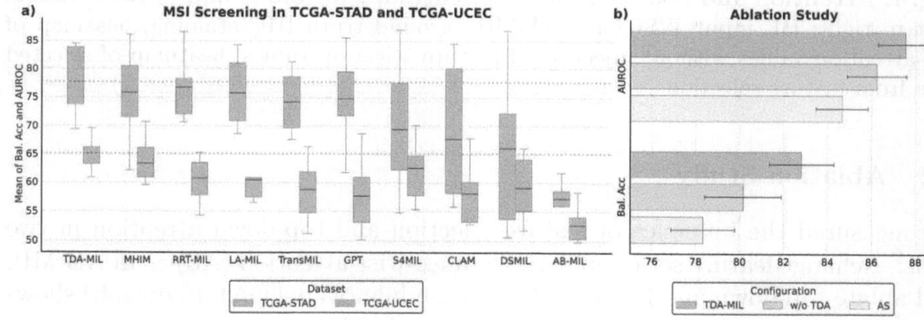

Fig. 3. Evaluation of MSI Screening on TCGA-STAD and TCGA-UCEC, and an Ablation Study. (a) MSI screening results are reported as mean of AUROC and balanced accuracy for each method; bars show patient-stratified cross-validation outcomes and dotted horizontal lines denote the average performance of TDA-MIL. **(b)** Ablation study: replacing the feature selection module with instance-wise attention (AS) and using pure self-attention, i.e., removing TDA (w/o TDA). Results are computed as the mean of the three tasks as in Table 1.

respectively; and TDA-MIL performs best overall. In MSI screening in colorectal cancer, TDA-MIL achieves a Bal. Acc of 86.45 with an improvement of 3.16 over the second best LA-MIL model and outperforms AB-MIL marginally in AUROC. However, for MSI prediction in stomach and uterine datasets RRT-MIL and MHIM are the second-best performing methods respectively, which are both outperformed by TDA-MIL. Overall, there is no single second-best baseline across all tasks, while TDA-MIL consistently surpasses the task-winning second-best approaches across various CPath tasks.

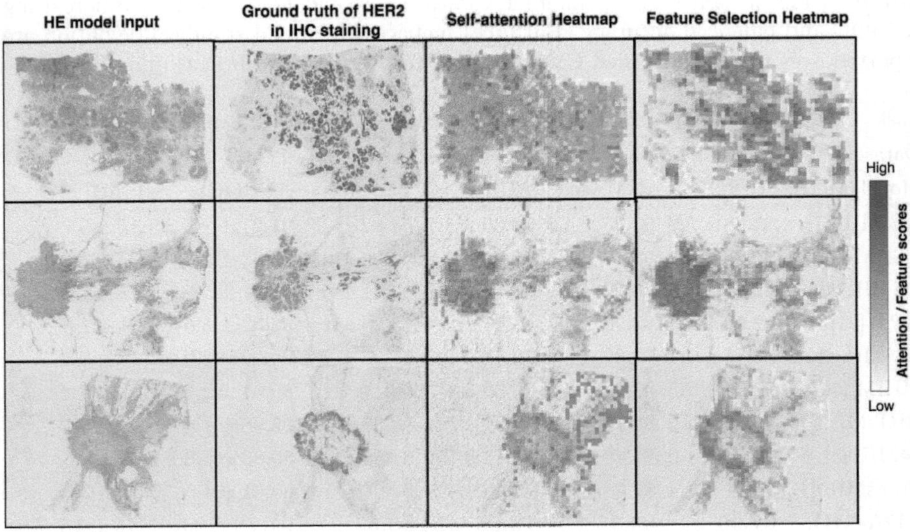

Fig. 4. Attention and feature selection heatmaps for molecular status HER2. Left to right: HE input WSI for TDA-MIL, ground truth IHC staining, heatmap of self-attention values without injection of feature selection values, heatmap of selected tiles from feature selection.

4.2 Ablation Study

We measured the influence of feature selection and top-down attention in two folds: exchange feature selection with instance-wise attention (AS) as in AB-MIL and ablate top-down and feature selection module (w/o TDA). Figure 3b) shows the results for exchanging the components or ablating components. Exchanging feature selection with instance-wise attention yields the worst performance across tasks. Ablating the TDA and feature selection part, i.e. using pure self-attention, decrease performance.

4.3 Qualitative Evaluation and Interpretability

Figure 3 shows an HE input for TDA-MIL, an IHC stained image of the same tissue for HER2, self-attention scores and task relevant tiles computed from the feature selection module. Samples in multi-stain are selected for clear IHC marker visibility from external ACROBAT dataset [11]. For the feature selection module, we visualized the normalized scores $\hat{x}_{i,\text{BU}}$ as defined by Eq. 3. It can be observed that more task-relevant areas are highlighted and less artifacts are visible in the heatmaps for selected features compared to self-attention values (Fig. 4).

5 Conclusion

In this work, we introduced TDA-MIL, a top-down attention-based approach that addresses a limitation of self-attention when used in MIL strategies for whole-slide image analysis. By integrating a feature selection module with self-attention in a two-step inference procedure, TDA-MIL learns a robust general representation and then refocuses the model on task-relevant patches, thereby offering enhanced context modeling leading to refined discriminative power. Extensive evaluations in CPath tasks demonstrate its broad applicability, highlighting the value of task-specific refinement and underscore the potential of TDA-MIL as a generalizable framework for challenging gigapixel image analyses.

Acknowledgements. This work was supported by the German Research Foundation (Deutsche Forschungsgemeinschaf, DFG) under project number 445703531. The authors gratefully acknowledge the computational and data resources provided by the Leibniz Supercomputing Centre (www.lrz.de).

Disclosure of Interests. The authors declare that they have no conflicts of interest related to this work.

References

1. Bandi, P., et al.: From detection of individual metastases to classification of lymph node status at the patient level: the CAMELYON17 challenge. IEEE Trans. Med. Imaging **38**(2), 550–560 (2019)
2. Network, C.G.A.: Comprehensive molecular characterization of human colon and rectal cancer. Nature **487**(7407), 330–337 (2012)
3. Chen, R.J., et al.: Towards a general-purpose foundation model for computational pathology. Nat. Med. **30**(3), 850–862 (2024). https://doi.org/10.1038/s41591-024-02857-3
4. Edwards, N.J., et al.: The CPTAC data portal: a resource for cancer proteomics research. J. Proteome Res. **14**(6), 2707–2713 (2015)
5. Fillioux, L., Boyd, J., Vakalopoulou, M., Cournède, P.H., Christodoulidis, S.: Structured state space models for multiple instance learning in digital pathology. In: Lecture Notes in Computer Science, pp. 594–604. Springer, Cham (2023)
6. Fourkioti, O., De Vries, M., Bakal, C.: CAMIL: context-aware multiple instance learning for cancer detection and subtyping in whole slide images. In: The Twelfth International Conference on Learning Representations (2024). https://openreview.net/forum?id=rzBskAEmoc
7. Ilse, M., Tomczak, J., Welling, M.: Attention-based deep multiple instance learning. In: Dy, J., Krause, A. (eds.) Proceedings of the 35th International Conference on Machine Learning. Proceedings of Machine Learning Research, vol. 80, pp. 2127–2136. PMLR (2018). https://proceedings.mlr.press/v80/ilse18a.html
8. Kang, M., Song, H., Park, S., Yoo, D., Pereira, S.: Benchmarking self-supervised learning on diverse pathology datasets. In: Proceedings of the IEEE/CVF Conference on Computer Vision and Pattern Recognition (CVPR), pp. 3344–3354 (2023)

9. Li, B., Li, Y., Eliceiri, K.W.: Dual-stream multiple instance learning network for whole slide image classification with self-supervised contrastive learning. In: Proceedings of the IEEE/CVF Conference on Computer Vision and Pattern Recognition, pp. 14318–14328 (2021)
10. Lu, M.Y., Williamson, D.F., Chen, T.Y., Chen, R.J., Barbieri, M., Mahmood, F.: Data-efficient and weakly supervised computational pathology on whole-slide images. Nat. Biomed. Eng. **5**(6), 555–570 (2021)
11. Rantalainen, M., Hartman, J.: Acrobat - a multi-stain breast cancer histological whole-slide-image data set from routine diagnostics for computational pathology (2023). https://doi.org/10.48723/W728-P041. https://snd.se/catalogue/dataset/2022-190-1/1
12. Reisenbüchler, D., Wagner, S.J., Boxberg, M., Peng, T.: Local attention graph-based transformer for multi-target genetic alteration prediction. In: Lecture Notes in Computer Science, pp. 377–386. Springer (2022). https://doi.org/10.1007/978-3-031-16434-7_37
13. Shao, Z., et al.: Transmil: transformer based correlated multiple instance learning for whole slide image classification. Adv. Neural. Inf. Process. Syst. **34**, 2136–2147 (2021)
14. Shi, B., Darrell, T., Wang, X.: Top-down visual attention from analysis by synthesis. In: Proceedings of the IEEE/CVF Conference on Computer Vision and Pattern Recognition, pp. 2102–2112 (2023)
15. Shi, B., Gai, S., Darrell, T., Wang, X.: Toast: transfer learning via attention steering. arXiv preprint arXiv:2305.15542 (2023)
16. Tang, W., Huang, S., Zhang, X., Zhou, F., Zhang, Y., Liu, B.: Multiple instance learning framework with masked hard instance mining for whole slide image classification. In: Proceedings of the IEEE/CVF International Conference on Computer Vision (ICCV), pp. 4078–4087 (2023)
17. Tang, W., Zhou, F., Huang, S., Zhu, X., Zhang, Y., Liu, B.: Feature re-embedding: towards foundation model-level performance in computational pathology. In: Proceedings of the IEEE/CVF Conference on Computer Vision and Pattern Recognition (CVPR), pp. 11343–11352 (2024)
18. Tourniaire, P., Ilie, M., Hofman, P., Ayache, N., Delingette, H.: Ms-clam: mixed supervision for the classification and localization of tumors in whole slide images. Med. Image Anal. **85**, 102763 (2023). https://doi.org/10.1016/j.media.2023.102763
19. Wagner, S.J., et al.: Transformer-based biomarker prediction from colorectal cancer histology: a large-scale multicentric study. Cancer Cell **41**(9), 1650–1661.e4 (2023). https://doi.org/10.1016/j.ccell.2023.08.002
20. Xu, F., et al.: Predicting axillary lymph node metastasis in early breast cancer using deep learning on primary tumor biopsy slides. Front. Oncol. 4133 (2021)
21. Zheng, Y., et al.: A graph-transformer for whole slide image classification. IEEE Trans. Med. Imaging **41**(11), 3003–3015 (2022)

Wavelet-Driven Decoupling and Physics-Informed Mapping Network for Accelerated Multi-parametric MR Imaging

Ruilong Dan[1], Kaicong Sun[1(✉)], Yichen Zhou[1], Minqiang Jia[1], Yuxuan Liu[1], Han Zhang[1], Xiaopeng Zong[1], and Dinggang Shen[1,2,3(✉)]

[1] School of Biomedical Engineering and State Key Laboratory of Advanced Medical Materials and Devices, ShanghaiTech University, Shanghai 201210, China
sunkc@shanghaitech.edu.cn

[2] Shanghai United Imaging Intelligence Co., Ltd., Shanghai 200230, China
Dinggang.Shen@gmail.com

[3] Shanghai Clinical Research and Trial Center, Shanghai 201210, China

Abstract. Multi-parametric magnetic resonance imaging (MRI) is an advanced MRI technique that can provide multiple quantitative maps simultaneously based on acquired multi-echo images. However, the lengthy scan time often limits its application. Accelerated multi-parametric MRI using deep learning is of great interest. The existing studies have two limitations: 1) inefficient use of the multi-echo information; 2) lack of physical prior for parametric mapping. To address these issues, in this work, we propose a novel decoupling-driven and physics-informed reconstruction network for accelerated multi-parametric MRI. Specifically, to better align and integrate multi-echo information, we propose a novel decoupling technique consisting of wavelet-driven decoupling module, contrastive and echo-dependent decoupling losses, such that the multi-echo features can be effectively decoupled into echo-dependent and echo-independent components. Only the echo-independent features are fused across multiple echoes. Besides, Bloch equations are incorporated as physical priors to guide the parametric mapping network. Experimental results on our in-house data (12-echo sequence) show that our method outperforms the state-of-the-art methods by 1.54% in average SSIM and 1.70 dB in average PSNR for 4× acceleration, which significantly advances the performance limitation for multi-parametric MRI. Our code is available at https://github.com/IDEARL23/WDPM-Net.

Keywords: Multi-parametric MRI · Quantitative MRI · Wavelet-driven · Feature decoupling · Physics-informed mapping

1 Introduction

Multi-parametric magnetic resonance imaging (MRI) is an advanced MRI technique that can provide intrinsic tissue properties like proton density (PD), T_1

J. C. Gee et al. (Eds.): MICCAI 2025, LNCS 15960, pp. 661–671, 2026.
https://doi.org/10.1007/978-3-032-04927-8_63

map, and T_2^* map simultaneously through a single scan [11,22] and is safer than the radioactive imaging modalities, such as computed tomograph (CT) and positron emission tomography (PET) [5]. However, the multi-echo nature of the acquisition sequence results in prolonged scan times, thereby highlighting the importance of accelerated acquisition in multi-parametric MRI.

Multi-parametric MRI pipelines usually contain two separate steps: multi-echo image reconstruction from under-sampled k-space data and desired maps or images estimation via signal modeling [4,22,23]. Current deep learning-based approaches for multi-parametric MR imaging mainly fall into two categories: (1) Two-step methods [2,24] and (2) One-step methods [1,6,9]. The former one first reconstructs multi-echo images from undersampled k-space [7,8], and then estimates parametric maps via empirical physical models [4,22,23]. Such framework suffers from error propagation due to the remaining artifacts left in the reconstructed images in step one. The latter one like MANTIS [9] directly maps k-space to parametric maps via unified networks, yet ignores intermediate supervision on the reconstructed images. The recent work SRM-Net [11] attempts joint optimization of the reconstruction and mapping networks. Although they have achieved noticeable performance improvement, they still have two main limitations: (1) the highly-coupled multi-echo information is inadequately exploited; and (2) no physical prior is integrated for parametric mapping. In summary, both categories struggle to balance multi-echo synergy, physics constraints, and co-optimization, highlighting unmet needs for effective multi-echo feature interaction mechanisms and physically constrained mapping architectures.

To this end, we introduce a novel Wavelet-driven Decoupling and Physics-informed Mapping Network (WDPM-Net) designed for efficient multi-echo feature interaction and physics-guided parametric mapping. ***First***, we decouple the echo-independent features (e.g., anatomical structures) from the echo-dependent features (e.g., contrast) by adaptively weighed wavelet transformation and two elaborately designed decoupling losses. The echo-independent features are aligned and fused to form consensus for robust reconstruction. ***Second***, we introduce a physics-informed mapping network (PIMN) by integrating analytically estimated maps as prior to guide the mapping network, ensuring more accurate and reliable parametric mapping. In summary, our contributions are mainly four-fold:

1. We propose a novel Decoupling-driven and Physics-informed Reconstruction Network (WDPM-Net) for accelerated multi-parametric MRI, effectively addressing the limitations of inefficient multi-echo exploration and overlooking of physical priors in deep learning-based multi-parametric MRI.
2. We introduce an effective feature decoupling scheme including dedicated Wavelet-driven network and elaborately designed contrastive and echo-dependent consistency losses, which enables more efficient multi-echo integration.
3. We have developed the first physics-informed parametric mapping network by combining data-driven prediction with physics-driven Bloch equations, improving the robustness and fidelity of the estimated parametric maps.

4. Experimental results on our in-house dataset (12-echo sequence) demonstrate that our method outperforms the state-of-the-art methods by 1.54% in average SSIM and 1.70 dB in average PSNR for 4× acceleration, significantly advancing the performance limitation for multi-parametric MRI.

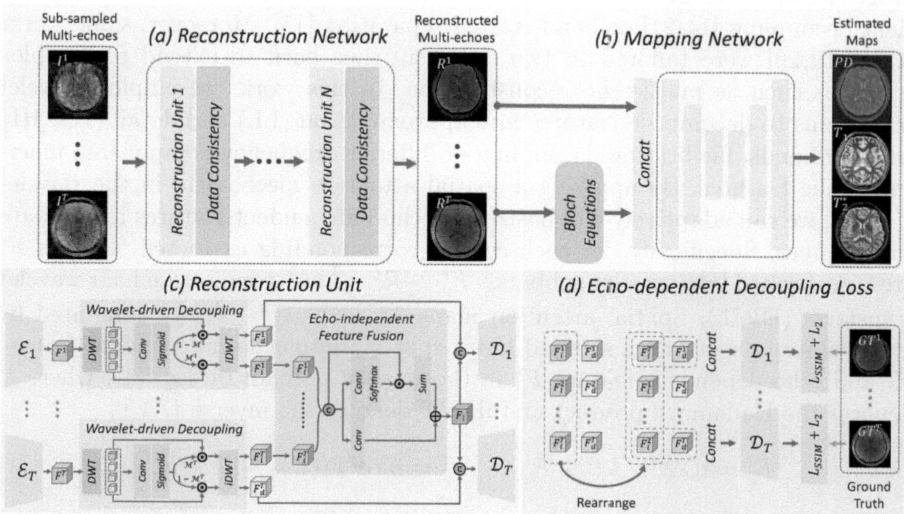

Fig. 1. The overall framework of the proposed WDPM-Net with (a) multi-echo reconstruction, (b) physics-informed parametric mapping in an end-to-end manner to accelerate multi-parametric MRI, (c) details of the reconstruction unit (RU), and (d) details of the echo-dependent decoupling loss. The reconstruction network consists of cascaded RUs, containing wavelet-driven decoupling and echo-independent feature fusion modules, to refine multi-echo MR reconstruction. The mapping network estimates the maps based on the reconstructed images under the guidance of Bloch equations.

2 Method

2.1 Overview of the Proposed Framework

Our framework comprises two main parts: (a) a multi-echo reconstruction network and (b) a parametric mapping network as shown in Fig. 1. The multi-echo reconstruction network consists of cascaded reconstruction units (RU) followed by a data consistency layer to reconstruct and refine the multi-echo images. Based on this, the physics-informed parametric mapping network generates the quantitative maps under the guidance of the initial estimation from multi-dimensional integration (MDI) and Bloch equations [22].

2.2 Wavelet-Driven Feature Decoupling and Fusion

Wavelet-Driven Decoupling Module. Feature decoupling aims to separate features into varied parts to finely control the information sharing adhered to domain priors and has presented promising performance in multi-modal feature fusion in MRI analysis [8,10,13,14]. Though there were attempts to disentangle features for MR reconstruction, their performance is limited by insufficient decoupling [14,21] or hand-crafted separation [15]. Moreover, some works [15,17,19,20] were tailored to two contrasts thus hard to extend to complex scenarios such as multi-echo reconstruction. In this work, we employ wavelet transforms to decompose features into approximation (LL) and detail (LH, HL, HH) subbands, facilitating decoupling of different frequency components inherent in the features. By applying a spatial attention mechanism in the wavelet domain, we can adaptively emphasize the echo-independent features in an adaptive manner. Specifically, for each echo t, corresponding extracted features F^t are transformed into wavelet subbands $F_w^t \in \mathcal{R}^{C\times\frac{H}{2}\times\frac{W}{2}}$ via discrete haar wavelet transform (DWT). Spatial attention maps $\mathcal{M}^t \in [0,1]^{4C\times\frac{H}{2}\times\frac{W}{2}}$, generated by a convolution layer and a sigmoid layer, split F_w^t into echo-independent feature F_i^t and echo-dependent feature F_d^t in the wavelet domain by Eq. (1), where $\odot$ denotes the Hadamard product and iDWT denotes the inverse DWT.

$$F_i^t = \text{iDWT}(\mathcal{M}^t \odot F_w^t), \quad F_d^t = \text{iDWT}((1-\mathcal{M}^t) \odot F_w^t) \tag{1}$$

Echo-Independent Feature Fusion. To obtain robust and consistent reconstruction across multi-echo images, we fuse echo-independent features $\{F_i^1, \ldots, F_i^T\}$ from T echoes. Specifically, we design a feature fusion module that adaptively fuses spatial structures across multiple echos. First, all features are concatenated and then processed by a convolutional layer to generate T spatial attention maps $\{\alpha^t\}_{t=1}^T$. These T maps are then normalized via softmax along the channel to adaptively weight the contribution of each echo's anatomical features. The weighted features are summed to produce an initial fused feature F_i^{init}, while a residual connection through a 1×1 convolution is added to F_i^{init}, yielding the final fused feature F_i that preserves consistent anatomical structures. The processes are formulated in Eqs. (2) and (3).

$$\{\alpha^t\}_{t=1}^T = \text{Softmax}\left(\text{Conv}\left(\text{Channel Concat}(F_i^1, F_i^2, \ldots, F_i^T)\right)\right) \tag{2}$$

$$F_i = \text{Conv}_{1\times1}\left(\text{Channel Concat}(F_i^1, F_i^2, \ldots, F_i^T)\right) + \sum_{t=1}^{T} \alpha^t \odot F_i^t \tag{3}$$

Echo-Dependent Decoupling Loss. To preserve the echo-dependent information of each echo image, e.g., contrast information, we propose an echo-dependent decoupling (ED) loss as depicted in Fig. 1(d). Specifically, the ED loss randomly rearranges the echo-independent features (i.e., F_i^1 to F_i^T) and constructs new T paired combinations. The re-ordered combinations are enforced to generate the same image as the corresponding ground-truth (GT) image. It is worth noting that our proposed ED loss has great scalability and can be extended to arbitrary amount of echo images. The ED loss is expressed as Eq. (4), where $\hat{I}^t$ is the t-th reconstructed echo image with λ_1 being a tunable hyperparameter.

$$\mathcal{L}_{\mathrm{ED}} = \sum_{t=1}^{T} [\|\hat{I}^t - \mathrm{GT}^t\|_2 + \lambda_1 L_{\mathrm{SSIM}}(\hat{I}^t, \mathrm{GT}^t)], \tag{4}$$

Contrastive Decoupling Loss. To explicitly perform decoupling and alignment for efficient fusion, a contrastive decoupling (CD) loss is designed to cluster the echo-independent features $\{F_i^t\}_{t=1}^T$ (i.e., positive pairs) and push the echo-dependent features $\{F_d^t\}_{t=1}^T$ apart (i.e., negative pairs). Specifically, all the $\{F_d^t\}_{t=1}^T$ and $\{F_i^t\}_{t=1}^T$ as well as the fused feature F_i are first flattened and normalized with L_2 norm. Thanks to the CD loss, we can regularize the feature space according to MR imaging priors, facilitating the decoupling between echo-independent and echo-dependent features while promoting alignment across echo-independent features. The total CD loss can be expressed as:

$$\mathcal{L}_{\mathrm{CD}} = \frac{1}{T(T-1)} \sum_{p \neq q} \cos(F_d^p, F_d^q) + \frac{1}{T} \sum_{t=1}^{T} \cos(F_i^t, F_d^t) - \frac{1}{T} \sum_{t=1}^{T} \cos(F_i^t, F_i) \tag{5}$$

2.3 Physics-Informed Network for Multi-parametric Mapping

Traditional parametric mapping is usually performed by Bloch equations, which are highly sensitive to the quality and fidelity of the reconstructed multi-echo MR images [11,22] and hinders the end-to-end optimization of reconstruction and mapping. Though there were attempts [11] introducing MLPs to imitate nonlinear mapping, the mapping performance is limited by the learning capacity of MLPs. In this work, we propose to apply more powerful network and integrate physical priors from Bloch equations [22] to improve the mapping performance.

Bloch Equation-Based Multi-parametric Mapping. The initial estimation of $\mathrm{T_1}$ map ($\mathrm{T_1}|_{\mathrm{init}}$) and $\mathrm{T_2^*}$ map ($\mathrm{T_2^*}|_{\mathrm{init}}$) can be calculated by Eq. (6), where TR_N ($N \in \{1,2\}$) is the repetition time. k is a physical parameter corresponding to repetition time, flip angle, and the acquired MR signal. $\mathrm{B_{1t}}$ is the transmission radio frequency field, $\Delta\mathrm{TE}$ is the difference of the echo times and ΔS is the corresponding signal difference. The above required parameters can be determined analytically. More details can be found in [22].

$$\mathrm{T_1}|_{\mathrm{TR}_N} = \frac{\mathrm{TR}_1}{|k_{\mathrm{TR}_N}| \cdot \mathrm{B_{1t}}}, \quad \mathrm{T_1}|_{\mathrm{init}} = \frac{\mathrm{T_1}|_{\mathrm{TR}_1} + \mathrm{T_1}|_{\mathrm{TR}_2}}{2}, \quad \mathrm{T_2^*}|_{\mathrm{init}} = \frac{-\Delta\mathrm{TE}}{\ln|\Delta S|} \tag{6}$$

Physics-Informed Mapping Network. Based on the reconstructed echo images and Eq. (6), we can yield the initial estimation of the maps $\mathbf{P}_{\mathrm{init}} = \{\mathrm{T_1}|_{\mathrm{init}}, \mathrm{T_2^*}|_{\mathrm{init}}\}$ by multi-echo fitting. These initial maps are concatenated with the reconstructed echo images $\mathbf{I}_{\mathrm{init}} = \{I_{\mathrm{init}}^t\}_{t=1}^T$ and processed by a UNet [16] as shown in Fig. 1(b). In this way, our PIMN bridges data-driven learning and physics-informed constraints for multi-parametric MRI, resulting in more robust and physically plausible mapping than solely data-driven or physics-driven estimations. Moreover, we can easily combine the PIMN with off-the-shelf multi-echo reconstruction methods for multi-parametric mapping as demonstrated in the experiments.

2.4 Loss Function

The total loss consists of three parts, including the multi-echo reconstruction loss in Eq. (7), feature decoupling loss in Eq. (8), and multi-parametric mapping loss in Eq. (9). We train our method in an end-to-end fashion, thus the total loss is summarized as Eq. (10), where λ_1, λ_2, λ_3, and λ_4 denote hyper-parameters used to balanced different losses, and are empirically set to 0.5, 1, 0.1, and 2, respectively.

$$\mathcal{L}_{\text{Recon}} = \|\mathbf{I}_{\text{final}} - \mathbf{I}_{\text{GT}}\|_2 + \lambda_1 L_{\text{SSIM}}(\mathbf{I}_{\text{final}}, \mathbf{I}_{\text{GT}}) \tag{7}$$

$$\mathcal{L}_{\text{Decouple}} = \mathcal{L}_{\text{ED}} + \lambda_2 \mathcal{L}_{\text{CD}} \tag{8}$$

$$\mathcal{L}_{\text{Map}} = \|\mathbf{P}_{\text{final}} - \mathbf{P}_{\text{GT}}\|_2 + \lambda_1 L_{\text{SSIM}}(\mathbf{P}_{\text{final}}, \mathbf{P}_{\text{GT}}) \tag{9}$$

$$\mathcal{L}_{\text{total}} = \mathcal{L}_{\text{Recon}} + \lambda_3 \mathcal{L}_{\text{Decoupling}} + \lambda_4 \mathcal{L}_{\text{map}} \tag{10}$$

3 Experiments

Dataset and Implementation Details. We evaluated our model on an in-house complex-valued dataset acquired by MULTIPLEX (MTP) [22] based on a 3T scanner (uMR 890). The MTP is a gradient echo pulse sequence (GRE) with following scanning parameters: $FA_1/FA_2 = 4° / 16°$, $TR_1/TR_2 = 37.2$ ms/40 ms, and six echo images per FA with TR_1 (TE=3.86 ms) and TR_2 (TE = 3.86/8.95/14.04/19.13/24.22 ms). Specifically, our dataset includes 42 subjects, with 12 echos per subject. The MRI scanning was based on a 64-channel head coil with a voxel size of $0.823 \times 0.823 \times 2\text{mm}^3$ and slice matrix size of $231 \times 272 \times 64$. To balance computation cost and reconstruction efficiency, we center-cropped the slices into 224×256 (50 slices per subject). The ground truth multi-coil multi-echo images and parametric maps are from vendor software. Our model inputs the coil-combined data using sensitivity maps calculated by ESPIRIT [18]. The 42 subjects were randomly split into subsets of 28, 4, and 10 subjects, corresponding to 1400, 200, and 500 slices for training, validation, and testing, respectively. To quantitatively assess performance, the following experiments apply two widely-used evaluation metrics, SSIM and PSNR. All our experiments were based on NVIDIA A100 GPUs (with 80 GB RAM) and the Pytorch framework. Moreover, we applied the AdamW optimizer and trained the model for 300 epochs with mini-batch size set to 2 and learning rate set to 1×10^{-3}. In this work, we utilized the encoder $\mathcal{E}$ and decoder $\mathcal{D}$ from MIMOUNet [3,12] to extract multi-stage features and applied our WD module

Table 1. Performance comparison of our model with existing methods on the dataset with equispaced sampling masks. The best results are in **bold**. AF: acceleration factor.

AF	Method	Reconstruction		PD		T1		T2*		Mapping	
		SSIM	PSNR	SSIM	PSNR	SSIM	PSNR	SSIM	PSNR	SSIM	PSNR
× 4	Zero Filling	0.5425	23.72	-	-	-	-	-	-	-	-
	MANTIS [9]	-	-	0.7092	22.74	0.7629	27.20	0.9005	30.98	0.7909	26.97
	SRM-Net [11]	0.8548	31.26	0.8153	22.96	0.7405	26.72	0.9047	31.46	0.8201	27.05
	JUST-Net [7]	0.9157	32.67	0.8575	23.32	0.7785	27.32	0.8790	28.76	0.8383	26.47
	Ours	**0.9311**	**34.37**	**0.8990**	**24.08**	**0.8088**	**28.34**	**0.9157**	**31.46**	**0.8745**	**27.96**
× 8	Zero Filling	0.4424	20.97	-	-	-	-	-	-	-	-
	MANTIS [9]	-	-	0.6827	20.90	0.7061	25.80	0.8821	29.59	0.7570	25.43
	SRM-Net [11]	0.8316	30.48	0.7979	22.06	0.7286	26.36	**0.8992**	**31.20**	0.8086	26.54
	JUST-Net [7]	0.8584	29.24	0.8329	22.63	0.7409	26.26	0.8630	29.58	0.8123	26.16
	Ours	**0.8766**	**30.57**	**0.8586**	**23.51**	**0.7566**	**27.07**	0.8939	29.95	**0.8364**	**26.84**

in each stage. Given the trade-off between training efficiency and reconstruction accuracy, RU's cascaded number N was set to 2.

Quantitative Comparison with State-of-the-Art Methods. Two representative MRI reconstruction methods are compared, including MANTIS [9] and SRM-Net [11]. Also, to assess the generalization capacity of our method, we extend a state-of-the-art (SOTA) multi-echo reconstruction method, JUST-Net [7], with our PIMN. The quantitative comparison with the existing SOTA methods on our dataset is summarized in Table 1, where our WDPM-Net consistently presents promising SSIM and PSNR under varied accelerating factors. For instance, our WDPM-Net outperforms the second best JUST-Net [7] by 3.62% in SSIM, for 4× acceleration. These quantitative results on varied AFs demonstrate that our WDPM-Net outperforms the existing benchmark methods significantly, which is highly attributed to the efficacy of our proposed decoupling-driven reconstruction and physics-informed mapping framework. Also, the performance of the JUST-Net equipped with our PIMN demonstrates the desired plug-and-play attributes of our PIMN.

Qualitative Comparison with State-of-the-Art Methods. Furthermore, we visually compared the investigated methods on the testing dataset and illustrated the results in Fig. 2. The first row presents the GT and the reconstructed images using different methods. The second row shows the corresponding error maps, and the third row shows the zoom-in views of the ROI. From the error maps, we can see that our method performs best among all the comparison methods on both multi-echo reconstruction and parametric mapping, respectively. Particularly, as evident in the regions pointed out by the yellow arrows, our reconstructed images contain more detailed and clearer tissue structures as the zoom-in views depict. Visual inspection coincides with the quantitative measures, revealing the clinical potential of our method.

Table 2. Ablation study with 4× acceleration and equispaced sampling for the three main components of our WDPM-Net, including the WD module, decoupling loss, and physics-informed mapping.

WD Module	Decoupling loss	Physics-Informed	Reconstruction		Mapping	
			SSIM	PSNR	SSIM	PSNR
			0.8942	33.13	0.8074	26.92
✓			0.9281	34.16	0.8379	27.30
✓	✓		0.9300	34.27	0.8570	27.69
✓	✓	✓	**0.9311**	**34.37**	**0.8745**	**27.96**

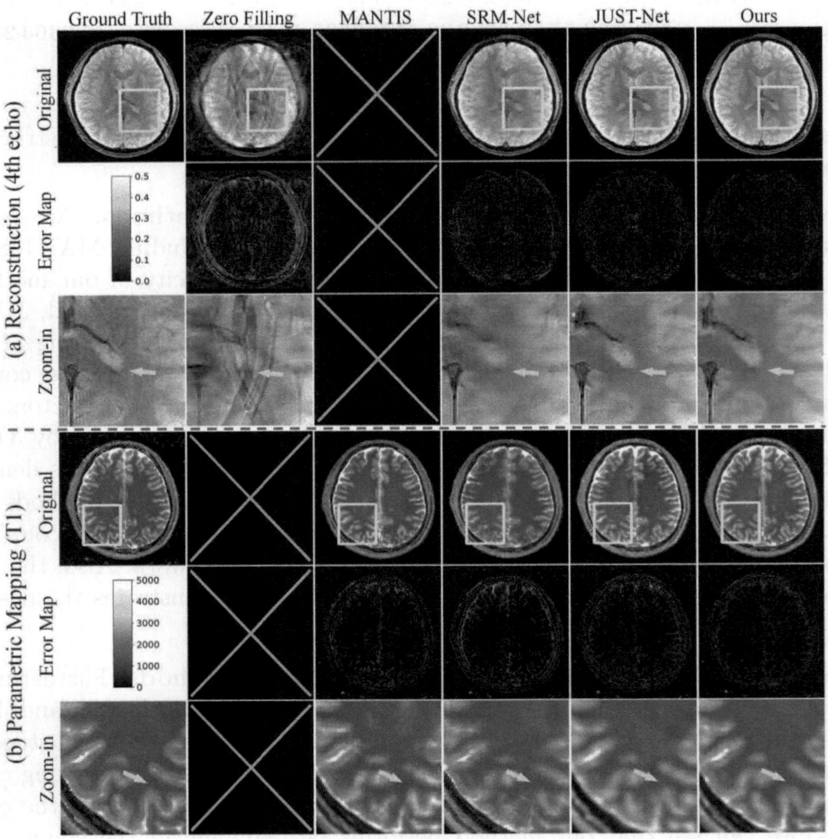

Fig. 2. Visual comparison of different methods on the test data with 4× equispaced sampling. The yellow boxes are shown in close-up views, and the reconstruction error maps of different methods are highlighted by the yellow arrows. The cross symbols indicate unavailable results. (Color figure online)

Ablation Study. In this section, we evaluated the efficacy of the three key components in WDPM-Net by gradually adding the components. The experimental results are in Table 2, where the models equipped with all components perform the best, indicating that the components can collaboratively enhance overall reconstruction and mapping performance.

4 Conclusion

In this paper, we have presented WDPM-Net, a novel decoupling-driven and physics-informed mapping network. Different from the existing multi-parametric mapping methods, we propose to decouple the features of multi-echo images into echo-independent and echo-dependent ones by a novel Wavelet-driven network and also elaborately designed decoupling losses. Thus, multi-echo information can be more efficiently exploited and fused. Furthermore, we propose to integrate Bloch equation as physical priors into the multi-parametric mapping network, which enhances the mapping robustness and reliability. Experimental results on our in-house dataset demonstrate that our method outperforms the state-of-the-art methods quantitatively and qualitatively, establishing a promising benchmark for accelerated multi-parametric MRI.

Acknowledgments. This work was supported in part by National Natural Science Foundation of China (grant numbers 82441023, U23A20295, 62131015, 62301318, 82394432), the China Ministry of Science and Technology (S20240085, STI2030-Major Projects-2022ZD0209000, STI2030-Major Projects-2022ZD0213100), Shanghai Municipal Central Guided Local Science and Technology Development Fund (No. YDZX20233100001001), The Key R&D Program of Guangdong Province, China (grant number 2023B0303040001), and HPC Platform of ShanghaiTech University.

Disclosure of Interests. The authors declare that they have no conflict of interest.

References

1. Cai, C., et al.: Single-shot T2 mapping using overlapping-echo detachment planar imaging and a deep convolutional neural network. Magn. Reson. Med. **80**(5), 2202–2214 (2018)
2. Chaudhari, A., Fang, Z., Hyung Lee, J., Gold, G., Hargreaves, B.: Deep learning super-resolution enables rapid simultaneous morphological and quantitative magnetic resonance imaging. In: Knoll, F., Maier, A., Rueckert, D. (eds.) MLMIR 2018. LNCS, vol. 11074, pp. 3–11. Springer, Cham (2018). https://doi.org/10.1007/978-3-030-00129-2_1
3. Cho, S.J., Ji, S.W., Hong, J.P., Jung, S.W., Ko, S.J.: Rethinking coarse-to-fine approach in single image deblurring. In: Proceedings of the IEEE/CVF International Conference on Computer Vision, pp. 4641–4650 (2021)
4. Cordes, D., Yang, Z., Zhuang, X., Sreenivasan, K., Mishra, V., Hua, L.H.: A new algebraic method for quantitative proton density mapping using multi-channel coil data. Med. Image Anal. **40**, 154–171 (2017)

5. Huang, H., et al.: MetaAD: metabolism-aware anomaly detection for Parkinson's disease in 3D[18] F-FDG PET. In: Medical Image Computing and Computer-Assisted Intervention – MICCAI, pp. 291–301. Springer (2024)
6. Jun, Y., Shin, H., Eo, T., Kim, T., Hwang, D.: Deep model-based magnetic resonance parameter mapping network (dopamine) for fast T1 mapping using variable flip angle method. Med. Image Anal. **70**, 102017 (2021)
7. Lee, J.H., et al.: Just-net: jointly unrolled cross-domain optimization based spatio-temporal reconstruction network for accelerated 3D myelin water imaging. Magn. Reson. Med. **91**(6), 2483–2497 (2024)
8. Lei, P., Fang, F., Zhang, G., Zeng, T.: Decomposition-based variational network for multi-contrast MRI super-resolution and reconstruction. In: IEEE/CVF Conference on Computer Vision and Pattern Recognition, pp. 21296–21306 (2023)
9. Liu, F., Feng, L., Kijowski, R.: Mantis: model-augmented neural network with incoherent k-space sampling for efficient MR parameter mapping. Magn. Reson. Med. **82**(1), 174–188 (2019)
10. Liu, X., Qiu, S., Zhou, M., Le, W., Li, Q., Wang, Y.: WFANet-DDCL: wavelet-based frequency attention network and dual domain consistency learning for 7T MRI synthesis from 3T MRI. IEEE Trans. Circuits Syst. Video Technol. **35**(6), 5617–5632 (2025)
11. Liu, Y., et al.: SRM-Net: joint sampling and reconstruction and mapping network for accelerated 3T brain multi-parametric MR imaging. IEEE Trans. Biomed. Eng. **72**(6), 1811–1824 (2025)
12. Liu, Z., Mao, H., Wu, C.Y., Feichtenhofer, C., Darrell, T., Xie, S.: A convnet for the 2020s. In: Proceedings of the IEEE/CVF International Conference on Computer Vision, pp. 11976–11986 (2022)
13. Mao, Y., Jiang, L., Chen, X., Li, C.: Disc-diff: disentangled conditional diffusion model for multi-contrast MRI super-resolution. In: Medical Image Computing and Computer-Assisted Intervention – MICCAI, pp. 387–397. Springer (2023)
14. Ouyang, J., Adeli, E., Pohl, K.M., Zhao, Q., Zaharchuk, G.: Representation disentanglement for multi-modal brain MRI analysis. In: Feragen, A., Sommer, S., Schnabel, J., Nielsen, M. (eds.) IPMI 2021. LNCS, vol. 12729, pp. 321–333. Springer, Cham (2021). https://doi.org/10.1007/978-3-030-78191-0_25
15. Qiu, Y., Zhang, H., Ma, Q., Yang, G., Lai, Z.: Multi-contrast MRI reconstruction based on frequency domain separation and cross-self-attention. IEEE Access **12**, 55062–55076 (2024)
16. Ronneberger, O., Fischer, P., Brox, T.: U-Net: convolutional networks for biomedical image segmentation. In: Navab, N., Hornegger, J., Wells, W.M., Frangi, A.F. (eds.) MICCAI 2015. LNCS, vol. 9351, pp. 234–241. Springer, Cham (2015). https://doi.org/10.1007/978-3-319-24574-4_28
17. Sun, K., Wang, Q., Shen, D.: Joint cross-attention network with deep modality prior for fast MRI reconstruction. IEEE Trans. Med. Imaging **43**(1), 558–569 (2024)
18. Uecker, M., et al.: ESPIRiT–an eigenvalue approach to autocalibrating parallel MRI: where sense meets grappa. Magn. Reson. Med. **71**(3), 990–1001 (2014)
19. Xiang, L., et al.: Deep-learning-based multi-modal fusion for fast MR reconstruction. IEEE Trans. Biomed. Eng. **66**(7), 2105–2114 (2019)
20. Yan, Y., Yang, T., Jiao, C., Yang, A., Miao, J.: Iwnext: an image-wavelet domain convnext-based network for self-supervised multi-contrast MRI reconstruction. Phys. Med. Biol. **69**(8), 085005 (2024)
21. Yan, Y., et al.: Cross-modal vertical federated learning for MRI reconstruction. IEEE J. Biomed. Health Inform. **28**(11), 6384–6394 (2024)

22. Ye, Y., Lyu, J., Hu, Y., Zhang, Z., Xu, J., Zhang, W.: Multi-parametric MR imaging with flexible design (multiplex). Magn. Reson. Med. **87**(2), 658–673 (2022)
23. Ye, Y., et al.: A multi-dimensional integration (MDI) strategy for MR T2* mapping. NMR Biomed. **34**(7), e4529 (2021)
24. Zhang, C., et al.: A unified model for reconstruction and R2* mapping of accelerated 7T data using the quantitative recurrent inference machine. Neuroimage **264**, 119680 (2022)

XFMamba: Cross-Fusion Mamba for Multi-view Medical Image Classification

Xiaoyu Zheng[1(✉)], Xu Chen[1], Shaogang Gong[2], Xavier Griffin[3], and Greg Slabaugh[1(✉)]

[1] Digital Environment Research Institute (DERI), Queen Mary University of London, London, UK
{xiaoyu.zheng,g.slabaugh}@qmul.ac.uk

[2] School of Electronic Engineering and Computer Science, Queen Mary University of London, London, UK

[3] Blizard Institute - Faculty of Medicine and Dentistry, Queen Mary University of London, London, UK

Abstract. Compared to single-view medical image classification, using multiple views can significantly enhance predictive accuracy as it can account for the complementarity of each view while leveraging correlations between views. Existing multi-view approaches typically employ separate convolutional or transformer branches combined with simplistic feature fusion strategies. However, these approaches inadvertently disregard essential cross-view correlations, leading to suboptimal classification performance, and suffer from challenges with limited receptive field (CNNs) or quadratic computational complexity (transformers). Inspired by state space sequence models, we propose *XFMamba*, a pure Mamba-based cross-fusion architecture to address the challenge of multi-view medical image classification. XFMamba introduces a novel two-stage fusion strategy, facilitating the learning of single-view features and their cross-view disparity. This mechanism captures spatially long-range dependencies in each view while enhancing seamless information transfer between views. Results on three public datasets, MURA, CheXpert, and DDSM, illustrate the effectiveness of our approach across diverse multi-view medical image classification tasks, showing that it outperforms existing convolution-based and transformer-based multi-view methods. Code is available at https://github.com/XZheng0427/XFMamba.

Keywords: Multi-view · Cross fusion · Medical image classification · State space models

1 Introduction

In many clinical applications, such as the detection of orthopedic fractures, radiologists analyze multiple views within a diagnostic examination. Typically each

J. C. Gee et al. (Eds.): MICCAI 2025, LNCS 15960, pp. 672–682, 2026.
https://doi.org/10.1007/978-3-032-04927-8_64

view provides complementary insights, but their correlation can also be leveraged to improve predictive accuracy. Numerous recent studies aim to develop multi-view networks inspired by the multi-view analysis performed by radiologists, leading to the creation of more robust and performant models [1,2].

Previous approaches in multi-view image analysis rely on Convolutional Neural Networks (CNN) or Transformers. While CNN-based approaches are recognized for their scalability and linear computational complexity in the number of image pixels, their reliance on local receptive fields introduces a bias toward local feature extraction, limiting their ability to directly capture long-range dependencies [3,4]. Moreover, CNNs often employ a weight-sharing mechanism across different view features of the input image, limiting their flexibility in adapting to unseen or low-quality medical images. To address the limited context understanding of CNNs, methods such as CVT [5] and MV-HFMD [24] utilize CNNs as the backbone for local feature extraction while incorporating cross-view attention mechanisms to transfer features between unregistered views. With the emergence of Transformers, transformer-based methods such as T-MVF [7] enhance visual modeling capabilities by leveraging global view features and dynamically adaptable weights. However, the attention mechanism in transformer-based methods results in quadratic computational complexity relative to input size and limitations on token size, posing substantial challenges to efficiency [8]. Building on the architecture of vision transformers, multi-view Swin Transformer approaches such as MV-Swin-T [6] and WT-MVSNet [9] aim to improve efficiency by reducing the dimensions or strides of processing windows. However, this optimization comes at the cost of limiting cross-view feature interactions.

To address the aforementioned limitations, Selective Structured State Space Models, also referred to as Mamba [8], have garnered increasing attention for their ability to selectively forget or propagate information, achieve global feature coverage, and utilize dynamic weights with linear computational complexity. Mamba has demonstrated exceptional performance in natural language processing [8] and computer vision tasks [11,17]. Furthermore, extensive research has investigated its potential in medical imaging, including image segmentation [10] and image classification [12]. Multi-view BI-Mamba [25] employs an early fusion mechanism to capture and integrate multi-view information. However, these recent studies incorporate Mamba as a plug-and-play module without a task-specific, in-depth design for multi-view medical imaging. Additionally, the exploration of Mamba in multi-view medical imaging tasks for cross-image fusion remains limited, particularly in architectural adaptations and fusion strategies.

Inspired by these advantages, we propose XFMamba, a pure Mamba network for multi-view cross-fusion, tailored to address the challenges of multi-view unregistered medical image classification. XFMamba integrates a four-stage encoder and two-stage fusion modules. The encoder effectively captures multi-scale features across multiple views. Furthermore, the two-stage fusion module is designed to exchange and align cross-view information and enhance multi-view integration. Comprehensive experiments conducted on the MURA musculoskeletal radiographs dataset [13], the CheXpert chest X-ray dataset [14], and the CBIS-DDSM mammography dataset [15] demonstrate that XFMamba outperforms state-of-the-art models across three varied clinical problems. The key contributions of this work can be summarized as follows:

1. Our cross-fusion work marks the successful application of state space models, specifically Mamba, in multi-view medical image classification.
2. We propose a two-stage, Mamba-based fusion mechanism designed to efficiently extract and seamlessly integrate information across multiple views.
3. Comprehensive evaluations across three varied datasets demonstrate the superior accuracy and efficiency of our method, establishing a new benchmark for Mamba's potential in the multi-view medical imaging domain.

2 Methods

2.1 Preliminaries

State Space Models. State Space Models (SSMs) [16] constitute a category of sequence-to-sequence modeling frameworks distinguished by their time-invariant dynamics, formally known as linear time-invariant (LTI) systems. SSMs, with their linear computational complexity, efficiently capture underlying dynamics by implicitly mapping to latent states, formally expressed as:

$$h'(t) = \mathbf{A}h(t) + \mathbf{B}x(t), y(t) = \mathbf{C}h(t) + \mathbf{D}x(t). \tag{1}$$

where, $x(t) \in \mathbb{R}$ denotes the input, $h(t) \in \mathbb{R}^N$ denotes the hidden state, and $y(t) \in \mathbb{R}$ denotes the output. $h'(t)$ refers to the time derivative of $h(t)$. Additionally, $\mathbf{A} \in \mathbb{R}^{N \times N}$ refers to the state matrix. $\mathbf{B} \in \mathbb{R}^{N \times 1}$, $\mathbf{C} \in \mathbb{R}^{1 \times N}$ and $\mathbf{D} \in \mathbb{R}$ denote the projection matrices.

Mamba [8] employs the zero-order hold (ZOH) discretization method to convert ordinary differential equations (ODEs) into discrete functions, making it particularly well-suited for deep learning applications. The discretization process applies a time step Δ to transform the continuous state parameters $\mathbf{A}$ and $\mathbf{B}$ into discrete parameters $\overline{\mathbf{A}}$ and $\overline{\mathbf{B}}$, which can be defined as:

$$\overline{\mathbf{A}} = exp(\Delta\mathbf{A}), \overline{\mathbf{B}} = (\Delta\mathbf{A})^{-1} exp(\Delta\mathbf{A} - I) \cdot \Delta\mathbf{B}. \tag{2}$$

Then, Eq. (1) takes the following form:

$$h_t = \overline{\mathbf{A}}h_{t-1} + \overline{\mathbf{B}}x_t, y_t = \mathbf{C}h_t + \mathbf{D}x_t \tag{3}$$

Equation (3) represents the fundamental operation within the SSM module.

2.2 Proposed Architecture

As shown in Fig. 1, the proposed approach consists of a four-stage multi-scale encoder and a two-stage cascade fusion module. For each input image, the encoder is comprised of four visual state space modules (VSSMs), incorporating downsampling operations and sequentially cascaded to extract hierarchical image features across multiple levels. Next, features extracted from the two branches are processed and integrated through a two-stage cascade fusion module, which consists of a Cross-View Swapping Mamba (CVSM) block for local feature cross-view correlation (shallow fusion) and a Multi-View Combination Mamba (MVCM) block for holistic feature multi-view correlation (deep fusion). Finally, the refined and merged features are fed into a classifier to produce the prediction.

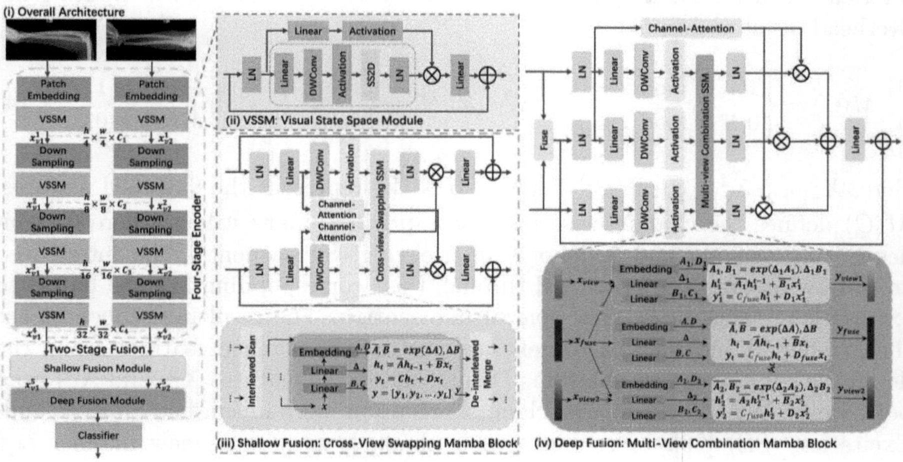

Fig. 1. XFMamba architecture. (i) The overall architecture is composed of a four-stage encoder and two-stage fusion module. (ii) Visual State Space Module (VSSM) for feature extraction. (iii) Cross-view swapping Mamba (CVSM) block for shallow fusion. (iv) Multi-view combination Mamba (MVCM) block for deep fusion.

XFMamba Encoder. The multi-scale encoder consists of two four-stage feature extraction branches that process two greyscale images with different views as input, defined as $X_{v1}, X_{v2} \in \mathbb{R}^{H\times W\times 1}$. The multi-scale encoder adopts VMamba as the backbone [17]. Inspired by ViT [18], it starts by partitioning the input into patches to produce feature maps, i.e., $X^1_{v1,v2} \in \mathbb{R}^{\frac{H}{4}\times\frac{W}{4}\times C_1}$. Next, the features are progressively processed through three stages of downsampling and VSSM, as shown in Fig. 1(i) and (ii), resulting in multi-scale feature representations, i.e., $X^2_{v1,v2} \in \mathbb{R}^{\frac{H}{8}\times\frac{W}{8}\times C_2}$, $X^3_{v1,v2} \in \mathbb{R}^{\frac{H}{16}\times\frac{W}{16}\times C_3}$, and $X^4_{v1,v2} \in \mathbb{R}^{\frac{H}{32}\times\frac{W}{32}\times C_4}$. We implement the VSSM using Selective Scan 2D (SS2D) modules proposed by VMamba [17]. As illustrated in Fig. 1(ii), the features are processed through a sequence comprising linear projection (Linear), depth-wise convolution (DWConv), and the Sigmoid Linear Unit (SiLU) activation function. Subsequently, an SS2D module with a multiplicative connection is utilized to capture spatially long-range (wider context) information.

Shallow Fusion: Cross-View Swapping Mamba. As illustrated in Fig. 1(iii), we introduce a CVSM block that leverages an interleaving mechanism to enhance cross-view features by incorporating information from the other view. The two input features are initially processed through linear layers and depth-wise convolutions, followed by SiLU activation, and are then passed into the cross-view interleaving SS2D. Specifically, the cross-view interleaving SS2D mechanism is firstly applied on X_{v1}^4 and X_{v2}^4, facilitating efficient feature interleaving across multiple views without incurring additional computational overhead, as shown in Eq. 4.

$$M(C) = \begin{cases} 1, \text{if } C \bmod 2 = 0, \\ 0, \text{if } C \bmod 2 = 1. \end{cases} \quad \widetilde{X}_{v1/v2}^5[:, C] = \begin{cases} \widetilde{X}_{v2/v1}^4[:, C], \text{if } M(C) = 1, \\ \widetilde{X}_{v1/v2}^4[:, C], \text{if } M(C) = 0. \end{cases} \tag{4}$$

Here, $\widetilde{X}_{v1/v2}^5, \widetilde{X}_{v1/v2}^4 \in \mathbb{R}^{B \times N \times C} (N = H \times W)$ represent the flattened features. $M(C)$ defines the channel-wise mask, comprising binary values where 0 signifies no interleaving and 1 denotes interleaving. Subsequently, the interleaved cross-view features are processed through their corresponding Mamba blocks to extract long-range cross-view spatial information. The resulting sequences are then reshaped and merged to generate the output scanned features using the de-interleaving merge mechanism. This results in the features X_{v1}^5 and X_{v2}^5 from both views. We adopt a shared channel-attention mechanism, i.e., Squeeze-and-Excitation (SE) [20], to reweight their channels in a cross-view manner and adaptively learn the inter-model relationships. Crucially, cross-multiplication is employed to process two channels instead of merely scaling each feature map by its own weights. Specifically, the attention derived from the global context of v_1 is used to scale the channels of v_2's feature map, and vice versa.

Deep Fusion: Multi-view Combination Mamba. In the previous CVSM block, features from two views interleave to enhance each other, capturing localized cross-view interactions. However, to generate a more holistic representation that integrates critical information from both views at a deeper level, we introduce the MVCM block. Unlike CVSM, which focuses on localized feature interleaving, MVCM performs a more comprehensive fusion by aggregating and refining the outputs of the CVSM block. This transition from shallow fusion (local feature mixing) to deep fusion (global integration) ensures that the final representation effectively captures higher-level fused features from both views, as illustrated in Fig. 1(iv). Specifically, the outputs X_{v1}^5 and X_{v2}^5 from the CVSM block are first fused as one of the inputs. Additionally, two separate view branches are introduced, each processing features from different views as input. Subsequently, the three branches are processed through linear layers, depth-wise convolution layers, and SiLU activation before being fed into the multi-view combination SSM. Within the multi-view combination SSM, the system matrix $\mathbf{C}_{fuse}$ information is exchanged across multiple selective scan modules to enhance multi-view feature fusion, inspired by [21, 22]. The Mamba selection mechanism generates the system matrices $\mathbf{B}$, $\mathbf{C}$ and Δ from linear projection layers, facilitating the context-awareness capability of the model. As specified in Eq. 3, the matrix $\mathbf{C}$ decodes information from the hidden state h_t to produce the output y_t. Therefore, we employ the matrix $\mathbf{C}_{fuse}$ of fused-branch to decode two separate view branches enhancing the model's multi-view context-awareness ability. The modified Eq. 3 is applicable to the multi-view combination SSM and is defined as follows:

$$\begin{aligned} y_{v1}^t &= \mathbf{C}_{fuse} h_{v1}^t + \mathbf{D}_{v1} x_{v1}^t, \quad y_{v1} = [y_{v1}^1, y_{v1}^2, \dots, y_{v1}^n], \\ y_{v2}^t &= \mathbf{C}_{fuse} h_{v2}^t + \mathbf{D}_{v2} x_{v2}^t, \quad y_{v2} = [y_{v2}^1, y_{v2}^2, \dots, y_{v2}^n], \\ y_{fuse}^t &= \mathbf{C}_{fuse} h_{fuse}^t + \mathbf{D}_{fuse} x_{fuse}^t, \quad y_{fuse} = [y_{fuse}^1, y_{fuse}^2, \dots, y_{fuse}^n]. \end{aligned} \tag{5}$$

Where, $x_{v1/v2/fuse}^t$ defines as the input at time step t, while $y_{v1/v2/fuse}$ represents the output of the multi-view combination SSM. After the multi-view combination SSM process, we propose a fuse-multiplication mechanism to compute channel attention on the normalized original fused view-specific features, adaptively learning the fused-view channel relationships to enhance the feature representation of a single view. Finally, additive fusion is applied to combine the three feature branches, resulting in a final holistic fused feature.

3 Experiments

3.1 Datasets and Evaluation Metrics

Datasets. To evaluate the effectiveness of our approach, we conducted experiments on the MURA [13], CheXpert [14], and CBIS-DDSM [15] datasets. The **MURA** dataset is the musculoskeletal abnormality detection dataset, which is a binary classification task and contains two view X-ray images for each patient. We have selected patients with two views and divided the patients into random subsets for training (35,185), validation (3,158) and testing (3,283). The **CheXpert** dataset comprises chest X-ray images annotated for 13 distinct observations with positive and negative labels. We selected patients that included both frontal and lateral views and randomly divided the samples into subsets for training (23,559), validation (3,926) and testing (3,928). The **CBIS-DDSM** is a mammography dataset with craniocaudal (CC) and mediolateral-oblique (MLO) views, which is a binary classification to predict benign versus malignant cases for each CC/MLO pair. The training, validation, and test splits comprise 857, 215, and 274 CC/MLO pair samples, respectively. We applied the cropped method for CBIS-DDSM, outlined by [23], adopting thresholding to position a fixed-size cropping window that captures the breast while minimizing the inclusion of an empty background. For all three datasets, we resized the images to 224×224 prior to inputting them into the models.

3.2 Experimental Details

Our XFMamba approach was implemented using PyTorch and trained on an NVIDIA A100 GPU. We adopted the Adam optimizer with the initial learning rate $1e^{-4}$ and weight decay $1e^{-5}$ to update the model parameters. The model was trained for 100 epochs with a batch size of 16. We initialized our XFMamba model with the pre-trained weights provided by VMamba [17] for the XFMamba feature encoder, constructing three variants of the model with different sizes: tiny (XFMamba-T), small (XFMamba-S), and base (XFMamba-B).

3.3 Comparison with State-of-the-Art Methods

To evaluate the performance, we compared the AUROC values of our XFMamba method with CVT [5], MVC-NET [2], MV-Swin-T [6], MV-HFMD [24], and BI-Mamba [25]. We follow the experimental protocol [5], which is optimized for the cross-entropy loss and repeated over four training runs. The quantitative results are presented in Table 1, showcasing the classification performance along with a comparison of model sizes across the three datasets. Our tiny model outperforms other methods on the MURA and CheXpert datasets while maintaining the second-fewest number of parameters. Additionally, our small and base methods achieve the highest AUROC values across all three datasets. This superiority arises from the capability of our proposed XFMamba model to efficiently extract features from each individual view and while leveraging cross-view features.

Table 1. AUROC and model size (millions of parameters) comparisons on MURA (2 classes), CheXpert (13 classes), and CBIS-DDSM (2 classes) datasets. The mean and standard deviation of AUROC over four training runs.

Method	Backbone	Params(M)	MURA	CheXpert	CBIS-DDSM
CVT	Resnet18	39.95	0.876 ± 0.003	0.909 ± 0.002	0.697 ± 0.002
BI-Mamba	ViM-s	**37.39**	0.878 ± 0.004	0.905 ± 0.002	0.611 ± 0.003
MV-HFMD	ViT-s	54.07	0.886 ± 0.002	0.909 ± 0.003	0.569 ± 0.002
MVC-NET	ResNet26	71.10	0.838 ± 0.003	0.892 ± 0.002	0.702 ± 0.003
MV-Swin-T	Swin-T	58.32	0.711 ± 0.002	0.882 ± 0.005	0.566 ± 0.002
XFMamba	VMamba-t	39.29	0.898 ± 0.002	0.917 ± 0.003	0.664 ± 0.002
XFMamba	VMamba-s	58.73	$\mathbf{0.910 \pm 0.003}$	0.918 ± 0.002	0.752 ± 0.004
XFMamba	VMamba-b	90.54	0.904 ± 0.004	$\mathbf{0.919 \pm 0.002}$	$\mathbf{0.761 \pm 0.003}$

Furthermore, Fig. 2(Left) compares the model computational complexity on the CBIS-DDSM dataset. The results indicate that our tiny model has the lowest FLOPs among the compared approaches while still achieving competitive performance. Figure 2(right) presents the qualitative results for GradCAM [26] heatmap examples in the CBIS-DDSM dataset against other methods. The white mask of the full mammogram image is the ROI of abnormalities. XFMamba's heatmaps are more precise and tend to focus on the ROI mask, whereas other models' heatmaps may highlight irrelevant areas. However, in failed cases, all models struggled to distinguish between abnormal and normal regions due to the limited visibility of the pathological structure.

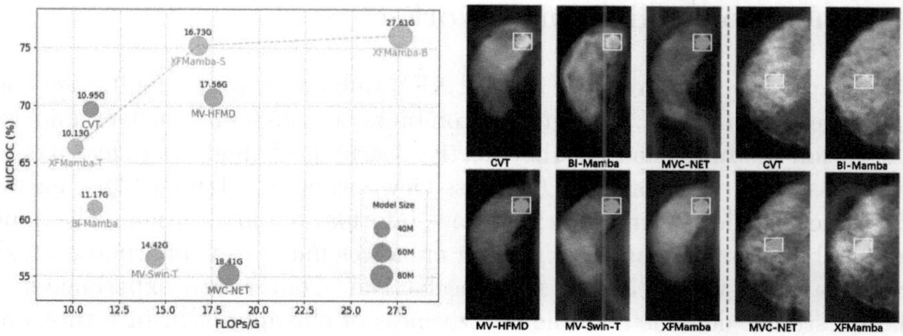

Fig. 2. (Left) Computational complexity comparison on CBIS-DDSM dataset. The size of each circle denotes the model size, i.e., parameters. (Right) Qualitative results for successful cases and failed cases (right of the orange dotted line) using different methods on the CBIS-DDSM dataset.

3.4 Ablation Study

As illustrated in Table 2, we performed an ablation study with XFMamba-S on the CheXpert dataset. Compared to the complete model, eliminating the CSVM-Block results in a 0.40% performance reduction. To evaluate the impact of removing the MVCM-Block, we test two simple fusion methods (addition and concatenation) reducing performance by 0.48% and 0.55%, respectively. Eliminating both blocks and employing late fusion for feature integration results in a 0.92% performance decline. Additionally, we compare our proposed cross-fusion model with both early-fusion and late-fusion models utilizing concatenation, finding that XFMamba outperforms them by 0.94% and 0.92%, respectively. To further assess the effectiveness of multiple views, we compare the lateral view, frontal view, and their combination, observing that incorporating multiple views improves performance by 1.03% and 2.17% compared to using only the lateral or frontal view, respectively.

Table 2. XFMamba ablation study on CheXpert.

Num.	Views	Fusion Type	CVSM-Block	MVCM-Block	AUROC↑
1	Lateral	no	✗	✗	0.8961
2	Frontal	no	✗	✗	0.9081
3	Frontal & Lateral	early-fusion	✗	✗	0.9090
4	Frontal & Lateral	late-fusion	✗	✗	0.9092
5	Frontal & Lateral	cross-fusion	✓	✗ (concat)	0.9129
6	Frontal & Lateral	cross-fusion	✓	✗ (add)	0.9136
7	Frontal & Lateral	cross-fusion	✗	✓	0.9144
8	Frontal & Lateral	cross-fusion	✓	✓	**0.9184**

4 Conclusion and Future Work

In this paper, we have proposed a novel XFMamba network for multi-view medical image classification. Our model combines a four-stage encoder and two-stage fusion module, enabling the effective learning of single-view features and their cross-view combination. Our cross-view swapping Mamba block enriches feature across views through cross-view information integration. The multi-view combination Mamba block further enhances multi-view information fusion through a multi-view selective scan mechanism. Comparison experiments and ablation studies demonstrate the effectiveness of our approach. In future work, our XFMamba network can serve as a baseline for multi-modality models that combine different imaging modalities. We also plan to explore clinical applications where multiple views are registered, or exhibit geometric relationships, leveraging our approach to enhance feature alignment and information fusion.

Acknowledgments. This work forms part of the research portfolio of the National Institute for Health Research Barts Biomedical Research Centre (NIHR203330). This work was also supported by the Engineering and Physical Sciences Research Council [grant number EP/Y009800/1], through Keystone project funding from Responsible Ai UK (KP0016). This research utilised Queen Mary's Andrena HPC facility, supported by QMUL Research-IT.

Disclosure of Interests. The authors have no competing interests to declare that are relevant to the content of this article.

References

1. Seeland, M., Mader, P.: Multi-view classification with convolutional neural networks. PLoS ONE **16**(1), e0245230 (2021)
2. Zhu, X., Feng, Q.: MVC-Net: multi-view chest radiograph classification network with deep fusion. In: IEEE 18th International Symposium on Biomedical Imaging (ISBI 2021), pp. 554–558. IEEE (2021)
3. Karthik, K., Kamath, S.S.: A deep neural network model for content-based medical image retrieval with multi-view classification. Vis. Comput. **37**(7), 1837–1850 (2021)
4. Yu, H., Yang, L.T., Zhang, Q., Armstrong, D., Deen, M.: Convolutional neural networks for medical image analysis: state-of-the-art, comparisons, improvement and perspectives. Neurocomputing **444**, 92–110 (2021)
5. van Tulder, G., Tong, Y., Marchiori, E.: Multi-view analysis of unregistered medical images using cross-view transformers. In: de Bruijne, M., et al. (eds.) MICCAI 2021. LNCS, vol. 12903, pp. 104–113. Springer, Cham (2021). https://doi.org/10.1007/978-3-030-87199-4_10
6. Sarker, S., Sarker, P., Bebis, G., Tavakkoli, A.: MV-Swin-T: mammogram classification with multi-view swin transformer. In: IEEE International Symposium on Biomedical Imaging (ISBI 2024). IEEE (2024)

7. Sun, Z., Jiang, H., Ma, L., Yu, Z., Xu, H.: Transformer based multi-view network for mammographic image classification. In: Wang, L., Dou, Q., Fletcher, P.T., Speidel, S., Li, S. (eds.) MICCAI 2022. LNCS, pp. 46–54. Springer, Cham (2022). https://doi.org/10.1007/978-3-031-16437-8_5
8. Gu, A., Dao, T.: Mamba: linear-time sequence modeling with selective state spaces. In: First Conference on Language Modeling (2024). https://openreview.net/forum?id=tEYskw1VY2
9. Liao, J., Ding, Y., Shavit, Y., et al.: WT-MVSNet: window-based transformers for multi-view stereo. Adv. Neural Inform. Process. Syst. **35**, 8564–8576 (2022)
10. Ma, J., Li, F., Wang, B.: U-Mamba: enhancing long-range dependency for biomedical image segmentation. arXiv preprint arXiv:2401.04722 (2024)
11. Zhu, L., Liao, B., Zhang, Q., Wang, X., Liu, W., Wang, X.: Vision mamba: efficient visual representation learning with bidirectional state space model. In: Proceedings of 41st International Conference on Machine Learning (ICML 2024), vol. 235, Article 2584, pp. 62429–62442. JMLR.org (2024)
12. Yue, Y., Li, Z.: MedMamba: vision mamba for medical image classification. arXiv preprint arXiv:2403.03849 (2024)
13. Rajpurkar, P., Irvin, J., Bagul, A., et al.: MURA: large dataset for abnormality detection in musculoskeletal radiographs. arXiv preprint arXiv:1712.06957 (2017)
14. Irvin, J., Rajpurkar, P., Ko, M., et al.: CheXpert: a large chest radiograph dataset with uncertainty labels and expert comparison. In: Proceedings of AAAI Conference on Artificial Intelligence, vol. 33, no. 1, pp. 590–597. AAAI Press (2019)
15. Lee, R.S., Gimenez, F., Hoogi, A., et al.: A curated mammography dataset for use in computer-aided detection and diagnosis research. Sci. Data **4**(1), 170177 (2017)
16. Gu, A., Goel, K., Ré, C.: Efficiently modeling long sequences with structured state spaces. In: Proceedings of 9th International Conference on Learning Representations (ICLR 2021), pp. 1–15. OpenReview (2021)
17. Liu, Y., et al.: VMamba: visual state space model. In: 38th Conference on Neural Information Processing Systems (NeurIPS 2024) (2024)
18. Dosovitskiy, A., et al.: An image is worth 16×16 words: transformers for image recognition at scale. In: International Conference on Learning Representations (ICLR 2021) (2021)
19. Li, Z., Pan, H., Zhang, K., et al.: MambaDFuse: a mamba-based dual-phase model for multi-modality image fusion. arXiv preprint arXiv:2404.08406 (2024)
20. Hu, J., Shen, L., Sun, G.: Squeeze-and-excitation networks. In: Proceedings of IEEE Conference on Computer Vision and Pattern Recognition (CVPR 2018), pp. 7132–7141. IEEE (2018)
21. Chen, C.F.R., Fan, Q., Panda, R.: CrossViT: cross-attention multi-scale vision transformer for image classification. In: Proceedings of IEEE/CVF International Conference on Computer Vision (ICCV 2021), pp. 357–366. IEEE (2021)
22. Wan, Z., Zhang, P., Wang, Y., et al.: Sigma: siamese mamba network for multi-modal semantic segmentation. In: IEEE/CVF Winter Conference on Applications of Computer Vision (WACV 2025), pp. 1734–1744. IEEE (2025)
23. Wu, N., Phang, J., Park, J., et al.: Deep neural networks improve radiologists' performance in breast cancer screening. IEEE Trans. Med. Imaging **39**(4), 1184–1194 (2019)
24. Black, S., Souvenir, R.: Multi-view classification using hybrid fusion and mutual distillation. In: Proceedings of IEEE/CVF Winter Conference on Applications of Computer Vision (WACV 2024), pp. 270–280. IEEE (2024)

25. Yang, Z., Zhang, J., Wang, G., et al.: Cardiovascular disease detection from multi-view chest X-rays with BI-Mamba. In: Linguraru, M.G., Dou, Q., Feragen, A., et al. (eds.) MICCAI 2024. LNCS, pp. 134–144. Springer, Cham (2024). https://doi.org/10.1007/978-3-031-72086-4_13
26. Selvaraju, R.R., Cogswell, M., Das, A., et al.: Grad-CAM: visual explanations from deep networks via gradient-based localization. In: IEEE International Conference on Computer Vision (ICCV 2017), pp. 618–626. IEEE (2017)

You Can Detect It: Fetal Biometric Estimation Using Ellipse Detection

Hongyuan Zhang[1,2], Haoyu Xie[2], Tingting Ye[3(✉)], and Songxiong Wu[4(✉)]

[1] CAIR, HKISI-CAS, Hong Kong SAR, China
[2] Medical AI Lab, Shenzhen University Medical School, Shenzhen, China
[3] Department of Ultrasound, Shenzhen University General Hospital and Shenzhen University Clinical Medical Academy, Shenzhen, China
ytty126@163.com
[4] Department of Radiology, Shenzhen University General Hospital and Shenzhen University Clinical Medical Academy, Shenzhen, China
wusongxiong@email.szu.edu.cn

Abstract. The cardiothoracic diameter ratio (CTR) biometric in four-chamber ultrasound plane is often measured for diagnosing congenital heart disease. However, due to the commonly existing artifacts like acoustic shadowing, manual measurement can be time-consuming and labor-intensive task, and may result in high measurements variability. Presently, one of the most popular approaches is segmentation-based methods, which utilize deep learning networks to segment the cardiac and thoracic regions. Then, the metric is calculated through an ellipse fitting scheme. This is inefficient, and requires additional post-processing. To tackle the above problems, in this paper, we therefore present a one-stage ellipse detection network, namely EllipseDet, which detects the cardiac and thoracic regions in ellipses, and then automatically calculates the CTR biometric in four-chamber view. In particular, we formulate the network that detects the center of each object as points and regresses the ellipses' parameters simultaneously. Besides, we propose a novel ellipse feature alignment module and Ellipse-IoU loss to further regulate the regression procedure. We have evaluated EllipseDet on a clinical echocardiogram dataset and the experimental results show that our proposed framework outperforms several state-of-the-art methods. As an open science, source code, images dataset and pre-trained weights are available at https://github.com/szuboy/FOCUS-dataset.

Keywords: Biometric Estimation · Ellipse Detection · Four-chamber Ultrasound View · Cardiothoracic Diameter Ratio

1 Introduction

Fetal congenital heart disease is one of the most common forms of birth defects worldwide, with an approximate incidence rate of 8% among newborn babies [20]. Four-chamber view of the fetal echocardiography is an important yet unique tool

J. C. Gee et al. (Eds.): MICCAI 2025, LNCS 15960, pp. 683–693 2026.
https://doi.org/10.1007/978-3-032-04927-8_65

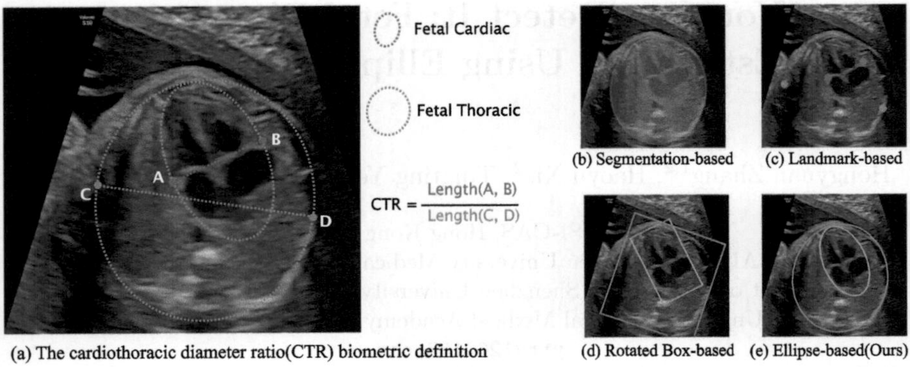

Fig. 1. (a) The definition of the CTR biometric in the four-chamber view and (b-e) the comparisons of object representation in different estimation pipelines.

for assessing the fetus for the presence of congenital heart disease [5]. In the four-chamber view, assessment and diagnosis are based on the cardiothoracic diameter ratio (CTR) [19] biometric, as shown in Fig. 1(a). Specifically, the CTR assesses heart size by comparing the maximum cardiac diameter to the thoracic diameter. However, due to the commonly existing acoustic shadowing artifact in fetal echocardiogram, the manual measurements is a labor-intensive and time-consuming process, and suffer from large inter-operator variance. Therefore, automatic and reliable fetal cardiac measurements are in high demand.

Recently, many deep learning methods have been proposed for fetal biometrics estimation [1,3,18], but they still cannot achieve satisfactory results due to the low quality of four-chamber view. Existing deep learning-based methods can be classified into three categories: segmentation-based methods, landmark-based methods and box-based methods, as shown in Fig. 1. Segmentation-based methods [12,15,18] are currently the most common approach, utilizing segmentation results as an intermediate processing step. However, these approaches make biometrics estimation heavily dependent on these results, leading to error accumulation across both stages [26]. Landmark-based methods [1], on the other hand, detect the landmarks directly on the image without relying on segmentation masks. Unfortunately, ambiguous boundaries caused by massive speckle noise and artifacts significantly hinder the effectiveness of landmark-based approaches. Besides, rotation object detection methods [3,23] can also be used as candidate solutions. However, traditional bounding box may perform poorly in representing such ball-shaped biomedical objects [16]. Thus, how to design a suitable method to represent the cardiac and thoracic regions is the key procedure for achieving precise biometric estimation.

In this paper, we present a one-stage detection framework, called EllipseDet, for fetal cardiac biometrics measurement in four-chamber scans. Unlike previous paradigms [1,15,18], our proposed method uses ellipses to represent the cardiac and thoracic regions and directly regresses the ellipses's parameters in one-stage without requiring additional post-processing, as shown in Fig. 1(e). We also introduce a novel feature alignment module and loss function tailored for detecting ellipse-shaped fetal cardiac and thoracic regions, making it suitable for practical medical object detection. **Overall, our contributions are listed as follows:**

- **Ellipse Representation:** We propose a simple ellipse representation for fetal biometrics estimation that better depicts fetal structures. In addition, we introduce a new ellipse feature alignment module and ellipse intersection over union (Ellipse-IoU) loss to further regulate the training procedure.
- **Optimized Biomedical Object Detection:** To the best of our knowledge, our EllipseDet is the first one-stage approach with ellipse representation and optimized Ellipse-IoU loss for fetal biometric estimation.
- **Superior Detection and Clinical Benefits:** The extensive experimental results demonstrate that our proposed method, EllipseDet, achieves superior detection performance and offers greater clinical benefits compared to other state-of-the-art techniques.
- **Public Fetal Cardiac Dataset:** We contribute the first publicly available fetal cardiac biometrics dataset, providing a foundation for future research.

2 EllipseDet for Fetal Biometrics Estimation

2.1 Architecture Overview

The overview of our EllipseDet is shown in Fig. 2. EllipseDet is a simple, unified network consisting of a backbone network [9], a feature pyramid network [13], and three task-specific heads. The backbone and feature pyramid network are designed to generate multi-scale feature maps for the subsequent head networks. And, the heatmap and offset heads are used to determine the central location, while the ellipse regression head determines the parameters of the ellipse.

Ellipse Representation. As shown in Fig. 1(e), given an ellipse object, we represent it as $E_i = (x_i, y_i, a_i, b_i, \theta_i)$ for the i-th object with class c_i, where (x_i, y_i) denotes the center point, a_i and b_i are the lengths of the semi-major and semi-minor axes, respectively, and θ_i corresponds to the tilt angle. The parameters $x_i, y_i, a_i, b_i, \theta_i \in \mathcal{R}^1$, where $\theta_i \in [-\frac{\pi}{2}, \frac{\pi}{2})$. Thus, we model the object as one center and three parameters.

Backbone Network. Following prior works [3,25], the ResNet50 [9] with Feature Pyramid Network (FPN) is adopted as the backbone network to produce multi-scale feature maps from input ultrasound image. We construct the pyramid with levels P_2 through P_4, where l indicates pyramid level and P_l has resolution 2^l lower than the input [13]. And, all pyramid levels have 256 channels.

Regression Head. As illustrated in Fig. 2, an ellipse object is decoupled into location and size, corresponding to the two branches in the regression head. Each branch has four convolutional layers and interacts with each other through a feature alignment module. However, since the output feature size is smaller than

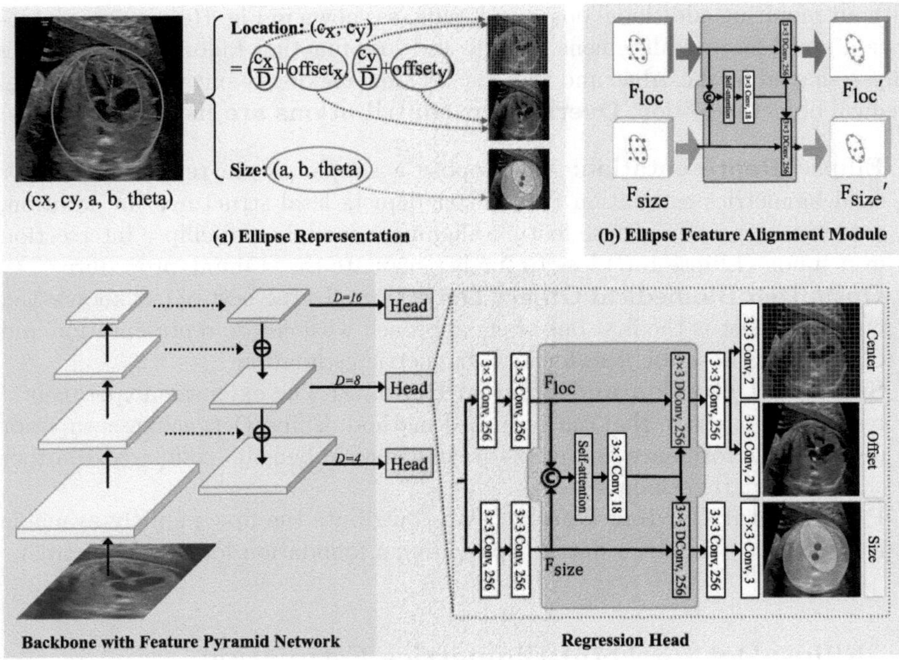

Fig. 2. The network structure of EllipseDet and the explanation of the modules: (a) Ellipse representation for thoracic and cardiac instances, (b) Visualization of the alignment between thoracic and cardiac features using the proposed ellipse feature alignment module.

that of the input image, we use heatmap-offset aggregation [3,28] to correct the discretization error from the output stride. Finally, there are three outputs in each head. Specifically, we predict the heatmap $\hat{Y} \in \mathcal{R}^{\frac{W}{D} \times \frac{H}{D} \times C}$, local offset $\hat{O} \in \mathcal{R}^{\frac{W}{D} \times \frac{H}{D} \times 2}$, and a size $\hat{S} \in \mathcal{R}^{\frac{W}{D} \times \frac{H}{D} \times 3}$, where D is the output stride, representing the down-sampling factor of the prediction, C is the number of object categories, and the three channels in size prediction represent the semi-major axis, semi-minor axis and angle, respectively. Following standard practice [10,28], the ground truth of the heatmap is modeled as a 2D Gaussian kernel:

$$Y = \exp(-\frac{(x - \tilde{x}_i)^2 + (y - \tilde{y}_i)^2}{2\sigma_p^2}) \quad (1)$$

where the $\tilde{x}_i = \lfloor \frac{x_i}{D} \rfloor$ and $\tilde{y}_i = \lfloor \frac{y_i}{D} \rfloor$ are the down-sampled target center point location and σ_p is the size-adaptive standard deviation [10]. The predicted heatmap is optimized using pixel-wise regression loss $\mathcal{L}_{heatmap}$ with focal loss [14]. Besides, the offset is calculated as $o = (\frac{x_i}{D} - \tilde{x}_i, \frac{y_i}{D} - \tilde{y}_i)$. The offset loss function $\mathcal{L}_{offset}$ is optimized with $\mathcal{L}_1$ loss. However, the training of the size regression branch is non-trivial. For an ellipse instance, its center and size are relevant and should be trained as a whole, rather than as independent regression tasks. To address this challenge, we introduce the Ellipse-IoU Loss, which is further discussed in the Sect. 2.3.

Ellipse Assembling and Biometric Estimation. During inference, we only assemble ellipse from predictions with the highest confidence scores across all FPN levels without any post-processing. Specifically, we first extract the peak in the heatmap for each category independently, and its location $(\hat{x}_i, \hat{y}_i)$ is taken as the center position of the cardiac or thoracic object. Meanwhile, the offset $(\delta x_i, \delta y_i)$ is obtained from $\hat{O}$. Then, the bounding ellipse is formed with center point $\hat{p}$ and size $\hat{s}$ as follows:

$$\hat{p} = ((\hat{x}_i + \delta x_i) * D, (\hat{y}_i + \delta y_i) * D) \qquad \hat{s} = \hat{S}_{\hat{x}_i, \hat{y}_i} = (\hat{a}_i, \hat{b}_i, \hat{\theta}_i) \tag{2}$$

Once we obtain the cardiac and thoracic parameters, we can calculate and estimate the CTR biometric as: $\text{CTR} = \hat{a}_{cardiac}/\hat{a}_{thoracic}$.

2.2 Ellipse Feature Alignment Module

Since conventional networks for rotated object detection have limited generalization to rotation variations, better feature alignment generally means better classification and regression in object detection [11,24]. Thus, as shown in Fig. 2, we introduce a simple yet effective ellipse feature alignment module (EFAM) in the regression head, enhancing the collaboration between location and size features during detection. It reintegrates the decoupled location and size features to provide shared offsets, which keeps task-specific features and mitigates inconsistent predictions. The detailed architecture of the module is presented in Fig. 2(b). We first use and concatenate the features $F_{\text{loc}} \in \mathcal{R}^{\frac{W}{D} \times \frac{H}{D} \times 256}$ and $F_{\text{size}} \in \mathcal{R}^{\frac{W}{D} \times \frac{H}{D} \times 256}$ of the two branches to form the task-interaction feature $F_{\text{loc-size}} \in \mathcal{R}^{\frac{W}{D} \times \frac{H}{D} \times 512}$, which contains location and size information. Then, we incorporate a self-attention module [21] into the EFAM to further improve ellipse feature learning. This operation allows for dynamic weighting of global rotated ellipse object features. Given the task-interaction feature $F_{\text{loc-size}}$, we compute the attention feature as follows:

$$F'_{\text{loc-size}} = \text{Attention}(F_{\text{loc-size}}) = \text{softmax}\left(\frac{QK^T}{\sqrt{d_k}}\right)V \tag{3}$$

where Q, K, V represent the queries, keys, and values of $F_{\text{loc-size}}$, respectively, and d_k is the dimension of the keys. After obtaining the feature $F'_{\text{loc-size}}$, we apply a 3×3 convolutional layer to output the offsets $F_{\text{offset}} \in \mathcal{R}^{\frac{W}{D} \times \frac{H}{D} \times 18}$. Using F_{offset}, we can adaptively align the features for location and size tasks:

$$F'_{\text{loc}} = f^{\text{loc}}_{\text{dcn}}(F_{\text{loc}}, F_{\text{offset}}) \qquad F'_{\text{size}} = f^{\text{size}}_{\text{dcn}}(F_{\text{size}}, F_{\text{offset}}) \tag{4}$$

Here, $f^{\text{loc}}_{\text{dcn}}(\cdot)$ and $f^{\text{size}}_{\text{dcn}}(\cdot)$ both the standard deformable convolutional operation [6]. Then, we can use the aligned features F'_{loc} and F'_{size} for regression.

2.3 Ellipse-IoU Loss

Compared with smooth-l_1 loss, the IoU loss considers the optimization as a whole. However, existing IoU losses [27] are box-based and not tailored for elliptical biomedical objects. In this work, we derive an differential algorithm to compute IoU for ellipse representation and achieve competitive performance.

As shown in Fig. 3, we first introduce the Ellipse-IoU loss based on the definition of IoU, instead of relying on the compromise method of the bounding rectangle box. This approach is more effective for accurately representing elliptical objects. Based on the ellipse representation [22], the Ellipse-IoU for a given instance is calculated as follows:

$$\text{Ellipse-IoU} = \frac{\int_0^{2\pi} \frac{1}{2}\min(r(\phi), \hat{r}(\phi))^2 d\phi}{\int_0^{2\pi} \frac{1}{2}\max(r(\phi), \hat{r}(\phi))^2 d\phi} \approx \lim_{n\to\infty} \frac{\sum_{i=1}^{N} \frac{1}{2}\min(r(\phi), \hat{r}(\phi))^2 \Delta\phi}{\sum_{i=1}^{N} \frac{1}{2}\max(r(\phi), \hat{r}(\phi))^2 \Delta\phi} \tag{5}$$

where ϕ is the angle with the x-axis, and $r(\phi)$ and $\hat{r}(\phi)$ are the axis lengths corresponding to the angle ϕ.[1] When N approaches infinity, the discrete form converges to the continuous form. We assume that the sampling is uniform, leading to $\phi = \frac{2\pi}{N}$, which further simplifies the expression:

$$\text{Ellipse-IoU} = \frac{\sum_{i=1}^{N} \min(r(\phi), \hat{r}(\phi))^2}{\sum_{i=1}^{N} \max(r(\phi), \hat{r}(\phi))^2} \tag{6}$$

Since the optimal Ellipse-IoU is always 1, the Ellipse-IoU loss can be defined as:

$$\text{Ellipse-IoU Loss} = 1 - \text{Ellipse-IoU} = 1 - \frac{\sum_{i=1}^{N} \min(r(\phi), \hat{r}(\phi))^2}{\sum_{i=1}^{N} \max(r(\phi), \hat{r}(\phi))^2} \tag{7}$$

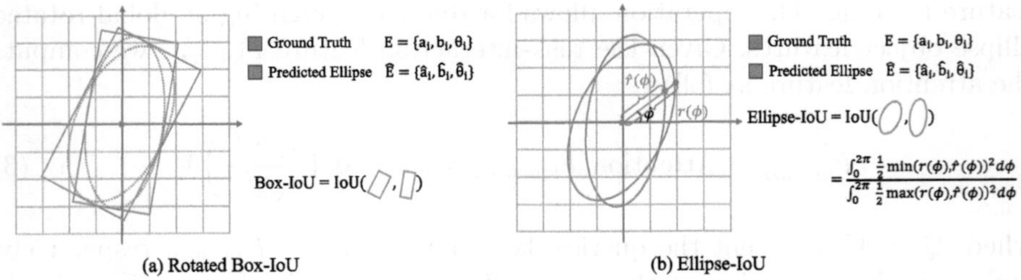

Fig. 3. Comparison of different Rotated Box-IoU [27] and Ellipse-IoU metrics.

[1] For any rotated ellipse instance (x, y, a, b, θ), the corresponding axis length r in ϕ can be expressed as: $r(\phi) = \sqrt{(a\cos(\phi-\theta))^2 + (b\sin(\phi-\theta))^2}$.

2.4 Final Training Loss Function

In this study, we do not normalize the scale and instead use the raw pixel coordinates directly, scaling the loss by a constant λ_{size}. The final loss is:

$$\mathcal{L}_{final} = \mathcal{L}_{heatmap} + \lambda_{offset}\mathcal{L}_{offset} + \lambda_{size}\mathcal{L}_{size} \tag{8}$$

where λ_{offset} and λ_{size} weights for the respective loss terms. To balance the Ellipse-IoU loss function and other loss functions, we set $\lambda_{size} = 10$ and $\lambda_{offset} = 1$ in all our experiments unless specified otherwise. In addition, we set the number of samples $N = 36$ for the Ellipse-IoU loss function by default.

Table 1. Experimental results between different methods across different metrics.

Methods		Cardiac Region		Thoracic Region		Biometric
		DSC (%) ↑	HD95 (pixel) ↓	DSC (%) ↑	HD95 (pixel) ↓	$\mathbf{P_{CTR}}$(%) ↑
U-Net [17]	[MICCAI2015]	89.12 ± 1.62	25.63 ± 4.24	93.96 ± 0.67	32.74 ± 4.68	90.23 ± 1.37
Rotated Faster RCNN [24]	[ICRSA2020]	86.60 ± 0.72	29.79 ± 1.28	91.00 ± 0.94	39.82 ± 4.60	88.30 ± 1.63
EllipseNet [3]	[MICCAI2022]	77.87 ± 2.48	48.37 ± 6.89	81.36 ± 3.24	70.46 ± 8.79	45.15 ± 15.12
Oriented RepPoints [11]	[CVPR2022]	60.00 ± 3.10	109.96 ± 10.06	69.42 ± 5.25	140.57 ± 27.50	61.94 ± 11.83
BiometryNet [1]	[MICCAI2022]	/	/	/	/	35.98 ± 12.61
E2EBM-Net [8]	[CBM2023]	/	/	/	/	31.45 ± 39.96
TransUNet [4]	[MedIA2024]	92.48 ± 0.36	16.76 ± 0.72	**96.48 ± 0.18**	16.26 ± 0.96	93.32 ± 1.38
EllipseDet(N = 36)	[OURS]	**93.18 ± 0.24**	**14.47 ± 0.96**	94.87 ± 0.84	**16.19 ± 0.86**	**94.41 ± 1.33**

3 Experiment and Results

3.1 Experiment Protocol

Fetal Cardiac Ultrasound Dataset: The dataset comprises 300 images, and is divided into 200 training sets, 50 validation sets and 50 test sets based on the patient-level. The annotation work involved the participation of 2 experts with more than 5-year clinical experience. It includes images of 217 subjects from two medical centers within the public FETAL_PLANE database [2] and incorporates ellipse annotations for both cardiac and thoracic across all cases.

Implementation Details: All the ultrasound images are resized 512 × 512 to before feeding into the network. Random left-right flipping, scaling, shifting and random Gaussian noise are added for data augmentation. Following previous studies [4], the segmentation model is trained using a combination of loss functions: Dice loss function and binary cross-entropy (BCE) loss function, with a BCE weight of 0.2. We trained all model with batch size 8 and initial learning rate 1.25×10^{-4} for 120 epochs. All experiments were conducted on a Linux Ubuntu 16.04 system, which was equipped with 2 NVIDIA TITAN RTX GPUs.

Evaluation Metrics: All experiments were evaluated using the Dice similarity coefficient (DSC) and 95th percentile Hausdorff distance (HD95). In addition, we presented the measurement precision of the CTR biometric. The CTR is formulated as $R = b_C/b_T$, where b_C represents the length of minor axis of the cardiac ellipse, b_T represents that of the thoracic ellipse [3]. Then, the CTR precision can be defined as: $P_{CTR} = (1 - \frac{|R_{\text{true}} - R_{\text{pred}}|}{R_{\text{true}}}) \times 100\%$, where R_{pred} and R_{true} denote the predicted CTR and the ground truth CTR, respectively.

3.2 Comparisons with State-of-the-Arts

To demonstrate the effectiveness of EllipseDet, we conducted experiments across three aspects: 1) Segmentation-based method: U-Net [17] and TransUNet [4]; 2) Landmark-based method: E2EBM-Net [8] and BiometryNet [1]; and 3) Detection-based method: EllipseNet [3], Oriented RepPoints [11] and Rotated Faster RCNN [24]. For U-Net and TransUNet, we employed the ellipse-fit post-processing method [7] to obtain final ellipse prediction results. Landmark-based methods use landmarks to directly compute CTR biometric, while detection-based methods utilize the inscribed ellipse of the rotated bounding box for evaluation metrics.

Quantitative Results: We listed the detection performance in Table 1. As expected, due to the lack of semantic information, the landmark-based E2EBM-Net [8] and BiometryNet [1] show poor biometric estimation performance ($P_{CTR} < 40\%$). Compared with the U-Net [17] and TransUNet [4] methods, our method does not require additional ellipse fitting post-processing methods and achieves state-of-the-arts results in almost all metrics (DSC $> 90\%$, $P_{CTR} > 90\%$). Besides, thanks to the optimization of the ellipse objects, EllipseDet also outperforms the other box-based methods in region segmentation and CTR estimation.

Qualitative Results: Figure 4 shows two cases illustrating the performance of different methods. As we can see, our EllipseDet aligns closely with true labels, effectively capturing thoracic and cardiac regions, while other methods struggle with boundaries. It also demonstrates superior accuracy in challenging scenarios.

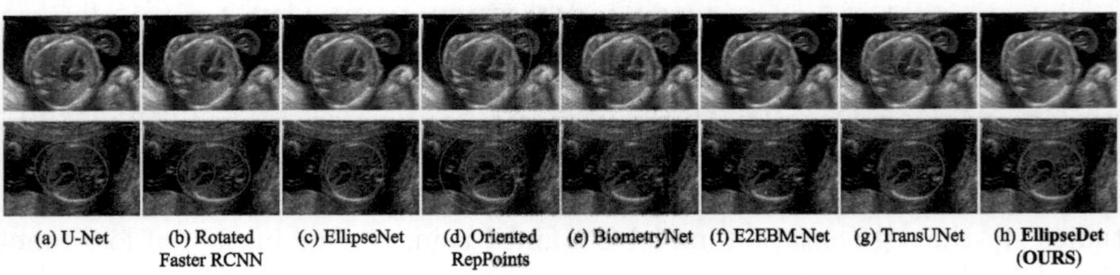

Fig. 4. Visualizations of different methods are shown, with red indicating true labels, cyan for predicted thoracic labels, and green for predicted cardiac labels. (Color figure online)

Table 2. Ablation studies of different components of our method on the dataset.

FPN	EFAM	Box-IoU	Ellipse-IoU		Cardiac Region		Thoracic Region		Biometric
			N = 36	N = 100	DSC (%) ↑	HD95 (pixel) ↓	DSC (%) ↑	HD95 (pixel) ↓	$\mathbf{P_{CTR}}$(%) ↑
					84.45 ± 1.62	31.11 ± 5.23	86.96 ± 3.78	63.11 ± 8.64	84.53 ± 6.46
✓					89.67 ± 0.50	24.17 ± 3.56	90.18 ± 1.95	24.80 ± 5.02	89.17 ± 1.73
✓	✓				90.19 ± 0.69	23.23 ± 1.47	90.94 ± 1.63	23.36 ± 4.55	92.10 ± 1.82
✓	✓	✓			91.24 ± 0.33	17.35 ± 1.20	92.02 ± 1.27	18.26 ± 2.61	93.32 ± 1.38
✓	✓		✓		93.18 ± 0.24	**14.47 ± 0.96**	94.87 ± 0.84	16.09 ± 0.86	94.41 ± 1.33
✓	✓			✓	**93.46 ± 0.22**	14.55 ± 0.97	**95.08 ± 0.77**	**15.60 ± 0.88**	**95.06 ± 1.45**

3.3 Ablation Studies

Table 2 shows ablation studies on the key components of our method with the fetal cardiac ultrasound dataset. The multi-scale features of the FPN can bring about a 5% performance improvement. Besides, using our proposed EFAM, the performance significantly improves by 3% in CTR estimation. Compared with the smooth-l_1 and Box-IoU loss, proposed Ellipse-IoU loss function brings continuous performance improvements in both segmentation and biometric estimation. Besides, the Ellipse-IoU loss function is not sensitive to the number of samples N. When $N = 36$ and $N = 100$, all metric performances are close (<1%).

4 Conclusion

This paper proposes an one-stage ellipse detection network for the cardiac and thoracic regions detection. In order to better represent the fetal cardiac and thoracic, two innovative modules (EFAM and Ellipse-IoU) are proposed for the optimization of elliptical objects. Extensive experimental results also show the effectiveness of our proposed EllipseDet method.

Disclosure of Interests. The authors have no competing interests to declare that are relevant to the content of this article.

References

1. Avisdris, N., et al.: Biometrynet: landmark-based fetal biometry estimation from standard ultrasound planes. In: International Conference on Medical Image Computing and Computer-Assisted Intervention, pp. 279–289. Springer (2022)
2. Burgos-Artizzu, X.P., et al.: Evaluation of deep convolutional neural networks for automatic classification of common maternal fetal ultrasound planes. Sci. Rep. **10**(1), 10200 (2020)
3. Chen, J., Zhang, Y., Wang, J., Zhou, X., He, Y., Zhang, T.: EllipseNet: anchor-free ellipse detection for automatic cardiac biometrics in fetal echocardiography. In: de Bruijne, M., et al. (eds.) MICCAI 2021. LNCS, vol. 12907, pp. 218–227. Springer, Cham (2021). https://doi.org/10.1007/978-3-030-87234-2_21

4. Chen, J., et al.: Transunet: rethinking the U-net architecture design for medical image segmentation through the lens of transformers. Med. Image Anal. **97**, 103280 (2024)
5. Copel, J.A., Pilu, G., Green, J., Hobbins, J.C., Kleinman, C.S.: Fetal echocardiographic screening for congenital heart disease: the importance of the four-chamber view. Am. J. Obstet. Gynecol. **157**(3), 648–655 (1987)
6. Dai, J., et al.: Deformable convolutional networks. In: Proceedings of the IEEE International Conference on Computer Vision, pp. 764–773 (2017)
7. Fitzgibbon, A., Pilu, M., Fisher, R.B.: Direct least square fitting of ellipses. IEEE Trans. Pattern Anal. Mach. Intell. **21**(5), 476–480 (1999)
8. Gao, Z., Tian, Z., Pu, B., Li, S., Li, K.: Deep endpoints focusing network under geometric constraints for end-to-end biometric measurement in fetal ultrasound images. Comput. Biol. Med. **165**, 107399 (2023)
9. He, K., Zhang, X., Ren, S., Sun, J.: Deep residual learning for image recognition. In: Proceedings of the IEEE Conference on Computer Vision and Pattern Recognition, pp. 770–778 (2016)
10. Law, H., Deng, J.: Cornernet: detecting objects as paired keypoints. In: Proceedings of the European Conference on Computer Vision (ECCV), pp. 734–750 (2018)
11. Li, W., Chen, Y., Hu, K., Zhu, J.: Oriented reppoints for aerial object detection. In: Proceedings of the IEEE/CVF Conference on Computer Vision and Pattern Recognition, pp. 1829–1838 (2022)
12. Liang, B., et al.: Automatic segmentation of 15 critical anatomical labels and measurements of cardiac axis and cardiothoracic ratio in fetal four chambers using nnu-netv2. BMC Med. Inform. Decis. Mak. **24**(1), 128 (2024)
13. Lin, T.Y., Dollár, P., Girshick, R., He, K., Hariharan, B., Belongie, S.: Feature pyramid networks for object detection. In: Proceedings of the IEEE Conference on Computer Vision and Pattern Recognition, pp. 2117–2125 (2017)
14. Lin, T.Y., Goyal, P., Girshick, R., He, K., Dollár, P.: Focal loss for dense object detection. In: Proceedings of the IEEE International Conference on Computer Vision, pp. 2980–2988 (2017)
15. Liu, L., Tang, D., Li, X., Ouyang, Y.: Automatic fetal ultrasound image segmentation of first trimester for measuring biometric parameters based on deep learning. Multimedia Tools Appl. **83**(9), 27283–27304 (2024)
16. Nguyen, E.H., et al.: Circle representation for medical object detection. IEEE Trans. Med. Imaging **41**(3), 746–754 (2021)
17. Ronneberger, O., Fischer, P., Brox, T.: U-Net: convolutional networks for biomedical image segmentation. In: Navab, N., Hornegger, J., Wells, W.M., Frangi, A.F. (eds.) MICCAI 2015. LNCS, vol. 9351, pp. 234–241. Springer, Cham (2015). https://doi.org/10.1007/978-3-319-24574-4_28
18. Sinclair, M., et al.: Human-level performance on automatic head biometrics in fetal ultrasound using fully convolutional neural networks. In: 2018 40th Annual International Conference of the IEEE Engineering in Medicine and Biology Society (EMBC), pp. 714–717. IEEE (2018)
19. Sylwestrzak, O., Respondek-Liberska, M.: Echocardiographic methods of fetal heart size assessment-heart to chest area ratio and transversal heart diameter. Prenatal Cardiol. **8**(1) (2018)
20. Van Der Linde, D., et al.: Birth prevalence of congenital heart disease worldwide: a systematic review and meta-analysis. J. Am. Coll. Cardiol. **58**(21), 2241–2247 (2011)
21. Vaswani, A., et al.: Attention is all you need. In: Advances in Neural Information Processing Systems, vol. 30 (2017)

22. Xie, E., et al.: Polarmask: single shot instance segmentation with polar representation. In: Proceedings of the IEEE/CVF Conference on Computer Vision and Pattern Recognition, pp. 12193–12202 (2020)
23. Xie, X., Cheng, G., Wang, J., Yao, X., Han, J.: Oriented R-CNN for object detection. In: Proceedings of the IEEE/CVF International Conference on Computer Vision, pp. 3520–3529 (2021)
24. Yang, S., Pei, Z., Zhou, F., Wang, G.: Rotated faster R-CNN for oriented object detection in aerial images. In: Proceedings of the 2020 3rd International Conference on Robot Systems and Applications, pp. 35–39 (2020)
25. Yang, X., Yan, J., Ming, Q., Wang, W., Zhang, X., Tian, Q.: Rethinking rotated object detection with Gaussian Wasserstein distance loss. In: International Conference on Machine Learning, pp. 11830–11841. PMLR (2021)
26. Zhao, L., Li, N., Tan, G., Chen, J., Li, S., Duan, M.: The end-to-end fetal head circumference detection and estimation in ultrasound images. IEEE/ACM Trans. Comput. Biol. Bioinform. (2022)
27. Zhou, D., et al.: IOU loss for 2D/3D object detection. In: 2019 International Conference on 3D Vision (3DV), pp. 85–94. IEEE (2019)
28. Zhou, X., Wang, D., Krähenbühl, P.: Objects as points. arXiv preprint arXiv:1904.07850 (2019)

Author Index

J. C. Gee et al. (Eds.): MICCAI 2025, LNCS 15960, pp. 695–699, 2026.
https://doi.org/10.1007/978-3-032-04927-8

MIX
Papier aus verantwortungsvollen Quellen
Paper from responsible sources
FSC® C105338

If you have any concerns about our products,
you can contact us on
ProductSafety@springernature.com

In case Publisher is established outside the EU,
the EU authorized representative is:
Springer Nature Customer Service Center GmbH
Europaplatz 3, 69115 Heidelberg, Germany

Printed by Libri Plureos GmbH
in Hamburg, Germany